Lecture Notes in Networks and Systems

Volume 472

The series "Lecture Notes in Networks and Systems" publishes the latest developments in Networks and Systems—quickly, informally and with high quality. Original research reported in proceedings and post-proceedings represents the core of LNNS.

Volumes published in LNNS embrace all aspects and subfields of, as well as new challenges in, Networks and Systems.

The series contains proceedings and edited volumes in systems and networks, spanning the areas of Cyber-Physical Systems, Autonomous Systems, Sensor Networks, Control Systems, Energy Systems, Automotive Systems, Biological Systems, Vehicular Networking and Connected Vehicles, Aerospace Systems, Automation, Manufacturing, Smart Grids, Nonlinear Systems, Power Systems, Robotics, Social Systems, Economic Systems and other. Of particular value to both the contributors and the readership are the short publication timeframe and the world-wide distribution and exposure which enable both a wide and rapid dissemination of research output.

The series covers the theory, applications, and perspectives on the state of the art and future developments relevant to systems and networks, decision making, control, complex processes and related areas, as embedded in the fields of interdisciplinary and applied sciences, engineering, computer science, physics, economics, social, and life sciences, as well as the paradigms and methodologies behind them.

Indexed by SCOPUS, INSPEC, WTI Frankfurt eG, zbMATH, SCImago.

All books published in the series are submitted for consideration in Web of Science.

For proposals from Asia please contact Aninda Bose (aninda.bose@springer.com).

More information about this series at https://link.springer.com/bookseries/15179

Isak Karabegović · Ahmed Kovačević ·
Sadko Mandžuka

Editors

New Technologies,
Development
and Application V

Set 1

 Springer

Editors
Isak Karabegović
Academy of Sciences and Arts of Bosnia
and Herzegovina
Sarajevo, Bosnia and Herzegovina

Ahmed Kovačević
Northampton Square
City University of London
London, UK

Sadko Mandžuka
Faculty of Traffic and Transport Sciences
University of Zagreb
Zagreb, Croatia

ISSN 2367-3370 ISSN 2367-3389 (electronic)
Lecture Notes in Networks and Systems
ISBN 978-3-031-05229-3 ISBN 978-3-031-05230-9 (eBook)
https://doi.org/10.1007/978-3-031-05230-9

This Springer imprint is published by the registered company Springer Nature Switzerland AG
The registered company address is: Gewerbestrasse 11, 6330 Cham, Switzerland

Interdisciplinary Research of New Technologies, their Development and Application

This book features papers focusing on the implementation of new and future technologies, which were presented at the International Conference on New Technologies, Development and Application, held at the Academy of Science and Arts of Bosnia and Herzegovina in Sarajevo on 23–25 June 2022. It covers a wide range of future technologies and technical disciplines, including complex systems such as Industry 4.0; patents in Industry 4.0; robotics; mechatronics systems; automation; manufacturing; cyber-physical and autonomous systems; sensors; networks; control, energy, renewable energy sources; automotive and biological systems; vehicular networking and connected vehicles; intelligent transport, effectiveness and logistics systems, smart grids, nonlinear systems, power, social and economic systems, education and IoT. Majority of organized conferences are usually focusing on a narrow part of the issues within a certain discipline while conferences such these are rare. There is a need to hold such conferences. The value of this conference is that various researchers, programmers, engineers and practitioners come to the same place where ideas and latest technology achievements are exchanged. Such events lead to the creation of new ideas, solutions and applications in the manufacturing processes of various technologies. New coexistence is emerging, horizons are expanding, and unexpected changes and analogies arise. Best solutions and applications in technologies are critically evaluated.

The first chapter covers mechanical design, Industry 4.0, robotics, cyber-physical systems, mechatronic systems, automation of production processes, 3D printing and advanced production and metallurgy. The first article is about service robots in COVID-19 pandemic. In order to control the pandemic, especially in medical institutions where patients are affected by the COVID-19 virus, it is necessary to treat patients with medicines, which can easily be achieved by the use of service robots. Article gives analysis of application of service robots for logistics during the COVID-19 pandemic and how it is related to Industry 4.0.

The second article presents development and implementation of digital twins, which allow the prediction of the behaviour of physical processes, services or systems and system optimization in a virtual, simulated environment, which is steadily increasing in the industrial environment. This article presents the development of a digital twin of a robotic cell by coupling state-of-the-art software environments. One of the articles covers the numerical investigation of aerodynamics of a naval surface combatant designed by the Office of Naval Research (ONR). As a more realistic ship, ONR Tumblehome hull was chosen instead of the generic frigate model SFS2 and ONRT model has been widely used for validation studies as a benchmark geometry. In one of the articles, the influence of capacity discharge percussion welding parameters on heat-affected zone properties of welded joints is analysed. One article aim is to analyse the use of hybrid powertrains in public transport by heavy vehicles. In the work, it was decided to investigate the various hybrid architectures normally used in vehicles and put on a multibody model to test optimal management systems of the engines present on the vehicle. Several articles discuss quality control. The last articles are about mathematics in engineering.

The second chapter covers computer science, information and communication technologies, Internet of things and cyber security. The first work is about energy-efficient AI systems based on memristive technology. Second article gives a method to optimize the geometric model, by defining a local sampling of the mesh, based on k-means method and minimization of the Hausdorff measure with a homologous model. The third article is about realization of single image super-resolution reconstruction based on wavelet transform and coupled dictionary. The one article provides digital transformation using artificial intelligence and machine learning on an electrical energy consumption case. One of the articles provides guidelines for a successful open data initiative implementation and presents key stakeholders with their engagement reasons, critical success factors and strategic themes for the designated area. In this chapter, some works are about COVID-19 such as the influence of COVID-19 pandemic on digitalization of medical services in Montenegro.

The third chapter is devoted to traffic and transport systems, logistics and intelligent systems. Chapter starts with application of big data sets and data science in transportation engineering. Big data sets originating from mobile telecommunication networks, besides their primary purpose within the telecom environment, are becoming more popular in other application areas such as transportation engineering. The second work describes the decision support system for safety management on the motorway section. The system is based on the estimation of the probability of traffic accidents on the motorway—crash potential. Based on this assessment, the system recommends active measures to reduce the likelihood of their actual occurrence. One article is about concept of road traffic noise monitoring in the function of environmental and health protection. Another article explains a new method for intelligent robot control, based on deep learning and reinforcement. The fundamental idea of this work is how the UAV equipped with a monocular camera can learn significant information about the object of interest in the context of its localization and navigation. One article explains real-time mobile robot perception based on deep learning detection model. One of the articles at the end of the chapter explains in an elegant and original manner an analysis of the level of pollution with pollutants eliminated in the atmosphere by the thermal engine of hybrid vehicles. The impact on the environment by the pollution with exhaust gases presented in the paper represents the results of some experiments obtained with measuring equipment, which accurately reproduces and records data of the level of the car pollutants measured at the output of the exhaust installation of the flue gas.

The fourth chapter is devoted to new technologies in the energy, fluids, power quality and advanced electrical power systems. The first article is about design and innovation in rotary positive displacement compressors. Rotary compressors have been in use for more than a century and are necessary for daily life in heating, ventilation, air-conditioning, refrigeration (HVAC-R) and air compression applications. Two new compressor designs will be highlighted here: the revolving vane compressor to reduce frictional losses and the coupled vane compressor to reduce material and fabrication costs, both of which having the potential to reduce the carbon footprint for their applications. The second article is about powering the smart parking system with photovoltaic solar panels. The paper describes how to make a small system (model) that contains all the necessary elements of the real system. By using the IoT solution, the hardware and software components of the system are created, and with the help of solar panels and its components, small power supply system is created. The third article is about decision-making model on new technologies for the production of mini hydropower plants in Bosnia and Herzegovina. Another article explains how to use support of artificial intelligence to complex hydro-energy systems. One work is about implementation of distributed information systems in solving problems of energy consumption monitoring.

The fifth chapter is devoted to new methods in the agriculture, ecology and chemical processes of a wide range of topics: electrochemical sensors based on molecularly imprinted polymers and different carbon materials for antibiotics detection, analysis of polyphenol content and antioxidant capacity of bear onion (allium ursinum) extracts obtained by different extraction techniques in selected solvents, consumption of cereals in Bosnia and Herzegovina - the health risk calculation, bioavailability of Co and Zn in some plant species depending on their concentration in the substrate.

The sixth chapter focuses on the field of geodesy, construction, new materials and sustainable innovation and others. The first article is about experimental statistical modelling of the pressing process of vibropressed concrete elements using Taguchi's method. The aim of the research is to optimize the vibropressing process, aiming to reduce the percentage of vibropressed concrete elements whose thickness and tensile strength do not fall within the tolerance range. The experimental statistical modelling method used is Taguchi's method, and the major factors influencing the vibropressing process that the authors chose after applying the dispersion analysis are the pressing pressure, the pressing time, the water/cement ratio and the type of cement used. Another article is about biocoagulants and bioflocculants in water and wastewater treatment technology. In this regard, a number of studies are currently being conducted to investigate different types of biocoagulants and bioflocculants, their performance and impact on health and environment. This article presents the advantages, limitations and challenges of employing biocoagulants and bioflocculants in water and wastewater treatment technology. One of the articles is about recycling of building materials as a principle of sustainable construction.

The seventh chapter covers economics, E-business, and entrepreneurships. The chapter starts with benefits and risks of applying Internet of bodies technology (IoB). Second article gives analysis of job insecurity and psychological safety in the workplace: evidence from Bosnia and Herzegovina. The study findings revealed differences in job security regarding career stage, professional position and contract type, while differences in psychological safety were confirmed in distinct career stages. One article is about perception of the role of private individuals participating in the exchange of goods and services through digital platforms in the collection of value-added tax revenues.

The whole content of this book is intended to a wide range of technical systems; different technical disciplines in order to apply the latest solutions and achievements in technologies and to improve manufacturing processes in all disciplines were systemic thinking which has a very important role in the successful understanding and building of human, natural and social systems. We hope this content will be the first in a series of publications that are intended to the development and implementation of new technologies in all industries.

Isak Karabegović

Contents

Contents

Electrical Engineering, Computer Science, Information and Communication Technologies, Control Systems

Intelligent Transport Systems, Logistics, Traffic Control

New Technologies in Agriculture, Ecology, Chemical Processes

New Technologies in Civil Engineering, Architecture, Construction

Economics, E-Business, Entrepreneurships

Contributors

Ivan Abramenko Kharkiv Petro Vasylenko National Technical University of Agriculture, Kharkiv, Ukraine

Delalić Adela University of Sarajevo, School of Economics and Business Sarajevo, Sarajevo, Bosnia and Herzegovina

Azra Ahmić Faculty of Economy, International University Travnik, Travnik, Bosnia and Herzegovina

Fatma Akcakoca Faculty of Agriculture, Ankara University, Ankara, Turkey

Amar Aladžuz School of Economics and Business Sarajevo, University of Sarajevo, Sarajevo, Bosnia and Herzegovina

Ismar Alagić TRA Tešanj Development Agency, Tešanj, Bosnia and Herzegovina;
Faculty of Mechanical Engineering, University of Zenica, Zenica, Bosnia and Herzegovina

Zijad Alibašić Center for Advanced Technologies in Sarajevo, Sarajevo, Bosnia and Herzegovina

Abdel Alibegović Faculty of Political Sciences, University of Sarajevo, Sarajevo, Bosnia and Herzegovina

Adi Alić School of Economics and Business Sarajevo, University of Sarajevo, Sarajevo, Bosnia and Herzegovina

Arnaut-Berilo Almira University of Sarajevo, School of Economics and Business Sarajevo, Sarajevo, Bosnia and Herzegovina

Aljinović Amanda Faculty of Electrical Engineering, Mechanical Engineering and Naval Architecture, University of Split, Split, Croatia

Bohdan Antypenko Faculty of Electronics and Information Technologies, Sumy State University, Sumy, Ukraine

Viktoriia Antypenko Faculty of Electronics and Information Technologies, Sumy State University, Sumy, Ukraine

Halit Apaydin Faculty of Agriculture, Ankara University, Ankara, Turkey

Melika Arifhodžić Faculty of Political Sciences, University of Sarajevo, Sarajevo, Bosnia and Herzegovina

Almira Arnaut-Berilo University of Sarajevo, School of Economics and Business Sarajevo, Sarajevo, Bosnia and Herzegovina

Elma Avdagić-Golub Faculty of Traffic and Communication, University of Sarajevo, Sarajevo, Bosnia and Herzegovina

Besim Balić Faculty of Forestry, University of Sarajevo, Sarajevo, Bosnia and Herzegovina

Kuan Thai Aw Nanyang Technological University, Singapore, Singapore

Seyfettin Bayraktar Department of Marine Engineering Operations, Yildiz Technical University, Istanbul, Turkey

Ernad Bešlagić Mechanical Engineering Faculty, University of Zenica, Zenica, Bosnia and Herzegovina

Muhamed Begović Faculty of Traffic and Communication, University of Sarajevo, Sarajevo, Bosnia and Herzegovina

Jasmin Bektešević Faculty of Mechanical Engineering, Department of Mathematics and Physics, University of Sarajevo, Sarajevo, Bosnia and Herzegovina

Rok Belšak Faculty of Mechanical Engineering, University of Maribor, Maribor, Slovenia

Denis Berberović School of Economics and Business Sarajevo, University of Sarajevo, Sarajevo, Bosnia and Herzegovina

Oleh Bevz Central Ukrainian National Technical University, Kropyvnytskyi, Ukraine

Mirha Bičo Ćar School of Economics and Business, University of Sarajevo, Sarajevo, Bosnia and Herzegovina

Vuk Bosković Faculty of Mechanical Engineering, Technical University of Munich, Munich, Germany

Lucija Brezočnik Intelligent Systems Laboratory, Faculty of Electrical Engineering and Computer Science, University of Maribor, Maribor, Slovenia

Miran Brezočnik Faculty of Mechanical Engineering, University of Maribor, Maribor, Slovenia

Biljana Buhavac Faculty of Civil Engineering, Department of Water Resources and Environmental Engineering, University of Sarajevo, Sarajevo, Bosnia and Herzegovina

Lucija Bukvić Faculty of Transport and Traffic Science, University of Zagreb, Zagreb, Croatia

Cătălin-Laurenţiu Bulgariu Faculty of Industrial Engineering and Robotics, Polytechnic University of Bucharest, Bucharest, Romania;
National Railway Company C.F.R. S.A., Bucharest, Romania

Stipo Buljan Federal Ministry of Energy, Mining and Industry, Mostar, Bosnia and Herzegovina

Piercarlo Cattani Department of Computer, Control and Management Engineering, University of Rome "La Sapienza", Rome, Italy

Giampiero Celenta MEID4 Academic Spin-Off of the University of Salerno, Fisciano, Italy

Volodymyr Chenchevoi Kremenchuk Mykhailo Ostrohradskyi National University, Kremenchuk, Ukraine

Bai Chuang School of Physics and Electronic Science, Changsha University of Science and Technology, Changsha, People's Republic of China

Vlad Ciprian Cupşan Universal Alloy Corporation Europe, Dumbrăviţa, Maramureş, Romania

Maria Curcio MEID4 Academic Spin-Off of the University of Salerno, Fisciano, Italy

Milica Daković-Tadić Faculty for International Economy, Finance and Business, University of DonjaGorica, Podgorica, Montenegro

Oleksandr Danyleiko Laser Systems and Advanced Technologies Department, National Technical University of Ukraine "Igor Sikorsky Kyiv Polytechnic Institute", Kyiv, Ukraine

Borna Dasović Faculty of Civil Engineering, Transportation Engineering and Architecture, University of Maribor, Maribor, Slovenia

Marco Claudio De Simone Department of Industrial Engineering, University of Salerno, Fisciano, Italy

Luka Dedić Promettis Ltd., Zagreb, Croatia

Remzo Dedić Department of Construction, Faculty of Mechanical Engineering, Computer Science and Electrical Engineering, University of Mostar, Mostar, Bosnia and Herzegovina

Adela Delalić School of Economics and Business Sarajevo, University of Sarajevo, Sarajevo, Bosnia and Herzegovina

Vesna Rašković Depalov Faculty of Technical Sciences, University of Novi Sad, Novi Sad, Serbia

Tome Dimovski Faculty of ICTs, "St Kliment Ohridski" University - Bitola, Bitola, North Macedonia

Milena Djukanovic Faculty of Electrical Engineering, University of Montenegro, Podgorica, Montenegro

Dmytro Lesyk Laser Systems and Advanced Technologies Department, National Technical University of Ukraine "Igor Sikorsky Kyiv Polytechnic Institute", Kyiv 03056, Ukraine;
New Technologies Research Centre, University of West Bohemia, Pilsen, Czech Republic;
Principles for Surface Engineering Department, G.V. Kurdyumov Institute for Metal Physics of the NAS of Ukraine, Kyiv 03142, Ukraine

Ali Dogrul Department of Naval Architecture and Marine Engineering, National Defense University, Turkish Naval Academy, Istanbul, Turkey

Koçi Doraci Universiteti Politeknik I Tiranes, Tirane, Albania

Branislav Dudić Faculty of Management, Comenius University in Bratislava, Bratislava, Slovakia;
Faculty of Economics and Engineering Management, University Business Academy, Novi Sad, Serbia

Vitaliy Dzhemelinkyi Laser Systems and Advanced Technologies Department, National Technical University of Ukraine "Igor Sikorsky Kyiv Polytechnic Institute", Kyiv, Ukraine

Edin Džiho Faculty of Mechanical Engineering Univerzitetski Kampus, Džemal Bijedić" University of Mostar, Mostar, Bosnia and Herzegovina

Alma Džubur Faculty of Civil Engineering, Department of Water Resources and Environmental Engineering, University of Sarajevo, Sarajevo, Bosnia and Herzegovina

Žana Džubur Faculty of Civil Engineering, "Dzemal Bijedic" University of Mostar, Mostar, Bosnia and Herzegovina

Matteo d'Amore Department of Pharmacy, University of Salerno, Fisciano, Italy

Apostol Eliza-Ioana Faculty of Industrial Engineering and Robotics, University Politehnica of Bucharest, Bucharest, Romania;
Scientific Researcher at INCAS - National Institute for Aerospace Research "Elie Carafoli", Bucharest, Romania

Maksim Fed Kremenchuk Mykhailo Ostrohradskyi National University, Kremenchuk, Ukraine

Hajar Feizi Faculty of Agriculture, University of Tabriz, Tabriz, Iran

Fahira Fejzić Čengić Faculty of Political Sciences, University of Sarajevo, Sarajevo, Bosnia and Herzegovina

Mirko Ficko Faculty of Mechanical Engineering, University of Maribor, Maribor, Slovenia

Iztok Fister Jr. Intelligent Systems Laboratory, Faculty of Electrical Engineering and Computer Science, University of Maribor, Maribor, Slovenia

Andrea Formato Department of Agricultural Science, University of Naples "Federico II", Portici, Naples, Italy

Faris Gačanović MGBH d.o.o. Sarajevo, Sarajevo, Bosnia and Herzegovina

Franjo Gilja Faculty of Mechanical Engineering, Computing and Electrical Engineering, University of Mostar, Mostar, Bosnia and Herzegovina

Dusan Golubovic Faculty of Mechanical Engineering, University of East Sarajevo, Sarajevo, Bosnia and Herzegovina

Janez Gotlih Faculty of Mechanical Engineering, University of Maribor, Maribor, Slovenia

Vincenzo Guercio Engineering School, Deim, University of Tuscia, Viterbo, Italy;
Engineering School, University of Tuscia, Viterbo, Italy

Domenico Guida Department of Industrial Engineering, University of Salerno, Fisciano, SA, Italy

Vasile Gusan Faculty of Industrial Engineering and Robotics, Polytechnic University of Bucharest, București, Romania;
Continental Automotive Systems, 8, Street Salzburg, Sibiu, Romania

Sabahudin Hadrovic Institute of Forestry, Belgrade, Serbia

Muhamed Hadžiabdić International University Sarajevo, Sarajevo, Bosnia and Herzegovina

Vahidin Hadžiabdić Faculty of Mechanical Engineering, Department of Mathematics and Physics, University of Sarajevo, Sarajevo, Bosnia and Herzegovina

Miodrag Hadžistević Faculty of Technical Sciences, Department of Production Engineering, University of Novi Sad, Novi Sad, Serbia

Svetlana Hadžić "Džemal Bijedić" University of Mostar, Mostar, Bosnia and Herzegovina

Hu Haitao School of Information and Electrical Engineering, China Agricultural University, Beijing, China

Jasmin Halilović University of Tuzla, Tuzla, Bosnia and Herzegovina; Faculty of Mechanical Engineering, University of Tuzla, Tuzla, Bosnia and Herzegovina

Wafaa M. Hikal Department of Biology, Faculty of Science, University of Tabuk, Tabuk 71491, Saudi Arabia; Environmental Research Division, Water Pollution Research Department, National Research Centre, Giza, Egypt

Ina Hodžić School of Economics and Business, University of Sarajevo, Sarajevo, Bosnia and Herzegovina

Milan Honner New Technologies Research Centre, University of West Bohemia, Pilsen, Czech Republic

Ilija Hristoski Faculty of Economics - Prilep, "St Kliment Ohridski" University - Bitola, Prilep, North Macedonia

Anes Hrnjić School of Economics and Business Sarajevo, University of Sarajevo, Sarajevo, Bosnia and Herzegovina

Matej Hruska New Technologies Research Centre, University of West Bohemia, Pilsen, Czech Republic

Ermin Husak Technical Faculty Bihać, Univesity of Bihać, Bihać, Bosnia and Herzegovina

Aida Husetić Faculty of Technical Engineering, University of Bihać, Bihać, Bosnia and Herzegovina

Ermin Huskić Department of Mechanical Design, Faculty of Mechanical Engineering, University of Sarajevo, Sarajevo, Bosnia and Herzegovina

Selma Husnić Faculty of Civil Engineering, Dzemal Bijedic University of Mostar, Mostar, Bosnia and Herzegovina

Želimir Husnić The Boeing Company, Philadelphia, USA

Kanita Imamović-Čizmić Department of Legal and Economic Sciences, Faculty of Law, University of Sarajevo, Sarajevo, Bosnia and Herzegovina

Aida Imamović Faculty of Metallurgy and Technology, University of Zenica, Zenica, Bosnia and Herzegovina

Gudelj Iris FOR FIVE dma, Sarajevo, Bosnia and Herzegovina

Safet Isić Faculty of Mechanical Engineering Univerzitetski Kampus, "Džemal Bijedić" University of Mostar, Mostar, Bosnia and Herzegovina

Merima Šahinagić Isović Faculty of Civil Engineering, Dzemal Bijedic University of Mostar, Mostar, Bosnia and Herzegovina

Dalila Ivanković Faculty of Civil Engineering, University Džemal Bijedić Mostar, Mostar, Bosnia and Herzegovina

Anđela Jaksic Stojanovic University of Donja Gorica, Podgorica, Montenegro

Anđela Jakšić-Stojanović Faculty of Culture and Tourism, University of Donja Gorica, Podgorica, Montenegro;
Faculty of International Economics, Finance and Business, University of Donja Gorica, Podgorica, Montenegro

Emina Japić Faculty of Technical Engineering, University of Bihać, Bihać, Bosnia and Herzegovina

Zlata Jelačić Department of Mechanics, Faculty of Mechanical Engineering, University of Sarajevo, Sarajevo, Bosnia and Herzegovina

Zijah Jelkić Veritas Automotive d.o.o., Sarajevo, Bosnia and Herzegovina

Đorđe Jevtić Faculty of Mechanical Engineering, Department of Production Engineering, University of Belgrade, Belgrade 35, Serbia

Aleksandar Jokić Faculty of Mechanical Engineering, Department of Production Engineering, University of Belgrade, Belgrade, Serbia

Marko Jovanović SMM Production Systems, Ltd., Maribor, Slovenia

Marinko Jurčević Faculty of Transport and Traffic Sciences, Univesity of Zagreb, Zagreb, Croatia

Liudmila Kamenska Ukrainian State University of Railway Transport, Kharkiv, Ukraine

Sergii Kamenskyi Radio Information Technologies LLC, Kharkiv, Ukraine

Enita Kapo Faculty of Political Sciences, University of Sarajevo, Sarajevo, Bosnia and Herzegovina

Edina Karabegović Technical Faculty Bihać, Univesity of Bihać, Bihać, Bosnia and Herzegovina

Isak Karabegović Academy of Sciences and Arts of Bosnia and Herzegovina, Sarajevo, Bosnia and Herzegovina

Sašo Karakatič Intelligent Systems Laboratory, Faculty of Electrical Engineering and Computer Science, University of Maribor, Maribor, Slovenia

Timi Karner Faculty of Mechanical Engineering, University of Maribor, Maribor, Slovenia

Ines Katić Križmančić 2 K ideja d.o.o., Zagreb, Croatia

Ivana Katnić Faculty of International Economics, Finance and Business, University of Donja Gorica, Podgorica, Montenegro

Uroš Klanšek Faculty of Civil Engineering, Transportation Engineering and Architecture, University of Maribor, Maribor, Slovenia

Sanela Klarić International Burch University, Sarajevo, Bosnia and Herzegovina

Nikola Knezović Faculty of Mechanical Engineering, Computing and Electrical Engineering, University of Mostar, Mostar, Bosnia and Herzegovina

Volodymyr Kopei Ivano-Frankivsk National Technical University of Oil and Gas, Ivano-Frankivsk, Ukraine

Selma Korac Faculty of Pharmacy, University of Sarajevo, Sarajevo, Bosnia and Herzegovina

Goran Kos Institut for Tourism, Zagreb, Croatia

Amel Kosovac Faculty of Traffic and Communication, University of Sarajevo, Sarajevo, Bosnia and Herzegovina

Zorana Kostić Faculty of Mechanical Engineering, University of Niš, Niš, Serbia

Pavel Kovač Faculty of Technical Sciences, University of Novi Sad, Novi Sad, Serbia

Džemal Kovačević University of Tuzla, Tuzla, Bosnia and Herzegovina; Faculty of Mechanical Engineering, University of Tuzla, Tuzla, Bosnia and Herzegovina

Amra Kožo School of Economics and Business, University of Sarajevo, Sarajevo, Bosnia and Herzegovina

Andrii Kuk Lviv Polytechnic National University, Lviv, Ukraine

Yaroslav Kusyi Lviv Polytechnic National University, Lviv, Ukraine

Vincenzo Laiola MEID4 Academic Spin-Off of the University of Salerno, Fisciano, Italy

Ana Lalevic Filipovic Faculty of Economics, University of Montenegro, Podgorica, Montenegro

Zorana Lanc Faculty of Technical Sciences, Department of Production Engineering, University of Novi Sad, Novi Sad, Serbia

Rosario La Regina MEID4 Academic Spin-Off of the University of Salerno, Fisciano, Italy

Ammar Lavić Faculty of Mechanical Engineering, "Džemal Bijedić" University of Mostar, Mostar, Bosnia and Herzegovina

Ivan Lazović Faculty of Management Zaječar, Megatrend University of Belgrade, Zaječar, Serbia

Nerma Lazović Faculty of Civil Engineering, Department of Water Resources and Environmental Engineering, University of Sarajevo, Sarajevo, Bosnia and Herzegovina

Samir Lemeš Polytechnic Faculty, University of Zenica, Zenica, Bosnia and Herzegovina

Serhii Leshchenko Central Ukrainian National Technical University, Kropyvnytskyi, Ukraine

Osman Lindov Faculty of Traffic and Communications, University of Sarajevo, Sarajevo, Bosnia and Herzegovina

Ahmet Lojo Faculty of Forestry, University of Sarajevo, Sarajevo, Bosnia and Herzegovina

Angelo Lorusso Department of Industrial Engineering, University of Salerno, Fisciano, SA, Italy

Darko Lovrec Faculty of Mechanical Engineering, University of Maribor, Maribor, Slovenia

Željko Lozančić Faculty of Civil Engineering, Department of Water Resources and Environmental Engineering, University of Sarajevo, Sarajevo, Bosnia and Herzegovina

Tetiana Lukan Ivano-Frankivsk National Technical University of Oil and Gas, Ivano-Frankivsk, Ukraine

Jovanka Lukic Faculty of Engineering, University of Kragujevac, Kragujevac, Serbia

Haris Lulić Institute of Metrology of Bosnia and Herzegovina, Sarajevo, Bosnia and Herzegovina

Mehmed Mahmić Technical Faculty Bihać, Univesity of Bihać, Bihać, Bosnia and Herzegovina

Sanjin Mahmutović Faculty of Political Sciences, University of Sarajevo, Sarajevo, Bosnia and Herzegovina

Bia Mandžuka Faculty of Transport and Traffic Sciences, Univesity of Zagreb, Zagreb, Croatia

Sadko Mandžuka Faculty of Traffic and Transport Sciences, Zagreb, Croatia; Department of ITS, University of Zagreb, Zagreb, Croatia

Marko Mančić Faculty of Mechanical Engineering, University of Niš, Niš, Serbia

Anna Marchenko Faculty of Electronics and Information Technologies, Sumy State University, Sumy, Ukraine

Carmine Maria Pappalardo Department of Industrial Engineering, University of Salerno, Fisciano, Italy

Mladineo Marko Faculty of Electrical Engineering, Mechanical Engineering and Naval Architecture, University of Split, Split, Croatia

Adnan Mašić Faculty of Mechanical Engineering, Department of Mathematics and Physics, University of Sarajevo, Sarajevo, Bosnia and Herzegovina

Adnan Mehonic Department of Electronic and Electrical Engineering, UCL, London, UK

Midhat Mehuljić Faculty of Mechanical Engineering, Department of Mathematics and Physics, University of Sarajevo, Sarajevo, Bosnia and Herzegovina

Shuli Mei College of Information and Electrical Engineering, China Agricultural University, Beijing, People's Republic of China

Orlić Merima University of Sarajevo, School of Economics and Business Sarajevo, Sarajevo, Bosnia and Herzegovina

Amina Milišić Faculty of Civil Engineering, Dzemal Bijedic University of Mostar, Mostar, Bosnia and Herzegovina

Zoran Miljković Faculty of Mechanical Engineering, Department of Production Engineering, University of Belgrade, Belgrade 35, Serbia

Peđa Milosavljević Faculty of Mechanical Engineering, University of Niš, Niš, Serbia

Dragoljub Mirjanic Academy of Sciences and Arts of the Republic of Srpska, Banja Luka, Bosnia and Herzegovina

Bijelić Mitar TNT Express INC, Franklin, WI, USA

Alexandra Mittelman Faculty of Management, Comenius University in Bratislava, Bratislava, Slovakia

Iuliana Moisescu Polytechnic University of Bucharest, Bucharest, Romania; Ministry of Culture, Bucharest, Romania

Radu Costin Moisescu Polytechnic University of Bucharest, Bucharest, Romania; State Office for Inventions and Trademarks, Bucharest, Romania

Valentyn Moiseenko Ukrainian State University of Railway Transport, Kharkiv, Ukraine

Naida Mujić Faculty of Electrical Engineering, University of Sarajevo, Sarajevo, Bosnia and Herzegovina

Admir Mulahusić Faculty of Civil Engineering, University of Sarajevo, Sarajevo, Bosnia and Herzegovina

Adis J. Muminovic Department of Mechanical Design, Faculty of Mechanical Engineering, University of Sarajevo, Sarajevo, Bosnia and Herzegovina

Marko Mumović Faculty of Mechanical Engineering, University of Montenegro, Podgorica, Montenegro

Edis Nasić University of Tuzla, Tuzla, Bosnia and Herzegovina

Edis Nasić Faculty of Mechanical Engineering, University of Tuzla, Tuzla, Bosnia and Herzegovina

Gheorghe Neamțu Faculty of Industrial Engineering and Robotics, University Politehnica of Bucharest, Bucharest, Romania

Gheorghe Ioan Pop University POLITEHNICA of Bucharest, Splaiul Independenței 313, Bucharest, Romania;
S.C. Universal Alloy Corporation Europe S.R.L., Dumbravița 244A, Maramures, Romania

Andrii Nekrasov Kremenchuk Mykhailo Ostrohradskyi National University, Kremenchuk, Ukraine

Yakiv Nemyrovskyi Central Ukrainian National Technical University, Kropyvnytskyi, Ukraine

Viktor Nenia Faculty of Electronics and Information Technologies, Sumy State University, Sumy, Ukraine

Emir Nezirić Faculty of Mechanical Engineering Univerzitetski Kampus, "Džemal Bijedić" University of Mostar, Mostar, Bosnia and Herzegovina

Gjeldum Nikola Faculty of Electrical Engineering, Mechanical Engineering and Naval Architecture, University of Split, Split, Croatia

Milijana Novovic Buric Faculty of Economics, University of Montenegro, Podgorica, Montenegro

Milijana Novović-Burić Faculty of Economics, University of Montenegro, Podgorica, Montenegro

Mirna Nožić Faculty of Mehanical Engineering, University "DžemalBijedić" of Mostar, Mostar, Bosnia and Herzegovina

Constantin-Dorin Olteanu Faculty of Industrial Engineering and Robotics, Polytechnic University of Bucharest, Bucharest, Romania;
Directorate for the Registration of Persons, Sibiu, Romania

Dženan Omeradžić Faculty of Technical Engineering, University of Bihać, Bihać, Bosnia and Herzegovina

Adnan Omerhodžić Faculty of Traffic and Communications, University of Sarajevo, Sarajevo, Bosnia and Herzegovina

Lejla M. Omerović Department of Textile Design and Management, University of Zagreb Faculty of Textile Technology, Zagreb, Croatia

Oleh Onysko Ivano-Frankivsk National Technical University of Oil and Gas, Ivano-Frankivsk, Ukraine

Kim Tiow Ooi Nanyang Technological University, Singapore, Singapore

Constantin Oprean Lucian Blaga University of Sibiu, Sibiu, România

Merima Orlić University of Sarajevo, School of Economics and Business Sarajevo, Sarajevo, Bosnia and Herzegovina

Carmine Maria Pappalardo Department of Industrial Engineering, University of Salerno, Fisciano, Italy

Katarina Pavlović Faculty for Project and Innovation Management, EDUCONS University, Belgrade, Serbia

Željka Pavlović Faculty of Textile Technology, Department of Textile Design and Management, University of Zagreb, Zagreb, Croatia

Sead Pašić Faculty of Mechanical Engineering Univerzitetski Kampus, Džemal Bijedić" University of Mostar, Mostar, Bosnia and Herzegovina

Ekrem Pehlic Faculty of Health Studies, University of Bihac, Bihac, Bosnia and Herzegovina

Belma Pehlivanović Faculty of Pharmacy, University of Sarajevo, Sarajevo, Bosnia and Herzegovina

Aleksandra Penjišević Faculty of Management, Union "Nikola Tesla" University, Sremski Karlovci, Serbia

Nedim Pervan Department of Mechanical Design, Faculty of Mechanical Engineering, University of Sarajevo, Sarajevo, Bosnia and Herzegovina

Antonija Petrov Faculty of Textile Technology, Department of Clothing Technology, University of Zagreb, Zagreb, Croatia

Milica Petrović Faculty of Mechanical Engineering, Department of Production Engineering, University of Belgrade, Belgrade, Serbia

Špela Pečnik Intelligent Systems Laboratory, Faculty of Electrical Engineering and Computer Science, University of Maribor, Maribor, Slovenia

Boran Pikula Faculty of Mechanical Engineering, University in Sarajevo, Sarajevo, Bosnia and Herzegovina

Oleksii Piskarev Kharkiv Petro Vasylenko National Technical University of Agriculture, Kharkiv, Ukraine

Vili Podgorelec Intelligent Systems Laboratory, Faculty of Electrical Engineering and Computer Science, University of Maribor, Maribor, Slovenia

Gregor Polančič Intelligent Systems Laboratory, Faculty of Electrical Engineering and Computer Science, University of Maribor, Maribor, Slovenia

Alina Bianca Pop Technical University of Cluj-Napoca, North University Center of Baia Mare, Baia Mare, Romania

Wei Qin College of Information and Electrical Engineering, China Agricultural University, Beijing, China

Luka Radunović Faculty of Electrical Engineering, University of Montenegro, Podgorica, Montenegro

Zijada Rahimić School of Economics and Business Sarajevo, University of Sarajevo, Sarajevo, Bosnia and Herzegovina

Zijada Rahimić School of Economics and Business, University of Sarajevo, Sarajevo, Bosnia and Herzegovina

Milena Rajić Faculty of Mechanical Engineering, University of Niš, Niš, Serbia

Sanja Raljević Jandrić Faculty of Political Sciences, University of Sarajevo, Sarajevo, Bosnia and Herzegovina

Bălaşa Raluca Faculty of Industrial Engineering and Robotics, University Politehnica of Bucharest, Bucharest, Romania;
Scientific Researcher at INCAS - National Institute for Aerospace Research "Elie Carafoli", Bucharest, Romania

Lejla Ramić Department of Legal and Economic Sciences, Faculty of Law, University of Sarajevo, Sarajevo, Bosnia and Herzegovina

Miloš Ranisavljev Faculty of Technical Sciences, Department of Production Engineering, University of Novi Sad, Novi Sad, Serbia

Zandra B. Rivera Engineering Research Institute, Universidad Privada San Juan Bautista, Lima, Peru

Raffaele Romano Department of Agricultural Science, University of Naples "Federico II", 80055 Portici, Naples, Italy

Flaviana Rotaru Faculty of Industrial Engineering and Robotics, Polytechnic University of Bucharest, Bucuresti, România;
ROHEALTH - Health and Bioeconomy Cluster, Bucharest, Romania

Hussein A. H. Said-Al Ahl Medicinal and Aromatic Plants Research Department, National Research Centre, Giza, Egypt

Ivana Salopek Čubrić Faculty of Textile Technology, Department of Textile Design and Management, University of Zagreb, Zagreb, Croatia

Merima Salčin Faculty of Civil Engineering, "Dzemal Bijedic" University of Mostar, Mostar, Bosnia and Herzegovina

Ahmedin Salčinović Federal Institute of Agropedology Sarajevo, Sarajevo, Bosnia and Herzegovina

Aida Sapcanin Faculty of Pharmacy, University of Sarajevo, Sarajevo, Bosnia and Herzegovina

Isad Saric Department of Mechanical Design, Faculty of Mechanical Engineering, University of Sarajevo, Sarajevo, Bosnia and Herzegovina

Isad Saric Department of Mechanical Design, Faculty of Mechanical Engineering, University of Sarajevo, Sarajevo, Bosnia and Herzegovina

Sarih Sarı Department of Naval Architecture and Marine Engineering, Yildiz Technical University, Istanbul, Turkey

Mohammad Taghi Sattari Faculty of Agriculture, University of Tabriz, Tabriz, Iran

Slavica Mačužić Saveljić Faculty of Engineering, University of Kragujevac, Kragujevac, Serbia

Borislav Savković Faculty of Technical Sciences, University of Novi Sad, Novi Sad, Serbia

Erjon Selmani Universiteti Politeknik I Tiranes, Tirane, Albania

Amra Serdarević Faculty of Civil Engineering, Department of Water Resources and Environmental Engineering, University of Sarajevo, Sarajevo, Bosnia and Herzegovina

Serhii Serhiienko Kremenchuk Mykhailo Ostrohradskyi National University, Kremenchuk, Ukraine

Sergii Shendryk Sumy National Agrarian University, Sumy, Ukraine

Vira Shendryk Sumy State University, Sumy, Ukraine

Ihor Shepelenko Central Ukrainian National Technical University, Kropyvnytskyi, Ukraine

Halima Sofradžija Faculty of Political Sciences, University of Sarajevo, Sarajevo, Bosnia and Herzegovina

Bohdan Solohub Lviv Polytechnic National University, Lviv, Ukraine

Evhen Solovykh Central Ukrainian National Technical University, Kropyvnytskyi, Ukraine

Borko Somborac Faculty of Management, Union "Nikola Tesla" University, Sremski Karlovci, Serbia

Denijal Sprečić University of Tuzla, Tuzla, Bosnia and Herzegovina; Faculty of Mechanical Engineering, University of Tuzla, Tuzla, Bosnia and Herzegovina

Svetlana Stevovic Academy of Sciences and Arts of the Republic of Srpska, Banja Luka, Bosnia and Herzegovina

Savo Stupar University of Sarajevo, School of Economics and Business, Sarajevo, Bosnia and Herzegovina

Nedim Suljić Faculty of Mining, Geology and Civil Engineering Tuzla, University of Tuzla, Tuzla, Bosnia and Herzegovina

Suad Sućeska Sarajevo, Bosnia and Herzegovina

Aida Sušić Smed Engineering BH d.o.o., Mostar, Bosnia and Herzegovina

Petrică Tertereanu Polytechnic University of Bucharest, Bucharest, Romania; County Police Inspectorate Vâlcea, Vâlcea County, Romania

Vito Tič Faculty of Mechanical Engineering, University of Maribor, Maribor, Slovenia

Angela Topić Faculty of Mechanical Engineering, Computing and Electrical Engineering, University of Mostar, Mostar, Bosnia and Herzegovina

Jusuf Topoljak Faculty of Civil Engineering, University of Sarajevo, Sarajevo, Bosnia and Herzegovina

Anes Torlaković School of Economics and Business, University of Sarajevo, Sarajevo, Bosnia and Herzegovina

Mirsad Trobradović Faculty of Mechanical Engineering, Department of Engines and Vehicles, University of Sarajevo, Sarajevo, Bosnia and Herzegovina

Nedim Tuno Faculty of Civil Engineering, University of Sarajevo, Sarajevo, Bosnia and Herzegovina

Sergii Tymchuk Kharkiv Petro Vasylenko National Technical University of Agriculture, Kharkiv, Ukraine

Faris Ustamujić Airbus DS GmbH, Taufkirchen, Germany

Kenan Varda Mechanical Engineering Faculty, University of Zenica, Zenica, Bosnia and Herzegovina

Ljiljan Veselinović School of Economics and Business, University of Sarajevo, Sarajevo, Bosnia and Herzegovina

Krešimir Vidović Faculty of Transport and Traffic Sciences, Univesity of Zagreb, Zagreb, Croatia

Francesco Villecco Department of Industrial Engineering, University of Salerno, Fisciano, Italy

Maja Tonec Vrančić Faculty of Transport and Traffic Science, University of Zagreb, Zagreb, Croatia

Grega Vrbančič Intelligent Systems Laboratory, Faculty of Electrical Engineering and Computer Science, University of Maribor, Maribor, Slovenia

Victor Vriukalo Ivano-Frankivsk National Technical University of Oil and Gas, Ivano-Frankivsk, Ukraine

Nenad Vujadinović Faculty of Arts, University of Donja Gorica, Podgorica, Montenegro

Miroslav Vujić Faculty of Transport and Traffic Sciences, Univesity of Zagreb, Zagreb, Croatia

Aleksandar Vujović Faculty of Mechanical Engineering, University of Montenegro, Podgorica, Montenegro

Dženan Vukotić Federal Institute of Agropedology Sarajevo, Sarajevo, Bosnia and Herzegovina

Nermina Zaimović-Uzunović Mechanical Engineering Faculty, University of Zenica, Zenica, Bosnia and Herzegovina

Bijelić Zdravko Institute in the Making "Logos", Novi Sad, Serbia

Milan Zeljković Faculty of Technical Sciences, Department of Production Engineering, University of Novi Sad, Novi Sad, Serbia

Min Zhao College of Information and Electrical Engineering, China Agricultural University, Beijing, China

Leiping Zhu College of Information and Electrical Engineering, China Agricultural University, Beijing, People's Republic of China

Mirha Bičo Ćar University of Sarajevo, School of Economics and Business, Sarajevo, Bosnia and Herzegovina

Fuad Ćatović Faculty of Civil Engineering, "Dzemal Bijedic" University of Mostar, Mostar, Bosnia and Herzegovina

Marko Ćećez Faculty of Civil Engineering, Dzemal Bijedic University of Mostar, Mostar, Bosnia and Herzegovina

Merima Činjarević School of Economics and Business Sarajevo, University of Sarajevo, Sarajevo, Bosnia and Herzegovina

Elvir Čizmić School of Economics and Business, University of Sarajevo, Sarajevo, Bosnia and Herzegovina

Alem Čolaković Faculty of Traffic and Communication, University of Sarajevo, Sarajevo, Bosnia and Herzegovina

Goran Čubrić Faculty of Textile Technology, Department of Clothing Technology, University of Zagreb, Zagreb, Croatia;
Department of Clothing Technology, University of Zagreb Faculty of Textile Technology, Zagreb, Croatia

Zoran Đikanović Faculty of International Economics, Finance and Business, University of Donja Gorica, Podgorica, Montenegro

Mirsad Donlagić Faculty of Technology, University of Tuzla, Tuzla, Bosnia and Herzegovina

Belma Đono Infobip, Sarajevo, Bosnia and Herzegovina

Himzo Đukić University of Mostar FSRE, Mostar, Bosnia and Herzegovina

Ammar Šabanović Faculty of Mechanical Engineering, "Džemal Bijedić" University of Mostar, Mostar, Bosnia and Herzegovina

Dina Šamić Municipality Centar, Sarajevo, Bosnia and Herzegovina

Milan Šekularac Faculty of Mechanical Engineering, University of Montenegro, Podgorica, Montenegro

Munira Šestić School of Economics and Business, University of Sarajevo, Sarajevo, Bosnia and Herzegovina

Mina Šibalić Faculty of Mechanical Engineering, University of Montenegro, Podgorica, Montenegro

Nikola Šibalić Faculty of Mechanical Engineering, University of Montenegro, Podgorica, Montenegro

Pero Škorput Faculty of Transport and Traffic Science, University of Zagreb, Zagreb, Croatia

Jasmina Pašagić Škrinjar Faculty of Transport and Traffic Science, University of Zagreb, Zagreb, Croatia

Marko Šoštarić Faculty of Traffic and Transport Sciences, Zagreb, Croatia

Damir Špago Faculty of Mechanical Engineering, "Džemal Bijedić" University of Mostar, Mostar, Bosnia and Herzegovina

Branko Štrbac Faculty of Technical Sciences, Department of Production Engineering, University of Novi Sad, Novi Sad, Serbia

Edin Šunje Faculty of Mechanical Engineering Univerzitetski Kampus, Džemal Bijedić" University of Mostar, Mostar, Bosnia and Herzegovina

Darko Šunjić Faculty of Mechanical Engineering, Computing and Electrical Engineering, University of Mostar, Mostar, Bosnia and Herzegovina

Suvada Šuvalija Faculty of Civil Engineering, Department of Water Resources and Environmental Engineering, University of Sarajevo, Sarajevo, Bosnia and Herzegovina

Lamija Šćeta School of Economics and Business Sarajevo, University of Sarajevo, Sarajevo, Bosnia and Herzegovina

Aurel Mihail Titu "Lucian Blaga" University of Sibiu, Sibiu, Romania

Crnjac Žižić Marina Faculty of Electrical Engineering, Mechanical Engineering and Naval Architecture, University of Split, Split, Croatia

Šaban Žuna Center for Advanced Technologies in Sarajevo, Sarajevo, Bosnia and Herzegovina

New Technologies in Mechanical Engineering, Metallurgy, Mechatronics, Robotics and Embedded Systems

Application of Service Robots for Logistics During the COVID-19 Pandemic Accelerates the Implementation of Industry 4.0

Isak Karabegović[1]([✉]), Edina Karabegović[2], Mehmed Mahmić[2], and Ermin Husak[2]

[1] Academy of Sciences and Arts of Bosnia and Herzegovina, St. Bistrik 7, 71000 Sarajevo, Bosnia and Herzegovina
isak1910@hotmail.com
[2] Technical Faculty, University of Bihać, IrfanaLjubijankića bb, 77 000 Bihać, Bosnia and Herzegovina

Abstract. The development of science leads to changes that come with the initiation and implementation of innovations in technologies that are implemented in production, thus leading to fundamental changes in the economic and social system of each society. Development of new technologies, including robotics & automation, intelligent sensors, 3D printers, cloud computing, radio frequency identification – RFID, etc., is the foundation of the fourth industrial revolution, i.e., Industry 4.0. The implementation of Industry 4.0, among other technologies, is mostly based on robotic technology, both industrial and service robots. Of all the service robots, we can single out the service robots for logistics, whose high implementation was caused by the COVID-19 pandemic. There are two reasons for the high trend in the application of service robots for logistics. One reason is the COVID-19 pandemic, where the use of service robots can meet one recommendation, and that is distance. In order to control the pandemic, especially in medical institutions where patients are affected by the COVID-19 virus, it is necessary to treat patients with medicines, which can easily be achieved by the use of service robots. Another reason for their application is global competition in the market where work must be done to improve quality, reduce production time, reduce production and maintenance costs, to replace workers on all monotonous and difficult work tasks. By applying service robots, we increase the implementation of Industry 4.0 and thus create a significantly higher level of added value, create more benefits and bring benefits to all people in the world.

Keywords: Service robots · Logistics · Automation · COVID-19 · Industry 4.0

1 Introduction

The development of technologies and their implementation in industry and the environment are characterized by radical and rapid changes in a positive sense. Changes come through the initiation and implementation of innovations in production technologies, which provides a new way of production and service sector thus leading to fundamental

I. Karabegović et al. (Eds.): NT 2022, LNNS 472, pp. 3–17, 2022.
https://doi.org/10.1007/978-3-031-05230-9_1

changes in the economic and social system of each society. The COVID-19 pandemic has enabled a possibility of implementing service robots for logistics, the reason being that service robots can replace people and fill the gap in the lack of skilled labor and the frequency and fatigue of work by handling simple transport jobs. During the pandemic, people have to keep their distance, especially in medical facilities where patients are infected with the COVID-19 virus. It is necessary that these patients are served with medicine, food, etc., which can be provided by service robots for logistics. The changes that are happening today in the industry and the environment in the world were called industrial revolution 4.0 at the World Economic Forum – WEF in Davos in 2016. Industry 4.0 is already being implemented worldwide, in other words, by introducing technologies into production processes such as:

new generations of digital technologies and infrastructure, artificial intelligence, machine learning, robotics, Internet, nanotechnology, genetic modification, new species and ways of energy and information storage, quantum computing, 3-D printing, genetic engineering and biotechnology, automation of production processes, data exchange and processing. The goal of Industry 4.0 is the networking of production processes, i.e., networking of smart industries and factories of the future, which is related to the concept of intelligent production [1–3].

Industry 4.0 relies on Cyber-Physical-Systems [4–9]. Analysis and discussion of Industry 4.0 has the task of bringing technological change closer to all those involved in its implementation, become aware of its impact, for the benefit of all. It is necessary to reach a platform that achieves cooperation between public and private [10–12]. Industry 4.0 has a lot of features, here are just a few:

- digitalization of industry, connection of machines, warehouses, logistics, and equipment (CPS-cybernetic systems),
- smart production systems, such as smart factories and smart products, but also logistics, marketing and smart services with a focus on individual customer needs,
- vertical integration within the organizational sector,
- horizontal integration through diverse business partners and supply chains, in terms of business and consumer integration, i.e. new business models,
- transformation of the value chain and product life cycle,
- accelerated development through increased flexibility and reduced costs of industrial processes,
- real-time production control and optimization.

A new wave of technological advancement and new digital technologies are benefiting the manufacturing industry by increasing productivity, revenue, employment, and investment. The implementation of Industry 4.0 brings the breadth and depth of change, i.e., it affects the transformation of entire production, management and administration systems [13, 14]. The implementation of Industry 4.0, and all the mentioned technologies forming its foundation, brings a wave of technological progress. In other words, they bring benefits to the processing industry through the growth of productivity, revenue, employment, and investment. Production processes become flexible and enable economical production of small series of products. Industry 4.0 leads to the creation and production of new products and services, as well as increased productivity and reduced costs, which affects economic growth and development. The consequences are structural changes, new jobs, growing demand for educated workers, but also the some job losses [15]. The implementation of innovations and patents in robotic technology are responsible for the development of robots, so that second-generation robots have been developed, both industrial and service robots. One of the reasons for the implementation of service robots is the integration of machines and devices with service robots, and the other is that their implementation automates production processes to become more flexible and easier to perform tasks [16].

2 Differences and Advantages of Industry 4.0 Compared to Industry 3.0

Industry 4.0 comes after Industry 3.0. The difference between these two industrial revolutions is reflected in the following: production processes in the third industrial revolution include manual labor in automated production, while in the fourth industrial revolution (Industry 4.0) intellectual labor is used instead of manual labor and automatic design, as shown in Fig. 1. The development and progress of new technologies, which are becoming more sophisticated every day, lead to a more orderly system as a whole in all segments of society [4, 5, 17].

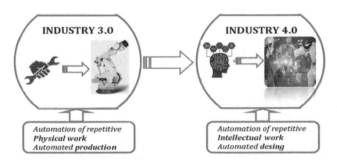

Fig. 1. Differences and advantages of Industry 4.0 compared to Industry 3.0

The Industry 4.0 platform has one common goal which is to match supply and demand in a very affordable way by providing the customer with a wide range of products without both parties interacting. The inclusion of digital technologies with other advanced technologies will transform the production processes in industrial production due to the fact that the price of this technology is continuously declining in the market and it is becoming more widespread in production processes. We must mention that robots are not intended to completely replace workers, but to work with them, so as to remove fences in production processes, which currently enclose industrial first-generation robots, as shown in Fig. 2 [4, 6, 9, 16].

During work, man can perform various tasks in all fields of work, and conduct very complex operations and analytical tasks. On the other hand, collaborative robot is easy to work with, performs monotonous repetitive operations, can handle hazardous substances, as well as lift heavy objects. The differences between workers and robots

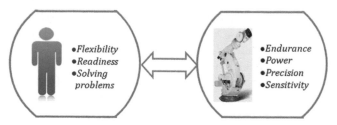

Fig. 2. Advantages and disadvantages between workers and industrial robots

are shown in Fig. 2.The implementation of basic technologies of Industry 4.0 is on the rise, as shown in Fig. 3, which requires flexible automation in all industries as well as reducing the time of production and sale of finished products with continuously high quality [4, 18–20].

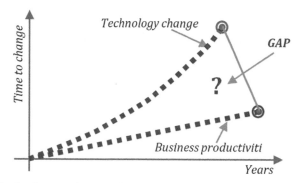

Fig. 3. Implementation of Industry 4.0 is enabling technological changes

The Industry 4.0 application in production processes provides technological improvements in terms of flexibility, accuracy, security and simplification of the use of new technologies. Technological changes occur by the exponential function, while business productivity takes place by the linear function, as shown. Based on Fig. 3, we see that there is a difference when it comes to technological change and productivity, and this difference can only be reduced by people with their agility.

3 Implementation of Professional Service Robots and Logistics Robots During the COVID-19 Virus Pandemic

In the 1990s, UNECE and the International Federation of Robotics (IFR) they adopted a system for classifying service robots. Earlier classifications of service robots for professional and personal needs were adopted by ISO. The classification of service robots was performed in two groups: *I – personal/home service robots* and *II – professional service robots*. Group I includes the following service robots: home service robots, fun service robots, handicap assistance robots, personal transport (AVG), security and surveillance, and other home robots.

Fig. 4. Case of implementation of professional service robots, and review of robots for logistics and manufacturing

Group II – *professional service robots* includes the following service robots: service robots for agriculture, livestock, forestry, spatial and mining systems, cleaning robots, construction robots, inspection and maintenance robots, logistics robots, medical service robots, robots for rescue and security, service robots for defense, underwater systems, mobile platforms, public relations robots and other service robots that are not specified, as shown in Fig. 4. The paper will offer a review and analysis of service robots for logistics and service robots in industrial production processes [5, 6, 16].

The implementation of robots in general brings improvements when it comes to production processes, especially second-generation robots, and some advantages are:

- significantly improved performance during performing operations,
- opportunities have been created to have flexible automation, and to automate tasks that have not been able to automate so far,
- in Industry 4.0 robots play an important role because their application allows companies to be specific to the global market,
- by applying collaborative robots, we can significantly improve the performance of non-ergonomic tasks,
- they are characterized by simplicity,
- reduced product life-cycle and increased product diversity require flexible automation, which will result in increased robot applications, etc.

Industry 4.0 will enable installation of next-generation industrial robots that will be aware, connected, process-responsive, and intelligent, as shown in Fig. 5.

Fig. 5. New-generation service robots for application in Industry 4.0

The diagram in Fig. 5 shows us that second-generation robots must be: connected, responsive, aware and intelligent. The advanced generation of robots must be equipped with: smart sensors to provide environmental information, communication devices with data exchange, a control system that allows autonomy or adaptation with external and internal control, a computer that can learn based on defined algorithms and make decisions. With robots conceived in this way, we are accelerating the application of Industry 4.0 in production. The application of new generation robots in production companies increases productivity, reduces production costs and achieves high product quality. An example of a production process with implemented new generation robots and new technologies on which Industry 4.0 is based is given in Fig. 6. All machines are interconnected, communicate, exchange data and only control can take place internally and externally, in other words schematically shows one intelligent production process [4].

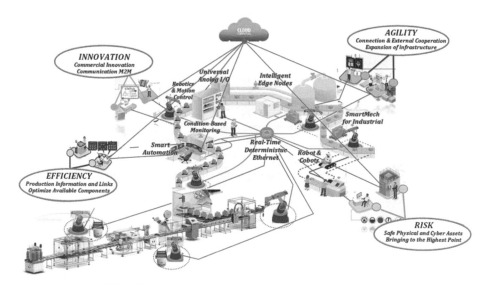

Fig. 6. Example of intelligent manufacturing process in industry using basic technologies of Industry 4.0 and new generation robots [4]

In the production process of any industry, we can conclude that the factors of efficiency, innovation, agility and risk are important, as shown in Fig. 6. Based on monitoring the production and operation of machines, we can provide preventive maintenance, to replace worn elements in time to avoid downtime. Adapt the organization of production itself by optimizing the capacity itself. It is necessary to provide internal internet for internal communication, as well as external communication, as well as to ensure the safe operation of the system. In the last ten years, there has been an increase in the number of innovations in all areas and all technologies in the world. These innovations in other technologies have been reflected in the development of robotic technology, so that second-generation robots have seen the light of day, which has reflected in the enormous trend of increased use of service robots for professional use. Figure 7 presents the trend of increased implementation of service robots for professional use. The statistical data were taken from the International Federation of Robotics (IFR), the United Nations Economic Commission for Europe (UNECE) and the Organization for Economic Cooperation and Development OECD) that have aggregate data coming from about 750 robot companies [21–30].

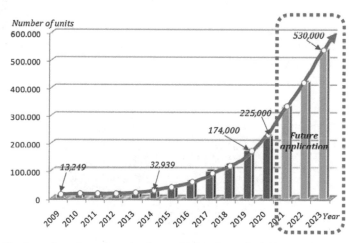

Fig. 7. Application of service robots in the world for the period 2009–2020, with forecasts of implementation until 2023 [21–30]

The trend of application of service robots for professional use is continuously increasing on annual basis. Based on the diagram shown in Fig. 7 we can conclude that the application takes place by exponential function. In 2009, 13.249 robot units were applied, and in just ten years this amount increased to 225.000 robot units, which is an increase of 179%. The growing trend is continuing, so that the use of about 530.000 units of service robots for professional use is predicted by 2023. This growing trend is due to the implementation of Industry 4.0 in production processes, which has a huge impact on the global economy, and the economic growth. Most countries in the world have hastily adopted Industry 4.0 implementation strategies. Those countries that have not yet adopted Industry 4.0 are currently working to adopt it in their production processes in order to achieve increased productivity and be competitive in the global market, i.e. have higher economic growth. The implementation of Industry 4.0 changes the paradigm of the production process, from centralized to decentralized or smart production process, as shown in Fig. 6. Industry 4.0 represents the computerization of production and the creation of smart factories of the future, where physical objects are integrated into the information network. Production systems are vertically networked with factory business processes, and horizontally connected to real-time manageable networks. In his paper, author K. Chukalovstates: *"The interaction of implemented systems, based on special software and user interface, which is integrated into digital networks, creates new functional systems for horizontal and vertical integration"* [31].

Implementation of Industry 4.0 is a driver and carrier of sustainable economic growth and development, given that the level of development and competitiveness of industry are correlated with the intensity of reindustrialization and the quality of the structure of the manufacturing industry. In order to obtain information where service robots are most used in professional services, we have analyzed the number of applied service robot units expressed as a percentage (%) in 2020 in all areas shown in Fig. 4. The analysis is shown in Fig. 8 [21].

The analysis of the application of service robots for professional services in 2020 has shown that service robots for logistics are the most used service robots in 2020, with about 47%. Such a large percentage of the application of service robots for logistics is due to the implementation of Industry 4.0, which aims at complete flexible automation of production processes, from raw materials to the finished product.

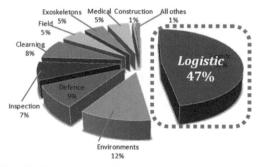

Fig. 8. Percentage of application of service robots for professional services in all areas in 2020 [21]

Service robots for logistics perform tasks that were originally developed for a strictly controlled and standardized production environment. Their application has been extended to warehousing, wholesale and e-commerce in service industries. The first use of automated guided vehicles (AGVs) was in structured factory premises and required special floors to guide movement. With the development of smart sensors they are used to move around loads, such as crates, shelves, and occasionally people. In other words, automated guided vehicles (AGVs) move independently with the avoidance of obstacles in the production process, warehouse, post office or airport.

Service robots for logistics – automated guided vehicles (AGV) are able to perform a number of tasks such as:

- *transport:* loading of pallets and bulk parts,
- *storage:* moving products from stretchable wrappers to docks or warehouses,
- *commissioning:* preparation of products for distribution to the customer,
- *delivery of parts:* towing trailer parts and materials to the place of consumption,
- *installation:* delivery of pallets for production processes,
- *assembly:* collecting parts for assembly and movement of products through production processes,
- *transfer:* materials, semi-finished products, products, etc.

Automated guided vehicles (AGVs) are applied to clinics, post offices, hotels, companies as couriers.

In order to prove the presented conclusions, we have conducted an analysis of the implementation of service robots for logistics in the last ten years, as shown in Fig. 9.

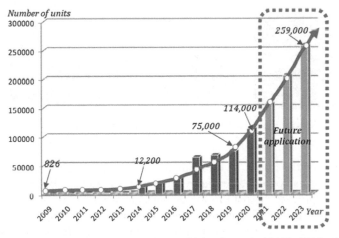

Fig. 9. Application of robots for logistics in the world for the period 2009–2020 with forecast of implementation until 2023 [21–30]

The trend of implementation of service robots for logistics in the world is exponential, as shown in Fig. 9. Their implementation was very low until 2014, when about 12.200 robots units were used. In the following years, the use of service robots for logistics grew rapidly, so that in 2020, about 114.000 robot units were applied. The increase in the trend of application of service robots for logistics can be attributed to the basic technologies on which Industry 4.0 is based and their implementation in production processes. It is estimated that the implementation of service robots for logistics will continue to grow in the coming years, so that implementation of about 259.000 units of service robots for logistics in production processes is expected by 2023. During the COVID-19 pandemic, workers in these sectors were often exposed to health risks, partly due to the way their work was organized. The development and implementation of service robots for logistics will no doubt try to address these risks, for example by limiting the need for physical intimacy between people, or by facilitating more efficient cleaning. With the threat of COVID-19 continuing, warehouse automation can be seen as an area of improved efficiency and safety, since it is a facility where a large number of automated mobile robots can operate and provide social distance and other security measures that workers may need. The COVID-19 pandemic has placed great emphasis on the implementation of service robots for logistics in many companies around the world, since service robots can fill the gap in the lack of skilled labor, and the frequency and fatigue of work by handling simple transport, delivery and routing jobs. Many companies, encouraged by the implementation of Industry 4.0 in production processes, have developed various designs of service robots for logistics. In order to provide a complete insight in the implementation of Industry 4.0 and the robotic technology, we will illustrate the solutions of several companies implementing service robots for logistics. One of the solutions is that of the company MiR – Mobile Industrial Robots which has developed various designs of collaborative service robot for logistics, a small part of which is shown in Fig. 10.

Fig. 10. Service robots of company "Mobile Industrial Robots – MiR" [32, 33]

The design of the robots of the company "Mobile Industrial Robots - MiR" are such that they are autonomous, reliably bypass all obstacles when performing tasks, independently optimize the path to the destination, when an obstacle is found on the road redirect their path. The new laser scanner technology allows them to visualize up to 360 degrees, and they are equipped with 3D cameras on the front of the robot, as well as two sensors on each corner for the right decision when performing a task. Robots use energy from batteries to move and perform tasks, and when they are discharged, they make their own decision about charging on a connector installed in a random space. Installed batteries provide the robot with operation up to 1000 cycles.The newest offer on the market provided by the MiR Company is the MiR1350 service robot that can carry a load of 1350 kg, which is convenient for transport in production processes when supplying raw materials to machines [32, 33]. Many companies are developing service robots for logistics. The number of implementations is increasing along with other core technologies of Industry 4.0. All companies tend to implement "smart transport", one of which is OMRON company. OMRON is the world's leading supplier of service robots for logistics, which can be installed in manufacturing processes in the industry. It is the first company to launch a commercial mobile robot on the market in 2013 [34]. OMRON has the largest installed base of mobile robots in production which can be applied in thousands of applications in multiple industries and are used to increase throughput, eliminate errors and improve material traceability.

Fig. 11. Service robots by company OMRON [34]

In June 2020, the company OMRON presented the design of a new robot HD-1500, which has the ability to carry a load of 1500 kg, and is suitable for work in the production process, as Fig. 11 shows [35]. The HD-1500 robot is flexible and provides many possibilities for automation in production, and is suitable for performing various tasks in operation. BionicHIVE is developing the SqUID service robot for complete warehouse automation. The SqUID robot can be adapted to an existing workspace without requiring additional investment. The complete solution can be adapted to any existing storage infrastructure, as shown in Fig. 12.

Fig. 12. Service robot "SqUID" by company BionicHIVE [35]

"SqUID" robots work autonomously and synchronized as one fleet of robots and all have three-dimensional movement in the workspace. Data analysis is done in real time which allows him to learn and create problems very well, especially in the warehouse. They are very easy to implement in warehouses, and their autonomy gives us the opportunity to plan properly in the coming period.

When we compare the potential benefits of BionicHIVE's solutions, they represent Industry 4.0 at its best [35]. In addition to these three, many other companies, encouraged by the implementation of Industry 4.0 in production processes, have developed various designs of service robots for logistics (over 750 companies in the world produce robots). In order to get more accurate picture of the implementation of Industry 4.0 and the robotic technology itself, we will illustrate the solutions of few logistics service robot companies. Due to paper limitations, we will provide only a few solutions, although each company deserves the same analysis. The solutions are shown in Fig. 13 [36].

Fig. 13. Different designs of collaborative service robots for logistics of several companies [36]

Second generation robots and we will single out logistics robots thanks to advanced technologies such as artificial intelligence, show advances in production automation, but many fears of labor shortages in the plants of manufacturing companies. Service robots are suitable for the service sector, logistics, maintenance, inspection and professional cleaning.

Many scientists and policy makers expect that the wave of automation or implementation of Industry 4.0 will soon disrupt the service sector. The prospect of a new generation of service robots, capable of serving and communicating with humans, would expand the limit of automation from industrial production tasks to service activities [37]. Finally, how the service robots, that are currently deployed, fit into the companies in terms of work organization, working conditions, health and safety, and the demand for skills. The implementation of service robots for logistics is increasing which leads to the automation of logistics processes in all areas, especially in warehouses and production processes, aiming to achieve smart production processes brought by Industry 4.0.

4   Conclusion

Industry 4.0 is already being introduced into production facilities that are present in the global market. The introduction of this technology in the work on which Industry 4.0 relies changes the approach to the organization of production, production processes and the way to satisfy the customer. It has led to a significant transformation of economic theory and the evolution of sustainable development paradigm. In 2019 we have witnessed the outburst of the COVID-19 pandemic, which continues today. One of the recommendations to control it is that people must keep their distance, which has placed great emphasis on the use of service robots in medical institutions where patients suffer from COVID-19 virus, and need to be served with medicines, food, etc. In other words, during the COVID-19 pandemic, there has been an increasing trend of application of service robots for logistics in all segments of society as well as in many companies around the world in production processes for transport, commissioning, parts delivery, transmission, and assembly. Service robots for logistics can fill the gap in the lack of skilled labor, and the frequency and fatigue of work by handling simple transport, delivery and routing tasks. The implementation of Industry 4.0 cannot be carried out without the implementation of both industrial and service robots. Industry 4.0 represents a change in the production paradigm and digitalization of the economy. It is overcoming the lack of resources and increased living standards, thus leading to a significant transformation of economic theory and the evolution of sustainable development paradigm. The trend of implementation of service robots in the world will increase in all segments of society, especially service robots for logistics used in medical institutions to control the COVID-19 pandemic. Another reason is the implementation of Industry 4.0, applied by all companies in developed countries in order to remain competitive in the global market. By implementing service robots, we increase the implementation of Industry 4.0 and thus create a significantly higher level of added value and employment.

References

1. McCann, P., Ortega-Argilés, R.: Smart specialization, regional growth and applications to European Union cohesion policy. Reg. Stud. **49**(8), 1291–1302 (2015)
2. Wang, S., Wan, J., Zhang, D., Li, D., Zhang, C.: Towards smart factory for industry 4.0: a self-organized multi-agent system with big data based feedback and coordination.Comput. Netw. **101**, 158–168 (2016)
3. Schwab, K.: The Fourth Industrial Revolution. World Economic Forum, Geneva, Switzerland (2016)
4. Karabegović, I., Kovačević, A., Banjanović-Mehmedović, L., Dašić, P.: Integrating Industry 4.0 in Business and Manufacturing. IGI Global, Hershey, PA, USA (2020). https://www.igi-global.com/book/handbook-research-integrating-industry-business/237834
5. Karabegović, I.: The role of industrial and service robots in the fourth industrial revolution. ACTE Technica Corviniensis-Bulletin of Engineering, University Politehnica Timisoara, Tome XI, Fascicule 2. April 2018. Hunedoara, Romania, pp. 11–16 (2018). http://acta.fih.upt.ro/pdf/2018-2/ACTA-2018-2-01.pdf
6. Karabegović, I., Karabegović, E., Mahmić, M., Husak, E.: The application of service robots for logistics in manufacturing processes. Adv. Prod. Eng. Manage. **10**(4), 185–194 (2015). https://www.apem-journal.org/Archives/2015/APEM10-4_185-194.pdf
7. Karabegović, I., Karabegović , E., Mahmić, M., Husak, E.: Implementation of Industry 4.0 and industrial robots in production processes. In: Karabegović, I. (ed.) New Technologies, Development and Application II 2019. Lecture Notes in Networks and Systems, vol. 76, pp. 96–102. Springer Nature, Switzerland, AG (2020)
8. Karabegović, I., Husak, E.: Industry 4.0 based on industrial and service robots with application in China. J. Mobility Vehicle **44**(4), 59–71 (2018)
9. Karabegović, I.: The role of industrial and service robots in fourth industrial revolution with focus on China. J. Eng. Archit. **5**(2), 110–117 (2017)
10. Mićič, V.: Industry 4.0 development conditions in the Republic of Serbia. J. Facta Univ. Econ. Org. **17**(2), 97–112 (2020). https://doi.org/10.22190/FUEO191112008M
11. WEF: The Future of Jobs Report 2018, pp. 6–44. World Economic Forum, Geneva, Switzerland (2018)
12. Rüßmann, M., et al.: Industry 4.0 The Future of Productivity and Growth in Manufacturing Industries, pp. 5–12. The Boston Consulting Group (2015)

13. Ecker, C.: Advantages and challenges for small manufactureres. Internationa Collaborative Robots, Workshop, Columbia, USA (2015)
14. Meniere, Y., Rudyk, I., Valdes, J.:Patents and the Fourth Industrial Revolution. Europeanpatent Office, Munich, Germany (2017)
15. Ostrgaard, E.: Collaborative Robot Technology and Applications. International Collaborative Robots, Workshop, Columbia (2015)
16. Karabegović, I., Husak, E.,Predrag, D.:The role of service robots in Industry 4.0 – smart automation of transport. Int. Sci. J. Industry 4.0 **4**(6), 290–292 (2019). https://stumejournals.com/journals/i4/2019/6/290
17. Crnjac, M., Veža, I., Banduka, N.: From concept to the introduction of Industry 4.0. Int. J. Ind. Eng. Manage. **8**(1), 21–30 (2017)
18. Kusmin, K.L.: Industry 4.0 Analytical Article, IFI8101 – Information Society Approaches and ICT Processes, School of Digital Technologies, Tallinn University, Estonia (2016)
19. Bunse, B., Kagermann, H., Wahlster, W.: Smart Manufacturing for the Future. Germany Trad & Invest, Berlin, Germany (2017)
20. Bechtold, J., Lauenstein, C., Kern, A., Bernhofer L.: Executive Summary: The Capgemini Consulting Industry 4.0 Framework. Capgemini Consulting, Paris, France (2017)
21. World Robotics 2020 Service Robots: The International Federation of Robotics. Statistical Department, Frankfurt am Main, Germany (2020). https://ifr.org/
22. World Robotics 2019 Service Robots: The International Federation of Robotics. Statistical Department, Frankfurt am Main, Germany (2019). https://ifr.org/
23. World Robotics 2018 Service Robots: The International Federation of Robotics. Statistical Department, Frankfurt am Main, Germany (2018). https://ifr.org/
24. World Robotics 2017 Service Robots: The International Federation of Robotics. Statistical Department, Frankfurt am Main, Germany (2017). https://ifr.org/
25. World Robotics 2016 Service Robots: The International Federation of Robotics. Statistical Department, Frankfurt am Main, Germany (2016). https://ifr.org/
26. World Robotics 2015 Service Robots: The International Federation of Robotics. Statistical Department, Frankfurt am Main, Germany (2015). https://ifr.org/
27. World Robotics 2014 Service Robots: The International Federation of Robotics. Statistical Department, Frankfurt am Main, Germany (2014). https://ifr.org/
28. World Robotics 2012 Service Robots: The International Federation of Robotics. Statistical Department, Frankfurt am Main, Germany (2012). https://ifr.org/
29. World Robotics 2011 Service Robots: The International Federation of Robotics. Statistical Department, Frankfurt am Main, Germany (2011). https://ifr.org/
30. World Robotics 2010 Service Robots: The International Federation of Robotics. Statistical Department, Frankfurt am Main, Germany (2010). https://ifr.org/
31. Chukalov, K.: Horizontal and vertical integration, as a requirement for cyber-physicalsystems in the context of industry 4.0. Industry 4.0 **2**(4), 155–157 (2017)
32. Optimized Internal Transportation of Heavy Loads and Pallets with MiR1000 and MiR500. Odense, Denmark. https://www.mobile-industrial-robots.com/. Accessed 6 Nov 2021
33. Diverseco Announce Appointment as Official Mobile Industrial Robots (MiR) Partner. https://diverseco.com.au/news/diverseco-mir-partnership/. Accessed 8 Nov 2021
34. Introducing the HD-1500: Our Strongest Mobile Robot. https://assets.omron.eu/downloads/brochure/en/v1/i854_hd-1500_leaflet_en.pdf. Accesed 8 Nov 2021
35. Autonomy: The SqUID allows us to plan the future. https://www.bionichive.com/. Accesed: 15 Nov 2021

36. https://www.google.com/search?q=service+robots+for+logistics+i&tbm=isch&ved=2ah
    UKEwja1YjGx4v0AhVVs6QKHSUnBBcQ2cCegQIABAA&oq=service+robots+for+log
    istics+i&gs_lcp=CgNpbWcQDDIHCCMQ7wMQJ1AAWABgwxoAHAAeACAAWOIAW
    OSAQExmAEAoAEBqgELZ3dzLXdpei1pbWfAAQE&sclient=img&ei=wY6KYZrxB
    tXmkgWlzpC4AQ&bih=757&biw=1821&rlz=1C1NHXL_hrBA708BA708#imgrc=94P-
    XI8uZvrhCM. Accesed 20 Nov 2021
37. Motteo, S.: JRC Technical Report, Automation and Robots in Services: Review of Data and
    Taxonomy. JRC Working Papers Series on Labour, Education and Technology (2020). https://
    ec.europa.eu/jrc

Design of a Digital Twin of a Robotic Cell for Product Quality Control

Janez Gotlih[1]([⊠]), Miran Brezočnik[1], Mirko Ficko[1], Marko Jovanović[2], Rok Belšak[1], and Timi Karner[1]

[1] Faculty of Mechanical Engineering, University of Maribor, Maribor, Slovenia
janez.gotlih@um.si
[2] SMM Production Systems, Ltd., Maribor, Slovenia

Abstract. Progress in automation is based on the development of methods that allow the construction of flexible and reconfigurable systems to perform tasks that need to be completed in the shortest possible time and with the required quality. In this sense, the development and implementation of digital twins, which allow the prediction of the behavior of physical processes, services or systems and system optimization in a virtual, simulated environment, is steadily increasing in the industrial environment. This article presents the development of a digital twin of a robotic cell by coupling state-of-the-art software environments. The individual parts of the digital twin system are presented and combined to form a functioning automated system. The operation of the virtual cell is verified by simulating a cycle consisting of transporting the product via conveyor belts through the safety door into the quality control cell, where inspection is performed using the UR5 robotic arm.

Keywords: Digital twin · Robotics · Quality control · Home appliance device

1 Introduction

Robotization is one of the most important trends in industries with mass production. In these environments, the preparation and precise design of automation systems is of great importance. The idea of digitally modeling physical objects and simulation of their behavior in real time was first introduced in 2002. At that time, this idea was referred to as the "mirrored spaces model", but over the years the term "digital twin" has become widely accepted [1, 2].

Today, it is estimated that in the near future, many companies will set up digital twin models that virtually simulate or predict the behavior of physical processes, services, or systems [3]. Setting up a digital twin in the design phase of a system makes it easier to test the effectiveness of the system so that potential problems in subsequent real-world implementations can be identified in the development phase [4–6]. Another way to use digital twins is to digitize systems that are already physically functioning to optimize or extend product flexibility in production at a later stage [7].

I. Karabegović et al. (Eds.): NT 2022, LNNS 472, pp. 18–29, 2022.
https://doi.org/10.1007/978-3-031-05230-9_2

Fig. 1. Integration of a digital twin into the project design

The transition from the initial idea, conceptual design and testing, implementation of corrections, virtual commissioning of the machine, and real commissioning to final production is supported by the digital twin, which contains all the details of the defined system (Fig. 1). The accuracy of such a virtual system depends on the accuracy of the description of the real components, where the kinematic and dynamic descriptions of the mechanical elements are implemented in a multi-physics environment [8, 9]. Thus, with the help of digitization, some corrections to critical elements can be predicted in advance.

Creating a detailed digital twin of an automated mechanical system is beyond the capabilities of a single software environment [10–13]. To integrate a multi-physics simulation with external control that enables automation, multiple software solutions must be coupled. A diagram of such a system with the individual software environments and communication channels is shown in Fig. 2.

In Fig. 2, SIMIT is the center of information processing where all software environments are interconnected [14]. By routing signals, the platform acts as a communication channel through which the different environments can communicate with each other [15]. Mechatronics concept designer (NX-MCD or MCD) is used to model and simulate complex electromechanical systems consisting of mechanical, electrical and automation assemblies, considering multi-physics [16]. MCD allows visualization of the physical system in a three-dimensional virtual environment. S7-PLCSIM Advanced simulates a programmable logic controller (PLC) and eliminates the need for a physical PLC [17]. TIA Portal enables the automation of the system, from the design of digital elements to the configuration of all devices [18] and also allows to set up a human–machine interface (HMI) for process control and supervision [19]. URSim allows the creation, simulation and testing of Universal Robot (UR) robotic arm systems in a three-dimensional virtual environment [20].

This article presents the development of a digital twin of a robotic cell by coupling state-of-the-art software environments. The individual parts of the digital twin system are presented and combined to form a functioning automated system. The operation of the virtual cell is verified by simulating a cycle consisting of transporting the product via conveyor belts through the safety door into the quality control cell, where quality control is performed using the UR5 robotic arm.

The paper is organized as follows: Sect. 2 presents the implementation of the different software environments. Section 3 presents the digital twin for the test cell. Section 4 concludes the paper.

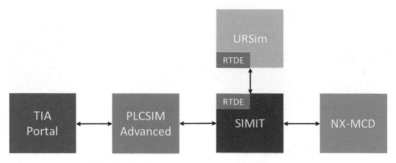

Fig. 2. Integration of software environments

2   Materials and Methods

In this article, the development of a digital twin for final inspection of kitchen ovens is presented. The quality control cell includes conveyor belts to transport the product, automatic safety doors, a clamping device for product positioning, and a robotic arm to perform the inspection process.

The digital twin is designed by using MCD, SIMIT, S7-PLCSIM Advanced, TIA Portal and URSim software environments. The fusion of all software environments enables the simulation of a control algorithm that can be set up, executed, and observed in real time in the TIA Portal environment. S7-PLCSIM Advanced controls the simulated hardware (servo motors, motor drives, sensors, etc.) in the SIMIT software environment, which allows accurate simulation of the physical devices in real time (motor responses, sensor responses, etc.). URSim is used to perform realistic offline programming of the robot arm. The graphical representation of the whole system is done in the MCD environment, where the physical properties of the elements are defined.

2.1   Implementation of the MCD environment

In the MCD environment, we have fully defined the quality control cell by modeling all transport, safety, and manipulation elements. The imported CAD models were defined as rigid bodies to which physical properties have been assigned. Joints, mechanical couplings, and actuators were used to define the behavior of each body and the interactions between them. Fixed, translational, and rotational joints were most used. Mechanical couplings were used to couple the rotational motion of the servo motor shaft and the linear motion of the clamp along the guide using the "motion profile" and "mechanical curve" functions. Actuators were added to the joints to allow the movement of rigid bodies. The speed control actuator determines the velocity of the motion of a rigid body, while the position control actuator determines the position limit for the motion of a rigid body. In some cases, user-defined behavior was implemented by runtime expressions and expression blocks. Collision sensors were used to define collision objects only for those objects that were expected to come into contact during the simulation. The use of collision sensors also enables the display of parameters and collision events that can be used to perform functions such as controlling operations, changing execution parameters, changing expression blocks, changing signal status, etc. Signals were added for communication with the SIMIT environment, which allows external control of the actuators based on a control algorithm and online sensor values.

2.2 Implementation of the S7-PLCSIM Advanced

We have used the S7-PLCSIM Advanced as a virtual controller to execute a control algorithm, as the S7-PLCSIM Advanced software environment allows simulating the operation of a physical industrial controller with all its functions. Such a simulation environment is useful when performing tests and developing an algorithm and makes it possible to perform tests without affecting the hardware and without having to purchase a controller in advance.

2.3 Implementation of the TIA Portal Environment

In the TIA portal environment, we have defined hardware components, topology and hierarchy of the components, and technology objects, developed a software algorithm to control the quality control cell according to the inspection protocol, and set up an HMI.

The hardware of the simulated system consists of a PLC and its central processing unit (CPU), a graphic HMI display and five motor drive units, that control two servo motors and three asynchronous motors. Figure 3 shows all hardware components interconnected via the Profinet protocol, which is an industrial communication protocol, based on industrial Ethernet, intended for fast communication between the industrial controller and the elements in the communication network and can be configured for real-time communication.

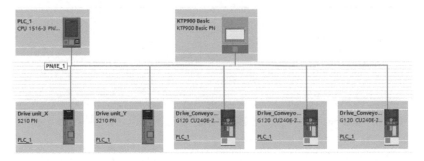

Fig. 3. Hardware components interconnected via the Profinet protocol

We have defined motor drives to be controlled in isochronous real time (IRT), which means that the cycle times of the controller and peripherals are synchronized. Components that communicate in IRT must be defined in the system topology. An exception is the HMI, for which no topology needs to be defined (Fig. 4).

For speed and position control of motor drives, we have used centralized control with technology objects, where all process control is done centrally on the PLC. Technology objects are functions that connect the motor drive control commands to the associated

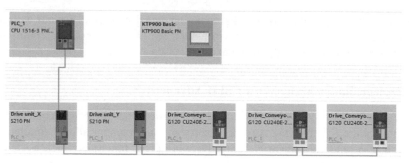

Fig. 4. Topology view of the system

motors between the TIA Portal software environment and S7-PLCSIM Advanced. We have created five technology objects, one for each motor drive. Each technology object needs to be configured for the type of movement, that it will perform and parameters like the component hardware, mechanics, dynamic properties, and control restrictions need to be specified (Fig. 5).

Different PROFIdrive telegrams may be used for control and communication between the PLC and the motor drive. We have used the Standard telegram 1 for speed control of conveyor belts, and the Standard telegram 105 for position control of servo motors of the clamp.

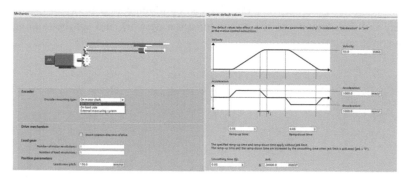

Fig. 5. Axis mechanics (left) and dynamic values of axis displacements (right)

One of the most common ways to describe a control algorithm is through flowcharts. We have defined the control algorithm for the final quality control by an algorithm consisting of a main program and two subroutines, a subroutine for the initial setup of the cell and product movements, and a subroutine for opening and closing the safety doors. The flowchart of the main program is shown in Fig. 6.

For the monitoring and control of the cell, we have developed the HMI through which the user can fully control the execution of the digital twin by passing the necessary signals from the control program to the HMI and visualizing them on the screen. Figure 7 on the left shows the main menu of the HMI, which includes buttons to start and stop the cycle, as well as sensor values for belt speed and product placement. On the right side

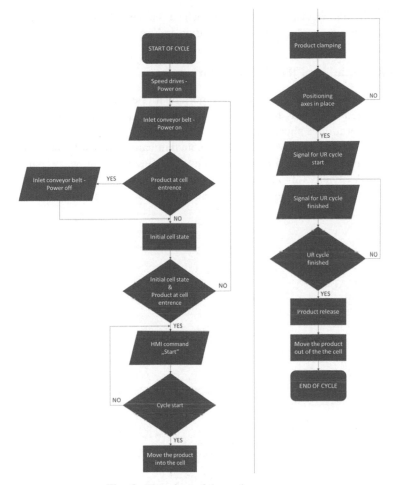

Fig. 6. Flowchart of the main program

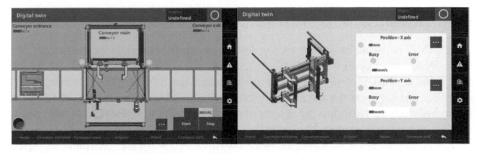

Fig. 7. HMI main menu (left) and clamp operation menu (right)

of Fig. 7 is the clamp visualization menu, which displays the desired position values, motion speed setting, and axis status (operation and error) of the clamping system.

2.4 Implementation of the URSim Environment

In the URSim environment we created an algorithm for performing motion and executing functions related to the UR robotic arm. An excerpt of the software algorithm for the robotic arm operation is shown in Fig. 8.

To define a virtual network that allows communication with software environments outside the virtual device in our case with the SIMIT environment, we prepared a virtual device in the VMware software environment. For communication, we created scripts that manipulate the input and output signals of the virtual controller of the simulated robotic arm.

Fig. 8. Excerpt of the software algorithm for the UR robotic arm

2.5 Implementation of the SIMIT Environment

In the SIMIT software environment, we defined motor drive models for the project. We have already mentioned that SIMIT also serves as a communication bridge between the graphical display in the MCD environment and the S7-PLCSIM Advance virtual controller and connects the URSim software environment with the system. In SIMIT it is also necessary to configure the project operating mode as asynchronous, synchronous or bus synchronous, whereby the latter is suitable to perform real-time simulations.

For communication with the S7-PLCSIM Advanced software environment, where the algorithm is loaded from the TIA Portal software environment, it is necessary to define the path to the project. We have chosen to connect SIMIT directly to the TIA

Portal project, because we run all software environments on the same computer. In this way, the data of the hardware defined in the project will be imported into SIMIT.

The modeling of the communication links between the TIA Portal and the MCD environment was performed using binary signals only. Figure 9 shows diagrams of the modeled communication between TIA Portal and MCD. Signals with green frames are defined as output signals connected to the input signal of another software environment marked with a red frame. In this way we have transferred the values of the limit switches and the signals for controlling the pneumatic cylinders.

To display the analogous values of the position of the robotic arm joints, we connected the output signals from the URSim software environment to the input signals of the MCD environment (Fig. 10). Thus, we achieved the mapping of the simulated movement of the robotic arm from the URSim to the MCD environment.

Fig. 9. Binary communication between MCD and TIA Portal

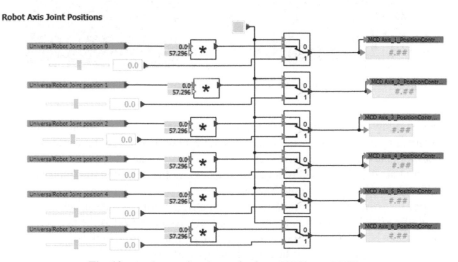

Fig. 10. Analogue signal transfer from URSim to MCD

3 Results

Verification of the commissioning of the quality control cell was performed in the digital twin to test the correctness of the control algorithm, the operation of the HMI, and to detect possible collisions. Figure 11 shows a virtual concept of the quality control cell on the left and the tested product, the kitchen oven, on the right.

At the beginning of the cycle, the control algorithm developed in TIA Portal performs the initialization of the cell with conveyor belts, entrance and exit safety doors, the clamping device, and the robotic arm. Using technology objects and motion blocks, the virtual power supply to the actuators is turned on and the input conveyor belt that transports the product to the entrance door is activated. When the product triggers the input position sensor, the conveyor stops, and the system waits for the input signal from the HMI to start the inspection cycle.

Manually starting the cycle triggers the opening of the cell entrance door, which is defined in the MCD environment as a displacement of the pneumatic cylinder or as a speed drive. After confirming the position of the opened door, the conveyor belts at the entrance and in the cell are started. When the product activates the position sensor in the cell at the approximate inspection location, both conveyors are stopped and the operation to close the entrance door is started. When the door is closed, the input conveyor delivering the new product to the entrance of the cell is started and the product positioning process begins.

Fig. 11. Virtual concept of the quality control cell (left) and the kitchen oven (right)

By starting the individual servo motors, the position of the product is precisely determined by means of pressure plates. First, the clamp moves to the working position and at the same time a pneumatic pressure plate starts to push the product to the desired position along the Y-axis. Then the clamping device closes the pressure plates to fix the position of the product along the X-axis.

After the positioning is complete, the simulated controller sends an activation signal through the SIMIT environment to the URSim environment to start the inspection cycle with the robot UR. An inspection of the operating knob, opening of the oven door, simulated galvanic inspection of the heating elements, and closing of the oven are performed. All robot movements are transferred from URSim to MCD, where a collision check is

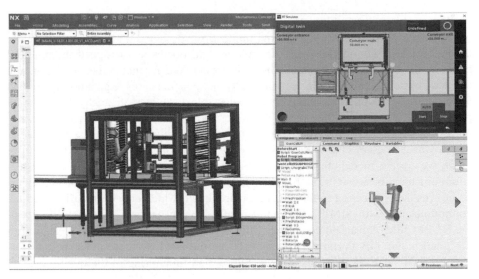

Fig. 12. Display of digital twin during simulation

performed. At the end of the cycle, the robot sends a signal to the controller for the completed cycle.

Once the inspection cycle is complete, the product is released by opening the clamps and retracting the simulated pneumatic cylinder to return the clamping device to its initial position. Then the exit door is opened, and the product is transported out of the cell by running the conveyor belts in the cell and at the cell exit. After the product leaves the cell, a new product is confirmed for entry.

Figure 12 shows the digital twin during operation. The window on the left side shows the graphical representation of the quality control cell in MCD. The upper right window shows the HMI with the main menu. The window in the lower right corner shows the URSim interface with the implementation of a motion algorithm.

4   Conclusion and Discussion

Increasingly sophisticated systems in the context of Industry 4.0 require sophisticated approaches to the development of industrial applications. With the help of digital twins, the development approach focuses primarily on digital knowledge that supports development in various phases, from individual components to structures to the virtual commissioning of sophisticated systems. The methods allow engineers to identify critical elements of the system at very early stages of development and work on them accordingly at the right time. Such a development approach enables the modularity of the system with the possibility of adding the desired functions. In addition to the digital system models, we also obtain the necessary control algorithms for managing industrial applications, which allows easier and faster implementation of software solutions in a real environment. At the same time, real commissioning is more predictable and faster compared to classical development methods.

In our project, we developed a digital twin for the commissioning of a robotic quality control cell. In the MCD software environment, we designed the cell with physically defined models. The models were defined as rigid bodies and assigned the required behavioral properties in the simulation environment. When running the simulation, the defined physical elements behave realistically, but in some cases (e.g., a poorly defined 3D model), the system may begin to behave unstable due to unrealistically defined mass properties and inertia of the model. We were able to solve this problem by manually correcting the inertia of the rigid bodies. For communication, we created a signal table, which was then connected to the SIMIT software environment. In the SIMIT environment, we created models that simulated the operation of motor drives and established communication between the software environments. The only problem we had to overcome was the connection to the URSim environment, which is not primarily listed for communication with SIMIT. The entire quality control process was controlled with the S7-PLCSIM Advanced controller, which cyclically executed a software algorithm designed in the TIA Portal environment. Using the HMI, we have developed a user-friendly graphical interface that allows control and monitoring of the quality control cycle execution.

Despite efforts to optimize the definition of some functions of the system, the processing complexity of the implementation of software environments in the simulation state is a major challenge for a standard personal computer. Using high computational accuracy of technology objects can also cause processor power to be exhausted during execution, causing execution cycles to fail or to be delayed. This was most evident in URSim - SIMIT - MCD RTDE (real time data exchange) communication, where the robot arm motion is executed in real time. The result was slower execution of the robot motion in the MCD environment as dictated by the URSim software environment, leading to motion errors.

Overall, the digital twin was successfully implemented and tested, and valuable insights into the usability of digital twins were gained. It was shown that with the right software environments, it is possible to simulate complex systems very accurately and optimize them for later implementation in real environments.

Acknowledgements. The authors thank the Slovenian Ministry of Higher Education, Science and Technology and the Slovenian Research Agency (Research Core Funding No. P2-0157) for financial support that made this work possible. The authors also acknowledge financial support from the ROBKONCEL project (OP20.03530).

References

1. Singh, M., Fuenmayor, E., Hinchy, E., Qiao, Y., Murray, N., Devine, D.: Digital twin: origin to future. Appl. Syst. Innov. **4**, 36 (2021). https://doi.org/10.3390/asi4020036
2. Shao, G., Helu, M.: Framework for a digital twin in manufacturing: Scope and requirements. Manuf. Lett. **24**, 105–107 (2020). https://doi.org/10.1016/j.mfglet.2020.04.004
3. Malakuti, S., Schlake, J., Ganz, C., Harper, K.E., Petersen, H.: Digital Twin: An Enabler for New Business Models (2019)
4. Jin, T., et al.: Triboelectric nanogenerator sensors for soft robotics aiming at digital twin applications. Nat. Commun. **11**(1), 5381 (2020). https://doi.org/10.1038/s41467-020-19059-3

5. Zheng, Y., Yang, S., Cheng, H.: An application framework of digital twin and its case study. J. Ambient Intell. Hum. Comput. **10**(3), 1141–1153 (2019). https://doi.org/10.1007/s12652-018-0911-3
6. Park, K.T., et al.: Design and implementation of a digital twin application for a connected micro smart factory. Int. J. Comput. Integr. Manuf. **32**(6), 596–614 (2019). https://doi.org/10.1080/0951192X.2019.1599439
7. Kuehner, K.J., Scheer, R., Strassburger, S.: Digital twin: finding common ground – a meta-review. Procedia CIRP **104**, 1227–1232 (2021). https://doi.org/10.1016/j.procir.2021.11.206
8. Tao, F., Cheng, J., Qi, Q., Zhang, M., Zhang, H., Sui, F.: Digital twin-driven product design, manufacturing and service with big data. Int. J. Adv. Manuf. Technol. **94**(9), 3563–3576 (2018) https://doi.org/10.1007/s00170-017-0233-1
9. Schroeder, G.N., Steinmetz, C., Pereira, C.E., Espindola, D.B.: Digital twin data modeling with AutomationML and a communication methodology for data exchange. IFAC-PapersOnLine **49**(30), 12–17 (2016). https://doi.org/10.1016/j.ifacol.2016.11.115
10. Liu, M., Fang, S., Dong, H., Xu, C.: Review of digital twin about concepts, technologies, and industrial applications. J. Manuf. Syst. **58**, 346–361 (2021). https://doi.org/10.1016/j.jmsy.2020.06.017
11. Adamenko, D., Kunnen, S., Nagarajah, A.: Comparative analysis of platforms for designing a digital twin. In: Ivanov, V., Trojanowska, J., Pavlenko, I., Zajac, J., Peraković, D. (eds.) Advances in Design, Simulation and Manufacturing III, pp. 3–12. Springer International Publishing, Cham (2020)
12. Lim, K.Y.H., Zheng, P., Chen, C.-H.: A state-of-the-art survey of digital twin: techniques, engineering product lifecycle management and business innovation perspectives. J. Intell. Manuf. **31**(6), 313–1337 (2020). https://doi.org/10.1007/s10845-019-01512-w
13. Phanden, R.K., Sharma, P., Dubey, A.: A review on simulation in digital twin for aerospace, manufacturing and robotics. Mater. Today: Proc. **38**, 174–178 (2021). https://doi.org/10.1016/j.matpr.2020.06.446
14. Virtual Commissioning and Operator Training with SIMIT: https://new.siemens.com/global/en/products/automation/industry-software/simit.html. Accessed 7 Dec 2021
15. Siemens PLM Software Website: Digital Twin: https://www.plm.automation.siemens.com/global/en/our-story/glossary/digital-twin/24465. Accessed 7 Dec 2021
16. Mechatronic Concept Design: Siemens Software: https://www.plm.automation.siemens.com/global/en/products/mechanical-design/mechatronic-concept-design.html. Accessed 7 Dec 2021
17. SIMATIC S7-PLCSIM Advanced V3.0 – Industry Support Siemens: https://support.industry.siemens.com/cs/document/109772889/trial-download-simatic-s7-plcsim-advanced-v3-0?dti=0&lc=en-WW. Accessed 7 Dec 2021
18. Totally Integrated Automation Portal: Automation Software – Siemens Global: https://new.siemens.com/global/en/products/automation/industry-software/automation-software/tia-portal.html. Accessed 7 Dec 2021
19. Martinez, S., et al.: A digital twin demonstrator to enable flexible manufacturing with robotics: a process supervision case study. Prod. Manuf. Res. **9**(1), 140–156 (2021). https://doi.org/10.1080/21693277.2021.1964405
20. Collaborative Robotic Automation: Cobots from Universal Robots: https://www.universal-robots.com/. Accessed 7 Dec 2021

Selective Surface Modification of Complexly Shaped Steel Parts by Robot-Assisted 3D Scanning Laser Hardening System

Dmytro Lesyk[1,3,4(✉)], Matej Hruska[2], Vitaliy Dzhemelinkyi[1], Oleksandr Danyleiko[1], and Milan Honner[2]

[1] Laser Systems and Advanced Technologies Department, National Technical University of Ukraine "Igor Sikorsky Kyiv Polytechnic Institute", Kyiv 03056, Ukraine
lesyk_d@ukr.net
[2] New Technologies Research Centre, University of West Bohemia, Pilsen, Czech Republic
[3] Principles for Surface Engineering Department, G.V. Kurdyumov Institute for Metal Physics of the NAS of Ukraine, Kyiv 03142, Ukraine
[4] Mechanical Engineering and Mechatronics Department, West Pomeranian University of Technology, Szczecin 71899, Poland

Abstract. Laser surface hardening is one of the most advanced surface modification techniques to increase the wear resistance of large-sized and complexly shaped metal products. In this study, the laser transformation hardening process for the high-quality surface treatment of the steel products is applied using a high-power disc laser with extremely good beam quality and three-dimensional (3D) scanning optics. The shaft AISI 1066 steel part was selectively processed by the robot-based laser hardening system to increase the surface hardness. At the same time, such a computer numerical control (CNC) laser system is ideal for remote surface treatment of complexly shaped metal products. The experimental tests with a solid-state disc laser of a maximum power of 5.3 kW were performed with a constant power strategy. Both plane and cylindrical areas on the shaft were hardened and compared. The results showed that the hardness values on the plane surfaces correlate well with the hardness values on the cylindrical surfaces. The hardening intensity was about 2.5 times higher than that of the unhardened carbon steel shaft.

Keywords: Shaft AISI 1066 steel part · Laser transformation hardening process · Disc laser · Robot-based 3D scanning · Surface hardness · Hardening intensity

1 Introduction

Currently, it is well-known that the use of expensive wear-resistant and corrosion-resistant materials is not economically rational to solve the problem of increasing the reliability and durability of specific metal products. To significantly improve the performance of structural metal parts are the most promising methods of surface hardening by changing the structure of the subsurface layer without changing the chemical composition and surface roughness of the processed surface [1, 2].

© The Author(s), under exclusive license to Springer Nature Switzerland AG 2022
I. Karabegović et al. (Eds.): NT 2022, LNNS 472, pp. 30–36, 2022.
https://doi.org/10.1007/978-3-031-05230-9_3

The method of increasing the durability of structural steel products due to selective surface hardening of their working areas using laser radiation is becoming extremely relevant and important in mechanical engineering. A stimulating factor in the development of the laser surface hardening techniques is the appearance on the market of modern powerful solid-state lasers (diode, fiber, and disc lasers) with a higher total level of technical and economic indicators as compared to the conventional lasers [3–6]. CO_2 lasers are used for decades as the standard tool for processing an enormous variety of materials. Compared to the radiation of CO_2 lasers, an important feature of solid-state lasers is that the laser radiation with a wavelength of about 1 μm interacts better with the surface of metals [4]. The continuous-wave lasers combined with scanning optics provide high processing performance, high uniformity of hardening, and allow processing of complexly shaped profiles.

By analogy with other types of hardening, the laser surface hardening of steels and alloys consists of an austenitic structure formation during high-speed heating and its subsequent transformation to martensite structure during cooling by absorbing and transferring of high concentration energy to a thin surface layer [7–9]. The heating and cooling rates are in a range of $10^8...10^{10}$ °C/s, which allow forming of special microstructures and mechanical properties of the selectively hardened surfaces. The layered hardened zone is formed in the near-surface layer [10]. The laser heat treatment (LHT) significantly improves the hardness and wear resistance of the materials [11–13]. In particular, Park et al. [11] studied the effect of the high-power diode-laser-assisted heat treatment on the fretting wear behavior of the AISI P20 steel. The surface hardness was increased by 128% after the LHT treatment of the mold steel. As a result, the coefficient of friction and wear volume magnitudes were smaller than in the untreated specimen. The wear resistance of the bearing steel was doubled by the CO_2 laser [12].

Recently, the surface properties and microstructures formed by the CO_2 [14], diode [15], and fiber [16] lasers are widely studied. Application of the diode lasers and fiber lasers combined with scanning optics result in deeper/wider hardening depth/width and homogeneous structure in the near-surface layers [15]. The high-power disc lasers with extremely good beam quality can be used for the surface treatment of complexly shaped metal products [17–19]. At the same time, the programmable focusing optics coupled with computer numerical control (CNC) robotic equipment allows performing remote hardening tasks within a three-dimensional (3D) workspace. A study made by Schuocker [18] revealed that the hardening depth was achieved up to 3 mm in the structural steel specimen with a thickness of 10 mm after the LHT treatment using a disc laser combined with scanner optics. The advantages of the 3D scanning laser hardening technique for the surface treatment of the complexly shaped steel parts were stated in the work [17].

Nevertheless, it should also be noted that the LHT parameters using the disc laser systems should be optimized for surface hardening of the carbon and tool steel parts. In this work, the steel shaft of the reduction gear was selectively hardened by a high-power disc laser to increase the surface hardness in the responsible working areas.

The aim of this work is to study the surface hardness and hardening intensity of the AISI 1066 steel shaft processed by the robot-based laser hardening system with the high-power disc laser and 3D scanning optics.

2 Experimental Procedures

The shaft AISI 1066 steel parts (285 mm × Ø25 mm) were applied to increase their durability and reliability (Fig. 1).

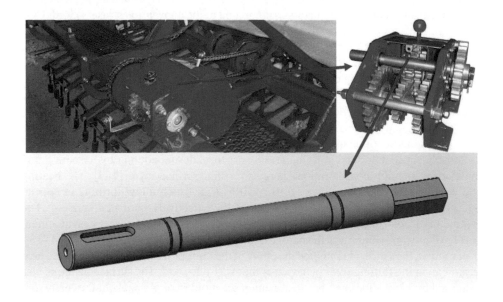

Fig. 1. 3D model of the shaft of the reduction gear

The studied shafts are used in the reduction gear of the seeders. The nominal chemical composition of the AISI 1066 carbon steel is given in Table 1.

Table 1. Chemical composition and initial surface macrohardness of the shaft AISI 1066 steel part in wt.%

Chemical composition (wt.%)									Initial HV_5
Name	C	Mn	Si	Cr	Cu	P	S	Fe	
AISI 1066	~0.6	~1.0	~0,25	≤0,25	≤0,2	≤0,035	≤0,035	Bal	~200

The laser heat treatment (LHT) of the cylindrical and plane (right side of the shaft, Fig. 1) surface areas on the shaft were performed using a Trudisk 8002 disk laser (a wavelength of laser radiation is 1.03 μm, the maximum laser power is 5.3 kW) and SCANLAB intelliWELD 30 FC V scanning optics (a focal length is 460 mm, the working distance is 488 mm) mounted in the industrial Fanuc M-710iC robot with numerical program control (CNC) as shown in Fig. 2.

The laser beam is fiber-delivered (200 μm in diameter) to the scanning head water-cooled collimator and then directed to the scanning head moving deflection mirrors. The scanning head is able to process an elliptical (cuboid-shaped) image field of dimensions 385×300 mm^2 ($220 \times 220 \times 140$ mm^2), the maximum laser beam deflection speed is 21.5 m/s. The scanning system collimation in Z-axis is allowed in the range of ±70 mm. The maximum laser power with a specified cooling is 8.0 kW. The scanning process was controlled by a RobotSync Unit or by SAMLight software [17].

The rotary positioner (axis B) was integrated into the laser hardening system for the processing of shaft parts (Fig. 2).

Fig. 2. Laser surface hardening of the AISI 1066 steel shaft

The LHT treatment was conducted in a range of laser power of 1.35...2.25 kW (Table 2). The obtained values of the energy density of the laser beam fall into the range between 4.5...7.5 J/mm^2. The scanning speed and processing time (shaft rotational speed or scanner feed rate) were constant. The laser tracks 10 mm wide on the shaft surface were produced at a spot size of 0.8 mm and a scanning speed of 20000 mm/s.

Table 2. Laser heat treatment parameters

Regime marking	Laser power, P (W)	Scanning speed, V (mm/s)	Processing speed, S (mm/s)	Energy density of laser beam, E (J/mm^2)
LHT1	1350	20000	11	~4.5
LHT2	1650			~5.5
LHT3	1950			~6.5
LHT4	2250			~7.5

The surface macrohardness of the shaft parts was measured using a GE MIC 20 TFT tester at a load on Vickers indenter of 5 kgf (50 N). In all cases, a total of ten measurements were conducted at different treated areas and the averaged values were reported. The scatter of the experimental data did not exceed by 5%. The intensity of surface hardening was evaluated by the HV magnitudes of surface hardness of the initial and hardened shaft by the following expression $I_{hard} = (HV_{hard} - HV_{in})/HV_{in} \cdot 100\%$.

3 Results

The surface macrohardness values of the shaft steel part are presented in Fig. 3. The LHT experiments showed that as compared to the unhardened areas (~200 HV_5), the surface hardness increased with ongoing LHT treatment. The surface hardness was enlarged by about 2.5 times after the LHT3 and LHT4 treatments. This is attributed to the formation of a fine-grained martensite structure. The observed surface hardness values and tendencies correlate well with the literature data [17, 18].

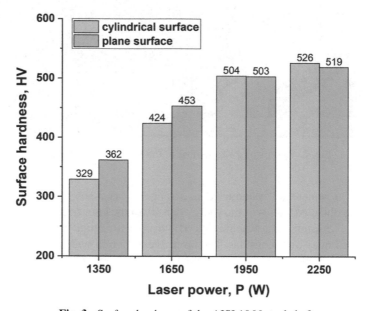

Fig. 3. Surface hardness of the AISI 1066 steel shaft

According to the results of hardening intensity (Fig. 4), it can be seen that the determined hardening intensity values of the LHT-processed areas were respectively increased by about 70%, 115%, 150%, and 155% after the LHT1, LHT2, LHT3, and LHT4 regimes.

It should also be noted that the HV magnitudes on the plane surfaces correlate well with the HV magnitudes on the cylindrical surfaces. Additionally, the hardness saturation (500...525 HV_5) is observed with a high laser power magnitude (1950...2250 W) (Fig. 3 and Fig. 4). As a result, it is recommended to apply the LHT3 regime to provide a hardened surface without the melting surface.

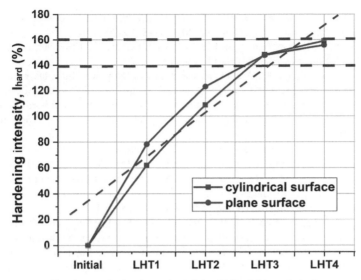

Fig. 4. Hardening intensity of the AISI 1066 steel shaft

4 Conclusion

The laser heat treatment (LHT) method is an advanced thermal surface treatment for improving the surface hardness and wear resistance of metal parts. The results of this study can be summarized as follows:

1. The shaft AISI 1066 steel parts were selectively treated by the robot-based laser hardening system consisted of the advanced high-power disc laser and scanning optics.
2. Based on the LHT experiments, the LHT3 regime ($P = 1950$ W) should be applied to carry out the laser surface hardening of both cylindrical and plane AISI 1066 steel surfaces.
3. The surface hardness magnitudes of the LHT3-hardened areas were increased by about 2.5 times as compared to the untreated specimen (~200 HV_5).
4. The study of the microstructure, hardening depth, and microhardness distribution in the near-surface layers of the LHT-hardened AISI 1066 steel shaft is planned in further research.

Acknowledgements. This work was supported by the Ministry of Education, Youth and Sports of the Czech Republic and the Ministry of Education and Science of Ukraine (Project No. 0122U002389).

References

1. Vollertsen, F., Partes, K., Meijer, J.: State of the art of laser hardening and cladding. In: Proc. Materials of Third Int. WLT-Conf. Lasers in Manuf., pp. 783–792 (2005)

2. Montealegre, M.A., Castro, G., Rey, P., Arias, J.L., Vázquez, P., González, M.: Surface treatments by laser technology. Contemp. Mater. **I–1**, 19–30 (2010)
3. Dinesh Babu, P., Balasubramanian, K.R., Buvanashekaran, G.: Laser surface hardening: a review. Int. J. Surf. Sci. Eng. **5**, 131–151 (2011)
4. Li, R., Jin, Y., Li, Z., Qi, K.: A comparative study of high-power diode laser and CO_2 laser surface hardening of AISI 1045 steel. J. Mater. Eng. Perform. **23**(9), 3085–3091 (2014)
5. Lee, K.-H., Choi, S.-W., Yoon, T.-J., Kang, C.-Y.: Microstructure and hardness of surface melting hardened zone of mold steel, SM45C using Yb:YAG disk laser. J. Weld. Join. **34**(1), 75–81 (2016)
6. Martínez, S., Lesyk, D., Lamikiz, A., Ukar, E., Dzhemelinsky, V.: Hardness simulation of over-tempered area during laser hardening treatment. Phys. Procedia **83**, 1357–1366 (2016)
7. Santhanakrishnan, S., Kong, F., Kovacevic, R.: An experimentally based thermo-kinetic phase transformation model for multi-pass laser heat treatment by using high power direct diode laser. The Int. J. Adv. Manuf. Technol. **64**(1–4), 219–238 (2013)
8. Lesyk, D., Martinez, S., Mordyuk, B., Dzhemelinskyi, V., Danyleiko, O.: Effects of the combined laser-ultrasonic surface hardening induced microstructure and phase state on mechanical properties of AISI D2 tool steel. In: Ivanov, V., Trojanowska, J., Machado, J., Liaposhchenko, O., Zajac, J., Pavlenko, I., Edl, M., Perakovic, D. (eds.) DSMIE 2019. LNME, pp. 188–198. Springer, Cham (2020). https://doi.org/10.1007/978-3-030-22365-6_19
9. Kennedy, E., Byrne, G., Collins, D.N.: A review of the use of high power diode lasers in surface hardening. J. Mater. Process. Technol. **155–156**, 1855–1860 (2004)
10. Chen, C., Zeng, X., Wang, Q., Lian, G., Huang, X., Wang, Y.: Statistical modelling and optimization of microhardness transition through depth of laser surface hardened AISI 1045 carbon steel. Opt. Laser Technol. **124**, 105976 (2020)
11. Park, C., Kim, J., Sim, A., Sohn, H., Jang, H., Chun, E.-J.: Influence of diode laser heat treatment and wear conditions on the fretting wear behavior of a mold steel. Wear **434–435**, 202961 (2019)
12. Lei, S., Liu, Q.K., Liu, Y.P., Li, H.: Wear behavior of laser-hardened GCr15 steel under lubricated sliding conditions. Materials Science Forum **628–629**, 697–702 (2009)
13. Pellizzari, M., De Flora, M.G.: Influence of laser hardening on the tribological properties of forged steel for hot rolls. Wear **271**(9–10), 2402–2411 (2011)
14. Orazi, L.: Experimental investigation on a novel approach for laser surface hardening modelling. Int. J. Mech. Mater. Eng. **16**(2), 1–10 (2021). https://doi.org/10.1186/s40712-020-001 24-0
15. Hagino, H., Shimizu, S., Ando, H., Kikuta, H.: Design of a computer-generated hologram for obtaining a uniform hardened profile by laser transformation hardening with a high-power diode laser. Precis. Eng. **34**(3), 446–452 (2010)
16. Tarchoun, B., El Ouafi, A., Chebak, A.: Experimental investigation of laser surface hardening of AISI 4340 steel using different laser scanning patterns. J. Mineral. Mater. Charact. Eng. **8**, 9–26 (2020)
17. Hruška, M., Vostřák, M., Smazalová, E., Svantner, M.: 3D scanning laser hardening. In: Proc. Materials of the 23rd Int. Conf.: Metal. Mater., Metal 2014, pp. 921–926. (2014)
18. Schuocker, D., Aichinger, J., Majer, R., Spitzer, O., Rau, A., Harrer, Th.: Improved laser hardening process with temperature control avoiding surface degradation. In: Proc. Materials of the 8th Int. Conf. on Photonic Technol., LANE 2014, pp. 1–5. (2014)
19. Hruška, M., Vostřák, M., Smazalová, E., Švantner, M.: Standard and scanning laser hardening procedure. In: Proc. Materials of the 22nd Int. Conf.: Metal. Mater., Metal 2013. (2013)

Ambient Light and Object Color Influence on the 3D Scanning Process

Kenan Varda[1] (iD), Ernad Bešlagić[1] (iD), Nermina Zaimović-Uzunović[1] (iD), and Samir Lemeš[2](✉) (iD)

[1] Mechanical Engineering Faculty, University of Zenica, Fakultetska 1, 72000 Zenica, Bosnia and Herzegovina
[2] Polytechnic Faculty, University of Zenica, Fakultetska 1, 72000 Zenica, Bosnia and Herzegovina
samir.lemes@unze.ba

Abstract. In 3D scanning of objects with different 3D scanners, two of the factors that most affect the quality of the results obtained are the lighting conditions and the colour of scanned objects. This paper presents the influence of light and colour of objects scanned using two different 3D scanners, RangeVision PRO and Artec Eva. To estimate the effect of these factors, 3D printed models of NACA aero profiles in two different colours, black and white and in four different scenarios of the object's brightness during scanning were used. The brightness values for all scenarios are shown numerically, and the number of registered triangular surfaces on the scans was taken as a relevant result. A rotary table was used in the scanning process. It enabled the scanning of models from all sides of the object by rotating for an angle of 360°.

Keywords: 3D scanning · 3D printing · Brightness influence · Colour influence

1 Introduction

As 3D scanning becomes more popular and widely used, more and more authors investigate the influence of different parameters' impact on the 3D scanning process.

Blanco and others in [1] dealt with how different light sources affect the quality of the digitized surface. The test was performed on a flat surface of EN AW 6082 aluminium. The study found that the influence of ambient light on the laser surface triangulation affects the quality of digitized point clouds, where sodium is present low-pressure luminaires performed best with minor cloud saturation points. Koseoglu et al. [2] examined the effect of light on intraoral scanning in which marked lighting conditions with room lighting at 1003 lx and light without illumination at 0 lx, using two scanning methods with blue and white light. By testing, they proved that lighting conditions are essential when scanning and recommended lighting conditions in-room lighting using blue light, i.e. RLB. Feng Li and others in [3] also examined the effect of light on scanning, using a scanner with structured light GOM ATOS III with light variation at 660, 280, and 0 lx (with varying ±20 lx). Experimental results, however, show that

I. Karabegović et al. (Eds.): NT 2022, LNNS 472, pp. 37–45, 2022.
https://doi.org/10.1007/978-3-031-05230-9_4

ambient light, in this case, does not contribute to some obvious systematic errors in terms of accuracy. Lemeš and Zaimović-Uzunović in [4] investigated a relationship between the ambient light intensity, the colour of a scanned surface, and the laser scanning quality on a low-cost 3D scanner. They concluded that the influence is especially strong when scanning glossy white, yellow and green surfaces. Kersten et al. showed that the evaluated systems achieved different accuracies for the test bodies in 3D compared to the references system and for determining parameters following the guideline VDI/VDE 2634 using stable reference bodies [5]. Revilla-León et al. in [6] observed significant differences in the trueness and precision values across different lighting conditions. Hsu et al. analyzed the 3D scanning process used to classify products in the production line [7]. They demonstrated that the typical lighting conditions presented in companies and industrial environments could affect target objects' classification and determine their positions. Singh and Nagla in [8] experimentally found that the variation in the measured distance by the 3D scanner in the dark is more consistent and accurate than the sensory information taken in high-intensity tube light/halogen bulbs and sunlight.

To perform this research, we considered that the 3D scanning process is performed on an object model that is interesting for this type of measurement. Objects with complex geometry that do not contain many elements of simple shapes are most often measured using a 3D scanner. We chose to scan the model of symmetrical NACA 0015 vertical wind turbine profile as a case study. The model was created for testing vertical wind turbines in a wind tunnel. The model's dimensions could not be easily acquired in reverse engineering, one of the main applications of 3D scanners. The length of the tendon profile for both blades is 60 mm. The blade models were made in SolidWorks software. Blade segments with a total height of 60 mm were modeled with the addition of a blade base to facilitate the processing of scans after scanning, as shown in Fig. 1.

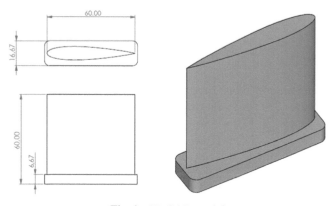

Fig. 1. 3D CAD model

After creating a 3D CAD model of benchmark parts, we used it to make the physical model using the 3D printer Ultimaker S5, which uses PLA material to print objects. One black and one white benchmark part were created using the same printing regime and the same material, assuming that the objects' geometry was identical (Fig. 2). It is important to emphasize that in the scanning process performed to examine the impact of the factors mentioned above, the deviations from dimensions and form were neglected, as they were not the primary targets of this research.

Fig. 2. 3D printed benchmark parts

2   3D Scanning and Measuring Equipement

We used two different scanners for 3D scanning. The RangeVision pro stationary scanner is a professional optical 3D scanner designed to scan complex objects. It is ideal for engineering when it comes to industrial design. This device has the exceptional precision provided by two 6 MP cameras. As it uses structured light technology, the professional RangeVision 3D scanner delivers high-quality scanning of any object, from jewellery to more oversized items, with an accuracy of up to 0.018 mm and a resolution of up to 0.04 mm. RangeVision PRO can quickly scan objects of different shapes and sizes by changing the scanner field (FOV) (Fig. 3).

Artec Eva is a lightweight handheld 3D scanner ideal for fast and high-quality textured and precise 3D models. Features such as lightness, speed and versatility, Eva is Artec's most popular scanner designed to scan medium and large objects (larger than 100 mm). Scanning is quick, recording accurate measurements in high resolution (Fig. 3).

Detailed specifications of both scanners are shown in Table 1.

As previously mentioned, we analyzed the influence of scanned objects' colour and ambient light on the scanned result's quality. A light intensity measuring device was used to measure illuminance. A generally accepted presumption is that brightness and light intensity are critical factors in 3D scanning. We used a multi-purpose device that can measure temperature, relative humidity, light and sound intensity to measure light intensity. It is an environmental meter type PCE-EM 882 (Fig. 4).

We used the standard LED bulbs to vary the light intensity. Their combination provided better illumination of the 3D scanner workspace.

Fig. 3. 3D scanners used for scanning (RangeVision PRO left, Artec EVA right)

Table 1. Detailed 3D scanners specifications

Specification	RangeVision PRO	Artec EVA
Dimensions of the scanning module (mm)	408 × 380 × 125	Starting from 100
3D resolution (mm)	0.25, 0.13, 0.05	Up to 0.2
Output formats	STL, OBJ, PLY, ASCII	STL, OBJ, PLY, BTX
3D accuracy (mm)	0.06 to 0.018	Up to 0.1
Scanning principle	Structured light	Structured light
Light source	LED (blue light)	Flash bulb/White 12 LED array
Scanning area (mm)	520 × 390 × 360; 280 × 210 × 200; 115 × 85 × 80	214 × 148 to 536 × 371
Scanning distance (mm)	900, 520, 260	400 – 1000

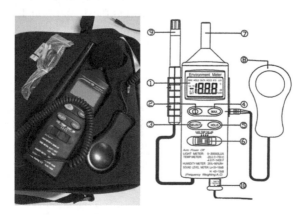

Fig. 4. PCE-EM 882 design and components

3   Scenarios and 3D Scanning Process

Before the scanning process, we first prepared a scanning room and checked the equipment conditions. In the preparation of four various scanning scenarios, the intensity of the light was varied. In the scanning scenarios marked with numbers 1 and 2, we used the standard room lighting. We used only half of the available lamps in scenario 1and all available room lamps were used in scenario 2. We used additional LED lamps in scenarios 3 and 4 to increase the brightness. The numerical value of illumination for each scenario is expressed in lux and shown in Table 2.

Table 2. Illumination values for 3D scanning scenarios

Name	Lux
Scenario 1	80 ÷ 90
Scenario 2	170 ÷ 200
Scenario 3	700 ÷ 720
Scenario 4	1290 ÷ 1310

3.1   RangeVision PRO Scanning Process

After defining the scenarios, we scanned the black and white blade profiles using the RangeVision PRO scanner. The scanner has three pairs of lenses for scanning objects of different sizes. For the benchmark part we used in this paper, we selcted a set of lens number 2. When selecting the mode of operation of this scanner, the options of the process itself are set in the Scan centre NG software. For this case, we selected the scanning using a rotary table on which a benchmark part is placed, which was calibrated using a predefined axis evaluation plate. Four scan positions were selected in the scan parameters, with a rotation of 90° for each of the positions, simplifying the level of scans to a mean value of 2 out of 3, and the projector's light type is cross.

The scanning process was controlled by software in the Scan centre NG environment, with changes in the lighting regime and the benchmark part in the scanning phase of the black object. Figure 5 shows one of the scenarios within the Scan centre NG software.

We scanned the black benchmark part analogous to scanning the white benchmark part in the same four scenarios. We noticed that the RangeVision PRO scanner does not correctly detect the black object on the rotary table in the preparation phase. We increased the exposure value under the 3D scanner options to avoid this negative phenomenon. However, by varying both the exposure values and the lighting of the room, in all four cases, the scanner failed to detect a single triangular surface on the black object. According to the manufacturer's instructions, scanning shiny objects is an additional aggravating circumstance with this scanner. The PLA material used by the 3D printer and from which the benchmark parts were made is shiny and reflective.

The number of registered triangular surfaces for the white and black benchmark part scanned on the RangeVision PRO is shown in Table 3.

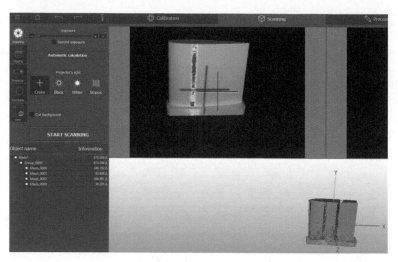

Fig. 5. Scan centre NG software environment in process of scanning

Table 3. Number of triangular surfaces registered in RangeVision 3D scanner scanning process

Number of triangular surfaces (RangeVision PRO)		
Scenario	White benchmark part	Black benchmark part
Scenario 1	815 264	0
Scenario 2	829 107	0
Scenario 3	849 683	0
Scenario 4	829 905	0

3.2   Artec EVA Scanning Process

Unlike the RangeVision PRO scanner, which requires setting more parameters and performing calibration, the Artec Eva scanner software interface is straightforward and the user can adjust only a few parameters. Before the scanning process in the Artec studio 15 software settings, the user has to select the scan display option, in this case "Geometry + Texture" and set the scanner scan speed to the maximum resolution of 16 frames or pictures per second. Pressing the "play" button on the Artec Eva scanner in the software opens an overview that allows orientation and shows that the scanner is directed towards the object to be scanned. The green scale from 400 to 1000 shows the distance of the scanner from the object in millimetres. To achieve the best scan results, it is necessary to keep the distance of the scanner from the scanned object in the middle of this scale.

Scanning with Artec Eve requires the user to hold the scanner in hand and move around the static scanned object. However, it is possible to use a turntable like the RangeVision PRO scanner, where the object is placed on a turntable and then rotated in our example by 360°. In this case, the scanner is static, and in this way, we tried to equalize conditions scanning with both scanners.

After the scan parameters were set, the same four scenarios were scanned as in the scan with the RangeVision PRO scanner and under the same ambient lighting. Eight scans were performed for each scenario and for both the white and the black benchmark part. Figure 6 shows one of the scanning scenarios using the Artec EVA scanner within the Artec studio 15 software environment.

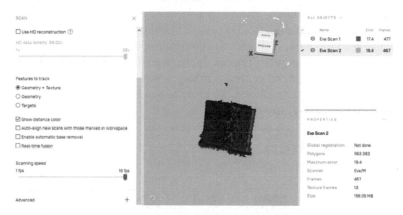

Fig. 6. Artec studio 15 software environment in process of scanning

After the scan, the right side of the screen provides information on the number of registered polygons during each scan. A polygon is a 3D mathematical construction made of three or more points that have their x, y and z coordinates in space. Table 4 shows an overview of the number of registered polygons for each white and black benchmark scenario.

Table 4. Number of polygons registered in Artec EVA scanner scanning process

Number of polygons (Artec EVA)		
Scenario	Black benchmark part	White benchmark part
Scenario 1	563 383	963 420
Scenario 2	519 043	1 053 072
Scenario 3	368 580	1 059 597
Scenario 4	430 677	1 068 856

4   Conclusion

In general, we can conclude that both used scanners have proven to be a good solution for fast and accurate obtaining of the geometry of a three-dimensional object. In the scanning process, the advantages and disadvantages of both scanners concerning both factors that were taken into account when designing this experiment are visible.

When scanning with a RangeVision PRO scanner, the results registered with the white benchmark part are reasonably uniform. They do not have significant discrepancies, which shows that different lighting does not significantly affect the quality and results of the scan. An interesting phenomenon occurred in scenario number 4 when the highest ambient lighting was available. There was a decrease in the registered elements for the white benchmark part, indicating that excessive lighting for such a building is unsuitable. It is clear that scenario number 3, with the illumination of 718 lx, gave the best results in a number of registered triangular surfaces. In the case of the black benchmark part, there was no registration of elements during scanning, which indicates that we have to avoid shiny black objects with this scanner. A graphical representation of the scan results for the RangeVision PRO is shown in Fig. 7.

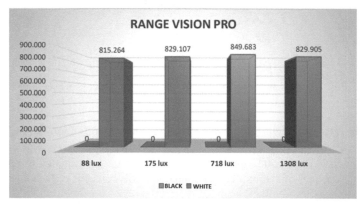

Fig. 7. Graphical representation of RangeVision PRO 3D scanning

With the Artec Eva scanner, scanning a black object was successful, which is a good indicator that the scanner is more suitable for scanning darker objects. The graphical representation of the results in Fig. 8 shows that for scanning a black object, the influence of ambient light is noticeable with a difference of 194 803 polygons, which is quite a large number. In this case, since the black object absorbs light, the number of polygons decreases with increasing light. In this example, the best illumination condition for scanning a black object is Scenario 1, with a low light usage level of 86.2 lx. The number of polygons registered at that level is 563 383, while the worst scan was shown in scenario 3 at 704 lx with 368 580 registered polygons. When scanning a white object, the Artec Eva scanner showed the registration of a more significant number of polygons compared to scanning a black object. The influence of ambient light is noticeable with a difference of 105 436 polygons, which is less than the difference with a black object.

The general recommendation would be that both scanners in a wide range of different lighting can be used for white objects. For black items, Artec EVA has proven to be a good solution. In the case of the RangeVision scanner, some future research could show how black objects can be treated so that scanning can be performed.

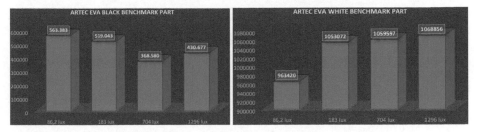

Fig. 8. Graphical representation of Artec EVA 3D scanning

References

1. Blanco, D., Fernández, P., Valiño, G., Rico, J.C., Rodríguez, A.:. Influence of ambient light on the repeatability of laser triangulation digitized point clouds when scanning EN AW 6082 flat faced features. In: AIP Conference Proceedings, vol. 1181, no. 1, pp. 509–520. American Institute of Physics (2009). https://doi.org/10.1063/1.3273669
2. Koseoglu, M., Kahramanoglu, E., Akin, H.: Evaluating the effect of ambient and scanning lights on the trueness of the intraoral scanner. J. Prosthodont. **30**, 811–816 (2021). https://doi.org/10.1111/jopr.13341
3. Li, F., Stoddart, D., Zwierzak, I.: A performance test for a fringe projection scanner in various ambient light conditions. Procedia CIRP **62**, 400–404 (2017). https://doi.org/10.1016/j.procir.2016.06.080
4. Lemeš, S., Zaimović-Uzunović, N.: Study of ambient light influence on laser 3D scanning. In: 7th International Conference on Industrial Tools and Material Processing Technologies ICIT & MPT, Ljubljana (2009)
5. Kersten, T.P., Lindstaedt, M., Starosta, D.: Comparative geometrical accuracy investigations of hand-held 3D scanning systems-an update. Int. Arch. Photogramm. Remote Sens. Spat. Inf. Sci. **42**(2), 487–494 (2018). https://doi.org/10.5194/isprs-archives-XLII-2-487-201
6. Revilla-León, M., Subramanian, S.G., Özcan, M., Krishnamurthy, V.R.: Clinical study of the influence of ambient light scanning conditions on the accuracy (trueness and precision) of an intraoral scanner. J. Prosthodont. **29**(2), 107–113 (2020). https://doi.org/10.1111/jopr.13135
7. Hsu, Q.-C., Ngo, N.-V., Ni, R.-H.: Development of a faster classification system for metal parts using machine vision under different lighting environments. The Int. J. Adv. Manuf. Technol. **100**(9–12), 3219–3235 (2018). https://doi.org/10.1007/s00170-018-2888-7
8. Singh, R., Nagla, K.S.: Error analysis of laser scanner for robust autonomous navigation of mobile robot in diverse illumination environment. World J. Eng. **15**(5), 626–632 (2018). https://doi.org/10.1108/wje-08-2017-0228

The Implementation Process and Factors that Influence the Quality of the Integration of Collaborative Robots in the Automotive Industry

Vasile Gusan[1,2] and Aurel Mihail Titu[3](✉)

[1] Faculty of Industrial Engineering and Robotics, Polytechnic University of Bucharest,
București, Romania
[2] Continental Automotive Systems, 8, Street Salzburg, Sibiu, Romania
[3] Lucian Blaga University of Sibiu, 10, Victoriei Street, Sibiu, Romania
mihail.titu@ulbsibiu.ro

Abstract. The automotive manufacturing process has the role of manufacturing the finished products or final assembly of the products intended for vehicles. In automotive, industrial organizations must produce as quickly as possible, at the highest possible quality and at the lowest possible cost in order to continue to secure a competitive place on the market. Due to this requirement, it was necessary to discover certain solutions through which: the manufacturing speed will increase and remain constant during manufacturing, the quality of the manufacturing process will be improved because a qualitative process will always lead to quality products, and manufacturing costs generated during process time to be reduced. These requirements led, in time, to the emergence of collaborative robots. If traditional robots were intended for particular applications, collaborative robots have much greater flexibility. The scientific paper presents, in an elegant manner, the process of integration of collaborative robots and the steps to be followed in order to transform manual processes, old or new, into manufacturing processes with collaborative robots. Certain factors that influence the quality of integration of collaborative robots are presented.

Keywords: Collaborative robot · Process · Automotive · Quality · Manufacturing

1 Introduction

The automotive industry is easily driven to total automation and digitization. Intelligent manufacturing facilitates product quality assurance and reduced manufacturing costs. By reducing manufacturing costs, industrial organizations aim to increase profits. Thus, collaborative robots are implemented on a large scale. Collaborative robots are a reliable solution to manufacture intelligently, qualitatively and at the lowest possible cost. Even if this aspect represents a slightly higher investment than a manufacturing flow without

© The Author(s), under exclusive license to Springer Nature Switzerland AG 2022
I. Karabegović et al. (Eds.): NT 2022, LNNS 472, pp. 46–57, 2022.
https://doi.org/10.1007/978-3-031-05230-9_5

collaborative robots, it has been scientifically and practically demonstrated that collaborative robots will pay off the investment over time. In terms of the acquisition cost, the price of collaborative robots is generally lower than the price of industrial robots. Due to this aspect, the amortization of the investment will take place faster. Also, the potential of collaborative robots is unlimited and they can be integrated into any industrial application and even integrated into production processes.

The integration of collaborative robots must be done following certain steps for the application to be a quality one. The quality of the application with collaborative robots can be influenced by several aspects such as the selection of the robot suitable for the application, programming, the use of grippers and supports to support the robots compliant for the application and many other aspects. A quality application will be effective and efficient, while a non-qualitative application may lead to multiple unplanned downtimes, premature failure of the robot or even its malfunction.

2   The Effect on the Quality of the Process After the Integration of the Collaborating Robots

One of the most used strategy today by each organization is continous quality improvement and control [1]. The quality of the process is determined by the process parameters. An important parameter of the manufacturing process is the handling of the product. This parameter, in most cases, cannot be measured automatically because it is about the effectiveness of the operator. The effectiveness of the operator refers to his ability to do things properly, following the procedures and work instructions.

Some factors that could affect the effectiveness of the operator are:

- Fatigue;
- Emotional state;
- Inattention;
- Lack of training;
- Lack of work experience;

These factors can also affect operators efficiency which has a direct impact on manufacturing flow productivity. For the operator to remain efficient all the non value activities must be eliminated [2].

These factors can influence the manufacturing process to be a variable process. A variable production process can easily come out of the necessary process parameters, and in this case, the quality of the products from the process is also variable.

Collaborative robots make it possible to obtain a larger number of compliant products. This is due to the guaranteed repeatability of Universal Robots, which can range from ±0.03 to ±0.11 mm. Repeatability is a property that varies depending on the type of robot used. Repeatability is the robot's ability to keep the same coordinates after performing 100 movements. It can be concluded that the collaborative robot is a very complex tool that can perform precise movements after multiple cycles. In automotive manufacturing flows, 40% of waste is caused by the mistakes of the worker or operator. To this percentage can be added some of the other 35% of scraps that are due to unknown causes, but which, certainly, some of them could be due to the mistakes of the worker. It can be estimated that the scrap due to man can reach somewhere around 75% of the total parts. These errors are generally due to the following causes:

- products dropped on the floor by the operator;
- products handled improperly;
- non-compliance with work instructions;
- loading and mixingthe wrong materials.

All these aspects can be due to the fatigue accumulated during working hours, the operator's inattention, the lack of knowledge of the operations on the respective manufacturing flow, the lack of interest of the operator, the emotional state, the stress of reaching the target and remaining efficient for many other reasons.

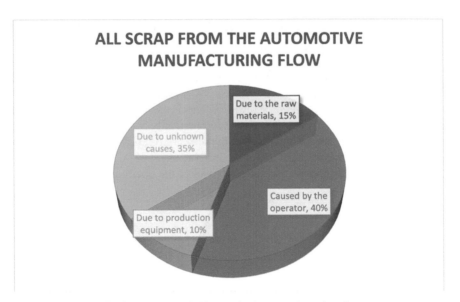

Fig. 1. Diagram of all scrap in the manufacturing flow

Following the implementation of collaborative robots on manufacturing flows, all these human errors can be eliminated. Due to this, the waste will be eliminated. If we consider that 40% of all manufactured products become scrap due to human error, the percentage of final products from the manufacturing flow will increase by 40%. This is the result of a cause-and-effect action. By eliminating the negative causes, the negative effects that lead to losses due to non-quality will also be eliminated. As it can be seen: quality is free while non-quality costs [3].

In the automotive field, respectively in the production of components and electronic products, the products dropped on the floor represent a great risk for the finished product. This situation can lead to the breakdown of electronic components that can affect the quality of the final product. The risk is even greater if the component cannot be detected following a circuit test or a final functionality test. This could reach the customer and generate complaints. Therefore, many automotive manufacturers give up the delivery of the product dropped on the floor and turn it into waste. However, the risk remains if the operator is not noticed by anyone and transfers the part further. This unfavorable situation could not happen if the product is handled by the collaborative robot. The only situation in which the robot could drop the product would be due to accidental opening of the gripper by man or certain defects of the gripper. If the product is not detected in the gripper, the collaborative robot can be programmed to stop automatically and display that the product has been lost. Thus, the risk of having products dropped on the ground is close to 0 if the collaborative robot will operate the production line.

Improper handling of products can lead to product defects. Due to this aspect, scraps may appear that cannot be recovered by the organization. For example, if the discussion is about the electronics industry, the product may be hit during its placement in the device, which would result in bending of the pins, cracking or breaking of certain components (Fig. 2).

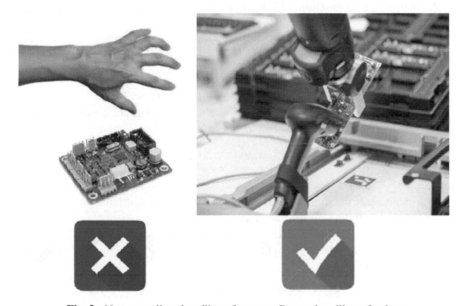

Fig. 2. Non-compliant handling of man vs. Proper handling of cobot

This could lead to irrecoverable scraps if they are detectable or complaints if they are not detectable. This can happen because of the operator:

- is tired;
- is careless;
- is disinterested;
- wants to be effective, but loses its effectiveness;
- is negatively affected emotionally;
- did not follow the work instructions.

Such situations could not happen if the products are handled by a collaborative robot. It rigorously maintains the steps, coordinates and speeds defined in the program. Scheduled steps are executed sequentially, and the decisions of the cobot can be made depending on the case are defined by the programmer.

To observe concretely and from a mathematical point of view what is the positive impact that collaborative robots bring on the quality of the process, a mathematical demonstration of a technological flow is required.

It will take into account a technological flow that must produce monthly: 50,000 products. The percentage of scrap is differentiated by cause and distributed according to Fig. 1. Each year, the production flow will have to produce:

$$P_A = P_L * 12 = 50000 * 12 = 600000 \text{ products/year;} \tag{1}$$

where:

P_A- represents the production for one year;
P_L- represents production for one month.

Of the one-year production, 5% of the products are scrap. Thus, the following can be distinguished:

$$R_A = P_A * 5\% = 600000 * 0.05 = 30000 \text{ scraps;} \tag{2}$$

where:

R_A- represents the waste from production for one year;

It is known that 45% of these scraps are due to non-compliant handling recognized as being due to the operator. It can be calculated that:

$$TR_{OR} = R_A * 45\% = 30000 * 0.40 = 12000 \text{ scraps/year;} \tag{3}$$

where:

TR_{OR}- represents the total waste recognized due to non-qualitative manipulations of the operator from production for one year;

Of the total scrap, 30% of scrap is due to unknown causes. However, it is hypothesized that at least 20% is also due to the operator, but these are not be recognized or identified. Thus, we can distinguish:

$$TR_{ON} = R_A * 20\% = 30000 * 0.2 = 6000 \text{ scraps/year;} \tag{4}$$

where:

TR_{ON}- represents the total of the unrecognized rejects due to non-qualitative manipulations of the operator coming from the production for one year;

In conclusion, it is possible to calculate the scrap per year by summing the recognized scrap with the unrecognized scrap. It can be calculated that:

$$TRA_{PO} = TR_{OR} + TR_{ON} = 12000 + 6000 = 18000 \text{ scraps/year;} \tag{5}$$

where:

TRA_{PO}- represents the total waste supposed to be due to non-qualitative manipulations of the operator from production for one year;

Fewer scraps would mean more compliant products delivered to the customer and increased profits over time. To observe concretely what the impact is in 5 years, based on the calculated mathematical data, a diagram can be made.

The diagram that concretely presents the situation is presented in Fig. 3.

All these scraps, whether it is the total scrap recognized due to non-qualitative manipulations of the operator, or the total scraps presumed to be due to non-qualitative manipulations of the operator are losses that the organization must bear financially.

Fig. 3. The number of rejects due to the operator and supposed to be due to the operator per year

where:

PF_{ON}- represents financial losses for one year assumed due to non-qualitative manipulations of the operator;

Thus, it will be possible to calculate the total financial losses per year assumed due to non-qualitative manipulations of the operator, as follows:

$$TPF_A = PF_{OR} + PF_{ON} = 1020000 + 510000 = 1530000 \text{ € lost/ year;} \quad (6)$$

where:

TPF_A- represents the total financial losses for one year due to non-qualitative manipulations of the operator, cumulating the percentage of the presumed losses;

Based on the calculated ones, a diagram can be generated, Fig. 4 to observe concretely what are the losses due to the non-quality that the organization has to face.

These financial losses could also be compounded by certain customer complaints. Even if the operator is trained in the way of working and is aware of the quality level that is required, it will not always be able to be controlled.

Fig. 4. Financial losses per year due to non-quality in Euro per year

Thus, if the tests do not detect 100% non-conformities on each product that could come from the manufacturing flow, which is generally impossible, the risk of sending non-compliant products to the customer could remain. Along with the risk of sending non-compliant products to the customer, there is also the risk of receiving complaints from him.

3 Factors that Influence the Quality of the Integration of Collaborative Robots

Many factors can influence the quality of collaborative robots. [4] Non-qualitative integration can affect the quality of operation of the manufacturing flow with collaborative robots. Non-qualitative integration can lead to:

- The accuracy of the collaborative robot during operation;
- Accidental downtimes of the collaborative robot during its normal operation;
- Premature failure of the wrists and motors of the collaborative robot;
- Damage to grippers and products;
- Injury of people working with the collaborative robot.

The main factors due to which a non-qualitative integration can take place are:

- Wrong choice of collaborative robot;
- Mounting the robot on surfaces with high vibrations;
- Wrong programming of the robot;
- The use of high speeds in areas of human interaction without their prior control.

If the collaborative robot does not work properly, it means that this aspect may be due to one of the main factors mentioned [5].

The collaborative robot must be chosen correctly according to its technical parameters. The criteria according to which the collaborative robot must be chosen are the necessary precision, the maximum payload, its speed and range.

Perfection has not yet been achieved in the world, and this applies to collaborative robots. Their maximum position deviations can vary between ±0.03 mm ±0.11 mm depending on the type of robot chosen. Due to this aspect, the collaborative robot should be chosen according to the precision imposed by the application. If the application requires accuracy of ±0.03 mm, and the purchased robot has an accuracy of ±0.11 mm, it can be said that the choice was wrong from the beginning and due to this aspect multiple problems will occur during operation.

In 2016, the maximum payload for a collaborative robot was 10 kg. Today, it can be much higher. [6] Collaborative robots remain limited depending on the maximum payload they can lift. This aspect also includes auxiliary gripping devices such as grippers. If the payload is exceeded, the collaborative robot will be able to stop during operation, its accuracy will be affected, and in time will occur even premature defects of its joints. For example, if the maximum payload that the robot can lift is 5 kg, but the integrator uses it, contrary to the manufacturer's instructions, to lift a mass greater than 5 kg, he will face the problems mentioned. The gripper must have collaborative properties to maintain a collaborative application [7] (Fig. 5).

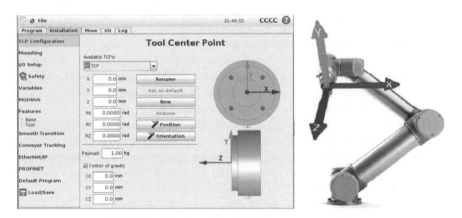

Fig. 5. Declaration of the Tool Center Point(TCP)

Due to the working speed, collaborative robots make it easier to achieve higher productivity [8]. The maximum speeds and accelerations that the collaborative robot can reach are mentioned in the brochures or data sheets received from the robot manufacturer. Many programs can simulate the movements of the collaborative robot so that it can be determined whether or not it will fit in the required cycle time. If the cycle time is shorter than the time when the collaborative robot can handle the product at maximum speed, it can be said that it has not been chosen properly or is not suitable for the application.

The range or extent to which the robot can operate is also a specific parameter of the collaborative robot. This represents the maximum distance that the collaborative robot can reach after stretching the joints. This distance can be increased by auxiliary devices that will be caught in the robot flange, but the mass of these devices must be taken into account when calculating payloads. If the robot cannot reach all the target areas, it is recommended to implement an additional robot to serve the affected areas.

Collaborative robots can be mounted in various planes/angles to serve the manufacturing flow. An important factor, however, is the rigidity of the supports on which they are mounted. If the collaborative robot is mounted on a support on which there are large vibrations and movements during operation, the accuracy of the robot may be affected and the robot may stop during operation. If the collaborative robot is planned to operate at high speeds, the support must be planned to be built as rigid as possible. The robot can be mounted on moving supports such as an electric axis, but it is important that it does not perform movements simultaneously with the robot, does not perform high-speed increases and sudden braking. The support must respect the dimensions of the robot flange so as to allow its orientation and fixation.

Collaborative robots can be programmed by novice or inexperienced users. [9] However, the quality of collaborative robot programming is a very important factor. Apparently, the interface is very human-friendly, easy to set up and program. However, it is very important that the person who programs the robot's movements must have the experience to use the robot's movements and set it up as efficiently as possible. If certain basic settings are not made correctly, the robot's accuracy, functionality, and time integrity of the robot's wrists may be affected. A simple example of such an aspect is the setting of the tool center point, known by the abbreviation TCP.

At the beginning of any collaborative robot programming application, the first step is to declare the robot in the soft tool center point (TCP-Tool center point). This point is especially important when starting any application. In this menu, the tool center point, its mass and the center point of the tool are declared. This setting can be found in a dedicated menu together with other global parameters [10].

Its declaration is made using x, y, z coordinates. The determination of this point is different depending on the application, for example, if it is desired to use a screwing application, the tip of the screwdriver would be the place where the center point of the tool will be found. In the case of a pick-up and placement application, the center point of the tool is located in the middle of the gripper grip area.

Also, the corresponding value of its weight must be changed in the program each time the robot lifts or leaves the part in the gripper. The center of gravity of the gripper must be declared in this area.

This value is important to declare because the collaborative robot uses a compensating force depending on the mass it will have to lift. Declaring this point incorrectly will facilitate the occurrence of positioning errors during the automatic operation of the collaborative robot, stops during operation and may even lead-in time to damage its arms (joints).

The tool center point can be determined by measuring from the robot flange. If a double gripper is to be used, several grippers or tools may be declared. For each tool, a tool center point must be declared in the robot program. The mass of the tool or gripper together with the center of gravity is also stated if they do not coincide.

There is an automatic tool center point determination system. The robot must be brought to the same point from four different positions. If the application is used accordingly, the robot will automatically calculate the center point of the tool and its center of gravity.

It is important to state the position in which the collaborative robot itself is positioned. Its base has the possibility of being mounted on a horizontal support, on the floor or ceiling or it can be mounted vertically, for example a rigid wall.

If the position is incorrectly declared, the robot will transmit this and its programming cannot be continued (Fig. 6).

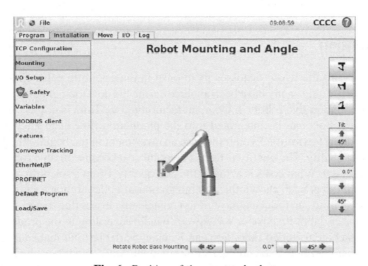

Fig. 6. Position of the mounted robot

Once these basic settings have been declared, the configurations must be saved. After saving the configurations, you can start its actual programming. You can load an empty program from the Robot Program menu, and then load an empty programming template in which you can start programming the robot. Programming the collaborative robot in a compliant way is a very important factor because its compliant programming will lead to a qualitative operation of it. If the programming is a quality one then the collaborative robot:

- will not stop during operation;
- no malfunctions will occur during operation;
- there will be no collisions;
- the robot will not malfunction over time;
- will work at maximum efficiency and effectiveness.

For an application to be collaborative, it must be implemented collaboratively. This refers to controlling the speed or functionality of the robot in common workspaces with humans. Even if the robot stops in a collision with the human, it could seriously hit the human before stopping if it is operating at high speeds. If the robot must operate at high speeds, it is recommended that perimeter scanners be used in areas accessed by humans so that the robot knows that humans are present and:

- Either to operate at a reduced speed while the operator is in the area and to change his speeds once he leaves the working perimeter;
- Either stop when the operator is in the area and restart when he leaves the perimeter.

Also, if the gripper is not collaborative, then the application will not be collaborative. If the robot uses, for example, a knife as a tool, then even if the robot is a collaborative one, the application will not be.

4 Conclusion

In conclusion, all the waste or losses mentioned in the scientific paper may be due to non-quality and could easily have been avoided during the development of the process by integrating collaborative robots in the manufacturing flow. Time is not wasted, because collaborative robots can be integrated into the production flow even after it has been made due to their flexibility. Even if they are an investment in the beginning, they are an investment in quality. The quality is free, it does not cost because in time robots will pay off the investment. What costs is actually the non-quality. From a scientific and practical point of view, it has been shown that the integration of collaborative robots into such a manufacturing flow can reduce almost, if not completely, losses due to non-quality. This is because the collaborative robot will always handle according to the product.

It can also be concluded that there can be many variables that make an application with collaborative robots to be qualitative or non-qualitative. What must be kept in mind is that a quality application with collaborative robots will add value, will be efficient and effective.

References

1. Godina, R., Matias, J.C., Azevedo, S.G.: Quality improvement with statistical process control in the automotive industry. Int. J. Ind. Eng. Manag. **7**(1), 1–8 (2016)
2. Hasta, A., Harwati, H.: Line balancing with reduced number of operator: a productivity improvement. IOP Conf. Ser.: Mater. Sci. Eng. **528**(1), 012060 (2019). https://doi.org/10.1088/1757-899X/528/1/012060
3. Kragic, D., Gustafson, J., Karaoguz, H., Jensfelt, P., Krug, R.: Interactive, collaborative robots: challenges and opportunities. In: IJCAI, pp. 18–25. (2018). https://doi.org/10.24963/ijcai.2018/3
4. Crosby, P.: Quality is Free. Ed. McGraw-Hill, New York, United States of America (1979)
5. Bragança, S., Costa, E., Castellucci, I., Arezes, P.M.: A Brief overview of the use of collaborative robots in industry 4.0: human role and safety. In: Arezes, P.M., et al. (eds.) Occupational and Environmental Safety and Health. SSDC, vol. 202, pp. 641–650. Springer, Cham (2019). https://doi.org/10.1007/978-3-030-14730-3_68

6. Sherwani, F., Asad, M.M., Ibrahim, B.S.K.K.: Collaborative robots and industrial revolution 4.0 (ir 4.0). In: 2020 International Conference on Emerging Trends in Smart Technologies (ICETST), pp. 1–5. IEEE (2020). https://doi.org/10.1109/ICETST49965.2020.9080724
7. Iqbal, Z., Pozzi, M., Prattichizzo, D., Salvietti, G.: Detachable Robotic Grippers for Human-Robot Collaboration. Front. Robot. AI **8**, 644532 (2021). https://doi.org/10.3389/frobt.2021. 644532
8. Bloss, R.: Collaborative robots are rapidly providing major improvements in productivity, safety, programing ease, portability and cost while addressing many new applications. Indus. Robot: An Int. J. **43**, 463–468 (2016). https://doi.org/10.1108/IR-05-2016-0148
9. Pieskä, S., Kaarela, J., Mäkelä, J.: Simulation and programming experiences of collaborative robots for small-scale manufacturing. In: 2018 2nd International Symposium on Small-scale Intelligent Manufacturing Systems (SIMS), pp. 1–4. IEEE (2018). https://doi.org/10.1109/ SIMS.2018.8355303
10. Schou, C., Andersen, R.S., Chrysostomou, D., Bøgh, S., Madsen, O.: Skill-based instruction of collaborative robots in industrial settings. Robot. Comput. Integr. Manuf. **53**, 72–80 (2018). https://doi.org/10.1016/j.rcim.2018.03.008

The Influence of Collaborative Robots on the Quality, Efficiency and Effectiveness of Automotive Manufacturing Flows

Aurel Mihail Titu[1]([✉]) and Vasile Gusan[2,3]

[1] Lucian Blaga University of Sibiu, 10 Victoriei Street, Sibiu, România
`mihail.titu@ulbsibiu.ro`
[2] Faculty of Industrial Engineering and Robotics, Polytechnic University of Bucharest, Bucharest, România
[3] Continental Automotive Systems, 8, Street Salzburg, Sibiu, România

Abstract. These days, industrial organizations are looking for multiple solutions so that they remain flexible and competitive in the market. This situation leads directly to the improvement of manufacturing flows by automation or their robotization. Quality, efficiency, and effectiveness are important aspects that any organization must take into account in order to remain competitive. Thus, the manufacturing processes with collaborative robots are becoming more and more common in the industry. This aspect is due not only to the lack of staff, but also to the multiple advantages collaborative robot bring. The current pandemic context further favors the implementation of these types of robots. Collaborative robots are becoming more and more a necessity today, due to the fact that they can easily mold to the process and improve it from several points of view. Collaborative robots bring an important impact in terms of quality, efficiency and effectiveness of manufacturing flows. The scientific paper presents in an elegant way the effect that collaborative robots bring, following the integration, on the manufacturing flows in automotive.

Keywords: Collaborative robot · Efficiency · Effectiveness · Automotive · Manufacturing

1 Introduction

The year 2020, a year in which the lives of many have changed, a year in which the world has stopped for a while and a year that will certainly not be easy to forget for many people. From then until now, quality, efficiency and effectiveness have remained performance indicators that directly affect the organization's profit. Industrial organizations have come to understand that to survive in the marketplace, it is important to be flexible. Flexibility is the ability of the organization to adapt quickly and easily to the market, customers or unforeseen situations. This flexibility has also led to changes in the production area of the companies. Industrial organizations are rapidly migrating to full automation of manufacturing flows. The goal is for them to be serviced by a

I. Karabegović et al. (Eds.): NT 2022, LNNS 472, pp. 58–67, 2022.
https://doi.org/10.1007/978-3-031-05230-9_6

single operator. This has led to the widespread deployment of collaborative robots. This is because collaborative robots are important tools that ensure product quality during manufacturing, improve the efficiency and effectiveness of manufacturing flows. By improving the quality, efficiency and effectiveness of production, collaborative robots facilitate the increase of the organization's profit.

2 The Concept of Efficiency and Effectiveness

The concept of efficiency can be defined as the way to achieve multiple goals with fewer or the same resources, be they financial or human.

The concept of effectiveness can be presented as the way to achieve the proposed objectives correctly.

According to the definitions, in order to concretize these terms in the production area, it can be said that efficiency is a concept that is based on productivity and profit, while efficiency is mainly focused on quality. [1] Given this analogy, it can be considered that manufacturing processes must be both efficient and effective.Digitization contributes to the efficiency of the process. [2].

An efficient but not effective technological process will be able to easily manufacture the number of products, with a required number of resources or even less. The product must be delivered to the customer according to plans, but their quality will not be compliant (Fig. 1).

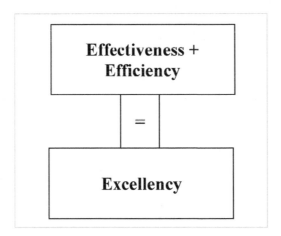

Fig. 1. Achieving excellence through efficiency and effectiveness

At the same time, an efficient but not efficient technological process will be able to manufacture the products without problems following the quality requirements, but it will not succeed in reaching the capacity desired by the customer and the resources invested will be more than originally planned.

Taking into account an existing technological process in automotive, it can be considered that efficiency refers to the speed and accuracy of manufacturing, and efficiency refers to maintaining the process in process parameters. Thus, an efficient technological process will be able to satisfy the customer's requirement in terms of capacity and if it is effective it can be considered a quality process, and a quality process is a stable process that will manufacture only quality products.

3   Collaborative Robots Are Quality Assurance Tools

Quality assurance is the tool by which a company ensures its competitiveness in the market and the customer's trust. This is presented as an existing concept in a qualitative system that comprises a whole set of systematically implemented and planned activities.[3].

The quality of the products can be maintained by preventing defects that may be present during processing or transport. Due to this aspect, especially for the production lines operated by the workers, multiple visual aids, work procedures and processes are made. In these documents it is presented:

- Working steps;
- Way of working;
- How to operate on equipment;
- How to handle subassemblies and products.

These procedures define an effective way in which activities on a manufacturing flow must be carried out. In case of any deviation, the risk of making one or more scraps increases. Production equipment and operators must be able to ensure the quality of processed products and subassemblies (Fig. 2).

Fig. 2. Collaborative robots are always on duty

At the same time, the manufacturing speed must be maintained to meet the customer's capacity requirements. A steady pace of work is difficult to maintain during a long shift of human resources. Even if the human resource manages to keep pace, over time, it can happen that due to accumulated fatigue the work procedures are not followed in a proper way. This aspect can lead to further rejection.

Collaborative robots are recommended to be implemented especially in manual processes. [4].

Everyone needs a vacation once in a while, can get sick or leave the company at any time. This aspect is being applied even to the production operator. He will have to be replaced by another worker who may be less experienced. Even if the worker is trained, there is a possibility for him to make mistakes because he does not know the manufacturing flow like an operator experienced with it.

Repetitive actions contribute much more to the appearance of scrap because the operator can be distracted after a while. The job becomes monotonous and boring, and due to this aspect, it will be difficult for the operator to remain effective.

Collaborative robots are resources that can manage the manufacturing flow, always respecting: working steps, working mode, equipment operation mode and subassembly handling mode. They can also maintain the same manufacturing speed, maintaining a steady pace over time. Collaborative robots cannot be distracted, tired and bored over time.

This situations may influence the managers to adopt in their organizations collaborative robots solutions. [5].

4 Production Efficiency by Implementing Collaborative Robots

Efficiency in a manufacturing flow is the way to get more products manufactured using fewer resources: time or monetary units.

This days, robots are being a driving force and the destination is automation. [6] A manufacturing flow without collaborative robots is one in which the manipulation is done by, in general, humans. Collaborative robots present the advantage to be more flexible and safer then industrial robots. [7] There are also manufacturing flows in which the products are handled by industrial robots or axes, but the comparison could not be feasible as it needs to be contained in an automatic cell. These cases are comparable because, in the case of working with workers and in the case of working with collaborative robots, the workspace will remain open. Both ways of working can be considered collaborative.

In a manufacturing flow, the resource is represented by time and monetary units. Industrial organizations want to manufacture as quickly as possible, ie in the shortest possible time, and as cheaply as possible, ie with as few monetary units as possible. Due to this aspect, especially in the production area, the resources to be reduced are manufacturing time and manufacturing cost.

Collaborative robots are a viable alternative in terms of reducing manufacturing time and manufacturing costs. [8] This aspect was practically demonstrated in an application with collaborative robots developed personally at Continental Sibiu in 2018.

In that application, both manufacturing flows were analyzed: with workers, and with collaborative robots, respectively. Basically, before the implementation of collaborative robots, workers managed to produce an average of 105 parts per hour.

Following the implementation of collaborative robots, the average production increased by 25 products per hour. Thus, it was possible to increase production by 24% from 105 products per hour to 130 products per hour. Collaborative workers and robots were the only variable aspects of the manufacturing flow, without an improvement of the process being made on the production equipment.

This was achieved by increasing the speed of handling the product and maintaining a constant pace of production. This would not have been possible in the case of working with operators, because its working rate is high at the beginning of the shift, and over time it decreases until the exit from the shift, while in the case of collaborative robots it is maintained 24 h for 7 days constantly.

The data comes from the manufacturing flow, where measurements were performed before and after implementation. As can be seen, the payback period of collaborative robots is very fast. [9] (Fig. 3).

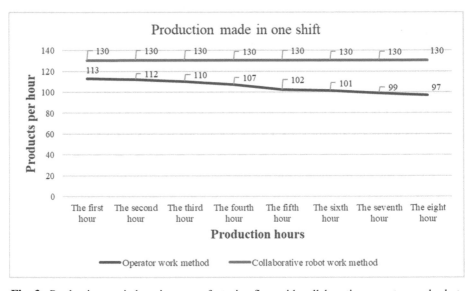

Fig. 3. Production carried out in a manufacturing flow with collaborative operators and robots

From the point of view of the parts produced on an shift to make a clear comparison from the point of view of production between the two situations, it is necessary to perform a calculation to see what the difference is between the two ways of working.

Thus, it is necessary to calculate the production for an 8-h shift of the way of working with collaborative operators and robots:

$$P_O = 113 + 112 + 110 + 107 + 102 + 101 + 99 + 97 = 841 \text{ products}; \quad (1)$$

where:

P_O- represents the production made with operators;

Respectively, for the way of working with collaborative robots:

$$P_C = 130 * 8 = 1.040 \text{ products made on an } 8 - \text{hour shift}; \quad (2)$$

where:

P_C- represents the production made with cobots or collaborative robots;

Difference in production:

$$D_F = P_C - P_O = 1.040 - 841 = 199 \text{ products;}$$ (3)

where:

D_F- represents the difference in terms of production made in addition to the way of working with operators;

The 199 products represent the difference between the two ways of working. This difference can extend exponentially to production in 3 shifts, respectively for one week, one month, one year. Depending on the experience of the operator working on the line, this can be even more variable. The measurements were performed with experienced operators. Also, it is observed, over time, that the performance of the operators decreases. This aspect will especially influence the production results in a 12-h shift.

The number of product items that go to the customer represents sales honorary orders, respectively profit. If the organization produces faster, it can deliver to the customer faster. If it always produces in time to deliver, situations may arise in which the organization must bear additional costs in order not to delay delivery because if the customer line stops due to lack of materials, the financial penalties are very high.

Such a product costs about 85 euros. Thus, the products can be transformed in cash. One-year impact on the production line can be calculated.

Input data:

$$C_P = 85 \text{ EURO;}$$

$$D_F = 199 \text{ products/shift;}$$

where:

C_P- represents the cost in EURO per product;

The organization works 24 h a day, 7 days a week. A shift is 8 h. The organization works 354 days a year.

$$P_Z = D_F * 3 = 199 * 3 = 597 \text{ products per day/three shifts;}$$ (4)

$$P_S = P_Z * 7 = 597 * 7 = 4.179 \text{ products per week;}$$ (5)

$$P_A = P_Z * 354 = 597 * 354 = 211.338 \text{ products per year;}$$ (6)

$$Dm_A = P_A * C_P = 211.338 * 85 = 17.963.730 \text{ EURO per year;}$$ (7)

where:

P_Z- represents the daily production;

P_S- represents the production made per week;

P_A- represents the production made per year;

Dm_A- represents the monetary difference obtained per year due to the implementation of the working mode with collaborative robots.

If there are similar manufacturing flows, depending on the number, this monetary difference can be further quantified.

The production result is directly impacted by the time a product is manufactured and packaged. Thus, it can be concluded that if production has increased, the time has been reduced due to collaborative robots.

If before they were produced in an 8-h shift, in working mode with operators, 841 products, and after the implementation of collaborative robots, the production increased to 1040 h. Thus, it can be considered that the manufacturing flow has been streamlined because, with the same resources in terms of time, it manages to achieve many more products which is the purpose of production. It can also be seen that monetary income has increased.

5   Effectiveness of a Manufacturing Flow with and Without Collaborative Robots

Even with safety limitations due to free human interactions, the working method with collaborative robots is a more efficient way. [10] The effectiveness of a manufacturing flow refers to the ability of the manufacturing flow to make compliant products. Compliant products are products that meet the customer's requirements, so they are quality products. We consider that the efficiency of a manufacturing flow can be measured by reporting the resulting products as qualitative in the process superimposed on the total number of products processed by the technological flow.

Thus, if we consider:

$$E_C = \frac{P_C}{T_P}; \tag{8}$$

where:

E_C – represents effectiveness;

P_C – represents the number of compliant products from the manufacturing flow;

T_P – represents the total number of products processed on the manufacturing flow. It cannot be greater than P_C.

Based on the formula, we calculated the efficiency of the manufacturing flow using both the working mode with operators and the working mode with collaborative robots. We have considered that 40% of all non-compliant products are due to the worker. Every year, 600,000 products are manufactured, of which 570,000 are compliant products delivered to the customer.

Thus, we calculated the total waste per year, the number from which we can calculate the total waste per year due to the way of working with the operator.

$T_{PA} = 600000$ scraps per year;

$P_{CMO} = 570000$ compliant products per year;

$$T_{RAO} = T_{PA} - P_{CMO} = 600.000 - 570.000 = 30.000 \text{ scraps per year;} \tag{9}$$

$$R_{DO} = T_{RA} * 40\% = 30.000 * 0,4 = 12.000 \text{ scraps per year;} \tag{10}$$

where:

T_{PA} – represents the products manufactured in a year in a way of working with operators;

P_{CMO} – represents the number of compliant products from the manufacturing flow in which the way of working with operators is used;

T_{RAO} – represents the total waste per year in a way of working with operators;

R_{DO} – represents the total waste per year due to the way of working with operators.

Following the data calculation, it is possible to calculate the effectiveness of the way of working with operators:

$$E_{CMO} = \frac{P_{CMO}}{T_{PA}} = \frac{570.000}{600.000} = 0,95; \tag{11}$$

We consider that in the case of working with operators, the effectiveness of the manufacturing flow is 95%.

A calculation of the effectiveness of the technological flow in the case of working with collaborative robots was also performed. It is known that all scraps due to the way of working with operators will disappear. Thus, a new total of scrap can be calculated per year.

$$T_{RAR} = T_{RAO} - R_{DO} = 30.000 - 12.000 = 18.000 \text{ scraps per year}; \tag{12}$$

where:

T_{RAR} – represents the total waste per year in a way of working with collaborative robots;

T_{RAO} – represents the total waste per year in a way of working with operators;

R_{DO} – represents the total waste per year due to the way of working with operators.

Due to this, it was possible to calculate a new number to represent the number of compliant products from the manufacturing flow with collaborative robots.

$$P_{CMR} = T_{PA} - T_{RAR} = 600.000 - 18.000 = 582.000 \text{ compliant products}; \tag{13}$$

where:

P_{CMR} – represents the number of compliant products from the manufacturing flow in which the way of working with collaborative robots is used;

T_{PA} – represents the products manufactured in a year in a way of working with operators;

T_{RAR} – represents the total waste per year in a way of working with collaborative robots.

Following the analytical calculation, it was possible to calculate the effectiveness of working with operators:

$$E_{CMO} = \frac{P_{CMO}}{T_{PA}} = \frac{582.000}{600.000} = 0,97; \tag{14}$$

It can be concluded that the way of working with collaborative robots is 2% more effectivethan the way of working with operators in this situation.

6   Conclusion

It can be concluded that a manufacturing process is important to be both efficient and effective so that it meets the customer's requirements in terms of capacity and quality. The concept of efficiency and the concept of efficiency facilitate the achievement of excellence.

Collaborative robots are tools for product quality assurance. They contribute to the increase in the number of products made in a proper way of a manufacturing flow. Collaborative robots do not leave jobs and keep working speedily.

Collaborative robots achieve after the implementation the efficiency of the manufacturing flow. They manage to increase the number of parts produced and manage to normalize a manufacturing flow with variable production. An efficient manufacturing flow will produce higher monetary income than an inefficient manufacturing flow. Due to this aspect, it can be considered that collaborative robots streamline not only the production but also the income of the organization.

The more efficient a manufacturing flow is, the better it will be to produce quality products. An efficient manufacturing flow is a flow capable of qualitatively processing the first products, saving time and money. The less efficient the manufacturing flow, the more it will generate losses for the industrial organization.

Collaborative robots bring real benefits to the organization by improving its production in terms of quality, increasing their efficiency and effectiveness and the profit of the industrial organization in which they were implemented.

References

1. Boehm, E.: Improving efficiency and effectiveness in an automotive R&D organization. Res. Technol. Manag. **55**(2), 18–25 (2012). https://doi.org/10.5437/08956308X5502011
2. Bevan, O., Freiman, M., Pasricha, K., Samandari, H., White, O.: Transforming Risk Efficiency and Effectiveness. McKinsey & Company (2019)
3. Oprean, C., Țîțu, M.: Managementul Calității în Economia și Organizația Bazată pe Cunoștințe, Ed. AGIR. Bucharest, Romania (2008)
4. Weckenborg, C., Kieckhäfer, K., Müller, C., Grunewald, M., Spengler, T.S.: Balancing of assembly lines with collaborative robots. Bus. Res. **13**(1), 93–132 (2019). https://doi.org/10.1007/s40685-019-0101-y
5. Simões, A.C., Soares, A.L., Barros, A.C.: Factors influencing the intention of managers to adopt collaborative robots (cobots) in manufacturing organizations. J. Eng. Tech. Manage. **57**, 101574 (2020). https://doi.org/10.1016/j.jengtecman.2020.101574
6. Galin, R., Meshcheryakov, R.: Automation and robotics in the context of Industry 4.0: the shift to collaborative robots. In: IOP Conference Series: Materials Science and Engineering, vol. 537(3), p. 032073. IOP Publishing (May 2019). https://doi.org/10.1088/1757-899X/537/3/032073
7. Ferraguti, F., Pertosa, A., Secchi, C., Fantuzzi, C., Bonfè, M.: A methodology for comparative analysis of collaborative robots for industry 4.0. In: 2019 Design, Automation & Test in Europe Conference & Exhibition (DATE), pp. 1070–1075. IEEE, (March 2019). https://doi.org/10.23919/DATE.2019.8714830

8. Sherwani, F., Asad, M. M., Ibrahim, B.S.K.K.: Collaborative robots and industrial revolution 4.0 (ir 4.0). In: 2020 International Conference on Emerging Trends in Smart Technologies (ICETST), pp. 1–5. IEEE (March 2020). https://doi.org/10.1109/ICETST49965.2020.908 0724
9. Zanchettin, A.M., Rocco, P., Chiappa, S., Rossi, R.: Towards an optimal avoidance strategy for collaborative robots. Rob. Comp.-Integ. Manuf. **59**, 47–55 (2019). https://doi.org/10.1016/j.rcim.2019.01.015
10. Raiola, G., Cardenas, C.A., Tadele, T.S., De Vries, T., Stramigioli, S.: Development of a safety-and energy-aware impedance controller for collaborative robots. IEEE Rob. Auto. Let. **3**(2), 1237–1244 (2018). https://doi.org/10.1109/LRA.2018.2795639

The Aerodynamic Wind Loads of a Naval Surface Combatant in Model Scale

Sarih Sarı[1][✉], Ali Dogrul[2], and Seyfettin Bayraktar[3]

[1] Department of Naval Architecture and Marine Engineering, Yildiz Technical University, 34349 Istanbul, Turkey
sarih@yildiz.edu.tr
[2] Department of Naval Architecture and Marine Engineering, National Defense University, Turkish Naval Academy, Istanbul, Turkey
[3] Department of Marine Engineering Operations, Yildiz Technical University, Istanbul, Turkey

Abstract. Design of the ships considering hydrodynamic and aerodynamic requirements is important especially for naval surface combatants. From the aerodynamic point of view, the superstructure and the flight deck, which is used for landing and take-off operations of aerial vehicles, have to be designed in terms of ship airwake and wind loads. The superstructure combined with the sea and weather conditions has a crucial effect on the flow characteristics such as the turbulence and vortices on the flight deck. This study covers the numerical investigation of aerodynamics of a naval surface combatant designed by the Office of Naval Research (ONR). As a more realistic ship, ONR Tumblehome hull was chosen instead of the generic frigate model SFS2 and ONRT model has been widely used for validation studies as a benchmark geometry. Numerical analyses were conducted by employing the k-ω turbulence model and solving the unsteady RANS equations. In the present study, the axial and tangential velocity distributions on the flight deck of ONRT were firstly validated in model scale. Following this, one more model scale geometry was generated and the aerodynamics of these vessels were investigated in headwind conditions. Thus, the scale effects on the aerodynamics of the ship were observed maintaining the dynamic similarity based on Reynolds number. Furthermore, the aerodynamic wind loads on the ONRT surface combatant vessel are presented for various wind-over-deck (WOD) angles in one model scale.

Keywords: Aerodynamics · CFD · ONRT · RANS · Ship airwake

1 Introduction

Ship design practices mostly focus on the hydrodynamic and structural design of the hulls. However, the wind loads on the ship have significant effects on the hydrodynamic and structural response of the ship. In addition, the wind loads are important on the aerodynamics of the ship. The aerodynamic behavior of the ships with a flight deck for fixed and/or rotary-wing aircraft has to be determined for safe flight operations. Within this, the aerodynamic flow field can be investigated in terms of ship airwake, ship-aircraft

I. Karabegović et al. (Eds.): NT 2022, LNNS 472, pp. 68–76, 2022.
https://doi.org/10.1007/978-3-031-05230-9_7

interaction. Also, the wind loads on the hull can be determined at different wind-over deck (WOD) angles which have an important role in the ship maneuvering.

The aerodynamics of the surface vessels can be investigated relying on experimental and numerical methods. The experimental methods are preferred in the wind tunnel setups and the ship superstructure is observed for different angles and velocities. The numerical methods are employed using commercial CFD solvers. Different turbulence models can be applied and detailed flow field analyses can be made rapidly. Several types of research have been made focusing on ship aerodynamics and ship airwake in recent years using SFS2 and ONRT hull forms.

Forrest and Owen [1] comprehensively investigated airwake on simple frigate shape-2 (SFS2) and Royal Navy Type 23 frigate. They compared the experimental results with the Detached Eddy Simulations (DES) results and managed to model high-scale turbulences. Yuan et al. [2] performed a validation study for SFS2 ship geometry for low-sea-state using OpenFOAM. Then, a Canadian patrol frigate was examined with delayed detached eddy simulations (DDES). The effect of the mast on helicopter operations was investigated numerically and experimentally. The axial and tangential velocity distributions above the flight deck at a constant height were obtained and a good agreement was achieved with the published other studies. Bardera et al. [3] addressed the airwake behind the SFS2 generic frigate both experimentally and numerically. Different bubble-shaped hangar geometries were investigated in terms of the flow characteristics on the flight deck. Experimental data obtained with particle image velocimetry (PIV) were compared with the numerical results. It was reported that the frigate design with an optimum hangar geometry is important for safe helicopter operations. Li et al. [4]. Performed numerical analyses focusing on the determination of the optimum hangar length for SFS and SFS2 ships with minimum recirculation zone. DDES turbulence model was employed to capture even small eddies and the optimum hangar geometry was obtained. Some other studies on the simulation of the airwake field employing different turbulence models. In this manner, Farish [5] assessed the ship airwake over SFS2 geometry in model scale. The model scale experiments were conducted utilizing hot-wire anemometry and PIV, and the results were compared with larger-scale wind tunnel measurements. The results were then compared with full-scale large eddy simulation (LES) analyses in terms of velocity distribution and turbulence intensity. It is found that the results agree well despite the difference in Reynolds number. Choi and Miklosovic [6] performed LES analyses to investigate the airwake on the flight deck of SFS2 geometry. The numerical results were compared with the wind tunnel data. It is seen that the numerical findings based on the moving mesh algorithm are in good agreement with the measurements. Nisham et al. [7] conducted numerical analyses using the DES method in their study. Full-scale SFS2 geometry was investigated in head wave conditions at various encounter frequencies and wind over deck (WOD) angles. The numerical results were obtained by performing analyses with and without the atmospheric boundary layer, and the results were introduced in terms of non-dimensional velocity distribution and ship motions. In addition, a detailed comparison of flow features around the ship deck was presented.

Krebill [8] focused on the interaction of ship motions and ship airwake in his thesis work. ONR Tumblehome was investigated using PIV and LDA techniques and the flow characteristics were examined in detail. It was found that the ship motions have a significant role in the velocity fluctuations. The study of Dooley et al. [9] focuses on the ship-helo coupled effect on the ship aerodynamics experimentally and numerically. They examined the effects of helicopter rotor on streamlines at full scale and model scale for ONR Tumblehome. Combined effects of motion, wave and turbulence were discussed for the airwake of the ONR Tumblehome at various speeds and sea states. In another study of Dooley et al. [10], the authors focused on the effects of waves and ship motions on the airwake of ONR Tumblehome with an operating helicopter in full-scale. Numerical analyses were performed in head wave conditions corresponding to sea state 3 and 6. ONRT bare hull and a generic helicopter based on S-60 Seahawk were used for the numerical investigation. Effects of the ship motions due to waves were observed in terms of velocity distribution over the flight deck in different sea states. The results were discussed with and without helicopter behind the ship. The paper of Dooley et al. [11] focused on the airwake of ONR Tumblehome at different Reynolds numbers numerically and experimentally. The numerical results were compared with the PIV results. Scale effects were also examined with multiple scales. Especially for low Reynolds numbers, grid dependence on the flow was emphasized.

In this study, the airwake of the benchmark model ONR tumblehome was investigated numerically using CFD method. The present study was focused on the flow around ONRT ship in water phase to mimic the experimental conditions. The analyses were carried out at $Re = 1x10^6$. The numerical approach was validated with the available experimental results in model scale. The validation study was conducted in terms of axial and tangential velocity distribution at two different locations at the aft region. Two different models were generated to observe the scale effects on ship aerodynamics. Finally, for the larger model, different wind-over deck angles were tested numerically and the effects of the wind direction on the ship airwake were investigated.

2   Theoretical Background

The numerical analyses were conducted using a commercial computational fluid dynamics (CFD) software solving unsteady Reynolds-Averaged Navier-Stokes (URANS) equations. The governing equations are the continuity (Eq. 1) and the momentum (Eq. 2) equations considering the flow is incompressible, Newtonian, isothermal and turbulent.

$$\frac{\partial U_i}{\partial x_i} = 0 \tag{1}$$

$$\frac{\partial U_i}{\partial t} + U_j \frac{\partial U_i}{\partial x_j} = -\frac{1}{\rho}\frac{\partial P}{\partial x_i} + \frac{\partial}{\partial x_j}\left[\upsilon\left(\frac{\partial U_i}{\partial x_j} + \frac{\partial U_j}{\partial x_i}\right)\right] - \frac{\overline{\partial u_i' u_j'}}{\partial x_j} \tag{2}$$

Here, ρ is the fluid density, U_i is the velocity vector, and P is the pressure. The last two terms belong to the viscous stress tensor while υ is the kinematic viscosity. Detailed information about the turbulence model (SST k-ω) and the equations can be found in Wilcox [12, 13].

3 Onrt Geometry

The present study focuses on the aerodynamic investigation of the well-known benchmark geometry ONRT model. The ship was designed by the Office of Naval Research (ONR) and the ship is the preliminary design of the Zumwalt Class US Navy destroyer, namely DTMB 5613. The ship has a large superstructure with a tumblehome hull. The perspective view of the 3-D model is given in Fig. 1. The main particulars of the ship are given in Table 1 at different scales.

Fig. 1. Perspective view of ONRT

Table 1. Main particulars of ONR tumblehome

Particular	Description	Full	Model	Model
λ	Scale factor	1	48.935	77
L_{WL} (m)	Length waterline	154.0	3.147	2.0
B_{WL} (m)	Beam waterline	18.78	0.384	0.244
T(m)	Draft	5.494	0.112	0.071
Δ	Displacement	8507 ton	72.6 kg	18.64 kg
C_B	Block coefficient	0.535	0.535	0.535

4 Numerical Results

4.1 Numerical Approach

Numerical analyses were conducted using a commercial CFD solver. The governing equations (URANS) were solved using the finite volume method. The computational domain was created neglecting the free surface effects similar to the wind tunnel experimental conditions. The computational domain was discretized by finite hexahedral volume elements using the trimmer mesh algorithm. Local mesh refinements were employed around the ship superstructure and flight deck. Prism layers were created to model the boundary layer flow accurately. Wall y^+ values were kept between 30 and 100 on the hull surface while the local wall y^+ values were under 1. The boundary conditions and the mesh structure applied on the computational domain were presented in Figs. 2 and 3. One may notice that the ship surfaces and the bottom surface were defined as no-slip

Fig. 2. Mesh structure and the local refinements

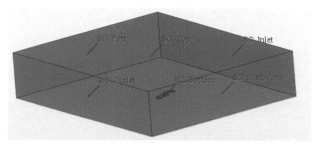

Fig. 3. Computational domain and the boundary conditions

wall while the outlet was defined as pressure outlet. Other surfaces including the inlet, top and side surfaces were set as velocity inlet.

The numerical setup was prepared to mimic the wind tunnel experiments. For this purpose, the wetted surface of the ship and the free surface were not modeled, instead, the flow region around the upper part of the ship was modeled. The numerical analyses were carried out in water fluid to mimic and validate the results of a recent study [11].

4.2 Validation and Scale Effect at Headwind Condition

The ONRT ship in different model scales ($\lambda_1 = 77, \lambda_2 = 48.935$) was investigated at headwind condition and the velocity distributions at two different sections were obtained in terms of axial and tangential velocities. Two vertical lines as shown in Fig. 4 were located at the aft and the obtained velocity distributions were validated with the experimental ones [11] for $\lambda_1 = 77$. The models are geometrically similar and dynamic similarity was maintained between the ships via Reynolds number. This means that the Reynolds numbers of these models are equal and it is considered as 10^6.

Fig. 4. The vertical lines located at the aft of the superstructure

The axial and tangential velocity distributions given in the Fig. 5 show that the numerical results are mostly in good agreement with the experimental results at two vertical lines. At $x/L = 0.708$, the axial velocities are very near to the experiments in both model scales. This means that the dynamic similarity of Reynolds works well in this flow region behind the superstructure and there is nearly no scale effect on the velocity field. The scale effect is again neglected in the tangential velocity distribution in the same vertical line. However, the vertical velocities do not match well with the experimental ones for $0.05 < z/L < 0.1$ at $x/L = 0.708$. This difference may be due to the lack of vortex modelling with RANS approach.

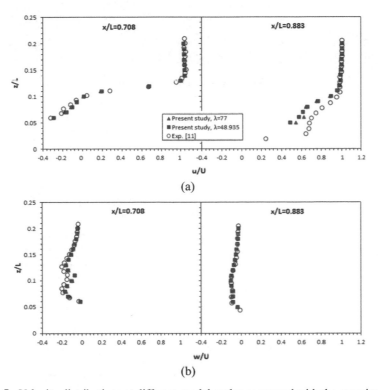

Fig. 5. Velocity distributions at different model scales compared with the experiments

At $x/L = 0.883$, the relative between the numerical and experimental results are quite larger in both velocity components. However, scale effects are not observed in the vertical line for the two velocity components. For $0 < z/L < 0.1$, the difference between two methods increases and this flow region is the reattachment zone where the large eddies are dominant in the flow. The streamlines following the superstructure causes a backward facing step flow that leads to high fluctuations in the velocity components.

4.3   Effects of WOD Angle for 1/48.935 Model

The effects of the wind over deck angle were investigated for $\lambda_2 = 48.935$. The numerical analyses were conducted between 90^0 and 180^0 covering the wind conditions between head and beam. Figure 6 shows the wind directions and angles. Here, 180^0 represents the headwind condition while 90^0 stands for the beam wind condition.

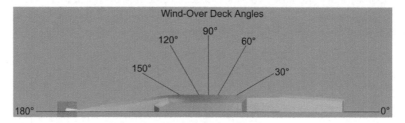

Fig. 6. WOD angle configuration

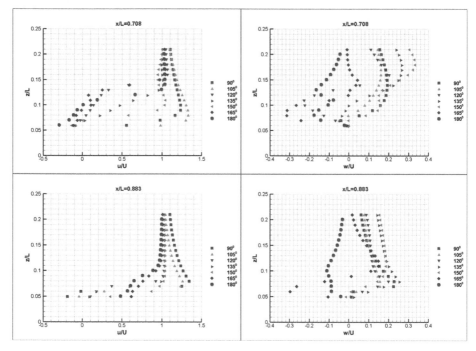

Fig. 7. Velocity distributions at different WOD angles

In Fig. 7, the velocity distributions at different wind-over deck angles were presented at two vertical lines behind the superstructure. At $x/L = 0.708$, the change in the axial velocity has an irregular trend because of the superstructure geometry. The velocity distribution is not changing compatible with the change in the wind angle. However, the beam wind conditions are not affected from the superstructure and the velocities are similar in 90^0 and 105^0. For the tangential velocity component, this similarity is valid between 90^0 and 120^0. At $x/L = 0.883$, the axial velocity distributions were less affected by the wind angle because of the open space in the flight deck. The velocities below the hangar bay have the same value in all angles while the tangential velocity is totally different from the headwind condition. Beam condition cover the angles between 90^0 and 120^0.

5 Conclusion

This paper focuses on the numerical investigation of the airwake in the flight deck of ONRT ship in different model scales and WOD angles. The numerical approach was first validated with the experiments for $\lambda = 77$ at headwind condition. Following this, the analyses were extended to $\lambda = 48.935$ for the angles between 90^0 and 180^0. The analyses showed that the numerical results were mostly in good agreement in two vertical lines behind the superstructure and the over the flight deck. The difference between two methods may be due to the lack of turbulence modelling with RANS approach. Also it is shown that no scale effects were observed in both velocity components. This may be due to the dynamic similarity satisfied by Reynolds number. The effects of wind angle is obvious in the bow-quartering angles. The angles around beam condition does not affect the velocity components much.

This study can be extended using different turbulence models (DES, LES) to observe the flow field in detail. In addition, the free surface effects can be modeled to observe the effects of ship motions.

References

1. Forrest, J.S., Owen, I.: An investigation of ship airwakes using Detached-Eddy Simulation. Comput. Fluids **39**(4), 656–673 (2010). https://doi.org/10.1016/j.compfluid.2009.11.002. Apr.
2. Yuan, W., Wall, A., Lee, R.: Combined numerical and experimental simulations of unsteady ship airwakes. Comput. Fluids **172**, 29–53 (2018). https://doi.org/10.1016/j.compfluid.2018.06.006. Aug.
3. Bardera, R., Matias-Garcia, J.C., Garcia-Magariño, A.: Aerodynamic optimization over helicopter flight-deck of a simplified frigate model using bubble-shaped hangars, Presented at the AIAA AVIATION 2020 FORUM (Jun 2020). https://doi.org/10.2514/6.2020-2960
4. Li, T., Wang, Y.-B., Zhao, N., Qin, N.: An investigation of ship airwake over the frigate afterbody. Int. J. Mod. Phys. B **34**(14 and 16), 2040069 (Apr 2020). https://doi.org/10.1142/S021797922040069X
5. Farish, D.: Large Eddy Simulations of Scaled Wind-Tunnel Experiments for Ship Airwake Analysis, MSc Thesis. The Pennsylvania State University. [Online] Available: https://etda.libraries.psu.edu/catalog/18161dmf5403 (2020). Accessed: 12 Nov 2020
6. Choi, J., Miklosovic, D.S.: LES simulations using the moving mesh method with comparison to experimental results for a periodic ship airwake. Presented at the AIAA AVIATION 2020 FORUM (2020). https://doi.org/10.2514/6.2020-2700
7. Nisham, A., Terziev, M., Tezdogan, T., Beard, T., Incecik, A.: Prediction of the aerodynamic behaviour of a full-scale naval ship in head waves using detached eddy simulation. Ocean Eng. **222**, 108583 (2021). https://doi.org/10.1016/j.oceaneng.2021.108583. Feb.

8. Krebill, A.: Effect of Ship Motion on Ship Airwake Aerodynamics, PhD Thesis, University of Iowa (2020). https://doi.org/10.17077/etd.005517

9. Dooley, G.M., Carrica, P., Martin, J., Krebill, A., Buchholz, J.: Effects of waves, motions and atmospheric turbulence on ship airwakes. Presented at the AIAA Scitech 2019 Forum. San Diego, California, USA (2019). https://doi.org/10.2514/6.2019-1328

10. Dooley, G., Ezequiel Martin, J., Buchholz, J.H.J., Carrica, P.M.: Ship airwakes in waves and motions and effects on helicopter operation. Comp. Fluids **208**, 104627 (Aug 2020). https://doi.org/10.1016/j.compfluid.2020.104627

11. Dooley, G.M., Krebill, A.F., Martin, J.E., Buchholz, J.H.J., Carrica, P.M.: Structure of a ship airwake at multiple scales. AIAA J. **58**(5), 2005–2013 (2020). https://doi.org/10.2514/1.J058994. Feb.

12. Wilcox, D.C.: Turbulence Modeling for CFD. 3rd edition. DCW Industries, La Cãnada, California, USA (2006)

13. Wilcox, D.C.: Formulation of the k-w turbulence model revisited. AIAA J. **46**(11), 2823–2838 (2008). https://doi.org/10.2514/1.36541

Multibody Modeling and Dynamical Analysis of a Fixed-Wing Aircraft

Maria Curcio[1], Carmine Maria Pappalardo[2(✉)], and Domenico Guida[2]

[1] MEID4 Academic Spin-Off of the University of Salerno, Via Giovanni Paolo II, 132, 84084 Fisciano, Italy
m.curcio16@studenti.unisa.it

[2] Department of Industrial Engineering, University of Salerno, Via Giovanni Paolo II, 132, 84084 Fisciano, Italy
cpappalardo@unisa.it

Abstract. This study is aimed at developing a simplified virtual model capable of simulating the dynamic behavior of a fixed-wing aircraft by applying the multibody approach. The case of an aircraft with simplified aerodynamics, an axial thrust, and without control surfaces is considered. After modeling the aerodynamic actions following a Lagrangian approach, the equations of motion are analytically derived and numerically implemented in a MATLAB computer code to develop a virtual model capable of simulating the dynamic behavior of the aircraft. Finally, the numerical results found are presented and a discussion on the numerical results is provided, paying attention to the Cessna 172 Skyhawk, which is considered as the case study.

Keywords: Fixed-wing aircraft · Longitudinal flight dynamics · Multibody simulation · Lagrangian mechanics · Cessna 172 skyhawk

1 Introduction

This section provides an introduction to the topics considered in the present paper. Unmanned Aerial Vehicle (UAV) is a type of aircraft that can be easily piloted either through a ground-based command station or through a pre-established flight plan. Due to their potential applications, their use is growing day by day [1, 2, 32], as they have the fundamental advantage of being able to perform even particularly complex operations, with lower economic and ethical costs [3–7]. Numerical-experimental methods are very suitable in cases like these [8–10]. Generally, in the field of flight dynamics, Newton's second law is mainly used to derive the equations of motion of an aircraft [11, 12]. Blended Newtonian and Lagrangian approaches are seldom used to carry out this type of analysis. However, once the equations of motionare obtained, it is possible to set up a dynamic simulation in a general-purpose software environment like MATLAB that is able to reproduce the dynamic behavior of the aircraft during its flight [13]. The fundamental purpose of this investigation is, therefore, to develop a simplified model that allows for the calculation of the equations of motion of an aircraft. This is obtained

© The Author(s), under exclusive license to Springer Nature Switzerland AG 2022
I. Karabegović et al. (Eds.): NT 2022, LNNS 472, pp. 77–84, 2022.
https://doi.org/10.1007/978-3-031-05230-9_8

through the integration between CAD (Computer-Aided Design) and MBD (Multi-Body Dynamics) methods [14, 15]. The multibody approach can be conveniently used to model the flight dynamics of an aircraft. First of all, it is necessary to model the system as a rigid body, studying the externally applied actions. Subsequently, all the applied actions are analyzed in their reference frame, and then projected in the main reference system used to carry out the analysis. This is done in this paper through symbolic and numeric computations performed in the MATLAB simulation environment, in which the equations of motions are symbolically derived and numerically solved. To achieve this goal, some preliminary hypotheses and assumptions are considered, like a simplified aerodynamic shape, an axial thrust, and the absence of control surfaces. The aircraft is modeled as a multibody system having three degrees of freedom in its longitudinal plane. To verify the consistency of the numerical results obtained from the dynamical simulations, the Cessna 172 Skyhawk is considered as the case study for our investigation, since the design data are free available on the internet or other sources [16, 17].

2 Multibody Dynamics and Flight Dynamics

2.1 Fundamentals of Multibody Dynamics

Articulated mechanical systems, as well as aerospace systems, are made up of a series of components held together by kinematic constraints [18, 19]. The kinematic and dynamic analysis, as well as the trajectory planning, are used to control the motion from one configuration to another [20, 21]. Lagrangian equations are a powerful tool for deriving the equations of motion of a complex multibody system [22]. Assuming the generalized coordinate vector denoted as $q = [q_1 \ q_2 \dots \ q_k \dots \ q_n]^T$, the set of Lagrange equation is given by:

$$\frac{d}{dt}\left(\frac{\partial T}{\partial \dot{q}_k}\right) - \frac{\partial T}{\partial q_k} + \frac{\partial U}{\partial q_k} = Q_k \tag{1}$$

where T is the system kinetic energy and U is the system potential energy. The Lagrangian component Q_k of a generic external force is given by:

$$Q_k = \sum_{j=1}^{N_a} \mathbf{F}_j \frac{\partial \mathbf{r}_j}{\partial q_k} \tag{2}$$

where $\mathbf{F}j = [F_{x,j} F_{y,j} F_{z,j}]^T$ denotes a generic external force applied on the generic rigid body j of the multibody system and N_a is the number of externally applied forces. By performing the analytical calculation mentioned above, a computational Lagrangian approach to derive the aircraft equations of motion is carried out in this paper through symbolic computation performed in MATLAB, leading to the system dynamical model.

2.2 Fundamentals of Flight Dynamics

Assuming, for simplicity, a flat earth surface, we need to introduce three reference frames to correctly develop the model of an aircraft. The first is the inertial earth-fixed reference system denoted as (O^E, X^E, Y^E, Z^E), with the origin conventionally placed on the surface of the earth and the Z^E axis pointing toward the center of the earth. In this work, the earth-fixed reference system is used as the global reference frame. The second reference system is the body-fixed reference system denoted as (G, X^B, Y^B, Z^B). It is used to determine the instantaneous position and orientation of the aircraft with respect to the earth-fixed axis. The point G is the center of mass of the airplane, the X^B axis is pointing toward the nose of the airplane, and the Z^B axis is orthogonal to X^B in order to have $X^B Z^B$ as plane of symmetry of the aircraft. The third reference frame is the wind-fixed reference system, and it is denoted as (G, X^W, Y^W, Z^W). This reference system is used to decompose the aerodynamic forces. The axis X^W points along the relative wind direction, while the Z^W axis, taken perpendicular to X^W, lies along the plane of symmetry of the airplane. Fundamental is the knowledge of the angle of attack α, as the aerodynamic actions depend mainly on it. It is defined as the angle measured between the longitudinal body axis and the direction of the relative wind. It is computed as follows:

$$\alpha = \tan^{-1}\left(\frac{w}{u}\right) \tag{3}$$

where w and u respectively represent the third and the first components of the aircraft velocity vector. Conventionally, the angle of attack is defined positive if the lift is directed in the foot-head direction of the pilot, negative vice versa. The main actions acting on an aircraft during its motion are of three types: the force of gravity F_G (acting in the earth-fixed axis system), the force of propulsion F_P (acting in the body-fixed axis system), and the force induced by the aerodynamic actions F_A (acting in the wind-fixed axis system). These externally applied actions are respectively modeled as follows:

$$\mathbf{F}_G = \begin{bmatrix} 0 \\ mg \end{bmatrix}_E, \quad \mathbf{F}_P = \begin{bmatrix} T_{p,x} \\ 0 \end{bmatrix}_B \approx \begin{bmatrix} T_p \\ 0 \end{bmatrix}_B, \quad \mathbf{F}_A = \begin{bmatrix} -D \\ -L \end{bmatrix}_W = \begin{bmatrix} -D\cos(\alpha) + L\sin(\alpha) \\ -D\sin(\alpha) - L\cos(\alpha) \end{bmatrix}_B \tag{4}$$

where m is the mass of the aircraft, g is the gravity acceleration, T_p is the propulsion or thrust force produced by the engine of the aircraft, L is the magnitude of the lift force, and D is the magnitude of the drag force. The aerodynamic actions can also be written as:

$$L = \frac{1}{2}\rho v_\infty^2 S C_L, \quad D = \frac{1}{2}\rho v_\infty^2 S C_D \tag{5}$$

where C_L and C_D are the lift and drag coefficients, respectively, ρ is the density of the air, v_∞ is the relative speed of the aircraft, and S is the aircraft reference surface [11, 12]. Finally, the net moment acting on the body about Y^B axis is called pitching moment and it is equal to:

$$M_m = \frac{1}{2}\rho v_\infty^2 S c C_m \tag{6}$$

where C_m is the pitching moment coefficient and c is the mean aerodynamic chord [11, 12]. In this investigation, for the simplified assumptions adopted above, the aerodynamic coefficients depend only on the angle of attack, while the other parameters are assumed to be constants.

3   Numerical Results and Discussion

3.1   Description of the Case Study

The aircraft is schematized as a rigid body with three degrees of freedom in its longitudinal plane, as shown in Fig. 1.

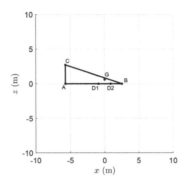

Fig. 1. System geometry schematization

The numerical results found using the system data reported in Table 1 are discussed below.

Table 1. System data

System data	Numerical value
Length	8.28 (m)
Height	2.72 (m)
Takeoff distance	290 (m)
Mean aerodynamic chord	1.47 (m)
Lift coefficient	0.147 (-)
Drag coefficient	0.019 (-)
Pitching moment coefficient	0 (-)
Moment of Inertia	1824.93 ($kg\ m^2$)

The system generalized coordinate vector is given by $\mathbf{q} = \begin{bmatrix} x\ y\ \theta \end{bmatrix}^T$, where x and y respectively represent the horizontal and vertical displacements of the aircraft center of mass, and θ is the aircraft angular displacement. The aircraft is schematized as a triangle, whose vertices are A, B, and C shown in Fig. 1. The points D_1 and D_2 shown in Fig. 1 correspond to the wheels of the aircraft. Finally, the point G is the center of mass of the aircraft. Considering the previously introduced reference frames, the position vectors of the relevant geometric points are derived by applying the fundamental formula of rigid kinematics and the main external actions are transposed in the global reference frame. Thus, one can write the system equations of motion as follows:

$$\mathbf{M\ddot{q}} = \mathbf{Q}_b \tag{7}$$

where $\mathbf{M}$ is the mass matrix, $\mathbf{\ddot{q}}$ is the generalized acceleration vector, and $\mathbf{Q}_b$ is the total body generalized force vector. The equations of motion of the aircraft system modeled in this paper represent a nonlinear set of Ordinary Differential Equations (ODEs) that can be readily implemented and solved by developing a specific computer code in the MATLAB simulation environment. By setting up a dynamic simulation in MATLAB, the initial conditions during the takeoff denoted with $\mathbf{q}_0 = \begin{bmatrix} x_0 & y_0 & \theta_0 \end{bmatrix}^T = \begin{bmatrix} 0 & 0 & 0 \end{bmatrix}^T$ and $\mathbf{\dot{q}}_0 = \begin{bmatrix} \dot{x}_0 & \dot{y}_0 & \dot{\theta}_0 \end{bmatrix}^T = \begin{bmatrix} 0 & 0 & 0 \end{bmatrix}^T$ are considered. A large set of numerical results is obtained, while only a sample of the numerical results is shown in Fig. 2.

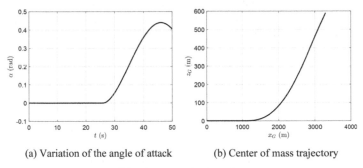

(a) Variation of the angle of attack   (b) Center of mass trajectory

Fig. 2. Numerical results

The resulting time law for the angle of attack is shown in Fig. 2a, while the resulting center of mass trajectory is shown in Fig. 2b.

3.2 Discussion

The results obtained from the simulation correctly reproduce the motion of the aircraft during the takeoff. A fundamental observation concerns the trend of the angle α. After an initial phase in which the angle of attack is zero, it begins to increase during the ground run. It can also be seen that, beyond a certain value, the angle α starts to decrease. This can be explained by considering that for the high value of the angle of attack the stall phenomenon occurs. This implies an abrupt reduction of the lift value and an increase of the drag value, leading to a loss of altitude of the aircraft [11]. To avoid this phenomenon, and to prevent an excessive increase of the angle of attack, it is advisable to act on the control system that regulates the aircraft control surfaces [12]. This issue will be fully addressed and solved in future investigations.

4   Conclusions and Future Work

The principal research areas of interest for the authors are multibody system dynamics of articulated mechanisms and machines, nonlinear control of robotic mechanical systems, and applied system identification of structural systems [23–28, 30, 31, 37]. In this research work, as simulators have been widely used by automotive manufacturers and researchers because they can reduce time and prototyping costs [31, 34, 38], a MAT-LAB computer program was developed by the authors to perform virtual simulations of the set of the multibody equations of motion analytically derived with the use of a Lagrangian approach. As future developments, the authors are planning to remove the simplifying hypotheses adopted in this research work, considering not only the local value of the angle of attack and transforming it as a continuous function byway of polynomial regression [35, 36], but also introducing the topic of aircraft control [33, 39–42]. By doing so, and employing the multibody system approach, a dynamic model having six degrees of freedom in space could be readily developed for the aircraft of interest in future works. As there is an increasing interest of companies in the aerospace sector [29], these important issues will be explored and addressed in future investigations.

References

1. Fahlstrom, P., Gleason, T.: Introduction to UAV Systems. John Wiley & Sons (2012)
2. De Simone, M.C., Guida, D.: Control design for an under-actuated UAV model. FMW Transactions **46**(4), 443–452 (2018)
3. Celenta, G., De Simone, M.C.: Retrofitting Techniques for Agricultural Machines, Lecture Notes in Networks and Systems. 128 LNNS, pp. 388–396 (2020)
4. Rivera, Z.B., De Simone, M.C., Guida, D.: Unmanned ground vehicle modelling in Gazebo/ROS-based environments. Machines **7**(2) (2019). art. no. 42
5. De Simone, M.C., Rivera, Z.B., Guida, D.: Obstacle avoidance system for unmanned ground vehicles by using ultrasonic sensors. Machines **6**(2) (2018). art. no. 18
6. Formato, A., Romano, R., Villecco, F.: A novel device for the soil sterilizing in sustainable agriculture. Lect. Notes Net. Sys. **233**, 858-865 (2021). https://doi.org/10.1007/978-3-030-75275-0_94
7. Salvati, L., d'Amore, M., Fiorentino, A., Pellegrino, A., Sena, P., Villecco, F.: On-road detection of driver fatigue and drowsiness during medium-distance journeys. Entropy **23**(2), 135 (2021)
8. Formato, A., Ianniello, D., Pellegrino, A., Villecco, F.: Vibration-based experimental identification of the elastic moduli using plate specimens of the olive tree. Machines **7**(2) (2019). https://doi.org/10.3390/machines7020046 art. no. 46
9. Sun, X., Liu, H., Song, W., Villecco, F.: Modeling of eddy current welding of rail: three-dimensional simulation. Entropy **22** (2020). art. no.947
10. Liguori, A., Formato, A., Pellegrino, A., Villecco, F.: Study of tank containers for foodstuffs. Machines **9** (2021). art. no. 44
11. Sinha, N.K., Ananthkrishnan, N.: Elementary Flight Dynamics with an Introduction to Bifurcation and Continuation Methods. CRC Press (2013)
12. Sinha, N.K., Ananthkrishnan, N.: Advanced Flight Dynamics with Elements of Flight Control. CRC Press (2017)
13. Yang, F., Yuan, X.G.: Matlab-based dynamic simulation of aircraft environmental control system. ActaSimulataSystematicaSinica **6** (2002)

14. Pappalardo, C.M., Lettieri, A., Guida, D.: Identification of a dynamical model of the latching mechanism of an aircraft hatch door using the numerical algorithms for subspace state-space system identification. IAENG Int. J. Appl. Math. **51**(2), 346–359 (2021)
15. Pappalardo, C.M., Manca, A., Guida, D.: A combined use of the multibody system approach and the finite element analysis for the structural redesign and the topology optimization of the latching component of an aircraft hatch door. IAENG Int. J. Appl. Math. **51**(1), 175–191 (2021)
16. Kim, C.W.: Scale Modeling of Cessna, p. 172 (2011)
17. Smith, R.: Cessna 172: A Pocket History. Amberley Publishing Limited (2010)
18. Sicilia, M., De Simone, M.C.: Development of an energy recovery device based on the dynamics of a semi-trailer. Lect. Notes Mech. Eng., 74–84 (2020)
19. Manrique-Escobar, C.A., Pappalardo, C.M., Guida, D.: A multibody system approach for the systematic development of a closed-chain kinematic model for two-wheeled vehicles. Machines **9**(11), 245 (2021)
20. Manrique-Escobar, C.A., Pappalardo, C.M., Guida, D.: On the analytical and computational methodologies for modelling two-wheeled vehicles within the multibody dynamics framework: a systematic literature review. J. Appl. Comp. Mech. **8**(1), 153–181 (2022)
21. Guida, R., De Simone, M.C., Dašić, P., Guida, D.: Modeling techniques for kinematic analysis of a six-axis robotic arm. IOP Conference Series: Materials Science and Engineering **568**(1) (2019). art. no. 012115
22. De Simone, M.C., Guida, D.: Modal coupling in presence of dry friction. Machines **6**(1) (2018). art. no. 8
23. Pappalardo, C.M., Lettieri, A., Guida, D.: A general multibody approach for the linear and nonlinear stability analysis of bicycle systems. Part I: Methods of constrained dynamics. J. Appl. Comp. Mech. **7**(2), 655–670 (2021)
24. Pappalardo, C.M., Lettieri, A., Guida, D.: A general multibody approach for the linear and nonlinear stability analysis of bicycle systems. Part II: Application to the Whipple-Carvallo bicycle model. J. Appl. Comp. Mech. **7**(2), 671–700 (2021)
25. Manrique Escobar, C.A., Pappalardo, C.M., Guida, D.: A parametric study of a deep reinforcement learning control system applied to the swing-up problem of the cart-pole. Appl. Sci. **10**(24), 9013 (2020)
26. Pappalardo, C.M., Guida, D.: Dynamic analysis and control design of kinematically-driven multibody mechanical systems. Eng. Lett. **28**(4), 1125–1144 (2020)
27. Ardila-Parra, S.A., Pappalardo, C., Estrada, O.A.G., Guida, D.: Finite element based redesign and optimization of aircraft structural components using composite materials. IAENG Int. J. Appl. Math. **50**(4), 1–18 (2020)
28. Pappalardo, C.M., Lombardi, N., Guida, D.: A model-based system engineering approach for the virtual prototyping of an electric vehicle of class l7. Eng. Lett. **28**(1), 215–234 (2020)
29. De Simone, M.C., Ventura, G., Lorusso, A., Guida, D.: Attitude controller design for microsatellites. Lect. Notes Net. Sys. **233**, 21–31 (2021)
30. De Simone, M.C., Guida, D.: Experimental investigation on structural vibrations by a new shaking table. Lect. Notes Mech. Eng., 819–831 (2020)
31. Colucci, F., De Simone, M.C., Guida, D.: TLD design and development for vibration mitigation in structures. Lect. Notes Net. Sys. **76**, 59–72 (2020)
32. Salvati, L., d'Amore, M., Fiorentino, A., Pellegrino, A., Sena, P., Villecco, F.: Development and testing of a methodology for the assessment of acceptability systems. Machines **8**(47) (2020)
33. Li, T., Kou, Z., Wu, J., Yahya, W., Villecco, F.: 4) Multipoint optimal minimum entropy deconvolution adjusted for automatic fault diagnosis of hoist bearing. Shock and Vibration **2021** (2021). art.no. 6614633

34. Liguori, A., Armentani, E., Bertocco, A., Formato, A., Pellegrino, A., Villecco, F.: Noise reduction in spur gear systems. Entropy **22**, 1306 (2020)
35. Dašić, P., Dašić, J., Antanasković, D., Pavićević, N.: Statistical Analysis and Modeling of Global Innovation Index (GII) of Serbia. In: Karabegović, I. (ed.) NT 2020. LNNS, vol. 128, pp. 515–521. Springer, Cham (2020). https://doi.org/10.1007/978-3-030-46817-0_59
36. Tošović, R.; Dašić, P., Ristović, I.: Sustainable use of metallic mineral resources of Serbia from an environmental perspective. Environ. Eng. Manage. J. (EEMJ) **15**(9), pp. 2075–2084 (Sept 2016). 1582–9596. https://doi.org/10.30638/eemj.2016.224
37. Formato, A., Ianniello, D., Romano, R., Pellegrino, A., Villecco, F.: Design and development of a new press for grape marc. Machines **7**(3), 51 (2019)
38. Villecco, F., Aquino, R.P., Calabrò, V., Corrente, M.I., Grasso, A., Naddeo, V.: Fuzzy-assisted ultrafiltration of wastewater from milk industries. In: Naddeo, V., Balakrishnan, M., Choo, K.H. (eds.) Frontiers in Water-Energy-Nexus—Nature-Based Solutions. Springer, Cham, Swiss, pp. 239–242 (2020)
39. Dašić, P., Dašić, J., Crvenković, B.: Applications of access control as a service for software security. Int. J. Indus. Eng. Manage. (IJIEM) **7**(3), pp. 111–116 (Sept 2016). 2217–2661
40. Dašić, P.: Response Surface Methodology: Selected Scientific-Professional Papers (in Serbian, Slovenian and Russian). SaTCIP Publisher Ltd., Vrnjačka Banja (2019). – 301 str. ISBN 978–86–6075–054–1
41. Dašić, P.: Reliability of Technical Systems: Selected Scientific-Professional Papers (in Serbian and Russian). SaTCIP Publisher Ltd., Vrnjačka Banja (2019). – 308 str. ISBN 978–86–6075–051–0
42. Dašić, P.: Scientific and Technological Trends: Selected Scientific-Professional Papers (in Serbian). SaTCIPPublesher Ltd., Vrnjačka Banja (2020). – 305 str. ISBN 978–86–6075–072–5

Improving the Performance of an Aerodynamic Profile by Testing in the Subsonic Wind Tunnel

Apostol Eliza-Ioana[1,2(✉)] and Bălașa Raluca[1,2]

[1] Faculty of Industrial Engineering and Robotics, University Politehnica of Bucharest, 313 Splaiul Independenței, 6 District, Bucharest, Romania
albuelizaioana@yahoo.com
[2] Scientific Researcher at INCAS - National Institute for Aerospace Research "Elie Carafoli", B-dul Iuliu Maniu no. 220, Bucharest, Romania

Abstract. The development of new technologies for commercial aviation involves significant risks to technologies, as these programs are often driven by fixed assumptions about the airline's future needs, while being subject to many technical uncertainties. The effect of these uncertainties is exacerbated by the fact that development programs are long and uncertainties continue to evolve depending on the aircraft. Unfortunately, the standard methods used to carry out all the activities related to the design, manufacture and testing of an aircraft are not sufficient to determine the performance of an aircraft. Thus, worldwide the activity of experimental aerodynamics is crucial in the development and modernization of civilian, military and spacecraft aircraft. The need to perform tests on models of complex phenomena in fluid mechanics, have required, since the late nineteenth century, the design and construction of specific experimental installations called wind tunnels. Thus, wind tunnels have been used with great success since the beginning of aviation as a tool to design new concepts in aerodynamics.

These works aim to improve the performance, accuracy and quality of testing, increase the competitiveness of similar installations on the world market, ensure the national capacity for research - development of new products of the aeronautical and defense industry.

Keywords: Aircraft · Technologies · Wind tunnel · Aviation

1 Introduction

The aerospace industry is very diverse, encompasses a wide range of commercial, industrial, military applications and covers a wide range of interests ranging from design and manufacturing to operation and maintenance of vehicles moving in the Earth's atmosphere or further into space.

Nowadays, every phase necessary for the design of aircraft consists in the improvement and optimization of each configuration. As the aerospace industry goes through these stages, the resources required for design, flight itself and related certifications are constantly growing.

© The Author(s), under exclusive license to Springer Nature Switzerland AG 2022
I. Karabegović et al. (Eds.): NT 2022, LNNS 472, pp. 85–93, 2022.
https://doi.org/10.1007/978-3-031-05230-9_9

Transport efficiency is proven by the diversification of means of transport and by the use of new means of communication. Air transport generates several specific features, such as speed, regularity, adaptability, accessibility, comfort and safety [1].

A special importance in this field is reflected in the safety of air transport because this absolutely necessary feature is influenced by certain technical, human, meteorological factors and certain factors related to air piracy.As we all know, the future of aviation is quite promising, so the progress of activities in space exploration and space travel become a reality. Aerospace engineering is a profession that uses the science of engineering in the development and study of aircraft, technology and spacecraft. It is one of the most important branches of engineering and has developed humanity to master the sky and space [6].

The measurement of the forces and the aerodynamic moments in the subsonic tunnel represents the purpose of this project through which the aim is to obtain the forces and moments produced by the air flow around a model with aerodynamic profile [3].

The objective of these measurements is to determine the estimated loads that will be reported on the actual scale of the wing for structural integrity and to improve its performance.

2   Aerodynamic Characteristics of an Airfoil

Experimental tests on aerodynamic models and airplanes have shown that air flowing along the surface of an aerodynamic model at various angles of attack creates areas along the aerodynamic surface where the pressure is negative or less than atmospheric pressure and areas where the pressure is positive or greater than atmospheric pressure. This negative pressure on the upper surface exerts a greater force on the aerodynamic model than that caused by the positive pressure caused by the air pressure on the surface of the lower wing.An aerodynamic profile is obtained by making a straight cross-section through an airplane wing, a helicopter propeller blade or a ship blade, a wind turbine or athe rotor of a hydraulic machine, etc. The shape of the aerodynamic profile is elongated in the direction offluid flow [7]. An aerodynamic profile is designed to ensure an optiOmal ratio between lift and resistance generated by its interaction with a fluid.The characteristic elements of an aerodynamic profile are sizes that define the shape, the generation mode and some functional aspects (Fig. 1).

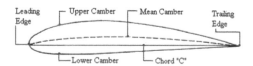

Fig. 1. Components of an airfoil

The aerodynamic profile is a cross-section through a wing and is an aerodynamically optimizedbody, in the sense of developing as high a load-bearing force as possible and a force as low as possible [2] (Fig. 2).

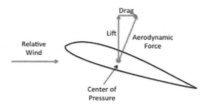

Fig. 2. Forces on an airfoil

Figure 3 shows the pressure distribution along an aerodynamic profile. The average pressure variation, for any given angle of attack, is referred to as the center of pressure (CP).

The center of pressure is defined as the average pressure variation for different angles of attack (CP).Through this center of pressure the aerodynamic force acts. The center of pressure moves forward when the angles of attack are large, and at low angles the center of pressure moves backward.The movement of the center of pressure is very important because it influences the air nets imposed on the wing structures. Aerodynamic balance and control of an aircraft are influenced by the center of pressure [2]. In this project we will study the aerodynamic characteristics of a model with the aerodynamic profile. This model is subjected to an air current and thus an aerodynamic force is exerted on it.

2.1 Model Design

The measurement of the forces and the aerodynamic moments in the subsonic tunnel represents the purpose of this project through which the aim is to obtain the forces and moments produced by the air flow around a model with aerodynamic profile [3].

The objective of these measurements is to determine the estimated loads that will be reported on the actual scale of the wing for structural integrity and to improve its performance.

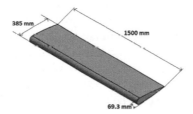

Fig. 3. Dimensions of the aerodynamic profile

These dimensions have been specially adopted to be able to insert inside the model an ERAD4000 type scanner for determining the pressure distribution on the model.

2.2 Experimental Program

This model will be mounted on the three supports of the external balance as shown in Fig. 5, this balance is located outside the subsonic wind tunnel and used to measure the loads transmitted from the aerodynamic model in the test section (three forces and three moments).The balance is also used to position the aerodynamic model in different positions at various angles of attack and rotation. Positioning angles are measured using absolute encoders (Fig. 4).

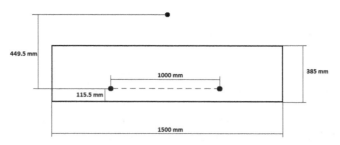

Fig. 4. The position of the balance support on the model

In order to be able to determine the pressure distribution on the basic model, 24 pressure taps were mounted through 6 different sections.

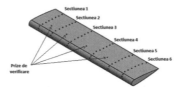

Fig. 5. Sections of the pressure sokets

The pressure taps must be very small in size, have no imperfections and be perpendicular to the aerodynamic surface. Pressure sensors are differential and the reference pressure is atmospheric pressure. The valve gates are connected by small tubes with a diameter of 1 mm to the existing pressure taps on the wing. These tubes are properly connected on the model in 6 different sections on the model [4].

We consider a system of reference axes with origin O at half the distance between the two legs of the balance, respectively in the middle of the model which is at a distance of 115.5 mm from the attack force of the model, as shown in Fig. 6.

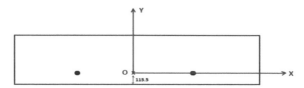

Fig. 6. Reference system

Thus the sections of the pressure taps will be placed at certain distances from the origin O as follows (Table 1):

Table 1. Position of the pressure taps

Section no.	X [mm]
Section 1	−625
Section 2	−437.5
Section 3	−62.5
Section 4	250
Section 5	562.5
Section 6	625

Measurements are made on a standard wing that is connected to a pressure sensor module.These modules have 48 gates each, 44 being connected to a pressure tap on the surface of the model, 2 to the atmospheric pressure (0.46) and 2 to the reference pressure (1.47).

The pressure tap is a hole attached to a sensor. These pressure taps must be small in size,have no imperfections and be perpendicular to the aerodynamic surface.Pressure sensors are differential, and the reference pressure is atmospheric pressure.The 48 protections of the scanner valve are connected by specific small tubes with a diameter of 1 mm to the existing pressuretaps on the wing.These tubes are properly connected to the standard wing so that they are proportionate to the pressure taps on the leading board, the flight board, the center body and the flaps.The pressure sensor, which has been specially designed for this purpose, consists of a membrane and a housing.

2.3 Forces and Moments Measurements

Before starting the data recording, a check of the pressure taps is performed. This consists in applying a pressure that helps to check the pressure taps if they are tight, if they are clogged or if they correspond.I applied the pressure with a pump on each pressure tap of the model, checking that they are tight and not strangled, 24 and 25 on the central body and 43 on the flight board.We have drawn up an execution plan which consists in making a table in Excel with the steps to be followed in order to measure.We had three slots that had to be mounted on the model in succession, each of them being measured at three different distances.

With the help of the program "Sequential reading of mechanical scanvalve V.3.22", an Excel file was opened in which are measured: lift, drag, pitch, roll, yawand side force [5] (Table 2).

Table 2. Forces measurements

L	D	Pitch	Roll	Yaw	Side force
166.5632	4.829634	−40.6661	−0.18788	0.260304	−5.25546
200.4611	4.884322	−41.6876	−0.23798	0.250096	−5.45412
231.073	5.53716	−42.148	−0.23798	0.234784	−5.512815
261.839	6.162654	−42.621	−0.23798	0.257752	−6.135885
289.2983	6.969302	−42.772	−0.22796	0.296032	−6.40227
314.6422	8.073316	−42.5556	−0.27305	0.436392	−6.9531
310.4319	15.43227	−38.0369	−0.27806	−1.86296	12.176955

With the help of the lift and drag we can calculate the aerodynamic coefficients as follows:

- lift coefficient and drag coefficient:

$$CL = 2 * L/\rho * V2 * S \text{ and } CD = 2 * D/\rho * V2 * S$$

where:
L - lift; D - drag; ρ - density; V - speed; S - surface;
Thus we obtain the following values for the load-bearing coefficient and the forward resistance coefficient (Table 3):

Table 3. CL and CD values

CL	CD	Angle of attack
1.102732	0.319746	5.28
1.327154	0.323367	7.28
1.52982	0.366588	9.28
1.733506	0.407999	11.28
1.915301	0.461403	13.27
2.08309	0.534494	15.28
2.055216	1.021694	17.28

Depending on the incident to which the model is subjected, we can graphically represent the polars, respectively the lift and the drag (Figs. 7 and 8):

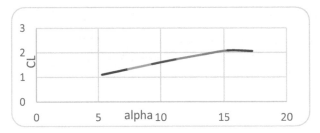

Fig. 7. Lift coefficient vs. Angle of atteck

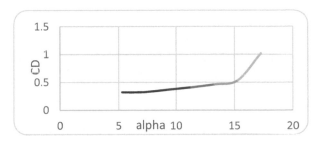

Fig. 8. Drag coefficient vs. Angle of attack

3 Pressure Distribution on the Airfoil

With the help of pressure taps mounted on the aerodynamic profile we can build the wing profile used for these measurements (Fig. 9):

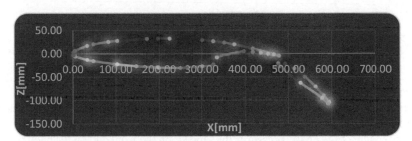

Fig. 9. The aerodynamic profile with pressure taps

Thus we can calculate the pressure in millibars on the model at different angles of incidence. With the help of the calculated pressure, we can graphically represent the pressure distribution on the aerodynamic profile (Figs. 10, 11 and 12).

It is observed that at different angles of incidence from the direction of the air nets, a force appears that tends to move the models in the direction of the air movement upwards. We also notice the formation of swirling areas on the extrados and soffit models.

From what has been shown so far, we can say that the total aerodynamic force arises due to:

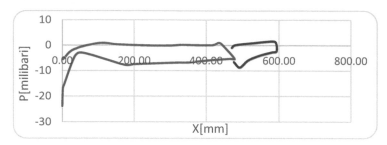

Fig. 10. Pressure distribution at the angle of incidence 4°

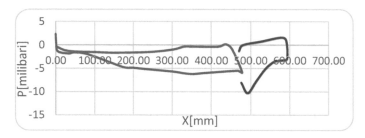

Fig. 11. Pressure distribution at the angle of incidence −8°

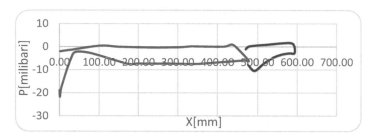

Fig. 12. Pressure distribution at the angle of incidence 0°

- the pressure difference that appears on the lower and upper part of the body, resulting in a force that tends to lift the body;
- the difference between the pressures that appear on the front and on the back the body;
- friction of the air with the surface of the body, resulting, together with the difference pressure from the previous point, a force that tends to rotate the plate, or otherwise said to tilt it in the direction of the movement of the air masses.

4   Conclusions

In relation to the requirements of aeronautics and astronautics, the construction of the respective facilities has acquired a very diversified multilateral development which has led to the establishment of criteria according to which certain classifications can be made.

In relation to the air entrainment mode, the wind tunnels are open circuit or closed circuit, and in relation to the operation mode they are with continuous or intermittent operation.

The main equipment for measuring the forces and aerodynamic moments that appear on the model due to the movement of air in the experimental area is the aerodynamic balance. It is able to measure according to its own reference system, the components of the torsor of the aerodynamic forces in the virtual center of the balance.

Every aircraft in flight has 6° of freedom, being free to make a translational movement along the three perpendicular axes and a rotational movement also around the three axes. The measurement of forces and moments in the wind tunnel aims to obtain their values on the 3 axes.

Related to this, we aim to improve the way data are processed and displayed. In order to make the testing more efficient, it is necessary to increase the data processing capacity. This is done by adding computing systems or improving current computing systems. Thus, the experimental data are available for analysis much faster, immediately after completion or even during testing as appropriate. In order to be able to interpret the data more easily, several screens will be placed where data related to the operating parameters of the tunnels and the measured aerodynamic sizes will be displayed.

The main objective of this paper is fundamental and applied research in order to improve the aerodynamic parameters resulting from the testing in the wind tunnel of an aircraft model, by addressing these specific aspects of quality assurance we want to achieve this goal and highlight the importance of scientific research in the aerospace field.

References

1. Wang, Y., Zhou, T.: Experimental study on high speed two-dimensional airfoil based on conventional high speed wind tunnel test. J. Phys.: Conf. Ser. **1626**, 012183 (2020) https://doi.org/10.1088/1742-6596/1626/1/012183
2. Soontornpasatch, T.: Computational study of low and high subsonic speed aerodynamic characteristics of the modified airfoil profile. IOP Conf. Ser.: Mater. Sci. Eng. **405**, 012002 (2018) https://doi.org/10.1088/1757-899X/405/1/012002
3. Makhija, D., et al.: Aerodynamic analysis of aircraft model using indigenously developed wind tunnel facility, IOP Conf. Ser.: Mater. Sci. Eng. **1206**, 012013 (2021)
4. Etkin, B., Reid, L.D.: Dynamic of Flight, Stability and Control. Third Eddition, John Wiley & Sons, INC, pp. 125–134 (1996)
5. Roskam, J.: Airplane Desing Part VI: Preliminary Calculation of Aerodynamic, Thrust and Power Characteristics, pp. 67–80. DARcorporation (2000)
6. Dmitriev, A.Y. et al.: Special aspects of quality assurance in the design, manufacture, testing of aerospace engineering products. IOP Conf. Ser.: Mater. Sci. Eng. **714**, 012006 (2020). https://doi.org/10.1088/1757-899X/714/1/012006
7. Kovrigin, E., Vasiliev, V.: Trends in the development of a digital quality management system in the aerospace industry. IOP Conf. Ser.: Mater. Sci. Eng. **868**, 012011 (2020). https://doi.org/10.1088/1757-899X/868/1/012011

The Influence of Lithium Concentration at the Surface of Al-Li Aerospace Alloy Extrusions on Dye Penetrant Inspection

Vlad Ciprian Cupșan[1,2]([✉]) and Aurel Mihail Titu[3]

[1] Polytechnic University of Bucharest, 313 Splaiul Independenței, Bucharest, Romania
vladcupsan@yahoo.com
[2] Universal Alloy Corporation Europe, Dumbrăvița, Maramureș, Romania
[3] Lucian Blaga University of Sibiu, 10 Victoriei Street, Sibiu, Romania
mihail.titu@ulbsibiu.ro

Abstract. This paper depicts the current processing of aerospace aluminium alloy Al-Li extruded material, with emphasis on nonconformities caused by the permanent fluorescent background during dye penetrant inspection. The process for dye penetrant inspection is described in full details to provide a clear image about each individual processing step. The second part includes a series of tests performed on aluminium alloy Al-Li type material during dye penetrant inspection, as part of the root cause identification process. The last part of the paper illustrates an innovative hypothesis related to the cause behind the permanent fluorescent background during dye penetrant inspection. The innovative hypothesis related to the influence of lithium concentration at the surface of the Al-Li aerospace alloy extrusions is tested by relevant trials and results are presented as part of the conclusions.

Keywords: Aluminum alloy · Al-Li · Dye penetrant · Fluorescent background · Lithium concentration

1 Introduction

In the current context of the commercial aircraft sector, the costs for each flight must be reduced continuously in order to allow an increase in competitiveness for each airline in service. The main driver for the cost of airline ticket prices is the fuel consumption of the aircraft during one flight cycle. There are many areas of improvement for the reduction of aircraft fuel usage, like: more efficient engines, smarter aerodynamic surfaces and aircraft weight reduction. The latter can be divided also in different areas of the aircraft, with the main one being the aerostructure of the aircraft. Being the heaviest part of the aircraft beside from the engines, it provides a lot of opportunities for decreasing weight. The aerostructure has a high complexity and at the same time a large diversity in singular components. Thus, the incremental improvements of the aircraft aerostructure for weight reduction went in the direction of identifying new lighter materials. It is only logical that largest components of the aerostructure have been taken in consideration for applying these lightweight materials. One example would be the crossbeams which are part of

I. Karabegović et al. (Eds.): NT 2022, LNNS 472, pp. 94–99, 2022.
https://doi.org/10.1007/978-3-031-05230-9_10

the passenger floor grid. These span over the width of the aircraft and are needed at each frame location for structural stability. Taking in consideration that composite materials were not an option for such components, aircraft designers went in the direction of innovative aluminum alloys [1]. The material which has been selected for the new types of aircraft such as the Airbus A350 is the aluminum-lithium alloys which are lighter than conventional aluminum alloys.

2   Dye Penetrant Inspection

Nondestructive testing, including dye penetrant inspection is a vital activity in the aerospace product manufacturing flows. Following the mechanical processing of aerospace materials it is necessary to identify any material failure, that can pose problems during product lifecycle after installation on the aircraft. For aluminum alloys, including aluminum-lithium alloys, the most used nondestructive inspection method is the dye penetrant inspection. In general, this inspection is performed following the mechanical processing of the material and before applying the surface treatment.

Dye penetrant inspection in the aerospace industry is considered to be a special process performed by highly trained personnel with clear and documented process steps. Beside from the preparation of the surface, where parts need to be degreased and etched to a certain depth, the dye penetrant includes the steps presented in Fig. 1 [2].

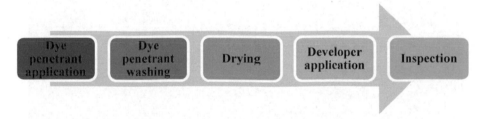

Fig. 1. Processing sequence performed during dye penetrant inspection

Each of the above mentioned process steps have requirements which have to be met with precision. For example, after penetrant application there are strict rules for penetrant dwell time. Also, during penetrant washing it is very important not to apply water with too much pressure, else there is the risk to remove penetrant from the defect areas.Still, there is a continous learning process related to dye penetrant inspection in conjunction with new aerospace materials and including the new demand rates from the aerospace industry. In this context the equipment used for dye penetrant inspection needs to evolve in order to cope with the unprecedented demand. Larger penetrant inspection lines are becoming more and more the standard in the industry. This poses new challenges in performing the process to the desired process parameters, as for example washing a production batch with hundreds of products takes much longer than performing the same activity for a few products. Out of this new mix of process, new materials and new equipment problems can arise which need to be analyzed and improved [3].

3 Permanent Fluorescent Background

Taking in consideration the combination of lightweight materials, such as aluminum-lithium alloys and the complexity and importance of the dye penetrant inspection is only logical that at some point issues start to arise [4]. In addition to the inherit complexity in melting and extruding, the aluminum-lithium alloys are very sensitive to process variations. During the subsequent processing of an aircraft component, by conventional machining, there are challenges with local overheating and metal smearing. This specificity in machining requires a lot of attention during the quality control inspections. Also it is necessary to dye penetrant inspect the product, to certify that it is according to the requirements and free of cracks. This makes the dye penetrant inspection vital to the airworthiness of such components. Following the mandatory precleaning process for the material, by means of chemical processing, the product must be inspected with dye penetrant as explained in the previous chapter [5].

It has been observed, generally in the industry, that the aluminum-lithium alloys tend to generate a light fluorescent background during dye penetrant inspection. This phenomenon was not considered important as it allowed the inspection of the product, under secure conditions. Still, there are cases in which some products showed more severe fluorescent background as the one presented in Fig. 2.

Fig. 2. Aerospace products presenting permanent fluorescent background

It was noticed that the fluorescent background could not be removed by regular solvent cleaning, in order to perform the wipe-off and reinspect the material. Taking in consideration these properties the defect was named: permanent fluorescent background. This type of defects is fairly new, with not much documentationbeing available related to this subject. Products affected by this problem, tend to show also brown staining in the impacted area when analyzed under white light, as presented in Fig. 3. Following a series of tests related to the defect, it has been concluded that it affects only aluminum-lithium alloys and not the conventional aluminum alloys, such as 2000 or 7000 series,which are regularly used for aerospace products [6]. Further observations revealed that the machined area did not present traces of the permanent fluorescent background, opening the discussion related to the surface of the extruded Al-Li profile.

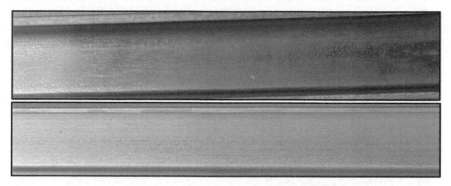

Fig. 3. Comparison between affected material and not affected material under white light

It was observed that some Al-Li materials behaved better, not presenting such severe fluorescent background, but rather the lighter version which does not affect the dye penetrant inspection. With all of the above taken in consideration the only logical area to look into is the material itself, more explicitly at the surface of the aluminum-lithium alloy extruded profile. Taking into account the differences between conventional aluminum alloys and the Al-Li type alloys the following hypothesis has been developed: the permanent fluorescent background is caused by lithium concentration at the surface of the extruded profile in conjunction with specific steps needed during penetrant inspection, such as dye penetrant application and washing. A potential violent reaction between the lithium compound and water vapor, entraps the penetrant at the surface of the extruded profile, creating also the brown staining [7].

4   Hypothesis Testing

In order to test the hypothesis presented previously, two aluminum-lithium alloy materials have been selected. One showed only the light but acceptable fluorescent background and the second materialpresented the severe permanent fluorescent background [8]. To test the lithium concentration at the surface of the extruded profiles, incremental chemical etching was performedto remove layers of approximatively 25 microns from the previous surface. In order to measure the lithium concentration the XRF measuring technique was used. Because the XRF technique measures at a depth between 0 and 50 microns, all values were considered to be in the middle of the interval. Due to this equipment limitation, the initial values for non-etched material are shown at 25 microns depth, in Fig. 4. Marked with shades of green are the samples from the material which in practice has developed the light fluorescent background and in shades of orange are the samples from the material that developed the severe permanent fluorescent background. It can be observed easily that the material most susceptible to severe permanent fluorescent background is the one showing a higher concentration of lithium at the surface in the 25 to 50 microns range. Contrary, the material which performs better under the same situation, presenting only a light fluorescent background has a gradient of Li concentration at the surface of the extruded profile. The gradient is starting from a 75 microns depth, where Lithium concentrations are similar in the range of 1.3–1.5%Li, to difference of 0.3%Li at 50 microns depth and a difference of 0.8%Li at 25 microns depth. By following the current gradient of lithium concentration in the Al-Li alloy which has better results during dye penetrant inspection, it is estimated that the lithium concentration exactly at the surface of the extruded profile tends towards zero.

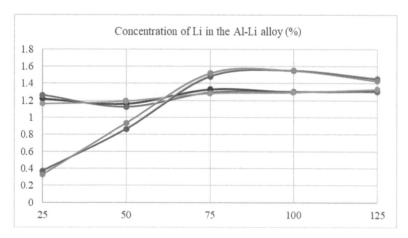

Fig. 4. Lithium concentration at the surface of the Al-Li extruded profiles

Therefore, the hypothesis is confirmed.The low lithium concentration at the surface of the Al-Li extruded profiles has a positive influence on the results following the dye penetrant processing, resulting in a light fluorescent background. On the other hand the profiles which have a higher concentration at the surface, after dye penetrant processing,show severe permanent fluorescent background. The tests performed confirm the hypothesis expressed and furthermore gives clear evidence in this regard.

5   Conclusion

The new type of defect called permanent fluorescent background,that interferes with the dye penetrant inspection, makes the processing of some aluminum-lithium alloy extruded profiles impossible. The severe permanent fluorescent background does not allow the identification of relevant indications, on the contrary hiding them. It is not recommended to continue the dye penetrant inspection when this defect is identified on the extruded profiles. Performing an extra deoxidizing cycle, including rinsing, allows the removal of the defect and the profiles can be re-inspected with dye penetrant. When the hypothesis was tested, the lithium concentration at the surface of the Al-Li extruded profiles has influenced the severity of the fluorescent background.Profiles having a low or near zero concentration at the surface have better results, considered acceptable for the penetrant inspection. Contrary, the profiles that had a constant lithium concentrationat the surface haveshown indications of severe permanent fluorescent background.

References

1. Li, J., et al.: Improvement of aluminum lithium alloy adhesion performance based on sandblasting techniques. Int. J. Adhesion and Adhesives **84**, 307–316 (2018)
2. Shipway, N.J., et al.: Automated defect detection for fluorescent penetrant inspection using random forest. NDT and E Int. **101**, 113–123 (2019)
3. Shipway, N.J., et al.: Using ResNets to perform automated defect detection for fluorescent penetrant inspection. NDT and E Int. **119**, 102400 (2021)
4. Karigiannis, J., et al.: Multi-robot system for automated fluorescent penetrant indication inspection with deep neural nets. Procedia Manufacturing **53**, 735–740 (2021)
5. Xiao, R., Zhang, X.: Problems and issues in laser beam welding of aluminum–lithium alloys. J. Manuf. Process. **16**, 166–175 (2014)
6. Zheng, J., Xie, W.F., Viens, M., Birglen, L., Mantegh, I.: Design of an advanced automatic inspection system for aircraft parts based on fluorescent penetrant inspection analysis. Insight - Non-Destructive Testing and Condi. Moni. **57**, 18–34 (2015)
7. Zolfaghari, A., Kolahan, F.: Reliability and sensitivity of visible liquid penetrant NDT for inspection of welded components. Materials Testing **59**, 290–294 (2017)
8. Endramawan, T., Sifa, A.: Non destructive test dye penetrant and ultrasonic onWelding SMAW butt joint with acceptance criteria ASMEStandard. Mater. Sci. Eng. **306**, 012122 (2018)

Breakdown of the Product Quality Assurance Flow Within the Advanced Product Quality Planning (APQP) Methodology in the Aerospace Industry

Aurel Mihail Titu[1,2(✉)] and Vlad Ciprian Cupșan[3]

[1] Lucian Blaga University of Sibiu, 10 Victoriei Street, Sibiu, Romania
mihail.titu@ulbsibiu.ro
[2] Polytechnic University of Bucharest, 313 SplaiulIndependenței, Bucharest, Romania
[3] Universal Alloy Corporation Europe, Dumbrăvița, Maramureș, Romania

Abstract. This paper presents a breakdown of the product quality assurance flow from the Advanced Product Quality Planning (APQP) methodology within the aerospace industry. It starts with the description of the current requirements for APQP deployment and continues with specific information related to APQP standardization in the aerospace industry, that includes also the AS 9145 aerospace standard. The paper gives step by step guidance on how to implement and use the product quality assurance flow as part of the aerospace APQP methodology. To highlight the contributions for the implementation of core APQP activities per the applicable requirements, the paper includes industry application models. The last part, the conclusions of the paper, provide a list of key points to be considered for any organization within the aerospace industry which aims to implement the Advanced Product Quality Planning (APQP) methodology based on the AS 9145 aerospace standard.

Keywords: Aerospace industry · APQP · AS 9145 · Product quality · Flow

1 Introduction

The aerospace industry has been at the forefront of innovation related to materials and processes, always pushing the boundaries in order to improve the customer experience and costs. Since the takeoff of the commercial aircraft sector, the aerospace industry has used the lessons learned by the military aircraft sector, thus implementing and using some aspects of standardization and statistical control. Based on this evolution and in conjunction with its specificity, the aerospace industry has incrementally improved the approaches applied during the design, plan and manufacturing of aircraft materials, components and assemblies. Mainly due to low number of aircraft build per year, until recently the aerospace industry has relied only on sporadic application of some advanced planning tools, but without linking the activities together. With the success seen in other industries, like automotive, the methodology for Advanced Product Quality Planning

© The Author(s), under exclusive license to Springer Nature Switzerland AG 2022
I. Karabegović et al. (Eds.): NT 2022, LNNS 472, pp. 100–107, 2022.
https://doi.org/10.1007/978-3-031-05230-9_11

has drawn the attention of the aerospace manufacturers. These companies under the International Aerospace Quality Group have started to develop guidance materials for the APQP methodology adapted to the aerospace industry.

2 APQP in the Aerospace Industry

2.1 APQP Standardization

The AS 9145 aerospace standard was released in 2016 by the International Aerospace Quality Group (IAQG) defining the principles for Advanced Product Quality Planning (APQP) and those of Production Part Approval Process (PPAP) in the aerospace industry [1]. The standard provides the requirements for the methodology, starting with the requirements for product development, from concept throughout the design phase, up to the production of the articles and until the feedback from the customer. PPAP approval is the confirmation from the customer that the design and production processes have demonstrate complete fulfillment of customer requirements.

This standard applies for the development of new products, but can be usedfor existing products that require major changes. The standard can be considered as a contractual requirement between customer and supplier, with the applicability being defined between the previously mentioned parties. The requirements specified in the AS 9145 aerospace standard are aimingthe improvement of product quality, product delivery, cost reduction and the fulfillment all customer requirements, ensuring customer satisfaction.The standardization of the requirements for design and manufacturing development and of the process to approve the production parts is meant to reduce or eliminate internal organizational requirements aligning them to those of the standard and of other organizations in the aerospace industry.

The standard is structured in 6 main chapters, with the first chapter defining the scope of the standard itself. Section 2 includes the relevant industry references, such as the AS 9100 aerospace standard for Quality Management Systems (QMS), but also a link to the Aerospace Advanced Product Quality Planning Manual from IAQG. Terms are defined in Sect. 3. The main chapter for the aerospace APQP standard is Sect. 4, which includes the general requirements, as well as the requirements for managing Advanced Product Quality Planning projects. This chapter also includes the requirements for each phase, starting with Phase 1 - Planning, followed by Phase 2 - Product design and development, Phase 3 - Process design and development, Phase 4 - Product and Process Validation and finishing with Phase 5 - On-Going Production, Use, and Post-Delivery Service. Chapter 5 includes all the industry requirements for the Production Part Approval Process (PPAP), which are: submission of PPAP file, PPAP disposition, customer approval or rejection of PPAP submission. Chapter 6 of the AS 9145 standard includes the notes. The standard also includes four annexes: Appendix A which includes the acronym log; Appendix B - Advanced Product Quality Planning phase activities, deliverables and outputs; Appendix C includes a Control Plan detailed explanation; Appendix D - Production Part Approval Process (PPAP) approval form. Figure 1 from the AS 9145 standard presents the APQP methodology divided by Phases and milestones. Also the standard includes two important tables, Table 1 which shows the acceptance criteria for process capability studies and Table 2 that includes the production part approval process (PPAP) file contents.

2.2   Breakdown of APQP

The aerospace APQP methodology as defined also by the AS 9145 aerospace standard is broken down as shown in Fig. 1 into phases, activities and deliverables.

Fig. 1. APQP breakdown according to AS 9145 aerospace standard

This being the classic way to see the APQP methodology in a similar way as a Quality Management System (QMS), consisting of procedures, work instructions and forms.Nonetheless, there isanother way to breakdown the aerospace APQP methodology, as described by the APQP standard and IAQG Manual for APQP [2], by specialized flows within the overall set of activities. In the APQP methodology for the aerospace industry the following flows can be identified as shown in Fig. 2.

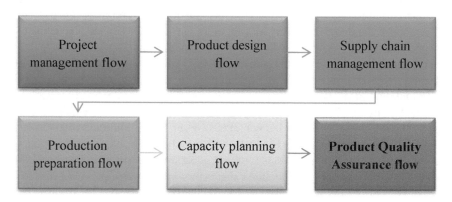

Fig. 2. APQP breakdown in specialized flows

Each flow has a specific goal and an area where the activities bring added value. For example, the Product design flow includes activities which are linked directly with product definition, from concept to the release of the design data. The capacity planning flow includes the planning activity and alsocapacity verification in Phase 4.

2.3 APQP Requirements for Deployment

The APQP methodology has already shown its benefits in the automotive industry, where it has allowed the creation of a standardized communication between customer and supplier. With this success story and the ever increasing demand of aircraft, despite the Covid-19 crisis, the aerospace industry has continued the deployment of APQP requirements. Main three actors in this deployment are the International Aerospace Quality Group (IAQG) and the two biggest aircraft producers, Airbus and The Boeing Company. IAQG is leading the development and release of aerospace standardized documents, such as the APQP standard and the aerospace APQP manual, created in collaboration with the IAQG member companies, as presented in Fig. 3.

Fig. 3. Location of IAQG member companies

On the other side, Airbus and The Boeing Company have started to deploy the requirements for applying APQP to all new work packages or transfers of work. This has been done by implementing contractual requirements for APQP deployment as a complete methodology, as well as for some APQP activities independently. Furthermore the two biggest aircraft manufacturers have started to require contractually the deployment of the methodology to all sub-tiers for each specific contract. It is expected that all aerospace contracts will include some requirements related to the deployment of the APQP methodology in the aerospace industry [3]. Companies from the industry which are operating under the context presented before, are executing the implementation of the APQP methodology into their own companies and quality systems. Taking in consideration the high number of companies, their sizes and the geographical location, gaps start to evolve related to the implementation. On one side there are companies in which management support for APQP is high and also resources are available, these have already finished the implementation of the methodology and are using it on a daily basis. On the other side there are the companies with limited resources which need to make wise decision related to the implementation, in order to be able to comply with the industry and customer requirements for APQP deploymentand at the same time use less resources for the implementation, by using already developed application models as the ones presented in the next chapter.

3   APQP Application Models for the Product Quality Assurance Flow

The aerospace APQP methodology is quite complex, it includes over 40 activities as described by the IAQG APQP aerospace manual, but still depending on the project and of the specificities of the product to be implemented with APQP, some of the activities bring less or none value. Thus, it is important to understand at the beginning of the project the necessary activities to complete the project successfully and with optimal resources. As explained before, the APQP activities are liked within specific flows throughout the APQP project lifecycle. One example of reducing the amount of APQP activities is when a product is produced only in house, in this case the supply chain management flow is not necessary and would consume resources, without adding any value to the project.

By screening the aerospace APQP methodology we can come to the conclusion that there is an APQP specific flow that needs to be always applied when executing an APQP project. The product quality assurance flow, which sometimes is called the APQP „Core" flow is always applicable to a certain extend. This flow is very similar to the automotive APQP flow [4], from which it actually originates, but due to the nature of the aerospace products and diversity needs to be applied in a different way in order to bring value to the aerospace APQP projects. The following part of this paper will show and explain aerospace APQP application of some activities within the product quality assurance flow.

3.1   Product Key Characteristics and Critical Items

One of the most important activities related to product quality, the identification of product special requirements is very different in the aerospace industry from two points of view. First due to the nature of the industry and the specificity of the aerospace design and development, which occurs very rarely, thus most of the products involved in Transfer of Work type projects, do not include already defined special requirements of the design data, as some of the drawings are from the 20th century. This reality shifts the responsibility for the definition of product special requirements to the organization which doing the manufacturing. Often the production organization does not have access to the design organization or to the design risk analysis. A second reason would be the complexity and safety concerns related to the aerospace products. This creates for each product being implemented with APQP a high number of characteristics which need to be analyzed. Both of the above mentioned reasons put a lot of strain on the APQP approach developed by the organization which is planned to do the manufacturing of the product and still comply with the requirements from APQP.

One way not do to this activity within the aerospace industry is to allocate the task of identifying product special requirements such as product key characteristics or critical items to a single person in the organization. Furthermore, if the responsible person does not have previous experience with similar products than the activity for sure will not bring any value to the project and most probably will derail the complete project. Such approaches are to be avoided.

A better approach would be for the organization to create a process for identification of product special requirements that includes the APQP pillars and fulfills the requirements of the IAQG APQP Checklists for this kind of activity [5]. An application model for such a process needs to involve a multifunctional team comprised of engineering, manufacturing engineering, quality, production and if available the customer or design organization representatives. The more experienced the multifunctional team is, the better the results for product special requirements identification process. A model to apply this is to define a person to lead this activity and facilitate the communication of the design data to all team members. The team leader must then collect all the feedback from the team in a structured way. A model for this activity is the one presented in Table 1, which is designed as a matrix and allows the identification and definition of key characteristics and critical items in an easy way, even for multiple part numbers from the same product family.

Table 1. Application model of product special requirements definition. Part 1.

Scope			
Work package	Aerospace WP	**Product family**	Painted profiles
Product list		**Product applicability matrix**	
Internal PN	**Customer PN**	**Characteristic 1**	**Characteristic 2**
Product A	Customer PN A	Applicable	N/A
Product B	Customer PN B	N/A	Applicable
Product C	Customer PN C	Applicable	Applicable
Preliminary identification of characteristics			
Description		Thickness deviation	Holes after painting
Nominal value		0	Pass
Lower Specification Limit		-0.2	n/a
Upper Specification Limit		0.2	n/a
Source		Drawing, area C5	Condition of Supply
Type of characteristic			
Type of feature		Key Characteristic	Critical Item
Multifunctional team members and approval			
Name	**Position**	**Signature**	
Member A	Program Manager		
Member B	Engineering		
Member C	Quality		
Member D	Production		

The above application model includes all relevant steps for a robust definition process for product special requirements. At the same time it allows the necessary traceability to the work package and product family, as well as including the product applicability matrix. A preliminary list is created at first in the „Preliminary identification of characteristics" where the feedback from the team is collected. Following the collection process, the next step according to the application model is to define the type of characteristic, by general agreement within the team. Last step being the approval of the list of special requirements, which will be used in the subsequent APQP activities.

3.2 Process Key Characteristics and Critical Items

Following the definition of the product special requirements and the development of the process flow diagram, the APQP product quality assurance process includes the definition of a process risk analisys which is the precursor of the process special requirements identification and definition. Here the automotive industry APQP approach is pretty straightforward allowing the definition of process special requirements directly from the Process Failure Modes and Effects Analysis (PFMEA). On the other hand the aerospace industry has other constraints which need to be taken into account. For example the aerospace industry due to its safety requirements has developed a lot of requirements for the manufacturing processes. One of the two main aircraft producers has actually included a process special requirements list for each of the process specifications. This definition, in general, helps the aerospace product manufacturers, but under certain conditions it can hide some process special requirements, as the Original Equipment Manufacturers (OEM) definition is generic and not linked with a specific product design.

Table 2. Application model of product special requirements definition. Part 2.

Scope						
Work package	Aerospace WP		**Product family**	Painted profiles		
Characteristic list			**Identification**	**Process applicability**		
Description	**LSL**	**Nominal**	**USL**	**Characteristic designation**	**Process step**	**Process step**
Concentration	6%	8%	10%	Critical Item	Applicable	N/A
Temperature	18	20	22	Key Characteristic	N/A	Applicable
				Process step	Machining	Painting
				Specification	Spec 1	Spec 2
Multifunctional team members and approval						
Name	**Position**		**Signature**			
Member A	Prog. Manager					
Member B	Engineering					
Member C	Quality					
Member D	Production					

As shown in Table 2, the model follows the necessary steps in order to comply with all the aerospace requirements for APQP, including the guidance and checklists provided by IAQG [6].

4   Conclusion

As described previously in the paper the implementation of certain aerospace APQP activities varies from organization to organization and does not allow the uniform approach. The application models included in this paper provide clear ways to incorporate and execute the current aerospace APQP requirements. These application models are related to critical activities that are part of the Product Quality Assurance flow, which in term is the most important flow from the APQP methodology. It is very important to execute product and process special requirements identification and definition to the highest accuracy possible, by following the guidelines included in this paper. Else, the complete set of aerospace APQP activities following the above mentioned elements, will be completely derailed. In such a case the complete project is at risk and potentially will lack in effectiveness due to a high number or product nonconformities, outside of the proposed project targets.

References

1. SAE International: Requirements for advanced product quality planning and production part approval process. AS **9145**, 2016–11 (2016)
2. IAQG: Aerospace APQP/PPAP Manual, SCMH, Section 7.2.3 (2017)
3. Titu, A.M., Cupşan, V.C.: Review of advanced product quality planning (APQP) in the aerospace industry, In: 11th Research/Expert Conference with International Participation „QUALITY 2019", B&H, Neum (2019). http://quality.unze.ba/zbornici/QUALITY%202019/013-Q19-011.pdf
4. EATON: Advanced Product Quality Planning (APQP) and Production Part Approval Process (PPAP), CQD-116, Rev 1 (2015)
5. Duwendag D.: APQP in der Luftfahrtindustrie nach DIN EN 9145:2019, Beuth Verlag (2021)
6. IAQG: Key Characteristic (KC) Traceability Form Guidance, SCMH, Section 7.2.17 (2020)

An Analysis of Heat Affected Zone at Percussion Welding of Austenitic Steel Wires

Edin Džiho[(⊠)], Sead Pašić, and Edin Šunje

Faculty of Mechanical Engineering Univerzitetski Kampus, Džemal Bijedić"
University of Mostar, 88104 Mostar, Bosnia and Herzegovina
edin.dziho@unmo.ba

Abstract. Capacity discharge percussion welding is nonconventional welding process where heat source is an electric arc obtained by discharging of capacitor bank. The arching time is very short, around 10 ms, and depends on a few parameters such as: capacity of capacitor bank, voltage, movement speed of welded pieces during welding. During arching a certain amount of material is melted and after that squeezed out by applied force used in welding process. Upon tese welding parameters depends many welded joint characteristics, quantity of squeezed material, heat affected zone properties, geometry of welded joint. All mentioned directly influence on mechanical properties of welded joints. In this paper is analysed the influence of capacity discharge percussion welding parameters on heat affected zone properies of welded joints.

Keywords: Percussion welding · Welding parameters · Heat affected zone · Microstructure · Mechanical properties of welded joint · Hardness

1 Introduction

All welding processes that use capacitor capacity as an energy source are considered capacitor discharge welding processes. There are a number of these processes, however, the most common are capacitor discharge percussion welding - CDPW, CD stud welding and capacitor discharge welding [1].

Capacitor discharge percussion welding is a butt welding method of the metal pieces under the influence of a mechanical load applied during or immediately after an electric arc burning. The arc is obtained by discharging the energy stored to the capacitors in order to melt the connecting surfaces [2].

A main characteristic of a capacitor discharge percussion welding compared to other capacitor discharge welding processes is a creation of the joints in solid phase during mutual moving of the pieces to be joined. This moving causes collisions of work pieces at the end of the process and provides a relatively constant electric arc [1, 2].

There are several welding process parameters, as most importantare determined:

- charging voltage of capacitors bank,
- capacity of capacitors bank,

© The Author(s), under exclusive license to Springer Nature Switzerland AG 2022
I. Karabegović et al. (Eds.): NT 2022, LNNS 472, pp. 108–113, 2022.
https://doi.org/10.1007/978-3-031-05230-9_12

- intensity of a pressing force of the wires during welding,
- velocity of the wires during welding.

Thus, by discharging the capacitor bank, a short term electric arc is generated and melts a certain amount of metal. After that, the pieces are mutually joined with certain force. This is the main reason why this process is commonly used for joining wires and similar shape products of small dimensions. Energy stored at capacitoer bank depends on capacity of capacitor bank and voltage charged at capacitor bank:.

$$W = \frac{1}{2}CU^2 \tag{1}$$

where:

W – stored energy at capacitor bank [J],
C – capacity of capacitor bank [F],
U – charged voltage of capacitor bank [V].

At percussion welding process welding time is very short, and process can be divided into three phases: melting of ignition tip, arcing and phase of pressing. During the phase of melting of ignition tip, ignition tip is heating and melting by Jules heat. During the phase of arcing, the arc is generated to melt working pieces, and then the pieces are conected by pressure force of spring. At the end, during the phase of pressing, the arc is extinct, melted material is squeezed from pool of melted material, until solidification.

In this paper were used the wires made of austenitic steel and the influence on welding parameters related to a mechanical properties of a welded joint was analysed.

2 Experimental Part

In experimental part of the work, the samples for welding were made of austenitic stainless steel Cr Ni 19 9. The samples were made of cold-drawn stainless Cr-Ni steels wire for TIG welding process, diameters1,0 mm and 1,6 mm. The chemical composition of the material of the experimental samples is given in Table 1.

Table 1. Shemical composition of the material

Wire mark in acconrdance to DIN	C_{max} [%]	Si [%]	Mn [%]	Cr [%]	Ni [%]
ER 308LSi	0,03	0,88	1,0–2,5	19,0–22,0	9,0–11,0

Welding of the experimental samples was done in three series, for different energies level, for both wire diameters, as shown in Table 2.

Table 2. Welding parametres of expermental samples

φ1,0 mm	Voltage [V]	φ1,6 mm	Voltage [V]
M1	45	V1	70
M2	60	V2	80
M3	75	V3	90

After welding, the metallographic analysis of the samples was performed. This analysis involved making macroslices of all welded specimens, and Fig. 1 shows two characteristic specimens. Then, characteristic sites for microstructure analysis were determined on the selected samples, as shown in Fig. 3 for the sample labeled M2. Samples were etched in Kalling reagent and observed at 750x magnification.

The presence of defects, such as cracks as well as metallic or non-metallic inclusions, was not found on any of the samples. The width of the welding zone varies depending on the welding parameters, as well as the final geometry of the welded joint. Insufficiently filled seam can be observed on certain samples welded with lower energy, and on the other hand, the presence of dripping is noticeable in some samples with higher energy (Fig. 1.a).

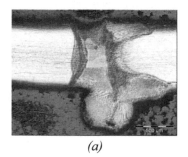
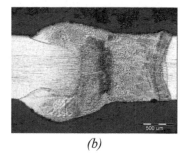

(a)   *(b)*

Fig. 1. Macroslices of charactersitic welded specimens [3]

Also, it is possible to observe the asymmetry of the left and right sides of the welded joint, which is due to the characteristic preparation of samples for welding. Namely, one of the wires is prepared by aligning the tip of the one wire, and the other has the tip of the wire at an angle of 50° ÷ 60°, depending on the welding parameters, Fig. 2.

Since the parent material is austenitic steel, the microstructure of the weld metal and HAZ is also austenitic. The microstructure of the parent material, which was not exposed to the heat, location 1, shows a cold-drawn austenitic microstructure with the presence of delta ferrite.

Fig. 2. Welded joint preparation [3]

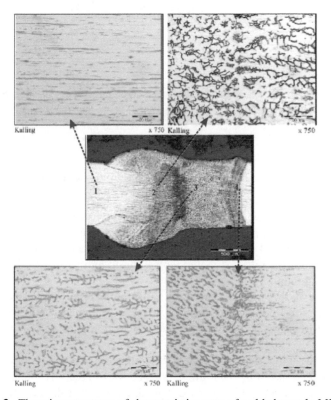

Fig. 3. The microstructures of characteristics zone of welded sample M2 [3]

The microstructure of the welded metal, locations 2 and 3, is also austenite, with a dendritic type. This shape of the structure is to be expected since this part of the welded joint is formed by crystallization of molten metal.

The fusion line is shown at location 4. A normal transition from the welded metal to the parent material is visible, with relatively narrow recrystallized zone. The width of the recrystallization zone varies depending on the applied energy level, which is expected.

2.1 Hardness Measurement Analysis

After the metallographic analysis, hardness measurements were performed on the same samples by the Vickers method HV1. Due to the very small dimensions of the samples, it was not possible to perform the test according to the standard prescribed conditions.

The test was not performed with the same step on all samples. Therefore, in addition to the hardness diagram, a drawing of the positions at which the measurement was made for the M2 sample was given.

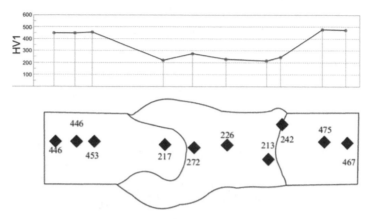

Fig. 4. Diagram of hardness measurements on a sample M2 [3]

By analyzing the obtained values of hardness measurements, it can be concluded that the hardness of the welded metal and the heat affected zone is less than the hardness of the parent material, Fig. 4. The reason is the fact that the parent material is cold-drawn wire, i.e. deformation-reinforced material,resulting in an increase in the hardness and strength of the material.

The decrease in the value of the hardness of the material follows the energy input during welding, so samples welded with lower energy input have a smaller decrease in hardness, while samples welded with higher energy input have a biger decrease in hardness.

It is also evident that there is a certain difference in the hardness of the weld metal and the parent material in the heat affected zone. More precisely, the hardness of the parent material in the heat affected zone is in all cases lower than the hardness of the weld metal. In addition, it is evident that the decrease in hardness in the HAZ is not the same on both sides of the weld. For the previously mentioned reasons, the HAZ is not symmetrical, but is wider on the wire that was connected to the positive pole (cathode) during welding. The above is in line with the fact that a higher temperature develops at the anode, which increases the HAZ.

3 Conclusion

Metallographic analysis has shown that all samples in the welding process form the same structure in the weld metal, ie cast dendritic microstructures. No cracks or any inclusions were found in any of the samples. Porosity is present in certain samples, mainly welded with higher values of energy input.

Microscopic examination showed that the transition from the parent material to the weld metal is much more emphasize on the wires that were connected to the negative pole (cathode) during welding, where a lower temperature develops. In all wires, at the positive pole (anode), this transition is much less emphasize.

Hardness analysis showed that in all samples there is a decrease in hardness in the weld metal and HAZ because the parent material is in the cold-drawn state. This reduction in hardness is biger with smaller diameter wires. The reason for this may be in the fact that the hardness of the parent material in wires of biger diameter is lower.

In all samples, the hardness in the HAZ is less than the hardness of the weld metal. Also, the width of the HAZ is not the same at the anode and cathode side. It is much wider at the anode side because the heating is more intense. The decrease in hardness and width of the HAZ is more emphasize in samples that are welded with higher drive energy [4].

References

1. Kaleko, D.M., Lebedev, V.K., Chvertko, N.A.: Processes Of Welding Using The Arc Discharge of the Capacitors, OPA Amsterdam (1999)
2. Kaleko, D.M., Moravsky, V.E., Chvertko, N.A.: Capacitor Discharge Percussion Welding. Naukova Dumka, Kyiv (1984)
3. Džiho, E.: Optimizacija perkusionog kondenzatorskog zavarivanja žica doktorska disertacija, Univerzitet „Džemal Bijedić" u Mostaru, Mašinski fakultet (2016)
4. Džiho, E., Pašić, S.: Influence of Welding Parameters on Mechanical Properties of Welded Joint at Capacity Discharge Welding Process, Welding conference "Zavarivanje 2020", Kladovo, Serbia (2021)

Influence of Delta Ferrite on Mechanical Properties of Nickel Free Austenitic Stainless Steels

Jasmin Halilović[(✉)], Denijal Sprečić, Edis Nasić, and Džemal Kovačević

University of Tuzla, 75000 Tuzla, Bosnia and Herzegovina
jasmin.halilovic@untz.ba

Abstract. This paper deals with the results of testing of mechanical properties of nickel free austenitic stainless steel depending on the content of delta ferrite. The appearance of delta ferrite is caused by different ratios of alphagen and gammagen alloying elements, and its appearance is greatly influenced by the method of production, as well as subsequent mechanical processing and heat treatment. Nickel free austenitic stainless steels were manufactured using an induction furnace by adding nitrided ferroalloy under atmospheric pressure. Mechanical properties investigation and microstructural analysis of forged and solution annealing ingots were performed for different nitrogen content. Analysis of microstructure was done by optical microscopy, whereas tensile tests were performed for characterization of mechanical properties.

Keywords: Nickel free austenitic stainless steels · Delta ferrite · Mechanical properties · Nitrogen

1 Introduction

Because of numerous advantages which are enabled by adding nitrogen, nickel free austenitic stainless steels alloyed with nitrogen have become the subject of intense research for a very short time [1]. Nitrogen is easily accessible and cheap and it is also a strong stabilizer of austenitic microstructure. Furthermore, nitrogen prevents allergic reactions caused by nickel and improves mechanical properties [1, 2]. Besides the advantages mentioned, there are numerous problems that have to be overcome during the production of nickel free austenitic stainless steels [2]. Several methods of production of these steels are applied, the main difference between them being the way of introduction of nitrogen into steel [3]. Nickel free austenitic stainless steelsalloyed with nitrogen are produced in induction furnaces, arc furnaces, AOD furnaces, plasma arc furnaces, pressure ESR process, pressure arc slag process and production by powder-metallurgy technologies [3–6]. In this paper we used an induction furnace for production of nickel free austenitic stainless steels and the introduction of nitrogen into steel was done by adding nitrided ferroalloy under atmospheric pressure. Solution alloying under atmospheric pressure is limited by nitrogen solubility borders in iron based alloys (up to 0,5%N), which requires bigger pressures if one wants to achieve bigger nitrogen content

I. Karabegović et al. (Eds.): NT 2022, LNNS 472, pp. 114–121, 2022.
https://doi.org/10.1007/978-3-031-05230-9_13

[1]. Using this method of production, it is difficult to get ingots with high chemical and physical homogenity. Because of that, treatment with hot plastic deformation is obligatory for further use in order to remove the funnel structure in the item [1, 7]. Also, in these conditions solidification of austenitic stainless steels is not a balanced process and you get a two-phase microstructure of austenite and delta ferrite [8]. Depending on the ratio ofalphagen and gammagen elements, solidification can begin with crystallization of delta ferrite or austenite [1, 8, 9]. Solidification of austenite stainless steels can occur in three ways [9]:

1. mode A (L → L + γ → γ),
2. mode AF (L → L + γ → L + γ + δ → γ + δ) or
3. mode FA (L → L + δ → L + δ + γ → γ + δ),

In this sequence, L is the liquid phase, γ isaustenite, and δ is delta ferrite. Which mode has begun is determined through the following criteria [8–10]:

1. mode A or AF ($Cr_{ekv}/Ni_{ekv} \leq 1,48$),
2. mode FA ($1,48 \leq Cr_{ekv}/Ni_{ekv} \leq 1,95$).

In this paper, the values of chrome and nickel equivalents are determined according to the ASTM A 800/A 800M – 01 standard (Standard Practice for Steel Casting, Austenitic Alloy, Estimating Ferrite Content). By analysing the chemical content and the microstructures of ingots without heat treatment, it has been established that a two-phase austenite and delta ferrite microstructure occurred, as well as a precipitate. With the aim of eliminating the precipitate and reducing the delta ferrite ratio, there has been subsequent heat treatment of solution annealing. This paper analyses the effects of chemical content and subsequent thermal processing on the occurrence of delta ferrite and precipitate, followed by the analysis of influence of delta ferrite on mechanical properties of nickel free austenitic stainless steel.

2 Experimental Work

The tested ingots (M1, M2 and M3 in Table 1), with different nitrogen content, were produced in an induction furnace. ARMCO iron produced in an open induction furnace was used as a gasket and it was later melted under slag. Solution alloying was done by adding ferroalloys (Cr, Mo, Mn, high nitrogen ferro-chrome) in different ratios in order to get suitable chemical content. The ingots obtained were subject to hot forging in order to reduce the thickness for 50–80%, then specimens of 95 × 35 ×25 mm were cut from ingot, solution annealed at 1100 °C for 60 min, and water quenched. The chemical content of the ingots obtained – austenitic stainless steels, is given in Table 1. Metallographic analysis of samples was done by optical microscopy (Optika XDS-3MET), and then the content of delta ferrite was evaluated by the application of software OptikaISView and ImageJ. The testing of mechanical characteristics was conducted on test pieces prepared in accordance with the EN 10002-1 standard – tensile testing (Fig. 1), at room temperature, using the Zwick/Roell material testing machine with the capacity of 30 kN.

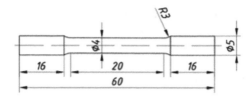

Fig. 1. Test piece geometry for the static tension test

Table 1. Composition of austenitic stainless steels (wt%)

Alloy	Cr	Mo	Mn	Si	Ni	N
A1	16,14	2,84	8,56	0,176	≤0,1	0,380
A2	17,6	2,81	8,44	0,205	≤0,1	0,420
A3	16,88	3,72	11,23	0,218	≤0,1	0,350

3   Results and Discussion

Since the Schaeffler diagram is not authoritative for the marking of microstructure after casting, a metallographic ingot/sample analysis without heat treatment was applied (hot forged). By analysing the microstructures of samples without heat treatment (shown in Table 2), it is clearly visible that a two-phase microstructure of austenite and delta ferrite was formed. The way in which the mentioned microstructure was formed, as mentioned before, is observed based on the relation of chrome and nickel equivalents. Table 3 shows the calculated relations betweenchrome and nickel equivalents and they are all lower than 1,48, which leads to the conclusion that solidification started according to the AF mode for all the ingots analysed. Also, besides the formation of a two-phase microstructure, it is visible that chromium nitrides have been formed in the form of precipitate on grain boundaries or in the form of lamellae in austenite grains, as well as carbides which precipitated on grain boundaries.

Precipitates occur in austenitic steels with nitrogen in bigger or smaller portions depending on chemical content and cooling conditions. What we have here are cooling conditions which are not balanced (as in the case of most industrial casting), where the austenite becomes oversaturated by carbon and chromium at temperatures below A1 (727 °C), which demands additional increase in the precipitate and delta ferrite ratio. The evaluation of delta ferrite and precipitate content was conducted by applying the ImageJ software – for the phase analysis of microstructure (Table 4).

The comparison of chemical analysis and the content of delta ferrite and precipitate shows that the A1 melt which has the lowest content of alphagen elements (Si, Cr, Mo), also has the lowest content of delta ferrite and precipitate, and the A3 melt which has the highest content of alphagen chemical elements, also has the highest content of delta ferrite and precipitate. It is also noticeable that melts with higher content of chromium and nitrogen have a higher content of chromium nitride.

Table 2. Microstructure of M1, M2 and M3 alloy

Alloy	Hot forged (950°C)	Solution annealed (1100°C)
A1		
A2		
A3		

Table 3. The calculated values between chrome and nickel equivalents and mode determination

Alloy	Cr_{ekv}/Ni_{ekv} $Cr_{ekv} = \%Cr + 1,5\%Si + 1,4\%Mo + \% Nb - 4,99$ $Ni_{ekv} = \%Ni + 30\%C + 0,5\%Mn + 26\%(N - 0,02) + 2,77$	Mode
A1	0,728	AF
A2	0,787	AF
A3	0,740	AF

Table 4. The evaluation of delta ferrite and precipitate content (%)

Alloy	A1	A2	A3
Delta ferrite + precipitates	10	11	19

With the aim of eliminating precipitate and a delta ferrite ratio, we did heat treatment of solution annealing. The aim of solution annealing is the presence of austenite microstructure at room temperature during quick cooling from high temperatures [1]. The cooling agent used was water. By heating the obtained samples we have a dissolution of delta ferrite into carbides and the sigma phase. Since the delta ferrite is rich in chromium, the carbides that are released are usually chromium carbides or, depending on the temperature, chromium nitrides can also be abstracted. The obtained microstructure of solution annealing melts is shown in Table 2. By applying an annealing temperature of 1100 °C, we can see that the A1 and A2 melts got an elongated austenite grain but that delta ferrite was not completely eliminated. Such high annealing temperature led to an increase in dissolution of nitrogen in delta ferrite and to a decreasein dissolution of nitrogen in austenite, which led to a formation of bigger greylayers of chromium nitride. With the A2 melt there is a bigger ratio of layers of chromium nitride because of greater content of chromium and nitrogen, or to put it differently, greater nitrogen content at annealing temperature of 1100 °C enables higher dissolution of nitrogen in delta ferrite. Table 2 shows that by increasing the content of manganese in the A3 melt a small amount of chromium nitride layers is formed because manganese supresses the creation of chromium nitride [11]. When it comes to the A3 melt, because of a great ratio of alphagen elements, the delta ferrite is not completely dissolved and it comes in higher ratios than in the previously analysed melts (A1 and A2). So it can be concluded that the annealing temperature of 1100 °C and the annealing time of 60 min can ensure the formation of an austenite microstructure with a low ratio of delta ferrite and a minimal ratio of precipitate (small eutectic carbides on the grain borders and chromium nitrides in the form of grey polygonal particles). Through the examination of mechanical characteristics it was necessary to determine the dependence of tensile strength and the elongation from the content of delta ferrite and precipitate. We examined the mechanical properties of melts with and without heat treatment of solution annealing. Table 5 shows the obtained values of tensile strengthand elongation.

Table 5. Results of the static test piece tensile test

Alloy	A1 hot forged	A2 hot forged	A3 hot forged	A1 solution annealed	A2 solution annealed	A3 solution annealed
Rm (MPa)	959	827	874	865	761	819
A (%)	56,2	37,2	34,8	50,9	42,7	62,1

By analysing Tables 1, 2, 4 and 5, it is visible that the melt (A1 – hot forged) with the lowest content of delta ferrite and precipitate also has the biggest tensile strength and elongation. The A3 melt – hot forged with the content of delta ferrite and precipitate of over 15% has a significant fall inelongation, and when it comes to tensile strength its fall is softer. Besides the influence of delta ferrite and precipitate on tensile strength and elongation, nitrogen and other prime alloying elements such as Cr, Mo and Mn have a strong influence on tensile characteristics from their direct connection with the microstructure. It can be concluded that even if a certain melt contains more nitrogen, it does not mean that the melt has to have greater tensile strength and elongation because the increase in the content of elements Cr, Mn, Mo asks for more nitrogen to form austenite bases. In other words, greater amounts of precipitate are formed and at the same time the austenite basis with nitrogen becomes poorer. All types of precipitate usually lead to a fall in elongation. For a better analysis of influence of microstructure and chemical content on tensile strength andelongation we have representative diagrams of dependence of tensile strength andelongation on the content of nitrogen and other chemical elements (Fig. 2).

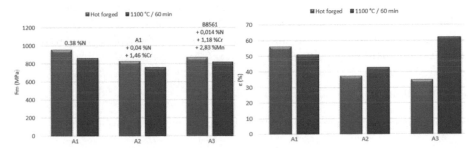

Fig. 2. Dependence of tensile strength and elongation on heat treatment and chemical content

At the annealing temperature of 1100 °C with melts A1 and A2 there has been a sudden fall in the value of tensile strength bacause of a significantly bigger austenite grain when compared to samples without heat treatment. Besides the fall of the value of tensile strength because of a bigger grain, there has also been a fall in tensile strength because of formation of bigger layers of chromium nitride. With the A3 melt bigger layers of carbides and nitrides were formed, which directly resulted in a decrease of value of tensile strength, but in a significantly smaller amount because of a bigger ratio of undissolved delta ferrite. All the things mentioned lead to a conclusion that the value of tensile strength grows with the decrease in the size of austenite grain and falls with the enlargement of austenite grain (if there is no delta ferrite ratio) and it also falls with the formation of layers of chromium nitride or massive carbides. Besides the things mentioned, by forming layers of delta ferrite or by enlarging grains of delta ferrite, there is an increase in the value of tensile strangth because of bigger percent of this phase in austenite basis. Besides the influence of appropriate phases and precipitate, as well as the size of the grain, chemical elements also significantly affect the value of tensile strength, primarily nitrogen.

It is visible from Fig. 2 that even if one increases the content of nitrogen in the A2 melt in relation to the A1 melt, there is no increase in tensile strength because other elements are not of the same content like the melt in question. The increased ratio of chromium in the A2 melt at annealing temperature of 1100 °C increased the dissolution of nitrogen in delta ferrite, which resulted in the formation of layers of chromium nitride and the fall in the value of tensile strength. One can see (Fig. 2.) that by increasing the ratio of the gammagen element manganese in the A3 melt, we supress the formation of chromium nitride and increase the value of tensile strength in that way. We have larger values of tensile strength for certain melts for samples which were not annealed because of bigger content of delta ferrite.

It is visible from Fig. 2 that elongation depends on the microstructure formed as well as the chemical content. Bigger ratio of alphagen elements reduces elongation while a bigger ratio of gammagen elements increases elongation. The most influential chemical element besides nitrogen is chromium because with its increase there is also a rise in the dissolution of nitrogen and the amount of formed chromium nitrides increases, which is clearly visible from Fig. 2. By reducing chromium content elongation increases because chromium nitride layers are reduced. Bigger content of nitrogen provides bigger elongation, while bigger content of manganese and molybdenum supresses the formation of chromium nitride. The smaller the density of the precipitate and the bigger the grain is, the bigger the elongation. Also, microstructures that contain smaller amounts of delta ferrite as a rule have bigger elongation.

4 Conclusions

Based on the analysis of the results obtained we can conclude the following:

- The choice of chemical content of a material is the first and very important step in the process of an alloy development. Not only does chemical content define the mechanical and chemical properties but also the dissolution of nitrogen, the stability of austenite and the tendency of formation of delta ferrite and precipitate.
- In order to prevent the occurrence of delta ferrite in steel, it is necessary to keep the content of elements which stabilize austenite, manganese and especially nitrogen as closer as possible to the upper limit of allowed content, while the content of elements that stabilise delta ferrite should be kept around the middle of the content allowed.
- By reducing the presence of alphagen elements, it will be more difficult for delta ferrites to form. In other words, its abstraction will be minimized and keeping chromium at minimal limits of 16,5% will reduce the formation of precipitates of chromium nitrides and chromium carbides, which directly reduce the values of tensile strength and elongation.
- The increase in delta ferrite content leads to the fall in the value of elongation.
- The smaller the density of the precipitate and the bigger the grain is, the bigger the elongation.
- The value of tensile strength grows with the reduction in the size of austenite grain and falls with the increase of austenite grain (if there is no delta ferrite ratio) as well as with the formation of layers of chromium nitrides or massive carbides. Beside this, by forming layers of delta ferrite and by enlarging grains of delta ferrite, there is a increase in the value of tensile strength.

References

1. Halilović, J.: Uticaj Parametara Nitriranja i Naknadne Termičke Obrade na Mikrostrukuru i Mehaničke Osobine Austenitnih Nehrđajućih Čelika Bez Nikla. Doctoral dissertation, Faculty of Mechanical Engineering, Tuzla (2019)

2. Wang, Q., Ren, Y., Yao, C., Yang, K., Misra, R.D.K.: Residual ferrite and relationship between composition and microstructure in high-nitrogen austenitic stainless steels. Metall. Mater. Trans. A **46**(12), 5537–5545 (2015). https://doi.org/10.1007/s11661-015-3160-5
3. Halilović, J.: Microstructure and mechanical properties of nickel free austenitic stainless steels produced by addition of nitrided ferroalloys during melting in induction furnace. J. Trends Dev. Mach. Assoc. Technol. **21**(1), 33–36 (2018)
4. Park, W., Jung, S.-M., Sasaki, J.: Fabrication of ultra high nitrogen austenitic stainless steel by NH3 solution nitriding. ISIJ Int. **50**(11), 1546–1551 (2010)
5. Li, H.-B., Jiang, Z.-H., Shen, M.-H., You, X.: High nitrogen austenitic stainless steels manufactured by nitrogen gas alloying and adding nitrided ferroalloys. J. Iron. Steel Res. Int. **14**(3), 63–68 (2007). https://doi.org/10.1016/S1006-706X(07)60045-4
6. Balachandran, G., Bhatia, M.L., Ballal, N.B., Krishna, R.P.: Processing nickel free high nitrogen austenitic stainless steels through conventional electroslag remelting process. ISIJ Int. **40**(5), 478–483 (2000)
7. Romu, J., Hännien, H.: High-Nitrogen Austenitic Stainless Steel Manufacturing Technologies. University of Technology, Helsinki (1993)
8. Tehovnik, F., Vodopivec, F., Kosec, L., Godec, M.: Hot ductility of austenite stainless steel with a solidification structure. Mater. Tehnol. **40** (2006)
9. Padilha, A.F., Tavares, C.F., Martorano, M.A.: Delta ferrite formation in austenitic stainless steel castings. Mater. Sci. **730**, 733–738 (2013)
10. Shankar, V., Gill, T.P.S., Mannan, S.L., Sundaresan, S.: Solidification Cracking in Austenitic Stainless Steel Welds, India (2003)
11. Menzel, J., Kirschner, W., Stein, G.: High nitrogen containing Ni-free austenitic steels for medical applications. ISIJ Int. **36**(7), 893–900 (1996)

Impact of Strain Rate on Plastic Properties of Metals

Stipo Buljan[1](✉) and Darko Šunjić[2]

[1] Federal Ministry of Energy, Mining and Industry, A. Šantića bb, B88000 Mostar,
Bosnia and Herzegovina
stipobuljan1@gmail.com
[2] Faculty of Mechanical Engineering, Computing and Electrical Engineering,
University of Mostar, Maticehrvatske bb, 88000 Mostar, Bosnia and Herzegovina

Abstract. The choice of technological factors and the clarification of the impact of strain rate at the impulse-high-speedprocessing of metals has a high degree of significance. Generally, there is no single opinion on the character of changes in the parameters of plasticity with the increase of the strain rate.

When testing carbon and alloy steels, aluminum, copper and its alloys, a decrease in plasticity was observed at high strain rate, which is explained by the existence of a "critical speed" for the mentioned materials at which the destruction occurs instantaneously. A metal that is essentially loaded with an impulse load shows that plastic deformation occurs by twinning and sliding within the grain. During this load, elastic-plastic waves occur in the workpiece, where the magnitude of the stress is related to the speed of propagation of the wave.

Keywords: Strain rate · High-speed processing · Deep drawing coefficient

1 Introduction

Metal forming processes are based on the plasticity property of metals. By direct or indirect application of external force, these processes process the metal –the workpiece into a semi-finished product or product of the desired shape and dimensions.

Rolling, forging, deep-drawing, bending, compression, extrusion, stamping are the most important processes of plastic treatment.

The need for work items of larger and more complex dimensions has led to the research of new ways of obtaining them, ie to unconventional processing procedures.

It was initially thought that studying high-speed loads was a new deformation mechanism that would be characteristic of this type of load. However, even with such a load, there are already three known deformation mechanisms that occur such as sliding, twinning and diffuse creep. A metal that is essentially loaded with an impulse load shows that plastic deformation occurs by twinning and sliding within the grain. During this load, elastic-plastic waves occur in the workpiece, where the magnitude of the stress is related to the speed of propagation of the wave.

I. Karabegović et al. (Eds.): NT 2022, LNNS 472, pp. 122–127, 2022.
https://doi.org/10.1007/978-3-031-05230-9_14

Some of the workpieces can only be obtained by applying new design technologies or non-conventional design technologies. In these design technologies, strain rates may be significantly lower than the deformity of conventional technologies, but it can be several tens of times larger. One of such methods is explosion design where we can practically produce forces, ie energy much higher than in the conventional way, but the degree of utilization will be lower and an additional problem will be how to cancel excess energy, take it outside but with taking care of the environment.

The processing covers a very wide range of diameters from 30 mm to 5000 mm and thickness from 0.5 mm to 30 mm [1].

In view of the absence of a single opinion on the character of changes to the plasticity parameters with the increase of strain rate, this paper will provide additional considerations on the impact of strain rate on plastic properties of metals.

2 Processability of Sheet Metal During Deep Drawing by Explosion in Water

The ability of some materials to be successfully processed by deep drawing is a test of the plastic properties of sheet metal when it is processed on presses. Many researchers have dealt with this issue and there are a large number of publications and papers with a large amount of data.

Methods for assessing the processability of materials in high-speed molding have not yet been sufficiently developed. Hydro explosive deep drawing can be used to assess the plastic capabilities of sheet metal in this group of high-speed processing.

The main factor that determines the ability of the material in deep drawing is the drawing coefficient oras it is also called in the literature, the degree of reshaping, because it determines how many times the diameter is larger than the diameter of the workpiece.

$$K = \frac{D}{d_1} = \frac{D}{1 - \varepsilon} = \frac{1}{m} = e^{\varphi} \tag{1}$$

When designing deep drawing technology, deformations can be expressed in the following ways:

$$\varepsilon = \frac{D - d_1}{D}; m = \frac{d_1}{D}; \varphi = \ln \frac{D}{d_1} \tag{2}$$

The degree of deformation can be related to other deformation expressions by a relation:

$$\varepsilon = \frac{D - d_1}{D} = 1 - m = \frac{K - 1}{K} = \frac{e^{\varphi} - 1}{e^{\varphi}} \tag{3}$$

The ratio of the diameter of the workpiece and the diameter of the platinum is denoted as the drawing ratio (m) and is calculated according to the form:

$$m = \frac{d_1}{D} = 1 - \varepsilon = \frac{1}{K} = \frac{1}{e^{\varphi}} \tag{4}$$

The main logarithmic deformation can be expressed by the following integral:

$$\varphi = \int_r^R \frac{d\rho}{\rho} = \ln \frac{R}{r_1} = \ln \frac{D}{d_1} \tag{5}$$

and if we put this together with other sizes we get the following pattern:

$$\varphi = \ln \frac{D}{d_1} = -\ln(1 - \varepsilon) = -\ln m - \ln k \tag{6}$$

In the above equations, D is the diameter of the platinum, and d_1 is the diameter of the deep-drawn workpiece. Another factor that is also very important byusing deep drawing is the relative thickness of the material and it can be expressed by following:

$$Sr = \frac{S}{D} = 100\% \tag{7}$$

s – thickness of platinum (mm)
D – diameter of platinum (mm)

When processing the material on the press, no matter how much its nominal force, it is impossible to transfer to the workpiece energy greater than that required for its deformation or energy required to obtain a certain shape. In such a case the energy required for shaping or the force of deep drawing depends mainly on the mechanical properties of the material, the geometry of the workpiece, the geometry of the tool, etc. Therefore, the ultimate drawing coefficient is determined on the basis of the mechanical characteristics of the material and does not depend on the force of the press.

In the case of deep drawing by explosion in water, different amounts of energy can be given to the same workpieces. In this case, the excess energy that is delivered to the workpiece is lost when the workpiece hits the edge of the mold. In that way, the energy of the explosive is not only spent on deforming the workpiece, but part of it passes into the kinetic energy of the workpiece. Due to the above, the ultimate drawing coefficient depends not only on the mechanical properties of the workpiece but also on the energy delivered to the workpiece or the amount of explosives [2, 3].

Another specificity of the deep drawing by explosion process is the high strain rate of the workpiece. While in the case of deep drawing on presses the drawing speeds are relatively low, in blasting they are proportional to the speeds of wave propagation on the material that is being processed.

During hydro-explosive drawing, at the initial moment, the values of velocity and acceleration in the material of the workpiece reach the highest values, especially in the part of the transitional rounding of the matrix, which causes an increase in deformations in that zone. From this narrow annular focus of deformation, the propagation of longitudinal waves begins, under the action of which the material is moved from adjacent zones to the deformation zone. Deformations spread towards the ends of the workpiece with the speed of propagation of longitudinal waves.

In such a development, the most critical is the initial moment when the crown of the workpiece has not yet begun to move, so the flow of metal in the meridional direction occurs only at the expense of thinning sheet metal on the transitional rounding of the matrix.

The movement of the ring begins when the longitudinal waves reach its free end. If the boundary deformations do not occur in the critical zone by then, the intensive movement of the ring begins and in that way the successful deep drawing of the workpiece is ensured. Otherwise, destruction occurs in the critical zone.

It follows from the above that the greater the amount of explosives, if the other factors are the same, the workpiece gets a higher initial speed and greater risk of destruction, and also increasing the diameter of the workpiece increases the ring and takes longer to start moving and thus create conditions for greater thinning of sheet metal in the critical zone. Having in mind all this, it can be concluded that the sheet metal holder has a very subtle role in deep drawing by explosion and the holding force should be determined so that it really is a sheet metal holder in hydro-explosive processing, but also does not help thin the material at the ratio of the matrix.

It has been experimentally proven that the limit ratio of deep drawing by explosion in water is lower than the limit ratio for standard methods, but it is good enough that this method can be successfully used especially in smaller production series and workpieces where large amounts of energy are required for processing, ie high nominal forces of machines.

3 Impact of Strain Rate on Plastic Properties of Metals

Increasing the strain rate affects the deformation resistance, and it increases with increasing the strain rate. At high speeds, the effect occurs due to the delay in the plastic strain rate compared to the speed of elastic deformation. At low strain rates, this effect is less significant [4]. When exploring the impact of the strain rate on cold deformation, most researchers observe the impact of speed on changes in the proportionality and drawing strength limits, and their relationship [5]. It has been proven that during cold plastic deformation, the ratio of drawing limit and drawing strength changes with increasing strain rate due to an increase in drawing limit [6].

The choice of impact factors and the clarification of the impact of the strain rate in pulsed - high-speed metal processing is of great practical importance. For now, there is generally no consensus on the nature of changes in plasticity parameters with increasing the strain rate. Primarily due to the small number of experiments performed in this type of processing, as well as due to the unique testing methodology.

When testing carbon and alloy steels, aluminum, copper and its alloys, a decrease in plasticity was observed at high strain rate, which is explained by the existence of a "critical speed" for the mentioned materials at which the destruction occurs instantaneously.

It was found that the critical speed for steel is in the range of 50 to 100 m/s and for aluminum 11 m/s. In addition to knowing the critical speed, it is very important to know the period of delay in the flow of material from the moment of loading. For plastic materials with strong flow surfaces, the delay period is a few milliseconds.

During the transition from static to dynamic testing with a speed of 3.8 m/s, a decrease in relative elongation was observed. By further increasing of the strain rate, there is a slight increase in the plasticity characteristics. Relative elongation of carbon and low-alloy steels in relation to static values increases by 26% at an impact speed of 300 m/s [7].

At a strain rate of 650 m/s, 1 × 18 H97 hardened steel specimens are destroyed without any traces of plastic deformation. This means that for this steel the mentioned strain raterepresents the critical speed. It has been shown experimentally for this steel that increasing the deformation speed to 200 m/s leads to an increase in the plasticity properties.

The largest increase in relative elongations compared to the static test is 80%. A further increase in the strain rate leads to a sharp drop in plastic properties. At speeds up to 500 m/s, the plastic properties are up to 2.5 times lower than static tests.

With hardened and tempered steel, the plasticity characteristics also increase up to a speed of 200 m/s. At this speed, the uniform elongation is 65% higher than in the static test, while at the speed of 500 m/s the plastic properties are reduced and are lower by 1.7 to 2.0 times lower than the static values [7].

Tests of the plastic properties of titanium alloy have shown that the plastic properties increase by an average of up to 20% for speeds up to 65 m/s compared to static characteristics. Further increase in speed negatively affects the plastic properties. At a speed of 350 m/s, the plasticity characteristics are 1.5 to 1.8 times lower than the static values [7].

For most of the tested metals, an increase in the value of plasticity properties was found in relation to static values with an increase in the strain rate. The maximum strain rate varies from metal to metal. Increasing the strain rate above these leads to a sharp drop in plastic properties and localization of plastic deformation.

The above statements mainly relate to pulsed shaping in the cold state. It is known that such designs can be done even when the material or workpiece is heated, and then these parameters would change.

4   Conclusion

Explosive materials are used in various technological operations of metal processing. Thanks to the high pressure, it is possible to get almost any amount of energy very quickly, easily and cheaply, and thus to process suitable materials that are difficult to process by other processes or are too expensive.

For different dimensions of the preparation, there are certain values of the strain rate and thus a certain amount of explosives that allows complete deep drawing. It follows that at a certain value of the deep drawing coefficient, the mentioned interval of the amount of explosives is minimal, and only one certain amount of explosives allows complete drawing of the workpiece. Drawing by explosion is not recommended for large drawing ratios.

The deformation state of the workpiece is very favorable, so that with the right choice of technological factors there is the lowest degree of deformation in the normal direction compared to other deep drawing processes. This indicates a high uniformity of deformation in the meridian cross section. Due to the high strain rate the structure in the workpiece is uniform, which is the result of simultaneous deformation of all zones along the cross section of the workpiece.

References

1. Lange, K.: Umformfechnik, vol. 4. Sprring- Verlag, Berlin, Heindeberg, New York, London, Paris, Tokio, Hong Kong, Barcelona, Budapest (1993)
2. Petjušev, V.G., Stepanov, G.V.: O postanovke visokoskorostnih mehaničeskih ispitanih materialov. Problemi Pročnosti (1972)
3. Buljan, S., Đukić, H.: Analise of limited possibilies of the level of deformation dependent on explosion wave. In: Proc. 2. International Conference "Business Systems Management - UPS-2001", ISBN 3-901509-06-02-x, str.33–37, Vienna – Mostar (2001)
4. Rešković, S.: Teorija Oblikovanja Deformiranjem. Sveučilište u Zagrebu Metalurški Fakultet, Sisak (2014)
5. Çetinarslan, C.S., Güzey, A.: Tensile Properties of Cold-Drawn Low-Carbon Steel Wires Under Different Process Parameters. Department of Mechanical Engineering, Faculty of Engineering and Architecture, Trakya University, Edirne, Turkey (2012)
6. Mihalikova, M.: Analysis of the influence of the loading rate on the mechanical properties of microalloyed steel. J. Met. Mater. Miner. svez. 17, br. 1, 25–28 (2007)
7. Beljajev, B.N.: Visokoskorostnaja Deformacija Metolov. Nauka i Tehnika (1976)

Studying the Mechanics of Low-Plastic Materials Surface Layer Processed by Deforming Broaching

Yakiv Nemyrovskyi, Ihor Shepelenko[✉], Evhen Solovykh, Oleh Bevz, and Serhii Leshchenko

Central Ukrainian National Technical University, 7 Universytetskyi Avenue, Kropyvnytskyi 25006, Ukraine
kntucpfzk@gmail.com

Abstract. A methods is proposed for modeling the stress-deformed state of low-plastic material surface layer, using the example of SCH20 cast iron, processed by deforming broaching. From the standpoint of the plasticity resource exhaustion, the analysis of the surface layer was carried out, which made it possible to establish the presence of plastic deformation local zone. The possibility of using a combined technology, including the operations of deforming broaching and finishing antifriction non-abrasive treatment, taking into account the influence of local plastic deformation zone, has been established.

Keywords: Deforming broaching · Plasticity · Modeling · Stress-deformed state · Local zone · Low-plastic material

1 Introduction

The operational properties of machine parts are significantly influenced by the physical and mechanical characteristics of the surface layer material formed after processing. They are assessed by the following main indicators: the degree of the surface layer hardening, the depth of hardening, residual stresses formed in the surface layer after treatment [1].

At present, in addition to the above parameters, to assess the surface layer quality of the processed part, a parameter is widely used that characterizes the defectiveness of the surface layer – the resource of the plasticity used [2]. Therefore, the parameters of plasticity are one of the important characteristics of the processed part surface layer quality. They determine its working capacity, especially under conditions of cyclic operational loads, for example, for cast iron liners of internal combustion engines (ICE) [3].

Moreover, the development of combined technologies, for example, combining operations of deforming broaching (DBR) and finishing antifriction non-abrasive treatment (FANT), makes it possible to create a working surface with predetermined operational properties. The development of such a technology requires a thorough study of surface treating process by DBR before applying an anti-friction coating. This issue is especially

© The Author(s), under exclusive license to Springer Nature Switzerland AG 2022
I. Karabegović et al. (Eds.): NT 2022, LNNS 472, pp. 128–134, 2022.
https://doi.org/10.1007/978-3-031-05230-9_15

relevant when processing low-plastic materials, for example, graphite-containing cast iron, the ability of which to plastically deform is limited by destruction [4].

An example of a combined technology use with the participation of DBR are steel parts, for example, piston pins of ICE, which are subjected to chemical-thermal operations (cementation, heat treatment). In this case, it is also very important to control the plasticity resource after DBR. Thus, the correct choice of the resource of plasticity used allows to obtain a working surface with an improved structure after chemical-thermal operations with a significant reduction in the cementation time.

2 Literature Review

The material plasticity, that is, the plastic deformation accumulated by the time of destruction, depends on a number of factors, among which, in addition to the nature of the material itself, the most important are the thermomechanical parameters of the processing process itself: the type of stress-deformed state (SDS) [2, 3].Deforming broaching is a process of a workpiece plastic forming and occurs at low temperatures and processing speeds. Therefore, the type of stress state has the greatest influence on the plasticity of cast iron in this process. The dependence of plasticity on the type of stress state is characterized by the plasticity diagram. Typically, the plasticity diagram is represented in the coordinates "stiffness coefficient (type) of the stress state – η and the accumulated deformation before fracture e_{ult}".

The authors of work [5] developed an original methods that allows deformation of cast iron samples in the region of high negative values of stiffness coefficient of the stress state. The implementation of this technique made it possible to construct a plasticity diagram for SCH20 cast iron, which is shown in Fig. 1 and represents a parabola section in the region of negative values.

As follows from Fig. 1, the plasticity of such a semi-fragile material, cast iron SCH20 at a value of $\eta = -4$ is very significant and reaches about 80%.

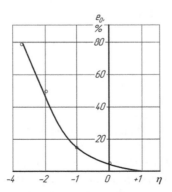

Fig. 1. Plasticity diagram of cast iron SCH20 [5]

To determine the resource of the plasticity used, various authors [6–8] proposed dependences (criteria), which are based on the principle that deformation occurs without destruction if the plasticity ψ satisfies the expression:

$$\psi \leq 1. \tag{1}$$

Among these dependencies, the Kolmogorov criterion is the most convenient for practical calculations in deforming broaching [9]:

$$\Psi = \int\limits_0^{e_0} \frac{de_0}{e_{ult}(\eta)}, \tag{2}$$

where e_0 – accumulated deformation,

$e_{ult}(\eta)$ – ultimate deformation of the material at the corresponding value of the stiffness coefficient of the current stress state, that is, the data that are determined by the plasticity diagram.

The value of η is determined based on the dependence:

$$\eta = \frac{3 \cdot \sigma}{\sigma_0}, \tag{3}$$

where σ – hydrostatic pressure,

σ_0 – stress intensity.

Taking into account the fact that products made of graphite-containing cast irons are classified as low-plastic materials, the study of their plasticity during DBR is especially relevant for the following reasons. First, it is necessary to determine the ultimate deformation in different zones of the deformation zone. This will make it possible to correctly establish such a technological parameter as the admissible value of residual plasticity, which, according to [2], has a very significant effect on the operational indicators of finished products. Secondly, according to work data [3], the value of the resource of plasticity used after the operation preceding the chemical-thermal treatment (cementation, nitriding) has a significant effect on the time minimizing of thermal operations, as well as on obtaining an improved structure after them. Therefore, the authors of works [10] recommend limiting the plastic deformation value ψ within:

$$\psi_{\text{max}} \leq [\psi], \tag{4}$$

where $[\psi] = (0.25 \div 0,3)\psi$.

As follows from the plasticity diagram (Fig. 1), plastic deformation of cast iron is possible only at negative values of η, therefore, an indispensable condition for DBR of a cast iron workpiece is the absence of plastic deformation on the outer surface of the workpiece, where, according to data [11], the stress state is close to biaxial stretching with a stiffness coefficient of the stress state $\eta = +2$. Therefore, plastic deformation should only occur in the material layer adjacent to the processed hole.

The aim of the work is to determine the SDS and the parameters necessary to determine the plasticity in the areas of the deformation zone in the layer adjacent to the processed hole. To achieve this goal, it is necessary to solve the following tasks:

– develop a method for studying SDS for a working layer of a cast iron workpiece adjacent to a deformable hole using the finite element method (FEM);

– investigate the SDS and the history of deformation for the layer adjacent to the processed hole;
– investigate the processed surface quality according to the resource parameter of plasticity used.

3 Research Methodology

Modeling of SDS during deforming broaching of a workpiece made of SCH20 cast iron was carried out using the Deform software package according to the scheme shown in Fig. 2.

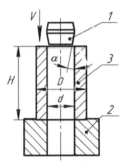

Fig. 2. Processing scheme: 1 – deforming element; 2 – base; 3 – studied sample

Data preparation for modeling the deforming broaching of the sleeve made of SCH20 cast iron using the Deform software package was carried out in several stages. Initially, the object of research was set: a sleeve made of SCH20 cast iron with the following dimensions: outer diameter $D = 55$ mm; inner diameter $d_{in} = 35$ mm; length H = 80 mm.

The working movement of the deforming element 2 (Fig. 2) was set by its monotonic longitudinal movement along the processed hole of the sleeve 3, which rests on a fixed base 2. The speed of the deforming element is $V = 0.5$ mm/s. The working cone angle of the deforming element α was 4°, respectively. The nominal tension on the deforming element is $a = 0.05$ mm per side.

The next stage is to determine the material properties for the studied object – sleeve made of SCH20 cast iron. To improve the accuracy of data calculating for the studied sample material, the experimentally obtained flow curve was used [4]. In addition to the flow curve, studied material properties were set by hardness $HB = 1.7$ GPa, Poisson's ratio $\mu = 0.27$ and Young's modulus $E = 1.6 \times 10^5$ MPa.

A window fragment of the Deform-3D program preprocessor is shown in Fig. 3.

To analyze the phenomena that take place on the surface of the processed hole, points P1, P2, P11 were set. Note that point P11 is removed from point P1 to a depth of 0.01 mm, and point P2 is located from point P1 at a depth of 0.25 mm (Fig. 4).

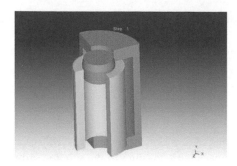

Fig. 3. A window fragment of the Deform-3D program preprocessor

Fig. 4. Fragment of the processed surface with points P1, P2 and P11 when modeling the DBR

4   Results

The deformation of a cast iron sleeve of the following dimensions (d_{in} = 35 mm; t_0/d_0 = 0.28), broaching modes (tension a = 0.1 mm) and tool geometry (α = 4°) were modeling. The presence of such sizes and modes provides critical contact pressures in the contact zone [3]. Let us consider the results of the SDS study of a workpiece from SCH20 cast iron when modeling according to the developed methods using the Deform software package.

Let's stop first on the study of the deformation parameters in the contact zone. The distribution of the deformation speeds intensity ξ_0 (Fig. 5) shows that the most intense deformation occurs at the treated surface itself at the beginning and end of the contact zone.

The distribution and components of the deformation speeds tensor have a similar character: radial – ξ_r and axial – ξ_z. This indicates a significant local symmetry of deformations in these areas.

The results of modeling the accumulated deformation e_o (Fig. 6) showed that its maximum region is not limited to the surface layer, but extends to a certain depth h, and this layer is close to the equally strengthened one.

The described results are confirmed by the previously obtained experimental data [12] on the presence of plastic deformation local zone – the so-called influx, located at the beginning of the contact zone and increasing the length of the contact zone. As for the transition zone of the contact area into the non-contact area, there also the modeling data show the presence of plastic deformation local zone.

The zone of local plastic deformation formed during the transition of the contact area with the non-contact zone behind it is characterized by the presence of stretching components of the stress tensor: σ_r, σ_z and σ_φ. The presence of stretching stresses accelerates the intensity of the plasticity resource exhaustion in the surface layer and leads to microfractures in the surface layer, the so-called peeling of the treated surface.

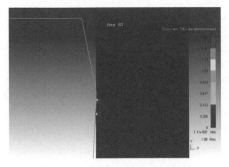 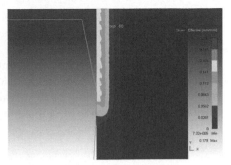

Fig. 5. The distribution field of the sample deformation speeds at the 60th step of deforming broaching simulation

Fig. 6. The distribution field of the accumulated deformation e_0 of the sample at the 60th step of deforming broaching simulation

The confirmation of peeling fact is shown in Fig. 7, which shows a photo of the deforming element after broaching under conditions of the plasticity resource exhaustion of the treated surface with peeling.

Fig. 7. The deforming element after broaching with the peeling of the treated surface: processed workpiece made of cast iron SCH20, *HB* 1.7 GPa; thickness $t_0/d_0 = 0.56$; angle $\alpha = 8°$; tension $a = 0.1$ mm; $\Sigma a = 0.4$; coolant – 5% emulsion

Microparticles of destroyed material are clearly visible on the working cone of the deforming element (Fig. 7). Obviously, such a surface cannot be coated.

The above material testifies to the need for a thorough study of the SDS results for the successful processing by DBR of cast iron products and allows us to draw the following conclusions.

5   Conclusions

The presented material made it possible to draw the following conclusions:

– to study the SDS of a product processed by DBR, a methods was developed for studying SDS by modeling the deforming broaching of the product made of SCH20 cast iron using the Deform software package;

- it was established that the most dangerous place from the position of the plastic resource exhaustion is the surface of the processed hole;
- a methods for using modeling data has been developed, which describes the accumulated deformation for calculating deformation hardening in a surface layer;
- when using a combined technology, including the DBR and FANT operations, improving the antifriction coating quality, which increases the operational resource of the processed part, is possible by eliminating the influence of the local plastic deformation zone formed after the contact area and affecting the plasticity resource exhaustion of the treated surface.

References

1. Suslov, A.G.: The Quality of Machine Parts Surface Layer, p. 320. Moscow (2000). [in Russian]
2. Rosenberg, A.M.: Mechanics of Plastic Deformation in the Processes of Cutting and Deforming Broaching, p. 320. Kiev (1990). [in Russian]
3. Rosenberg, O.A.: Technological Mechanics of Deforming Broaching, p. 203. Voronezh (2001). [in Russian]
4. Shepelenko, I., Tsekhanov, Y., Nemyrovskyi, Y., Eremin, P., Bevz, O.: Plasticity studies during deformation under conditions of significant negative values of the stiffness coefficient of the stress state. In: Karabegović, I. (ed.) New Technologies, Development and Application IV. NT 2021. Lecture Notes in Networks and Systems, vol. 233, pp. 215–223. Springer, Cham (2021). https://doi.org/10.1007/978-3-030-75275-0_25
5. I Shepelenko Y Tsekhanov Y Nemyrovskyi V Gutsul S Mahopets 2021 Compression mechanics of cylindrical samples with radial deformation limitation V Ivanov I Pavlenko O Liaposhchenko J Machado M Edl Eds Advances in Design, Simulation and Manufacturing IV DSMIE 2021 Proceedings of the 4th International Conference on Design, Simulation, Manufacturing: The Innovation Exchange, DSMIE-2021, June 8–11, 2021, Lviv, Ukraine – Volume 2: Mechanical and Chemical Engineering Lviv Ukraine 2021 06 08 2021 06 11 Lecture Notes in Mechanical Engineering Springer Cham 53 62 https://doi.org/10.1007/978-3-030-77823-1_6
6. Smelyanskiy, V.M.: Mechanics of parts hardening by surface plastic deformation, p. 300. Moscow (2002). [in Russian]
7. Del, G.D.: Technological Mechanics, p. 174. Moscow (1978). [in Russian]
8. Grushko, A.V.: Maps of Materials in Cold Working by Pressure, p. 348. Vinnitsa (2015). [in Russian]
9. Ogorodnikov, V.A., Kiritsa, I.Yu., Muzyichuk, V.I.: Plasticity diagrams and peculiarities of their construction. Improving pressure treatment processes and equipment in metallurgy and mechanical engineering, pp. 251–255 (2006). [in Russian]
10. Kolmogorov, V.L.: Mechanics of metals pressure processing, p. 688. Moscow (1986). [in Russian]
11. Rosenberg, O.A.: Technological mechanics of deforming broaching, p. 203. Voronezh (2001). [in Russian]
12. SF Studenets' PM Eryomin OV Chernyavs'kyi 2015 The influence of deformation conditions on the structure and hardening of cast iron surface layer in machining with combined carbide broaches J. Superhard Mater. 37 4 282 288 https://doi.org/10.3103/S1063457615040085

Development of the Technique for Designing Rational Routes of the Functional Surfaces Processing of Products

Yaroslav Kusyi[1(✉)], Oleh Onysko[2], Andrii Kuk[1], Bohdan Solohub[1], and Volodymyr Kopei[2]

[1] Lviv Polytechnic National University, Lviv, Ukraine
jarkym@ukr.net
[2] Ivano-Frankivsk National Technical University of Oil and Gas, Ivano-Frankivsk, Ukraine

Abstract. The quality of the products is a wide concept, which is closely related with the development and design of the stage parts and stages of the Product Life Cycle. Therefore rational technological processes of products machining provide regulated machining accuracy, quality of surface layers of the products, their operational characteristics and reliability indicators. Traditional techniques for planning a rational route of treatment of product surfaces only ensured to form restricted number of quality parameters of products. Developed technique of planning a rational route of treatment of products provides relationships between material homogeneity of product and technological methods of surfaces treatments using the LM-hardness method. This technique is realized for steel products including for shaft 6E4-2717.00.00.01. The values of the Weibull homogeneity coefficients (m) during machining of shaft 6E4-2717.00.00.01 increasefrom 6.12–11.46 to 198.23–344.59. At that the material constants A_m in the technological chain "input blank – output product" change from 0.814 to 0.966 during machining of shafts 6E4-2717.00.00.01 by cutting and abrasive methods of processing.

Keywords: Technological process · Technological inheritability · LM-hardness method · Object-oriented design · Functionally-oriented design

1 Introduction

The achieving of the necessary parameters of accuracy, surface layers quality of products is provided by rational technological processes planning of manufacturing products. Technological processes of manufacturing products are realized by means of the cutting methods, surface plastic deformation, heat treatment and coating. The optimal structure of technological routes of functional surfaces processing is determined by providing the minimum technological cost of manufacturing machine products from one side, and achieving the required operational characteristics and reliability indicators in accordance with the operating conditions of machine products from another ones [1–3].

During machining of unconjugated surfaces of mechanical engineering products, the rational structure of technological routes of their processing is formed according to the

© The Author(s), under exclusive license to Springer Nature Switzerland AG 2022
I. Karabegović et al. (Eds.): NT 2022, LNNS 472, pp. 135–143, 2022.
https://doi.org/10.1007/978-3-031-05230-9_16

principle of object-oriented design of technological processes. This principle is provided by a variety of processing methods in compliance with the regulated requirements for the accuracy and quality of the surface layer at a minimum technological cost. Processing methods are chosen according to the method of blank production, its overall dimensions and accuracy, material properties, technical condition of technological equipment etc.

For functional, conjugate surfaces of mechanical engineering products, the rational structure of technological routes of their processing is formed according to the principle of functionally-oriented design of technological processes. The main criterion is to provide the necessary operational characteristics and reliability indicators of mechanical engineering parts in compliance with the regulated requirements for accuracy and surface layers quality of the products [4, 5], achieved by machining [6–8] and assembly [9–11].

Therefore, the development of technique for designing a rational structure of technological routes for processing of the functional, conjugate surfaces of products is an important task for modern mechanical engineering.

2 Literature Review

Two traditional variants for planning a rational route of treatment of product surfaces are used in mechanical engineering practice [12, 13].

The first variant is based on providing of the technical requirements for the accuracy and quality of executive surfaces using typical technologicalroutes for their treatment. This method of the treatment route planning for executive surfaces of the product is preliminaries. It requires a high qualification of the technologist and doesn't take into account the conditions of further operation for a particular product. The first variant of the planning for a rational route of a product surfaces treatment is implemented in the preliminary calculation of the production type and form of its organization [1, 12, 13].

The calculation method for determining of the surface treatment route is based on the calculation of the general refinement. Possible methods of surface treatment are determined by the refinement coefficient ε_i, which takes into account the inheritability of product properties from the previous to the current technological step [1]:

$$\varepsilon_i = [TOp_{i-1}]/[TOp_i], \tag{1}$$

where $[TOp_{i-1}]$, $[TOp_i]$ are the values of the limiting parameter for the technological operation by machining of the product's certain surface, respectively, at the previous and current technological steps.

In mechanical engineering practice, the machining accuracy of a certain surface of the product serves the most important criterion for evaluating the refinement [1]:

$$\varepsilon_i = T_{i-1}/T_i, \tag{2}$$

where T_{i-1}, T_i are, respectively, the tolerances of the size provided at the previous and current technological steps.

The design refinement ε_c is calculated by the accuracy criterion [1]:

$$\varepsilon_c = T_{bl.}/T_{prod.}, \tag{3}$$

where $T_{bl.}$ is the tolerance of the blank for the accuracy of the selected method of product blank obtaining; $T_{prod.}$ is the tolerance of the product according to the requirements of the design documentation.

The number of technological processing methods n is calculated by [1]:

$$n = \ell g(\varepsilon_c)/0.46. \tag{4}$$

The trueness checking of the number of the technological treatments n is determined by:

$$\prod_{i=1}^{n} \varepsilon_i \geq \varepsilon_c. \tag{5}$$

The accuracy of the product surfaces regulated by the designer must be consistent with the roughness parameters Ra or Rz [12, 13].

However, for some products, requirements on the quality of the surface layer of the executive surfaces of machine products are higher than on the accuracy one.

In this case, the criterion for estimating the design refinement is regulated in the design documentation by the roughness parameters Ra or Rz [1]:

$$\varepsilon = Ra(Rz)_{-1}/Ra(Rz), \tag{6}$$

where $Ra\,(Rz)_{i-1}$, $Ra\,(Rz)_i$ are respectively, the surface roughness parameters Ra or Rz at the previous and current manufacturing steps.

The estimated refinement of ε_c is calculated by [1]:

$$\varepsilon_c = Ra(Rz)_{bl.}/Ra(Rz)_{prod.}, \tag{7}$$

where $Ra\,(Rz)_{bl}$ is, respectively, the surface roughness of the workpiece and the parameter $Ra\,(Rz)_{prod.}$.

It is especially important for a certain group of products to provide the operational characteristics of their functional surfaces given the catastrophic consequences of their failures during operation. Therefore, the system principle of providing and support the life cycle of these products and machines requires consideration at the stage of technological preparation of production in accordance with operating conditions indicators of wear resistance, fatigue strength, etc., which are determined by reliability regulated parameters, including reliability and durability [14, 15].

During machining a layer of metal – allowance is removed from the workpiece surface after each technological step. It contains comprehensive information about the quality of the surface layer, the size of the defective layer of metal to be removed, the total deviation of the shape and relative position of the surfaces and the error of processing.

The minimum allowance of the workpiece material removed during cutting is determined by [1]:

– for sequential treatment of opposite surfaces (one-sided allowance):

$$z_{\min i} = R_{Z\,i\text{-}1} + T_{i-1} + \rho_{i-1} + \varepsilon_i, \tag{8}$$

– for parallel treatment of opposite surfaces (bilateral allowance):

$$2z_{\min i} = 2 \cdot \left[R_{Z\,i-1} + T_{i-1} + \rho_{i-1} + \varepsilon_i \right], \tag{9}$$

– for processing of external (internal) cylindrical surfaces of a detail (bilateral allowance):

$$2z_{\min i} = 2 \cdot \left[R_{Z\,i-1} + T_{i-1} + \sqrt{\rho_{i-1}^2 + \varepsilon_i^2} \right], \tag{10}$$

where $R_{z_{i-1}}$ is the roughness achieved at the previous transition, μm (mm); T_{i-1} is the depth of the defective layer, which was obtained at the previous transition, μm (mm);

ρ_{i-1} is a spatial deviation of the workpiece obtained at the previous transition, μm, (mm);

ε_I is aninstallation error at the performed technological transition, μm (mm).

During machining and formation of the main characteristics of the products material the information on the inheritability of their properties is required to analyse, systematize and process, which are a set of indicators of product quality [4, 5].

3 Research Methodology

3.1 Theoretical Bases

Theoretical approaches, methods of computer modelling and statistical processing and analysis of operational information with the research of failure physics and development of methods to provide operational characteristics and reliability of products are implemented from a single synergistic position to assess spontaneous formation of structures and their behaviour using the inheritability criterion [14, 15].

During the manufacturing of a product the technological operation is a subsystem for the technological process, but the technological step is a subsystem for certain technological operation [1].

The investigation of the process of technological inheritability of material properties and product quality parameters provides to improve the performance and reliability of machine parts using two conditions: establishing the causes of physical phenomena during the products manufacturing; capabilities of technological operations and processes management [1, 14, 15].

Mathematical dependences (8)–(10) describe the processes of technological inheritability of quantitative indicators of the product, in particular, product tolerances, surface quality parameters etc. At the same time, the material of the workpiece is the main carrier of information about the inheritability of properties along with many parameters of the processed surfaces [1].

The material parameters change during products manufacture by means of the influence of technological factors. In mechanical manufacturing in solving applied problems, the method of LM-hardness, providing the high efficiency, is used [16–18]. An important feature of the LM-hardness method is the determination of the degree of scattering of the characteristics of the product material mechanical properties on the destroyed samples after service life at different levels of stress-strain state. This method is easiest to implement in mechanical manufacturing, using hardness as a mechanical characteristic, the defined value of which is used to indirectly assess the properties of product material [1, 14, 15].

Homogeneity is an important parameter that integrally describes the certain state of the product material during processing of the obtained results after hardness measurements. It is analysed by the Weibull coefficient (m) [16–18]:

$$m = \frac{d(n)}{2,30259 \cdot S(\ell g(H))},$$
(11)

where $d(n)$ is a parameter that depends on the number of measurements n [16–18];

$$S(\ell g(H)) = \sqrt{\frac{1}{n-1} \cdot \sum_{i=1}^{n} \left(\ell g(H_i) - \overline{\ell g(H)}\right)^2}, \overline{\ell g(H)} = \frac{1}{n} \cdot \sum_{i=1}^{n} \ell g(H_i).$$
(12)

The calculated refinement ε_c is estimated by [1]:

$$\varepsilon_c = m_{prod.}/m_{bl.}$$
(13)

where $m_{prod.}$, $m_{bl.}$ are respectively, the coefficients of homogeneity for the blank and product.

Refinement for the i-th technological step at known values of the scattering characteristics of the material hardness is represented by a mathematical dependence [1]:

$$\varepsilon_i = m_i/m_{i-1},$$
(14)

where m_i, m_{i-1} are respectively Weibull homogeneity coefficients at the current and previous technological steps.

3.2 Experimental Research Methodology

The rational technological route planning for machining a workpiece of product using experimental studies is implemented for the shaft 6E4-2717.00.00.01, manufactured by "Transsystem" enterprise (Lviv, Ukraine) (see Fig. 1) [1].

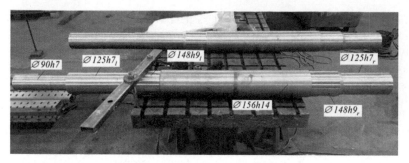

Fig. 1. Shaft 6E4-2717.00.00.01

Shaft 6E4-2717.00.00.01 is the basic product of the drive drums of the belt conveyor. TS 1850. Material of the product is steel 40X GOST 4543-71 (European equivalents are 37Cr4KD, 41CrS4). Round stock is a type of blank. Overall dimensions of the blank are Ø 160 × 2105 mm, its weight is 332.24 kg, weight of the product is 243 kg and material using coefficient– 0.73 [1].

It is established from the analysis of the service functions of the shaft 6E4–2717.00.00.01 (see Fig. 1) that the processing accuracy of the shaft necks is a criterion for determining the calculated refinement using the principle of object-oriented design [1, 4, 5].

Establishment of the methods of shaft surfaces treatment by means of the criterion of accuracy and technological routes planning of their treatment was realized using relations (2)–(5).

Homogeneity of the processed material at this stage of experimental research using the principle of functionally-oriented designanalysed by means of current values of Weibull homogeneity coefficient (m) using relations (11)–(12) [14–16].

Hardness was measured using a portable hardness tester TD-42 madeby "Ultracon"-company according given to the large sizes and weight of the blank. It was used the standard Brinell method of hardness measurement [1].

The constant of material $A_m$is calculated by [1]:

$$A_m = \left(m + 2\big/4m + 4\right)^{1/m}. \tag{15}$$

The hardness of the workpiece was measured in axial sections under the location of the executive, conjugate surfaces of the shaft – necks in the size Ø 90h7, Ø 125h7 (from the left), Ø 148h9 (from the left), Ø 148h9 (from the right), Ø 125h7 (from the right) and the free surface in the size Ø 156h14 (see Fig. 1). Two series of experiments included 30–35 hardness measurements for each section of the workpiece.

The results of the research are presented in Fig. 2, where (IT) – laws of the calculated refinement using criterion of the processing accuracy, (m) – laws of the calculated refinement using the homogeneity criterion of the product surfaces material. It is established that during machining a steel shaft for both methods there is a decrease in the values of intermediate refinements and tolerances in the transition from previous (rough) to finish (final) processing methods in the technological chain "workpiece surface – product surface". Improving the accuracy of workpiece machining using the principle of object-oriented design is characterized by reducing of the size tolerances. At the same time, the improving the accuracy of workpiece machining after the next technological step using the criterion of material homogeneity is provided by increasing the values of the Weibull homogeneity coefficients (m). This pattern characterizes the increasing for the homogeneity of the material structure of product and reduces the tendency to its technological damageability from the workpiece to the product (see Fig. 2, a).

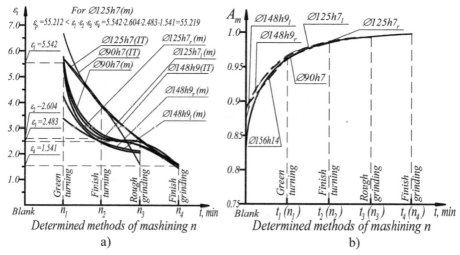

Fig. 2. Change of intermediate refinements after technological steps using criterions of accuracy and material homogeneity a) and material constants b) for processing routes of a shaft surfaces according to technological inheritability (t_1 (n_1), t_2 (n_2),..., t_k (n_k) – time spans of 1, 2, k technological steps; (IT), (m) – change of intermediate refinements using the processing accuracy and the homogeneity coefficient of the material)

For the proposed method of developing a technological route, the change of calculated refinements is smoother than for the traditional ones. The law of change of calculated refinements from green turning methods to rough grinding, when the main allowance is removed and the surface formation is provided, is similar to exponential. This allows us to use the mathematical apparatus of Markov chains to model the physical process in further investigations.

Besides, the improving (refinement) of the material constant A_m occurs using the principle of the technological inheritability of product parameters after each technological step in the technological chain "input workpiece – the final product". The value of A_m varies within narrow limits during machining of the workpiece of the shaft (see Fig. 2, b) for the selected processing method of the product executive surface. At the same time, more scattering takes place for workpieces and rough technological treatments with stabilization of constant values for finishing ones. This indicates about a higher homogeneity of the material structure compared to rough methods of the product surface treatment. Besides, the known laws of the choice of technological treatments for the principle of object-oriented design are confirmed: at each subsequent transition the surface treatment is usually increased by 1–3 accuracy grades for rough machining,

1–2 accuracy grades – for semi-rough machining and 1 accuracy grade – for machining finishing (see Fig. 2, a, b).

4 Conclusion

The main conclusions have been drawn based on the research results.

1. Technological inheritability of material properties during machining of steel products using the principle of functionally-oriented design provides the transformation of structurally inhomogeneous material of blanks into structurally homogeneous material of finished products. Thus the values of the Weibull homogeneity coefficients (m) increase from 6.123–11.457 (for the starting workpiece) to 198.227–344.588 (after different grinding methods).
2. The material constants A_m in the technological chain "input blank - output product" change from 0.814 to 0.966 during machining of shafts 6E4–2717.00.00.01 by cutting and abrasive methods of machining. In this case, the higher scattering of the values of the constants takes place for the workpiece and rough technological transitions with the stabilization of the values of the constants for finishing. This indicates a higher homogeneity of the structure after finishing compared to the rough methods of surface treatment and providing recommendations for the rational design of the route of treatment for the executive surfaces of the products.

References

1. Kusiy, Ya.: Scientific and applied bases of technological inheritability of quality parameters for providing of operational characteristics of products: Thesis of Doctor of technical sciences, Ukraine, Lviv (2021). (in Ukrainian)
2. Yoshimura, M.: System design optimization for product manufacturing. Concurr. Eng. **15**(4), 329–343 (2007). https://doi.org/10.1177/1063293x07083087
3. Kopei, V.B., Onysko, O.R., Panchuk, V.G.: Principles of development of product lifecycle management system for threaded connections based on the Python programming language. J. Phys. Conf. Ser. **1426** (2020). https://iopscience.iop.org/article/10.1088/1742-6596/1426/1/012034/pdf. https://doi.org/10.1088/1742-6596/1426/1/012034
4. Stupnytskyy, V.: Features of functionally-oriented engineering technologies in concurrent environment. Int. J. Eng. Res. Technol. (IJERT) **2**(9), 1181–1186 (2013)
5. Stupnytskyy, V., Hrytsay, I.: Computer-aided conception for planning and researching of the functional-oriented manufacturing process. In: Tonkonogyi, V., et al. (eds.) InterPartner 2019. LNME, pp. 309–320. Springer, Cham (2020). https://doi.org/10.1007/978-3-030-40724-7_32
6. Pryhorovska, T., Ropyak, L.: Machining error influnce on stress state of conical thread joint details. In: Proceedings of the International Conference on Advanced Optoelectronics and Lasers, CAOL, 2019-September: art. no. 9019544, pp. 493–497. (2019). https://doi.org/10.1109/CAOL46282.2019.9019544
7. Onysko, O., Panchuk, V., Kopei, V., Havryliv, Y., Schuliar, I.: Investigation of the influence of the cutter-tool rake angle on the accuracy of the conical helix in the tapered thread machining. J. Phys: Conf. Ser. **1781**(1), 012028 (2021). https://doi.org/10.1088/1742-6596/1781/1/012028

8. Ropyak, L.Ya, Vytvytskyi, V.S., Velychkovych, A.S., Pryhorovska, T.O., Shovkoplias, M.V.: Study on grinding mode effect on external conical thread quality. IOP Conf. Ser. Mater. Sci. Eng. **1018**(1), 012014 (2021). https://doi.org/10.1088/1757-899X/1018/1/012014

9. Ropyak, L.Y., Pryhorovska, T.O., Levchuk, K.H.: Analysis of materials and modern technologies for PDC drill bit manufacturing. Progress Phys. Metals **21**(2), 274–301. (2020). https://doi.org/10.15407/ufm.21.02.274

10. Kopei, V., Onysko, O., Panchuk, V., Pituley, L., Schuliar, I.: Influence of working height of a thread profile on quality indicators of the drill-string tool-joint. In: Tonkonogyi, V., Ivanov, V., Trojanowska, J., Oborskyi, G., Pavlenko, I. (eds.) InterPartner 2021. LNME, pp. 395–404. Springer, Cham (2022). https://doi.org/10.1007/978-3-030-91327-4_39

11. Bazaluk, O., Velychkovych, A., Ropyak, L., Pashechko, M., Pryhorovska, T., Lozynskyi, V.: Influence of heavy weight drill pipe material and drill bit manufacturing errors on stress state of steel blades. Energies **14**(14), art. no. 4198 (2021). https://doi.org/10.3390/en14144198

12. Bagge, M.: Process planning for precision manufacturing. An approach based on methodological studies: Doctoral thesis, Sweden, Stockholm (2014)

13. Gupta, D.P., Gopalakrishnan, B., Chaudhari, S.A., Jalali, S.: Development of an integrated model for process planning and parameter selection for machining processes. Int. J. Prod. Res. **49**(21), 6301–6319 (2011)

14. Kusyi, Ya.M., Kuk, A.M.: Investigation of the technological damageability of castings at the stage of design and technological preparation of the machine Life Cycle. J. Phys. Conf. Ser. **1426** (2020). https://iopscience.iop.org/article/10.1088/1742-6596/1426/1/012034/pdf. https://doi.org/10.1088/1742-6596/1426/1/012034

15. Kusyi, Ya., Stupnytskyy, V.: Optimization of the technological process based on analysis of technological damageability of casting. In: Advances in Design, Simulation and Manufacturing III. Proceedings of the 3rd International Conference on Design, Simulation, Manufacturing. Volume 1: Manufacturing and Materials Engineering, pp. 276–284, Ukraine, Kharkiv (2020). https://doi.org/10.1007/978-3-030-50794-7_27

16. Lebedev, A.A., Muzyka, N.R., Volchek, N.L.: A new method of assesment of material degradation during its operating time. Zaliznychnyi Transp. Ukrainy **5**, 30–33 (2003)

17. Lebedev, A.A., Makovetskii, I.V., Muzyka, N.R., Volchek, N.L., Shvets, V.P.: Assessment of damage level in materials by the scatter of elastic characteristics and static strength. Strength Mater. **38**, 109–116 (2006)

18. Muzyka, N.R., Shvets, V.P., Boiko, A.V.: Procedure and instruments for the material damage assessment by the LM-hardness method on the in-service scratching of structure element surfaces. Strength Mater. **52**(3), 432–439 (2020). https://doi.org/10.1007/s11223-020-00195-6

Investigation of the Influence of tapered Thread Pitch Deviation on the Drill-String Tool-Joint Fatigue Life

Volodymyr Kopei[1], Oleh Onysko[1,3(✉)], Yaroslav Kusyi[2], Victor Vriukalo[1], and Tetiana Lukan[1]

[1] Ivano-Frankivsk National Technical University of Oil and Gas, Ivano-Frankivsk, Ukraine
oleh.onysko@nung.edu.ua
[2] Lviv Polytechnic National University, Lviv, Ukraine
[3] Computerized Mechanical Engineering Department, Ivano-Frankivsk National Technical University of Oil and Gas, Karpatska, 15, Ivano-Frankivsk 76000, Ukraine

Abstract. Pipes and drill string elements are connected by tapered threads on the pin and box. These connections are called string-grid tool-joint. Such connections are made by lathe. Therefore, their accuracy largely depends on the kinematics of the lathe, as well as the accuracy of the profile of the cutters and their geometric parameters. One of the basic parameters of thread accuracy is its pitch. An important indicator for a tapered threaded connection is the accuracy of the lead angle and pitch. There for these two parameters and their influence on strength are the objects in this research. Increasing the thread pitch in the direction of the larger diameter of the pin taper can lead to a decrease in the fatigue strength of the joint.

Keywords: Drill-string tool-joint · Lathe machining · Inclination angle of the tapered thread · Pin and box · FEA · Fatigue

1 Introduction

The drilling of the oil and gas well requires maximum improvement of drilling machine equipment, among which drill string pipes are the most widely used component. The most complex and responsible part of the drill-string pipe is the tool-joint – the connector of pipes and other elements of the drill string. It is intended for mechanical fastening of drill pipes among themselves and for ensuring tightness. These requirements largely depend on the accuracy of the tapered thread on both parts of the drill-string pipe - pin and box. One of the most important influence parameter of the thread accuracy is pitch accuracy. Since the tapered threads are produced by means of machining on lathes, the accuracy thus largely depends on the accuracy of lathe tool kinematic. Among the parameters that affect the efficiency of the pipes and the accuracy of the cut is also the accuracy of the profile and pitch diameter.Therefore, the comprehensive identification of problems of precision threading, which arise in the process of their manufacture is associated with both theoretical studies of the kinematics of lathe machining, and with studies of the fatigue strength of the pipe thread joint in general.

I. Karabegović et al. (Eds.): NT 2022, LNNS 472, pp. 144–154, 2022.
https://doi.org/10.1007/978-3-031-05230-9_17

2   Literature Review

To increase the reliability of connections of elements of drill pipe columns, a number of methods are used, among which the most common are design [1–3] technological [4–6] and operational methods [7–10]. In [11] the influence of errors in the manufacture of modern drill bits on the stress state of the components of the drill string is researched. Paper [12] shows the dependence of the lathe machining process depending on the static and kinematic geometric parameters of the tool cutter. The research [13] deals with modelling of the wear resistance of the tool depending on machining parameter: cutting force in the turning process. In paper [14], the results of researches of the vertex turning kinematics parameters of cylindrical thread are submitted. The experimental investigations of the cutting process using work pieces from hard-machining high-alloy chromium steels are carefully considered in the work [15]. In [16] the generalized kinematic model of lathe machining operations is presented. That model can be applied for turning planning but not for accuracy analyze. The researches of the accuracy of the cylindrical thread obtained by using of the machining are shown in [17]. The article [18] theoretically investigates of the kinematics of the surface forming of the tapered thread by lathe machining but don't studies the influence of tool parameters on thread pitch accuracy. The analysis of the stress distribution on the thread teeth are shown in [19]. But the accuracy investigation is not presented in it. Experiments of a design are shown in article [20]. The geometry in multi-point thread turning is studied in [21]. But the accuracy of thread pitch is not investigated in it. The loading analysis on the thread teeth in cylindrical pipe thread connection are studied in paper [22]. Analytical study of kinematic rake angles of cutting edge of lathe tool for tapered thread manufacturing is obtained in [23]. In paper [24] using the cutting force prediction and using a validated mechanistic force model, the energy consumption in turning can be estimated but the pitch thread accuracy is not researched in [24]. Paper [24] is about computerized simulation and investigations of non-symmetric thread profiles but without pitch accuracy researches. In [25] the modulated tool path machining for threading applications is investigated well. In article [26] it is recommended to change the tolerance limits of this flank angle from (29.5°, 30.5°) to (30°, 30.5°) but not summered the pitch tolerance. The accuracy of the tapered thread profile is studied in [27].In [28] it is proved that the optimal function of the variable pitch of the box or pin can significantly improve the fatigue strength of the connection.

3   Research Methodology of the Lathe Machining Kinematics of Thread

Methodologically that study relies on the theory of circular vector functions [18, 23]. The parametric model of the cutting process whose parameters are the kinematic geometric parameters of the cutter is offered in work [29]. According to this theory, the radius vector $\vec{r}$ describes the movement of an arbitrary point M along a helical curve with a constant pitch on the tapered surface using this equation [18]:

$$\vec{r} = a(\vartheta) \cdot \vec{e}(\vartheta) + \vec{k} P\vartheta/2\pi \tag{1}$$

where:

P – thread pitch,
ϑ – rotate angle arbitrary point M relatively thread axis,
$a(\vartheta)$ – the tapered radius at arbitrary point M,
$\vec{e}(\vartheta)$ i $\vec{g}(\vartheta)$ – couple of the circular vector functions of unit valuesituated relatively axis OX form angles η i $\eta + 90°$ (Fig. 1).

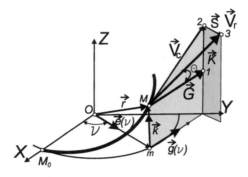

Fig. 1. Scheme of the moving an arbitrary point M along a helical curve

The circular vector function $\vec{g}(\vartheta)$ can be expressed as follows [14, 15]:

$$\vec{g}(\vartheta) = \vec{e}(\vartheta + \pi/2)$$

On the other hand, since this vector lies on the tangent line to the circuit in which the vector $\vec{e}(\vartheta)$ is placed radially, then [18, 23]:

$$\vec{g}(\vartheta) = \frac{d\vec{e}(\vartheta)}{d\vartheta}$$

The differential equation that defines the derivative of the position vector $\vec{r}$ with respect to its rotate angle υ around the axis Ox is offered in article [29]:

$$\frac{d\vec{r}}{d\vartheta} = \left(\frac{P tg(\varphi)}{2\pi}\right) \cdot \vec{e}(\vartheta) + \left(P \cdot tg(\varphi) \cdot \frac{\vartheta}{2\pi} + \left(\frac{d_3}{2} - \Delta\right)\right) \cdot \vec{g}(\vartheta) + \vec{k} \cdot P/2\pi$$

$$(2)$$

On the Fig. 1 the vector $\vec{S}$ (between points 2 and 3) corresponds to the first addend of the formula 2, the vector $\vec{G}$ (between points M and 1) corresponds to the second addend of the same equation and the vector $\vec{K}$ (between points 1 and 2) corresponds to the third addend of the formula.

$$\frac{d\vec{r}}{d\vartheta} = \vec{V}_r = \vec{S} + \vec{G} + \vec{K}.$$

$$(3)$$

Vector $\vec{V}_r$ is tangent line to helical curve at a point M but it isn't correspond summary speed of the its motion because its value is defined as a change of the position vector long with respect to a change in rotate angle υ around the axis of the thread. So: – Eq. 2 does not describe the speed of the motion of the point M; – because of the vector $\vec{S}$ in formula 3 lies individually for every arbitrary point on the helical curve it is impossible to provide it by the machine tool.

3.1 Theoretical Determining of the Angle of Inclination of A Conical Helix with a Constant Pitch (Standard Inclination Angle)

Based on Eqs. (1)–(3) and using Fig. 1 the formula of the inclination angle is [18, 29]:

$$\theta = arctg \frac{p}{\sqrt{(r + p\upsilon tg(\beta))^2 + (p\upsilon tg(\beta))^2}}$$

where β – half conical angle.

Because according to the expressions $\vartheta = \frac{h}{p}, p = \frac{P}{2\pi}$, then after substituting these expressions into the Eq. (4):

$$\theta = arctg \frac{P}{\sqrt{(P \cdot tg(\beta))^2 + \pi^2(2r + 2h \cdot tg(\beta))^2}} \tag{4}$$

The efficiency of the cutting process depends on the magnitude of the static rake angle γ_a at the corner of the cutter as well as the rake angles at other points of the cutting edge (Fig. 2).

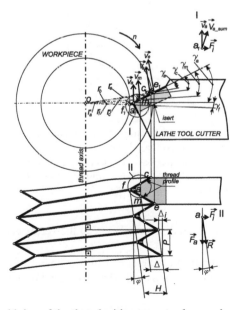

Fig. 2. Scheme of machining of the thread with a taper angle φ and constant pitch P by lathe

The nominal value of the rake angle at an arbitrary point M of the cutting edge of the threaded cutter can be determined using Fig. 2 by equation (Fig. 2, Fig. 3):

$$\gamma_m = arcsin\left(\frac{r_a}{r_m}sin(180° - \gamma_a)\right)$$

where:

r_m – the radius for the point M is determined by the formula:
$r_m = r_c - \Delta = \frac{d_3}{2} + Ptg\varphi \cdot \frac{\vartheta}{2\pi} + b - \Delta,$
r_a – the radius for point A is determined by the formula:
$r_a = \frac{d_3}{2} + Ptg\varphi \cdot \frac{\vartheta}{2\pi} + b + f - H.$

The equation of the speed of point M (arbitrary point of the cutting edge) is obtained in [29]:

$$\frac{d\overrightarrow{r}_M}{dt} = 2\pi n\left[\left(\frac{Ptg(\varphi)}{2\pi}\right)\overrightarrow{e}(\vartheta) + \left(P \cdot tg(\varphi)\frac{\vartheta}{2\pi} + \left(\frac{d_3}{2} - \Delta + b\right)\right)\overrightarrow{g}(\vartheta) + \overrightarrow{k}\frac{P}{2\pi}\right]$$

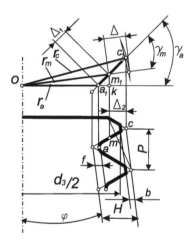

Fig. 3. Scheme for determining the parameters of placement of an arbitrary point of the cutting edge of the cutter for turning a tapered thread. The symbols indicate: $d3$ - outer diameter of the tapered thread on the side of the smaller base; b - section of the top of the thread; f is the cut of the cavity of the cut; Δ is the radial distance of an arbitrary point M of the cutting edge from the outer vertex of the output triangle;

Δ_I is the difference of radial distances between the vertex and arbitrary point of the cutting.

Therefore, the determination of the speed of movement of an arbitrary point on a helical line with a constant pitch on a conical surface can be represented as follows:

$$\frac{d\overrightarrow{r}_M}{dt} = \left|\overrightarrow{S_m}\right| \cdot \overrightarrow{e}(\vartheta) + \left|\overrightarrow{V_m}\right| \cdot \overrightarrow{g}(\vartheta) + \left|\overrightarrow{V}_{s\,axial}\right| \cdot \overrightarrow{k}$$

The constituent vectors of this expression are determined by formulas:

$$\overrightarrow{S_m} = (Ptg(\varphi)n) \cdot \overrightarrow{e}(\vartheta) \tag{5}$$

$$\overrightarrow{V_m} = 2\pi n\left(P \cdot tg(\varphi)n \cdot t + \left(\frac{d_3}{2} - \Delta + b\right)\right) \cdot \overrightarrow{g}(\vartheta) \tag{6}$$

$$\overrightarrow{V_{s\,axial}} = \overrightarrow{k}\,Pn \tag{7}$$

However, the machine is not able to provide the vector $\overrightarrow{S_m}$, only $\overrightarrow{S_a}$ in every points of the carbide insert cutting edge, and: $\left|\overrightarrow{S_a}\right| = \left|\overrightarrow{S_m}\right|$. Transverse motion with speed $\overrightarrow{V}_{s\,axial}$ is on the schema on Fig. 4.

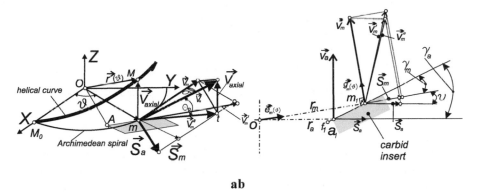

ab

Fig. 4. Position of the angle $\odot_p$, (a) and vectors $\overrightarrow{V_m}$, $\overrightarrow{V'_m}$ i $\overrightarrow{V''_m}$ (a, b)

So the real inclination angle of the tapered thread θ_p can be calculated based on Fig. 4 by the formula:

$$\theta_p = arctg\left(\frac{\left|\overrightarrow{V}_{s\,axial}\right|}{\left|\overrightarrow{V}''_m\right|}\right) \tag{8}$$

Therefore, based on [29] and on Eq. (6), the real inclination angle of the tapered thread should be calculated by the formula:

$$\theta_p = arctg\left(\frac{Pn}{\sqrt{(\left|\overrightarrow{V_m}\right| - ptg(\varphi)n \cdot \sin(\gamma_a - \gamma_m))^2 + (Ptg(\varphi)n \cdot \cos(\gamma_a - \gamma_m))^2}}\right) \tag{9}$$

4   Theoretical Investigation of the Real Inclination Angle and Pitch Deviation of the Tapered Thread

The axial (pitch)deviation of an arbitrary point m of the cutting edge of the cutter from a given trajectory in the real process of turning conical lock threads can be defined as a function of the difference between the theoretical θ and the real angle of rise of the thread θp. On Fig. 5 the scheme of formation of the axial (pitch) deviation of the theoretical model relative to the real conical helix is illustrated.

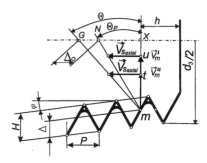

Fig. 5. The scheme of formation of the axial (pitch) deviation on the real and the theoretical tapered helix thread

At the point m as in Fig. 5 two vectors $\vec{V}''_m$ and $\vec{V}'_m$ are plotted. At their ends, i.e. at points t and u, the longitudinal feed vectors Vs_{axial} are placed. The axial (pitch) deviation can be calculated as the difference between the segments (GX) and (NX) by the formula:

$$\Delta_o = \left(\frac{d_3}{2} + h \cdot tg(\varphi) - \Delta\right) \cdot \left(tg(\theta) - tg(\theta_p)\right) \tag{10}$$

where:

θ – standard inclination angle, determined by using of dependence (8);
θ_p – predict thread inclination angle, determined by using of dependence (9);
d_3 – minor diameter of the tool-joint tapered thread according to Standard API 7;
h – distance from pin end to determined thread turn;
φ – cone angle of tapered thread;
Δ – distance from major diameter to definite point m.

5   Methodology of the Investigation of the Pitch Deviation on the Tool-Joint Fatigue Life

In order to study the effect of pitch irregularity, which is caused by technological factors, on fatigue strength a finite element analysis was performed. The parametric model of the tool joint developed by the authors [30] was used. The model was developed in Abaqus/CAE. Drilling tool joint ZN-80 GOST 5286 with Z-66 GOST 28487 thread (2

3/8 REG API Spec. 7 equivalent) was selected for research. The material of the parts is SAE-4140 steel with a Young's modulus $E = 201$ GPa, a Poisson's ratio $v = 0.33$, a yield strength $\sigma_y = 965$ MPa, a ultimate tensile strength $\sigma_t = 1076$ MPa, a endurance limit $\sigma_{-1} = -572$ MPa). Material plasticity and friction are simulated. To simulate the make-up of an axisymmetric model of a threaded connection, the axial elongation ("bolt load") of the box shoulder [31] by the value $\Delta = 0.25$ mm (Fig. 6) is performed.

The calculations of the fatigue life N using Brown-Miller strain-life equation [32] and fe-safe 6.5 software are performed. At the left end of the pin (Fig. 6) the pressure L acts. It simulates the external tensile load. We consider the fatigue loading cycle $L_{min} = 0$ MPa, $L_{max} = 350$ MPa.

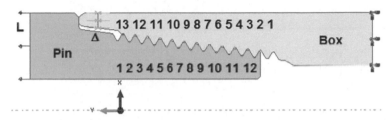

Fig. 6. Model of the ZN80 tool joint with numbered thread roots of the pin and box

The box pitch is standard and constant. The pin pitch is standard at the end of the pin and varies along the y-axis. The axial coordinate of the root with number i of the pin thread can be determined as follows:

$$y(i) = -P \cdot i + d(12 - i),$$

where i is the number of the thread root,

P is the thread pitch,

d is the pitch error along the y-axis. Three values are selected for research: 0 mm (constant pitch), +0.01 mm (pitch increases), −0.01 mm (pitch decreases).

6 Results of the Investigation of the Pitch Deviation on the Tool-Joint Fatigue Life

The results are shown in Figs. 7, 8. In the standard tool joint (Figs. 7a, 7b), the smallest lg$N$ values are observed in the first ($i = 1$, $i = 2$) thread root of the pin (3.68, 2.2) and the box (4.37, 4.19).

The value $d = + 0.01$ mm (Figs. 7c, 7d) leads to a decrease in lg$N$ in the first roots of the pin thread (3.21, 2.73) (Fig. 7a) and an increase in lg$N$ in the first roots of the box thread (4.5, 4.43) (Fig. 7b). The unevenness of the lg$N$ distribution in the pin thread increases (Fig. 7a).

The value $d = -0.01$ mm (Figs. 7e, 7f) leads to an increase in lg$N$ in the first roots of the pin thread (Fig. 8a) (3.6, 3.46) and a decrease in lg$N$ in the first roots of the box thread (4.24, 3.89) (Fig. 8b). The values in the last roots of the pin thread also decrease (Fig. 7a). The unevenness of the lg$N$ distribution in the pin thread decreases (Fig. 7a).

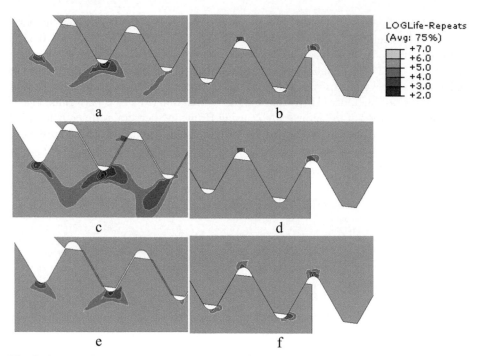

Fig. 7. lg*N* distribution: a, c, e – left side; b, d, f – right side; a, b – *d* = 0 mm; c, d – *d* = + 0.01 mm; e, f – *d* = −0.01 mm.

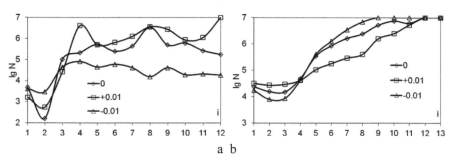

Fig. 8. Values of lg*N* in roots *i* of pin (a) and box (b) threads

7 Conclusion

1. Studies have shown the difference between the tapered thread pitch predicted by standard and the tapered thread pitch predicted by lathe machining possibility.

2. Increasing the thread pitch in the direction of the larger diameter of the pin taper can lead to a decrease in the fatigue strength of the joint. A particularly dangerous situation is when the pitch errors of the pin and the box are summed up. To prevent this, the authors recommend making the pin thread with *d* < 0 mm and *d* > −0.01 mm.

References

1. Vlasiy, O., Mazurenko, V., Ropyak, L., Rogal, O.: Improving the aluminum drill pipes stability by optimizing the shape of protector thickening. Eastern-Europ. J. Enterprise Technol. 1(7–85), 25–31 (2017). https://doi.org/10.15587/1729-4061.2017.65718
2. Ropyak, L.Ya., Pryhorovska, T.O., LevchukK, H.: Analysis of materials and modern technologies for PDC drill bit manufacturing. Progress Phys. Metals 21(2), 274–301 (2020). https://doi.org/10.15407/ufm.21.02.274
3. Grydzhuk, J., Chudyk, I., Velychkovych, A., Andrusyak, A.: Analytical estimation of inertial properties of the curved rotating section in a drill string. Eastern-Europ. J. Enterprise Technol. 1(7–97), 6–14 (2019). https://doi.org/10.15587/1729-4061.2019.154827
4. Pryhorovska, T., Ropyak, L.: Machining error influnce on stress state of conical thread joint details. In: Proceedings of the International Conference on Advanced Optoelectronics and Lasers, CAOL, 2019-September 9019544, pp. 493–497 (2019). https://doi.org/10.1109/CAOL46282.2019.9019544
5. Shatskyi, I.P., Perepichka, V.V., Ropyak, L.Y.: On the influence of facing on strength of solids with surface defects. MetallofizikaiNoveishieTekhnologii 42(1), 69–76 (2020). https://doi.org/10.15407/mfint.42.01.0069
6. Ropyak, L.Ya., Vytvytskyi, V.S., Velychkovych, A.S., Pryhorovska, T.O., Shovkoplias, M.V.: Study on grinding mode effect on external conical thread quality, IOP Conf. Ser: Mater Sci. Eng. 1018, 012014 (2021). https://doi.org/10.1088/1757-899X/1018/1/012014
7. Panevnik, D.A., Velichkovich, A.S.: Assessment of the stressed state of the casing of the above-bit hydroelevator. NeftyanoeKhozyaystvo - Oil Industry 2017(1), 70–73 (2017)
8. Velychkovych, A., Petryk, I., Ropyak, L.: Analytical study of operational properties of a plate shock absorber of a sucker-rod string. Shock and Vibration 3292713 (2020). https://doi.org/10.1155/2020/3292713
9. Shatskyi, I., Ropyak, L., Velychkovych, A.: Model of contact interaction in threaded joint equipped with spring-loaded collet. Eng. Solid Mech. 8(4), 301–312 (2020). https://doi.org/10.5267/j.esm.2020.4.002
10. Skitsa, L., Yatsyshyn, T., Liakh, M., Sydorenko, O.: Ways to improve safety of pumping-circulatory system of a drilling rig. Mining Mineral Deposits 12(3), 71–79 (2018). https://doi.org/10.15407/mining12.03.071
11. Bazaluk, O., Velychkovych, A., Ropyak, L., Pashechko, M., Pryhorovska, T., Lozynskyi, V.: Influence of heavy weight drill pipe material and drill bit manufacturing errors on stress state of steel blades. Energies 14(14), 4198 (2021). https://doi.org/10.3390/en14144198
12. Baizeau, T., Campocasso, S., Fromentin, G., Rossi, F., Poulachon, G.: Effect of rake angle on strain field during orthogonal cutting of hardened steel with c-BN tools. In: 15th CIRP Conference on Modelling of Machining Operations. Procedia CIRP 31, pp. 166–171 (2015)
13. Zhang, G., Guo, C.: Modeling flank wear progression based on cutting force and energy prediction in turning process. Procedia Manuf. 5, 536–545 (2016). https://doi.org/10.1016/j.promfg.2016.08.044
14. Zanger, F., Sellmeier, V., Klose, J., Bartkowiak, M., Schulze, V.: Comparison of modeling methods to determine cutting tool profile for conventional and synchronized whirling. ProcediaCIRP 58, 222–227 (2017). https://doi.org/10.1016/j.procir.2017.03.216
15. Wang, J., Mathew, Ph., Li, X., Huang, C., Zhu, H.: Experimental study on cutting characteristics for buttress thread turning of 13%Cr stainless steel. Ad-vances in Materials Processing IX in Key Engineering Materials 443, pp. 262–267 (2010)
16. Khoshdarregi, M.R., Altintas, Y.: Generalized modeling of chip geometry and cutting forces in multi-point thread turning. Int. J. Mach. Tools Manuf. 98, 21–32 (2015). https://doi.org/10.1016/j.ijmachtools.2015.08.005

17. Slătineanu, L., et al.: Requirements in designing a device for experimental investigation of threading accuracy. MATEC Web of Conference 112 01005/ (2017). https://doi.org/10.1051/matecconf/201711201005
18. Medvid, I., Onysko, O., Panchuk, V., Pituley, L., Schuliar, I.: Kinematics of the tapered thread machining by lathe: analytical study. In: Tonkonogyi, V., et al. (eds.) InterPartner 2020. LNME, pp. 555–565. Springer, Cham (2021). https://doi.org/10.1007/978-3-030-68014-5_54
19. Wang, Y., Xia, B., Wang, Z., Chai, C.: Model of a new joint thread for a drilling tool and its stress analysis used in a slim borehole. Mech. Sci. **7**, 189–200 (2016). https://doi.org/10.5194/ms-7-189-2016
20. Saruev, A.L., Saruev, L.A., Vasenin, S.S.: Drill pipe threaded nipple connection design development. IOP Conf. Series: Earth Environ. Sci. **27**(1), 012056 (2015). https://doi.org/10.1088/1755-1315/27/1/012056
21. Xu, H., Shi, T., Zhang, Z., Shi, B.: Loading and contact stress analysis on the thread teeth in tubing and casing premium threaded connection. Math. Problems Eng. **2014**, 287076 (2014). https://www.hindawi.com/journals/mpe/2014/287076/
22. Chen, S., An, Q., Zhang, Y., Gao, L., Li, Q.: Loading analysis on the thread teeth in cylindrical pipe thread connection. J. Pressure Vessel Technol. **3**(132), 0312021–0312028 (2010)
23. Onysko, O., Kopei, V., Panchuk, V., Medvid, I., Lukan, T.: Analytical study of kinematic rake angles of cutting edge of lathe tool for tapered thread manufacturing. In: Tonkonogyi, V., et al. (eds.) InterPartner 2019. LNME, pp. 236–245. Springer, Cham (2020). https://doi.org/10.1007/978-3-030-40724-7_24
24. Fromentin, G., Döbbeler, B., Lung, D.: Computerized simulation of interference in thread milling of non-symmetric thread profiles. In: 15th CIRP Conference on Modelling of Machining Operations. Procedia, CIRP 31, 2015, pp. 496–501 (2015)
25. Berglind, L., Ziegert, J.: Modulated Tool Path (MTP) Machining for Threading Applications. In: 43rd Proceedings of the North American Manufacturing Research Institution of SMEProcedia Manufacturing, vol. 1, pp 546–555 (2015)
26. Kopei, V., Onysko, O., Odosii, Z., Pituley, L., Goroshko, A.: Investigation of the influence of tapered thread profile accuracy on the mechanical stress, fatigue safety factor and contact pressure. In: Karabegović, I. (ed.) NT 2021. LNNS, vol. 233, pp. 177–185. Springer, Cham (2021). https://doi.org/10.1007/978-3-030-75275-0_21
27. Onysko, O., Medvid, I., Panchuk, V., Rodic, V., Barz, C.: Geometric modeling of lathe cutters for turning high-precision stainless steel tapered threads. In: Ivanov, V., Trojanowska, J., Pavlenko, I., Zajac, J., Peraković, D. (eds.) DSMIE 2021. LNME, pp. 472–480. Springer, Cham (2021). https://doi.org/10.1007/978-3-030-77719-7_47
28. Kopei, V.B., Onysko, O.R., Panchuk, A.G., Dzhus, A.P., Protsiuk, V.R.: Improving the fatigue life of the tool-joint of drill pipes by optimizing the variable pitch of the box thread. IOP Conf. Ser.: Mater. Sci. Eng. **1166**, 012017 (2021). https://doi.org/10.1088/1757-899X/1166/1/012017
29. Onysko, O., Panchuk, V., Kopey, V., Havryliv, Y., Sculiar, I.: Investigation of the influence of the cutter-toolrake angle on the accuracy of the conical helix in the tapered thread machining, International Conference of Applied Sciences. J. Phys: Conf. Ser. **1781**(2021), 012028 (2021). https://doi.org/10.1088/1742-6596/1781/1/012028
30. vkopey/ThreadsAbaqus: Abaqus/CAE python scripts for modelling threaded connections of oil and gas equipment. https://github.com/vkopey/ThreadsAbaqus
31. Hoffman, E.L.: Finite element analysis of sucker rod couplings with guidelines for improving fatigue life: Sandia report Sandia National Laboratories, p. 66 (1997)
32. Kandil, F.A., Brown, M.W., Miller, K.J.: Biaxial low cycle fatigue fracture of 316 stainless steel at elevated temperatures. In: International Conference on Mechanical Behaviour and Nuclear Applications of Stainless Steel at Elevated Temperatures, pp. 203–210. The Metals Society, London (1982)

Stress and Deformation Analysis of Different Bolt Models in Finite Element Analysis

Aida Sušić[1], Emir Nezirić[2(✉)], Safet Isić[2], and Edin Šunje[2]

[1] Smed Engineering BH d.o.o., Mostar, Bosnia and Herzegovina
[2] Faculty of Mechanical Engineering Univerzitetski Kampus, Džemal Bijedić" University of Mostar, Sjeverni logor b.b., 88104 Mostar, Bosnia and Herzegovina
emir.neziric@unmo.ba

Abstract. Finite Element Method (FEM) is a commonly used numerical method for solving engineering problems by cutting the structure into several elements connected to nodes. A set of algebraic equations is solved (instead of multiple partial differential equations) to obtain the desired results. The method is usually applied through different computer software on the structure models. It is required more time to solve large models with a lot of elements. To reduce solution time, it is required to simplify some of the parts of the structure and use some approximations. Some of the usual components in structures are bolts, which could be simplified by different models. The proposition of the simple structural model is created in FEM software named ANSYS is created and analyzed with different bolt models. Stress and deformation results are compared to the analytical solution for the proposed model. It is shown that the best results for deformation are obtained for bolt modelled as line body, and the best result for the stress in bolts are obtained for a solid body model of a bolt and nut. For all variants of the bolt model, is shown that it gives larger values of stress and deformations than analytical calculations, which increases the safety factor of a structure. Usage of the right approximation of structure parts in FEM can decrease solution time and still give an appropriate result for deformation and stress.

Keywords: Finite element method · Bolt · Stress · Deformation

1 Introduction

Calculations of the stress and deformation of bolted connections for practical use are regulated by international standards for different fields of engineering. For example, bolted joints of steel structures are regulated by EN 1993-1-8 [1] and pressure vessels with bolted domed ends are regulated by EN 13445-3 [2]. Most of the standards are focused on the definition of how to handle some safety coefficients, standardized calculation factors for the increase or decrease of applied loads and procedures for calculations. Calculations of the bolted connection for the general load are well known and defined in a lot of the textbooks thatare focused on the mechanical components design [3–5] or steel structures design [6–8].

© The Author(s), under exclusive license to Springer Nature Switzerland AG 2022
I. Karabegović et al. (Eds.): NT 2022, LNNS 472, pp. 155–162, 2022.
https://doi.org/10.1007/978-3-031-05230-9_18

The Finite Element Method (FEM) for modelling and analysis of bolted connection-sis discussed in various researches. Montgomery J. presented the possibility of modelling of bolts in FEM software Ansys [9], where it was compared its impact on the gap caused by preloading force on the bolt. It was concluded that understanding the results is more important than the value of the results for the person who is reading the results.

Grizejda R. in [10] have compared a simplified bolt model with the use of the FEM and compared it to the solid model of the bolt to obtain results for the stiffness of the bolts. It has been concluded that a simplified model could be used as a substitute for the solid bolt model if an adequate preload distribution was used.

In [11] authors hadanalyzed various preloading conditions of bolted connections modelled by FEM, where is concluded that contact deformation in bolted connection is decreasing with the distance from bolt location.

In [12] authors have recreated a test made by Shi G. et al. in FEM software Abaqus. They have concluded that with the sold model of bolts it is possible to have almost the same results obtained by the experimental tests.

McCarthy et al. in [13] have compared FEM results to experimental results of a single bolted connection. Results have shown that strain values for the modelled bolted connection have good agreement with experimental results, but bending and axial stiffnes were too high.

In [14] have analyzed four different FEM models of bolts in structure, where is concluded that a solid model of a bolt gives the best results and could be modeled to analyze the structures.

Masaly et al. in [15] have analyzed a bolted connection subjected to cyclic loading, where solid bolts were used in the FEM model. It is concluded that experimental and FEM results have a good agreement, but analysis with a solid model of a bolt is time-consuming and that it is necesary to use proper simplifications of bolted connections.

In this paper, it would be analyze the different types of bolted connection modelled in FEM software Ansys. Results obtained by FEM would be compared to analytically obtained results for deformation and stress in the connection objects.

2   Analytical Analysis of Deformation and Stress in Bolted Conection

2.1   Analytical Calculation of Bolted Connection Deformations

The simplest bolted connection constist of the two steel plates (i.e. flanges, beam-to-column connection, motor baseplate to stand) connected with a bolt through holes, as shown in Fig. 1. Bolt connection could have additional components such as shims.

Deformation of the plates caused by the bolt preloading could be calculated as

$$\lambda = \frac{F_0}{k_c} \tag{1}$$

where:

F_0 – preloading force.

k_c – plates stifness.

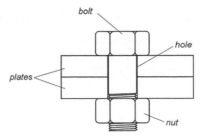

Fig. 1. Simple bolt connection

Plates stifness could be calculated as [3]

$$k_c = \frac{A_c E_c}{g} \qquad (2)$$

where:

A_c – effective area.

E_c – elasticity modulus of plates.

g – thickness of plates (lenght between bolt head and nut).

Effective area i approximately equal to the average area of the shaded gray surface on Fig. 2.

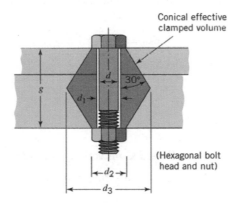

Fig. 2. Effective area of bolted connection [3]

Effective area could be calculated by [3]

$$A_c = \frac{\pi}{4}\left[\left(\frac{d_3 + d_2}{2}\right)^2 - d_1^2\right] \qquad (3)$$

where

$d_1 \approx d$ (for small clearances).

$d_2 = 1{,}5d$ (for standard hexagonal bolts).

$d_3 = d_2 + g \tan 30°$

2.2   Analytical Calculation of Bolt Stress In Bolted Connection Example

As a working load, it would be used applied force on the bolted connection example, shown on Fig. 3. Bolt is inserted in a hole with diameter 1 mm larger than the bolt diameter, where plates are overlaping. Overhung plate is loaded with force F on its free end.

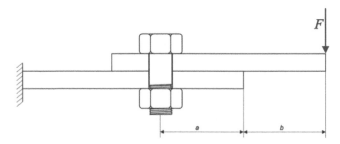

Fig. 3. Example of a bolted connection for analysis

Force in the bolt could be calculated from the proportion:

$$F_b a = Fb \Rightarrow F_b = \frac{b}{a} F \tag{4}$$

Total force in the bolt is calculated as sum of the preloading force and the governing applied force.

$$F_{tot} = F_0 + F_b \tag{5}$$

Stress in the bolt could be calculated as

$$\sigma = \frac{F_b}{A} \tag{6}$$

where A is the bolt section area.

2.3   Analytical Analysis of An Bolted Connection Example

As an example, there would be used a connection with folowing data.

- Bolt M10
- $F_0 = 1$ kN
- $F = 1$ kN
- $g = 20$ mm (2*10 mm)
- $E_c = 200$ Gpa
- $a = 50$ mm
- $b = 50$ mm

With the given data, the plates deformation is calulated by using the Eqs. (1)–(3). Value of the calculated plates deformation is 0,41 μm.

Stress in the bolt was calculated for a analyzed bolted connection by using Eqs. (4)–(6), and its value is 38,24 MPa.

3   Fem Analysis of the Bolted Connection

The analyzed bolted connection is modelled and analyzed in software ANSYS, where the three types of bold model is used for analysis:

- line body
- solid body (bolt and nut as two bodies)
- solid body (bolt and nut as one body)

Plates were created as solid objects, where contacts between plates were set as *frictional* with 0,2 friction coefficient. For a line body, contact between line body as bolt and plates was created by using *joint* as connection between line body end and contact surface between nut/head and the same area on the plates. For the solid body, contact between nut/head and plates was designed as *frictional* with friction coefficient 0,2.

Deformation of plates for different models of bolt was shown on Fig. 4.

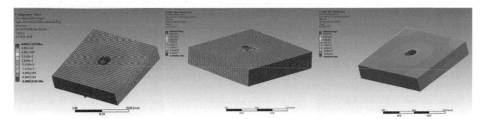

Fig. 4. Deformation for different bolt models: line body, solid body (bolt and nut as two bodies) and solid body (bolt and nut as one body)

Total deformation comparable to the analytical results could be obtained from the sum of minimum and maximum deformation obtained from the FEM analysis, since it would be equal to the total deformation through the plates thickness.

Total stress in the bolt for different models of bolt was shown on Fig. 2.

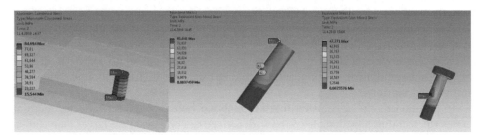

Fig. 5. Stress for different bolt models: line body, solid body (bolt and nut as two bodies) and solid body (bolt and nut as one body)

4   Results and Discussion

Results obtained from the analytical calculations and FEM analysis are shown in Tables 1 and 2.

Table 1. Deformation of the plates caused by preload in bolt

Bolt model	Deformation [μm]	Deviation from the analytical solution [%]
Analytical solution	0,41	–
Line body	0,44	7,32
Solid body (bolt and nut as two bodies)	0,86	102,44
Solid body (bolt and nut as one body)	0,61	48,78

Table 2. Stress in bolt caused by preload in bolt

Bolt model	Stress [μm]	Deviation from the analytical solution [%]
Analytical solution	38,24	–
Line body	84,69	121,46
Solid body (bolt and nut as two bodies)	43,00	12,44
Solid body (bolt and nut as one body)	47,27	23,61

As it could be seen, line body gives the best results for the deformation of plates with difference of 7%. Result closest to the analytical values for the stress in the bolt was stress obtained by solid body bolt model with bolt and nut as two bodies. Its deviation from the analytical result is 12,44%.

As it could be seen on the Fig. 5, stress of all bolt models on the middle of the bolt span does not have much mutual difference in value, and it has range of 42–47 MPa for all FEM bolt models. It could be seen that stress of the bolts for all models is larger up to 20% from the analytical solution, and it could be used as valid for the bolted connection strength calculations.

In addition to the conclusion that all analysed bolt models are valid for analysis comes the fact that all calculated values (deformation and stress) are larger than analytical solution, so it would be safer to use FEM calculations than analytical calculation procedure. This should be confirmed by trials on some more complicated structures, and confirm it with experimental testing.

5 Conclusions

In this paper was conducted a comparative analysis of different bolt FEM model with analytical solutions for the bolted connection. After the conducted analysis, conclusions could be made as follows:

- Analytical analysis of the bolted connection is well documented and covered by multiple international standards.
- Numerical analysis of the bolted connection could be conducted by different types of bolt models: line body, solid body, bonded connection, etc.
- Line body bolt model is most suitable for deformation analysis, since it would give you appropriate results which are close enough to analytical results, but with possibility to consider the local effects.
- Solid body with bolt made by two parts would be suitable to analyze bolted connections according to stress of the bolts, but the time required for this analysis would be much larger than the line body bolt model.

References

1. EN 1993-1-8, Eurocode 3: Design of steel structures – Part 1–8: Design of joints, European Comitee for Standardization (2005)
2. EN 13445-3, Unfired pressure vessels – Part 3: Design, European Comitee for Standardization (2005)
3. Juvinall, R.C., Marshek, K.M.: Fundamentals of Machine Component Design. Willey, Hoboken (2017)
4. Ugural, A.C.: Mechanical Design of Machine Components. CRC Press, Boca Raton (2015)
5. Schmid, S.R., Hamrock, B.J., Jacobson, B.O.: Fundamentals of Machine Elements. CRC Press, Boca Raton (2014)
6. Boracchini, A.: Design and Analysis of Connections in Steel structures, Ernst and Sohn, Berlin, Germany (2018)
7. McKenzie, W.M.C.: Design of Structural Elements to Eurocodes. Palgrave Macmillan, New York (2013)
8. Pearsons, J.D., et al.: Manual for the Design of Steelworks Building Structures to Eurocode 3, The Institution of Structural Engineers, London, GB (2010)
9. Montgomery, J.: Methods for modeling bolts in the bolted joint, in: Proceedings of the ANSYS 2002 User's Conference, Pittsburgh (2002)
10. Grzejda, R.: Modelling bolted joints using a simplified bolt model. J. Mech. Transp. Eng. **69**(1), 29–37 (2017). https://doi.org/10.21008/j.2449-920X.2017.69.1.03
11. Piscan, I., Predincea, N., Pop, N.: Finite element analysis of the bolted joint. Proc. Manuf. Syst. **5**(3), 167–172 (2010)
12. Krolo, P., Grandić, D., Bulić, M.: The guidelines for modelling the preloading bolts in the structural connection using finite element methods. J. Comput. Eng. **2016**, 8 (2016). Article ID 4724312. doi: https://doi.org/10.1155/2016/4724312
13. McCarthy, M.A, McCarthy, C.T., Lawlor, V.P., Stanley, W.F.: Three-dimensional finite element analysis of single-bolt, single-lap composite bolted joints: part I—model development and validation. Composite Stuct. **71**, 140–158 (2005). https://doi.org/10.1016/j.compstruct.2004.09.024

14. Kim, J., Yoon, J.C., Kang, B.S.: Finite element analysis and modeling of structure with bolted joints. Appl. Math. Model. **31**, 895–911 (2007). doi. https://doi.org/10.1016/j.apm.2006.03.020

15. Masaly, E., El-Heweity, M., Abou-Elfath, H., Osman, M.: Finite element analysis of beam-to-column joints in steel frames under cyclic loading. Alexandria Eng. J. **50**(1), 91–104 (2011). https://doi.org/10.1016/j.aej.2011.01.012

Bim Modeling in Mechanical Engineering

Ermin Huskić, Adis J. Muminovic$^{(\boxtimes)}$, Isad Saric, and Nedim Pervan

Department of Mechanical Design, Faculty of Mechanical Engineering, University of Sarajevo, Sarajevo, Bosnia and Herzegovina
adis.muminovic@mef.unsa.ba

Abstract. Building information modeling (BIM) is a process based on so called intelligent three dimensional (3D) models. These models incorporate not only dimensional properties of the product. They incorporate all information's about functional or descriptive properties. There is significant difference between computer aided design (CAD) modeling and BIM modeling. Goal of this paper is to describe and highlight that differences. Through the paper it is described when, where and why BIM modeling needs to be used. Idea behind BIM modeling principles is described in detail. In addition, advantages and disadvantages of BIM modeling is highlighted. Potential possibilities of using BIM modeling in the field of mechanical engineering have been presented.

Keywords: Building information modeling (BIM) · Computer aided design (CAD) · Mechanical engineering · MEP

1 Introduction

BIM modeling is a new and advance 3D modeling methodology which is still in its developing stage. It is developed firstly for architectural and civil engineering purposes with a goal for easy and fast creations of quality 3D models. These modeling principles can be used in mechanical engineering also, especially in the area of Mechanical, Electrical and Pluming (MEP) for design of heating, ventilation, and air conditioning (HVAC) systems.First appearance of BIM modeling can be connected to 1970s. Due to limited technological development in that period they did not have significant development until these days. Today, BIM modeling is main modeling principles in the area of architecture and civil engineering.

A cross-disciplinary analysis of BIM and mechanical engineering is very well done in paper [1]. In conclusions of paper [1] impact of BIM on mechanical engineering is highlighted and 11 research opportunities were identified. In addition, paper [1] provides directions for studies where research is focused on the integration of mechanical engineering systems in BIM workflows and on the extension of BIM capability to model future mechanical engineering systems. Paper [2] gives overview of essential technologies, discusses their intended purpose, and gives outline of the currently achieved functionality in the area of BIM modeling with a special focus on mechanical engineering. In paper [3], current trends, benefits, possible risks, and future challenges of BIM for

© The Author(s), under exclusive license to Springer Nature Switzerland AG 2022
I. Karabegović et al. (Eds.): NT 2022, LNNS 472, pp. 163–175, 2022.
https://doi.org/10.1007/978-3-031-05230-9_19

the Architecture, Engineering and Construction (AEC) industry are discussed. The findings of this study provide useful information for AEC industry practitioners considering implementing BIM technology in their projects. From above mentioned papers it can be seen that there was some interest in research about BIM modeling in mechanical engineering few years ago, but it is hard to find more papers from last year's regarding importance and influence of BIM modeling in mechanical engineering. It is still unknown will BIM modeling stay just for development of Heating, Ventilation, and Air Conditioning (HAVC) systems or does it have potential to be used in other fields of mechanical engineering.

2 2. CAD OR BIM

2.1 2.1. CAD Modeling

CAD modeling is a process of computer representation of products, parts, buildings, machines, etc., using drawings, different types of 3D models and different materials [4]. CAD models have integrated dimensional and material information's but they do not have any descriptive information's like, for example, information's about who is the owner of that product, what is input power needed for product to operate, what kind of energy input product needs, what is designed life time of a product, etc. Main goal of CAD is to help engineers during design process [5]. CAD is not only 2D or 3D drawing tool, it can help engineers to solve many different problems in the area of mechanical engineering. Problems like stress analysis, dynamic simulations, design optimization, fluid flow, heat transfer, etc., [6–10]. Today, developments of new products in all industries is impossible without some form of CAD or its tools. First appearance of CAD can be linked up to the 1960s. At the beginning, CAD was only connected to Computer Aided Drafting. Design and modeling came later. It was used only for drawings using first developed computers. First CAD 3D models was wired models developed in 1970s and first solid models was created using first personal computers (PC) in 1980s.

There is a lot of reasons why CAD modeling is essential today in any form of product development and design, some of them can be summarized in a form of:

1. **Productivity.** Using CAD 3D models, it is possible to visualize and analyze functionality and design of developed products. Number of errors in design and development process can be drastically decreased and because of that, productivity is drastically increased. It is much cheaper if the error is noticed during design process in comparison if the error is noticed during manufacturing process. In addition, time from concept idea to fully functional virtual prototype is drastically reduced in comparison to the process of development of real prototype. Flexibility of design process are increased using CAD models. It is easy to change any part of product during design stage. All changes will be automatically updated in all parts of product development, like drawings or CAM.
2. **Quality of product and design process.** This is main advantage which CAD brings in the process of design and development. Quality of product development and design is drastically increased using CAD modeling. Better products can be designed and developed, with small amount of errors. Also, errors can be found and fixed more

quickly and easily. A lot of different analysis can be carried out for new product before its production, for example structural stress analysis, thermal analysis, flow simulation, optimization, etc.

3. **Communications.** Communications between engineers in design process are much better because they can share 3D models wary easily and quickly using the internet from any place around the world. In addition, communication in whole product life cycle management (PLM) if much better if CAD models are used. For example, communication between CAD as design field and CAM as manufacturing field is very easily, basically it is a so called one click communication.

4. **Development of database.** Using CAD modeling it is easy to create database of standard parts and assemblies. This standard parts can be easily and quickly used infinitive amount of times in all design projects.

2.2 BIM Modeling

According to the definition given by American National Committee BIM is digital representation of physical and functional characteristics of real model. BIM represent knowledge integration from real model in to digital model through whole life cycle of product. On this way, all necessary information's about the product are always available for all necessary decision making. BIM is initially developed for civil and architectural fields, but it's become more and more connected to all other fields which is connected to building engineering. BIM can be considered as advance modeling methodology with a goal to create models which are not only geometric representation of a model, they incorporate all necessary information's which can be used through whole life cycle of product. Using this types of models, it is easy for engineers or stakeholders to make important decisions during design and development of product or during product exploitation. Concept of BIM can be traced back to 1970s, but first official presentation of BIM concept is done 1986 by Robert Aisha. The idea is then called Building Model. Commercial use of BIM happens when AUTODESK company take everything about BIM development. First software which can be considered as BIM software was ArchiCAD from 1987. At the beginning ArchiCAD was standard CAD software, latter it becomes more BIM oriented software. Today, there is a lot of different BIM software's, some of most commonly used are: ArchiCAD, AUTODESK Revit, MagicCAD, Tekla Structures, VectorWorks, Trimble SketchUp, etc.

As early mentioned BIM is firstly developed for architecture and civil engineering. Today it is used in all other disciplines connected to buildings engineering. BIM is centered around information's. All descriptive or quantitative information's which are associated with a model can be easily exchanged between different disciplines. All peoples who work with a model can add new information's to the model or use information's which are already associated with it. This use of information is so called information workflow inside BIM (Fig. 1).

BIM modeling concept implies that all information's from all life cycles of products are consisted inside one model. One model is used through whole life of a product. Life of a product starts with concept design and detail design. Selected concepts are developed inside detail design and then analyzed using different analysis method depending on the type of the product. Last step in product design and development is the step where

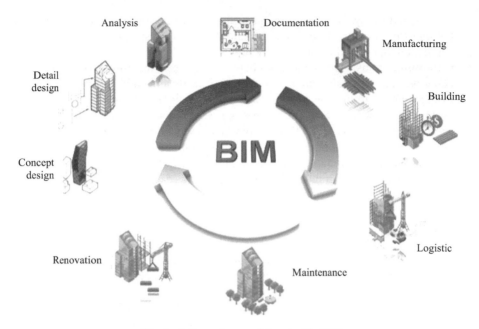

Fig. 1. Information workflow inside BIM

technical documentation is prepared. This is the step where standard CAD modeling stops in a form that additional information's are added to the model. CAD model can be used in CAM for manufacturing, but initial CAD model is not changed during manufacturing in any case.

In BIM modeling concept, all information's regarding manufacturing can be added directly to the model in a form of descriptive or quantitative information's. Same concept is used for all other steps inside life cycle of a model: building, logistic, maintains and renovation. Idea of BIM is that developed model fallow product through its whole life. For example, if house is taken as an example, idea of BIM modeling is that model of a house should be stored in some kind of database inside city government official offices and servers and any change to the real house should be implemented to the model. Also using model of a house renovations and maintenance will be much faster and cheaper. Final goal of a BIM model is to develop a model of whole city using this methodology, including traffic, water supply system, gas system, electric system, etc. Maintenance and management of the towns and cities will be much easier if this types of models are developed.

2.3 Why BIM

Why mechanical engineers should learn and be familiar with concept of BIM? Answer to this question is very simple. Building of a skyscraper or a house is a complex engineering process which includes a lot of mechanical engineering systems inside. Engineering systems like HVAC system, water supply, steel construction, facade system construction, windows and doors, etc. Windows and doors are great example where mechanical engineers should transfer from CAD to BIM concept. Architects and civil engineers want to use already developed models of windows and doors inside its architectural projects to save time and money, they cannot do that if the whole project is a BIM project and doors and windows are developed as CAD projects. Mechanical engineers inside doors and windows manufacturing company should use BIM modeling for design, development and manufacturing of its product. With a sell of real product (door or window) they sell the BIM model also. Or, even better solution is to offer BIM models of its products (doors and windows) for free with their online catalog. If the architect uses their BIM model inside its projects they will probably buy the real product also.

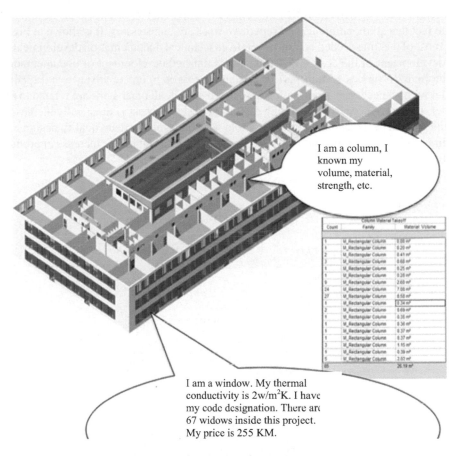

Fig. 2. Illustrative models in BIM project

Idea of BIM modeling is to make a model "alive" and "intelligent", model should "speak" about yourself. If a model of a screw is taken as an example. In standard CAD modeling, model of a screw will have geometry and material property. In BIM modeling model of a screw can have properties about maximal yield strength, maximal allowed torque for fastening, price of a screw, possible manufacturing companies listed, how much screws are in the project, etc. In civil engineering, BIM model of a wall or column have additional properties associated with it. Model of a wall "speaks" about its own properties. It can "tell" engineers about its thermal properties, number of windows associated with it, material used for construction, etc. Windows can "speak" about its thermal conductivity, prices, manufacturing company, expected life, code designation, number of glass plates inside, number of walls inside the window profile, etc. Illustrative examples are shown at Fig. 2.

3 Advantages of Bim

Main advantages of CAD and BIM modeling in comparison to the classic 2D design lies in the fact that all monotonous and repetitive work are unnecessary. It is shown at Fig. 3 that most of the time needed is transferred from technical documentation development to the development of the design.When design is finished development of documentation is automatized process. Main part of design process is to create advance and quality model with all necessary information associated with it, all other steps are related to the model. Using CAD or BIM modeling quality of design process is drastically increased. Engineers and designers can put their whole working time to create quality design and quality model. This results with cost and time saving with significant increase in product quality.

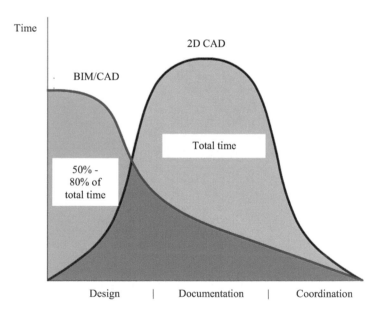

Fig. 3. Difference between 2D CAD and BIM/CAD project development processes

If CAD modeling is considered as 3D, BIM modeling can be considered as even 7D. Where 4D, 5D, 6D and 7D means following:

– 4D are time information's about the project and its elements. All models inside the projects will have information's about the time when they need to be installed during building process.
– 5D can be considered as information's about the prices of all elements installed inside the project. All 3D models carry with them information's about the prices. Using this information's, it is easy to calculate the price of whole project.
– 6D are information's about the energy consumptions of all elements and systems inside the project. All 3D models carry with them information about the quantity and type of energy they need. Using this information's, it is easy to calculate the energy consumption of whole project (for example a building or a house).
– 7D are information's about maintenance. All 3D models have integrated information's about its maintenance needs. Using this information's, it is easy to follow maintenance plans and service life of whole project.

Main advantages of BIM strategy can be summarized as:

– 20% lower building cost
– 33% lower cost of building maintenance,
– 47% - 65% less errors inside the project, noticed during building process,
– 44% - 59% increase of project quality,
– 35% - 43% less risk,
– 34% - 40% better building infrastructure,
– 50% less environmental pollution,
– 50% less time for project development.

4   Bim in Mechanical Engineering, Where to Use It

As it is already mentioned, BIM modeling is developed firstly for architectural and civil engineering. According to that, it should be used in all fields of mechanical engineering connected to buildings engineering. Parametric BIM models enables engineers to integrate knowledge from mechanical engineering into the project of a building. Usually, BIM software's have special section for mechanical and electrical engineering, it is called MEP (Mechanical, Electrical, Plumbing) section. Today, BIM is used for development of HVAC (Heating, Ventilation, Air Conditioning) systems and systems for water supply and drains. These systems must be developed by mechanical engineers. Also, all 3D models for these systems (pumps, heat exchangers, pipes, valves, ventilators, radiators, etc.) must be developed using BIM strategy by mechanical engineers. For example, for architects it is important that they have BIM models of radiators if they want to include them in their projects. This is the reason why all manufacturers of all building equipment for HVAC systems need to have developed online catalogs of their products in the form of BIM models. HVAC systems and systems for water supply and drains are most important fields of mechanical engineering where BIM modeling is used today, with its full capacity. Similar systems are systems for fluid processing industry. This is a field where BIM modeling can be used in a same way as for HVAC systems. Example of BIM model for HVAC and fluid processing systems is shown at Fig. 4.

Fig. 4. Example of BIM model for HVAC and fluid processing systems

Other fields of mechanical engineering where BIM modeling should be used, but it is not used at all (or it is used only partially) are:

1. **Steel structures.** BIM modeling in this fields are not fully developed and used. Usually BIM software's have special add in for this, which needs to be additionally installed. Principe of modeling is simple, in the first step, analytical model of structure needs to be defined. In second step vertical columns, horizontal beams and internal branches needs to be defined. In third step section profiles and type of connections are defined. Using developed analytical model, with associated sections and connection, it is easy to carry out numerical stress analysis of the structure. Example of metal structure developed in Autodesk Revit software are shown at Fig. 5. Autodesk Revit is one of the most used software's for BIM.

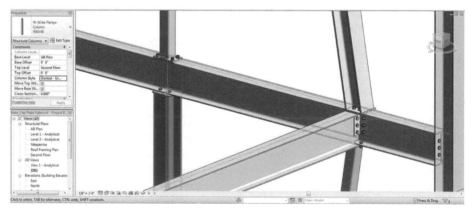

Fig.5. Steel structure developed in Revit software

Connections between steel elements are easily selected and numerically analyzed. It is necessary to select type of elements which needs to be connected and type of connection (bolts or welds). 3D model of connection is automatically generated. Also, numerical analysis for selected loads are automatically carried out. Two rows bolted connection of IPE 300 beams is shown at Fig. 6.

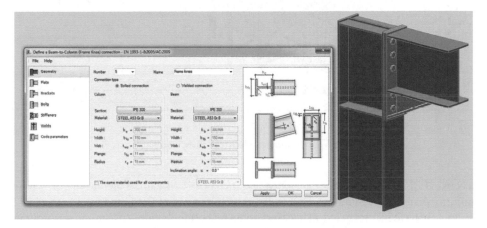

Fig. 6. Two rows bolted connection of IPE 300 beams

2. **Furniture design and development.** Furniture is basic part of every internal building planning. If an architect wants to insert some type of furniture in its BIM project that models must be BIM models. Product designers, industrial designers and mechanical engineers develop their models as some kind of polygonal or CAD models, this models can't be used inside BIM projects.

Also, if BIM models are developed and used for furniture design, that enables architect or users to additionally select best solution of that furniture for its project design. BIM model of furniture can have different configurations and different sizes integrated all in one model. Also, BIM model can have integrated additional information's regarding price of the furniture, delivery time, etc. Figure 7 shows simple model of kitchen table with chairs. Using this model user can select dimensions of the table (height, width, length) or number of chairs. All other dimensions are automatically generated. For selected parameters user will automatically get the price of the table and its delivery time.For furniture manufacturers it is recommended to create BIM models of all their products, and to offer them in online catalogs. If an architect uses their model inside its project, user will probably buy that product.

4. Lifts, cranes and other mechanical equipment for building engineering. Building process must be planned in detail and BIM modeling is a great tool for that. Building planning incorporate a lot of mechanical equipment which will be used during building

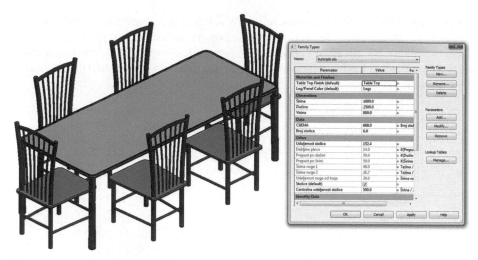

Fig. 7. Furniture modeling in BIM software

process. If that equipment is developed as BIM models it can be incorporate in whole building planning BIM model. Especially in the cases where BIM models of mechanical equipment have incorporated all necessary data for better planning, usage and positioning of that equipment. Figure 8 shows BIM model of a crane Liebherr-Tower Luffing-357HCL1832. This model is fully parametrized BIM model of a crane. User can import this model in a BIM project, chose position, height and load capacity, all other parts of the crane are automatically selected and generated.

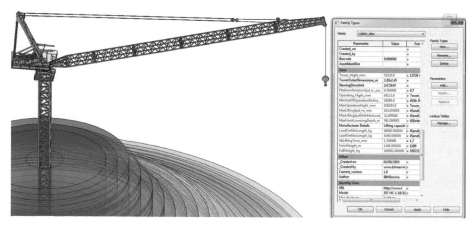

Fig. 8. Crane modeling in BIM software

Also, all other, non-dimensional calculations are automatically calculated (power of electric motors, diameters of ropes, size of hook, etc. Using this BIM model, mechanical engineers are incorporate knowledge into the model and user (building planer) can use that model and select what parameters are necessary for their building planning case. Same case is with all types of lifts. There is a lot of different types of lifts, or same lifts with different working characteristic's like speed or load capacity. For civil engineers it will be great if they can have already developed BIM models of lifts so they can insert it inside their projects and chose necessary parameters.

5. **Doors and windows.** Doors and windows are usually designed by architects or product designers and developed by mechanical engineers. Mechanical engineers usually develop doors and windows using CAD models. CAD models are great for manufacturing, but they cannot be used inside BIM projects. Doors and windows are great examples of product with similar design but with a lot of different parameters and dimensions. This is the reason why they are great for BIM modeling. BIM software's have integrated function for doors and windows import. Using this functions, it is possible to select doors or windows from the catalog. Selected and imported doors and windows probably will be purchased by end user. Because of this, for doors and windows manufacturers it is necessary to have developed their products in a form of BIM models. Except dimensional parameters, BIM models of doors and windows have additional parameters like heat transfer coefficients, solar heat gain, etc. These parameters can be used for automatic calculation of heating and cooling of a rooms and buildings. Example of doors and windows imported in BIM project on a wall can be seen at Fig. 9. It can be noticed that there is a lot of descriptive parameters associated with a BIM model.

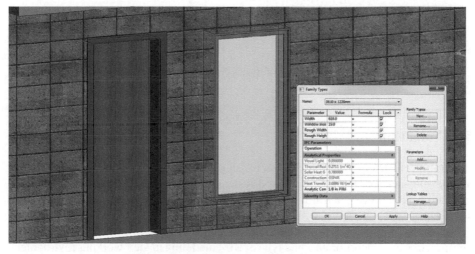

Fig. 9. Door and window in BIM software

6. **Stairs and fences.** Stairs and fences are also developed and manufactured by mechanical engineers. Here is the same case as with furniture and doors and windows. Again, architects can only use already developed BIM models and insert them in their projects. Mechanical engineers must develop such BIM models so they can be additionally adjusted for needed design and dimensions. This models will require more time to develop, but that will save a lot of time for architect during import, adjust and use of that model. Some examples of developed BIM models of fences are shown at Fig. 10.

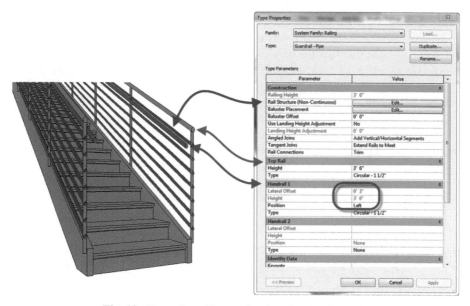

Fig. 10. Examples of fences developed as BIM models

5 Conclusion

Integration of BIM modeling in mechanical engineering is justified. In the field of design and development of HVAC systems and systems for water supply and drain it is not just justified it is necessary and it must be used. BIM modeling can be considered as next step in project development, with a special focus of buildings engineering, from concept design up through whole life of a building. All engineering fields connected to buildings engineering must incorporate some type of BIM modeling in their working environment.

Main focus of BIM is to integrate knowledge into the model. Developed BIM models have integrated all necessary information's regarding the product which is represented by the model. Mechanical engineers must be familiar with this concept and manufactures of all mechanical equipment which will be used for building engineering must incorporate BIM modeling into their working cycles. Architects and civil engineers must have available all BIM models of doors, windows, stairs, fences, sanitary equipment, furniture, etc., in online catalogs for free download and integration into their products. In addition, mechanical engineers must have available all BIM models of HVAC and water supply equipment for automatic integration in their projects also. They can develop these models by themselves or download it from online catalogs.

More detail explanation about development process of BIM models will be discussed in future research. Development of these models (so called families) are done using some of familiar functions from CAD modeling (Extrusion, Blend, Sweep, etc.). Modeling process for BIM and CAD are very similar. Transfer from one methodology to another is easy process.

References

1. Adanić, L., de Oliveira, S.G., Tibaut, A.: BIM and mechanicalengineering - a cross-disciplinaryanalysis. Sustainability **13**, 4108 (2021). https://doi.org/10.3390/su13084108
2. Svetel, I., Jaric, M., Budimir, N.: BIM: Promises and Reality. Spatium **2014**(32) (2014). https://doi.org/10.2298/SPAT1432034S
3. Azhar, S.: Building Information Modeling (BIM): Trends, Benefits, Risks, and Challenges for the AEC Industry. Leadership Manage. Eng. **11**(3) (2011). https://doi.org/10.1061/(ASCE)LM.1943-5630.0000127
4. Smajic, J., Muminovic, A.J., Saric, I., Muminovic, A.: Development and Design of a Machine for Hybrid Manufacturing. In: Karabegović I. (eds) New Technologies, Development and Application IV. NT 2021. Lecture Notes in Networks and Systems, vol. 233. Springer, Cham (2021) .https://doi.org/10.1007/978-3-030-75275-0_15
5. Saric, I., Smajic, J., Muminovic, A.J.: Integrated development and design of gears reduction driveIn: Karabegović, I. (eds.) New Technologies, Development and Application III. NT 2020. Lecture Notes in Networks and Systems, vol 128. Springer, Cham (2020)
6. Muminovic, A.J., Colic, M., Mesic, E., Saric, I.: Innovativedesign of spur gear tooth with infill structure. Bull. Polish Acad. Sci. Tech. Sci. **68**(3), 477–483 (2020)
7. Ma, Y., Niu, W., Luo, Z., Yin, F., Huanga, T.: Static and dynamic performance evaluation of a 3-DOF spindle head using CAD–CAE integration methodology, Robot. Comput.-Integrat. Manuf. **41** (2016). https://doi.org/10.1016/j.rcim.2016.02.006
8. Muminovic, A.J., Muminovic, A., Mesic, E., Saric, I., Pervan, N.: Spur gear tooth topology optimization: finding optimal shell thickness for spur gear tooth produced using additive manufacturing. TEM J. **8**(3), 788–794 (2019)
9. Biedrzycki, M.: Design of a hot runner injection mold with the use of CAD/CAE systems, Master Thesis, Faculty of Production Engineering (FMIE), The Institute of Manufacturing Processes (FMIE/IoMP) (2021)
10. Agarwal, A., Marumo, R., Letsatsi, M.T.: Numerical analysis of tube flow with helical insert using nano fluid to enhance heat transfer. Mater. Today Proc. **47** (2021). https://doi.org/10.1016/j.matpr.2021.05.281

Influence of Multiple Tool Geometry on Drawing Force

Mirna Nožić[1](✉) and Himzo Đukić[2]

[1] Faculty of Mehanical Engineering, University "DžemalBijedić" of Mostar, Univerzitetski kampus, 88 104 Mostar, Bosnia and Herzegovina
mirna.nozic@unmo.ba
[2] University of Mostar FSRE, Mostar, Bosnia and Herzegovina

Abstract. The paper presents experimental results of research on the influence of the geometry of a multistage tool on the drawing force. The results refer to six different five-step tools that are dimensioned in three different ways. The tools were tested in production conditions on horizontal mechanical presses on two different products with different total logarithmic degree of deformation.

Keywords: Multistage tools · Dimensioning of multistage tools · Drawing with wall thickness reduction · Tool geometry · Drawing force · Redistribution of total logarithmic degree of deformation

1 Introduction

In the process of deep drawing with the reduction of wall thickness, multi-stage tools are used. The executive parts of each multi-stage tool are: a puller and a set of extraction rings. The number of rings in a multi-stage tool is usually from two to six, so depending on that number, we are talking about: two-stage, three-stage, four-stage, five-stage or six-stage tools [1, 2]. The rings are arranged in the carrier according to the diameter from the largest to the smallest, which corresponds to the diameter of the extracted work piece. One workpiececan be obtained by drawing on different multistage tools with the same diameter of the rear ring in the tool. In Fig. 1.a. a 3D view of a multi-stage tool with five extraction rings (1–5) and an extractor 7 is given. Cooling rings are placed between the second and third and third and fourth rings. Figure 1.b shows a cross-section of the drawing ring.

The geometry of the drawing ring is determined by the diameter Ødi and the cone angle αi. The problem of dimensioning a multistage tool involves determining the diameter and angle of the cone rings in the tool. For the designer, this is a complex task, because it needs to be solved in such a way as to ensure the extraction of the workpiece of the given dimensions, without disturbing the stability of the drawing process.

When dimensioning multistage tools, the starting point is the total load, which is expressed in terms of the total logarithmic degree of deformation φ_u. This load is redistributed to the rings in the multistage tool. Ways of redistributing the total load to the rings in a multistage tool give different models for dimensioning multistage tools.

I. Karabegović et al. (Eds.): NT 2022, LNNS 472, pp. 176–183, 2022.
https://doi.org/10.1007/978-3-031-05230-9_20

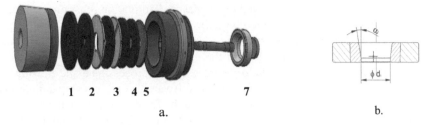

Fig. 1. 3D view of a multistage tool and cross section of the drawing ring

In the literature [3–10], these models are elaborated in detail, so they will only be listed here. In production practice, the dimensions of the rings in a multistage tool are mainly determined by pull-out tests until the given dimensions of the workpiece are reached. The problem comes down to determining the dimensions ($\emptyset d_i$, α_i) of the rings in the tool, which will ensure the drawing of the given workpiece without cracking. This approach usually does not take into account the correct redistribution of the load within the tool, which leads to the fact that some rings are more loaded, more worn and have to be changed more often. In order to achieve stability of the drawing process and reduce delays due to frequent changes of rings in the tool, models for dimensioning multistage tools have been developed, which start from the total tool load and define the way of its redistribution to rings in the tool.

2   Theoretical Methods for Determining The Drawing Force

The procedure for determining stress and strain force by approximate methods of plasticity theory has been discussed in detail in the literature [11, 12]. The expressions for the calculation of stress and deformation forces are derived by: Method of integration of approximate differential equations of equilibrium and plasticity, Method of deformation work, Method of upper estimation and Method of sliding lines. In the domestic literature, the expression for the pulling force obtained by the method of integration of approximate differential equations of equilibrium and plasticity is accepted. In papers [12–14], expressions for the calculation of the deformation force during drawing on two-stage tools are given.With the increase in the number of rings in the multistage tool, the problem of calculating the deformation force becomes more complicated. In papers [2, 4], expressions for the calculation of the drawing force during drawing on multistage tools with: three, four, five and more rings are given.

3   Experimental Research of Drawing Force

Experimental research was conducted on five-step tools for making two workpieces with different total degrees of deformation. Six five-step tools were investigated, whose dimensions and geometry are the result of different ways of dimensioning. For each work item, experiments were performed with a tool used in production practice and with tools dimensioned according to theoretical models: the model of equal load of all rings in a multistage tool (model 1) and the model of deformation separation (model 2). Experimental research was performed on horizontal mechanical presses in production conditions. The registration of the drawing force was performed using measuring tapes glued to the extractors. The signal from the measuring tapes was transmitted to the Spider 8 measuring device with the help of cables, and the processing of the drawing force diagram was performed using the accompanying software.Cylindrical preparations made of CuZn28 brass were used for experimental research. The dimensions of the preparation for the first work item are: outer diameter d = 14.2 mm, height h = 22 mm and bottom thickness s = 3.5 mm. For the second work item, the preparation had the following dimensions: outer diameter d = 12 mm, height h = 15 mm and bottom thickness s = 3.4 mm.

3.1 Sizing of Tools for Experimental Research

The load of the tools used in the experimental investigations was monitored via the value of the logarithmic degree of deformation on each ring in the tool.For five-stage tools for the first workpiece, dimensioned in different ways (production, according to model 1 and model 2), the values of the angles of the drawing rings and the logarithmic degrees of deformation are given in Table 1.

Table 1. Ring angles and logarithmic degrees of deformation of the investigated five-stage tools

Five-level tools for the first working subject								
	Production tools			Tool dimensioned according to model 1			Tool dimensioned according to model 2		
Reg num rings	exp. tag	$\varphi_u = 1{,}73$		exp. tag	$\varphi_u = 1{,}73$		exp. tag	$\varphi_u = 1{,}73$	
	E1			E2			E3		
	φ_i	φ_i/φ_u	α_i	φ_i	φ_i/φ_u	α_i	φ_i	φ_i/φ_u	α_i
1	0.251	0.145	6°	0.490	0.283	18°	0.351	0.203	10°
2	0.515	0.298	8°	0.372	0.215	12°	0.465	0.269	18°
3	0.389	0.225	10°	0.290	0.168	10°	0.339	0.196	10°
4	0.353	0.204	8°	0.353	0.204	8°	0.353	0.204	8°
5	0.224	0.129	8°	0.224	0.129	8°	0.224	0.129	8°

The values of the angles of the drawing rings and the logarithmic degrees of deformation for five-stage tools for the second workpiece are given in Table 2.

Table 2. Ring angles and logarithmic degrees of deformation of the investigated five-stage tools

Five-level tools for second working subject									
	Production tools			Tool dimensioned according to model 1			Tool dimensioned according to model 2		
Reg. num. rings	exp. tag E4	$\varphi_u = 1{,}79$		exp. tag E5	$\varphi_u = 1{,}79$		exp. tag E6	$\varphi_u = 1{,}79$	
	φ_i	φ_i/φ_u	α_i	φ_i	φ_i/φ_u	α_i	φ_i	φ_i/φ_u	α_i
1	0.486	0.272	15°	0.507	0.283	18°	0.486	0.272	16°
2	0.313	0.175	8°	0.384	0.215	12°	0.435	0.243	16°
3	0.321	0.179	8°	0.329	0.184	10°	0.330	0.184	10°
4	0.305	0.170	8°	0.308	0.172	8°	0.283	0.158	8°
5	0.364	0.203	8°	0.261	0.146	8°	0.254	0.142	8°

4 Results of Experimental Research

Experimental research of six five-stage tools in production conditions gave images of their total load. For all drawings, at least ten diagrams of the dependence of the deformation force on time were recorded. Multiplying the time with the average deformation rate, diagrams of the dependence of the force on the drawing stroke were obtained. Figure 2 shows selected recorded pull-out force diagrams for all six five-stage tools.

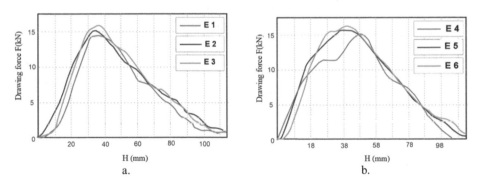

Fig. 2. Pull-out dependence diagrams for all six five-stage tools

The values of the pull-out force at the characteristic points (1–7) were read from the recorded diagrams for all experiments. Measuring points for deformation of the bottom of the workpiece are marked with points 1–5, while points 6 and 7 mark measuring points for deformation of the casing. Table 3 shows the mean values of the drawing force for experimental studies E1–E6.

Table 3. Drawing force values

	When deforming the bottom					When deforming the casing	
exp. tag	Measuring points					Measuring points	
	1	2	3	4	5	6	7
E1	0,18	1,85	6,55	12,50	14,52	8,11	3,32
E2	0,41	3,83	8,76	12,72	15,11	7,82	3,54
E3	0,45	3,32	8,89	13,73	15,53	8,55	3,93
E4	1.21	5.94	9.61	10.73	12.70	6.82	1.92
E5	0.18	5.21	12.11	13.84	14.96	9.79	4.42
E6	0.23	4.9	13.30	14.72	15.11	9.19	3.85

Drawing force values F (kN)

5   Analysis of Experimental Results

Based on the values of logarithmic degrees of deformation for all rings in the investigated five - stage tools (Table 1 and Table 2), the load analysis of five - stage tools was performed.

Using additivity as a property of logarithmic deformation, diagrams were made with the total load of each ring in all investigated five - step tools. Figure 3a shows a diagram of the total load of the tool rings in experiments E1–E3, and Fig. 3b shows a diagram of the total load of the tool rings in experiments E4–E6.

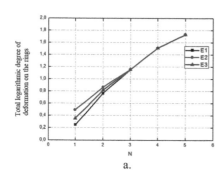
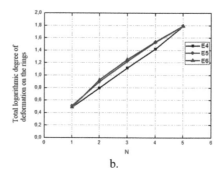

a.   b.

Fig. 3. Dependence diagrams of the total logarithmic degree of deformation by rings in all five-stage tools

Based on the values of the pull-out force from Table 3, diagrams of the dependence of the pull-out force when deforming the bottom of the workpiece on the ordinal number of rings in the tool were made. The force growth diagram for experiments E1–E3 is given in Fig. 4a, and for experiments E4-E6 in Fig. 4b.

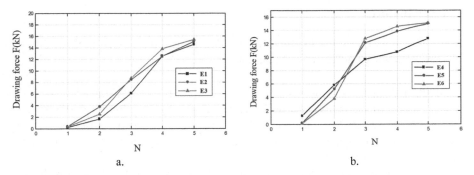

Fig. 4. Diagram of the dependence of the pulling force during the deformation of the bottom of the workpiece on the ordinal number of the ring in the investigated five-stage tools.

The value of the angle of the drawing ring depends on the logarithmic degree of deformation. Figure 5a shows that the contact length between the workpiece and the drawing ring (l) can be calculated by the expression:

$$l = \frac{s}{\sin \alpha} \tag{1}$$

where: s is the thickness of the wall that is reduced on the ring, α is the angle of the drawing ring.

In experimental research, values of the drawing ring angle in the range from 6° to 18° were used. In the diagram in Fig. 5b the relationship between contact length and thickness s (l/s) for angles ranging from 6° to 18° is shown. It can be seen from the diagram that the contact length is 6° l = 9.57 s, and for 18° l = 3.23 s.

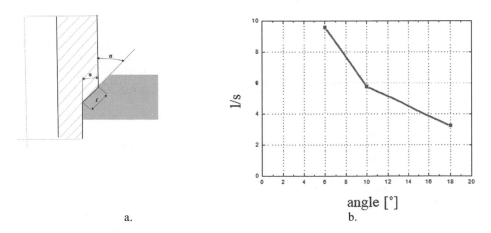

Fig. 5. Influence of angle on the length of contact between the workpiece and the drawing ring

6 Conclusion

Based on the recorded total load of all five-stage tools dimensioned in three different ways, the following conclusions can be made:

- The total pulling force is the lowest for both workpieces for tools dimensioned according to production conditions (Fig. 2). Analyzing the pulling force according to this model, it can be seen that a sudden increase in force occurs on the fifth ring in experiment E4 (Fig. 2b), which led to frequent production delays due to the replacement of the fourth and fifth rings;
- Pull-out force diagrams according to model 1 in all tools have approximately the same load increase on all rings, which results in a reduction in the number of production delays due to the replacement of worn rings;
- Analyzing the influence of the angles of the drawing rings, it can be concluded that the contact length significantly affects the friction force during drawing, so when constructing multistage tools, care must be taken that it is approximately the same on all drawing rings.

This requirement is relatively difficult to achieve in practice, because the value of the angle depends on the degree of deformation, and the contact length depends on the thickness of the reduction s.

References

1. Nožić, M., Đukić, H.: Istraživanje ukupnog opterećenja višestupanjskih alata, MATRIB 2015, nternational coference on materials, tribology, recycling, Vela Luka, Hrvatska (2015)
2. Nožić, M., Đukić, H., Šunjić, D.: Comparison of theoretical, experimental and numerical methods for process analysis for deep drawing with reduction of wall thickness. In: 13th International Conference on Accomplishements in Mechanical and Industrial Engineering, DEMI 2017, Banja Luka (2017)
3. Đukić, H., Nožić, M.: Novi pristup dimenzioniranju kod dubokog izvlačenja. In: MATRIB 2013, International Conference on Materials, Tribology, Recycling, VelaLuka, Croatia (2013)
4. Nožić, M., Đukić, H.: Eksperimentalna validacija analitičkih modela za proračun sile vučenja na višestepenim alatima. In: MATRIB 2014 International Coference on Materials, Tribology, Recycling, Vela Luka (2014)
5. Nožić, M., Đukić, H.: Novi pristup dimenzionisanju višestepenih alata. In: XXXI Savjetovanje proizvodnog mašinstva Srbije, Kragujevac (2006)
6. Nožić, M., Đukić, H.: Influence of total strain distribution on force drawing in plus ring tools. In: MATRIB 2004, International Conference, Vela Luka, Croatia (2004)
7. Nožić, M., Đukić, H.: Eksperimentalno određivanje pojedinačnog i ukupnog opterećenja petostepenog alata. In: 4th Međunarodna naučna konferencija o proizvodnom inženjerstvu "Development and modernization of production" RIM 2003, Bihać (2003)
8. Nožić, M., Đukić, H.: Single and total load of rings in plus ring tools. In: 7th International Research/Expert Conference"Trends in the Development of Machinery and Associated Technology" TMT 2003, Barcelona-Spain (2003)
9. Nožić, M., Đukić, H.: Influential parameters on single and total load of plus ring tools, 8. Savjetovanje o materijalima, tehnologijama, trenju i trošenju MATRIB 2003, Vela Luka, Croatia (2003)

10. Đukić H., Nožić M.: Model za dimenzionisanje višestepenih alata 5.Međunarodni naučno-stručni skup "Tendencije u razvoju mašinskih konstrukcija i tehnologija" TMT 2000, Zenica, (2000)

11. Nožić M., Đukić H: Eksperimentalna provjera opšteg modela za dimenzionisanje višestepenih alata, 5.Međunarodni naučno-stručni skup "Tendencije u razvoju mašinskih konstrukcija i tehnologija" TMT 2000, Zenica ,(2000)

12. Popović P.: Istraživanje i razvoj metoda projektovanja i proračuna tehnologije i izrade cilindričnih elemenata metodom dubokog izvlačenja sa redukcijom debljine zida omotača po visini, Završni elaborat, Naučnosistraživački projekat, Mostar, (1984)

13. Lee, C., Hong, S.: Curvature area prediction for the deep drawing-ironing process of a cylindrical cup using finite element method and regression analysis. J. Mech. Sci. Technol. 32(12), 5913–5918 (2018). https://doi.org/10.1007/s12206-018-1142-4

14. D. Adamović, V. Mandić, M. Živković,Z. Gulipija, M. Stefanović, M. Topalović, S. Aleksandrović,Numerical modeling of ironing process, Journal for Technology of Plasticity, Vol. 38 (2013), Number 2

Experimental Determination of Influence of Cooling Parameters on Injection Molded Part Dimensional Stability

Edin Šunje[(⊠)] and Edin Džiho

Faculty of Mechanical Engineering Univerzitetski Kampus, Džemal Bijedić" University of Mostar, 88104 Mostar, Bosnia and Herzegovina
edin.sunje@unmo.ba

Abstract. Injection molding is one of the commonly used methods for plastics processing. The process of injection molding consists of a sequence of stages and the time it takes to complete all the stages referred to as cycle time of injection molding process. The process of injection molding comprises the steps of closing the mold, injection of molten plastics within the closed mold, cooling the molten plastic, opening, and ejecting the product. The process of cooling is the most important stage of the injection molding process, it takes up to 75% of the total cycle time. The effects of coolant flow, melt temperature, cooling time and holding pressure on injection molded part quality has been analyzed. Physical and numerical experiment has been conducted and the most influential parameters has been determined. The material used in experiment is polypropylene Sabic PP 412MN40.

Keywords: Polymer · Injection molding · Cooling · Coolant flow regimes · Mold temperature · Melt temperature · Holding pressure

1 Introduction

Injection molding method was designed world of plastics part today. This method is possible to produce wide range of products, very simple such as DVD case up to very complex, such as the cars instrument panels. Product list is endless. Therefore, injection molding has found application in almost every sphere of our daily life, such as the car industry, aerospace industry, ski industry etc. As mentioned above, the cooling phase takes up to 75% of total process cycle time. In addition, the process of cooling has a huge impact on the quality of the molded parts. If at the same time it is considered that it is usually a serial or mass production, it can be concluded that the enormous need is necessary for reducing the time the cooling process. L-E Rannar and others [1–4] conducted research and analyzed the possibilities to improve the cooling system. The research was based on the development of custom cooling system, where the geometry of the cooling system follows the geometry of the molded part. Results of these studies showed that is possible to reduce the cooling time to 80% [5]. by changing the geometry

© The Author(s), under exclusive license to Springer Nature Switzerland AG 2022
I. Karabegović et al. (Eds.): NT 2022, LNNS 472, pp. 184–193, 2022.
https://doi.org/10.1007/978-3-031-05230-9_21

of the cooling system and optimization of other parameters It is important to note that the part warpage ratio increase by cooling time decrease. Authors Rahul Vashisht and Arjun Kapila [6] conducted research and analyzed possibility of applying other coolant which aim better properties than water. They concluded that it was possible to decrease the cycle time using the same process parameters. Recent research conducted at Ball State University in Indiana by Nathan March and Rex Cannes [7] have examined the possibility of reducing the cooling time variating the coolant flow regime and showed that it is possible to reduce the cooling time. The presented results can be seen that no significant reflection on the quality of the part surface as well as the mechanical properties of the pieces. It has been shown that increase of coolant flow leads to decrease of part warpage. Goal of this research is to determine the effects of convective cooling to the dimensional stability of the molded part.

2   Experimental Part

In previous research [8] the experiment has been conducted on influence of mesh type on injection molding results reliability. As the further research in this paper has been analyzed the influence of coolant regimes on injection molded part dimensional stability. The experimental research has been conducted on angle bracket which purpose is to connect aluminum and PVC extruded profiles. The material the bracket is made of is Sabic PP 412MN40.

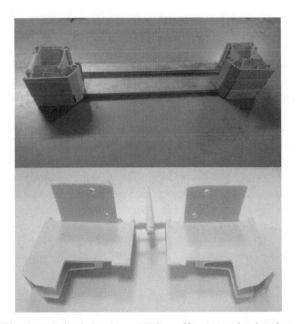

Fig. 1. Detail of aluminum-PVC profile connection brackets

In order to compare both physical and simulation experiment results, it was necessary to measure coolant pressure at the mold coolant inlet and outlet connection. The pressure was measured using pressure transducer HBM P8AP with nominal pressure of 20 bar and accuracy of 0.3. On the same locations, the thermocouple K type has been used for coolant temperature measurement. Beside above mentioned, mold temperature has been also measured at the nearest possible location to the mold cavity.

Measuring the temperature inside the mold is very difficult and requires additional mold machining in order to install the sensor inside the cavity. Temperature and pressure measuring probe is shown on Fig. 2.

Fig. 2. Temperature and pressure measuring probe

Temperature and pressure measuring probe has been designed in a way that coolant media achieves direct contact to thermocouple, so the potential heat conduction losses has been eliminated. The measuring chain is shown in Fig. 2.

The experiment has been conducted at Arburg 320D 500–210 machine, with technical characteristics given in Table 1. Part geometry required such orientation inside the mold so that the most of bracket is positioned in the ejection side of the mold. Consequently, most of heat exchange has been occurred in ejection side of the mold.

Taguchi orthogonal array L9 has been used for DOE for both physical and simulation experiment. The aim of experiment was to maximize cavity mass and therefore to achieve optimum part dimensional stability. According to Taguchi's method signal to noise ratio (S/N ratio) was used to maximize cavity mass. Signal represent desirableeffect for target value whereas noise represent undesirable effect for target value. Larger is better for S/N ratio is used to maximize cavity mass [7].

$$(S/N)_j = -10\log\left[\frac{1}{n}\sum\nolimits_{i=1}^{n}\left[1/y_{ij}^2\right]\right], i = 1..n \tag{1}$$

where:

n - number of repetitions,

y - observed quality characteristic.

Coolant pressure, that has been recalculated to flow, melt temperature, cooling time and holding pressure have been varied on 3 levels. Holding pressure has been varied because of its known influence on molded part dimensional stability. Varied parameters are given in Table 2.

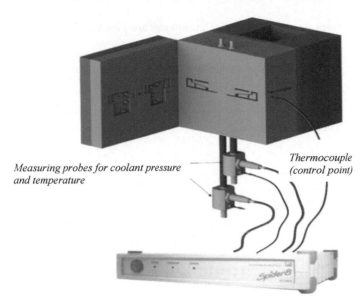

Fig. 3. Measuring chain

Table 1. Arburg 320D 500–210 machine characteristics

Parameter	Value	Unit
Clamping force	50	t
Injection rate	80	cm^3/s
Injection pressure	2050	bar
Screw diameter	30	mm
Shot size	90	g
Max volume	106	cm^3
Hydraulic response	0.2	s

Table 2. Values of varied injection molding parameters

Level	Flow [l/min]	Melt temperature [°C]	Cooling time [s]	Holding pressure [bar]
1	9	190	30	25
2	10	210	32	30
3	11	230	34	35

Other, constant, injection molding parameters are given in Table 3.

Table 3. Values of constant injection molding parameters

Parameter	Value	Unit
Injection pressure	500	[bar]
Injection rate	52/45	[%]
Switchover (V/P)	Automatic	
Injection time	1	[s]
Holding pressure time	5	[s]
Mold open time	7	[s]
Cycle time	43–47	[s]
Clamping force	50	[t]

On the other side, numerical model has been established according to recommendation from previous research [8]. Mesh type, size and mapping has been adapted to the model where the balance between mesh quality and hardware resources established. Numerical analyses conducted under the same conditions as physical experiment. The results of physical and numerical experiments are given in Tables 4 and 5, respectively. Part mass has been taken as a signal response. After determination of optimum parameters variant, dimensional measuring control has been conducted. Mitutoyo micrometers with a measuring range of 0–25 mm and 50–70 mm has been used for dimensional control process. The goal of experiment was to determine the influence of cooling parameters on injection molded part dimensional stability.

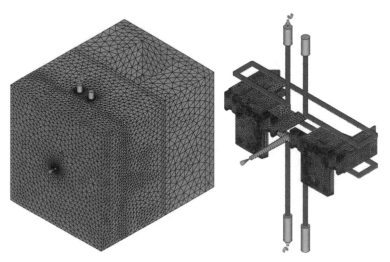

Fig. 4. Numerical model

The results of physical experiment are given in Table 4.

Table 4. Results of physical experiment

Expnb	Varied parameters				Mass						
	Q [l/min]	Tr [°C]	Tcool [s]	PN [bar]	M1 [g]	M2 [g]	M3 [g]	M4 [g]	M5 [g]	M6 [g]	S/N
E1	9	195	30	30	29,8	30,2	29,6	30,1	30,3	30,2	29,5512
E2	9	210	32	32	29,9	30,1	29,7	30,1	30,3	30,3	29,5610
E3	9	230	34	34	30,5	30,4	30,5	30,4	29,4	30,0	29,5979
E4	10	195	34	32	29,8	29,7	30,1	29,9	29,9	30,0	29,5132
E5	10	210	30	34	30,2	30,1	29,9	30	29,8	30,0	29,5422
E6	10	230	32	30	30,9	30,8	30,5	30,3	30,6	30,4	29,7091
E7	11	195	32	34	29,6	29,9	29,7	29,9	29,8	30,0	29,4889
E8	11	210	34	30	30,6	30,6	29,8	30,2	30,3	30,3	29,6278
E9	11	230	30	32	30,8	30,6	30,6	30,4	30,7	30,5	29,7142

The results of simulation experiment are given in Table 5.

Table 5. Results of simulation experiment

Exp. nb	Varied parameters					
	Q[l/min]	T_{melt}[°C]	t_{cool}[s]	P_N[bar]	m [g]	S/N
E1	9	195	30	30	28,41	29,0694
E2	9	210	32	32	28,24	29,0173
E3	9	230	34	34	28,09	28,9710
E4	10	195	34	32	28,46	29,0847
E5	10	210	30	34	28,17	28,9957
E6	10	230	32	30	28,06	28,9618
E7	11	195	32	34	28,44	29,0786
E8	11	210	34	30	28,27	29,0265
E9	11	230	30	32	28,02	28,9494

3   Results and Discussion

After statistical analyses has been conducted and determination of influential parameters an optimum variant has been chosen. The results graphical interpretation is given in

Fig. 5. Experimental results shows that the most desirable effect for quality characteristics has the combination of parameters rowE9 from Table 4, where coolant flow and melt temperature were on the third level, cooling time at first level and holding pressure at the second level.

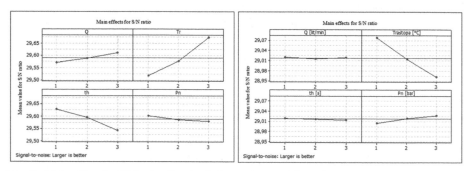

Fig. 5. Main effect for S/N value for physical experiment (left) and simulation experiment (right)

Physical and simulation experiment has shown that melt temperature is the most influential parameter, while coolant flow is the third most influential parameter. In physical experiment cooling time takes the second place, while holding pressure takes 4th place. In simulation experiment holding pressure is the 2nd most influential parameter, where cooling time has the smallest impact. The results for mean value and the rank of influential parameters for SN ratio are given in Table 6 and Table 7.

Table 6. Mean value and influential parameters for SN ratio (physical experiment)

Level	Flow	Melt temperature	Cooling time	Holding pressure
1	29,57	29,52	29,63	29,60
2	29,59	29,58	29,60	29,59
3	29,61	29,67	29,54	29,58
Delta	0,04	0,16	0,09	0,02
Rank	3	1	2	4

The explanation for the holding pressure impact deviation on part quality in physical and simulation experiment comes from mold design. Observed part is deep, that requires mold side-action elements as a part of designing solution. Any additional elements increase the mold complexity and makes it more expensive. Instead of side-action elements mold has been designed from several fixed parts so the machining was possible. During the experiment conduction it has been noticed that the increasing the value of the holding pressure causes material flow between mold segments and parting surfaces between injection and ejection mold side. For above mentioned reasons, holding pressure desirable effect could not have been achieved. This effect could not be performed

Table 7. Mean value and influential parameters for SN ratio (simulation experiment experiment)

Level	Flow	Melt temperature	Cooling time	Holding pressure
1	29,02	29,08	29,02	29,00
2	29,01	29,01	29,02	29,02
3	29,02	28,96	29,02	29,03
Delta	0,01	0,12	0,00	0,02
Rank	3	1	4	2

Table 8. Dimensional control results

Nb.	Physical experiment								Sim. exp.
	Nominal value	M1 [mm]	M2 [mm]	M3 [mm]	M4 [mm]	M5 [mm]	M6 [mm]	M_{avg} [mm]	M_{sim} [mm]
1	70,00	69,76	69,88	69,83	69,83	69,87	69,81	69,84	69,13
2	14,30	14,14	14,11	14,17	14,17	14,18	14,14	14,14	14,14
3	22,30	20,69	20,41	20,56	20,67	20,67	20,67	20,68	21,35
4	22,30	22,29	22,22	22,10	22,15	22,22	22,16	22,18	22,14

in simulation experiment. The results shows that the coolant flow parameter has a linear characteristic in physical experiment while in simulation experiment there is no significant difference between parameters level. Both experiments have shown that coolant flow has the most desirable effect on its third level.

Transient cooling analysis has been conducted. A lot of parameters have influence on transient cooling analysis. Beside of process parameters it is necessary to know material thermic properties, mold temperature at production startup, cooling media temperature, ambiental temperature, mold tool plate surface machining quality, mold opening and closing time, cooling channels roughness and coefficient of effectiveness. After above mentioned parameters set up, an analysis has been performed. In Fig. 6 are given measuring results of temperature for the mold control point, while in Fig. 7 shows the result of performed numerical analyses.

The results shows that the numerical analysis result for the mold temperature has significant matching in trend and intensity with the measured ones. The result deviation is <1 °C for the maximum mold temperature value and <2 °C for the minimum mold temperature values. The simulation result accuracy could be raised by increasing the mesh elements and the number of time steps for calculation of the heat flux. Increasing the number of heat flux calculation time steps would lead to increasing of total calculation time and hardware recourses. Measuring control has been performed and the results has been shown in the Table 8. Dimensional control has been performed on four measures on four locations on both cavities.

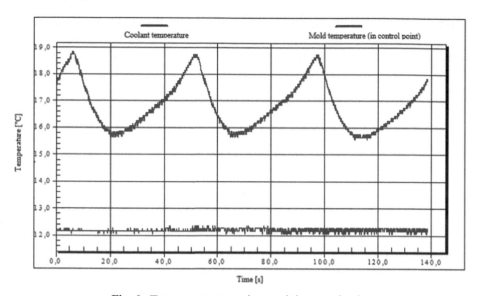

Fig. 6. Temperature measuring result in control point

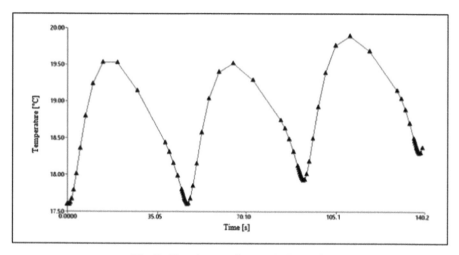

Fig. 7. Transient cooling analysis results

After the dimensional control and statistical analyses, it can be concluded that the average dimensional deviation of both experiments is less than ±3%.

4   Conclusion

Experimental determination of influence of cooling parameters on injection molded part dimensional stability has been conducted in this paper. Numerical model was established, and the results compared with the experimental results. After the results discussion, it can be concluded as follows:

- Numerical calculations give the high level of results reliability.
- The numerical analyzes results depends on mesh type and size, input parameters and boundary conditions.
- It is possible to reduce injection cycle time by choosing proper combination of cooling parameters.
- Increasing the coolant flow has a positive impact to injection molded part quality, as well as on injection molding cycle time.
- Melt temperature, cooling time, coolant flow, respectively, are the most influential parameters on injection molded part dimensional stability.

References

1. Rännar, L.-E.: Efficient cooling of FFF injection molding tools with conformal cooling channels - an introductory analysis. In: 1st International Conference on Advanced Research in Virtual and Rapid Prototyping Leiria, Portugal (2003)
2. Rännar, L.-E., Glad, A., Gustafson, C.-G.: Efficient cooling with tool inserts manufactured by electron beam melting. Rapid Prototyping J. **13**(3), 128–135 (2007). ISSN: 1355-2546
3. Rännar, L.-E., Gustafson, C.-G.: An investigation of optimal process settings in injection molding using inserts manufactured by rapid tooling. J. Manuf. Sci. Eng. (2008)
4. Rännar, L.-E., Gustafson, C.-G.: Effective injection molding with rapid tooling and optimization. Int. J. Adv. Manuf. Technol. (2008)
5. Masood, S.H., Trang, N.N.: Thermal analysis of conformal cooling channels in injection molding. In: Proceedings of the 3rd BSME-ASME International Conference on Thermal Engineering, Dhaka, Bangladesh (2006). C
6. Vahisht, R., Kapila, A.: A comparative study of coolants based on the cooling time of injection molding. Int. J. Emerging Technol. Adv. Eng. **4**(6), 830–834 (2014)
7. Marsh, N., Kanu, R.: The influence of liquid coolant flow regimes on the quality of injection molded plastics parts. In: IAJC-ASEE International Conference (2011)
8. Šunje, E.: Influence of mesh type on injection molding process simulation results reliability. Int. J. Eng. Technol. IJET-IJENS **17**(01), 32–41 (2017)

Estimation of Shrinkage and Warpage in Glass-Fiber Reinforced Polyamidehinge

Edin Šunje and Edin Džiho[✉]

Faculty of Mechanical Engineering Univerzitetski Kampus, Džemal Bijedić" University of Mostar, 88104 Mostar, Bosnia and Herzegovina
edin.sunje@unmo.ba

Abstract. The estimation of shrinkage and warpage is key input parameter for proper injection mold design. Shrinkage is the property of polymeric material, and it cannot be eliminated. If the shrinkage is uniform, part will be smaller but not warped, on the other side if the shrinkage is differential it will lead to part warpage. There are several reasons for part warpage: part geometry, injection pressure and temperature, mold cooling, material property. Especially, it is challenging to predict warpage in fiber reinforced materials, due to its differential shrinkage in flow direction and perpendicular to flow direction. In this study, numerical analysis is used to predict warpage on plastic hinge made of glass fiber reinforced polyamide. The result verification has been conducted through simulation dimensional control comparison with the corresponding measured dimensions of the real part.

Keywords: Polyamide 6 · Glass-fiber · Shrinkage · Warpage · Injection molding · Injection mold

1 Introduction

Dimensions of injection molded part are primarily determined in relation to dimensions of mold core and cavity, and secondary in relation to all other process parameters. It is obvious that the part dimensions are determined by dimensions of core and cavity. Experience has shown that dimensional deviations can be adjusted by changing the injection molding process parameters. Shrinkage can be defined as a geometric change in size, part will not suffer any distortion it only becomes smaller. This kind of shrinkage is called uniform shrinkage and can be easily compensated by part scaling.

Commonly, shrinkage is not uniformed that lead to part warpage, change in shape and dimensions. If the material is anisotropic, like fiber reinforced materials, it is not easy to meet dimensional requirements due to out of plane distortionor residual stress [1]. By adding filler, which are usually rigid, in material it is possible to reduce the shrinkage, but the viscosity increases. The higher viscosity requires higher processing temperature and that affect material pvT data.

According to Wang, Chang, and Hsu [2] the main cause of warpage can be classified in five groups: pressure (runner design, packing pressure and packing time), temperature (melt temperature, mold temperature and cooling system), part insert (cooling difference

© The Author(s), under exclusive license to Springer Nature Switzerland AG 2022
I. Karabegović et al. (Eds.): NT 2022, LNNS 472, pp. 194–201, 2022.
https://doi.org/10.1007/978-3-031-05230-9_22

and stiffness difference), part design (thickness distribution, reinforcing structure and material properties), fiber orientation which depends on melt advancement and thickness distribution.

Nowadays, numerical simulations are used to predict magnitude of shrinkage and warpage. Conventional trial-error methods are expensive, time consuming and costly what classifies it as not acceptable [3]. The goal of this study is to establish reliable numerical model, simulate process, estimate shrinkage and warpage, and verify study results. An experimental study has been conducted on hinge made of PA6GF30 material.

2 Simulation Background

First phase in injection molding process is the injection phase. As soon as the melted plastics is injected into the mold, tin boundary layer of plastics start to cool, while the melt inner layers propagate trough mold cavity. Melt flow can be treated as Generalized Newtonian Fluid. It can be mathematically described as follows [4]:

$$\frac{\partial \rho}{\partial t} + \nabla \cdot \rho \boldsymbol{u} = 0 \tag{1}$$

$$\frac{\partial}{\partial t}(\rho \boldsymbol{u}) + \nabla \cdot (\rho \boldsymbol{u}\boldsymbol{u} - \boldsymbol{\sigma}) = \rho g \tag{2}$$

$$\boldsymbol{\sigma} = -p\boldsymbol{I} + \eta(\nabla \boldsymbol{u} + \nabla \boldsymbol{u}^T) \tag{3}$$

$$\rho C_p \left(\frac{\partial T}{\partial t} + \boldsymbol{u}\nabla T \right) = \nabla(k\nabla T) + \eta \dot{\gamma}^2 \tag{4}$$

where:

$\boldsymbol{u}$ – velocity vector.
T – the temperature.
t – the time.
p – pressure.
σ – total stress tensor.
ρ – the density.
η – the viscosity.
k – thermal conductivity.
C_p – the specific heat.
$\dot{\gamma}$ – the share rate.

Modified-Cross model has been used to describe viscosity of melt:

$$\eta(T, \dot{\Upsilon}) = \frac{\eta_0(T)}{1 + (\eta_0 \dot{\Upsilon}/\tau^*)^{1-n}} \tag{5}$$

where n is power low index, η_0 the zero shear viscosity, τ^* parameter that describes transition regions between zero share rate and power law region of viscosity curve.

After the filling phase is over it is necessary to apply additional pressure to compensate in-mold shrinkage of material due to itscooling. It means more material must be injected in mold cavity. Holding pressure intensity is commonly up to 80% of injection pressure, and the holding pressure time is determined by gate material freezing. Modified Tait equation is used to describe behavior of material in packing phase [5].

$$V(T, P) = V_0(T)\left[1 - C \cdot ln\left(1 + \frac{P}{B(T)}\right)\right] + V_t(T, P) \tag{6}$$

where:

$$V_0(T) = \begin{cases} b_1 + b_1 T, T > T_t, melt\ state \\ b_{1s} + b_{2s}T, T \le T_t, solid\ state \end{cases}$$

$$B(T) = \begin{cases} b_{3m}\exp(-b_{4m}T), T > T_t, melt\ state \\ b_{3s}\exp(-b_{4s}T), T \ll T_t, olid\ state \end{cases}$$

$$V_t(P, T) = \begin{cases} 0, T > T_t, melt\ state \\ b_7 + \exp(b_8 T - b_9 P), T \le T_t, solid\ state \end{cases}$$

$$T \equiv T - b_5$$

transition temperature: $T_t \equiv b_5 + b_6 P$
for amorphous polymer, $b_{1m} = b_{1s}$
for crystalline polymers, $b_{1m} > b_{1s}$

A cooling phase starts as soon as material reaches into the relatively cold mold.Beside the cooling phase takes up to 80% of the total cycle time, proper cooling iscrucialstep to meet the part dimensional requirements. Mold cooling is the 3-dimensional transient heat transfer phenomena. Heat transfer can be described with three-dimensional Poisson equation [6].

$$\rho C_p \frac{\partial T}{\partial t} = k\left(\frac{\partial^2 T}{\partial x^2} + \frac{\partial^2 T}{\partial y^2} + \frac{\partial^2 T}{\partial z^2}\right) \tag{8}$$

where:

T – the temperature.
t – the time.
x, y and z–the coordinates.
ρ – the density.
C_p – the specific heat.
k – the thermal conductivity.

3   Case Study

Simulation of injection molding process of plastics hinge made of PA6GF30 material has been performed.Simulation has been conducted under the same process parameters like in production. Part is given in Fig. 1. along with established numerical model.

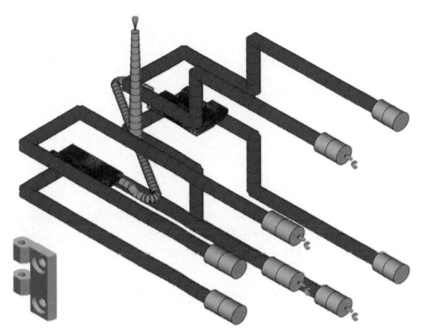

Fig. 1. Numerical model

Simulation set up is iterative process, where is necessary to determine proper mesh type and size. Generally, mesh size should be decreased up to the moment when there is no significant impact on the simulation result. Part is produced on Arburg 270S machine. Production process variables are given in Table 1.

Table 1. Injection molding parameters

Parameter	Value	Unit
Injection pressure	700	[bar]
Injection time	1.02	[s]
Switchover (V/P)	Automatic	
Holding pressure	450/400	[bar]
Holding pressure time	5	[s]
Melt temperature	290	[°C]
Cycle time	33	[s]
Cooling media temperature	60	[°C]

Material used in experiment is Akromid PA6GF30byAkro-Plastic GmbH. Cross-WLF model has been used to define viscosity model. Shear rate -viscosity diagram is shown on Fig. 2.

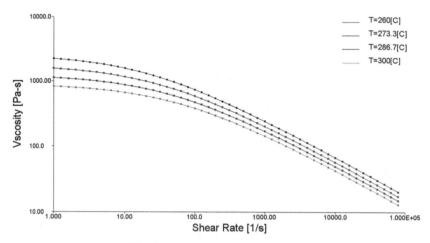

Fig. 2. Shear rate-viscosity diagram

A modified Tait model has been used to describe pvT behavior of material (Table 2).

Table 2. Injection 2-domain modified Tait pVT model coefficients

Coefficient	Value	Unit
b_5	479.13	[K]
b_6	2.25e−07	[K/Pa]
b_{1m}	0.0008128	[m^3/kg]
b_{2m}	4.445e−07	[m^3/kg-K]
b_{3m}	1.34295e+08	[Pa]
b_{4m}	0.004316	[1/K]
b_{1s}	0.0007717	[m^3/kg]
b_{2s}	2.026e−07	[m^3/kg-K]
b_{3s}	1.93874e+08	[Pa]
b_{4s}	0.0004161	[1/K]
b_7	4.107e−05	[m^3/kg]
b_8	0.04609	[1/K]
b_9	1.309e−08	[1/Pa]

After analyses has been conducted, warped part has been exported to CAD software where is the dimensional control of the warped part performed. All measured results are given in Table 3.

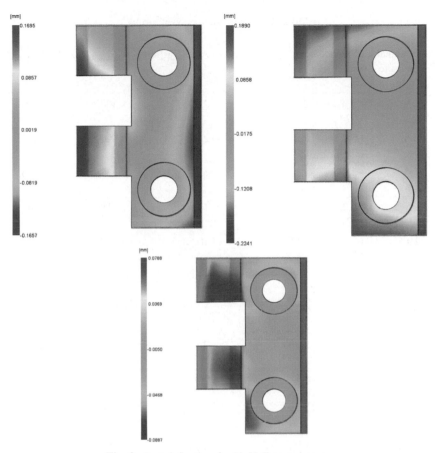

Fig. 3. Part deflection for X, Y, Z components

Measuring protocol has not been presented according to classified data policy, but only results. Based on conducted research, the simulation results have shown all results fitted into the tolerance and has shown high accuracy level. Each measure has been taken on 4 places in simulation and the real part and the average value is presented. Also, absolute measure deviation has been calculated and expressed in percentage (Fig. 3).

Table 3. Dimensional control results

Nb.	Nominal measure [mm]	Tolerance [mm]	Avg$_{measured}$ [mm]	Avg$_{simul}$ [mm]	$\Delta_{measured}$ [%]	Δ_{Simul} [%]
1	2.5	+0.07; −0.07	2.535	2.505	1.400	0.200
2	2.5	+0.07; −0.07	2.52	2.51	0.800	0.400
3	1.5	+0.07; −0.07	1.55	1.4875	3.333	0.833
4	11	+0.22; −0.22	10.855	10.9075	1.318	0.841
5	10.5	+0.22; −0.22	10.525	10.5825	0.238	0.786
6	2.5	+0.07; −0.07	2.515	2.505	0.600	0.200
7	5.3	+0.12; −0.12	5.325	5.3775	0.472	1.462
8	10	+0.10; -0.10	9.995	9.9175	0.050	0.825
9	9	+0.18; −0.18	9.1	9.03	1.111	0.333
10	5.5	+0.12; −0.12	5.55	5.515	0.909	0.273
11	6	+0.10; −0.10	5.955	5.9675	0.750	0.542
12	5	+0.12; −0.12	5.065	4.985	1.300	0.300
13	20	+0.26; −0.26	20.055	19.955	0.275	0.225
14	10	+0.10; 0.00	10.055	10.0625	0.550	0.625
15	9.88	+0.05; −0.05	9.845	9.89	0.354	0.101
16	20	+0.05; −0.05	20.01	19.97	0.050	0.150
17	12.5	+0.10; −0.10	12.525	12.465	0.200	0.280
18	25	+0.10; −0.10	25.025	25.025	0.100	0.100

(*continued*)

Table 3. (*continued*)

Nb.	Nominal measure [mm]	Tolerance [mm]	Avg$_{measured}$ [mm]	Avg$_{simul}$ [mm]	$\Delta_{measured}$ [%]	Δ_{Simul} [%]
19	40	+0.00; −0.20	39.995	39.935	0.012	0.162

As it can be seen from the table above all dimensions fits into the tolerance interval. Maximum measure deviation for the physical experiment is 3.33%, while in simulation experiment maximum dimension deviation is 1.46%.

4 Conclusion

An experimental verification of the shrinkage and warpage estimation has been conducted in this paper. The research has shown high level of results reliability from the simulation experiment. Deviation of measured and simulation results is less than 2%. It can be concluded that is possible to successfully predict magnitude of shrinkage and warpage and prove valuable input information for mold designers. Simulation results also depends on rheological, thermal ant pvT properties of materials that must be precisely defined. Proper type of mesh type and mesh size should be chosen and the boundary condition set.

References

1. Malloy, R.A.: Plastic Part Design for Injection Molding, ISBN 978-3-446-40468-7 (2010)
2. Wang, M.-L., Chang, R.-Y., (David) Hsu, C.-H.: Molding Simulation: Theory and Practice, ISBN: 978-1-56990-619-4 (2018)
3. Šunje, E.: Influence of mesh type on injection molding process simulation results reliability. Int. J. Eng. Technol. IJET-IJENS **17**(01)
4. Azdast, T., Behravesh, A.H., Mazaheri, K., Darvishi, M.M.: Numerical Simulation and Experimental Validation of Residual Stress Induced Constrained Shrinkage of Injection Molded Parts, Polimery 2008, 53, nr 4 (2008)
5. Yang, W.H., Peng, A., Liu, L., Hsu, D.C., Chang, R.Y.: Integrated numerical simulation of injection molding using true 3d approach. In: ANTEC 2004
6. Paclt, R.: Cooling/heating system of the injection molds. J. Technol. Plasticity **36**(2), (2011)

Dependence of Basic Geometric Characteristics of Polymer Components on Parameters of Injection Molding and Gate Position

Edis Nasić[(✉)], Denijal Sprečić, Jasmin Halilović, and Džemal Kovačević

Faculty of Mechanical Engineering, University of Tuzla, Ulica Urfeta Vejzagića 4, 75000 Tuzla, Bosnia and Herzegovina
edis.nasic@untz.ba

Abstract. Dimensional shrinkage of components is a very common occurrence in the injection molding process. This phenomenon leads to deviations, often to deformations that affect the final shape, dimensions and quality of the final component. It is difficult to completely avoid this phenomenon. In relation to the material used,component shape and construction tool, there are several influential factors. This paper analyzes the influence of melt temperature, holding pressure, injection velocity and gate position on geometrical characteristics - dimensional shrinkage. The material used to make the components is low density polyethylene (LDPE). The results of the research showed that by changing the holding pressure, injection velocity, melt temperature and gate position, dimensional shrinkage can be affected.

Keywords: Shrinkage · Deformations · Molding parameters · Injection molding · Gate position

1 Introduction

Injection molding is a process where desired shape of a component is formed by injecting melted polymer material under high pressure into the tool cavity. The injection process consists of the injection phase, holding pressure and cooling. Each phase affects the characteristics of the components. Shrinkage and dimension change of a part is a characteristic occurrence in the process of injection molding. It occurs because the density of a material differs from processing temperature to the temperature of the environment. Dimensional change of a product can be up to 20–25% according to its volume, measured on the melting temperature and the temperature of the environment [1]. Unlike amorphous materials, crystalline materials are especially prone to thermal dimensional change. When a crystalline material is cooled under the transition temperature, the molecules are arranged in a more orderly way forming crystallites. On the other hand, the microstructure of amorphous materials does not change with the change of phase. This leads to the fact that crystalline materials have a greater difference in specific volume between their melting and solid (crystalline) phase, as well as greater

© The Author(s), under exclusive license to Springer Nature Switzerland AG 2022
I. Karabegović et al. (Eds.): NT 2022, LNNS 472, pp. 202–208, 2022.
https://doi.org/10.1007/978-3-031-05230-9_23

dimensional variations [2–4]. Shrinkage is affected by the crystalline structure formed, density, cooling rate, process parameters, geometry (complexity) of a product, thickness of the wall, gate position, direction of solution flow etc. [5]. Shrinkage basically takes place in two phases [6]. The first phase occurs immediately after injection when the amount of shrinkage is the greatest. The second phase is subsequent shrinkage, which lasts a couple of hours longer where the changes are minimal. Generally, it cannot be completely avoided but the amount of shrinkage can be affected by the choice of process parameters and gate position.

2 Experimental Work

The components were produced on the injection machine ARBURG ROUNDER 221E/170R. The material used isLDPE, which is crystalline polymers. The main characteristics of the LDPEmaterial arethe following: the structure is linear with a higher number of side branches, the density ranges from 0,91–0,94 (g/cm^3), the crystallization degree is up to 50 to 60 (%), crystallization temperature is 105–115 (C^0), it has lower solidity and hardness [7, 8]. Geometric characteristics of the part, sprue, runnerand the gate, are shown on Fig. 1.

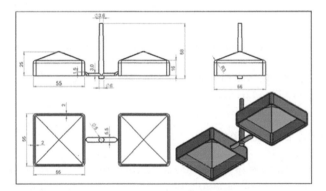

Fig. 1. Dimensions and 3D model of a component

The parameters and levels of variation used in experiments of injection molding are given in Table 1. The experiments were conducted according to L9 plan with three repetitions for each combination of parameters.

Figure 2 a) and b), shows profiles of holding pressure and areas of shrinkage measuring.

The expressions used to determine dimensional characteristics are:

$$D_{mu} = \frac{S_{wk1} - S_{w1}}{S_{wk1}} \cdot 100\% \tag{1}$$

$$D_{mst} = \frac{S_{wk2} - S_{w2}}{S_{wk2}} \cdot 100\% \tag{2}$$

Table 1. Parameters used in experiments of injection molding

Injection parametar	Unit	Level 1	Level 2	Level 3
Melt temperature (Tm)	°C	180	190	-
Holding pressure (hydraulic) (Phn)	bar	35	45	55
Injection velocity (Vi)	mm/s	30	33	36

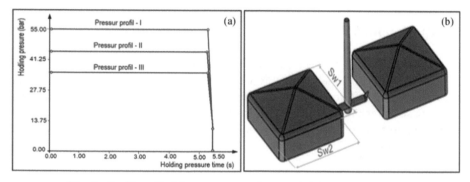

Fig. 2. Profiles of holding pressure and areas of shrinkage measuring

D_{mu} – shows the dimension at gate position, D_{mst} – shows the dimension at flow side,

S_{wk1}, S_{wk2} – shows the real dimension of the tool, S_{w1} i S_{w2} – shows the dimensions of a product measured in chosen areas.

3 Results and Discussion

Average values of shrinkage in chosen areas are given in Table 2, a) and b) Fig. 3, a) and b), shows the main effects of variation for parameters: melt temperature (Tm), holding pressure (Phn) and injection velocity (Vi). The graph of the main effects clearly shows that the increase, the holding pressure and the injection velocity result in linear decrease in shrinkage. However, the amount of shrinkage increases with the increase of melt temperature. The nature of influence of parameters of injection molding is identical in flow direction as well as at gate position.

Table 2. Shrinkage of average values (by percent)

a) Gate position			
No	Avg. va (%)	No	Avg. va (%)
1	2.59	10	2.77
2	2.52	11	2.60

(continued)

Table 2. (*continued*)

a) Gate position

No	Avg. va (%)	No	Avg. va (%)
3	2.41	12	2.50
4	2.45	13	2.57
5	2.31	14	2.39
6	2.43	15	2.50
7	2.26	16	2.35
8	2.31	17	2.41
9	2.21	18	2.34

b) Flow direction

No	Avg. va (%)	No	Avg. va (%)
1	2.82	10	2.95
2	2.64	11	2.78
3	2.51	12	2.60
4	2.58	13	2.71
5	2.42	14	2.53
6	2.56	15	2.65
7	2.38	16	2.47
8	2.47	17	2.53
9	2.35	18	2.45

The results of analysis variance (ANOVA) are shown in Table 3, a) and b). According to the results, ANOVA parameters that have a significant influence on the shrinkage for observed areas are: holding pressure (p = 0,000), melt temperature (p = 0,007 and 0.018) and injection velocity (p = 0,044).

Table 4 shows the average values of shrinkage for chosen areas based on the results from Table 2.

According to the results shown in Table 4, it is visible that amounts of shrinkage are higher in flow direction when compared to the gate position. Figure 4 shows the shrinkage in observed areas where the difference mentioned can be observed.

The gate position area is exposed to higher and longer holding pressure, both in the injection phase as well as in the packaging phase. Higher holding pressures have a direct influence on the decrease of shrinkage of a observed area. A higher amount of shrinkage in flow direction is connected to the orientation of the molecules of the material. When a component is ejected from the tool, there is a bigger intermolecular attraction in the material during cooling, which results in greater shrinkage, especially in flow direction.

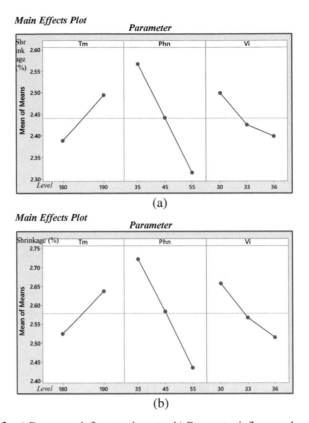

Fig. 3. a) Parameter influence character b) Parameter influence character.

The results obtained, showed in Fig. 3, and Table 3, show that the greatest influence on shrinkage comes from the holding pressure, followed by melt temperature and injection velocity. From influence main effects plots it is visible that the value of shrinkage decreases with the increase in pressure in two areas. Higher holding pressure ensures better packaging of a material. During the holding phase, the product is exposed to higher and longer lasting of pressure until gate is solidificated (seald), with the addition of a new quantity of material with the increase in density [4], which results in smaller amounts of shrinkage.

Injection velocity has a similar effect. When injecting a tool, injection velocity affects the formation of a hardened layer near the wall, which further affects the viscosity and molecule orientation. Higher speed prevents premature hardening of a material because of increased viscous heating, with the decrease in viscosity. This ensures later gate sealing and enables a higher possibility of affecting a product with the holding pressure. The mentioned might be a reason for less shrinkage with the increase of injection velocity.

The increase in temperature observed in the (pvt) diagram increases the difference of specific volume from the melted to the hardened state on applied pressure and decrease in density of part. This might be a reason for greater shrinkage with the increase in temperature for the material used. Also, higher melt temperatures during injection are a

Table 3. Medium schrinkage analysis variance

a) Gate position

Source	DF	Seq SS	Contribution	Adj SS	Adj MS	F-Value	P-Value
Tm	1	0.05014	15.25%	0.05014	0.050139	10.60	0.007
Phn	2	0.19001	57.78%	0.19001	0.095006	20.08	0.000
Vi	2	0.03193	9.71%	0.03193	0.015967	3.37	0.069
Error	12	0.05678	17.27%	0.05678	0.004732		
Total	17	0.32887	100.00%				

b) Flow direction

Source	DF	Seq SS	Contribution	Adj SS	Adj MS	F-Value	P-Value
Tm	1	0.05705	12.46%	0.05705	0.057047	7.43	0.018
Phn	2	0.24570	53.69%	0.24570	0.122851	15.99	0.000
Vi	2	0.06272	13.70%	0.06272	0.031359	4.08	0.044
Error	12	0.09219	20.14%	0.09219	0.007683		
Total	17	0.45766	100.00%				

($\Lambda = 0.05$ - Factor of significance)

Table. 4. Average values of shrinkage of chosen areas

Material	Gate position	Flow direction
LDPE	2,44 (%)	2,58 (%)

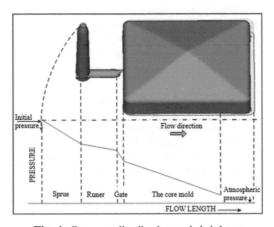

Fig. 4. Pressure distribution and shrinkage

result of the ejection of a component in higher temperatures. This results in the increase in intermolecular attraction in the material during component cooling, as well as higher amounts of shrinkage.

4   Conclusions

- It is possible to affect dimensional shrinkage of polymer components bymodification of parameters of injection molding and gate position.
- Dimensional shrinkage of the areas observed is smaller with the increase of holding pressure and injection velocity.
- Dimensional shrinkage of the areas observed is greater with the increase of melt temperature.
- Gate position affects the amounts of shrinkage. Greater shrinkage occurs in the flow direction theh in direction perpendicular. Flow direction is defined by gate position.
- When it comes to the range of parameter variation used in experiments of injection molding, holding pressure is a parameter which affects the shrinkage of observed areas the most, with a ratio of 57,78% and 53.69% in the total variation. The cumulative ratio of melt temperature and injection velocity in the total variation for areas observed is 24,96% and 26,16%.

References

1. PDF help dokument, Cmold20001/document/dg_doc/develop/books/dg/dgtoc.htm
2. António José Vilela Pontes;Doktorska Disertacija,Shrinkage and ejection forcesin injection moulded products, Universidade do Minho 2002
3. Patricija Sonia Ferraz Ferreita Alves; Doktorska Disertacija, Shrinkage and Warpage Behaviour on Injection Moulding Parts, Universidade do Minho 2008
4. Shoemaker, J.: Moldflow Design Guide A Resource for Plastics Engineers First Edition Moldflow Corporation Framingham, Massachusetts. Moldflow Corporation, U.S.A (2006)
5. Jachowicz, T., Gajdoš, I., Krasinskyi, V.: Research on the content and filler type on injection shrinkage. Adv. Sci. Technol. Res. J. 8(23), September 2014
6. Dangel, R.: Injection Moulds for Beginners. Hanser Publishers, Munich (2016)
7. Goodship, V.: ARBURG Practical Guide to Injection Moulding, 2nd Edition, First Published in 2017
8. Fischer, J.M.: Handbook of Molded Part Shrinkage and Warpage, Plastics Design Library 2003
9. Kumar Lal, S., Vasudevan, H.: Optimization of Injection Moulding Process Parameters in the Moulding of Low Density Polyethylene (LDPE). Int. J. Eng. Res. Dev. 7(5), 35–39 (2013)
10. Alam, M.M., Kumar, D.: Reducing shrinkage in plastic injection moulding using taguchi method in tata magic head light. Int. J. Sci. Res. (IJSR) 2(2), 107–110 (2013). India Online ISSN: 2319–7064
11. Vashishta, R., Kapilaa, A.: Analyzing effects of different gates on component and molding parameters. Int. J. Current Eng. Technol. 4(4), August 2014

Development of 3D Printed Transfemoral Prosthetic Leg with Actuated Joints

Zlata Jelačić[1]([✉]), Faris Ustamujić[2], Remzo Dedić[3], and Želimir Husnić[4]

[1] Department of Mechanics, Faculty of Mechanical Engineering, University of Sarajevo,
Vilsonovo šetalište 9, 71000 Sarajevo, Bosnia and Herzegovina
jelacic@mef.unsa.ba
[2] Airbus DS GmbH, Taufkirchen, Germany
[3] Department of Construction, Faculty of Mechanical Engineering, Computer Science and
Electrical Engineering, University of Mostar, 11000 Mostar, Bosnia and Herzegovina
[4] The Boeing Company, Philadelphia, USA

Abstract. The largest number of amputees use the so-called passive prosthetic devices, meaning that they do not have an external power source. These devices are using elastic or some other potential energy stored during a gait sequence and released in the next sequence. That causes problems in the case of movements requiring more power to perform, like climbing the stairs. The lack of muscles in the case of lower leg amputation makes it impossible to perform without the external power source.in order to overcome this problem, prosthetic devices for lower leg amputations must be externally powered in its knee and ankle joints. This paper describes the concept of a new type of 3D printed above-knee prosthesis with hydraulic system is integrated in the lower leg area of the above-knee prosthetic leg.

Keywords: Above-knee prosthetic leg · Rehabilitation robotics · Hydraulic control system · Wearable robotics

1 Introduction

Although there has been a significant development in new technologies during the last decade or so, commercial prosthetic devices largely remain passive, meaning they do not have an external power source. However, many movements related to the lower extremities, like climbing the stairs and walking up the steeper slopes, need significant power. That power comes from the leg muscles, but in the case they are not available, their function should be provided by an external power source. The goal is therefore to integrate the power system in the prosthetic leg housing.

Note: The Boeing Company and Airbus DS GmbH are not associated with this paper.

I. Karabegović et al. (Eds.): NT 2022, LNNS 472, pp. 209–219, 2022.
https://doi.org/10.1007/978-3-031-05230-9_24

The reason that the vast majority of the current prosthetic devices is passive is the fact that the integration of the external power would drastically increase the weight of the prosthetic leg. The biggest trade-off is therefore always the one between the user comfort and the natural way of moving.

The new generation of prosthetic devices are computer-controller and play in on the improvement of comfort. They tend to optimize the movement of the prosthetic device in the way they decrease the step deviations and lower the metabolic energy consumption of the user. However, most of the prosthetic devices focus on the gait part on flat ground or small slopes. The phases that are controlled are the standing and swinging phase together with the first step. The goal is to get as close as possible to natural gait. Users of such prosthetic devices are able to walk at different speeds on flat or inclined ground, but cannot climb the stairs in a natural way. To do that there has to be an external power source which will provide the necessary energy to lift the body.

While climbing the stairs, the body is being lifted from one stair to another. During this phase only one leg touches the stair and lifts the body. At the same time the other leg is in swing phase. During the climbing phase, the entire body load is on one legand the moments in the knee and ankle joints exceed body weight by several times. In the case of above knee amputation, the lack of muscles makes it impossible to produce these forces, and passive prostheses which only use stored energy, are unable to perform such activities. Therefore in order to perform high demanding power activities and movements, the joints in the knee and ankle need to be powered.

Our prosthetic leg is designed in such a way that it is able to perform the main leg movements in the sagittal plane autonomously. Therefore the knee and ankle joints are powered in order to obtain the required movements of the human leg. Powered knee enables overcoming large forces that occur during loading response. Previous research has shown that it does not suffice to only power the knee joint. The ankle joint plays an important role in the swing phase, especially during the stair ascent, and needs to be powered as well [1]. Powered ankle joint enables dorsal and plantar flexion movement of the entire foot. It also provides better stabilization of the knee and the entire prosthesis and the power needed in push-off phase.

The hydraulic power system concept is designed to enable characteristic movements of the prosthesis in sagittal plane while walking and ascending stairs. We are concentrated on testing the prosthesis on stair ascent, being the most complicated type of movement regarding energy requirements. The main components of the hydraulic system are two hydraulic actuators. One hydraulic cylinder is for powering the knee and the other for powering the ankle joint. The hydraulic system also includes the power pack unit and the accompanying hydraulic installation (Fig. 1). In order to keep the cost down, it was decided to choose off-the-shelf hydraulic power pack unit. It it consists of all the needed hydraulic installation components (electrical motor, hydraulic pump, reservoir, appropriate valves, connections etc.) integrated in one unit.

Fig. 1. Previous model of the transfemoral prosthetic leg with Endolite housing and separate drives for knee and ankle joint

2 New Concept Design of the Transfemoral Prosthetic Leg

The new 3D printed prosthesis structure is required to support the new, hydraulic actuator concept as well as the human using it. Its development was summed up in well-defined steps, leading to a step-by-step approach.

This approach consisted of:

- designing a new concept of structure,
- manufacturing the designed structural parts out of wood as basis material to accelerate the process and get a feeling for the functionality and feasibility of the concept,
- testing of the functionality of components and the entire concept with a fully equipped prosthetic structure
- based on observations, implementing design optimization and manufacturing the parts subsequently with 3D printers to achieve more complex shapes with integrated bracketry concepts to accommodate the actuator system

This paper will describe the actions and observation covered in the last step listed above. Previous steps have been covered in separate publications, see [1, 2].

As mentioned above, the new structural concept was examined with a fully equipped rapidly manufactured prosthetic structure. Analysis of the observation led to implementing design optimization and manufacturing the parts subsequently with 3D printer technology. Such approach enabled achieving more complex shapes with integrated bracketry concepts to accommodate the actuator system.

In order to accommodate the prosthetic concept to various height of population with amputations, the structural concept incorporates a tubular adapter, with adaptable length, positioned in the lower leg of the above-knee prosthesis. This resembles most of the classical prosthesis solutions. However, the location of the tube is different as it is now placed at roughly the middle of the lower leg structure. See Fig. 2 for an illustration.

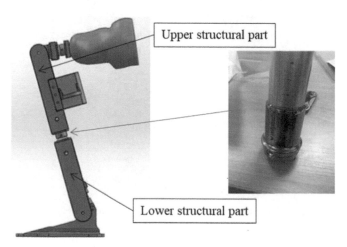

Fig. 2. Uniprox A3 tube adapter and its location in prosthesis concept between its lower and upper structural part.

Uniprox A3 tube adapter is available with a maximum length of 400 mm. As such it provides for a considerable adaptability lengthwise in a rather straightforward way by cutting it to the needed size. The interface on both sides of the tubular adapter is conceived via an adapter as manufactured by the company Otto Bock, see Fig. 3.

Fig. 3. Otto Bock adapter for tubular interface connection in lower leg.

This adapter is connected by four standardised screws to the remainder of the structure, i.e. both to the upper as well as the lower structural part of the lower leg prosthesis. The cross-sectional area has been sized as 54 x 54 mm to enable proper load transfer and sufficient safety margins considering the loads expected at this stage of design.

2.1 Design of the Lower Structural Part

The lower structural part of the prosthesis (Fig. 2) is designed with the aid of Solidworks and is depicted in detail in Fig. 4. This is the so-called "equipped" model, i.e. the cylinder actuator as well as the foot model are presented in the integrated mode.

Fig. 4. Lower structural part of the prosthesis with the connected foot model.

In Fig. 4 the interface holes for the Otto Bock adapter can be seen, on top of which the Uniprox tubular adapter is attached. Such a connection also enables the possibility of angular alignment besides the adaptation of the tube lengthwise.

2.2 Design of the Upper Structural Part

The upper structural part has an inverted interface concept, leading to simplicity in design. This is illustrated in Fig. 5, where upper actuator cylinder housing is visible including the bolted interface to the hydraulic pump and electrical motor. Also indicated in the figure is the knee rotational axis interface.

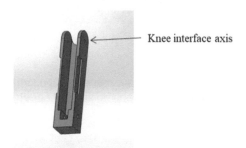

Knee interface axis

Fig. 5. Upper structural part of the prosthetic leg

This part has a demanding design given the boundary conditions of the cylinder movement in both extended and retracted position. At the same time, the knee joint must demonstrate the necessary rotational freedom in the correct angular segment. The final design concept chosen is shown in Fig. 6. The cut-outs and chamfers visible are implemented for proper cylinder head movement. At the bottom of the knee joint structure, there are four bolt interfaces, providing for the classical upper leg attachment interface via Otto Bock adapters.

Fig. 6. Knee joint structural design concept.

In previous description of the upper structural prosthesis part, the bolted interface towards the bracket holding electrical motors is mentioned. This bracket is illustrated in the Fig. 7. The dimensions are chosen such that closed fit is achieved when motor cylinders are positioned.

Fig. 7. Bracket design for electrical motors.

Integrating all the upper prosthesis structural components, i.e. upper structural prosthesis part, knee-joint and bracket, leads to the structural model as illustrated in Fig. 8.

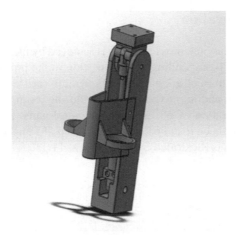

Fig. 8. Integrated view of the upper structural part model.

2.3 Manufacturing of Structural Parts

Following the design of the lower leg prosthesis structural components, as described in Section A, these parts were printed with the Zortrax M200 Plus printer.

Due to the operational printing volume available, it was not possible to print the upper and lower prosthesis structural parts in one shot, hence, they were split in two pieces each, which were then subsequently bonded with a suitable adhesive. An example of printed parts and adhesive bonding is given in Fig. 9.

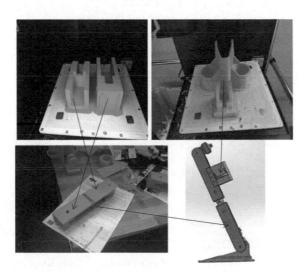

Fig. 9. 3D printed structural prosthesis parts and adhesive bonding of the lower prosthesis structural part.

3   Prosthesis Integration

The integration of all rotational moving joints is achieved by means of bolts with metric sizes of M6, M8 and M10 depending on the location in the prosthesis. The screw and bolt interfaces for the Otto Bock adapters have been reinforced with helicoils to facilitate greater pull-out strength and improved reusability during testing.

Following the integration of all movable structural parts, the foot model and the hydraulic actuators have been installed. As the final step, the upper leg part was installed by an orthopaedist. This is illustrated by Figs. 10 and 11.

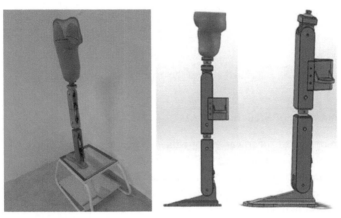

Fig. 10. Equipped 3D printed structural model of a lower leg prosthesis as mounted on an upper leg (bended knee position) – prototype vs. CAD model.

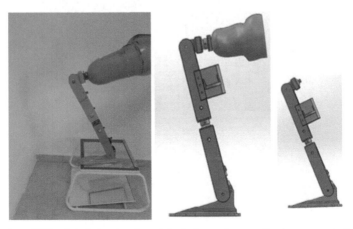

Fig. 11. Equipped 3D printed structural model of a lower leg prosthesis as mounted on an upper leg (straightened knee position) – prototype vs. CAD model.

4   Installation of Actuator and Control Equipment

As the final step the actuator equipment is installed. This includes electrical motors, hydraulic pumps and oil reservoir, all connected via flexible hoses. An illustration of the fully equipped functional prosthesis prototype can be seen in Fig. 12 [3, 4].

To close the control loop, accelerometers and motion sensors have been positioned and connected to the electrical control system. This makes the structural prosthesis model ready for functional testing phase.

Fig. 12.   Appearance of the 3D printed prototype prosthesis with hydraulic system and metal foot.

5   Model Validation and Next Steps

The developed control unit uses the specially designed control algorithm. It serves to control the joints of the prosthetic leg separately. The goal is to let the joints follow the desired trajectories (Fig. 13). Set of control sensors form an important part of the control unit. They are included in the feedback control loop to allow the reference tracking of the prosthesis (Fig. 14).

Fig. 13.   Control unit of the prosthetic leg

The sensors of the control unit, which are based on accelerometer and gyroscope, are also used for measurements. One set of three sensors is placed on the prosthetic leg and can directly measure the reference angles in the knee, ankle and metatarsophalangeal joint. The angle in the hip joint, here also referred to as thigh angle, is measured as the angle between the vertical axis and the position of the sensor mounted on the socket of the prosthetic leg.

Fig. 14. Control/measurement sensors

In order to autonomously test the efficacy of the prosthetic leg, the prosthetic sensors are used. Three sensors are placed on the prosthesis and three on the sound subject or the sound leg of an amputee. The sensors serve to map the angles of the leg and more precisely the angles of the knee, ankle and the foot. To facilitate the placement of sensors on the prosthesis and on the leg of a healthy subject, plastic cages were made for each sensor individually through 3D prototyping. The same has been done for the electronics. The prosthetic controller determines the joint inputs required at each instant in time to make the prosthesis track a commanded motion [5–7] (Fig. 15).

Fig. 15. SmartLeg prototype in real-time control testing

6 Discussion and Conclusion

Today, prosthetics strives to enable people to live as normal a life as possible, to play sports, etc. Demands are constantly increasing, and it only could be achieved with new solutions.

This paper aimed to offer a high quality and inexpensive solution that would allow climbing stairs in a natural way, in addition to the usual everyday activities. Previous prototypes have shown the feasibility of this goal. The prototype and the described integrated power system show an improvement of the previous ones based on the experimental output.

References

1. Jelačić, Z., Dedić, R., Džindo, H.: Active above-knee prosthesis. Elsevier Academic Press (2020). ISBN 978-0-12-818683-1
2. Jelačić, Z.: Contribution to dynamic modelling and control of rehabilitation robotics through development of active hydraulic above-knee prosthesis. Ph.D. dissertation, Dept. Mech. Eng., University of Sarajevo, Sarajevo, Bosnia and Herzegovina (2018)
3. Jelačić, Z., Dedić, R.: Real time control of above-knee prosthesis with powered knee and ankle joints. In: Karabegović, I. (ed.) NT 2019. LNNS, vol. 76, pp. 278–284. Springer, Cham (2020). https://doi.org/10.1007/978-3-030-18072-0_33
4. Jelačić, Z., Dedić, R.: Modelling and reference tracking of the robotic above-knee prosthetic leg with actuated knee and ankle joints during stair climbing. Heal. Technol. **10**(1), 119–134 (2019). https://doi.org/10.1007/s12553-019-00383-8
5. Dedić, R., Husnić, Ž, Ustamujić, F., Jelačić, Z.: Development of the concept of the integrated hydraulic system of the knee prosthesis. In: Karabegović, I. (ed.) NT 2021. LNNS, vol. 233, pp. 88–94. Springer, Cham (2021). https://doi.org/10.1007/978-3-030-75275-0_11
6. Jelačić, Z., Dedić, R.: Reference tracking of the robotic above-knee prosthetic leg with actuated knee and ankle joints using robust control. In: Badnjevic, A., Škrbić, R., Gurbeta Pokvić, L. (eds.) CMBEBIH 2019. IP, vol. 73, pp. 287–292. Springer, Cham (2020). https://doi.org/10.1007/978-3-030-17971-7_44
7. Jelačić, Z.: Comparison between the kinematic behaviour of different prototypes of prosthetic leg with actuated knee and ankle joints. In: Badnjevic, A., Gurbeta-Pokvić, L. (eds.) CMBEBIH 2021. IP, vol. 84, pp. 97–106. Springer, Cham (2021). https://doi.org/10.1007/978-3-030-73909-6_11
8. Dedić, R., et al.: IOP Conf. Ser.: Mater. Sci. Eng. 1208 012017 (2021). https://doi.org/10.1088/1757-899X/1208/1/012017
9. Husnić, Ž., et al.: IOP Conf. Ser.: Mater. Sci. Eng. 1208 012008 (2021). https://doi.org/10.1088/1757-899X/1208/1/012008

Process of Rapid Prototyping Using Wind Turbine as an Object of Experimental Research

Mina Šibalić[✉], Nikola Šibalić, Milan Šekularac, and Aleksandar Vujović

Faculty of Mechanical Engineering, University of Montenegro, Podgorica, Montenegro
minasibalic@edu.ucg.ac.me

Abstract. Techniques that manufacture parts by additive methods (by gradually adding solid material) are classified as rapid prototyping. Rapidprototyping has revolutionized prototyping with a key difference from conventional methods – supstractive methods. Overall process of new product development includes the prototype as an integral part of product engineering design. The prototype is a preliminary version of the final product and its purpose is to test or analyze the principles of operation as well as design evaluation. Using 3D printing (3DP) technology for prototyping significantly reduces production time and costs. Previously faced limitations by engineers, regarding tools which are available to them, are overcame with use of 3D printers. In this paper, the focus will be on CAD design, rapid prototyping (RP) of small wind turbine. Also, the paper generates prototypes from real material, using 3DP technologies and laboratory tests, that is determination of power on the turbine shaft.

Keywords: Prototype · Wind turbine · Rapid prototyping · 3DP

1 Introduction

Unknown three-dimensional shapes can be replicated in a much faster and easier way with the help of the concept of digitalization, which includes collecting a series of points that describe a certain surface and translate them into digital form.

Prototype is a word of Greek origin, derived from the word prototipos (proto-, type) and means the first design or something from which it was copied and developed [1]. Based on the meaning of the word, a prototype is defined as an original copy, which is later copied and developed. The concept of reversible engineering is seen as a systematic approach in analyzing existing parts and can be used either as an analysis of the design to observe and access the mechanisms of operation of the device. In a narrower sense, RE is defined as the duplication of an existing part or finished product without the aid of a drawing, technical documentation [2]. By using RP, it is possible to experiment with physical objects, reducing price and needed time up to 90%, while the time to place products on the market is significantly shorter.

The beginnings of RP were in the 1980s. The American company 3D Systems can be considered a pioneer of this technology. Charles Hull applied for a patent for a process that allows melting photopolymers using UV light stereolithography (SLA) and it has

I. Karabegović et al. (Eds.): NT 2022, LNNS 472, pp. 220–226, 2022.
https://doi.org/10.1007/978-3-031-05230-9_25

made it possible to replicate the same models quickly and easily. The first use of 3D printing in medicine was in 1999, when the bladder was printed, in 2000 the human kidney was printed for the first time, then the first artificial leg was printed in 2009, which enabled the personalization of this type of prosthesis later. Further improvement of 3D printing, in 2011 printing of 14k gold began [3].

2 Product Development

In order for a product to be ready for launch, it is necessary that the process from idea to realization goes through certain stages of development.

The product development process consists of the following steps:

1. Creating an idea,
2. Planning,
3. Conception,
4. Development,
5. Testing,
6. Final product design.

Prototyping plays a significant role in the creation of a new product, allowing allpotential errors to be removed before the start of production of the finished product. How a prototype is used to develop a new product or part, made for the purpose of performing a certain experiment, i.e. testing, it is necessary to define the key differences between the prototype and the finished product. The main differences are the material, processes and quality and accuracy of production [4].

Material: For the production of prototypes, the material that is used has sufficiently similar properties to the material intended for the final product, and the manufacturing process is much cheaper.

Processes: The production of prototypes is of an individual character, which is why certain flexible processes are chosen for their production. Subsequent production of final products is more demanding and uses more expensive processing processes.

Quality and accuracy of production: Prototyping forgives certain mistakes in the production of details, while for the final product it is necessary that these details are accurate and precise.

3 Rapid Prototyping

A series of independent strategies that reduce time production which is required to create full-size three-dimensional shapes most generally describe rapid prototyping. In order to obtain files, CAD software are used. Fast prototyping hardware automates production to circumvent the traditional way associated with making models and figures; these include various 3DP techniques and CNC carving.

Traditional methods have certain weaknesses, which limit efficiency and production. Clay modeling and foam board construction are traditional methods that often lack durability and strength. Another aggravating circumstance is that they are not always repeatable, for example, when making complex objects by carving, the artist will not be able to be impeccably precise every time.

The methods used for rapid prototyping can use all types of materials, which leads to solving problems that have arisen with traditional methods. All rapid prototyping technologies (additive methods) apply layer by layer of material in the form of cross-sections of the model in the x-y plane along the z axis and thus create a model.

All RP techniques have same approach. It consists of [4]:

- Object modeling in 3D CAD software,
- The drawn model is converted to STL (STereoLitography) file. The file represents a list of triangular surfaces,
- Certain software converts a virtual model into a series of cross-sections, in layers from 0.05 mm to 0.3 mm thickness,
- When printing, the machine uses this model to form a layer-by-layer model.
- Ease of use, etc.
    Disadvantages of 3D printing technologies:

Due to its good features, 3D printing technology is increasingly used. Some of advantages are [5]:

- Low cost of materials,
- There is no strict requirement against the generation of structural support,
- Direct production of color prototypes,
- High production speed,
- Possibility of installation in an office environment,
- Maximum utilization of construction material,
- Ease of use, etc.

In addition to the mentioned advantages, it is necessary to pay attention to the disadvantages, which may be fewer, but in certain cases they are very important, for example when choosing a process.

- Limited functionality of parts,
- Accuracy of the part,
- Poor finishing,
- Limiting the choice of materials,
- Power outage, etc.

4   CAD/RP System Integration and Practical Application

By reducing production time, the cost is reduced, which is the goal of reversible engineering and rapid prototyping, through computerized model generation based on existing physical models. RE collects product data and creates their CAD model, after which, with the help of RP technology, a physical model is obtained in a simple way.

Based on previous research and experience, the shapes of the blades given in Fig. 1 in the form of a function were adopted:

$$y_i = k^{i-1} \cdot x^{\frac{1}{k}} \tag{1}$$

For prototype $i = 1, 2, 3$; adopted value $k = 3$.

While the overall dimensions of the front surface of the wind turbine are $H = 10.50$ cm and $D = 17$ cm.

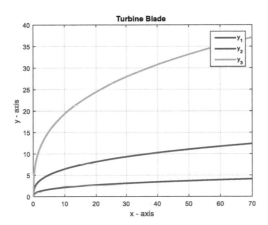

Fig. 1. Adopted blade shape for Prototype $i = 1, 2, 3$

In the software package CREO Parametrics, which is a three-dimensional computer program for computer and technical construction and design in production processes of a wide range, designed three different CAD models (Prototypes) of blades for small vertical wind turbine.

Figure 2 shows the shapes of Prototypes 1, 2 and 3, as well as a blade carrier that is designed so that blade prototypes can be replaces quickly and easily.

Figure 3 shows the Prototype 3 assembly, which was designed in CREO. Generating the initial three-dimensional CAD model is a technique of pre-processing by triangular approximation into the appropriate pattern following with a printing process on a 3D printer.

The 3D printing process is one of the techniques of RP and is very common, thanks to the quality of production, relatively low cost of production and currently available printing material. The 3DP process is based on "printing" – successive application of cross-sections of layers of material used for printing, the appropriate thickness, converted directly from 3D models. In this way, a physical object is created (Fig. 4). The 3D model is made of ABS or acrylonitrile butadiene sitren thermoplastic polymer which is very resistant to external influences. ABS has a wide application when it comes to production and processing. Most standard plastic processing machines use this polymer. The melting temperature is 210–270 °C, while the tool temperature is from 40 °C to 70 °C.

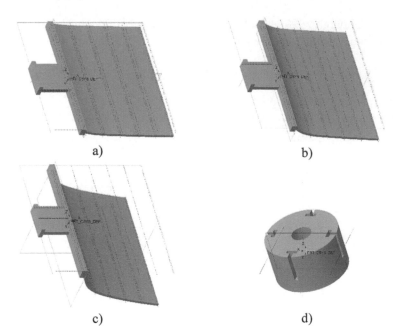

Fig. 2. Generating prototype geometry: a) For $i = 1$, Prototype 1 blade, b) For $i = 2$, Prototype 2 blade, c) For $i = 3$, Prototype 3 blade, d) blade carrier

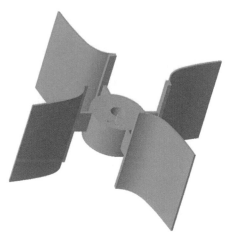

Fig. 3. Overview of the Prototype 3, solid model assembly

Fig. 4. 3D model made of real material using 3DP technology

5   Wind Energy

Wind is an inexhaustible ecological source of energy and a significant resource in electricity production. In addition to environmental benefits, environmental protection, which produces energy with the help of wind turbines, primary wind energy is also free. According to the mode of operation, a small wind turbine is generally the same as a large wind turbine [6].

Testing of the obtained prototypes was done in experimental conditions. The experiment was performed with the aero tunnel which gives an air flow speed of $v = 5$ m/s at the exit. We were able to measure the power of the turbine with the help of the construction shown in Fig. 5.

Fig. 5. Experimental varification Prototype 1, 2 and 3

The calculation procedure of the obtained power and torque of the turbine is given by the expression [7]:

$$P_t = \omega M_t \rightarrow M_t = \frac{Pt}{\omega} \tag{2}$$

where: P_t – turbine power, M_t – turbine torque, ω – angular velocity,

$$P_w = \frac{1}{2} \cdot \rho \cdot A \cdot v3 \tag{3}$$

where: P_w – the power that the wind has when iy hits the turbine,
  For values: $\rho = 1.1250$ kg/m^3,

$$\eta h = \frac{P_t}{P_w} \rightarrow P_t P_w \eta h \tag{4}$$

where: ηh – hydraulic efficiency,

$$A = D \cdot H \tag{5}$$

where: A – frontal surface that gives torque to the turbine,
  For adopted values: $\omega = 0.2$ rad/s; $\eta h = 0.15$, follows $P_w = 1.2551$ W, $P_t = 0.1883$ W and $M_t = 0.9413$ Nm.

6 Conclusion

The reduction in the cost of the product, as well as the time of its production, has greatly influenced the application of new technologies for the purpose of prototyping the given products. Reducing production time also reduces their cost, which is the goal of RE and RP. 3DP has optimized the production of the most complex products.

Wind is a form of clean, renewable energy resource. Wind turbines have a supplementary role for the application of non-renewable resources in energy.

By analizing different solutions (prototypes) of wind turbines, a decision can be made as to which solution is the most suitable. RP allows for a short time to make models on which comparative analysis is performed. Research has shown that Prototype 3 gives the best results, which need to be further improved through optimizing the geometry of the prototype. This type of wind turbine could be used as an aid to generate energy together with solar panels.

References

1. Ilustrated Dictionary Oxford, Mladinska knjiga Beograd (2008)
2. Topčić, A., Cerjaković, E.: Izrada prototipa, Univerzitet u Tuzli, Mašinski fakultet u Tuzli (2014). god
3. History of 3D Technology in the Last Three Decades. https://blogs.3ds.com/india/history-of-3d-technology/. Pristupljeno 10, August 2021
4. Upcraft, S., Fletcher, R.: The rapid prototyping technologies. Assem. Autom. **23**(4), 318–330 (2003). https://doi.org/10.1108/01445150310698634
5. 3D štampa Maketa – Osnovne prednosti. Solfins doo. Pristupljeno 20. Nov 2021. https://solfins.com/en_US/blog/3d-print-scan-6/post/3d-stampa-maketa-osnovne-prednosti-305
6. Ivanović, D.: Obnovljivi izvori energije, Podgorica (2015)
7. Manwell, J.F., McGowan, J.G.: Wind Energy Explained, Theory, Design and Application, Second Edition, USA

Realization of the Robotic ARM/Plotter

Milena Djukanovic[1]([✉]), Luka Radunović[1], Vuk Bosković[2], and Bai Chuang[3]

[1] Faculty of Electrical Engineering, University of Montenegro, 81000 Podgorica, Montenegro
`milenadj@ucg.ac.me`
[2] Faculty of Mechanical Engineering, Technical University of Munich, 85748 Munich, Germany
[3] School of Physics and Electronic Science, Changsha University of Science and Technology, Changsha, People's Republic of China

Abstract. Nowadays, when distance is not a very important category, signature is a rare thing that cannot be easily transported. At the same time, signature is a repetitive action that in total takes a lot of our time for different kind of everyday activities. In this regard, the aim of this paper and product presented here is development of a machine/system that successfully and authentically simulates human handwriting. The unique properties allow user to be next to or far away from the machine, while having his writings stored on his own account, which makes an efficient time save.

Keywords: Simulation of human handwriting · Plotter · Easy portability · Repetitive action · Time saving

1 Introduction

Technological advancements of today allow robotics and its components to replace humans in doing various tasks. Automatization of the processes became available both in the industry and in the private use. People have always striven to do every day routine tasks as quick as possible, so they could have time left to focus on more important work. A solution they came up to is assistive technologies and robotics. Robot, as an electro-mechanical system which can autonomously perform a given function or a task using the program, or controlled by human command, has a big variety of usage, starting from small housework and finishing with complex tasks, such as precise drawing, cutting, welding and assembling parts. The intelligence that one robot possesses directly depends on its components and program itself and its adaptability to unpredictable situations [1].

Major advantages of robotics over humans are more efficient economy, precision, speed and infinite number of repetitions of a given task [1, 2]. They differ, from autonomous robotic hands, used in industry, which have many different movements and functions, to simpler mechatronic systems or devices that work in automation.

The main focus of this paper is on robotic components that perform repetitive tasks. Thus, this paper is focused on the development of robot arm/plotter which would perform routine hand-executed functions, such as writing, drawing, engraving, but its primary purpose is signing of documents when the user is not directly next to them [2]. Before

© The Author(s), under exclusive license to Springer Nature Switzerland AG 2022
I. Karabegović et al. (Eds.): NT 2022, LNNS 472, pp. 227–235, 2022.
https://doi.org/10.1007/978-3-031-05230-9_26

the realization we have prepared the conceptual solution and programming part of the robotic arm [3, 4].

For this new system of signing to work, two people are needed. The one that needs the signature and the one that needs to sign. First person in need of the signature must position his paper on the machine bed, fixing it and using the clamps that are embedded on the machine bed as well. Paper does not need to be put parallel to the machine clamps and x-y axis, because the software will automatically fix the parameters and rotate the signature as needed. This leaves second person only in need of a smartphone device and an internet connection, while location of this person is not of any importance.

For the security reasons, and for the time optimisation of the whole system, database is added. Its functions are not only to save the signature of the second person in process, which could be used multiple times (second person would only need to position his presaved signature), but to save digital key of his smartphone as well, making the system almost 100% secured [5].

2 Materials and Methods

In order to have a semi-automated machine that works in real time, this system must be divided into smaller parts-subsystems. System is divided in three, internet connected subsystems, with distance not being an important factor for it to work (Fig. 1).

1. Smartphone app
2. Web server
3. Plotter Machine

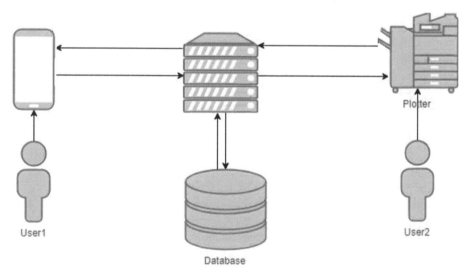

Fig. 1. Subsystem relation: User1 using smartphone application, User2 sending commands using Plotter, Database encrypting and saving data.

2.1 Smartphone Application

Smartphone application is made in Android Studio.

User's frontend consists of two main functions:

1. Signature saving
2. Signature positioning

1. - User draws his signature on Android screen using his fingers or smartphone pen, in any direction that he wants, and clicks the "save button". Saved signature can be changed just by redrawing it with a new one, and the old one gets replaced with a new one and erased from the whole database system.

2 - User gets picture of plotter's working table when he requests it through the application and positions his signature by dragging it across the screen. User 2, operator of the plotter machine may request User1 to sign the document that is on the working platform as well by clicking the button on the plotter, and sending notification with picture in it (Fig. 2).

Fig. 2. Smartphone application model: this is the second main function-signature positioning. Red ellipse is showing the line, place for the signature on the document. User1 then, after reading and checking authenticity of the document, drags and places his digital signature from tab on the left side to the signature line.

Backend is communicating with web server only when post/get request is made by user (changing signature or sending command for drawing the same on the plotter). When the first function is done - saving of the signature, application sends screenshot of the signature to the web server which stores it in database as jpeg file. After the second function starts - signature positioning, User1 requests picture of the plotter's working table from the web server. Picture is first saved in database and only then sent to User1. Then User1 positions his signature and sends a post request to the web server by clicking the "send button" (X and Y positions of the starting point are sent and stored) [6].

2.2 Web Server

Web server used in this system is home-based raspberry pi 3 LAMP (Linux, Apache, MySQL, PHP) server with the ability to store and send data in real time [5, 6].

2.3 Plotter Machine

Plotter is consisted of mechanical elements (Fig. 3)

1. Frame
2. Linear rails
3. Ball screw with ball screw nut
4. Ball screw end support

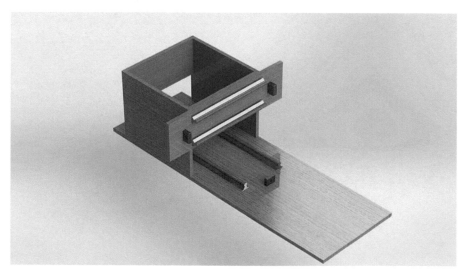

Fig. 3. Frame of the plotter: 3D model modelled in SolidWorks software

Electrical elements:

1. Microcontroller with wireless connection, connected to internet
2. Stepper motor drivers
3. Stepper motors
4. Limit switches

Microcontroller has two modes. In one mode it operates as a communication device, sending and receiving files to/from web server. In the second mode it converts received signature image to G-code and sends them to the stepper motor drivers. It uses signature position data (X and Y position of the starting point) sent by the User1 and adds them to the G-code [7] (Fig. 4).

Fig. 4. Prototype of the plotter machine with dimensions matching models' designed dimensions, made both with material commonly found in the laboratory and with industry-standard electro-mechanical elements.

3 Results

3.1 Plotter Test

As the plotter was controlled by G-code, it was possible to import motorcycle vector image file and allow the software - smartphone application to convert it into G-code

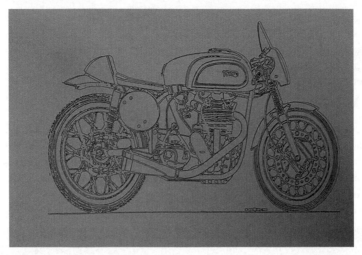

Fig. 5. Plotter drawing of motorcycle on the A4 format paper, using regular pen as its' working head.

after dragging it on the paper, as shown in the second main function of the application. This picture is just showing the precision of developed machine, which can be significantly increased and upgraded just by implementing the same procedures and electro-mechanical elements into more rigid, stable housing, which would reduce the vibrations that are caused by stepper motors and friction between pen and the paper [5, 7, 8] (Fig. 5).

3.2 Calculations and Plotting Precision

To take drawing and writing precision to a higher level that allows signature authenticity and adequate equipment is needed. This depends both from mechanical and electrical components [8, 9] (Fig. 6).

Parts of the electro-mechanical system that affect precision are:

1. Spindle geometry - thread pitch
2. Stepper motor specifications - number of steps per revolution
3. Stepper motor driver – micro-step value

In this project, the following components are used:

1. Spindle 1605

   a. The main characteristic in precision calculation is the thread pitch of the spindle. In this case its value is 5 mm. This value signifies the length of translational movement per one spindle revolution.

$$P = 5\,mm$$

Fig. 6. 3D model of stepper motor assembly made in SolidWorks, it consists of stepper motor, shaft coupling, precise ball screw and ball screw bearing with housing.

2. Stepper motor - NEMA 17

    a. The key parameter in motor choice is how many steps can it make per one revolution of 360 degrees. NEMA 17 is able to deliver 200 steps in one revolution in its lowest precision state, determined by stepper motor driver.

$$n = 200$$

3. Stepper motor driver - TB6600

    a. As microcontroller's electronics is not able to deliver specified amount of power to the stepper motor itself, stepper motor driver is required. It operates at 12 V and drives sufficient current to the motors depending on the program that microcontroller's board proceeds. Except its main power delivery function, the driver has the ability to dose the current flowing to the motor winding, while dosing the magnetic force between its electro-magnets represented in percentage of dosed power, thus enabling shorter steps. Every motor has so-called step modes stated in its specifications. TB6600 driver is able to split one step into 32 micro-steps. So the step ratio is 1/32 [10, 11] (Fig. 7).

$$sr = 1/32$$

Calculation of the smallest possible theoretical movement gives a figure about the best performance of a machine. The drawing resolution is inversely proportional to the

Fig. 7. TB6600 stepper motor driver

number of steps. If the step is shorter, this makes the theoretical resolution of the machine bigger [9, 10]. This brings us to the exact calculation:

$$Mm = P/n * sr = 5/200 * 1/32 = 0.00078125 \text{ mm} = 0.7\mu\text{m}$$

4 Conclusion

This value is only theoretical electro-mechanical minimum value that cannot be achieved in real life using regular wooden flame that is holding all linear mechanics in plotter system, as it is soft material that allows little structure deforming, decreasing stability and precision. For the industrial machine, one that chases microns, proper metal, epoxy-granite structure must be calculated and designed, not allowing negative effects produces by vibrations and inertia. Main topic for future research will be optimising this mechanical frame, conducing software cad/cae multi-iterating process and comparing rigidity and precision results with machine that is made by final iteration plans.

Acknowledgments. This research was funded by bilateral project "Research and application of service robots for non-production services based on microcontroller's monitoring" supported by Ministry of Education, Science, Culture and Sports in Montengro and Ministry of Science and Technology in PR Republic of China.

References

1. Williams, G.: Title of the article: NC robotics: build your own workshop bot. McGraw-Hill Year 2003, Volume Tab robotics, ISBN: 0071418288
2. Cekus, D., Skrobek, D., Zając, T.: A Dynamic Analysis of an Industrial CNC Plotter. In: 22nd International Conference. Engineering Mechanics 2016. Svratka, Czech Republic, 9–12 May 2016

3. Djukanovic, M., Grujicic, R., Radunovic, L., Boskovic, V.: Conceptual solution of the robotic Arm/Plotter. In: Karabegović, I. (eds.) New Technologies, Development and Application. NT 2018. Lecture Notes in Networks and Systems, vol 42, pp. 170–179. Springer, Cham (2019). ISBN: 978-3-319-90892-2

4. Djukanovic, M., Grujicic, R., Radunovic, L., Boskovic, V.: Programming of the Robotic Arm/Plotter System. In: Avdaković, S. (eds.) Advanced Technologies, Systems, and Applications III. Lecture Notes in Networks and Systems, vol 60, pp. 342–354. Springer, Cham (2019). ISBN: 978-3-030-02576-2

5. Santhosh Krishna, B.V., Oviya, J., Gowri, S., Varshini, M.: Cloud robotics in industry using Raspberry Pi

6. Publisher: IEEE, Published in: 2016 Second International Conference on Science Technology Engineering and Management (ICONSTEM). Date of Conference: 30–31 March 2016. ISBN: 978-1-5090-17060-5

7. Mattson, M.: programming: principles and applications. Cengage Learning (2009)

8. Xu, X.W., Newman, S.T.: Making machine tools more open, interoperable and intelligent—a review of the technologies. Comput. Ind. 57(2), 141–152 (2006)

9. Hashim, N.S.: (2012). Design of mini machine (Doctoral dissertation, UMP)

10. Jayachandraiah, B., Krishna, O.V., Khan, P.A., Reddy, R.A.: Fabrication of Low Cost 3-Axis (2014)

11. Router. International Journal of Engineering Science Invention, 3(6), 01–10

12. Overby, A.: CNC Machining Handbook: Building, Programming, and Implementation, May 2010

13. Publisher: [1] McGraw-Hill, Inc. Professional Book Group 11 West 19th Street New York, NYUnited States. ISBN:978-0-07-162301-8

14. Espalin, D., Muse, D.W., MacDonald, E., Wicker, R.B.: 3D Printing multifunctionality: structures with electronics. Int. J. Adv. Manuf. Technol. 72(5–8), 963–978 (2014). https.//doi.org/10.1007/s00170-014-5717-7

Dynamic Analysis of a Hybrid Heavy-Vehicle

Marco Claudio De Simone[1(✉)], Vincenzo Laiola[2], Zandra B. Rivera[3],
and Domenico Guida[1]

[1] Department of Industrial Engineering, University of Salerno, Via Giovanni Paolo II, 132,
84084 Fisciano, Italy
{mdesimone,guida}@unisa.it
[2] MEID4 Academic Spin-Off of the University of Salerno, Via Giovanni Paolo II, 132,
84084 Fisciano, Italy
[3] Engineering Research Institute, Universidad Privada San Juan Bautista, Av. José Antonio
Lavalle N° 302-304, Lima 15067, Peru

Abstract. The aim of this work is to analyze the use of hybrid powertrains in public transport by heavy vehicles. In the last years, the increasing demand for sustainable mobility has stimulated the use of electrified powertrains also in the heavy-duty transport sector. Numerous European regulations prohibit non-electrified vehicles from entering historical centers, in highly congested urban areas and airports. The use of an electric motor coupled to an Internal Combustion Engine (ICE) in a bus, whose motion is characterized by continuous transients, especially in the case of vehicles dedicated to urban transport where the distance between two successive stops is reduced. In the work, it was decided to investigate the various hybrid architectures normally used in vehicles and put on a multibody model to test optimal management systems of the engines present on the vehicle.

Keywords: Multibody · Dinamics · Heavy-vehicle · Hybrid · Simscape

1 Introduction

The architecture of a hybrid vehicle refers to the different ways of connecting the powertrain (IC engine and electric machines) to the output shaft and the power divider [1]. With the term power-split HEV Device (PSD), it is usually indicated the device generally consists of one or multiple planetary gearboxes dedicated to the purpose. Using the planetary gear in a PSD allows replacing the components present on a traditional vehicle (gearbox, alternator, starter motor) with two electric machines with two electric motors [2, 3].

2   Materials and Methods

The PSD is a combinatorial planetary gear, reported in Fig. 1, which combines mechanical powers. The power generated by the ICE is divided into two paths: mechanical and electrical [4]. In the mechanical path, the power of the IC engine is transferred from the carrier to the planetary ring, which is connected to the vehicle axle. In the electric path, the power not transferred in the first path is transformed into alternating electric current(AC) via the MG1 [5].

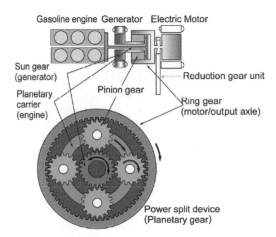

Fig. 1. Mechanical connections between PSD and motors

This electricity is first converted into direct current and then used to recharge the battery or transferred to the MG2 traction unit in electricity [6, 7]. To analyze the relationships between the angular velocities of the components of the power-split device it is possible to apply the Willis formula, reported in (1).

$$\varepsilon_0 = \frac{\omega_R - \omega_C}{\omega_S - \omega_C} = -\frac{z_S}{z_R} \tag{1}$$

where z_S and z_R represent the sun gear and ring-gear-tooth number. For the dynamic study of gearing it is possible to write the power and torque balance on the group by introducing the mechanical efficiency η_m:

$$\begin{cases} P_{ICE}\eta_m = P_S + P_R \\ T_{ICE}\eta_m = T_S + T_R \end{cases}$$

The task of the Power Split Device is to distribute and stabilize the rotation speed of the IC engine with the highly variable rotational speeds of the wheels, which depend on the driving conditions [8]. At the same time trying to give the driver a driving experience similar to a traditional vehicle with a manual or robotic gearbox, to adjust the speed of the combustion engine the PSD changes the rotation speed of the MG2 to reconciles the speed difference between wheels, MCI, and MG1 [9]. By analyzing the energy scheme of the vehicle, it is possible to understand that the possible configurations, in terms of power flows, are essentially two:

1) During normal driving conditions, the IC engine works at low and constant rpm, which coincides with the maximum efficiency condition [10]. This operation is guaranteed by asking the generator to absorb a small and constant torque.
2) When strong accelerations are needed, to prevent the thermal engine from increasing rpm and lowering the efficiency, the generator demands higher torques. In this way, the angular speed of the engine and the MG1, connected to the solar gear, will be lower. By decreasing the angular speed of the sun gear with a control system, the device gives the possibility to redirect the mechanical power to the planetary gear connected to the wheels [11].

When the vehicle is moving, in this case, the PSD allows different driving strategies only ICE, only electric, or combined). The choice depends on the strategy used to have the best efficiency [12, 13]. A further distinction is possible depending on whether the route is urban or extra-urban. Instead, when the vehicle is braking, the internal combustion engine does not act on the wheels for traction or is switched off. The MG2 motor is used as a power source thanks to regenerative braking. Regenerative braking converts the otherwise wasted mechanical energy generated by a vehicle's deceleration into electricity and stores it in the batteries [14] (Fig. 2).

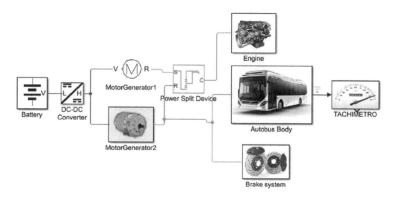

Fig. 2. SimScape multibody multidomain model

Energy recovery takes place by reversing the rotation of the electric motor and producing a resistance on the wheels acting as a brake [15]. Instead of converting the current into motion, the motor acts as an alternator and converts the kinetic energy back into electric current. If additional stopping power is required, friction brakes (e.g. disc brakes) are applied automatically. In order to study the dynamic behavior of a hybrid vehicle, it is necessary to model not only the vehicle but also all the components that make up the powertrain [16, 17]. Their cooperation will allow to predict the behavior and to evaluate the values of torque and mechanical power that the motor will generate/need, the consumption of fuel, the electrical energy requested by the battery drain, and the use of mechanical brakes. In order to study such a complex system, it is not possible to proceed simply by employing mathematical equations whose resolution would become complicated to manage [18]. Many software allows developing detailed models starting from three-dimensional geometries or integrating multidomain models (electrical, mechanical, hydraulic, etc.) into a single system. Such a solution allows not only to investigate the kinematical and dynamical behavior but allow to carry out an energy management investigation in order to easily examine its behavior in different operating conditions [19]. Such an approach is called Model Base Design and allows to create, for each organ of the power train, autonomous submodels that faithfully describe its behavior, and by interfacing these submodels with each other, respecting the real interconnections, it is possible to represent the drive train under exam. Such an approach, therefore, allows a modular description of the system in which each submodel is characterized by input and output variables [20]. An important advantage offered by this procedure is also represented by the possibility of being able to intervene on the various submodels without altering the structure of the overall system. Therefore, following such an approach, a Simscape multibody model of the hybrid vehicles was developed in Simulink, in particular, by using the Simscape Driveline, Multibody, Electrical, and Mechanical Simscape's libraries [21]. The goal of this investigation is to predict the performance and energy consumption in function of the type of route to be covered, from the capacity of the vehicle and the actual crowding, and the imagined propulsion. For the study, it was decided to use the data of a Volvo 7900 bus of 12 m in length for modeling the vehicle. The brake system of our bus is operated by 4 electronically controlled disc brakes. The combined braking function is provided, the electric motor acts as a retarder and harvests the braking energy. The disc brake is the most used type of braking system on the market, more expensive than a drum brake but much more performing [22]. The disk is the fundamental part of the system, its task is to degrade the kinetic energy into thermal energy, for this reason, one of its fundamental characteristics is the ability to disperse heat. Also very important is the ability to withstand considerable mechanical stress, especially at medium and high temperatures, the non-deformability to always allow a flat contact between the pad and the disc, the isolation from vibrations, and abrasion resistance. Disc brakes, usually positioned inside the wheel, are operated by a hydraulic system (braking system) which, by varying the pressure of the particular liquid inside it, gives the pads a thrust directly proportional to the force exerted on the pedal [23] (Table 1).

Table 1. Volvo 7900 data

Parameters	Values	Parameters	Values
Length (m)	12	Front area (m^2)	6.825
Weight (kg)	10450	Drag coefficient	0.55
Fully loaded weight (kg)	19000	Wheel radius (m)	0.56
Number of wheels per axle	2	Number of disc brakes	4

The internal combustion engine plays a fundamental role in the study of hybrid vehicles. Its management is closely linked to fuel consumption and the consequent polluting emissions [24]. The parameters that describe the behavior of an internal combustion engine are numerous, however, it is possible to consider a limited set of them which still allows a simplified model to be obtained. This model allows, although in an approximate manner, a concise but fairly realistic description of its operating conditions. With a high degree of abstraction, it is possible, considering a limited number of quantities, to obtain a simplified model, which allows describing in a synthetic but quite realistic way its operating conditions [25]. For the simulation, the internal combustion engine (diesel) "OM 936" (Euro 6) produced by Mercedes for hybrid buses was used.

The second energy converter present in the reference vehicle is the brushless motor, a direct current electric motor having a permanent magnet (MG) rotor. It can operate in two different ways: by converting the electrical power taken from the battery into mechanical power useful to the axle or, by converting the mechanical power received by the wheel axle (during deceleration) or from the IC engine in electric power to recharge the batteries. For this reason, the component will be indicated more properly Motor-Generator in the following paragraphs. Motor-Generator 1 is the electric machine connected to the sun gear of the power split device, designed to convert mechanical energy from the engine into electrical energy [26]. The model used for the MG1 is the SRI 300 produced by Dana TM4 for cars and heavy vehicles, with a maximum power of 70 kW and maximum torque of 400 Nm. The Motor-Generator 2 is the electric machine connected to the planetary wheel (Ring) of the power split device. It is responsible for the traction of the vehicle and also for converting the kinetic energy during vehicle braking into electrical energy, thanks to regenerative braking. The model used for the MG2 is the P211 produced by Magtec for hybrid and electric buses, with a maximum power of 200 kW and maximum torque of 2000 Nm. For our bus, it was decided to use a lithium-ion accumulator with a capacity of 400 Ah. The rechargeable lithium batteries operating at room temperature represent the electrochemical storage technology on which the greatest investments are focused. Lithium-ion batteries can just as well be found in consumer electronics (telephones, laptops) as in electric cars [27].

3   Results

The simulation was conducted considering the vehicle fully loaded, with passengers on board and a total mass of 19 tons. The bus performs a WLTP Class 2 speed profile cyclically for 15480 s (equal to 4 h and 30 min), starting with a full tank and 100% charged battery.

In Fig. 3a the trend of bus speeds on a 30-min extract and the speed demand in km/h between the required speed and the real speed of the vehicle.

In this model, the electric motor was mainly used for traction, while when the error is greater than 1 km/h. At the end of the simulation, the bus traveled 150.6 km in 15480 s and consumed 11.94 L of diesel. The bus will therefore travel 12.62 km for every liter of diesel consumed. As regards electricity consumption, the model provides that the battery is recharged both through a generator (MG1) powered by the thermal engine and, to a small extent, by the use of regenerative braking through the electric motor

(MG2), which during braking it acts as a generator by reversing the direction of rotation of the motor. The battery consumption, after 15480 s, starting from a capacity of 150 Ah it arrives at the end of the simulation with a capacity of 101.2 Ah. Figure 3b shows the trends of the mechanical (electric motor) and electric power expressed in kW, on a 30-min extract. As far as mechanical power is concerned, we can see how power is nearly alwaysconstant to maintain the number of revolutions of the diesel engine at low and constant speeds for maximum engine efficiency.

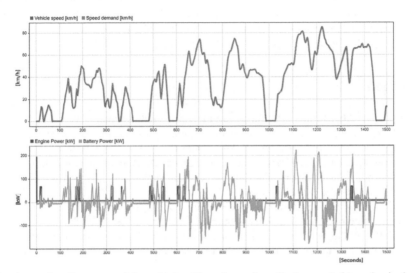

Fig. 3. a) speed performance of the vehicle with real speed profile (orange), b)mechanical power (red) and electrical power (blue) generated.

For braking, the powersplit hybrid bus uses a combined braking system. Figure 4 shows the saturation of the disc brakes, i.e. how much they are used during motion. The saturation is a percentage value where 0 means that the brakes are not used while when the value is equal to 1 it means that the maximum braking force is applied to the brakes.

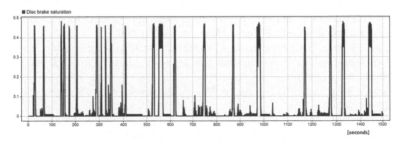

Fig. 4. Disc brake saturation

4 Conclusion

The objective of this work is the analysis of a heavy vehicle with hybrid powersplit architecture for public transport on wheels. The heart of this type of architecture is constituted by the power split device, which combines mechanical powers deriving from different sources, electric and the combustion engine, operating at different speeds and transmitting them to a single output. This solution allows to benefit from the reduced consumption and emissions of a serial hybrid but at the same time use both the engine, thermal and electric, in the event of significant climbs or accelerations, characteristic of a parallel hybrid. The multi-domain model of the hybrid bus was created in the Simulink environment to analyze consumption and performance.

References

1. Pappalardo, C.M., Lettieri, A., Guida, D.: Identification of a dynamical model of the latching mechanism of an aircraft hatch door using the numerical algorithms for subspace state-space system identification. IAENG Int. J. Appl. Math. **51**(2), 346–359 (2021)
2. Manrique-Escobar, C.A., Pappalardo, C.M., Guida, D.: A multibody system approach for the systematic development of a closed-chain kinematic model for two-wheeled vehicles. Machines **9**(11), 245 (2021)
3. Huang, J., et al.: Sizing optimization research considering mass effect of hybrid energy storage system in electric vehicles. J. Energy Storage **48**, art. no. 103892 (2022)
4. Dašić, P.: Reliability of technical systems: Selected scientific-professional papers (in Serbian and Russian). Vrnjačka Banja: SaTCIP Publisher Ltd., 2019. – 308 str. ISBN 978-86-6075-051-0
5. Gao, Z., et al.: Comprehensive powertrain modeling for heavy-duty applications: A study of plug-in hybrid electric bus. Energy Conversion Manage. **252**, art. no. 115071 (2022)
6. Dašić, P.: Scientific and technological trends: selected scientific-professional papers (in Serbian). Vrnjačka Banja: SaTCIPPublesher Ltd. (2020), – 305 str. ISBN 978-86-6075-072-5
7. Pappalardo, C.M., Manca, A., Guida, D.: A combined use of the multibody system approach and the finite element analysis for the structural redesign and the topology optimization of the latching component of an aircraft hatch door. IAENG Int. J. Appl. Math. **51**(1), 175–191 (2021)
8. Formato, A., Ianniello, D., Romano, R., Pellegrino, A., Villecco, F.: Design and development of a new press for grape marc. Machines **7**(3), 51 (2019)
9. Janovec, M., Čerňan, J., Škultéty, F., Novák, A.: Design of batteries for a hybrid propulsion system of a training aircraft. Energies **15**(1), art. no. 49 (2022)
10. Manrique Escobar, C.A., Pappalardo, C.M., Guida, D.: A parametric study of a deep reinforcement learning control system applied to the swing-up problem of the cart-pole. Appl. Sci. **10**(24), 9013 (2020)
11. Hergart, C.: Sustainable Transportation, Energy, Environment, and Sustainability, pp. 7–38 (2022)
12. Pappalardo, C.M., Guida, D.: Dynamic analysis and control design of kinematically-driven multibody mechanical systems. Eng. Lett. **28**(4), 1125–1144 (2020)
13. Patel, D.K., Singh, D., Singh, B.: A comparative analysis for impact of distributed generations with electric vehicles planning, Sustainable Energy Technologies and Assessments, 52, art. no. 101840 (2022)

14. Ramya, K.C., Ramani, J.G., Sridevi, A., Rai, R.S., Shirley, D.R.A.: Analysis of the Different Types of Electric Motors Used in Electric Vehicles, EAI/Springer Innovations in Communication and Computing, pp. 43–57 (2022)
15. Manrique-Escobar, C.A., Pappalardo, C.M., Guida, D.: On the analytical and computational methodologies for modelling two-wheeled vehicles within the multibody dynamics framework: a systematic literature review. J. Appl. Comput. Mech. 8(1), 153–181 (2022)
16. Formato, A., Ianniello, D., Pellegrino, A., Villecco, F.: Vibration-based experimental identification of the elastic moduli using plate specimens of the olive tree. Machines, vol. 7, no. 2, art. no. 46 (2019). https://doi.org/10.3390/machines7020046
17. Sun, X., Liu, H., Song, W., Villecco, F.: Modeling of eddy current welding of rail: three-dimensional simulation. Entropy, vol. 22, art. no. 947 (2020)
18. Liguori, A., Armentani, E., Bertocco, A., Formato, A., Pellegrino, A., Villecco, F.: NoiseReduction in spur gear systems. Entropy 22, 1306 (2020)
19. Ardila-Parra, S.A., Pappalardo, C., Estrada, O.A.G., Guida, D.: Finite element based redesign and optimization of aircraft structural components using composite materials. IAENG Int. J. Appl. Math. 50(4), 1–18 (2020)
20. Salvati, L., d'Amore, M., Fiorentino, A., Pellegrino, A., Sena, P., Villecco, F.: On-road detection of driver fatigue and drowsiness during medium-distance journeys. Entropy 23(2), 135 (2021)
21. Liguori, A., Formato, A., Pellegrino, A., Villecco, F.: Study of Tank Containers for Foodstuffs. Machines, vol. 9, art. no. 44 (2021)
22. Dašić, P.: Response surface methodology: Selected scientific-professional papers (in Serbian, Slovenian and Russian). Vrnjačka Banja: SaTCIP Publisher Ltd., 2019. – 301 str. ISBN 978-86-6075-054-1
23. Li, T., Kou, Z., Wu, J., Yahya, W., Villecco,F.: Multipoint Optimal Minimum Entropy Deconvolution Adjusted for Automatic Fault Diagnosis of Hoist Bearing. Shock and Vibration, v. 2021, art.no. 6614633 (2021)
24. Pappalardo, C.M., Lombardi, N., Guida, D.: A model-based system engineering approach for the virtual prototyping of an electric vehicle of class l7. Eng. Lett. 28(1), 215–234 (2020)
25. Formato, A., Romano, R., Villecco, F.: A novel device for the soil sterilizing in sustainable agriculture. In: Karabegović, I. (ed.) NT 2021. LNNS, vol. 233, pp. 858–865. Springer, Cham (2021). https://doi.org/10.1007/978-3-030-75275-0_94
26. Dašić, P., Dašić, J., Antanasković, D., Pavićević, N.: Statistical analysis and modeling of global innovation index (GII) of Serbia. In: Karabegović, I. (ed.) NT 2020. LNNS, vol. 128, pp. 515–521. Springer, Cham (2020). https://doi.org/10.1007/978-3-030-46817-0_59
27. Tošović, R., Dašić, P., Ristović, I.: Sustainable use of metallic mineral resources of Serbia from an environmental perspective. Environ. Eng. Manage. J. (EEMJ) 15(9), 2075–2084 (2016). ISSN 1582-9596. https://doi.org/10.30638/eemj.2016.224

Dynamic Analysis and Attitude Control of a Minisatellite

Rosario La Regina[1], Carmine Maria Pappalardo[2(✉)], and Domenico Guida[2]

[1] MEID4 Academic Spin-Off of the University of Salerno, Via Giovanni Paolo II, 132, 84084 Fisciano, Italy
[2] Department of Industrial Engineering, University of Salerno, Via Giovanni Paolo II, 132, 84084 Fisciano, Italy
{cpappalardo,guida}@unisa.it

Abstract. In the last decade, the interest of many companies to undertake missions aimed at space research has increased dramatically. Currently, the focus is more and more on the large family of miniaturized satellites. This paper aims to define the main components of a minisatellite. Additionally, the torques for attitude stabilization are evaluated. The implementation system identified herein consists of magnetorquers and three reaction wheels. The simulations carried out to assess their behavior were performed in a virtual environment using SIMSCAPE-MULTIBODY. Its use has made it possible to develop control systems for the satellite under study. Finally, an effective proportional-derivative (PD) feedback controller was chosen and tuned through numerical experiments.

Keywords: Minisatellite · Magnetorquers · Reaction wheels · Attitude control · Multibody

1 Introduction

This paper aims to design a satellite capable of surviving in LEO(Low Earth Orbit) and interfacing with its magnetic field. Minisatellites are more launched at an altitude of 400–600 (km), and descend over several months to a few years, in line with policies against the accumulation of debris in space. The minisatellite designed herein is thought of using COTS (Commercial Off-the-Shelf components), to keep costs low, and include in the hypothetical design only the minimum components for its proper operation. Cube-Sat is a mini-satellites classification, whose standard shape is $10 \times 10 \times 10$ (cm), with a maximum weight of 1 (kg), defined as 1U. These units can pair together to create more complex nanosatellites, up to a composition of 12U [1]. The CubeSat standard was developed by Twiggs (Stanford University's Space Systems Development Laboratory) and Puig-Suri (California Polytechnic State University) in 1999, and the first CubeSat launched (AAU CubeSat) into orbit was in 2003 [2]. After assembly, the Cube-Sats are placed in orbit using a mechanism called MRFOD (Morehead-Roma FemtoSat Orbital Deployer) or the more common P-POD (Poly Picosatellite Orbital Deployer) system [3–5].The most important aspect in the analysis of minisatellite design is the

I. Karabegović et al. (Eds.): NT 2022, LNNS 472, pp. 244–251, 2022.
https://doi.org/10.1007/978-3-031-05230-9_28

method of attitude control, which also allows for proper antenna pointing. The term attitude of a spacecraft means its orientation in space. The study of this aspect can be divided into three parts: determination, prediction, and control.In this article, the focus is only on control as it is the most critical one. The methods of controlling the attitude of a spacecraft are divided into two macro classes: passive and active. The first class includes those techniques that use the natural dynamics of the vehicle to satisfy orientation objectives. The second class includes methods based on the use of actuators, such as reaction wheels and thrusters. Actuators are very restricted by the environment and the low power usually available in space [6]. Some typologies of actuators can not use at all, such as direct current (DC) brush motors that do not work correctly in the vacuum, or shape memory alloys (SMA),which need cooling and have very low efficiency [7]. Pneumatic and hydraulic devices are also rarely considered, the former because of their high air consumption and low efficiency, the latter because of the need for heavy ancillary equipment. The actuators considered for most applications are limited to electric motors, electromagnetic and piezoelectric actuators [8]. Actuation systems are managed by control algorithms that receive data on generic body position from attitude determination sensors. Among the algorithms defined in the field of nanosatellites, the traditional algorithm is the proportional-integral-derivative (PID) controller [9, 10]. These minisatellites are becoming very popular for numerousapplications, such as extending internet access to the entire population of Earth or performing tasks previously performed by other satellites [11, 12].

2 System Description

The CubeSat consists of many components, some of which are essential for its proper operation. In this section, the focus is on the major components required to handle the basic tasks. The parts considered are the Antenna System, the Communication System, the Electric Power System (EPS), the Structure, the Onboard Computer, the Actuators, and the Sensors [13]. The antenna system consists of one or more spring-loaded ribbon antennas that deploy as they exit the P-POD. It operates on VHF/UHF frequencies. Once the antenna system is defined, the next step is to choose the communication system that allows the transmission and reception of information. The EPS consists of solar panels, batteries, and regulators. The mechanism of operation is simple, it performs in sequence a conversion from solar to electrical energy, storing it and then regulating the distribution of direct current to the other subsystems. When the vehicle is exposed to sunlight, part of the converted current will be directly used, and the rest will be stored. Conversely, if sunlight is not present, power is taken from the batteries [14]. The chosen structure is based on the geometry proposed by the CubeSat standard, that is, the 2-Unit CubeSat. Depending on the manufacturer, there is sometimes a switch in the structure, which allows the circuit to be opened during the launch phase when it is inserted into the P-POD, and then to close when it is put into orbit [15]. The choice of the on-board computer is directly linked to the control system. To have a compact and versatile solution, it was decided to use Cube ADCS from CubeSpace. The whole system is assembled using PC 104 cards compatible with the CubeSat standard. Before moving on to the chosen implementation system, an overview of the possible choices on the market is presented. The Attitude

Determination and Control System (ADCS) deflects, stabilizes, points, and rotates a satellite in the desired orientation despite external or internal disturbances. Typical for satellites in Low Earth Orbit, disturbance torques arise from aerodynamic and solar forces and magnetic moments. They are caused by poor wiring of the onboard electronic subsystems or current flowing in the solar panels between the connections of the various cells. There are also disturbances generated by solar radiation whose values depend on the absorption and specular reflection coefficients of the outer surfaces of the satellite.These torques are about two orders of magnitude smaller than the aerodynamic ones when in low orbit, and their influence can sometimes be neglected. To summarize, the list of ADCS actuators is: permanent magnets, magnetic torsions, impulse wheels, reaction wheels, propulsion systems, gravity gradient arms, and moment control gyroscopes [16]. Among these, the most common are explained as follows. Magnetic actuators use magnetic coils or bars to generate torques to control the satellite attitude and angular velocities. Magnetic torques can also be used to manage angular momentum build-up on reaction wheels. The torques generated are low and depend on the magnetic variation of the magnetic field of the Earth.Reaction wheels are used for each axle of the spacecraft to compensate for external disturbances and implement various commanded attitude maneuvers [17]. External torque disturbances can lead to the accumulation of wheel angular momentum and ultimate saturation [18]. Increasing angular momentum to saturation levels requires a desaturation strategy called momentum dumping. This is achieved using the magnetorquers and the thrusters. However, the torque delivered by each wheel does not depend on external factors. In this case, it is chosen to use a hybrid solution consisting of magnetorquers and reaction wheels. The system is assembled considering a three-axis stabilization requirement, so each axis is associated with a reaction wheel. By using the CubeWheel Small, the maximum torque that can be produced by each wheel is 0.23 (N*mm). A magnetometer is implemented for the attitude determination phase and can be connected directly to the board.The assumptions mentioned above are used for developing the dynamical model proposed in this paper.

3   Dynamical Model

To describe the motion of the satellite, a proper reference system must be defined. The reference system chosen is the orbit reference coordinate (ORC) frame shown in Fig. 1.

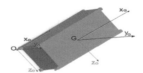

Fig. 1. Satellite orbit reference system

It is centered in the center of mass of the satellite and has the z-axis pointing towards the center of the Earth, the x-axis pointing in the direction of motion tangent to the orbit, and the y-axis normal to the orbital plane. The Euler angles, following the Tait-Bryan sequence, can be used to evaluate theattitude of the satellite. The roll, pitch, and yaw angles describe the transition, from the initial to the final configuration, of the satellite body. The roll angle θ describes rotations around the x-axis, the pitch angle ψ describes spins around the y-axis, and the yaw angle φ describes turns around the z-axis. Thus, the RPY (Roll, Pitch, and Yaw) sequence is used.Currently, virtual prototyping of the system is not yet fully defined. Thus, to carry out a first dynamical simulation, it is considered appropriate to use the system maximum mass prescribed by the CubeSat standard, which is equal to 2.66 (kg) [19]. Taking into consideration all bodies with the same density, the center of mass is calculated using SIMSCAPE-MULTIBODY [20]. The position of the center of masswith respect to the fixed triad represented in Fig. 1 is given by the vector [0.099 (m), 0.051 (m), 0.052 (m)]. The system is considered as a rigid body constrained at the center of mass by a gimbal joint. The dynamical model of the system under study has only three degrees of freedom, identified by the roll, pitch, and yaw angular displacements of the satellite chassis, and is given by $\mathbf{M\ddot{q}} = \mathbf{Q}_b$, where $\mathbf{M}$ is the system mass matrix, $\ddot{q}$ is the generalized acceleration vector, and $\mathbf{Q}_b$ is the total body generalized force vector.

4   Results and Discussion

Defined the case study, the simulation in inverse dynamics is carried out considering the environment as zero gravity and in the absence of any drag. This simulation is meant to calculate feedforward torques by imposing three cubic laws of motion, one for each degree of freedom of the system, that allow the system to move from an initial configuration, given by the coordinate vector [0 (deg), 0 (deg), 0 (deg)], to a final configuration, given bythe coordinate vector [45 (deg), 60 (deg), 30 (deg)], using a simulation time of 150(s) [21]. In Fig. 2, the results obtained from the inverse dynamic are represented.

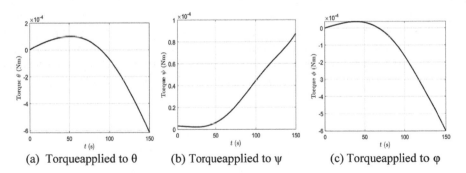

(a) Torqueapplied to θ (b) Torqueapplied to ψ (c) Torqueapplied to φ

Fig. 2. Torque in feedforward

The next step is to calculate the forward dynamics using the same system. Also, an appropriate damping coefficient equal to 0.0003 (Nm/deg/s)is inserted as a disturbance element to simulate the drag [22]. Figure 3 and Fig. 4 show the system response when feedforward torque is applied in input to the multibody system.

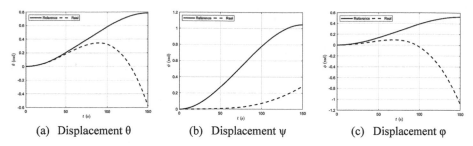

Fig. 3. Angular displacement without the PDcontroller

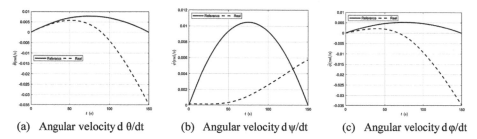

Fig. 4. Angular velocity without the PD controller

The analysis of the results obtained reveals the need to include a feedback controller for each of the three trajectories that the satellite must track. This need arises from the fact that the trajectory imposed to the system in inverse dynamics and that exhibited by the system in forward dynamics do not coincide. It is possible to evaluate an error function that can be calculated as the difference between the reference trajectory and the real one [23]. This work proposes the use of a feedback controller to counteract this error. In particular, the PD controller is chosen, whose starting parameters are computed with Ziegler-Nichols criterion, and then iteratively improved [24, 25]. Thus, the proportional and derivative parametersare tuned for each trajectory to be observed.In particular, these are: [0.5, 0.02] for θ; [0.7, 0.07] for ψ; and [0.35, 0.05] for φ. Using three separate controllers, the corresponding numerical results are shown in Fig. 5 and in Fig. 6.

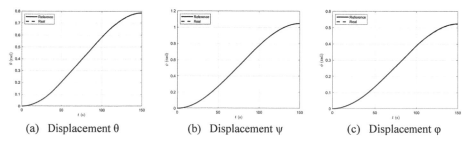

Fig. 5. Angular displacement with the PD controller

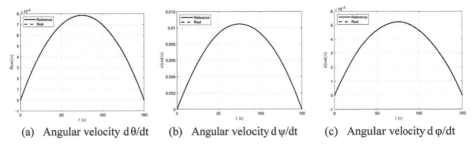

(a) Angular velocity d θ/dt   (b) Angular velocity d ψ/dt   (c) Angular velocity d φ/dt

Fig. 6. Angular velocity with the PD controller

Finally, the totalcontrol torques obtained as the sum of the feedforward and feedbackaction terms are shown in Fig. 7.

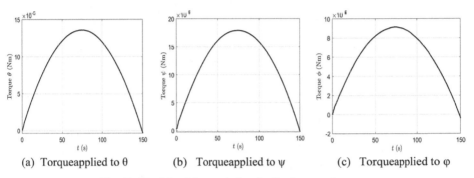

(a) Torqueapplied to θ   (b) Torqueapplied to ψ   (c) Torqueapplied to φ

Fig. 7. Total feed forward plus feedback control torque

5   Conclusions and Future Work

The authors already covered several areas of interest related to this work, which are multibody system dynamics applied to machines and mechanisms, nonlinear control, and system identification of structural systems [28–30]. The starting point of this work aims to highlight the key aspects and critical issues of attitude control of a minisatellite. The essential components for the survival of the minisatellite in LEO have been outlined, with particular attention to the actuator, which made possible the attitude control, the final goal of the current research. The theoretical introduction to satellite construction can be defined as the first step for future virtual prototyping [26]. Future research work will be devoted to further refining the simulation by introducing more disturbances and improving the control algorithm proposed in this paper to avoid errors. To this end, the authors plan to include a controller based on neural networks [27].

References

1. Dašić, P.: Scientific and technological trends: Selected scientific-professional papers (in Serbian). Vrnjačka Banja: SaTCIP Publesher Ltd., 2020. – 305 str. ISBN 978-86-6075-072-5

2. Dašić, P., Dašić, J., Antanasković, D., Pavićević, N.: Statistical analysis and modeling of global innovation index (GII) of Serbia. In: Karabegović, I. (ed.) NT 2020. LNNS, vol. 128, pp. 515–521. Springer, Cham (2020). https://doi.org/10.1007/978-3-030-46817-0_59

3. Formato, A., Ianniello, D., Pellegrino, A., Villecco, F.: Vibration-based experimental identification of the elastic moduli using plate specimens of the olive tree, Machines 7(2), art. no. 46 (2019). https://doi.org/10.3390/machines7020046

4. De Simone, M.C., Guida, D.: Experimental investigation on structural vibrations by a new shaking table, Lecture Notes in Mechanical Engineering, pp. 819–831 (2020)

5. Liguori, A., Formato, A., Pellegrino, A., Villecco, F.: Study of Tank Containers for Foodstuffs. Machines, vol. 9, art. no. 44 (2021)

6. Tošović, R., Dašić, P., Ristović, I.: Sustainable use of metallic mineral resources of Serbia from an environmental perspective. Environ. Eng. Manage. J. (EEMJ) 15(9), 2075–2084 (2016). ISSN 1582-9596. https://doi.org/10.30638/eemj.2016.224

7. Sun, X., Liu, H., Song, W., Villecco, F.: Modeling of eddy current welding of rail: three-dimensional simulation. Entropy, vol 22, art. no.947 (2020)

8. Dašić, P.: Response surface methodology: selected scientific-professional papers (in Serbian, Slovenian and Russian). Vrnjačka Banja: SaTCIP Publisher Ltd., 2019. – 301 str. ISBN 978-86-6075-054-1

9. Dašić, P.: Reliability of technical systems: selected scientific-professional papers (in Serbian and Russian). Vrnjačka Banja: SaTCIP Publisher Ltd. (2019). – 308 str. ISBN 978-86-6075-051-0

10. Pappalardo, C.M., Guida, D.: Dynamic analysis and control design of kinematically-driven multibody mechanical systems. Eng. Lett. 28(4), 1125–1144 (2020)

11. Dašić, P., Dašić, J., Crvenković, B.: Applications of access control as a service for software security. Int. J. Ind. Eng. Manage. (IJIEM) 7(3), 111–116 (2016). ISSN 2217-2661

12. Celenta, G., De Simone, M.C.: Retrofitting techniques for agricultural machines. Lecture Notes in Networks and Systems, vol. 128. LNNS, pp. 388–396 (2020)

13. De Simone, M.C., Rivera, Z.B., Guida, D.: Obstacle avoidance system for unmanned ground vehicles by using ultrasonic sensors. Machines 6(2), art. no. 18, (2018)

14. Sicilia, M., De Simone, M.C.: Development of an Energy Recovery Device Based on the Dynamics of a Semi-trailer. Lecture Notes in Mechanical Engineering, pp. 74–84 (2020)

15. Colucci, F., De Simone, M.C., Guida, D.: TLD design and development for vibration mitigation in structures. Lecture Notes in Networks and Systems, vol. 76, pp. 59–72 (2020). Kyd

16. Rivera, Z.B., De Simone, M.C., Guida, D.: Unmanned ground vehicle modelling in Gazebo/ROS-based environments. Machines 7 (2), art. no. 42 (2019)

17. Liguori, A., Armentani, E., Bertocco, A., Formato, A., Pellegrino, A., Villecco, F.: Noise reduction in spur gear systems. Entropy 22, 130 (2020)

18. De Simone, M.C., Guida, D.: Modal coupling in presence of dry friction, *Machines*, 6 (1), art. no. 8, (2018)

19. L. Salvati, M. d'Amore, A. Fiorentino, A. Pellegrino, P Sena, F. Villecco (2020) Development and Testing of a Methodology for the Assessment of Acceptability Systems. Machines, vol. 8, no. 47

20. De Simone, M.C., Guida, D.: Control design for an under-actuated UAV model. FME Trans. 46(4), 443–452 (2018)

21. Guida, R., De Simone, M.C., Dašić, P., Guida, D.: Modeling techniques for kinematic analysis of a six-axis robotic arm. IOP Conf. Ser. Mater. Sci. Eng. 568(1), art. no. 012115 (2019)

22. De Simone, M.C., Ventura, G., Lorusso, A., Guida, D.: Attitude controller design for microsatellites. In: Lecture Notes in Networks and Systems, 233, pp. 21-31. (2021). doi: https://doi.org/10.1007/978-3-030-75275-0_2

23. Li, T., Kou, Z., Wu, J., Yahya, W., Villecco, F.: Multipoint Optimal Minimum Entropy Decon-volution Adjusted for Automatic Fault Diagnosis of Hoist Bearing, Shock and Vibration, vol. 2021, art. no. 6614633 (2021)
24. Villecco, F., Aquino, R.P., Calabrò, V., Corrente, M.I., Grasso, A., Naddeo,V.: Fuzzy-assisted ultrafiltration of wastewater from milk industries. In: Naddeo, V., Balakrishnan, M., Choo, K.H. (eds.) Frontiers in Water-Energy-Nexus—Nature-Based Solutions, pp. 239–242. Springer, Cham (2020)
25. Salvati, L., d'Amore, M., Fiorentino, A., Pellegrino, A., Sena, P., Villecco, F.: On-road detection of driver fatigue and drowsiness during medium-distance journeys. Entropy **23**(2), 135 (2021)
26. Pappalardo, C.M., Lombardi, N., Guida, D.: A model-based system engineering approach for the virtual prototyping of an electric vehicle of class l7. Eng. Lett. **28**(1), 215–234 (2020)
27. Manrique Escobar, C.A., Pappalardo, C.M., Guida, D.: A parametric study of a deep rein-forcement learning control system applied to the swing-up problem of the cart-pole. Appl. Sci. **10**(24), 9013 (2020)
28. Pappalardo, C.M., Manca, A., Guida, D.: A Combined use of the multibody system approach and the finite element analysis for the structural redesign and the topology optimization of the latching component of an aircraft hatch door. IAENG Int. J. Appl. Math. **51**(1), 175–191 (2021)
29. Formato, A., Ianniello, D., Romano, R., Pellegrino, A., Villecco, F.: Design and Development of a New Press for Grape Marc. Machines **7**(3), 51 (2019)
30. Formato, A., Romano, R., Villecco, F.: A novel device for the soil sterilizing in sustainable agriculture. In: Karabegović, I. (ed.) NT 2021. LNNS, vol. 233, pp. 858–865. Springer, Cham (2021). https://doi.org/10.1007/978-3-030-75275-0_94

Controller Software Optimization in Adaptive Extreme Automation Systems

Sergii Tymchuk[1], Ivan Abramenko[1], Vira Shendryk[2(✉)], Sergii Shendryk[3], and Oleksii Piskarev[1]

[1] Kharkiv Petro Vasylenko National Technical University of Agriculture, Kharkiv, Ukraine
[2] Sumy State University, Sumy, Ukraine
v.shendryk@cs.sumdu.edu.ua
[3] Sumy National Agrarian University, Sumy, Ukraine

Abstract. The paper is devoted to the analysis of algorithms for finding optimal solutions for nonlinear functions with several variables. Often microprocessors in process automation systems monitor and perform calculations to find the extremum of such functions in order to generate a control signal that will be transmitted to the executive. The algorithm of formation of control signals in the adaptive microprocessor system of automation of technological process having the extreme function of the purpose is improved. The following is proposed to ensure high control efficiency with a minimum amount of measurement information. Algorithm for changing the size of a simplex while maintaining its regularity, taking into account the sign of the criterion function at the search stage. The number of steps in the observation phase in which at least one of the previous vertices remains intact. A computational experiment confirmed the effectiveness. Modeling of the process of finding the extremum of the criterion function of the two control influences showed the high efficiency of the proposed algorithm for the development of control microcontroller software.

Keywords: Software · Extreme control system · Control algorithm · Sequential simplex method · Modelling

1 Introduction

To increase the efficiency of production activities, automation is subject to complex technological processes, the management of which is to optimize the time-drifting extreme function of several variables. Information support of the relevant automation systems consists of several stages: obtaining the appropriate measurement information, information processing in the microprocessor controller, and the issuance of current control signals to the actuators. These operations must be performed in real-time and therefore the optimization of control algorithms and the search for optimal values of the extreme control function at a particular time is a very important task. The criterion of time requires finding optimal solutions in the shortest possible calculation time, so it is necessary to improve optimization algorithms, adapt them to the design capabilities of microcontrollers.

I. Karabegović et al. (Eds.): NT 2022, LNNS 472, pp. 252–259, 2022.
https://doi.org/10.1007/978-3-031-05230-9_29

Algorithms for determining the maximum function of several control variables can be divided into three groups depending on what characteristics of the criterion function are used in them to select the size and direction of the search step [1, 2]. The first group consists of zero-order algorithms or direct search methods that use only direct calculations of the values of the function. The second group consists of first-order algorithms that use, in addition to the values of the criterion function and the values of its first derivatives. Finally, the third group includes second-order algorithms based on the calculation of the values of the criterion function, as well as its first and second derivatives.

Based on the analysis of static and dynamic properties of control objects, the appearance of the surface of criterion functions, as well as based on the requirement of maximum simplicity of software implementation in control algorithms should use direct multi-dimensional search methods. All these methods are based on the organization of the iterative process of a sequential approach to the extremum of the unimodal function of several variables within a given region of uncertainty [3].

2 Literature Review

A review of the literature has shown that currently there are several different zero-order algorithms for determining the extremum of nonlinear functions of several variables [2–5].

In the method of direct search (Hook-Jeeves) first set some starting point $[x_1, ...x_n]$, then explore the surroundings of this point and find the direction where the greatest change in the criterion function. In the selected direction, the movement is carried out until there is a change in the values of the function, then a new survey is conducted around, etc. In general, this algorithm is characterized by ease of implementation of computational procedures, but also low accuracy.

In the method of coordinate descent (Gauss-Seidel) coordinate vectors are used as search directions. In this case, the search is conducted in the vector field of directions $\overline{d}_j$ $(j = 1, \cdots , n)$, where n - the number of controls. This method has a disadvantage, which is the low speed of finding the optimum and low accuracy of finding the result.

More advanced, but also more complex is the method of rotating coordinates (Rosenbrock method), which consists in organizing the rotation of the coordinate system so that at each iteration one of the coordinates corresponds to the direction of the fastest change in the criterion function. This method also requires more computing power than the previous ones.

A common disadvantage of these methods is that in the case of "ravine", strongly curved isolines of the objective function, these methods may be unable to provide progress to the optimum [6, 7]. Thus, the calculation is "stuck" in the "ravine".

Therefore, the use of the deformed polyhedron method (Nelder-Mead complex-simplex method), which has the property of adapting to the topography of the criterion function, is considered more promising. This method consists in constructing a polyhedron for the criterion function $f(x)$ in k-dimensional space (a prerequisite is k $\leq$ 6, otherwise the speed and computational efficiency of the algorithm are lost), which contains a k + 1 vertex. In each of the vertices, the values of the function $f(x)$ are calculated, and the minimum of these values is determined, as well as the corresponding vertex. Then a new polyhedron is built, and the process is repeated.

In practice, a modification of the polyhedron deformation method, the sequential simplex method (SSM), has proved its worth [6–8]. This modification can significantly reduce computation time. The idea of this method for the case of the function of two independent variables is as follows. In the plane of arguments, the initial simplex is constructed, formed by three points that do not lie on one line (any triangle). Further movement of the simplex in space occurs by mirroring the vertices that have the minimum value of the criterion function. At the heart of this movement is the assertion that the direction of the gradient of the criterion function is on average close to the direction from the worst vertex through the center of gravity of the opposite face. Reaching the extremum, the simplex begins to rotate around the vertex that best meets the criterion of efficiency. This looping phenomenon for the case of two variables is that the newly obtained vertex of the last calculated simplex is excluded from consideration and returned to the formation of the previous simplex, this step can be used to determine the end of the search process.

The experience of using the sequential simplex method shows the following positive aspects [7–9]:

- search with the use of SSM does not require complex calculations, all its stages are clearly formalized;
- SSM algorithm combines trial and working search steps, which allows each calculation of the function to approach its extremum;
- the direction of further movement depends on the ratio of the values of the objective function in the vertices of the simplex, which requires only the establishment of the ranks of these values;
- the simplicity of consideration of restrictions - the top of a simplex which does not satisfy restrictions, is simply rejected;
- high efficiency when searching in difficult conditions when calculating the extremum of the "ravine" objective function.

The use of SSM to optimize the control function in specific automated systems has led to a number of refinements [10, 11]. The first modifications were to introduce various ways to change the size and shape of the simplex, which increased the speed of the simplex at the beginning of the search, as well as the accuracy of the extremum at the final stage. The result of further research were algorithms in which the speed of the simplex changed depending on the success of the previous step of the search. These modifications of SSM, due to their developed adaptive properties, have proven themselves to optimize complex functions, but the use of irregular simplex, as well as the dependence of the search step on measurement errors, reduce the effectiveness of these methods in the drift of the target function due to interference.

Thus, when the extremum drift of the criterion function depends on the characteristics and the presence of a high level of obstacles, to optimize them, it is advisable to use SSM in its modification, using only regular (equilateral) simplex or regular simplex, the size of which varies according to pre-known law. However, when using a strictly constant simplex, one can expect either an increase in search time (exit of the system to the extremum zone) when choosing a small simplex size, or unreasonably large loss losses (accidental risk in the vicinity of the target function extremum) with a large simplex size. Known algorithms SSM with a natural change in the size of the simplex while maintaining its regularity involve pre-setting the rule of its change and the total number of steps, which is impossible in conditions of a priori uncertainty.

The analysis shows the need for additional research on the possibility of using the SSM method to localize the extremum of the criterion function of the automation system in real-time with noisy measurements of control effects.

3 Research Methodology

To ensure high control efficiency with a minimum amount of measurement information, it is proposed to use an algorithm to change the size of the simplex while maintaining its regularity, taking into account the sign of the criterion function at the search stage and the number of steps at the observation stage.

Features of this algorithm are:

- after subtracting the values of the criterion function in the vertices of the original simplex analysis of their signs is performed and, if all values are positive (i.e., the system is in the range of acceptable modes), further movement occurs with reduced simplex size;
- when entering the region of extremum, which is determined by reaching a certain number of steps, which remains intact at least one previous vertex of the simplex, the size of the simplex is reduced to the value at which the extremum drift is monitored;
- at any decrease in the size of the simplex, the new test vertices lying on the edge opposite the remaining vertex, with the maximum value of the criterion function, are determined by linear approximation of the values of the function in the corresponding vertex of the base to decrease.

We give an example of practical use of the proposed algorithm for the criterion function of two parameters $\varnothing = f(X_1, X_2)$ [12]. First, you need to determine the parameters: the coordinates of the initial simplex, and the size of the edges at different stages of the search L_0, L_1, L_2.

The range of possible values of control variables can be determined based on the allowable modes of operation of the automation object.

The coordinates of the center of the initial simplex in natural units are determined by the formula.

$$\tilde{X}_{0,i} = \frac{X_{i,\max} + X_{i,\min}}{2}, \tag{1}$$

where $X_{i,\max}, X_{i,\min}$ - limit values of the i-control variable, $i = 1, 2$;
$\tilde{X}_{0,i}$ - coordinate of the center of gravity for the same variable.
We introduce the rationing of variables.

$$x_i = \frac{2X_i - X_{i,\max} - X_{i,\min}}{X_{i,\max} - X_{i,\min}}, \tag{2}$$

where X_i is the current value of the control variable in natural units.

Then the coordinates of the center of the original simplex in relative units according to (2) are zero.

The length of the edges of the initial simplex can be found from the expression for L_0.

$$L_0 \leq L = \frac{\overleftrightarrow{d}}{2+k}\sqrt{\frac{k(k+1)}{2}}, \tag{3}$$

where $\overleftrightarrow{d}$ - the range of change of control variables, equal to the difference of the limit values, expressed in relative units; k - the dimension of the simplex.

Using the properties of an equilateral triangle, the coordinates of the vertices of the original simplex with the center at the coordinate point can be found from the relations given in Table 1.

The magnitude of the edge of the simplex L_1 is found based on static characteristics $\Phi = f(X_1)$ and $\Phi = f(X_2)$, based on the range of changes in control effects, at which $\Phi > 1$.

Table 1. Initial matrix of control effects of SSM

№ vertices	Management influence	
	X_1	X_2
1	$X_{1,1} = 0$	$X_{2,1} = \frac{L_0}{2\sqrt{3}}$
2	$X_{1,2} = -\frac{L_0}{2}$	$X_{2,2} = -\frac{L_0}{\sqrt{3}}$
3	$X_{1,2} = \frac{L_0}{2}$	$X_{2,3} = -\frac{L_0}{\sqrt{3}}$

The required extremum tracking accuracy is provided by the appropriate rib size L_2, due to the ratio.

$$L_2 = \delta\sqrt{\frac{2k}{k+1}}, \tag{4}$$

where δ is the permissible error in determining the extremum point of the criterion function.

4 Results

To implement the proposed approach, an appropriate program of calculations in the MATLAB environment was developed.

During the simulation, the coordinates of the original simplex were assumed to be equal to: $L_0 = 0,4$; $L_1 = 0,2$; $L_2 = 0,06$; $X_{0,1} = [0; 0,128]$; $X_{0,2} = [-0,105; -0,06]$; $X_{0,3} = [-0,105; -0,06]$.

In addition, a component of measurement noise was added to the measurement signals, which was set following the uniform distribution law.

The choice of the sampling interval was made by Kotelnikov's theorem taking into account the margin factor, according to the method described in [13, 14], based on comparing the amplitudes of the harmonics of the maximum spectrum frequency in different components of the measured signal. Its value was $\Delta t = 2.4691\mathrm{e}{-04}\ s$.

Modeling of the process of finding the extremum of the criterion function of the two control influences showed the high efficiency of the proposed algorithm for the development of control microcontroller software. Changing controls during the search process is shown in Fig. 1.

From the analysis of Fig. 1 we can conclude that according to the appearance of the surface of the criterion function in the range of possible values of the control variables of the conditions of entry into the zone of permissible modes occurs in the first step. Therefore, after determining the values of the function in the three vertices of the initial simplex is the reduction of the edge of the simplex to the value $L_1 = 0, 2$.

The further search process with the edge occurs by reflecting the worst vertex of the simplex relative to the opposite face and continues 8 times. The next reduction in

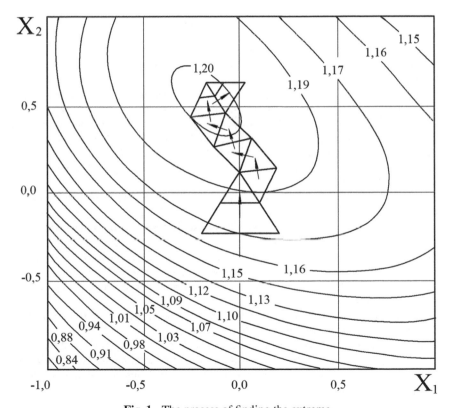

Fig. 1. The process of finding the extreme

the size of the edge of the simplex occurs to the value $L_2 = 0,06$. Thanks to what it is possible to trace criterion function in the field of an extremum for $n = 2$ steps.

5 Conclusion

The analysis revealed the most effective method of finding the extremum of the function of the two control variables. The selection and adaptation of an effective optimization method were performed to implement it later in the software of the control microcontroller of the process automation system. It was necessary to take into account that the microcontroller performs real-time control based on the information obtained from the measurement of the current state of the controlled variables. Therefore, the criteria for determining the effectiveness of the optimization method were low computing power, speed, sufficient accuracy with the existing noise of the control signals. To confirm the efficiency of the computational experiment was performed. Calculations showed that the use of a two-channel automatic optimization system using a modified simplex method can increase the control efficiency by an average of 7%. Specific numerical ratios of process characteristics depend on the nature and magnitude of the perturbations.

References

1. Vlasov, K.P., Anashkin, A.S.: Theory of automatic control. S.- Pb. St. Petersburg Mining Institute, 103 p. (2003)
2. Khalil Hassan, K.: Nonlinear Systems, USA, Pearson Education Limited, 560 p. (2013)
3. Winter, R.B.: Optimal control. Boston – Basel – Berlin, Burkhauser, 504 p., (2000)
4. Roy, V.F.: Synthesis of the algorithm for managing complex electricity consumers Lighting Engineering and Power Engineering, № 2, KNAMG, Kharkiv, pp. 74–78 (2009)
5. Vasiliev, F.P.: Optimization methods. Moskow, Factorial, 824 p. (2002)
6. Attetkov, A.B., Galkin, S.V., Zarubin, V.S.: Optimization methods. Moskow, Publ. N.E. Bauman MSTU, 440 p. (2003)
7. Clarke, F.: Necessary Conditions in Dynamic Optimization. Memoirs Amer. Math. Soc. **173**(816), 113 +(2005)
8. Dykhta, V., Lyapunov-Krotov, A.: Inequality and sufficient conditions in optimal control. J. Math. Sci. **121**(2), 2156–2177 (2004)
9. Vasiliev, O.V., Arguchintsev, A.V., Terletskiy, V.A.: Optimization methods for systems with lumped and distributed parameters based on admissible variations. Transactions of the 12th Baikal Int. conf. "Optimization methods and their applications." Plenary. Report, Irkutsk, pp. 52–68 (2001)
10. Srochko, V.A.: Modernization of gradient type methods in optimal control problems. Izv. Universities. Mathematics **12**, 66–78 (2002)
11. Aksyonov, E.P.: Optimal Decision Methods: textbook. Perm, IPC "Prokrost", 90 p. (2016)
12. Abramenko, I.G., Fyong, L.M., Vlasov, K.P.: Method of determining the efficiency criterion for systems of automatic optimization by flotation separation processes. Vestn. Kharkiv. Polytechnic Inst. **61**, 118–119 (1999)

13. Abramenko, I.G., Bovchalyuk, S.Y., Fomenko, V.O.: Determination of sampling intervals of time series of measurements of technological process parameters in ASC TP. Problems of energy supply and energy saving in agro-industrial complex of Ukraine: Bulletin of KhNTUSG. Professional edition, issue 196, Kharkiv, pp. 56–58 (2018)

14. Tymchuk, S., Abramenko, I., Zahumenna, K., Shendryk, S., Shendryk, V.: Determination of the sampling interval of time series of measurements for automation systems. In: Karabegović, I. (ed.) NT 2020. LNNS, vol. 128, pp. 478–483. Springer, Cham (2020). https://doi.org/10.1007/978-3-030-46817-0_55

The Establishment of an Advanced Brush Model for Simulation of Vehicle Dynamics

Zijah Jelkić[1]([✉]) and Boran Pikula[2]

[1] Veritas Automotive d.o.o., Rajlovaćka b.b., 71000 Sarajevo, Bosnia and Herzegovina
Zijah.Jelkic@veritas-ag.de
[2] Faculty of Mechanical Engineering, University in Sarajevo,
71000 Sarajevo, Bosnia and Herzegovina

Abstract. During many years of experiment, there have been used different tire models from purely experimental models with different approximation mathematical expressions, via physical models that take into account the important parameters of the tires and the tire contact patch, to FEM. However, these models can be inadequate for numerical simulations due to large CPU time consumption. In this paper, the Brush model and the proposed advanced Brush model is processed to define the forces of adhesion between the tire and the road, which is sufficiently simple and fast for simulations with sufficient accuracy. The advanced Brush model is defined by the most realistic contact pressure between the tire and the road which changes as a function of the force generated and the vertical load on the tire.

Keywords: Adhesion · Contact pressure · Vehicle dynamic · Simulation

1 Introduction

Modern vehicles are safer and at the same time more comfortable than vehicles manufactured several decades ago. The reason for this is the use of various electronic systems and their design using modern methods of analysis and simulation models. Therefore, modeling and simulation of the behavior of the tire becomes very important. Many simulation models can describe the formation of forces between the tire and road [1]. These models can be very simple, while some are very complicated and these are used to describe the interaction between the tire and the road rather than the definition of vehicle dynamics. As the tire is the only point of contact between the vehicle and the road, it is very important to properly define the forces created between the tire and the road. When modeling the vehicle dynamics, a simple and at the same time accurate model is required to describe vehicle behavior in dynamic conditions. One of the most known models, to define the force between the tires and the road, is the model of Pacejka "Magic" formula [2, 3] which has been created on the base of numerous experimental results. The main disadvantage of the Pacejka model is the need for a large number of parameters to achieve the desired accuracy and is not intended for dynamic conditions.

I. Karabegović et al. (Eds.): NT 2022, LNNS 472, pp. 260–273, 2022.
https://doi.org/10.1007/978-3-031-05230-9_30

To eliminate some of these deficiencies, the SWIFT model (Short Wave Length Intermediate Frequency model) was developed, which is based on the "Magic" formula [4, 5]. The SWIFT model belongs into dynamic models and is used for the analysis of the driving comfort, simulations that include systems that can surpass the dynamics of tires such as ABS, ESP, and analysis of the vibration in the range of 10–25 Hz.

TMeasy model is one of the semi-empirical models, where the emergence of forces and moments defines based on experimental results and observations of the relations of force and slip [6]. Semi-empirical models represent a compromise between the applicability, complexity, and efficiency in computing time on one hand, and the accuracy and visibility on the other hand. In semi-empirical models, the contact patch is flat, and the forces and moments are approximated with appropriate mathematical equations. The parameters of this equation are determined based on the results of experimental measurements. TMeasy doesn't offer as high accuracy as the "Magic" formula, but this model requires fewer parameters to define the forces and moments.

TreadSim model belongs to the group of the complex physical model [2, 6]. This model was developed to research other characteristics that cannot be present by analytical models. Some of these features are pressure distribution, the friction coefficient depending on the speed and pressure, the tire stiffness, longitudinal bending of the tire carcass. This model assumes that the tire is mounted on a rigid rim and is rolling on a flat surface without bumps. Forces and moments are defined simulating the deformation of tread elements running over the contact patch. It is assumed that the contact patch is a rectangular shape, the tire deformation depends on the vertical load, the tire pressure, and the speed which is determined empirically. The pressure distribution is assumed to be and trapezoid shape where one side of the trapezoid is a curve [7]. The shape of the contact pressure can be changed by using two parameters.

LuGre also belongs to the dynamic models, in which the forces and moments generated between the tire and the road are determined based on the friction parameter, which is defined in space and time, and the coordinate along with the contact patch [8–10]. In this model, it is assumed that the distribution of the contact pressure is an asymmetric trapezoidal shape. The main feature of this model is that it takes into account the hysteresis in the stress-deformation curve, where in the other models the change is linear. The model has a compact mathematical form of which has a relatively simple analytical solution for the quasi-stationary conditions. Other advantages of this model are computational efficiency and simple parameterization of the model to a model based on the physical characteristics of the tire. There are several extended versions of the model that take into account the variable sliding speed along the contact surface, combined longitudinal and lateral slip [8, 10].

Nowadays the FEM (Finite Element Method) models are a lot attractive [11, 12]. Among them is the FTire model (Flexible ring Tire model). It is used for analyzing ride comfort, handling, and road load prediction. FTire model is strictly based on physical appearance and characteristics between the tire and the road. Even though certain simplifications are unavoidable, this clean mechanical, thermo-dynamical, and tribological structure of the model guarantees a consistent and plausible model behavior. This model takes into account the most relevant sources, up to very high frequencies and extremely short wavelengths.

In addition to the empirical, complex physical, and FEM models, there can be added a group of simple physical models. The Dugoff, Fiala, and Brush models can be categorized in that group [2, 13, 14]. These models are among the first developed tire models for the definition of forces and moments. The main advantage of this model compared to other models is its simplicity. This simplicity of design is the key reason for the application of these models to use for the simulation of vehicle dynamics. The Brush model is perhaps the most interesting because it represents the most realistic characteristics of the tire and its interaction with the road. The disadvantage of this model is that the original form is designed for quasi-stationary conditions and is not suitable for dynamic conditions. The most important parameter of this model is the contact pressure distribution, which in the classic model is assumed as a symmetric parabolic.

2   The Brush Model

The Brush model is based on the assumption that between the carcass and the road are small elastic rubber parts that are attached to the carcass. That small elements represent the treads, and they represent the total elasticity of the pneumatic [2, 15]. Every tread can be deformed independently, and all that together reminds of a brush. The Brush model describes the formation of the forces and moments between the contact patch and the road so that the contact patch is divided into adhesion zone and slide zone.If the peripheral velocity, which is the product of the angular velocity of the tire ω and the dynamic radius of the tire r_d, is different from the translational velocity of the tire V_x, then longitudinal slip during braking process will appear [2]:

$$S_s = \frac{V_x - \omega r_d}{V_x} \tag{1}$$

or in the case of applying drive torque to the tire as the following:

$$S_s = \frac{\omega r_d - V_x}{\omega r_d} \tag{2}$$

The longitudinal and lateral slips can be defined as [2]:

$$\sigma_x = \frac{S_S}{1 + S_S}, \tag{3}$$

$$\sigma_y = \frac{tg\alpha}{1 + S_S}, \tag{4}$$

where S_s is the longitudinal slip and α is the slip angle.

The total, i.e. combined slip can be calculated as:

$$\sigma = \sqrt{\sigma_x^2 + \sigma_y^2}. \tag{5}$$

2.1 The Definition of Forces with the Classic Brush Model

The main characteristic of the classic Brush model is that the distribution of the vertical load F_z, respectively the contact pressure p, is symmetrical and parabolic between the tire and the road, and it does not change through the width of the contact patch. For simplicity, in many cases, the 2nd order parabola is used to describe the contact pressure in the classic Brush model [2, 15, 16]. The second assumption is the shape of the contact patch. In reality, the shape of the contact patch is rectangular, with rounded edges [16]. In this paper, it will be assumed that the shape of the contact patch is rectangular with sharp edges. This assumption won't have a big influence on the results because the total contact area is almost the same, but it greatly simplifies the calculations.

2.1.1 Pure Longitudinal Force

In the adhesion zone, the force depends on the tire stiffness and longitudinal elastic deformation, while in the sliding zone, the force depends on the contact pressure and the coefficient of friction, that is presented in the Fig. 1.

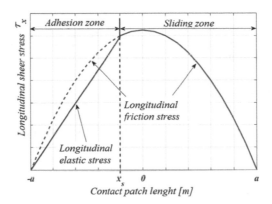

Fig. 1. Shear stress in the individual zones

The total longitudinal force can be expressed based on the generated shear stress on the contact patch between the tire and the road [16]. In the adhesion zone, shear stress occurs as the result of the tread deformation. The deformation changes linearly and depends on the deformation of the tread elements. In the same time, the shear stress in the sliding zone can be defined based on the contact pressure $p(x)$ and the coefficient of friction μ_x toward the x-axis. Because the distribution of the contact pressure is parabolic, the contact pressure can be described with the equation:

$$p(x) = \frac{3}{4} \frac{F_z}{wa} \left[1 - \left(\frac{x}{a}\right)^2 \right], \tag{6}$$

where F_z – is the vertical load, w – is the width of the contact patch, a = ½ of the contact patch length, x – longitudinal coordinate.

Finally, the total longitudinal force will be defined as:

$$F_x = \int_{-a}^{x_s} \frac{C_s S_s}{2a^2}(x+a)dx + \int_{x_s}^{a} \mu_x \frac{3}{4}\frac{F_z}{a}\left[1-\left(\frac{x}{a}\right)^2\right]dx, \tag{7}$$

were x_s is the point of separation of the two zones, C_s longitudinal tire stiffness and μ_x longitudinal friction coefficient. The point of separation x_s is where the line of the elastic stress and the line friction stress are intersecting (Fig. 1). Then the point of separation can be found equating the elastic stress and the friction stress:

$$\frac{C_s S_s}{w 2 a^2}(x+a) = \mu_x \frac{3}{4}\frac{F_z}{wa}\left[1-\left(\frac{x}{a}\right)^2\right]. \tag{8}$$

The solution of Eq. (8) gives the point of separation x_s:

$$x_s = a - \frac{2}{3}\frac{C_s S_s}{\mu_x F_z}a. \tag{9}$$

By introduction of a relative parameter:

$$x_n = \frac{x_s}{a}, \tag{10}$$

the final equation for the longitudinal force F_x can be now obtained from the Eq. (7) which depends on the tire stiffness C_s, longitudinal slip S_s, longitudinal friction coefficient μ_x, vertical load F_z and the relative point of separation x_n:

$$F_x = \frac{C_s S_s}{4}(x_n + 1)^2 + \mu_x F_z\left(\frac{1}{2} - \frac{3}{4}x_n + \frac{1}{4}x_n^3\right). \tag{11}$$

The longitudinal force, in the case when whole contact patch is sliding, can be calculated as:

$$F_x = \mu_x F_z. \tag{12}$$

2.1.2 Pure Lateral Force

The lateral force can be defined the same way as the longitudinal, based on the shear stresses in the zones of adhesion and sliding:

$$F_y = \int_{-a}^{x_s} \frac{C_\alpha S_\alpha}{2a^2}(x+a)dx + \int_{x_s}^{a} \mu_y \frac{3}{4}\frac{F_z}{a}\left[1-\left(\frac{x}{a}\right)^2\right]dx. \tag{13}$$

while the point of separation can be defined as:

$$x_n = 1 - \frac{2}{3}\frac{C_\alpha S_\alpha}{\mu_y F_z}. \tag{14}$$

The final equation for the lateral force F_y can be now obtained from the Eq. (13) which depends on the tire stiffness C_α, lateral slip S_α, lateral friction coefficient μ_y, vertical load F_z and the relative point of separation x_n:

$$F_y = \frac{C_\alpha S_\alpha}{4}(x_n + 1)^2 + \mu_y F_z \left(\frac{1}{2} - \frac{3}{4}x_n + \frac{1}{4}x_n^3\right). \tag{15}$$

Similar to the longitudinal force, in the case when whole contact patch is sliding, can be calculated as:

$$F_y = \mu_y F_z. \tag{16}$$

2.1.3 Combined Slip

In the case of a combined slip, it is necessary to define the resulting force F_r which is the result of the appearance of longitudinal and lateral slip. As the lateral and longitudinal coefficients of friction are equal, as well as the lateral and longitudinal tire stiffness, the resulting force F_r can be defined as:

$$F_r = \frac{C_e \sigma}{4}(x_n + 1)^2 + \mu_e F_z \left(\frac{1}{2} - \frac{3}{4}x_n + \frac{1}{4}x_n^3\right). \tag{17}$$

The relative point of separation is defined using the equivalent tire stiffness and equivalent coefficient of friction and using the combined slip. The non-dimensional length of the sliding zone is defined as previous based on the relative point of separation. The individual components, F_x and F_y, can be obtained based on the theoretical longitudinal and lateral slip:

$$F_x = \frac{\sigma_x}{\sigma}F_r, \tag{18}$$

$$F_y = \frac{\sigma_y}{\sigma}F_r. \tag{19}$$

3 Definition of the Forces with the Advanced Brush Model

During rolling of the tirein real conditions the pressure distribution is not symmetrical over the contact patch. That is shown in the Fig. 2.

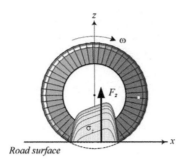

Fig. 2. Contact pressure distribution for rolling tire [16]

Such pressure distribution can be mathematically described as:

$$p(x, y) = \frac{49}{36} \frac{F_z}{wl} \left[1 - \left(\frac{x}{a} \right)^6 \right] \left[1 - \left(\frac{y}{b} \right)^6 \right] \left(1 - K^x \frac{x}{a} \right). \tag{20}$$

In Eq. (20) the parameter K^x represents the asymmetry coefficient that can be described with the equation:

$$K^x = tan\,\Theta = \frac{F_x}{F_z}. \tag{21}$$

The angle Θ represents the slope of the tangent at the origin of the resultant force F_r as it shown in the Fig. 3 a). The influence of the asymmetric coefficient on the contact pressure distribution is illustrated in the Fig. 3 b).

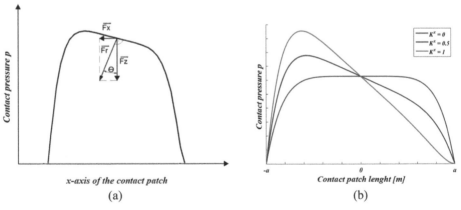

(a)   (b)

Fig. 3. Contact pressure distribution in case of formed longitudinal force Fxa) and illustration of the contact pressure change with the change of the asymmetric coefficient K^x b)

In the case of pure lateral force, the equation for the contact pressure will be slightly different, with the asymmetric coefficient for the lateral direction K^y:

$$p(x, y) = \frac{49}{36} \frac{F_z}{wl} \left[1 - \left(\frac{x}{a} \right)^6 \right] \left[1 - \left(\frac{y}{b} \right)^6 \right] \left(1 - K^y \frac{y}{b} \right). \tag{22}$$

The asymmetric coefficient for the lateral direction can be defined as:

$$K^y = tan\,\Theta = \frac{F_y}{F_z}. \tag{23}$$

The distributions of the contact pressure on the contact patch are illustrated in the Fig. 4, based on the Eqs. (20) and (22) and as such will be used for the advanced Brush model.

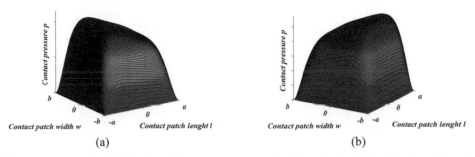

Fig. 4. Asymmetric distribution of the contact pressure in the longitudinal direction a) and lateral direction b)

3.1 Pure Longitudinal Force

The process of defining the pure longitudinal force is the same as for the classic Brush model. The contact patch is divided into two zones, the adhesion zone, and the slide zone Fig. 5. The main difference is the equation for the contact pressure. Based on that the longitudinal force can be defined as:

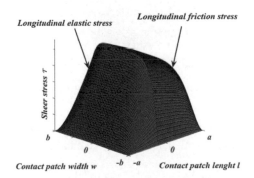

Fig. 5. Longitudinal sheer stress in particular zones

$$F_x = \int_{-b}^{b}\int_{-a}^{x_s} \frac{7}{3}\frac{C_S S_S}{wl^2}(x+a)\left[1-\left(\frac{y}{b}\right)^6\right]dxdy + \int_{-b}^{b}\int_{x_s}^{a}\frac{49}{36}\frac{F_z}{wl}\left[1-\left(\frac{x}{a}\right)^6\right]\left[1-\left(\frac{y}{b}\right)^6\right]\left(1-K^x\frac{x}{a}\right)dxdy \tag{24}$$

Because of the complexity, the equation for defining the point of separation is a polynomial of the sixth order:

$$K^x x_n^6 - (K^x + 1)x_n^5 + (K^x + 1)x_n^4 + (K^x + 1)x_n^3 + (K^x + 1)x_n^2 + (K^x + 1)x_n + 1 - \frac{6}{7}\frac{C_S S_S}{\mu_x F_z} = 0 \quad (25)$$

As such it's impossible to solve analytically but instead, there must be used numerical methods to find the solution of the Eq. (25).

The final equation of the longitudinal force can be written as:

$$F_x = \frac{C_S S_S}{4}(x_n + 1)^2 + \frac{7}{12}\mu_x F_z \cdot \left(\frac{6}{7} - \frac{3}{8}K^x - x_n + \frac{K^x}{2}x_n^2 + \frac{1}{7}x_n^7 - \frac{K^x}{8}x_n^8\right). \quad (26)$$

From Eq. (26) it can be conclude the longitudinal force depends on the longitudinal tire stiffness, longitudinal slip, longitudinal coefficient of friction, vertical load, and in this case from the asymmetric coefficient K^x. As in the classic Brush model, it is necessary to define the non-dimensional length of the slide zone to define when the Eq. (26) will be used.

3.2 Pure Lateral Force

In the case of pure lateral force, the Eq. (22) for the contact pressure will be used. The lateral shear stress in the particular zones is presented in the Fig. 6.

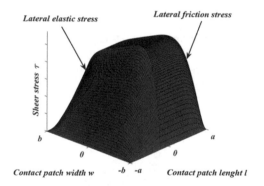

Lateral elastic stress   Lateral friction stress

Sheer stress τ

b

0   0

Contact patch width w   -b -a   Contact patch lenght l

Fig. 6. Lateral sheer stress in the particular zones

Because there is no longitudinal force, the contact pressure along the length of the contact patch will be symmetrical. The lateral force can be defined:

$$F_y = \int_{-b}^{b}\int_{-a}^{x_s} \frac{7}{3}\frac{C_\alpha S_\alpha}{wl^2}(x + a)\left[1 - \left(\frac{y}{b}\right)^6\right]\left(1 - K^y\frac{y}{b}\right)dxdy$$

$$+ \int_{-b}^{b}\int_{x_s}^{a} \mu_y \frac{49}{36}\frac{F_z}{wl}\left[1 - \left(\frac{x}{a}\right)^6\right]\left[1 - \left(\frac{y}{b}\right)^6\right]\left(1 - K^y\frac{y}{b}\right)dxdy \quad (27)$$

The relative point of separation therefore will be:

$$x_n^5 - x_n^4 + x_n^3 - x_n^2 + x_n - 1 + \frac{6}{7} \frac{C_\alpha S_\alpha}{\mu_y F_z} = 0. \tag{28}$$

Based on Eq. (27), the final equation for the lateral force will be:

$$F_y = \frac{C_\alpha S_\alpha}{4} (x_n + 1)^2 + \frac{7}{12} \mu_y F_z \left(\frac{6}{7} - x_n + \frac{1}{7} x_n^7 \right). \tag{29}$$

3.3 Combined Slip

For the combined or total slip, the same assumption for the equality of the lateral and longitudinal tire stiffness and coefficient of friction will be taken. Because of the longitudinal and lateral presence at the same time, the contact pressure will be as illustrated in the Fig. 7.

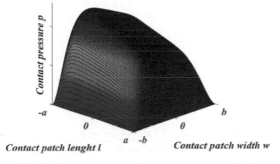

Fig. 7. Contact pressure in case of combined slip

The equation to describe the contact pressure will be:

$$p(x, y) = \frac{49}{36} \frac{F_z}{wl} \left[1 - \left(\frac{x}{a} \right)^6 \right] \left[1 - \left(\frac{y}{b} \right)^6 \right] \left(1 - K^x \frac{x}{a} \right) \left(1 - K^y \frac{y}{b} \right) \tag{30}$$

Based on the assumed contact pressure, the resultant force will be:

$$F_r = \frac{C_e \sigma}{4} (x_n + 1)^2 + \frac{7}{12} \mu_e F_z \left(\frac{6}{7} - \frac{3}{8} K^x - x_n + \frac{K^x}{2} x_n^2 + \frac{1}{7} x_n^7 - \frac{K^x}{8} x_n^8 \right) \tag{31}$$

The relative point of separation will be the solution of the Eq. (32):

$$K^x x_n^6 - (K^x + 1) x_n^5 + (K^x + 1) x_n^4 + (K^x + 1) x_n^3 + (K^x + 1) x_n^2 + (K^x + 1) x_n + 1 - \frac{6}{7} \frac{C_e \sigma}{\mu_e F_z} = 0. \tag{32}$$

Finally, the lateral and longitudinal forces can be separated based on the combined slip:

$$F_x = \frac{\sigma_x}{\sigma} F_r, \tag{33}$$

$$F_y = \frac{\sigma_y}{\sigma} F_r. \tag{34}$$

4 Results

The experimental results for the longitudinal and lateral forces, for different values of the vertical load, are shown in Fig. 8. For the classic Brush model Fig. 8a) and advanced Brush model Fig. 8b). The calculated values of the longitudinal force with the classic Brush model are shown in Fig. 8a). There is a very good matching with the experimental results. However, the calculated values with the advanced Brush model are shown in Fig. 8b), but they come with some deviation from the experimental results. The reason for this can be that the experimental results were measured in ideal laboratory conditions (quasi-stationary conditions). The advanced Brush model is designed strictly for dynamic conditions, where the classic Brush model is more for quasi-stationary conditions usable.

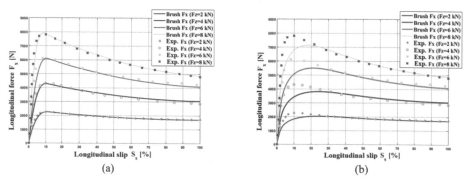

Fig. 8. Longitudinal force for the classic Brush model with the experimental results a) and for the advanced Brush model b) in function of longitudinal slip

The values of the longitudinal force F_x for the vertical load $F_z = 8$ kN are illustrated in the Fig. 9 for the case when the asymmetric coefficient takes the value $K^x = 0$ and $K^x = 1$. When $K^x = 0$, then the contact pressure distribution is symmetrical which describes

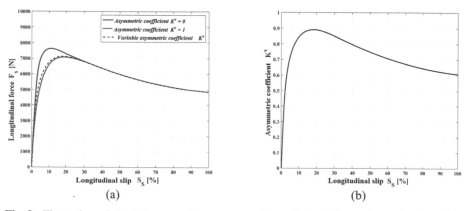

Fig. 9. Illustration of the influence of the asymmetric coefficient K^x on the results a) and the change of the asymmetric coefficient K^x during the change of longitudinal slip

the tire rolling in quasi-stationary conditions. In case when $K^x = 1$, then the vertical load is equal to the lateral force. Between these two boundaries is the curve when the asymmetric coefficient changes from min to max during the simulation.

The lateral force calculated with the classic Brush model is matching the experimental results for lower vertical loads, on higher values of vertical load it begins to deviate (Fig. 10 a)). The reason for that can be the lack of data for the experimental results. For the advanced Brush method, the deviation is even bigger, for the same reason as in the case of the longitudinal force (Fig. 10 b)).

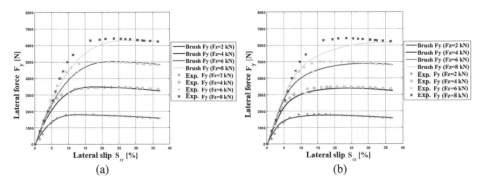

Fig. 10. Lateral force calculated with the classic Brush model a) and with the advanced Brush model b) in function of the lateral slip

Based on the defined interaction between the longitudinal and lateral forces for the classic and advanced Brush model, the forces can be calculated in the case of a combined slip. In this paper, 4 cases of combined slip were discussed, for the slip angles of $\alpha = 2°, 4°, 6°, 8°$.

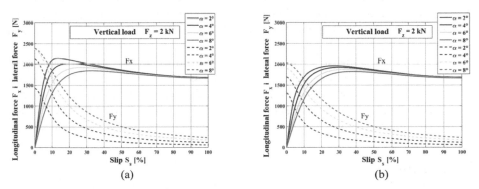

Fig. 11. Dependence of the longitudinal and lateral forces in function of the slip and slip angle for the classic a) and advanced b) Brush model

Based on the Fig. 11 and Fig. 12 it can be concluded that by rising the slip angle the lateral force begins to rise and the longitudinal force begins to decrease. In the case of the classic Brush model the longitudinal and lateral forces are closer to the friction circle (Fig. 12 a)) than in the case of the advanced Brush model (Fig. 12 b)).

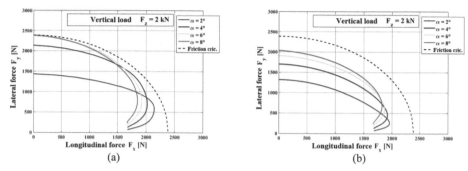

Fig. 12. Lateral force vs. longitudinal force in case of the classic a) and advance b) Brush model

5   Conclusion

The most influential parameter for the Brush model is the distribution of the contact pressure on the contact patch of the tire. The change of the contact pressure with the change of the load, respectively with the change of the formatted grip forces, was illustrated in a relatively simple way. Such assumption was introduced in the advanced Brush model with the asymmetric coefficient which depends on the vertical load and the formatted force.

This model is suitable for describing the formation of forces between the tire and the road in dynamic and extreme conditions which are not presented in the original Brush model. The advanced Brush model considers the dynamic changes between the load transitions in form of the asymmetric coefficient that was added in the model. This coefficient represents the more realistic impact of the load changes between the tire and the road.

The presented model is relatively light for simulations, accurate and is not time consuming. From that reason, it can be used for advances simulation of vehicle dynamics in different conditions and not just to keep track of the relations between the tire and the road. Also, it can be easily implemented in software like Matlab/SIMULINK to simulate the whole vehicle dynamics in situations like braking on surfaces with different traction.

References

1. Raymond, M., Brach, R.: Matthew brach: tire models for vehicle dynamic simulation and accident reconstruction. SAE, 1–2 (2009). ISSN 0148–7191
2. Pacejka, H.: Tyre and Vehicle Dynamics, Elsevier, 3rd edn. ,59–202 (2012). ISBN-13: 978–0–08-097016-5

3. Lu, C.-Y., Shih, M.-C.: Application of the pacejka magic formula tyre model on a study of a hydraulic anti-lock braking system for a light motorcycle vehicle system dynamics. Int. J. Vehicle Mech. Mobility **41**(6), 431–448 (2010). https://doi.org/10.1080/004231105123313 83848

4. Schmeitz, A.J.C., Besselink, I.J.M., de Hoogh, J., Nijmeijer, H.: Extending the Magic Formula and SWIFT tyre models for inflation pressure changes. Researchgate, 1–3 (2005). ISSN 0083-5560

5. Pauwelussen, J.P., Gootjes, L., Schroder, C., Kohne, K.-U., Jansen, S., Schmeitz, A.: Full vehicle ABS braking using the SWIFT rigid ring tyre model. Control Eng. Practice **11**(2) (2003). ISSN: 0967-0661:199-207

6. Uil, R.T.: Tyre models for steady-state vehicle handling analysis. Eindhoven University of Technology, DCT **2007**(142), 11–36 (2007)

7. Sjahdanulirwan, M., Yang, Q.: Prediction of tyre-road friction with an inverted-boat shaped pressure distribution. Int. J. Vehicle Mech. Mobility **24**, 145–161 (2007)

8. Deur, J., Ivanović, V., Troulis, M., Miano, C., Hrovat, D., Asgari, J.: Extensions of the LuGre tyre friction model related to variable slip speed along the contact patch length, Vehicle System Dynamics. Int. J. Veh. Mech. Mobility **43**, 508–524 (2011). https://doi.org/10.1080/004231 10500229808

9. Yamashita, H., Matsutani, Y., Sugiyama, H.: Longitudinal tire dynamics model for transient braking analysis: ANCF-LuGre tire model. J. Comput. Nonlinear Dyn. **10**(3), 1–11 (2015). CND-14–1109

10. Velenis, E., Tsiotras, P., Canudas-de-Wit, C., Sorine, M.: Dynamic tyre friction models for combined longitudinal and lateral vehicle motion, Vehicle System Dynamics. Int. J. Veh. Mech. Mobility **43**, 9–18 (2011). inria-00000921

11. Gipser, M.: FTire – the tire simulation model for all applications related to vehicle dynamics. Vehicle Syst. Dyn. Int. J. Veh. Mech. Mobility **45,** 139–151 (2008) https://doi.org/10.1080/ 00423110801899960

12. Tonuk, E., Samim Unlusoy, Y.: Prediction of automobile tire cornerong force characteristics by finite element modeling and analysis. Comput. Struct. **79**(13) (2001).https://doi.org/10. 1016/S0045-7949(01)00022-0:1219-1232

13. Rajamani, R.: Vehicle dynamics and control, 2nd edn. Springer (2012). ISSN 0941-5122,355-396

14. Kiebre, R.: Contribution to the modelling of aircraft tyre-road interaction, Universit´e de Haute Alsace - Mulhouse, tel-00601774, p. 8–29 (2011)

15. Svendenius, J: Tire Modeling and Friction Estimation, Department of Automatic Control Lund University, pp. 39–66 (2007). ISSN 0280-5316

16. Jazar, R.: Vehicle Dynamics – Theory and Application, 1st edn., pp. 95–161. Springer (2008). ISBN 978-0-387-7424-3

Innovative Burn Risk Assessment for Products with Metal Surfaces

Zorana Lanc$^{(\boxtimes)}$, Milan Zeljković, Miodrag Hadžistević, Branko Štrbac, and Miloš Ranisavljev

Faculty of Technical Sciences, Department of Production Engineering, University of Novi Sad, Trg Dositeja Obradovića, 21000 Novi Sad, Serbia
zoranalanc@uns.ac.rs

Abstract. Innovative burn risk assessment presented in this paper deals with hot metal surfaces of the product, which can be found at workplaces as well as households. Presented risk assessment enables quantification of the risk of burning by determinating the severity of burns depending on the degree of burns and the size of the hot surface. By using the IR camera for temperature measurement, it is possible to assess the risk of burning across the entire surface of the product and provide an IR image with the accurate location of the burn risk area. This is the most important advantage of innovative burn risk assessment and it is presented in this paper on a simple example of an iron.

Keywords: Burns · Risk assessment · Product · Metal surface · IR camera

1 Introduction

Knowledge of human thermal response to contact with hot solid surfaces has practical application in the design and assessment of products such as cookers, irons, toasters, kettles, and many others [1]. It should be considered that hot surfaces can cause burns and therefore they should be correctly assessed in order to eliminate or minimize the risk of burning. This is achieved by assessing the risk of burning caused by hot surfaces of products, according to the international standard EN ISO 13732–1: 2010. The standard has a couple of disadvantages. First, the standard only deals with first-degree burns, lacking the ability to assess the risk of second-and third-degree burn. Second, the standard doesn't take into account the size of the surface, one of the two most important factors for determinating the severity of the burn, besides the depth of damaged tissue. Thirdly, the standard doesn't quantify the risk of burning, but only provides its qualitative assessment. This paper presents innovative burn risk assessment for products, specifically with metal surfaces. The reason for this is that in order to determine the depth of burn, or as it is called degree of burn, two major factors have to be known: temperature and contact period. The exact temperature and contact period at which burn will occur and at which degree strongly depends on the type of material. Many studies have been conducted on live tissues using pigs, mouses, and rats to determine the depth of burn in regards to mentioned factors, where hot metals or hot water was predominantly used as the heat

© The Author(s), under exclusive license to Springer Nature Switzerland AG 2022
I. Karabegović et al. (Eds.): NT 2022, LNNS 472, pp. 274–282, 2022.
https://doi.org/10.1007/978-3-031-05230-9_31

source. Having collected all relevant data from these studies authors published their findings in research paper [2], explaining in detail how they determined temperature and contact period for first-, second-and third-degree burn arising from hot metal. In order to quantify the risk of burning three risk factors have to be known: the probability, frequency, and severity of the injury. The severity of the burn, as mentioned, depends on the burn degree and the size of the burn, which can be linked to the size of the hot surface in the way explained in mentioned research paper [2]. Knowing all the necessary information from previous work, such as burn depth and the size of hot metal surface it is possible to determine the severity of the burn and therefore quantify the risk of burning arising from the hot metal surfaces of products. Innovative burn risk assessment uses an IR camera for temperature measurement, enabling assessment across the entire surface of the product, unlike thermocouples suggested by the standard. This is the most important advantage of innovative burn risk assessment and it is demonstrated in this paper on the simple example of an iron.

2 Burn Risk Assessment

This chapter explains the differences between burn risk assessment according to EN ISO 13732-1 and innovative burn risk assessment for products with metal surfaces [3].

2.1 Burn Risk Assessment According to EN ISO 13732-1

In order to assess the risk of burning if the unprotected human skin comes into contact with hot surfaces, according to EN ISO 13732-1, the following procedure should be carried out:

1. identification of hot, touchable surfaces;
2. task analysis;
3. measurement of the surface temperature;
4. choice of applicable burn threshold value;
5. comparison of the surface temperature and the burn threshold;
6. determination of the risk of burning;
7. repetition of the assessment.

Identification of hot, touchable surfaces requires gathering information such as: accessibility of the surfaces; rough estimation of the surfaces temperatures; materials of which the surfaces consist; all operating conditions of the, etc.

In the task analysis information concerning the use of the product should be collected. In this step the following information is obtained: surfaces that are, or which may be, touched; intentional or unintentional touching; duration of contact with surfaces; persons of interest; probability of unintentional touching; frequency of intentional touching; actual range of power/temperature settings of the product during use.

Measurements of surface temperatures should be carried out on those parts of the product where contact of the skin with the surface can occur, by means of an electrical thermometer with a contact sensor made of metal and insignificant heat capacity. The accuracy of the instrument should be at least $\pm$ 1 °C in the range up to 50 °C and at least $\pm$ 2 °C in the range above 50 °C.

Choice of applicable burn threshold value depends on information concerning the contact period and surface material that are gathered in previous steps. From the results of the task analysis it can be deduced whether contact of the skin with a hot surface can occur unintentionally or intentionally and the category of person who comes or who might come into contact with the hot surface. The standard provides a table for recommended duration of contact period in regards to common situations for intentional and unintentional contact, age and physical abilities.

When it comes to selection of the burn threshold two most important information should be gathered: contact period and type of material. Using the contact period and the material of which the surface consists, the burn threshold value is either a value spread for short contact periods or a certain value for a longer contact period.

After selecting burn threshold value, the measured surface temperature should be compared with the burn threshold value. The following results are possible: the surface temperature is above the burn threshold; the surface temperature lies inside the burn threshold spreads or the surface temperature is below the burn threshold.

If the measured surface temperature is above the burn threshold, cutaneous injury upon contact with the hot surface is to be expected, i.e. there is a risk of burning. The risk of burning is all the greater: the higher is the measured surface temperature; the longer the period the surface temperature exceeds the burn threshold; the less the risk of burning is known to the person liable to be burned (e.g. children); the smaller the chance for counter-reaction; the more accessible the hot surface; the higher the contact risk in accordance with the intended use; the more frequently the contact is likely to occur; the smaller can be expected the previous knowledge of the user concerning safe handling of the product.

If the measured surface temperature lies inside the value spreads, cutaneous injury may or may not occur. This corresponds to the remaining uncertainty of the burn threshold specification. There is still a certain risk of burning which can be qualified similarly as in the case where measured surface temperature is above the burn threshold.

If the measured temperature lies below the burn threshold, the skin will not normally suffer injury. There is in general no risk of burning.

The assessment of the risk of burning should be carried out for all hot surfaces of the product which will be or can be touched during use. The assessment should be repeated, if the construction of the product is changed, the range of power/temperature settings of the product changes, the use of the product changes, or there is a change in any other circumstance which might lead to a different result in the assessment of the risk of burning.

2.2 Innovative Burn Risk Assessment for Products with Metal Surfaces

In order to assess the risk of burning according to innovative burn risk assessment for products with metal surfaces, the following procedure should be carried out:

1. identification of the hot metal surface(s) of the product;
2. task analysis;

3. IR temperature measurement of the hot metal surface(s) of the product;
4. burn degree;
5. severity of the burn;
6. determination of the risk of burning;
7. IR image(s) of the burn risk area;
8. repetition of the assessment.

Identification of the hot metal surface(s) of the product requires gathering the following information: type of metal of which the surfaces consist; operating conditions of the product especially the worst case, i.e. the case with the highest surface temperatures and the rough estimation of the size of the hot metal surface from the Table 2.

In the task analysis all necessary information concerning the use of the product should be collected. In this step the following information is obtained: metal surfaces that are, or which may be, touched; intentional or unintentional touching; duration of contact with hot metal surfaces; probability of unintentional touching; frequency of intentional touching. Information regarding intentional and unintentional contact as well as the duration of a contact period can be extracted from the standard. Innovative burn risk assessment is not applicable for children.

Measurements of surface temperatures should be carried out by means of an electrical thermometer with a contact sensor made of metal and insignificant heat capacity and an IR camera. The accuracy of the thermometer should be at least $\pm$ 1 °C in the range up to 50 °C and at least $\pm$ 2 °C in the range above 50 °C. As for the IR camera it is recommended to use the ones with uncooled infrared FPA detector. The accuracy of the IR camera should be at least $\pm$ 1 °C in the range up to 200 °C and at least $\pm$ 2 °C in the range above 200 °C.

After measuring the highest temperature and having the information on duration of a contact period the burn degree can be determined from Table 1.

Table 1. Burn degree in dependence of temperature and contact period for metals [2]

Contact period [s]	Temperature [°C]				
	50 - 60	60 - 70	70 - 80	80 - 100	100
1					
2 - 5					
5 - 12					
12 - 20					
20 - 60					
60 - 300					
300					

In the Table 1 burn degree are marked by colors. Green color represents combination of temperature of hot metal surface and contact period that lead to first-degree burn. Yellow color represents combination of temperature of hot metal surface and contact period that lead to second-degree burn. Red color represents combination of temperature of hot metal surface and contact period that lead to third-degree burn. Blue color represents safe combination of temperature of hot metal surface and contact period in regards to burns.

Knowing the approximate size of the hot metal surface as well as the burn degree, the severity of the burn can be determined from the Table 2.

Table 2. Burn severity risk matrix

Burn degree	Size of the hot metal surface [m²]			
	< 0.0346	0.0346 – 0.0865	0.0865 - 0.173	> 0.173
First-degree	minor	minor	minor	minor
Second-degree	minor	minor	moderate	major
Third-degree	minor	moderate	major	major

The risk of burning can be quantified using Kinney, AUVA, PILZ, or GURAD-MASER risk assessment. The mentioned risk assessments are suggested for the quantification of the risk of burning in two cases: when only one person uses the product and when more than one person is using the product at the same time. In the first case is recommended to use Kinney or AUVA risk assessment, whereas in the second case the PILZ and the GUARDMASTER risk assessment are more applicable.

From the previous step two scenarios are possible:

– there is no risk of burning, in which case the risk assessment is finished or
– there is risk of burning and IR image of a burn risk area should be presented.

Every IR camera has its program that enables finer adjustment of an IR images. One of the functions is creating isothermal areas on the IR images, meaning if the risk assessment from the previous step results in different levels of risks of burning, they can be easily shown on IR images in certain color in correspondence to temperature range. In this way isothermal areas on the IR images show the exact location of the burn risk area as well as its size.

The assessment should be repeated, if the construction of the product is changed, the range of power/temperature settings of the product changes, the use of the product changes, or there is a change in any other circumstance which might lead to a different result in the assessment of the risk of burning.

3   Comparison of Innovative Burn Risk Assessment and Burn Risk Assessment According to EN ISO 13732-1

The standard EN ISO 13732-1 describes the procedure for assessing the risk of burning on a simple example of an iron. For that reason, the authors decided to conduct innovative burn risk assessment on the iron with similar technical details and to show all the strengths of the proposed burn risk assessment. Both burn risk assessments deal only with hot metal surfaces of the iron.

3.1 Assessment of Risk of Burning for Metal Surfaces of the Iron According to EN ISO 13732-1

A flat iron is used to demonstrate the assessment of the risk of burning according to EN ISO 13732-1. On a flat iron three areas with different surface temperatures, different risks of burning and different possibilities of the application of protective measures can be distinguished: the soleplate, the handle and the intermediate area.

In this paper, we are dealing with hot metal surfaces of the product, therefore only the burn risk assessment for the soleplate will be explained in more detail below.

Following the risk assessment procedure according to Sect. 2.1 the following information regarding soleplate is collected.

a) Product information: accessibility - easily touchable; temperature estimation - very hot; surface material - steel, operating conditions - three selectable power stages.

b) Task analysis: surface which is or may be touched - soleplate surface; intentional or unintentional touching - unintentional; persons who contact or may contact - adults and children; duration of contact – 0.5 s for healthy adults, 4 s for children, 15 s for very young children; probability of unintentional touching – low during operation, higher during vertical storage of the flat iron and during vertical storage medium to high for people, who are not aware of the risk of burning, like young children; frequency of intentional touching - zero; actual range of power/temperature settings - maximum selectable power stage.

c) Measurement of surface temperature: the maximum measured temperature of the soleplate when the flat iron is operated at maximum power is 250 °C.

d) Choice of applicable burn threshold: the burn threshold spread for a smooth surface made of steel and a contact period of 0.5 s is 67 °C to 73 °C. These values apply if only adults have access to the flat iron. If also children have access burn spreads of 58 °C to 63 °C (4 s contact period) respective 55 °C to 59 °C (15 s contact period) apply.

e) Comparison and conclusion: the measured surface temperature is far above the applicable burn threshold spreads. Cutaneous injury upon contact with the hot soleplate is to be expected. This applies both for contact of adults and for contact of children with the soleplate surface.

f) Result of risk assessment: the temperature of the soleplate far exceeds the burn thresholds. A risk of burning therefore exists. The probability of contact with the heated soleplate is low when the flat iron is used by experienced adults at home or in the workplace. When inexperienced young children have access to a flat iron in the home the probability of a contact with the heated soleplate is medium-to-high. When the heated soleplate is touched by the unprotected skin severe injuries can be expected to occur. Altogether, the risk of burning is low-to-medium when the flat iron is used by experienced adults. The burning risk is medium-to-high when inexperienced children have access to a heated flat iron.

3.2   Innovative Assessment of Risk of Burning for Metal Surfaces of the Iron

A soleplate of a flat iron is used to demonstrate the innovative assessment of the risk of burning for metal surfaces of the product. Following the innovative risk assessment procedure according to Sect. 2.2 the following information regarding soleplate is collected.

a) Product information: accessibility - easily touchable; type of metal - steel, operating conditions - three selectable power stages; maximum operating temperature is 200 °C; size of the soleplate is less than 0.0346 m^2.

b) Task analysis: surface which is or may be touched - soleplate surface; intentional or unintentional touching - unintentional; persons of interest - adults; duration of contact – 0.5 s for healthy adults; probability of unintentional touching – low during operation, higher during vertical storage of the flat iron; frequency of unintentional touching - depends on situation, in households the frequency is weekly, wherein the case of professional usage at workplace is daily; frequency of intentional touching - zero.

c) Measurement of surface temperature: the maximum measured temperature of the soleplate when the flat iron is operated at maximum power is ~ 210 °C (Fig. 1).

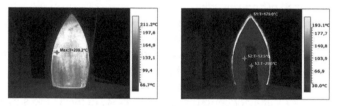

Fig. 1. IR images of the soleplate of an iron

d) Burn degree: since the innovative burn risk assessment deals only with temperatures up to 100 °C it is safe to say that when it comes to higher temperatures of 200 °C and greater, the third-degree burn occurs upon immediate contact with the hot metal surface.

e) Severity of the burn: as it can be seen from Table 2 for small size hot surfaces (<0.0346 m^2) there is only possibility of minor burn.

f) Risk assessment according to Kinney [4]: the probability of contact with hot soleplate of iron is unusual but possible (3); although there is only a risk of a minor burn, mostly because of the size of the hot soleplate, in worst case scenario the medical treatment by a doctor is needed and therefore the severity of an injury according to Kinney method is 2; in case of occasional (weekly) use of iron the frequency is 3, whereas in case of a regular (daily) use is 6. By multiplying the factors, two different levels of risk are obtained - negligible risk (18) in the case of weekly use of iron and low risk (36) in the case of daily use of iron.

g) IR image of burn risk area: there are three isotherms on Fig. 2. Yellow isotherm shows temperature from 66.7 °C to 101 °C, gray isotherm from 101 °C to 150 °C and red isotherm from 150 °C to 210 °C. As it can be seen from Fig. 2. the edges of the soleplate are the safest when it comes to risk of burning, where top of the iron and the bottom part have the highest risk of burning.

Fig. 2. IR image of the burn risk area

h) Results of innovative burn risk assessment: there are two levels of burn risk - negligible risk, where no action is required in case of weekly use of iron and low risk, where there is no need for additional activities in the management of the operation, but more cost-effective solution or improvement without additional investment should be considered. The probability of contact with the heated soleplate is low when the flat iron is used by experienced adults at home or in the workplace, and the risk is grater on the top and the bottom of the soleplate than on the edges of the soleplate. When the heated soleplate is touched by the unprotected skin minor burns can be expected to occur. Altogether, the risk of burning is negligible-to-low when the flat iron is used by experienced adults.

4 Conclusion

Presented innovative burn risk assessment for products with metal surfaces in this paper has the following advantages: quantification of the risk of burning, taking into account the size of the hot metal surface as well as the burn degree and visual interpretation of exact location and size of the burn risk area. The disadvantage of innovative burn risk assessment is that it can be used only for metal surfaces and it is not applicable for children, therefore further work is needed.

References

1. Parsons, K.: Human Thermal Environments: The Effects of Hot, Moderate, and Cold Environments on Human Health, Comfort, and Performance, 3rd edn. Taylor & Francis Group, Boca Raton (2014)

2. Lanc, Z., Zeljković, M., Hadžistević, M., Štrbac, B., Živković, A.: Granični uslovi za nastanak kontaktnih termičkih opekotina izazvanih vrućim metalnim površinama mašina. In: Proceedings: 13 International Scientific Conference ETIKUM, pp. 263–266, Novi Sad, Serbia (2021)
3. ISO 13732-1:2006 - Ergonomics of the thermal environment — Methods for the assessment of human responses to contact with surfaces — Part 1: Hot surfaces (2006)
4. Stanković, M., Stanković, V.: Comparative analysis of methods for risk assessment - 'Kinney' and 'Auva.' Saf. Eng. 3(3), 129–136 (2013)

Investigation of Properties of Tungsten Heavy Alloys for Special Purposes

Šaban Žuna[1]($\boxtimes$), Zijad Alibašić[1]($\boxtimes$), and Aida Imamović[2]($\boxtimes$)

[1] Center for Advanced Technologies in Sarajevo, St. Vilsonovo šetalište 9, 71 000 Sarajevo, Bosnia and Herzegovina
{saban.z,zijad.a}@cnt.ba
[2] Faculty of Metallurgy and Technology, University of Zenica, Travnička Cesta 1, 72 000 Zenica, Bosnia and Herzegovina
aida.imamovic@unze.ba

Abstract. Tungsten heavy alloys have wide applications where hardness, high density, high wear, and high-temperature resistance are required. A special use for tungsten heavy alloys are in military sectors. Within the advanced technologies in the production of tungsten heavy alloys, and own results of tungsten heavy alloys for special purposes are presented.

In this paper, two different tungsten alloys and their specifics from the aspect of chemical, mechanical and metallographic results are analyzed.

Keywords: Tungsten alloys · Haevy metals · Special purpose

1 Introduction

Tungsten heavy alloys have a wide range of applications and are an excellent material for gamma and X-ray insulation, i.e. they are used for radiation protection plates, for vibration dampers [1]. These alloys are also used for various products, such as bullet, shells, shrapnel head, grenades, rocket components, tank panzers, cannons, firearms, etc.

Tungsten alloys are a typical class of two-phase composites consisting of spherical tungsten based particles surrounded by a ductile phase of a lower melting point matrix consisting of Ni, Fe, Cu and Co [2], with very high melting point, a high density, a low coefficient of expansion and a high modulus of elasticity.

This paper presents recent results of research into the microstructure and mechanical properties of tungsten alloys for special purpose.

2 Basic Characteristics of Tungsten Heavy Alloys

According to ASTM B777-15 (2013) tungsten alloys are classified into four classes according to the nominal tungsten content. Table 1 shows the requirements for the composition (tungsten percentage), density and hardness of tungsten alloys.

I. Karabegović et al. (Eds.): NT 2022, LNNS 472, pp. 283–288, 2022.
https://doi.org/10.1007/978-3-031-05230-9_32

Table 1. Composition (tungsten percentage), density and hardness for tungsten alloys per ASTM B777-15

Class	W (mas.%)	Density (g/cm^3)	Hardness, HRC maks
1	90	16,85–17,30	32
2	92,5	17,15–17,85	33
3	95	17,75–18,35	34
4	97	18,25–18,85	35

Note: The hardness values given in the table apply to untreated (sintered or annealed) materials. For machined materials, hardness values can reach up to 49 HRC

Among the factors that most affect the mechanical properties of tungsten alloys are, in addition to the chemical composition, the production process [3] and the parameters temperature/time [4], particle size and shape [5], grain growth [6], the interface between particle/matrix distribution [7].

Aappropriate ratio of Ni and Coisrequired to increase the mechanical properties, in order to achieve appropriate values of tensile strength, ductility and hardness.

The addition of Mo reduces the concentration of W in the matrix durings intering, resulting in the finer structure of the alloys, as well as higher tensile strength and ductility [8].

The ratio of Ni and Co from 2 to 9 is required to increase the mechanical properties [2]. Such alloys have a high initial strength Rm, ductility and hardness, for example 91W-6Ni-3Co has Rm = 960 MPa, from A = 40%, hardness = 31 HRC in the initial, sintered state, so subsequent processing could increase the strength without maintaining a sufficient level of ductility.

The use of these alloys is limited to military applications, where a high level of mechanical properties is required.

3   Research Results

Microstructure and production parameters are keyfactors influencing the properties of heavy alloys.

Two samples of tungsten alloys were tested, in order to determine which quality class the materials of the samples belong to, as well as the microstructural and mechanical properties of these samples.

This paper presents the results of microstricture research on the example of two characteristic tungsten alloys, namely 91W-6Ni-1, 8Fe-1Co -sample marked W-I-91 and 93W-5Ni-1, 6Fe-0, 3Co-sample marked W-T-93.

3.1   Results of Chemical Analysis

For samples W-I-91 and W-T-93, chemical analysis tests were performed in the Kemal Kapetanović Institute, University of Zenica, and the results are presented in Tables 2 and 3.

Table 2. Chemical analysis of the W-I-91 sample [9]

Sample label	Content, %							
	Ni	Fe	Co	Si	Cr	Mn	S	W
W-I-91	6,00	1,81	1,05	0,11	0,05	0,03	0,002	rest

Table 3. Chemical analysis of the W-T-93 sample [9]

Sample label	Content, %							
	Ni	Fe	Co	Si	Cr	Mn	S	W
W-T-93	4,99	1,58	0,3	–	<0,002	0,04	-	rest

The chemical composition for the observed hard metal alloys corresponds to class 1 and class 2 according to ASTM B777-15, but with a two-component binder based on Ni-Co, i.e. W93-Ni6-Co1 or W91-Ni6-Co3 [9].

The results of basic mechanical properties for samples W-I-91 and W-T-93, which were obtained in the Institute "Kemal Kapetanović" University of Zenica, and relate to tensile strength, hardness and elongation are presented in Table 4.

Table 4. Mechanical properties for samples W-I-91 and W-T-93 [9]

Sample	Tensile strength, MPa	Elongation, %	Hardness, HRC
Sample W-I-91 (91W-6Ni-1,8Fe-1Co)	1485	10	48–50
SampleW-T-93 (93W-5Ni-1,6Fe-0,3Co)	900	18	37–40

The samples of the tested alloys W-I-91 and W-T-93 do not have a big difference in the hardness in the sintered state and with the application of advanced processing technologies the mechanical properties can be further improved.

3.2 Microstructure Test Results

Metallographic analysis of samples W-I-91 and W-T-93 was also performed in the Kemal Kapetanović Institute of the University of Zenica. Samples were analyzed in the transverse and longitudinal directions, Figs. 1, 2, 3, 4, 5, 6, 7 and 8. Sintered microstructure of the matrix of samples with round tung sten-based particles was observed in both samples.

The size of the tungsten-based particles in sample W-I-91 are uniform with average size about 30 μm.

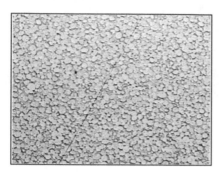

Fig. 1. W-I-91 sample microstructure, transversely, 150x [9]

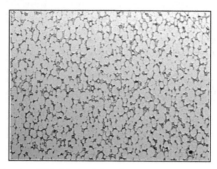

Fig. 2. W-I-91 sample microstructure, longitudinally, 150x [9]

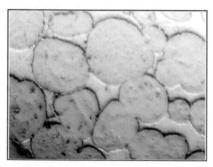

Fig. 3. W-I-91 sample microstructure, transversely, 1500x [9]

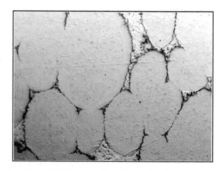

Fig. 4. W-I-91 sample microstructure, longitudinally, 1500x [9]

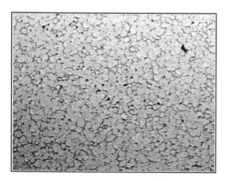

Fig. 5. W-T-93 sample microstructure, transversely,150x [9]

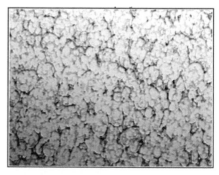

Fig. 6. W-T-93 sample microstructure, longitudinally, 150x [9]

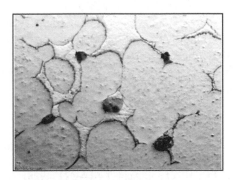

Fig. 7. W-T-93 sample microstructure, transversely, 1500x [9]

Fig. 8. W-T-93 sample microstructure, longitudinally, 1500x [9]

In the sample W-T-93, tungsten-based particles are significantly larger, and the size of individual particles exceeds 60 μm.

The results indicated:

- that the volume fraction of tungsten particles in sample W-I-91 is higher than in sample W-T-93,
- sample W-I-91 is characterized by tungsten-based particles that are more spherical in type than in sample W-T-93
- observation of sample W-I-91 in longitudinal section shows a certain degree of grain deformation, so it can be concluded that this sample suffered a certain degree of plastic deformation after sintering, and that it has less residual porosity (black parts) compared to sample W-T-93.

A higher level of mechanical properties of the sample materials W-I-91 could be achieved only by using powders of smaller granulation or correcting the chemical composition (for example replacing Fe with Co), otherwise by reducing the grain size with cold plastic deformation.

Additional increase in strength and hardness for W-T-93 can be achieved by post-sinter thermo mechanical treatment (extrusion) and rotary forging (swaging) and aging.

4 Conclusion

- The aim of alloying tungsten is to improve its chemical, physical, and mechanical properties combining the useful properties of tungsten heavy alloys.
- Sample W-I-91 (91W-6Ni-1,8Fe-1Co) has a tensile strength of 1485 MPa, a hardness of 48–50 HRC and an elongation of 10%.
- Sample W-T-93 (93W-5Ni-1,6Fe-0,3Co) has a tensile strength of 900 MPa, a hardness of 18 HRC and an elongation of 18%.
- Metallographic tests of samples W-I-91 and W-T-93 show significantly more spherical and finer tungsten-based particles comparising with W-I-91 alloy.

- Mechanical properties can be adjusted to some extent with chemical composition corrections, primarily with Ni/Fe and Fe/Co ratio, and additionally with post-sinter thermo mechanical treatment and plastic deformation.

References

1. Reiser, J., Rieth, M., Dafferner, B., Hoffmann, A.: Tungsten foil laminate for structural divertor applications - basics and outlook. J. Nucl. Mater. **423**(1–3), 1–8 (2012)
2. Caldwell, S.G.: Tungsten heavy alloys. Powder Metal Technol. Appl. **7**, 914–921 (1998)
3. German, R.M., Bose, A., Mani, S.S.: Sintering time and atmosphere influences on the microstructure and mechanical properties of tungsten heavy alloys. Metall. Trans. A **23**(1), 211–219 (1992)
4. Guo, W., Liu, J., Yang, J., Li, S.: Effect of initial temperature on dynamic recrystallization of tungsten and matrix within adiabatic shear band of tungsten heavy alloy. Mater. Sci. Eng., A **528**(19–20), 6248–6252 (2011)
5. German, R.M., Bose, A.: Fabrication of intermetallic matrix composites. Mater. Sci. Eng., A **107**, 107–116 (1989)
6. Lu, P., German, R.M.: Multiple grain growth events in liquid phase sintering. J. Mater. Sci. **36**(14), 3385–3394 (2001)
7. Demirskyi, D., Borodianska, H., Agrawal, D., Ragulya, A., Sakka, Y., Vasylkiv, O.: Peculiarities of the neck growth process during initial stage of spark-plasma, microwave and conventional sintering of WC spheres. J. Alloy. Compd. **523**, 1–10 (2012)
8. Lu, P., Xu, W., Yi, W., German, R.M.: Porosity effect on densification and shape distortion in liquid phase sintering. Mater. Sci. Eng., A **318**(1–2), 111–121 (2001)
9. Žuna, Š., Alibašić, Z., Nuhić, I.: Prethodno istraživanje legura tvrdih metala za KEP, Centar za napredne tehnologije u Sarajevu, pp. 28–39 (2021)

Complexity Measurement for Assembly of Personalized Products

Crnjac Žižić Marina[(✉)], Aljinović Amanda, Gjeldum Nikola, and Mladineo Marko

Faculty of Electrical Engineering, Mechanical Engineering and Naval Architecture, University of Split, Split, Croatia
mcrnjac@fesb.hr

Abstract. Today, industries need processes that rapidly meet customer needs and preferences. The assembly process is a breaking point that brings many challenges especially when a customer decides about the configuration of the product. This causes a high variety of information in the system and its uncertainty. Since complexity caused by personalized products affects assembly performances, research in recent years has focused on approaches to complexity measurement, such as information theory, axiomatic design and heuristic approaches. This paper illustrates the example of the application of the complexity measurement on the assembly process for personalized products. There is a discussion on the link between complexity measurement and process performances and the role of complexity measurement for managing processes. The research on complexity measurement had made it possible to detect the maximum complexity on the workstation to make changes if necessary. The importance of new technologies that enable the application of complexity measurement in the real industrial environment is emphasized.

Keywords: Complexity measurement · Information theory · Assembly · Personalized products · Internet of Things

1 Introduction

The general development of complexity theory in the literature can be observed through several approaches, thus creating an appropriate framework for the division of devel oped measures of complexity regardless of what complexity refers to. The complexity measures are developed to solve different problems in production or assembly systems. For example, there are the complexities of product [1, 2] or product family design [3, 4], the complexity of the entire production system [5, 6], the complexity of layout [7, 8] and many other problems. It is important how the authors define the complexity they study. The effort is made to apply complexity theory to real-world examples in practice. However, there is much more to be done, to use the information obtained by observing the complexities that appear in production or assembly systems for managing these systems. The paper structure is as follows. The second section of the paper presents a classification of complexity theories. The third section describes the complexity indicator and properties that the indicator satisfies. The fourth section discusses the role of complexity measurement for managing processes and the fifth is the conclusion.

I. Karabegović et al. (Eds.): NT 2022, LNNS 472, pp. 289–296, 2022.
https://doi.org/10.1007/978-3-031-05230-9_33

2   Development of Complexity Measures

Two important classifications in complexity theories emerge from the literature review. The first is related to the approaches for measurement of complexity in production systems, and the second is related to the classification of complexity depending on the domain.

Approaches for the measurement of complexity in the production systems are presented in Fig. 1. The information theory is given by Shannon. It is based on the theory of probability and statistics, [9]. Measuring the amount of information is obtained by calculating their entropy. Entropy is used to quantify the uncertainty that characterizes production or uses the information needed to describe the system components or their state. It can also be said that entropy serves to quantify the lack of knowledge about the system. Entropy can represent the natural tendency of the system to come to a state of complete chaos, which is a consequence of its complete disorganization. Kolmogorov's entropy is belonging to the theory of information and provides a calculation of the complexity rate by taking an analysis of the state of system components at different time intervals and considering the average increase in the complexity rate of a random set of measured data, [10]. This provides a measure of system dynamics that represents the evolution of the system over time. Nonlinear dynamics theory and chaos theory provide the basis for interpreting the nonlinearity, instability and unreliability which characterize systems with increasing complexity [11]. El Maraghy et al. have developed coding system where each complexity code represents the information needed to describe different areas of interest in production [12]. Other theories include axiomatic design and questionnaire research. In axiomatic design, Suh [13] defines complexity as a measure of unreliability in meeting functional requirements.

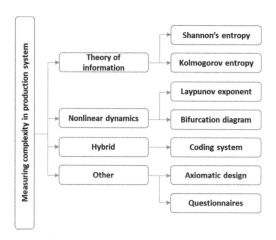

Fig. 1. Approaches for measurement of complexity in production system [14]

Domains of complexity theories are presented in Fig. 2, where the functional domain relies on axiomatic design mentioned before (depends on time), and the physical domain is divided into the static and dynamic domains. The static domain (also called structural) deals with structures, configurations, number of different components (related to the product, system…) and its interdependencies and interconnections. Static complexity can be viewed as a function of the structure of the system, connective patterns, variety of components and the strengths of interactions [15]. The dynamic domain concerns the uncertainty of the behaviour of the observed system over a period of time.

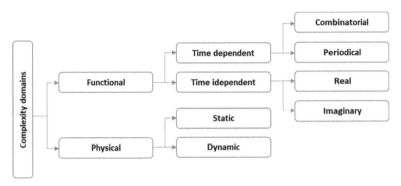

Fig. 2. Domains of complexity [13]

3 Complexity Indicator for Assembly of Personalized Products

3.1 Description of the Complexity Indicator for Assembly Workstation

The complexity measurement for the assembly of personalized products provides an opportunity to detect possible problems in the observed assembly processes. The complexity indicator developed by [16], serves to visualize the distribution of complexity at each assembly station, in order to do the allocation of workers so that quantities of assembled products can be completed and delivered on time. Shannon's entropy [9] was used to develop a complexity indicator. The expression for Shannon's entropy is:

$$H = -\sum_{s=1}^{N} P_s log_2 P_s \tag{1}$$

where:
N – is the number of states of the system.
P_s – is the probability that the observed system will be in the state s.
To adapt the expression (1), the proportion of time w_i^u spent on the workstation u for particular part, subassembly or product i during the assembly process is used instead of P_s defined above. A new expression for complexity indicator for the specific productis:

$$H_{i,u} = -\sum_{i\in PR} w_i^u log_2 w_i^u \tag{2}$$

where:

w_i^u – is the proportion of time spent on the workstation u for particular part, subassembly or product i within the assembly process or in warehouse, calculated as:

$$w_i^u = \frac{\chi_i^u}{\sum_{i \in PR} \chi_i^u} \tag{3}$$

The complexity indicator for assembly workstation represents the sum of these indicators for each product:

$$\sum_{i=1}^{N} H_i = -\sum_{i=1}^{N} \sum_{i \in PR} w_i^u log_2 w_i^u \tag{4}$$

3.2   The Complexity Metric: An Example of Pen Assembly

The assembly of two different pens is done for demonstration of the complexity indicator. Figure 3 shows pens and its differences so the first is called product "A" and the second is called product "B".

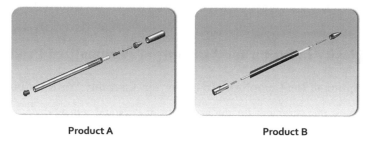

Product A   Product B

Fig. 3. Example of products for assembly

The assembly is done on the four workstations (WS), where two warehouses (W) supply workstations with necessary parts. The graph representation of assembly is shown in Fig. 4.

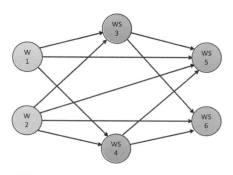

Fig. 4. Graph representation of assembly

The part flow in the assembly system is governed by the type of parts being produced and workstation capabilities. The data about the connections between workstations, the quantities of parts for product A and B, time for assembly or for delivery are in Table 1.

Table 1. Input data for the example of assembly

Relations between workstations	Quantity of parts for product A	Quantity of parts for product B	Time for assembly/delivery for product A	Time for assembly/delivery for product B
1–3	2	2	5	5
1–4	2		5	
1–5	1		5	
2–3	1	2	5	5
2–4		2		5
2–5	1		5	
2–6		1		5
3–5	1		10	
3–6		1		13
4–5	1		2	
4–6		1		2
5	1		11	
6		1		10

The data about the total time spent on the specific product in warehouses and on workstations is given in Table 2. That data serves to calculate the proportions of time for each product as the input data for the complexity indicator. In the last column of Table 2, the entropies for warehouses and workstations are calculated.

Deshmuk [15] discusses how the static complexity increases as the number of variety of components in the system and their interaction increases, which can be applied in this case. The assembly of one product is done to show the results of the complexity indicator, Table 3. Compared to the results of the complexity indicator in Table 2, it is visible when the number of parts increases, the complexity increases.

Table 2. Data about the complexity indicator when product A and B are assembled

WS	Assembly timefor product A [s]	Assembly time for product B [s]	Proportion of time product A	Proportion of time product B	Complexity indicator – Entropy [bit]
1	136,36	58,18	0,70	0,30	0,88
2	54,55	145,45	0,27	0,73	0,85
3	54,55	75,64	0,42	0,58	0,98
4	10,91	11,64	0,48	0,52	1
5	60		1		0
6		58,18		1	0

Table 3. Data about the complexity indicator when only product A is assembled

WS	Assembly time for product A [s]	Proportion of time product A [s]	Complexity indicator – Entropy [bit]
1	136,36	1	0
2	54,55	1	0
3	54,55	1	0
4	10,91	1	0
5	60	1	0

4   Complexity Measurement in Context of Industry 4.0

The performances of production system not only depend on the type of parts and the resources, but also on the control policy used [15]. This is the starting point for development of any complexity measure. The main aim is to use that measure in the control policies or managing production or assembly processes. As Industry 4.0 focuses on the smart products and processes, the digitalization is crucial po. With the development of complexity theories mentioned in Sect. 2, develops the Internet of things. This technology enables real-time data collection (i.e. in the assembly system) which is used for calculation of complexity with the complexity indicator. Today, widely used technology for track and tracing is the radio frequency identification (RFID) technology [17]. In theprevious case for assembly of pens, the RFID technology can be used to collect the data about the assembly time necessary at each workstation. With collection of this data, the complexities at each workstation can be calculated. The calculation of complexity indicators can be integrated in the existing information systems in the company so the person who manages the processes can use it for decision making.

5   Conclusion

The greatest value of defining a particular complexity measure is in finding the information on how and to what extent a particular complexity affects the performances in production or assembly processes. This is why future research will be focused on the connection of developed indicator with performances such as assembly cycle time or total assembly costs. To achieve this, it is inevitable to use new technologies characteristic for Industry 4.0, which enable the collection of necessary data (i.e. assembly time) used for complexity calculation which in the end serves for managing processes to achieve the best performances.

References

1. Alkan, B., Vera, D., Ahmad, B.: A method to assess assembly complexity of industrial products in early stage design phase. IEEE Access **6**, 989–999 (2017)
2. Modrak, V., Berdnar, S.: Using axiomatic design and entropy to measure complexity in mass customization. In: 9th International Conference on Axiomatic Design, vol. 34, pp. 87–92 (2015)
3. Park, K., Kremer, G.O.: Assessment of static complexity in design and manufactruing of a product family and its impact on manufacturing performance. Int. J. Prod. Econom. **169**, 215–232 (2015)
4. Wang, H., Zhu, X., Wang, H., Hu, S., Lin, Z., Chen, G.: Multi-objective optimization of product varety and manufacturing complexity in mixed-model assembly lines. J. Manuf. Syst. **30**, 16–27 (2011)
5. Samy, S., ElMaraghy, H.: A model for measuring complexity of automated and hybrid assembly systems. Int. J. Adv. Manuf. Technol. 813–833 (2012). https://doi.org/10.1007/s00170-011-3844-y
6. Schoettl, F., Paefgen, M., Lindemann, U.: Approach for measuring change-induced complexity based on the production architecture. In: Proceedings of the 4th CIRP Conference on Manufacturing Systems, pp. 1–6 (2014)
7. Espinoza, V., ElMaraghy, H., AlGeddaway, T., Samy, S.: Assessing the structural complexity of manufacturing systems layout. In: Proceedings of the 4th CIRP Conference on Assembly Technologies and Systems, pp. 65–70 (2012)
8. ElMaraghy, H., AlGeddawy, T., Samy, S., Espinoza, V.: A model for assessing the layout structural complexity of manufacturing systems. J. Manuf. Syst. **33**, 51–64 (2014)
9. Shannon, C.: A mathematical theory of communication. Bell Syst. Tech. J. **27**(3), 379–423 (1948)
10. Frizelle, G., Suhov, Y.: The measurement of complexity in production and other commercial systems. Proc. Royal Soc. London A: Math. Phys. Eng. Sci. **464**(2098), 2649–2668 (2008)
11. Kellert, S.: In the Wake of Chaos. The University of Chicago Press, Chicago (1993)
12. ElMaraghy, H., Kuzgunkaya, O., Urbanic, R.: Manufacturing systems configuration complexity. CIRP Ann. Manuf. Technol. **54**(1), 445–450 (2005)
13. Suh, N.: Complexity: Theory and Applications. University Press, Oxford (2005)
14. Efthymiou, K., Mourtzis, D., Pagoropoulos, A., Papakostas, N., Chryssolouris, G.: Manufacturing systems complexity analysis methods review. Int. J. Comput. Integr. Manuf. **29**, 1025–1044 (2016)
15. Deshmukh, A., Talavage, J., Barash, M.: Complexity in manufacturing systems, Part 1: analysis of. IIE Trans. **30**, 645–655 (1998)

16. Crnjac, M.: Decision support system modelling for personalized product assembly managing, Split: PhD thesis (2021)
17. Aljinovic, A., Gjeldum, N., Crnjac, M., Mladineo, M.: Analysis of performances of RFID systems for assembly line purposes in learning factory. In: Mechanical Technologies and Structural Materials 2019, Split, Croatia (2019)

Cellulose Filters in Industrial Processes and Compatibility with Ionic Liquids

Darko Lovrec[✉] and Vito Tič

Faculty of Mechanical Engineering, University of Maribor, Smetanova 17, 2000 Maribor, Slovenia
darko.lovrec@um.si

Abstract. Ionic liquids, combinations of ions and cations, offer many combinations of these. Therefore, there are many possibilities for the use of these liquids in different areas of industry, as they can be tailored to a specific purpose of use. Due certain special, excellent properties, they are already found today in various fields of industrial technical use, where they are declared as a high-tech fluid.

However, in many industrial processes, the liquid used must be filtered before use or during use. In the case of ionic liquids, the question arises as to the compatibility of the ionic liquid used with the material used as the filter element. In the presented case, we focused on the compatibility of ionic fluids used in hydraulic systems as hydraulic fluid and commonly used filter materials.

Keywords: Industrial processes · Cellulose filters · Ionic liquids · Hydraulic · Compatibility

1 Introduction

Filtration of gases or fluids is used in various fields of technology. This involves filtering the air, either ambient, compressed, sterile, steam filtration, or filtering of various technical gases. Filtration is even more common when using different fluids, either in connection with nutrients or in a variety of technological processes. Filters are used in the purification or filtration of water and soft drinks, in the dairy industry, in winemaking, in breweries, pharmacy, the food industry and elsewhere.

A wide variety of materials are used in these areas as filter materials. They can be made of polyester materials, fibreglass, or other materials. Very often we also encounter cellulose filters, especially when it comes to filtering different media in different technological processes. Cellulose filters are also used in cases where it is necessary to filter various lubricating oils [1].

In the case of using oil in the technical process, it is necessary to clean the oil in different cases. So, it is necessary to clean new or used oil in the transfer from storage to machine, or vice versa, to clean oils used in test runs, to remove the different contaminants after repair work, or maintain the required cleanliness level of lubricating oil. The latter is especially true for the commonly used hydraulic mineral based oils used within today's modern hydraulic systems. Filter cartridges of appropriate fineness are used for this

I. Karabegović et al. (Eds.): NT 2022, LNNS 472, pp. 297–307, 2022.
https://doi.org/10.1007/978-3-031-05230-9_34

purpose. In the case of hydraulic systems we use filter cartridges of appropriate fineness for this purpose, depending on the type and purpose of using the filter, e.g. a filter suitable for use on a pressure line, on a return line, or for bypass filtration [2].

At the same time, it is necessary to distinguish between surface filters and depth filters. This is the same in all cases of filtration of liquid lubricants, where we use mainly mineral oils, for example, in motor vehicles, lubrication of gears and bearings, in the case of hydrostatic guides, or within hydraulic systems [3].

2   Effective Filtering

Effective filtration of each medium used in technological processes and systems is undoubtedly a prerequisite for the long service life of all components, and indirectly the entire machine or device, or the purity of the medium is a prerequisite for use in a particular technical process [2].

Where we filter, depends on the technological process and the place of protection of the components. Based on this, we distinguish between filters at the media entry into the system, after the media entry into the system, or before a certain component in the system that requires higher cleanliness of the media. However, it can be filtered at the exit of the system into the environment, or before leaving the system back into the reservoir, for example, in the lubricating oil tank, so that it is so clean for reuse – circulation in the system [3].

By-pass filtration is an often used method of offline filtering, and is often fixed permanently to an application where the current filtration is insufficient. This is especially the case in harsh environments or under demanding workloads. This type of filtration is equally effective as a "quick cleanse" to alleviate dirt levels or resolve unexpected water contamination. In doing so, you should know what the target cleanliness of the system is (as mentioned, 3 microns for hydraulic oils in servo systems and 6 microns for gear lubrication oils), and if you have persistent moisture problems. In both cases, it is always best to determine the source of any contamination.

Effective filtering is always important, even in the case of "seemingly clean systems", which, on the other hand, may mask some problem that needs to be addressed first. Therefore, it is appropriate to determine the appropriate ISO cleanliness level in an appropriate way of oil sampling – the appropriate place for sampling and the method of sampling.

2.1   Cellulose Filter

Cellulose filters (usually in the form of a filter paper) are generally considered to be a versatile and diverse microfiltration agent that works by trapping particles in a random matrix of cellulosic fibres. Cellulose filter papers can be divided into quantitative or qualitative, depending on their use.

In addition, according to the manufacturers, cellulose filters are generally excellent in terms of separation effect, and they are considered economical filters because they can be removed in an environmentally friendly way; they are harmless from a medical point of view, and, thus, to the user, and they are reliable and very effective [4].

Experts dealing with filtration efficiency often recommend the use of cellulose depth filters to achieve efficiency. Cellulose depth filtration is a relatively old but reliable technology. In this case, each metal vessel holds the vital filtration cartridge tightly (made of tightly packed, long fibre cellulose); this allows the oil to travel up the centre and push downslowly through the entire thickness of the cellulose media. This "depth" matrix is used to trap suspendedparticles, separating them from their carrying fluid. Working by adsorption and absorption, it has theunique filtration ability to trap particles and moisture [5].

Deep filtration is most used in applications where exceptional lubricant cleanliness is required. For this purpose, and to achieve the best results, certain filter manufacturers recommend the use of deep cellulose type filters, which, with proper operation and design of the filter, can achieve a filtration rate of up to 3-micron particles. In the case of hydraulic drive technology, this level of cleanliness is suitable for high-tech servo drive systems.

2.2   Filter Materials for Lubricating Systems

Filters, the building blocks of any lubrication system designed to filter lubricating oils, and, thus, hydraulic oils or other types of hydraulic fluids, consist of a filter housing, a filter cartridge (a core of a filter), i.e. that part of the filter element which performs the actual filtration and/or dewatering function. Filter cartridge consist of several pleated filtration and support layers, which are placed as a cylinder around or inside the stabilising support tube. These mesh packs are sealed by the end-caps. Regardless of the type of filter, flow direction through the filter elements is from out to in. Depending on the filter material, the filter mesh pack is encased in an additional outer plastic sleeve. As an example, the construction of a typical filter element is illustrated in Fig. 1.

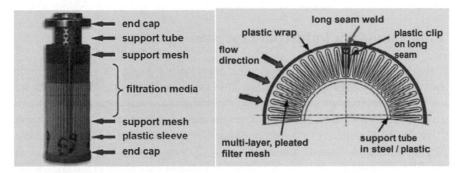

Fig. 1. Typical filter construction (left) and filter cartridge cross-section (right) (source: Hydac, Betamicron 4) [6].

The type of hydraulic fluid inside your machine plays a significant role in determining which filter cartridge you should install. This includes not only the fineness of the filter, the size of the filter and its efficiency (beta factor), but also the strength of the filter (depending on the installation location and the pressure present there), as well as the filter material used.

The latter must be suitable for the lubricants and various hydraulic fluids used – compatibility with lubricants and hydraulic fluids: Hydraulic oils H to HLPD types according to DIN 51524, other lubrication oils according to DIN 51517, API, ACEA, DIN 51515, ISO 674, Compressor oils according to DIN 51506, Biodegradable operating fluids according to VDMA 24568, Environment friendly hydraulic fluids HETG, HEES, HEPG type, Fire-resistant fluids HFA, HFB, HFC and HFD type…

Given the wide range of different types of lubricants and hydraulic fluids, it is also necessary to pay attention to the type of material used for the filter element and its compatibility with the fluid. Often are used glass fibres, synthetic fibres, metal fibres, stainless steel wire mesh, and cellulose fibres. For the removal of emulsified or free water, the filter material (e.g. glass fibre) is supplemented with water/humidity absorber material.

For lubrication systems in general there are a huge range of filter cartridges on the market now. Each has its advantages, efficiency and its own characteristics. For example, the cellulose depth filters are considered for cleaning oil slowly and continuously, and so can be more effective for a constant permanent installation. The cellulose depthfiltrationremoves dirt and water in the one same application, a finer level of filtration to remove particulate downto 3 microns can be achieved, it is possible to remove water, emulsified or free. Additionally, they are "kind" to the planet, a natural, sustainable product.

All the above applies to known media, known lubricants and known types of hydraulic fluids. In the case of a new type of fluids, which areused increasingly in industrial processes, it is necessary to check the compatibility of known filter materials with the used fluid and their impact on filterability and the ability to filter the fluid.

Such a situation arises in the case of the use of new, high-tech ionic liquids., which are usedincreasingly in various fields of technology, in a wide variety of processes and for a wide variety of purposes.

3   Ionic Liquids in Industrial Processes

Ionic liquids, a combination of a particular anion with a particular cation, are molten organic salts with very low melting points. They are a new class of liquid high-tech materials with extraordinary properties, among them nanostructural organisation. The wide field of applications of these materials develops towards a truly cross-sectional technology, like the development of Nanotechnology 20 years ago. Based on their unconventional properties Ionic liquids allow fundamentally new approaches to technical challenges: They have the potential to open doors to radical innovations in different areas of technology.

Looking back in the history of Ionic liquids their industrial application played a role as soon as 1934 (the dissolution and chemical modification of cellulose), and the first large scale industrial application was implemented by the Eastman Chemical Company more than fifteen years ago [7] Their unusual properties triggered the imagination and creativity of physicists, chemists, engineers, and even medical scientists from the very beginning, to invent new technologies in a large number of potential application fields [8].

Table 1 shows ionic liquid applications which are in a pilot stage or already commercialised. As can be seen, these applications cover many diverse technical fields, and even megatrends like mobility, health, and the green economy.

The Table lists implemented ionic liquid applications based on publicly available data and approved personal communications of the author with the industry in almost every technical field like, e.g., solvents, energy, catalysts, electrolytes, Nanotechnology, chemicals, electronics, paper and pulp, textiles, pharmaceuticals, biotech, nutrition, health, personal care, metal processing, oil & gas,the automobile industry, and all the way to the area of hydraulic drives and systems.

So, Ionicliquids applied in the fields of lubricants [9–11], heat transfer and storage fluids [12–14], heating, ventilation, air conditioning (HVAC) [15–18], sealing fluids [19], cutting and drilling fluids [20, 21], pressure transmission fluids (hydraulics) [22] and generally as operating liquids [23–26] in process machines are summarised as ionic engineering fluids. Thus, it is evident that ionic liquids, due to their excellent lubricating and other properties, are also used as lubricants and as hydraulic fluids.

Table 1. Ionic liquids' application areas [7]

Company	Process	IL is…	Scale
Air Products	Storage of hazardous gases	Solvent	Commercial
Arkema	Fluorination	Solvent	Pilot
BASF	Acid Scavenging (BASIL™)	Auxiliary	Commercial
BASF	Antistatic polymers	Performance Additive	Commercial
BASF	Azeotrope cleavage	Auxiliary	Pilot
BASF	Cellulose reshaping	Solvent	Pilot
BASF	Extractive distillation	Extractant	Pilot
BASF	Aluminum plating	Electrolyte	Pilot
BP	Aromatics alkylation	Activating Solvent	Pilot
Central Glass Co	Coupling reaction	Solvent	Commercial
Chevron Philips	Olefin oligomerization	Catalyst	Pilot
Eastman	Rearrangement (epoxybutene)	Catalyst	Commercial
Eli Lilly	Ether cleavage	Catalyst /Reagent	Pilot
Evonik-Degussa	Lithium ion batteries (Creavis)	Electrolyte	Pilot
Evonik-Degussa	Paint additive (Compatibilizer)	Performance Additive	Commercial

(*continued*)

Table 1. (*continued*)

Company	Process	IL is…	Scale
G24i	Dye sensitized solar cell	Electrolyte	Commercial
G2E	DSSC Energy glass	Electrolyte	Commercial
IFP	Olefin dimerization (Difasol process)	Activating Solvent	Pilot
IoLiTec /Wandres	Cleaning fluid	Performance Additive	Commercial
Klüber Lubrication	Wind turbine lubrication	Performance Additive	Pilot
Linde	Gas compression	Operating Fluid	Commercial
Novasina	Gas sensor	Electrolyte	Commercial
Panasonic	Supercapacitor	Electrolyte	Commercial
Petro China	Alkylation ("Ionicylation")	Catalyst	Commercial
Petronas	Mercury removal from natural gas	Reagent	Commercial
Pionics	Lithium ion batteries	Electrolyte	Commercial
Proionic	Carbonate based IL synthesis (CBILS)	Product	Commercial
Proionic /Mettop	High temperature cooling	Operating Fluid	Commercial
Proionic/University of Maribor/HAWE	Hydraulic drive in metalurgy	Hydraulic fluid	Commercial
…			

Although the application of ionic liquids as hydraulic fluids has been proposed frequently by many people working in the field for many years, only an almost negligible number of patent applications disclosing substantial inventive details has been submitted so far, and only a very limited number of publications are available. Obviously, the complex technical parameters, Standards, and regulatory affairs to be matched for a successful implementation into existing hydraulic devices is challenging.

In recent years, ionic liquids have found their use within hydraulic systems – due to their non-flammability, environmental friendliness, and other excellent properties, such as viscosity adjustment, high viscosity index value and consequent wide temperature range, flow at very low temperatures (up to −50 °C), practically zero vapour pressure…. however, they are already in industrial use [27]. They are currently in use in hydraulic systems operating under special, very demanding operating conditions. In these systems, however, as with all other hydraulic systems operating with conventional hydraulic fluids, the fluid must be filtered during system operation. This raises the question of the compatibility of ionic liquids with conventional filter materials, as well as the permeability of the liquid through the filter.

4 Filtration Ability of Different Liquid Lubricants

Since the fluid in a hydraulic system also acts as a lubricant which reduces the wear of the hydraulic elements, the reduction of solid particles in the circulation is of the utmost importance. This is particularly applicable when clearances in the hydraulic system components are small, as, then, a high degree of fluid cleanliness is required for trouble-free operation and reaching the expected useful lifetimes of the components. Contaminants are removed by filters.

The ability of the hydraulic fluid to flow through fine filter elements without their clogging is called the filterability, or filtration capacity. The filtration capacity refers to the commonly used mineral hydraulic oils, and is assessed by the laboratory test method according to Standard ISO 13357. The first part of that Standard covers the testing in the presence of water in oil, and the second part the testing without it. For other liquids, in principle, the test could turn out to be inapplicable, because the testing filter's membranes are perhaps incompatible with the tested fluid.

4.1 Filterability Test According to Standard ISO 13357

The filterability test is based on determining the ability to flow through the filter material in accordance with the conditions prescribed in the ISO 13357-1:2017 Standard (Petroleum products – Determination of the filterability of lubricating oils, Procedure for oils in the presence of water) and the ISO 13357-2:2017 Standard (Petroleum products – Determination of the filterability of lubricating oils, Part 2: Procedure for dry oils) [28, 29]. The Standards prescribe the preparation of the sample, preparing the filtering device, providing conditions for performing the test, and the implementation of the test itself,as well giving the filterability results based on the calculation of the measured parameters in two stages– Stage I filterability and Stage II filterability.

The test has to be performed three times. The test assumes the use of a sample of mineral oil without the presence of water and in the presence of water (330 ml ± 5 ml of the tested sample and 66 ml ± 0.02 ml of water of degree 3 according to ISO 3696). A cellulose membrane filter from mixed cellulose esters of 47 mm diameter and pore size of 0.8 mm is prescribed as the filter material for this filterability test.

The Stage I filterability is the ratio in percentage between 240 ml of oil and the filtered volume in the time theoretically required for filtering 240 ml, if there were no membrane clogging. The Stage II filterability is the ratio in percentage between the flow near the beginning of filtration and the flow within the range between 200 ml and 300 ml of filtered volume. The results of Stage I or Stage II filterability are given as "pass", i.e. satisfies, or "fail", i.e. does not satisfy, in relation to the limit value 50. The average of the measured values may be added in parentheses to the results above 50. If the filtration time exceeds 7,200 s (2 h), "not filterable" is given as the result. When determining the filterability without the presence of water, the filtration process is identical, only the described process of addingwater and sample heating in the oven is omitted.

As mentioned, the filterability test is intendedprimarily for testing mineral oils, but can also be used for other types of hydraulic fluids that are essentially water-free, e.g. synthetic HFD types of hydraulic fluid.

4.2 Filterability of Different Types of Hydraulic Fluids

Different types of hydraulic fluids, including ionic liquid, were used to test the filterability of the hydraulic fluid according to the Standard method, which prescribes the use of a cellulose filter. The samples ofmineral hydraulic oil (Hydrolubric VG 46), fire resistant hydraulic fluids of the HFDU type, on top of which they are faster biodegradable hydraulic fluids (Quintolubric 888-46, and Quintolubric 888-68, with two different viscosity grades of 46 cStand 68 cSt), as well as two samples of ionic liquids, IL-EMIM-EtSO4 and IL-17PI045.

As an example of the results of the filterability test, Table 2 shows the values for the synthetic highly flammable hydraulic fluid type HFDU –Quintolubric, for two different viscosity grades, VG 46 and VG 68. As expected, Quintolubric 888-46 and Quintolubric 888-68 had good filterability, as in both cases the filterability in Stage I and in Stage II was considerably higher than 50. The results are shown in Table 2.

Table 2. Filterability of Quintolubric 888-46 and Quintolubric 888-68

Standard ISO 13357-1 and ISO 13357-2	Filterability		
Sample	Stage I	Stage II	Result
Quintolubric 888-46; without water	98.6	98.7	Pass
Quintolubric 888-46; 0.2% of water	105.6	105.6	Pass
Quintolubric 888-68; without water	99.9	99.8	Pass
Quintolubric 888-68; 0.2% of water	100.6	100.6	Pass

In the described way, it was also intended to determine the filterability of both samples of ionic liquids. Standards ISO 13357-1 and ISO 13357-2 state that, in addition to mineral oils, the test method can also be used for other liquids, but the latter are maybe incompatible with membrane cellulose filters.

Before testing the filterability with a non-standard hydraulic fluid, in our case with both ionic liquids, a simple testwas performed for this purpose. For this purpose, the filter element was covered with 3 ml of ionic liquid, and the effect of the liquid on the filter material was observedafter 5 min. The results of the simple test of the compatibility of both ionic liquids with a cellulose filter are shown in Fig. 2.

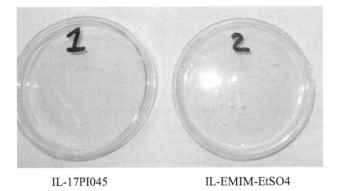

IL-17PI045   IL-EMIM-EtSO4

Fig. 2. Compatibility of ionic liquids and mineral oil with a cellulose filter element

In the present case, the concern about the possible incompatibility of ionic liquids with the cellulose filter proved to be true: Both ionic liquids samples deformed the cellulose filter membrane and the filterability determination by the method according to the ISO 13357 Standard was impossible. The filter element covered with IL-17PI045 started to decompose immediately after contact with the liquid, and was dissolved (Fig. 2 left). The sample of IL-EMIM-EtSO4 ionic liquid deformed the filter element into a glassy circle (Fig. 2 right).

Within this test it was proved that the cellulose filter paper was incompatible with the ionic liquid IL-17PI045, and that it was conditionally compatible with the ionic liquid EMIM-EtSO4. According to the results when testing the filterability in conformity with the ISO 13357 Standard and compatibility with the cellulose filter paper discussed at this point, it can be concluded that the ionic liquids IL-17PI045 and EMIM-EtSO4 (at least those two) are incompatible with the cellulose filter paper.

Therefore, the use of cellulose filter elements in hydraulic systems, other lubricating systems or within processes where ionic liquids need to be filtered, is not recommendable with those two liquids. This finding is a sufficient warning that, in these cases, before use, we check the compatibility of a particular type of ionic liquid with cellulose-based filter material individually. The filterabilities through filter elements from other materials, for example, glass fibres would, as found in another study and in practical applications using ionic liquids, caused were no problems [30].

5   Conclusion

Filtration of various liquids is very often present within the framework of various technological processes. In some cases, filtration is even necessary. These certainly include a wide variety of lubrication systems, as well as all hydraulic systems. In hydraulic systems, hydraulic mineral oil is the most commonly used hydraulic fluid, but it is not necessary.

A wide variety of filter materials are used for filtration. Cellulose filters used in various applications are considered to be environmentally friendly and very effective, especially when it comes to deep filtration. That is why they are also found in lubrication and hydraulic systems. In addition, it should be noted that the filtration ability is determined by a standard method, which provides for the use of cellulose filter paper, and was developedpreferably for media based on mineral oils.

In the case of different liquids that need to be filtered and the ability to filter is determined by this method, there may be problems with the compatibility of the tested liquid with the cellulose filter. This also proved to be the case when the filtration ability was determined by the standard method (using a cellulose filter) for ionic liquids, which are already used as high-tech fluids for lubricating fluids in various systems, or in hydraulic systems as a working medium. Therefore, it is recommended to test a certain ionic liquid for compatibility with a cellulose filter, or to use another type of filter material before use. In this case, the filterability cannot be tested in full compliance with the Standard for determining the filterability of a liquid.

References

1. N.N. Why cellulose depth filtration remains one of the best oil cleaning techniques. © 2021 Kleenoil (2021). https://upisecke.za.net/Kleenoil_filters.pdf
2. Blok, P.: The Management of Oil Contamination, Koppen & Lethem, ISBN 90-9008458-4, p. 328 (1994)
3. N.N. Filter Handbook, Hydac International, Brochure No.: E7011-3-11-16, p. 22 (2016)
4. N.N. Why Cellulose Filter Aids? J. Rettenmaier&Soehne GmbH + Co KG. https://www.jrs.eu/jrs_en/fiber-solutions/bu-filtration/why-organic-filter-aids/ (2022)
5. N.N. KLEENOIL, Filterelements. https://www.kleenoilpanolin.com/en/kleenoil/filter/, 2018 © KLEENOIL PANOLIN AG (2018)
6. N.N. HYDAC, Filter elements, Data Sheet No. E 7.200. 11/03.12 (2012)
7. Kalb, S.R.: Ionic liquids – A New Generation of High-Tech Engineering Liquids, International conference Fluid Power 2015, Maribor. University of Maribor. Proceedings. ISBN 978-961-248-491-0, pp. 49–77 (2015)
8. Kalb, R.S.: Toward industrialization of ionic liquids. In: Shiflett, M.B. (ed.) Commercial Applications of Ionic Liquids. GCST, pp. 261–282. Springer, Cham (2020). https://doi.org/10.1007/978-3-030-35245-5_11
9. Uerdingen, M., Anastas, P.: Handbook of Green Chemistry **6**, 203–219 (2010)
10. Kondo,Y., Koyama,T., Sasaki, S.: Tribological Properties of Ionic Liquids, Ionic Liquids – New Aspects for the Future, InTech Open, pp. 127–141 (2013)
11. Zhou, F., Liang, Y., Liu, W.: Chem. Soc. Rev. No. 38, pp. 2590–2599 (2009)
12. Beck, M., Schmidt, C., Ahrenberg, M., Schick, C., Kragl, U., Kessler, O.: Ionic liquids as new quenching media for aluminium alloys and steels. HTM J. Heat Treatm. Mat. **70**, 73–80 (2015)
13. Paul, T.C., Morshed, A.K.M.M., Fox, E.B., Visser, A.E., Bridges, N.J., Khan, J.A.: Exp. Thermal Fluid Sci. **59**, 88–95 (2014)
14. López-González, D., Valverde, J., Sánchez, P., Sanchez-Silva, L.: Energy, no. 54, pp. 240–250 (2013)
15. Zheng, D., Dong, L., Huang, W., Wu, X., Nie, N.: Renew. Sustain. Energy Rev. **37**, 47–68 (2014)

16. Schneider, M.C., Zehnacker, O., Schneider, R., Seiler, M.: Working fluid for absorption heat pumps, Patent EP2636715, 11 September 2013 (2013)
17. Roemich, C., et al.: Ionic liquids as sorption cooling media, Cleantech Conference & Showcase, 18–21 June, Santa Clara, CA, United States (2012)
18. Dinnage, P., Kalb, R.: Liquid sorbent, method of using a liquid sorbent, and device for sorbing a gas. Patent US20110247494, 13 October 2011 (2011)
19. Hilgers, C., Uerdingen,M., Wagner, M., Wasserscheid, P., Schlucker, E.: Patent WO2006087333, 15 February 2006 (2006)
20. Kalb, R., Hofstaetter, H.: Method of treating a borehole and drilling fluid, Patent WO2010106115, 23 September 2010 (2010)
21. Olivares, M.D.B., Ballesta, A.E.J., Molina, J.S.: Protic ionic liquids, Patent ES2373298, 02 February 2012 (2012)
22. Regueira, T., Lugo, L., Fernandez, J.: Ionic liquids as hydraulic fluids: comparison of several properties with those of conventional oils. Lubr. Sci. 26(7–8), 488–499 (2014)
23. Predel, T.C., Schluecker, E.: IonischeFlüssigkeitenalsBetriebsflüssigkeitenin der Pumpentechnik. ChemieIngenieur Technik 79(9), 1415 (2007)
24. Schluecker, E., Blendinger, S., Ismaier, A., Ruschel, A., Predel, T.: Strömungsanalyse mit Hilfe von High-Speed-Kameras. ChemieIngenieur Technik 80(12), 1753–1758 (2008)
25. Schluecker, E., Wasserscheid, P.: ionic liquids in mechanical engineering. ChemieIngenieur Technik 83(9), 1476–1484 (2011)
26. Almbauer, R., Kalb, R., Kirchberger R., Klammer, J.: Method and apparatus for operating steam cycling process with lubricated expander for recovery of waste heat from internal combustion engines, Patent WO2011151029, 5 July 2012 (2012)
27. Lovrec, D.: Ionic liquids – the path to the first industrial application, International Conference Fluid Power 2021, conference proceedings, University of Maribor, University Press. 2021, pp. 211–224. https://doi.org/10.18690/978-961-286-513-9.17,(2021)
28. ISO 13357-1:2017, Petroleum products - Determination of the filterability of lubricating oils – Part 1: Procedure for oils in the presence of water (2017)
29. ISO 13357-2:2017, Petroleum products – Determination of the filterability of lubricating oils – Part 2: Procedure for dry oils (2017)
30. Lovrec, D., Tič, V.: A new approach for long-term testing of new hydraulic fluids. In: Karabegović, I. (ed.) NT 2021. LNNS, vol. 233, pp. 788–801. Springer, Cham (2021). https://doi.org/10.1007/978-3-030-75275-0_87

Modelling and Development of Integrated Hydraulic System of the Transfemoral Prosthetic Leg

Zlata Jelačić[1]([✉]), Želimir Husnić[2], Remzo Dedić[3], and Faris Ustamujić[4]

[1] Department of Mechanics, Faculty of Mechanical Engineering, University of Sarajevo, Vilsonovo šetalište 9, 71000 Sarajevo, Bosnia and Herzegovina
jelacic@mef.unsa.ba
[2] The Boeing Company, Philadelphia, USA
[3] Department of Construction, Faculty of Mechanical Engineering, Computer Science and Electrical Engineering, University of Mostar, 11000 Mostar, Bosnia and Herzegovina
[4] Airbus DS GmbH, Taufkirchen, Germany

Abstract. Most prosthetic devices are using elastic or some other potential energy stored during a gait sequence and released in the next sequence. That causes problems in the case movements require more power to be executed, like climbing up the stairs. In the case of transfemoral amputation, the user lacks most of the leg muscles.Hence,in order to overcome this problem, prosthetic devices for transfemoral leg amputations must be externally powered in its main joints, the knee and ankle joints. This paper describes the concept of a new type of 3D printed above-knee prosthesis with integrated hydraulic system in the lower leg area.

Keywords: Above-knee prosthetic leg · Rehabilitation robotics · Hydraulic system

1 Introduction

The energetically passive devices, which form the vast majority, and are unable to restore the full mobility of the lower-limb amputees. The reason behind this constatation is the fact that the integration of the external power would drastically increase the weight of the prosthetic leg. The biggest trade-off is therefore always the one between the user comfort and the natural way of moving.

Many movements related to the lower extremities, like climbing up the stairs and walking up the steeper slopes, need significant power. That power comes from the leg muscles, but in the case they are not available, their function should be provided by an external power source. The goal is therefore to integrate the power system in the prosthetic leg housing.

While climbing the stairs, the body is being lifted from one stair to another. During this phase only one leg touches the stair and lifts the body. At the same time the other

Note: The Boeing Company and Airbus DS GmbH are not associated with this paper.

I. Karabegović et al. (Eds.): NT 2022, LNNS 472, pp. 308–316, 2022.
https://doi.org/10.1007/978-3-031-05230-9_35

leg is in swing phase. During the climbing phase, the entire body load is on one leg and the moments in the knee and ankle joints exceed body weight by several times. In the case of above knee amputation, the lack of muscles makes it impossible to produce these forces, and passive prostheses which only use stored energy, are unable to perform such activities. Therefore in order to perform high demanding power activities and movements, the joints in the knee and ankle need to be powered.

Our prosthetic leg is designed in such a way that it is able to perform the main leg movements in the sagittal plane autonomously. Therefore the knee and ankle joints are powered in order to obtain the required movements of the human leg. Powered knee enables overcoming large forces that occur during loading response. Previous research has shown that it does not suffice to only power the knee joint. The ankle joint plays an important role in the swing phase, especially during the stair ascent, and needs to be powered as well [1]. Powered ankle joint enables dorsal and plantar flexion movement of the entire foot. It also provides better stabilization of the knee and the entire prosthesis and the power needed in push-off phase.

The hydraulic power system concept is designed to enable characteristic movements of the prosthesis in sagittal plane while walking and ascending stairs. We are concentrated on testing the prosthesis on stair ascent, being the most complicated type of movement regarding energy requirements. The main components of the hydraulic system are two hydraulic actuators. One hydraulic cylinder is for powering the knee and the other for powering the ankle joint. The hydraulic system also includes the power pack unit and the accompanying hydraulic installation (Fig. 1). In order to keep the cost down, it was decided to choose off-the-shelf hydraulic power pack unit. It consists of all the needed hydraulic installation components (electrical motor, hydraulic pump, reservoir, appropriate valves, connections etc.) integrated in one unit.

2   New Concept Design of the Transfemoral Prosthetic Leg

The prosthesis structure is required to support the novel, hydraulic actuator concept as well as the human using it. Its development was summed up in well-defined steps, leading to a step-by-step approach.

This approach consisted of:

- designing a novel concept of structure,
- manufacturing the designed structural parts out of wood as basis material to accelerate the process and get a feeling for the functionality and feasibility of the concept,
- testing of the functionality of components and the entire concept with a fully equipped prosthetic structure
- based on observations, implementing design optimization and manufacturing the parts subsequently with 3D printers to achieve more complex shapes with integrated bracketry concepts to accommodate the actuator system

This paper will describe the actions and observation covered in the last step listed above. Previous steps have been covered in a separate publication, see [1, 2].

As mentioned above, the new structural concept was examined with a fully equipped rapidly manufactured prosthetic structure. Analysis of the observation led to implementing design optimization and manufacturing the parts subsequently with 3D printer technology. Such approach enabled achieving more complex shapes with integrated bracketry concepts to accommodate the actuator system.

In order to accommodate the prosthetic concept to various height of population with amputations, the structural concept incorporates a tubular adapter, with adaptable length, positioned in the lower leg of the above-knee prosthesis. This resembles most of the classical prosthesis solutions. However, the location of the tube is different as it is now placed at roughly the middle of the lower leg structure. See Fig. 1 for an illustration.

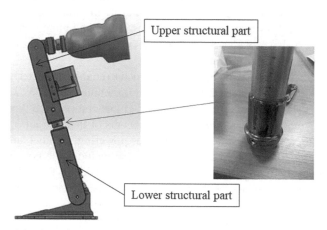

Fig. 1. Uniprox A3 tube adapter and its location in prosthesis concept between its lower and upper structural part.

3 Hydraulic Power System

In previous research [3], modified prostheses with on one hand linear actuator in the knee joint and on the hand in knee and ankle joint, were tested experimentally under laboratory conditions. Theobtained diagrams showed that the prosthesis behaved much better with an additional, separately actuated hydraulic linear actuator in the ankle than in the case when it was missing [4].

The biomechanical analysis helped to find the most efficient position for the installation of the hydraulic cylinders. The stroke and other dimensions were defined as well [5].

Next step was to solve the way of supplying the cylinder with oil and select all the necessary hydraulic components to control and regulate the oil flow.

Biomechanical analysis and experiments have shown that the movement of the linear actuators in the knee and ankle joints are not equal, regarding the movement time and stroke.It was hence decided to separate the control of the linear actuators in the knee

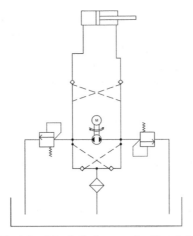

Fig. 2. Hydraulic scheme of the joint drive in the knee and ankle joints of the above-knee prosthetic leg.

and ankle. This is in accordance with the biomechanics of the human leg. The knee and ankle are controlled by different groups of muscles (Fig. 2).

In the previous versions of the prosthetic leg, the questions of housing the power unit had not been answered yet. The goal was to first design a working prototype before optimizing the design. After achieving this goal, it was time to design a new prototype. The design is made in such a way that it can incorporate the power unit into the housing of the lower leg of the above-knee prosthesis.

In order to keep the weight as low as possible, it was decided to print the leg parts. During design process, special attention needed to be given to the hydraulic connection

Fig. 3. Appearance of the 3D printed prototype prosthesis with hydraulic system and metal foot.

parts and hoses. After the completion of the housing, the power unit, consisting of pumps and electric motors needed to be installed. They were connected to small plates on the sides of the lower leg, and the oil tank is installed a little higher on the front side of the lower leg. The final first version of the 3D printed prosthetic leg prototype with integrated control and power system is shown in Fig. 3.

3.1 Hydraulic System Modelling and Simulation

Modelling of the Hydraulics System was performed separately for the System with Cylinder of 25 mm piston stroke for the ankle and for the Hydraulic System with the Cylinder of 35 mm piston stroke for the knee.

The Hydraulics System modelling and simulation was performed in the MatLab-Simulink. The Model of Hydraulics System with cylinder is shown in Fig. 4.

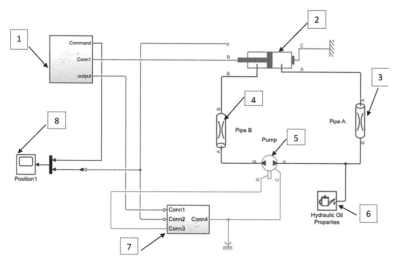

Fig. 4. Hydraulic system model.

Figure 4. Legend

1. System Inputs, Load and Command Signal
2. Double Acting Cylinder
3. Tubing connected to Cylinder Port A
4. Tubing connected to Cylinder Port B
5. Hydraulics fixed displacement pump
6. Hydraulic Fluid Properties
7. Blocks for subtraction, gain, saturation and Pump angular velocity source
8. Scope to display signals during simulation

The Model is a conceptual representation of the prosthesis hydraulics systems described herein.

The Hydraulics Cylinder Loads used for the Model simulation were in range of the Load corresponding to person of approx. 90 kg weight. Simulation with other loads could be performed as required.

The Model and Simulation provide insight into how the modelled hydraulics system will behave. The simulation provide data that would help to fully understand the hydraulics systems characteristics and performance.

It should be noticed that simulations are conducted within the limits of operation of the physical prosthesis hydraulics systems. The relevant characteristics of the system that is modelled are obtained and documented.

3.2 Ankle Hydraulics System

Simulation of the Model of Hydraulics System with 25 mm cylinder piston stroke is performed using Simulink with results explained herein. During simulation the piston at full extend position is consider as starting setting. The load of 2379 N was applied at the piston fully extended position.

The pressure in the piston side (chamber A) declined, at the same time the piston rod side (chamber B) pressure increased. As results of pressure change the piston motion begin. The piston displacement is shown in Fig. 5.

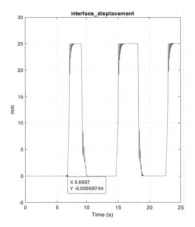

Fig. 5. Piston displacement.

The pressure in the cylinder chambers A and B are shown in Fig. 6.

Simulation was continued when the pressure in the cylinder chamber A was reduced close to zero, and pressure in chamber B was increased to 1.65 bar. As results, the piston moved from fully extended to fully retracted position. During simulation the piston was kept at fully retracted position 1.67 s.

The piston motion from retracted position to extend direction begin when the pressure in the cylinder chamber A increased, and pressure in chamber B declined.

During the cylinder piston extending from retracted to extend position, the force reach value of 2287 N maximum. The pressure in the cylinder chamber A was increased to 74 bar maximum.

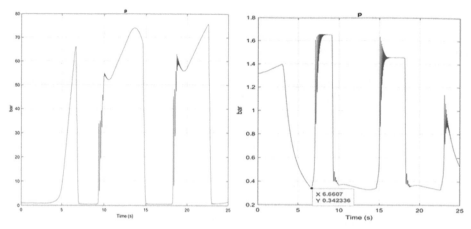

Fig. 6. Hydraulic cylinder chambers A and B pressure in the ankle joint.

The piston travel from fully extended to fully retracted position in 0.76 s and from retracted to extended position in 0.91 s.

3.3   Knee Hydraulics System

Simulation of the Model of Hydraulics System with 35 mm cylinder piston stroke is performed using Simulink with results explained herein.During simulation the piston at full extend position is consider as starting setting. The pressure in the piston side (chamber A) declined, at the same time the piston rod side (chamber B) pressure increased. As results, the piston motion begins. The piston displacement in retracted and extended direction are shown in Fig. 7.

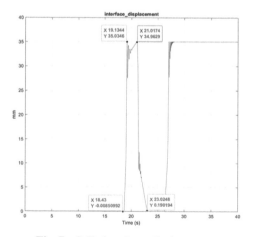

Fig. 7. Cylinder piston displacement.

The pressure in the cylinder chambers A and B are shown in Fig. 8.

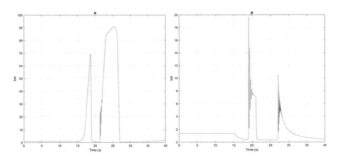

Fig. 8. Hydraulic cylinder chambers A and B pressure in the knee joint.

As results of described pressure change the piston moved from fully extended to fully retracted position. During simulation the piston was kept at fully retracted position 1.89 s.

The piston motion from retracted position to extend direction begin when the pressure in the cylinder chamber A increased, and pressure in chamber B declined. During simulation the piston was kept at fully extended position 3.32 s.

During the cylinder piston extending from retracted to extend position, the load reaches maximum value of 2814 N. The pressure in the cylinder chamber A was increased to 91 bar maximum.

The piston travel from fully extended to fully retracted position in 0.7 s and from retracted to extended position in 2.0 s.

4 Model Validation and Next Steps

Hydraulic model validation would be required to determine the degree to which a model and simulation are accurate representations of the phenomena and the prosthesis behaviour from the perspective of intended use.

Once the process of the model validation is completed, simulation would offer insight to possible failure modes and most likely points of failure. It would improve the prosthesis hydraulics performance, quality and work life.

Next step could be to consider modelling the hydraulics system with one pump and two servo valves. Simulation of different model and comparison of data, performance would provide facts essential to make decision which option would be the best choice for the prosthesis described herein.

The designed prototype has to be tested autonomously first, meaning that the experiments should show the degree of real-time tracking between the prosthetic leg and the input signals from the healthy leg. After this phase the tests with amputees in laboratory environment could then be performed.

Another important step is the development of the rehabilitation program through which the end-users would learn how to use the prosthetic leg as optimal as possible.

5 Discussion and Conclusion

Today, prosthetics strives to enable people to live as normal a life as possible, to play sports, etc. Demands are constantly increasing, and it only could be achieved with new solutions.

This paper aimed to offer a high quality and inexpensive solution that would allow climbing stairs in a natural way, in addition to the usual everyday activities. Previous prototypes have shown the feasibility of this goal. The prototype and the described integrated power system show an improvement of the previous ones based on the experimental output.

References

1. Jelačić, Z., Dedić, R., Džindo, H.: Activeabove-kneeprosthesis. Elsevier Academic Press, ISBN 978–0–12–818683–1 (2020)
2. Jelačić, Z., Dedić, R.: Real time control of above-knee prosthesis with powered knee and ankle joints. In: Karabegović, I. (ed.) NT 2019. LNNS, vol. 76, pp. 278–284. Springer, Cham (2020). https://doi.org/10.1007/978-3-030-18072-0_33
3. Jelačić, Z., Dedić, R.: Modelling and reference tracking of the robotic above-knee prosthetic leg with actuated knee and ankle joints during stair climbing. Heal. Technol. **10**(1), 119–134 (2019). https://doi.org/10.1007/s12553-019-00383-8
4. Dedić, R., Husnić, Ž, Ustamujić, F., Jelačić, Z.: Development of the concept of the integrated hydraulic system of the knee prosthesis. In: Karabegović, I. (ed.) NT 2021. LNNS, vol. 233, pp. 88–94. Springer, Cham (2021). https://doi.org/10.1007/978-3-030-75275-0_11
5. Jelačić, Z., Dedić, R.: Reference tracking of the robotic above-knee prosthetic leg with actuated knee and ankle joints using robust control. In: Badnjevic, A., Škrbić, R., Gurbeta Pokvić, L. (eds.) CMBEBIH 2019. IP, vol. 73, pp. 287–292. Springer, Cham (2020). https://doi.org/10.1007/978-3-030-17971-7_44

Comparison of Optimization Methods
in Optimization of Two Bar Structure

Ermin Husak[⊠] and Mehmed Mahmić

Technical Faculty, University of Bihać, IrfanaLjubijankića bb,
77 000 Bihać, Bosnia and Herzegovina
erminhusak@gmail.com

Abstract. In this paper, the analysis of the results obtained by optimizing the two bar structure is given. The minimum mass of the structure was chosen as an objective function. The constraint placed on the structure is the permissible stress in each bar. The optimization of the two bar structure was done with four methods: nonlinear programming, genetic algorithm, particle swarm optimization and ant colony optimization. The best results were obtained by particle swarm optimization.

Keywords: Optimization · Structure · Bar · Optimization methods

1 Introduction

Optimal design is a modern approach to design where optimization is an unavoidable phase of this process. Optimal design involves determining the optimal geometric parameters for a set objective function. The optimization process is carried out on the basis of some of the developed optimization methods. The most important classical group of optimization methods is mathematical programming. These methods are based on calculating the gradients of the objective and constraint function, so they are often called gradient methods. The modern group of optimization methods is represented by heuristic methods. These methods do not have a strictly defined way of finding the optimal solution but are based on a random approach. In this paper, we will take a simple example of an optimal structure design with two bars and compare the obtained optimal solutions with nonlinear programming NP, particle swarm optimization PSO, genetic algorithm GA and ant colony optimization ACO. The given example is often used for the purpose of comparing the results which can be seen in the literature [1–3].

2 Optimal Design of Two bar Truss

A structure with two bars connected in a node and loaded with a double value force is given, as shown in the Fig. 1. It is necessary to find the diameter of the tube d which has a constant value of the wall thickness $s = 2.4$ mm and the height of the structure h in order for the system to have a minimum mass. The maximum permissible stress is σ_p

© The Author(s), under exclusive license to Springer Nature Switzerland AG 2022
I. Karabegović et al. (Eds.): NT 2022, LNNS 472, pp. 317–321, 2022.
https://doi.org/10.1007/978-3-031-05230-9_36

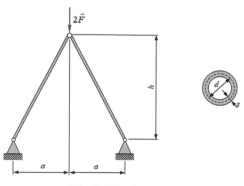

Fig. 1. Two bar truss

= 413.68 MPa. Other quantities that have constant values in this problem are: force F = 147 kN, a = 762 mm and average material density ρ = 8303.97 kg/m³.

The objective function is defined in the following form:

$$min \ f(d, h) = 2\rho\pi ds\sqrt{a^2 + h^2} \tag{1}$$

or when constant values are included, the following is obtained:

$$min \ f(d, h) = 132,525d\sqrt{0,58 + h^2} \tag{2}$$

while the constraints, which in this example refer to stresses in the bars, can be defined analytically or numerically. Since this is a simple example, it is simpler to define the constraint by an analytical function, while for more complex problems are used numerical methods. The stress in the bars must not exceed the permissible value and this is expressed in a form of inequality constraint (3).

$$\frac{F \cdot \sqrt{a^2 + h^2}}{d \cdot \pi \cdot s \cdot h} \leq \sigma_p \tag{3}$$

or when constant values are inserted in inequality, it is obtained

$$\frac{44,531\sqrt{762^2 + (h \cdot 10^3)^2}}{d \cdot h \cdot 10^6} - 1 \leq 0 \tag{4}$$

Note that the values of the variables in the above equations are in meters.

3   Results and Discussion

The previous example of a two bar structure was optimized with four methods: nonlinear programming, genetic algorithm for constrained optimization, particle swarm optimization, and ant colony optimization. All optimizations were performed in the MATLAB software package. In nonlinear programming optimization, the objective function is defined separately from the constraint functions. The optimization was completed in 11

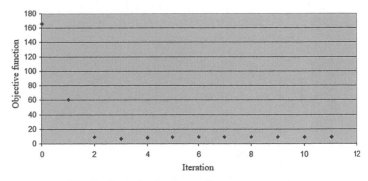

Fig. 2. Optimization by nonlinear programming

iterations, which can be seen in the Fig. 2. All optimization results for all four methods are given in Table 1.

Figure 3 shows the optimization flow of a genetic algorithm in constrained optimization. As with nonlinear programming, the objective function is defined separately from the constraint function. Unlike genetic algorithms without constraints, where in most cases the condition for stopping optimization is the number of generations, this is not the case with constraint optimization. Figure 3 shows that optimization has stopped in the eighth generation. It must be emphasized that certain parameters of heuristic methods, such as population size, mode of crossing and mutations, can significantly affect the number of generations and the time of optimization.

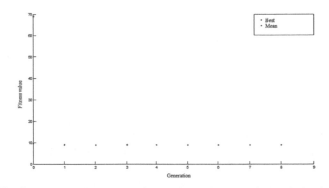

Fig. 3. Optimization by genetic algorithms in constraint optimization

Figure 4 shows the particle swarm optimization flow. Unlike the previous two optimization cases, the objective function in particle swarm optimization and ant colony optimization is defined over the penalty function. The condition for stopping the particle swarm optimization process is the number of generations.

Figure 5 shows diagram of ant colony optimization. Since with this method, each ant must pass a given number of rounds, the condition for stopping is the number of rounds [4–6].

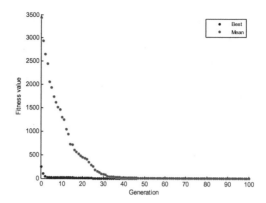

Fig. 4. Particle swarm optimization results

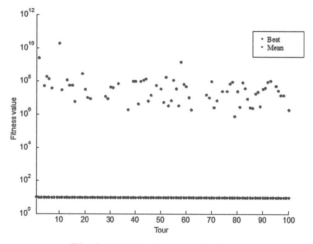

Fig. 5. Optimization by ant colony.

After the results are obtained by optimizing the structure with two bars, a comparison of the optimization methods can be made. For comparison, those results were selected whose objective function had the highest goodness, i.e. those where the mass was the lowest of certain optimizations, not taking into account population size or optimization time.In the following Table 1, the best results of each method are given.

Table 1. Obtained values of variables and objective function

	NP	GA	PSO	ACO
d *(mm)*	63	62,3	62,9	63,3
h *(mm)*	762	779,4	761,9	753,5
$F_{Obj.}$ *(kg)*	8,9938	8,9962	8,9937	8,9952

When the results are compared, it can be seen that the best value of the objective function was given by particle swarm optimization. Of course, the difference between the objective functions, i.e. the mass of the structure obtained by different methods, is not significant. The purpose of using the different optimizations was only to determine which optimization method would give the best result.

4 Conclusion

Optimization as a domain of science has become important in all areas of engineering and technology. The specific area of its use is also the optimal design. Over time, various optimization methods have been developed. The basic division of these methods is into gradient and non-gradient, i.e. heuristic methods. Depending on the problem to be solved, some methods have certain advantages over other methods. Heuristic methods are considered better for non-deterministic problems because of their heuristic way of searching feasible region. The two bar structure is an example that is often used in constraint optimization for the purpose of comparing the obtained values of the objective function. In this paper, the optimization of this structure is done with four methods, where the best solution of the objective function is obtained by particle swarm optimization.

References

1. Arora, J.: Introduction to Optimal Design. McGraw-Hill Book Company, New York (1989)
2. Singiresu, S.R.: Engineering Optimization, Theory and Practice, 4th edn. Published by John Wiley & Sons Inc, Hoboken, New Jersey (2009)
3. Jongbin, I., Jungsun, P.: Stochastic structural optimization using particle swarm optimization, surrogate models and Bayesian statistics. Chin. J. Aeronaut. **26**(1), 112–121 (2013)
4. Husak, E., Mahmić, M.: Size and topology optimization of structures. In: Karabegović, I. (ed.) New Technologies, Development and Application IV 2021. Lecture Notes in Networks and Systems, vol. 128, pp. 62–67 (2021)
5. Husak, E., Karabegović, I., Isić Safet, Karabegović, E.: Application of new optimization methods in structural design. In: Proceedings: 2nd International Scientific Conference on Engineering MAT 2012, Antalya, Turkey, pp. 140–143 (2012)
6. Jovanović, M., Husak, E.: Optimization Based on Simulation of Ants Colony, New Technologies, Development and Applications II. Lecture Notes in Networks and Systems, vol. 76. Springer, 2020. pp. 96–10 (2019). https://doi.org/10.1007/978-3-030-18072-0_36

Web-Based Remote Control and Monitoring of Pneumatic Workstation

Vito Tič[(⊠)] and Darko Lovrec

Faculty of Mechanical Engineering, University of Maribor, Smetanova 17,
2000 Maribor, Slovenia
vito.tic@um.si

Abstract. Automation systems based on programmable logic controllers are more and more present in every industrial machinery, especially in rapidly evolving and growing Industry 4.0 systems. Modern state-of-the-art automation systems should also have a possibility of remote control and monitoring of the process.

Aim of the research and development work was to investigate possibilities of remote control and monitoring of automation systems based on industrial PLC equipment without any need for additional software. The paper presents a remote control and monitoring solution on a laboratory pneumatic workstation using Beckhoffsoftware and hardware equipment.

Keywords: Pneumatic · Workstation · Remote control · Monitoring · PLC · Beckhoff

1 Introduction

Automation systems are present in modern industrial environment at every step, whether it is a smallor a large manufacturing company. Automation enables the production of a wide variety of products with high productivity, efficiency and low cost. The most common automation systems found in industrial environments are: mechanical, electrical, electronic, pneumatic systems and hydraulic systems.

Each of the systems has its advantages and disadvantages. The most optimal solution to each automation problem is obtained by combining individual systems, of which we use those that have the most advantages in the field of use and thus take advantage of the best features of each technique. The fact is that electronic control systems have so many advantages in this technique that no other system can replace them.

Modern industrial automation is therefore not possible without the use of industrial programmable logic controllers (PLCs) that allow the control of other systems. Certainly one of the biggest advantages of PLCs is their high flexibility and productivity.

With the introduction of Industry 4.0 we are experiencing the rapid development of PLCs and industrial communications technology along with widespread use of internet. Although there have been several solutions available for remote control and monitoring of PLC operation, these systems usually need specific and special software to access the PLC from remote location [1–3]. Therefore, our study focused on developing remote

© The Author(s), under exclusive license to Springer Nature Switzerland AG 2022
I. Karabegović et al. (Eds.): NT 2022, LNNS 472, pp. 322–328, 2022.
https://doi.org/10.1007/978-3-031-05230-9_37

control and monitoring of PLC-automated process without the need for any additional software on remote machine from where we are accessing the PLC and automation process. For this purpose, a previously developed laboratory pneumatic workstation was used [4–6], which is entirely based on real industrial components and represents an automated workstation based on the use of pneumatic systems.

2 Development and Design of Pneumatic Workstation

The workstation is based on the pre-designed work cycle. Workpieces are loaded into a storage magazine, from where a two-way cylinder pushes them to the pick-up point. The transfer of the workpiece from the pick-up point to the workplace is carried out by a 3-axis pneumatic manipulator, which can perform different movements of the workpiece. The workpiece is picked with the help of pneumatic gripper which also has possibility of 180° rotation. After the workpiece is moved from the pick-up point to the workplace, it is clamped into its position using one-way pneumatic gripper.

The main two production processes performed on the workpiece are drilling and stamping. Depending on the chosen control loop, only one of the selected operations can be performed, or both can be performed. Due to the available rotary movement on the manipulator gripper, both processes can be repeated on the other side of the workpiece. After production process is finished, the manipulator transfers the workpiece to an appropriate container – there are three available containers for finished pieces, depending on the production process chosen by the user.

In this way, our designed workstation provides a high level of flexibility in design of control system; from the simplest control cycles, where only three axes are managed: the user manually inserts the workpiece into the workplace, followed by clamping, drilling, stamping, un-clamping and manual removal of the workpiece; to very complex control cycles, where all 9 available axes are operated: ejection of the workpiece from the container, pick-up and transfer to the workplace, both machining operations, clamping and turning of the workpiece by 180°, re-clamping and repeating the machining operations and transfer to an appropriate container in respect to its current fill capacity.

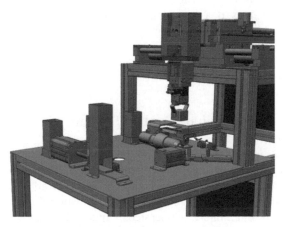

Fig. 1. 3D model of pneumatic workstation.

The basic concept of the workstation and its operation was further developed and conceived in SolidWorks, as presented in Fig. 1. Also, a simulation of movement through the work process was created, making it easier to reconcile the individual movements of all 9 axes.

While designing the pneumatic workstation, we tried to use and display as many different components as possible. Therefore, several different types of limit switches are used on the workstation in regard to the adequacy of their use. Thus, positions of the controlled axes are detected by mechanical, inductive, reed and pressure switches, each of which have their specific properties that enables an appropriate indication of the position of individual axis. Namely, all states cannot be detected only with one type of limit switch, for example: it is practically impossible to detect whether the gripper has completed the process of clamping the workpiece with a mechanical limit switch. That is why pressure switches were used, which, in addition to the adjustable switching point, also show the current pressure in line.

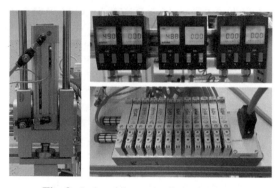

Fig. 2. Industrial pneumatic components

The presented nine pneumatic axes are controlled by the modern Festo VTUG valve terminal, as presented in Fig. 2. The valves can be controlled by triggering individual 24VDC coils on the valves, or via a special communication interface that can be added to the valve terminal and enables control via ProfiNet connection. Therefore, physical connections were carried out in such a way that the valves can be triggered either via digital outputs on the PLC, or via ProfiNet industrial communication between the valve terminal and thePLC.

3   Design and Simulation of Control System

The entire workstation was also designed in the Automation Studio, which enables us to design a pneumatic scheme together with an electrical scheme and a PLC program, as shown in Fig. 3. Further on, it allows all connection of all listed sub-system and simultaneous simulation of their operation. The pneumatic scheme of the workstation consists of 7 two-way and 2 one-way operated cylinders. In addition to the common bi-stable 3/2 and 5/2 directional valves, a 5/3 mono-stable valve with closed center position was also used for the Y axis of the manipulator, which enables us to stop the axis in its middle position. All cylinders also have one-way chokes valves on the discharge side, which allow us to adjust the speed of each axis.

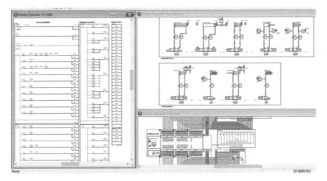

Fig. 3. Simulation of control system in Automation Studio

The design of the pneumatic control system was followed by the design of an electrical scheme of the control system. In the electrical scheme, a programmable logic controller (PLC) is added as a central element and its digital inputs and outputs are connected to individual elements. All elements of the electrical scheme are connected to the pneumatic scheme and form a functioning simulation system.

The third part of the simulation was the PLC program. The Automation Studio enables us to simultaneously simulate the operation of the pneumatic system, electrical circuit and control program. In this way, we created a program for one working cycle of the workstation and tested it.

During the simulation program displays the control states of pneumatic valves, pressures in pneumatic lines, movements of cylinders, activity of the limit switches, electrical signals on control circuit and execution of PLC program. It also has an option on running the simulation step-by-step, where user can look more closely at individual electrical, pneumatic and logic signals.

4 Beckhoff Soft-PLC Control System

Beckhoff Soft-PLC CX-5140 is used for automation and control of the pneumatic workstation. TwinCAT 3 is used for programming the CX-5140 in Visual Studio Shell, which combines the configuration and programming of the PLC. As shown in Fig. 4, TwinCAT 3 offers us 2 different text programming languages: Instruction List (IL) and Structured Text (ST); as well as 4 different graphical programming languages: Function Block Diagram (FBD), Ladder Logic Diagram (LD), Continous Function Chart (CFC) and Sequential Function Chart (SFC).

Due to the different programming languages available, it is important to consider some aspects of using and choosing a language before deciding on a particular type of programming language for our application. There are also some general guidelines for selection:

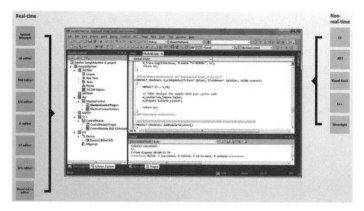

Fig. 4. Programming in BeckhoffTwinCAT 3

- Generally accepted programming language: LAD.
- Easy end user maintenance: SFC.
- Admission in Europe: IL or ST.
- Program execution speed: IL or ST.
- Use of digital I/O and basic processing: LAD or FB.
- Ease of later code change: LAD.
- Ease of use for new engineers: ST.
- Ease of use of complex mathematical operations: ST.
- Repetitive procedures that are interrupted or carried out simultaneously: SFC.

Initially, the sub-programs to control one pneumatic cylinder were designed and developed, followed by further development of whole automation process. The Sequential Function Chart (SFC) programming language was used for the most parts of the programming, due to:

- possibility of modular program construction,
- easy application programming with repetitive functions,
- transparency of the program,
- ease of final maintenance,
- programming speed,
- simple conversion from a draft state flow chart,
- good detection of programming errors.

5   Remote Control and Monitoring

Remote control and monitoring of the PLC automation process (pneumatic workstation) is based odTwinCAT 3 PLC HMI Web, which is a HTML5-based web client for displaying the visualization. The greatest advantage of the presented solution is that the remote control can be accessed form any device that has an up-to-date browser, such as personal computer, tablet PC or mobile phone, regardless of the platform and with no additional software needed to be installed. Remote control can also be accessed and opened simultaneously on several different devices.

Figure 5 presents web-based visualization of the remote control and monitoring of the pneumatic workstation. At the top of the screen the user has an option for manual control of the workstation, where initialization, reset and E-Stop can be triggered. Additionally, there are buttons to manually engage only a single operation, e.g. drill, stamp or drill & stamp procedure.

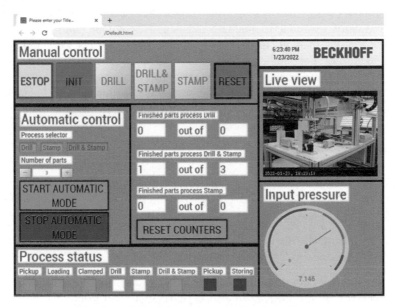

Fig. 5. Web-based remote control and monitoring

In the middle-left section there are controls for automatic control of the pneumatic workstation. The user can select between three different processes and enters the number of parts to be manufactured. The process is started and stopped with two button, as shown in Fig. 5.

Middle part of the screen shows current stack of different produced workpieces in one of three exiting containers, whereas bottom part of the screen displays current status of the machine and the manufacturing process.

Right side of the screen consists of input pressure monitoring and live view from a camera, which is incorporated into the web-based visualization.

6   Conclusion

The paper presents the development of automation process on pneumatic workstation, which was designed entirely using industrial components. The automation is based on state-of-the-art Beckhoffcontrol system, which can be controlled completely remotely from any up-to-date device, such as PC, tablet or mobile phone, without installing and using any additional software.

Presented remote control and monitoring solution can have many benefits for small and large scale factories, especially in these pandemic times, when many of companies have faced with interrupted production abilities due to shortage of personnel and/or materials. Having 24/7 remote access to their equipment to control, monitor and troubleshoot the process, could help them reduce operational costs and minimize downtime.

On the other hand, the presented laboratory remote controlled workstation canalso serve as a didactic tool in learning how to design, implement and control modern pneumatic systems, as well as to maintain and diagnose common faults in such systems.

Without a doubt, there are also a lot of possibilities for further work and upgrades. First and foremost the pneumatic workstation should be upgraded with safety equipment, i.e. safety light curtains.Further on, a self-diagnosis and error report system can be developed and implemented. Last but not least, the workstation can also be upgraded to communicate with other machines using Industrial Internet of Things (IIoT) concepts.

References

1. Shyr, W.-J., Su, T.-J., Lin, C.-M.: Development of remote monitoring and a control system based on PLC and WebAccess for learning mechatronics. Int. J. Adv. Rob. Syst. (2013). https://doi.org/10.5772/55363
2. Qasim, I., Anwar, M.W., Azam, F., Tufail, H., Butt, W.H., Zafar, M.N.: A model-driven mobile HMI framework (MMHF) for industrial control systems. IEEE Access **8**, 10827–10846 (2020). https://doi.org/10.1109/ACCESS.2020.2965259
3. Mhetraskar, S.S., Namekar, S.A., Holmukhe, R.M., Tamke, S.M.: Industrial automation using PLC, HMI and its protocols based on real time data for analysis. Int. J. Adv. Res. Eng. Technol. (IJARET) **11**(10) (2020). https://doi.org/10.34218/IJARET.11.10.2020.129
4. Krošel, A.: Načrtovanje, izdelava in krmiljenjenamenskegapnevmatičnegamanipulatorja. Diplomskodelo. Maribor (2016)
5. Tič, V., Krošel, A.: Razvojdelovnepostaje za didaktičnenamenenačrtovanja, implementiranja in krmiljenjapnevmatskihsistemov. Ventil : revija za fluidnotehniko in avtomatizacijo **23**(1), 48–52 (2017)
6. Krošel, A.: Krmiljenjepnevmatskedelovnepostaje z vodilomProfinet in krmilnikomBeckhoff. Magistrskodelo. Maribor (2019)

Research on Modeling the Technological Processing of Typographic Film

Alina Bianca Pop[1], Gheoghe Ioan Pop[2,3], Constantin Oprean[4],
and Aurel Mihail Titu[4(✉)]

[1] Technical University of Cluj-Napoca, North University Center of Baia Mare, 62A, Victor Babeș Street, Baia Mare, Romania
[2] University POLITEHNICA of Bucharest, SplaiulIndependenței 313, București, Romania
[3] S.C. Universal Alloy Corporation Europe S.R.L. Dumbravița 244A, Maramures, Romania
[4] Lucian Blaga University of Sibiu, 10 Victoriei Street, Sibiu, Romania
mihail.titu@ulbsibiu.ro

Abstract. This paper was developed based on a study conducted within a company whose main object of activity is the realization of a wide range of printing products and the provision of services in this field. The research starts with a SWOT analysis of the company, followed by the presentation of the preparation of the printing film and the elaboration of the factorial experiment strategy applied in order to model and partially optimize the technological process of the printing film. Using a known software application, the experimental data obtained based on the adopted influencing factors were performed - the temperature in the pools of substances, the level of the developer in the pool and the level of the fixer in the pool, having as objective the quality of the raster point and the lack of stains on film. The last part of the paper highlights our own points of view and the conclusions drawn from the elaboration of the study.

Keywords: Scientific research · Experimental statistical modeling · Strategies in the field of quality · Processing · Typographic film

1 Introduction

The company in which the study was conducted has as main object of activity the realization of a wide range of printing products and the provision of services in this field, printing of newspapers, books, leaflets, flyers, calendars, diaries, catalogs, leaflets, leaflets, posters. One of the long-term objectives planned by the company is to achieve an efficient quality management system, regarding the quality policy. The activity of the company is structured within functional compartments, each of them having a well-defined role in the general gear of the company [1].

For strategic and marketing management, a very commonly used tool is SWOT analysis. The opportunities offered by the environment are closely related to the resources and cultural environment of the organization, which, if not exploited, remain only potential opportunities. In analyzing the organization's resources, the assessment should not be made from an internal perspective, but in relation to the characteristics imposed by the environment and competition (Table 1).

I. Karabegović et al. (Eds.): NT 2022, LNNS 472, pp. 329–336, 2022.
https://doi.org/10.1007/978-3-031-05230-9_38

Table 1. SWOT analysis of the company

Strengths:	Weaknesses:
The organization is ISO certified; The partners are ISO certified; 6 employees are TESA accredited; Participates in fairs with typographic themes; It has a good organization of processes; Internal organization by sections; It is equipped with high circulation equipment; Uses quality technologies; Employees with work experience; Young employees with assertion desires; Experience in the field; Management quality.	The building is too small; Lack of new technologies and equipment; Lack of marketing staff; Location of the printing house; Weak market reputation; Inefficient strategic orientation; Higher costs than the competition.
Opportunities:	Threats:
Construction of a new headquarters, on the newly acquired land; Bringing new equipment, according to the new technologies in the field; Manifestation of regressive tendencies in competing organizations, amid the global economic crisis.	Competition on the local and national market; Increase in prices by suppliers; Vulnerability to the economic crisis, which has arisen domestically and internationally; Unfavorable changes in the euro exchange rate and trade policies.

2   Research Strategy in the Technological Process of Typographic Film Processing

The first impact in the execution process takes place within the printing preparation department, between the client and the organization, after signing the contract. This department is equipped with high-performance computers, monitors capable of working at optimal resolutions for printing, an Imagesetter, for the process of image distillation and film development [2, 3] in order to transmit information to the editing department, from where on plates, will the information is transmitted to the printing presses [4, 5].

After the computer prepares the material provided by the customer for printing, the operator transmits the information in the form of a file of type * ps to the machine called Imagesetter, which burns a film at a high resolution, and develops it [6]. The operator from the printing preparation department then sends the developed film to the editing department, which will transpose the information from the film, on printing plates, which will then be fixed in the printing machine [7]. From here, the printer will print the information on the plates, on sheets of paper [8].

The application of factorial experiments involves the planning of tests, and this is done according to an "optimal plan", given by an optimal ratio between measurements and information obtained. With the help of the desired model, we then have the accuracy of the estimates. For this we must concretely and completely define the problem we have to solve, ie to specify the function of objective interest. In the technological process of processing the film for printing, the objective function that we want to improve is the quality of the raster point and the lack of stains in the image fixed on the film, after development, adjusting the level of substances in the pool (developer and fixer). and the temperature at which they must be. In the card of the department for the preparation of the printing it is desired to optimize the process of processing the typographic film. From the computer where the print job is being prepared, the operator transmits the information to a computer with a *.ps file. The information is usually a leaflet, a poster, flyers, etc., in a word polychrome. For information in a single color - black, for example - the information is transmitted on the track. Polychromes, however, need maximum precision and quality, so we transmit the information on film [9]. The quality of the print and therefore the customer's satisfaction depends on the way we process a film. When the file reaches the computer that is going to transmit the information, it already has the *.ps extension, which means that the image has been decomposed into the four basic colors (blue, magenta, yellow, black) that are used in any printing house. From the overlap of the four colors, four film shots, results the polychromies.

3   Factorial Experiment

Before the operator can transmit the file to the machine called Imagesetter, which performs the process of film data writing and development, it must follow some parameters: film size, substance temperature, developer level and fixer level [10, 11]. We thus obtain the following input data, of the influencing factors:

- $x1$ - the temperature in the pools of substances, which is the same for both the developer and the fixative: maximum 34 °C (+), minimum 33 °C (−);
- $x2$ - pool developer level: maximum 12 cm (+), minimum 11 cm (−);
- $x3$ - the level of the fixative in the pool: maximum 12 cm (+), minimum 11 cm (-)

The objective Y function pursued, in this case is the quality of the raster point; lack of stains in the image fixed on the film.

4   Results Interpretation

In the following, an interpretation of the results obtained based on the factorial experiment approach was used using data processing tools [12].

In this sense, the starting point was the elaboration of the histograms presented in the following figures on temperature, developer and fixative.

In Fig. 1 we notice that the temperature variable has a normal distribution, between the given values, which are very close. In Fig. 2 we notice the normal distribution of the data, distribution offered by the very close values.

In Fig. 3 the distribution is a normal one, in the form of a bell. In the analyzed histograms, we notice that the values are close, they all have a normal distribution. This

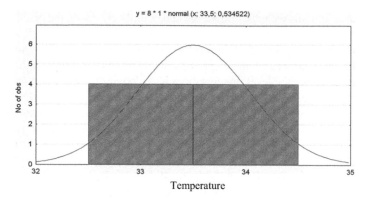

Fig. 1. Temperature histogram

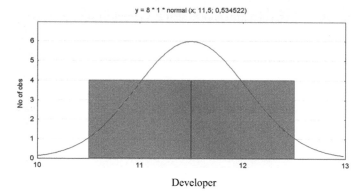

Fig. 2. Histogram for the developer

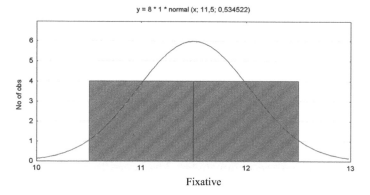

Fig. 3. Histogram for the fixative

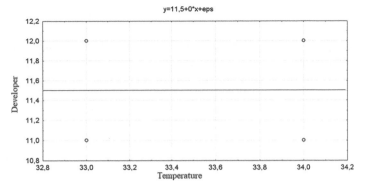

Fig. 4. Correlation between temperature and developer values

draws attention to the fact that the interaction between these variables will have to be considered in order to find out the optimal values.

Next, the data dispersion will be analyzed. In Fig. 4, each circle represents a subject. The value of the temperature is on the x-axis, and the value of the developer is on the y-axis. We notice that the circles are far from the red line in the middle, this fact suggests that there is not a very close connection between the two variables.

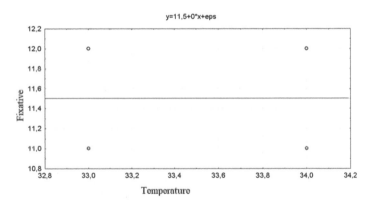

Fig. 5. Correlation between temperature and fixative values

In Fig. 5 we notice the same fact as in the case of Fig. 4. The circles are far from the red line, in the middle. The value of the fixative, represented on the y axis is not under the influence of the temperature value, x axis.

In the dispersion diagram in Fig. 6 we notice that we have not drawn any line, the circles on each axis determine the relationship between the three variables, which seem to correlate on the diagram. We do not observe any independent relationship of one of the variables.

In Fig. 7 we notice a slight increase towards level 6 of the objective function, when the temperature is high and the level of the fixator is low. Which denotes in these conditions the appearance of stains.

In Figs. 8 and 9 we notice the colored network surfaces according to the dispersion mode of the variables.

Influenced by the level of substances. At the maximum level, we will not have stains or a more erased raster.

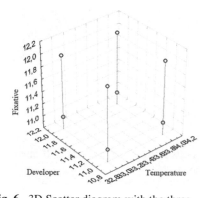

Fig. 6. 3D Scatter diagram with the three variables

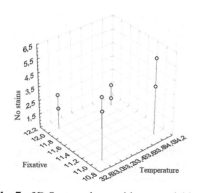

Fig. 7. 3D Scatter chart, with two variables related to the objective function

z=-227,88+46,063*x-90,73*y-0,434*x*x-1,5*x*y+6,01*y*y

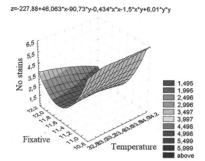

Fig. 8. 3D graph - two variables reported with the objective function

z=-134,698+9,222*x+18,415*y-1,271*x*x+1,5*x*y-1,583*y*y

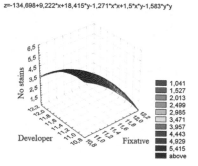

Fig. 9. 3D graph, two variables related to the objective function

In Fig. 10 we can see a correlation between the two variables, the results are symbolically colored in colors of different intensities.

In the final model (Fig. 11) the maximum values for developer and fixer were obtained, which provide the quality of the developed film.

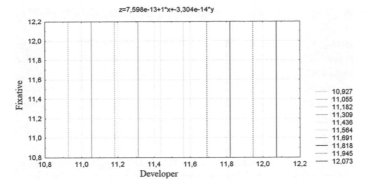

Fig. 10. 3D graphic, developer and fixer variables

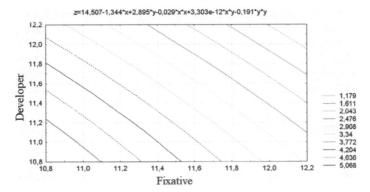

Fig. 11. 3D graphic, fixer and developer variables

5   Conclusions

Following this experimental research, it can be seen that in order to obtain the desired quality of the raster point and the absence of stains, the level of the developer must be at a maximum level of 12 cm, the level of the fixer also at 12 cm, and the optimum temperature is 34 °C.

If the developer has a lower level, then the raster point will be much too weak, likewise, if the fixer has a level lower than 12 cm, the films will have white spots.

The temperature influences the quality of the developed films, only if one of the substances has a level other than 12 cm.

It can also be seen that the fixer may have a slightly lower level, up to a minimum of 11.5 cm, but the developer is not allowed to descend below the level of 12 cm in the pool.

References

1. ColladoAgudo, J., de Leaniz, P.M.G., Crespo, A.H., Gómez-López, R.: Quality-certified hotels: the role of certification bodies on the formation of customer behavioral intentions. Sustainability 13(22), 12733 (2021). https://doi.org/10.3390/su132212733
2. Zhao, T., Hu, L., Zhang, Y., Fang, J.: Super-resolution network with information distillation and multi-scale attention for medical CT image. Sensors 21, 6870. (2021). https://doi.org/10.3390/s21206870
3. Dong, C., Loy, C.C., He, K., Tang, X.: Image super-resolution using deep convolutional networks. IEEE Trans. Pattern Anal. Mach. Intell. 38, 295–307 (2015)
4. Neamțu, G.V., Mohora, C., Anania, D.F., Dobrotă, D.: Research regarding the increase of durability of flexible die made from 50CrMo4 used in the typographic industry. Metals 11(6), 996 (2021)
5. Kipphan, H.: Handbook of Print Media Technologies and Production Methods, vol. 1, pp. 773–901. Springer, Berlin/Heidelberg, Germany. ISBN 978-3-540-29900-4. (2001)
6. Zhang, C., et al.: The effect of substrate biasing during dc magnetron sputtering on the quality of VO2 thin films and their insulator–metal transition behavior. Materials 12, 2160. Materials 13, 5132 (2020). https://doi.org/10.3390/ma13225132
7. Lim, B., Son, S., Kim, H., Nah, S., Lee, K.M.: Enhanced deep residual networks for single image super-resolution. In Proceedings of the IEEE Conference on Computer Vision and Pattern Recognition Workshops (CVPRW 2017), Honolulu, HI, USA, 21–26 July 2017, pp. 1132–1140 (2017)
8. Skrzetuska, E., Michalak, D., Krucińska, I.: Design and analysis of electrodes for electrostimulation (TENS) using the technique of film printing and embroidery in textiles. Sensors 21(14), 4789 (2021). https://doi.org/10.3390/s21144789
9. Balbas, D.Q., Sánchez-Rodríguez, E., Ramírez, Z., Corn, Á.H.: Interdisciplinary study of a Mexican 16th-century polychrome maize stem, paper, and colorín wood sculpture. Heritage 4, 1538–1553 (2021). https://doi.org/10.3390/heritage4030085
10. Țîțu, M., Oprean, C., Boroiu, A.: Cercetareaexperimentală aplicatăîncreştereacalităţii produselorşiserviciilor, ColecţiaPrelucrareaDatelorExperimentale, Editura AGIR, ISBN 978-973-720-362-5, 684 pages, Bucharest (2011)
11. Oprean, C., Țîțu, M.:Cercetareaexperimentalăşiprelucrareadatelor. Partea a II-a. ColecţiaPrelucrareaDatelorExperimentale, EdituraUniversităţii "Lucian Blaga" din Sibiu, ISBN (10) 973-739-296-5, ISBN (13) 978-973-739-296-1, ISBN (10) 973-739-298-1, ISBN (13) 978-973-739-298-5, p. 535 Sibiu (2007)
12. Țîțu, M., Oprean, C.: Cercetareaexperimentalăşiprelucrareadatelor. Partea I. ColecţiaPrelucrareaDatelorExperimentale, EdituraUniversităţii "Lucian Blaga" din Sibiu, ISBN (10) 973-739-296-5, ISBN (13) 978-973-739-296-1, ISBN (10) 973-739-297-3, ISBN (13) 978-973-739-297-8, p. 415, Sibiu (2006)

Ergonomic Analysis of Driver Postures on Electric Scooter

Slavica Mačužić Saveljić[(✉)] and Jovanka Lukic

Faculty of Engineering, University of Kragujevac, Sestre Janjic 6, 34000 Kragujevac, Serbia
s.macuzic@kg.ac.rs

Abstract. Micromobility includes light vehicles such as electric scooters, bicycles, electric bicycles. It is becoming more and more common in cities, both because it is easier to apply and to reach a certain place faster. Micro vehicles produce a small amount of exhaust gases and emit low noise levels. They are characterized by the fact that they represent the best mode of transport on short distances. Their use reduces the load on the lumbar spine in relation to motor vehicles, but still has a measurable impact on oscillatory comfort. In this paper an electric scooter in the CatiaV5 R18 software package was modelled and RULA (Rapid Upper Limb Assessment) package was used for human body load investigations. The influence of the anthropometric data of ten digital human models (five male and five female) on human body load were analysed. The body load was examined in driving conditions. The results showed that the obtained values of body load were different between subjects as well as their body parts.

Keywords: Micromobility · Load · Driver · CatiaV5R18 · RULA

1 Introduction

Micromobility is a term associated with the rapid evolution of light vehicles. "Micro vehicles" are becoming more and more represented for both private and business purposes. According to [1], the term "micro" can refer to vehicles that are usually less than 500 kg, but also to short-distance travel. Such vehicles have a low adverse impact on the environment [2].

Micromobility defines the use of micro vehicles or vehicles with a mass not exceeding 350 kg and a speed not exceeding 45 km/h. For this reason, these vehicles are safety for pedestrians as opposed to motor vehicles. The practicality of application at these vehicles is significant. They are suitable for fast and short trips, especially in urban areas. Assessment of discomfort or body position in driving condition is performed using digital human models and methods RULA, REBA etc. [3]. The RULA method, introduced in 1993, assesses static muscle activity and the force acting on the upper extremities. It is designed to quickly assess the load on the musculoskeletal system due to the position of the neck, waist and upper extremities. Positions of individual body segments are investigated and the more there is deviation from the neutral posture the higher will be the score of each body part. The risk is calculated into a score of 1 (low) to

I. Karabegović et al. (Eds.): NT 2022, LNNS 472, pp. 337–344, 2022.
https://doi.org/10.1007/978-3-031-05230-9_39

7 (high). In research [4], an assessment was made for the load on the upper extremities, more precisely in the neck area, while sitting in a car seat. RULA analysis was used to assess risk factors for seating position in the car. In the CATIA software package, the car seat was modelled and a manikin was placed, on which research was later done. RULA analysis for car seats without side backrests, the load on the neck ligaments and other structures of the cervical spine was established. The second analysis, which was also done with the RULA analysis, but with the installation of side neck rests, indicates better results and a more comfortable ride. In the second analysis, the bending angle in the left side of the door decreased from grade 3 to grade 1. The final score of the RULA analysis was 2, indicating a reduction in muscle fatigue on the left side of the neck. Paper [5] describes a test of the comfort of bus drivers. A group of five men with different anthropometric data was used. Driver's body anthropometrics measurements and seat dimensions are modelled in the CATIAV5 software package. The Human body model is adjusted according to the real-time driving position. RULA analysis was used for the analysis of sitting comfort. The results of the RULA analysis showed that greater seating comfort was observed in men of medium height compared to those of higher height. Therefore, for further research, it was necessary to modify the seat of the bus in order to suit a larger population in driving condition. RULA analysis for ergonomic design of car seats was also used in paper [6]. The RULA analysis was used to calculate the risk of musculoskeletal loading within the upper limbs and neck, based on variables such as weight, distance and frequency. In this paper, the RULA analysis gave a final score of 4, which indicates that the current sitting position causes a load on the muscles and upper limbs, and that it is necessary to change such a sitting position. After that, the improvement and optimization of the seats was done, and the RULA analysis was done again. The grade value of the RULA analysis was 2. It indicates lower loads than the previous one. A score of 2 means that sitting in this position does not cause additional loads on the upper extremities of the body. The result of this indicates that the driver is in the best seating position without the risk of straining the body.

Based on the final results of the RULA analysis, four levels of action have been proposed. Each of them shows the level of intervention needed to reduce the risk of injury [7, 8]:

- Action level 1: posture is acceptable
- Action level 2: further investigation is needed and changes may be needed
- Action level 3: investigation and changes are required soon
- Action level 4: investigation and changes are required immediately.

In this paper, digital human models were used to determine the optimalposition of the human body in driving condition for different subjects. Ten subjects, five male and five female, with different body heights were analysed. The driving conditions of all subjects were tested on a virtual model of an electric scooter Hover-1 Alpha [9]. It is important to examine the loads on human body parts, because this is a new vehicle for individual transport. The maximum speed is 29 km/h, and the test conditions are adjusted to the driving modes.

2   Methods

A quick assessment of the upper extremities analyses the upper limbs of the mannequin based on the anthropometric data of the created mannequin in the CatiaV5 R18 software package. RULA is a method for researching the human factor, i.e. biomechanical and postural load of the whole body. It is intended to be used as part of a broader study of ergonomics. For this research, certain parts of the model's body were adjusted according to the construction and application of the electric scooter. The first task was to create an electric scooter and then place the mannequin in the appropriate riding position. Figure 1 shows the appearance of a modelled electric scooter and mannequin.

a) Male mannequin   b) Female mannequin

Fig. 1. Modelled electric scooter with mannequins

Anthropometric data, of ten subjects (male and female), shown in Table 1 and Table 2. The choice of the respondents had to satisfy the condition of the technical characteristics of the scooter, that the maximum weight is 120 kg.

Table 1. Anthropometric data of male subjects

	Height (mm)	Arm length (mm)	Torso width (mm)	Foot width (mm)
Subjects	1772.8 ± 14.87	608.8 ± 5.26	332.2 ± 10.08	74.2 ± 1.30

Table 2. Anthropometric data of female subjects

	Height (mm)	Arm length (mm)	Torso width (mm)	Foot width (mm)
Subjects	1677.8 ± 7.19	573.2 ± 9.70	293.2 ± 8.01	68.8 ± 2.28

The digital human model was placed on an electric scooter. RULA analysis was done for the human model in the driving position. Upper body posture was assessed using the RULA score list. The range of motion for each body part is divided into segments. Number 1 represents the minimal risk of loading a certain part of the body. Higher numbers indicate a higher risk factor that causes a load on the body. Exposure levels RULA analysis is divided into four categories (Table 3).

Table 3. Categories RULA analysis [10]

RULA level	0	1	2	3
RULA score	1–2	3–4	5–6	7
Risk level	Negligible	Low	Medium	High

The RULA method was developed to identify muscular efforts associated with various factors, which lead to the appearance of fatigue, discomfort. The working principle of the analysis consists of the following. The human body is divided into two groups - postures. Posture A includes: shoulders, elbows and wrists, and posture B includes: neck, torso and legs. The score of these two groups (Score A posture and Score B posture), together with muscles and force, gives two new scores, the result of which gives the total score (Grand score) that describes the given body posture (Fig. 2).

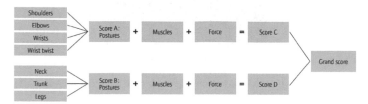

Fig. 2. Method of work RULA analysis [11]

3   Results

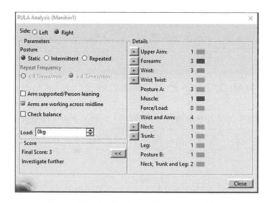

Fig. 3. Method of work RULA analysis

In this paper, an electric scooter was modelled and various anthropometric data of ten models was used in order to study the musculoskeletal load of the body. Estimation of body load was done using RULA analysis. An example of the obtained results shown on Fig. 3.

The green colour of the final score indicates that such a position of the model is acceptable if is not repeated for a long period. The yellow colour indicates that further examination of that position is necessary, and that it is desirable to change such a position. The colour orange requires further examination and change of position, which is useful to do as soon as possible. The red colour represents the greatest load that requires a quick change of such a position.

In the Table 4, the coefficient of determination (R^2) is given as a function of body parts dimension for the male subjects.

Table 4. Determination coefficients for the male population in static posture

Coefficient of determination (R^2)

	Upper arm	Forearm
Height (mm)	0.67	0.66
Arm length (mm)	0.56	0.52
Torso width (mm)	0.69	0.10
Foot width (mm)	0.70	0.12

The upper arm has the highest correlation (0.70) with the width of the foot. Also, a good correlation (0.69) exists between the upper arm and the width torso. A moderate correlation (0.66) exists between forearm and body height. Better values of the coefficient of determination (R^2) are noticed for the female population compared to the male population, for certain parts of the body (Table 5).

Table 5. Determination coefficients for the female population in static posture

Coefficient of determination (R^2)

	Upper arm	Forearm
Height (mm)	0.72	0.73
Arm length (mm)	0.85	0.53
Torso width (mm)	0.44	0.19
Foot width (mm)	0.23	0.09

The highest correlation (0.85) was observed between load index of the upper arm and arm length. The correlation between foot width and forearm is very small, so there is no correlation in that part of the body. The correlation with the width of the foot is small. These values were obtained for the static load of the body. For the same anthropometric characteristics of the body, a correlation was determined when it comes to intermittent body load. Figure 4 shows a comparative representation of the coefficients of determination for body height and upper arm load, in the case of static body load and intermittent, for the male population.

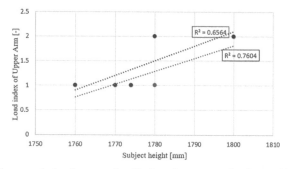

Fig. 4. Comparative correlation between load index of upper arm load vs. subject height for male population; red colour-static posture, blue colour-intermittent posture

Coefficient of determination (R^2) between load index of forearm load vs. subject height for male population, for the case of static posture had a value of 0.52, while in the case of intermittent posture 0.55. With intermittent body load for the male population, a better correlation (0.76) was obtained between body height and upper arm load, as well as upper arm load and arm length (0.75).

In contrast to the static load on the body, the correlation between forearm load and chest width and foot width is significantly higher, and is 0.70 (Table 6).

Table 6. Correlation for the male population during intermittent load

Coefficient of determination (R^2)		
	Upper arm	Forearm
Height (mm)	0.76	0.52
Arm length (mm)	0.75	0.56
Torso width (mm)	0.01	0.70
Foot width (mm)	0.007	0.70

Table 7 shows there is no correlation between the upper arm and the width torso and the width of the feet, when it comes to the male population.

In the case of the male population, there is a better correlation between upper arm load and height compared to female population. There is no correlation between upper arm load and torso width as well as foot width in both populations (Table 7).

The best correlation (0.78) was observed between the forearm length of the arm length for the female population. Figure 5 shows a comparative representation of the coefficient of determination (R^2) for body height and upper arm load for the female population.

Table 7. Correlation for the female population during intermittent load

Coefficient of determination (R^2)

	Upper arm	Forearm
Height (mm)	0.75	0.45
Arm length (mm)	0.20	0.78
Torso width (mm)	0.0002	0.58
Foot width (mm)	0.04	0.10

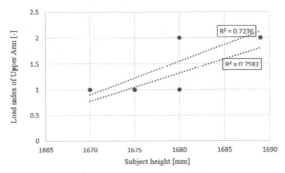

Fig. 5. Comparative correlation (R^2) between the subject height and load index of upper arm for female population; red colour-intermittent posture, purple colour-static posture

From the Fig. 5 it can be concluded that there is a higher correlation (0.7582) when the case is intermittent body load. Figure 6 shows a comparative correlation of subject height and forearm for the female population.

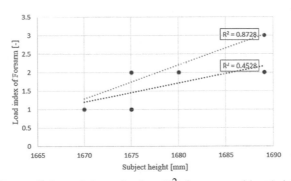

Fig. 6. Comparative coefficient of determination (R^2) between subject height and load index forearm for female population; red colour-intermittent posture, purple colour-static posture

It is noticed that there is a better correlation at static load on the body. However, the load on the forearm is higher at static load on the body compared to intermittent load.

4   Conclusion

The aim of this study was to determine the load on the upper body while riding an electric scooter. Testing was performed in the Catia software package using RULA analysis. Ten digital human models (five male and five female) were analyzed. Their different anthropometric characteristics were taken into account. The obtained results show that there is a correlation between body height and upper body load, but it is more pronounced in female than in male subjects. The highest load of the upper arm (3) in the male population has a subject height of 1800 mm, while the same load was recorded in the female population at a height of 1689 mm. Women with a height of 1670 mm and men with a height of 1760 mm had the lowest load on the forearm (1). The results show that, when using a given electric scooter, with increasing height, the load on the upper torso increases in both populations. By applying the RULA analysis, it is possible to determine which anthropometric dimensions are best suited for the use of a given electric scooter, with the least load on individual parts of the body.

Acknowledgements. This research supported by the Ministry of Education, Science and Technological Development of Republic of Serbia through Grant TR35041.

References

1. Dediu, H.: Where does the Word Micromobility come From? Micromobility industries, https://micromobility.io/blog/2019/8/1/where-does-the-word-micromobility-come-from. Accessed 26 Nov 2019
2. https://assets.ey.com/content/dam/ey-sites/ey-com/en_gl/topics/automotive-and-transport ation/automotive-transportation-pdfs/ey-micromobility-moving-cities-into-a-sustainable-fut ure.pdf, pristupljeno 19.11.2021
3. McAtamney, L., Corlett, E.N.: RULA: a survey method for the investigation of workrelated upper limb disorders. Appl. Ergon. **24**(2), 91–99 (1993)
4. Huat, N.L., Shafizal, M., Dullah, A.R., Khalil, S.N.: RULA analysis of the headrest head supporter on car seat. In: Proceedings of Mechanical Engineering Research Day: 288–290 (2020)
5. Gowtham, S., et al.: Seating comfort analysis: a virtual ergonomics study of bus drivers in private transportation. In: 3rd International Conference on Advances in Mechanical Engineering, ICAME (2020)
6. Mat, S., Abdullah, M.A., Dullah, A.R., Shamsudin, S.A.,Hussin, M.F.: Car seat design using RULA analysis. In: Proceedings of Mechanical Engineering Research Day (2017)
7. Kee, D., Karwowski, W.: Acomparison of three observational techniques for assessing postural loads in industry. Int. J. OccupSaf Ergon. **13**, 3–14 (2007)
8. Hoy, J., Mubarak, N., Nelson, S., de Landas, M.S., Mag-nusson, M., Okunribido, O., Pope, M.: Whole body vibration and posture as risk factors for low back pain among forklift truck drivers, J. Sound Vib. **284**, 933–946 (2005)
9. https://www.amazon.com/Hover-1-Electric-Scooter-Display-High-Grip/dp/B09J93CL44. Accessed 28 Dec 2021
10. Ansari, N.A., Sheikh, M.J.: Evaluation of work posture by RULA and REBA. Case Study IOSR J. Mech. Civil Eng. **11**, 18–23 (2014)
11. Malchaire, J., Gauthy, R., Piette, A., Strambi, F.: A classification of methods for assessing and/or preventing the risks of musculoskeletal disorders, European Trade Union Institute (2011). ISBN 978–2–87452–223–9

Mechanism Design for a Low-Cost Automatic Breathing Applications for Developing Countries

Marco Claudio De Simone[1]([✉]), Giampiero Celenta[2], Zandra B. Rivera[3], and Domenico Guida[1]

[1] Department of Industrial Engineering, University of Salerno, Via Giovanni Paolo II, 132, 84084 Fisciano, Italy
{mdesimone,guida}@unisa.it
[2] MEID4 Academic Spin-Off of the University of Salerno, Via Giovanni Paolo II, 132, 84084 Fisciano, Italy
[3] Engineering Research Institute, Universidad Privada San Juan Bautista, Av. José Antonio Lavalle N° 302-304, Lima 15067, Perú

Abstract. The aim of this paper is to present a design activity conducted for developing a new device for automatic breathing applications for developing countries using, to generate the necessary airflow, an Ambu (Auxiliary Manual Breathing Unit), a self-expanding bag for manual ventilation, used for support respiratory activity.

The idea is to create a mechanism to automate the device, through the use of open-source components and/or in any case easily available, and make it usable for intensive care applications for sedated patients or to support ventilatory activity in patients in cardiac arrest. The requirements of the device are essentially two: high reliability guaranteed by the automatic and/or mechanical transmission and adjustment mechanism of the airflow and the characteristic breathing times, all made accessible at a low cost.

Keywords: Multibody · Dinamics · Ventilation device · Ambu · Simscape

1 Introduction

Mechanical ventilators are one of the essential elements in ICU to treat respiratory insufficiency caused by the different variants of the virus, among other diseases [1]. According to WHO figures, the total number of people diagnosed positive exceeds 318 million globally, with Europe and the Americas on top of the list with 116 million each. The total number of deaths exceeds 5 million worldwide and both regions have 1.7 and 2.4 million deaths respectively [2]. These figures give a rough idea of the number of mechanical ventilators needed to cope with this pandemic. The problem of the lack of mechanical ventilators is still latent since there are still massive infections. Globally, the vaccinated population is estimated at 49.8%, who remain susceptible to infection but are not expected to require intensive care on a massive scale. However, the other half of the population is still at risk of developing bronchopulmonary infectious conditions that could become complicated and require admission to the ICU [3].

© The Author(s), under exclusive license to Springer Nature Switzerland AG 2022
I. Karabegović et al. (Eds.): NT 2022, LNNS 472, pp. 345–352, 2022.
https://doi.org/10.1007/978-3-031-05230-9_40

2   Materials and Methods

Breathing is the physiological process, fundamental for life, which allows the human organism to withdraw from the air the oxygen necessary for the survival of organs and tissues, and, at the same time, to dispose of the carbon dioxide generated by the activity of cells [4]. The mostly unconscious phenomenon, breathing consists of two moments that alternate cyclically after a pause: the inspiratory moment (or inspiration), through which the human organism enters the atmospheric air into the lungs and takes oxygen from this, exchanging it with anhydride carbon dioxide, and the expiratory moment (or exhalation), by which the organism expels the atmospheric air from the lungs by now poor in oxygen but rich in carbon dioxide.

Breathing is cyclic and consists of three phases:

Inspiration (I): During this phase, air is introduced into the lungs. Inspiration occurs thanks to the contraction of the intercostal muscles and the diaphragm, this contraction causes an increase in lung volume and a decrease in intrapleural pressure: the result is an aspiration of air into the lungs. For this dynamic to be activated, the pressure in the lungs must become lower than atmospheric pressure.

Exhalation (E): During this phase, the air is expelled from the lungs. The chest volume decreases, the lungs are compressed and the air expelled. The efficiency of the exhalation is above all passive, unlike the inhalation, and is due to the elastic return of the lung tissue.

Pause (P): Between one exhalation and the next inhalation there may be a pause of variable duration.

Lungs are the site of gaseous exchanges that allow the circulating blood to come into contact with the gases of the air and eliminate CO_2 and take on O_2 (semipermeable membrane of the alveoli). To this end, in an average adult, there is an entry of ambient air into the respiratory tract of about 500 ml. for each respiratory act (tidal volume) and about 12 acts per minute for an adult, and 23–40 per minute for newborns [5]. These values increase under exertion or due to illness. For each breath, the inhalation lasts about half of the exhalation: I over E ratio of 1: 2.

In mechanical ventilation instead of expanding due to negative pressure, the lung expands thanks to the active introduction of air under pressure and therefore the physiological principle of the negative intrathoracic pressure is subverted and circulation problems (reduction of venous return) and circulation problems can be created. baro trauma on the delicate cells that form the alveolar semipermeable membrane. This concept explains the need to always try to mechanically ventilate with the lowest possible pressures [6].

Therefore, mechanical ventilation is a form of instrumental therapy which, through a mechanical ventilator, supports the patient with severe respiratory insufficiency, allowing him to ventilate adequately and maintaining normal gas exchanges between the lungs and the environment.

The ventilators have a closed loop system for which a Tidal Volume is established that is administered to the patient, the Tidal Volume received by the patient (called TVe) is recorded, which is analyzed and compared with the Tidal Volume inspiratory (TVi),

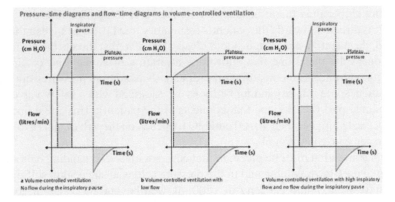

Fig. 1. Pressure diagrams vs. time and volume flow vs time in volume controlled ventilation

adjusting the output to the next breath. The User Interface is a touch monitor for setting the ventilation where parameters, limits, and alarms are established via software.

Generally, at least three parameters are shown, two of which are measured (pressure and flow), while the third value (volume) is derived, since the parameters measured directly by the ventilator are pressure, flow, time, temperature, and humidity.The volume is determined based on the analysis of the flow over time through a valve whose area is known; care must be taken because, in the event of a system leak, the calculated volume does not appear reliable.In the case of volume-controlled ventilation (VCV) the independent variable is the volume that is administered at constant flow, while the pressures vary in terms of peak, plateau, and PEEP [7]. The flow curve has a characteristic square wave linked to the initial need for high flows to overcome the resistance of the airways (mechanical and natural); a constant flow follows which then suddenly drops to zero until the expiratory phase (see Fig. 1).

2.1 The Ambu-Auxiliary Manual Breathing Unit

The AMBU self-expanding bag is used to provide manual ventilatory support for those patients with insufficient lung ventilation.They can be connected or not to external sources of O_2 and are usually used: during cardiopulmonary resuscitation (CPR), to support ventilation and oxygenation in acute patients, in intra and extra-hospital transport, and also to perform therapeutic hyperinflation maneuvers [8]. AMBU balloons are composed of a compressible unit (the self-expanding balloon) at the ends of which there are two one-way valves: one allows the air to enter the balloon, the other conveys the air to the outside, avoiding the phenomenon of rebreathing (inhalation of exhaled air). At the proximal end, the AMBU balloons are equipped with a universal fitting (15 mm) for connection to masks, endotracheal tubes, HME filters, catheter mounts, and tracheostomy tubes. O_2, whose concentrations depend on the source used and the presence or absence of reservoirs, can be connected to the connector located in the distal portion of the balloon [9]. To manually ventilate an intubated or tracheostomized user, the balloon must be connected to the HME-mount complex or directly to the cannula or tube (in emergency situations). Then the balloon is compressed, thus generating inside it a pressure higher than the atmospheric one; the valve system ensures that the compressed air is directed towards the patient's respiratory tree [10]. Then when the balloon is released, the negative pressure inside it closes the valve that directs the airflow to the user and opens the distal one that allows the aspiration of ambient air or air, and O_2 inside. With reference to the insufflation pressure, to avoid high pressures during manual ventilation, some self-expanding balloons are equipped with a safety valve, located in the proximal part (i.e. the one that connects to the patient): This valve opens when pressure is created higher than 40–60 cmH2O (depends on the valve), thus avoiding that ventilation can cause barotrauma.

Particular attention must be given, in maneuvers with self-expanding balloon, to the evaluation of the blown volume and to the pressure achieved since adult self-expanding balloons have a maximum capacity of 1600 ml, well beyond the value of about 500–600 ml equal to tidal volume to be delivered to the patient. This makes it clear that the self-expanding balloon must never be fully pressed, but it is sufficient to compress it with one hand to deliver the correct volume [11].

The functional requirements required of the retrofitted device are shown in Table 1.

Table 1. Functional requirements required of the retrofitted device

Parameters		Pediatric age	Adults
Breath frequency	(per min)	30 ÷ 40	12 ÷ 16
Inspiratory time	(sec)	0.4 ÷ 0.5	1,2 ÷ 1,4
Ratio breathing times	$\left(^{t_i}/_{t_e}\right)$	1:1,5 ÷ 1:2	1:2 ÷ 1:4
Inspiratory flow	(l/min)	2 ÷ 3	24
Volume per inspiratory act	(ml)	18 ÷ 24	500
	(ml/kg)	6 ÷ 8	6 ÷ 8

2.2 Mathematical Model

For the calculation of the necessary compression, the shape of the AMBU was approximated to a cylindrical tank with hemispherical bottoms (see Fig. 2). The goal is to identify the volume between the external envelope and a secant plane placed at h.

The volume variation of the spherical cap, which is a function of the advancement of the secant plane for the hemispherical bottoms and of the tank radius, can be calculated using the following equation:

$$\Delta V_{sphere} = \pi h(t)^2 \left(r - \frac{h(t)}{3} \right)$$

The volume variation for the cylindrical body, on the other hand, can be expressed as follows:

$$\Delta V_{cylinder} = L \left(r^2 \cos^{-1} \left(\frac{r - h(t)}{r} \right) + (h(t) - r) \sqrt{2rh(t) - h(t)^2} \right)$$

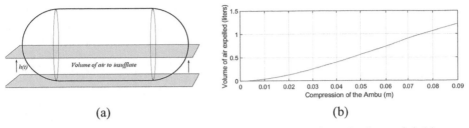

(a)   (b)

Fig. 2. a) Cylindrical tank with hemispherical bottoms and b) estimated volume of air blown as the position of the secant plane varies.

Having indicated with r and L respectively the curvature and the length of the cylindrical portion of the tank. Instead, with $h(t)$, it is indicated the progress of the cursor that comes into contact with the AMBU by compressing it [12, 13].

From the estimate made, considering the AMBU as a cylindrical tank with hemispherical bottoms, it is possible to evaluate the compression useful for delivering the 0.5 L to be blown into the patient. In Fig. 2b, the trend of the volume that can be dispensed as the progress of the cursor changes is diagrammed. From the graphs, it is possible to estimate that a stroke of the compression system of about 5 cm is required to deliver the required volume of gas [14].

2.3   Design of the Laws of Motion and the Camsprofile

At the basis of the specific constructive and operating schemes of the system are the project specifications regarding both the characteristics of the flows to be obtained from the machine, and the versatility of the energy source to be used for operation.It is required that the output flow of the balloon can be delivered following specific time laws [15]. In particular, four different delivery laws are required which share the same maximum compression value of the balloon and the same cycle time.

The trajectory chosen to define the laws of motion is the cycloidal one, which compared to other possible laws that present a discontinuity in acceleration in the initial and final instants, and therefore undefined (or infinite) jerk values. The numerical activity was conducted in Matlab software [16, 17].

To satisfy these requirements, a system based on four side-by-side cams was chosen, one for each hourly rate required for the flow.

As an example, Fig. 3 shows the development of the geometrical shape of the cam starting from the laws of motion for the pusher for the third characteristic ratio [18].

To allow the four cams to act alternately on the tappet, the cam group has been made sliding along the rotation axis of the crankshaft. In Fig. 4 are reported the motion laws requested and the and the flow generated for the same case-study [19, 20]. The concepts reported in Fig. 5, represent two views of the system under development. The machine was conceived by imagining an unclosed seat for the balloon and a box-like structure that isolates the moving parts necessary for operation from the outside. The translation of the cam group along the axis is carried out by means of a space cam system. The operator acts on the knob connected to the cam, the rotation generates the translation of

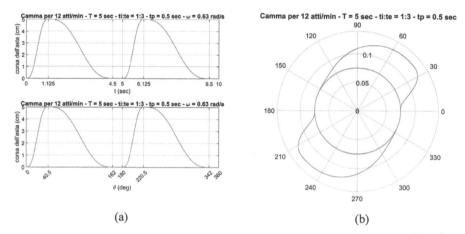

Fig. 3. a) Cycloidal motion laws and b) geometric shape of the cam for the ratio $\frac{t_i}{t_e} = \frac{1}{3}$.

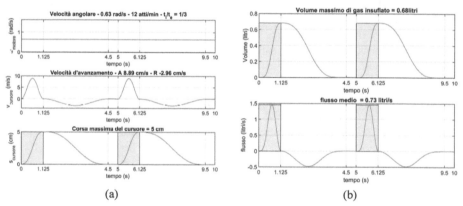

Fig. 4. a) Cycloidal motion laws and b) air flow estimation for the ratio of the times $\frac{t_i}{t_e} = \frac{1}{3}$.

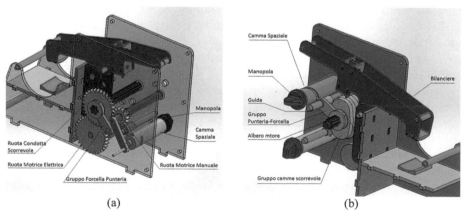

Fig. 5. Concept of the machine: a) detail of the crank; b) detail of the knobs for adjusting the relationship between breathing times and manual or motorized operation.

the tappet-fork group along the appropriate guide and therefore, of the cam group. It is required that the machine can operate on two different energy sources [21]. Operation in ordinary conditions must be guaranteed by an electric motor but, it must be possible to convert the electromechanical power supply to manual [22].

3 Conclusion

In this paper, a new device for automatic breathing applications for developing countries was presented using, to generate the necessary air flow, an Ambu normally used to temporarily support respiratory activity [23, 24]. It was decided to make the device more reliable and easy to maintain, using a cam motion transmission system of different geometry for the compression of the balloon to achieve the different breathing ratios required. From the numerical study carried out it is clear that the correct choice allow to generate the necessary flow of air for the patient in the appropriate conditions required.

References

1. Dašić, P., Dašić, J., Antanasković, D., Pavićević, N.: Statistical analysis and modelling of global innovation index (GII) of Serbia. In: Karabegović, I. (ed.) NT 2020. LNNS, vol. 128, pp. 515–521. Springer, Cham (2020). https://doi.org/10.1007/978-3-030-46817-0_59
2. Gulla, K.M., Kabra, S.K., Lodha, R.: Feasibility of pediatric non-invasive respiratory support in low- and middle-income countries. Indian Pediatr. **58**(11), 1077–1084 (2021). https://doi.org/10.1007/s13312-021-2377-1
3. Manrique-Escobar, C.A., Pappalardo, C.M., Guida, D.: A multibody system approach for the systematic development of a closed-chain kinematic model for two-wheeled vehicles. Machines 9(11), 245 (2021)
4. Nachiappan, N., Koo, J.M., Chockalingam, N., Scott, T.E.: A low-cost field ventilator: an urgent global need. Health Sci. Rep. 4(3)(e349) (2021)
5. Tošović, R., Dašić, P., Ristović, I.: Sustainable use of metallic mineral resources of Serbia from an environmental perspective. Environ. Eng. Manage. J. (EEMJ) 15(9), 2075–2084 (2016)
6. Pappalardo, C.M., Lettieri, A., Guida, D.: Identification of a dynamical model of the latching mechanism of an aircraft hatch door using the numerical algorithms for subspace state-space system identification. IAENG Int. J. Appl. Math. 51(2), 346–359 (2021)
7. Dada, S., et al.: Experiences with implementation of continuous positive airway pressure for neonates and infants in low-resource settings: a scoping review. PLoS ONE, 16 (6 June 2021), art. no. e0252718 (2021)
8. Pappalardo, C.M., Lettieri, A., Guida, D.: A general multibody approach for the linear and nonlinear stability analysis of bicycle systems. Part I: Meth. Constr. Dyn. J. Appl. Comput. Mech. 7(2), 655–670 (2021)
9. Dašić, P., Dašić, J., Crvenković, B.: Applications of access control as a service for software security. Int. J. Ind. Eng. Manage. (IJIEM) 7(3), 111–116 (2016)
10. Dašić, P.: Response surface methodology: Selected scientific-professional papers (in Serbian, Slovenian and Russian). Vrnjačka Banja: SaTCIP Publisher Ltd., 2019. – 301 str. ISBN 978–86–6075–054–1 (2019)
11. Pappalardo, C.M., Lettieri, A., Guida, D.: A general multibody approach for the linear and nonlinear stability analysis of bicycle systems. Part II: Appl. Whipple-Carvallo Bicycle Model J. Appl. Comput. Mech. 7(2), 671–700 (2021)

12. Dašić, P.: Reliability of technical systems: Selected scientific-professional papers (in Serbian and Russian). Vrnjačka Banja: SaTCIP Publisher Ltd., 2019. – 308 str. ISBN 978–86–6075–051–0 (2019)
13. Formato, A., Ianniello, D., Romano, R., Pellegrino, A., Villecco, F.: Design and Development of a New Press for Grape Marc., Machines, vol. 7, no. 3, p. 51 (2019)
14. Dašić, P.: Scientific and technological trends: Selected scientific-professional papers (in Serbian). Vrnjačka Banja: SaTCIP Publesher Ltd., 2020. – 305 str. ISBN 978–86–6075–072–5 (2020)
15. Formato, A., Ianniello, D., Pellegrino, A., Villecco, F.: Vibration-based experimental identification of the elastic moduli using plate specimens of the olive tree, Machines, vol. 7, no. 2, art. no. 46 (2019). https://doi.org/10.3390/machines7020046
16. Sun, X., Liu, H., Song, W., Villecco, F.: Modeling of Eddy Current Welding of Rail: Three-Dimensional Simulation, Entropy, vol. 22, art. no. 947 (2020)
17. Liguori, A., Armentani, E., Bertocco, A., Formato, A., Pellegrino, A., Villecco, F.: Noise reduction in spur gear systems. Entropy **22**, 1306 (2020)
18. Pappalardo, C.M., Manca, A., Guida, D.: A combined use of the multibody system approach and the finite element analysis for the structural redesign and the topology optimization of the latching component of an aircraft hatch door. IAENG Int. J. Appl. Math. **51**(1), 175–191 (2021)
19. Manrique Escobar, C.A., Pappalardo, C.M., Guida, D.: A parametric study of a deep reinforcement learning control system applied to the swing-up problem of the cart-pole. Appl. Sci. **10**(24), 9013 (2020)
20. Pappalardo, C.M., Guida, D.: Dynamic analysis and control design of kinematically-driven multibody mechanical systems. Eng. Lett. **28**(4), 1125–1144 (2020)
21. Salvati, L., d'Amore, M., Fiorentino, A., Pellegrino, A., Sena, P., Villecco, F.: Development and Testing of a Methodology for the Assessment of Acceptability Systems. Machines, vol. 8, no. 47 (2020)
22. Pappalardo, C.M., Lombardi, N., Guida, D.: A model-based system engineering approach for the virtual prototyping of an electric vehicle of class l7. Eng. Lett. **28**(1), 215–234 (2020)
23. Salvati, L., d'Amore, M., Fiorentino, A., Pellegrino, A., Sena, P., Villecco, F.: On-road detection of driver fatigue and drowsiness during medium-distance journeys. Entropy **23**(2), 135 (2021)
24. Formato, A., Romano, R., Villecco, F.: A novel device for the soil sterilizing in sustainable agriculture. In: Karabegović, I. (ed.) NT 2021. LNNS, vol. 233, pp. 858–865. Springer, Cham (2021). https://doi.org/10.1007/978-3-030-75275-0_94

On the Possibility of Designing a New Piston Scheme with Two Rings

Erjon Selmani[✉] and Koçi Doraci

Universiteti Politeknik I Tiranes, Tirane, Albania
eselmani@fim.edu.al

Abstract. The traditional scheme of three-ring automovive pistons, have been changed towards a new scheme with two piston rings. The main focus of the analysis were the inter-ring gas leakages, but in addition to this, the oil consumption and friction losses were also investigated. A piston with only two rings is smaller and hence lighter, and with reduced manufacturing costs. According to results, the simple elimination of a ring leads to the increase of the blow-by, oil consumption and the friction, but if tighter tolerances are applied, optimistic results in terms of oil consumption can be obtained with a slight increasein the blow-by and friction.

Keywords: Piston rings · Blow-by · Oil consumption · Internal combustion engines

1 Introduction

Internal combustion engines has undergone several improvements in the latest years, one of them is the trend of downsizing. On the other side, this trend has been accompanied by a general reduction in fuel consumption, pollutant emissions and frictions.Among the several elements in engines, blow by is an indication of the quality of tightness of the ringpack. Several studies has been performed on the topic of the blow by [1–4], but in all the cases the piston consisted of a three-ring pack. In [5] physical tests were made on a two ring piston, but the focus was the oil consumption of the engine. This article proposes an analysis of the blow-by gases as well as the consumption of lubricating oil by an engine to which a change has been applied. Specifically, the second ring has been eliminated and some geometric and mechanical parameters have been modified. All the geometrical data, together with the related working parameters of the engine were written in ©Ricardo RINGPAK for the solution.

2 Approach for Ring and Gas Dynamics. Oil Consumption Mechanisms

Piston rings are curved beams with an end gap, typically there are three rings, two compression rings and one oil ring Fig. 1. On the ring section are applied all the forces and moments, Fig. 2.

© The Author(s), under exclusive license to Springer Nature Switzerland AG 2022
I. Karabegović et al. (Eds.): NT 2022, LNNS 472, pp. 353–359, 2022.
https://doi.org/10.1007/978-3-031-05230-9_41

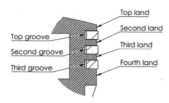

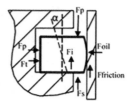

Fig. 1. Piston and ring area   **Fig. 2.** Forces on the ring

In order to compute gas flow from the combustion chamber and the related oil flow, the pressure distribution in each zone must be calculated for each crank angle. On this purpose, the ideal gas Eq. (1) is coupled with the continuity Eq. (2). The mass flow through the gap and through ring-groove clearance can be modeled using the orifice-flow model for a laminar and compressible flow [6]. The ring has three degree of freedom, and hence, the Newton's second law of equilibrium can be applied in axial and radial direction and in the twist motion of the ring, Eqs. (3–5) can be written:

$$\left(\frac{dp}{dt}v + \frac{dv}{dt}p\right) = \left(\frac{dm}{dt}\right)RT + mR\left(\frac{dT}{dt}\right) \tag{1}$$

$$\frac{dm}{dt} = Q_{in} - Q_{out} \tag{2}$$

$$m_r\frac{d^2x}{dt^2} = F\_Pressure - F\_inertia - F\_squeese - F_{friction} \tag{3}$$

$$m_r\frac{d^2y}{dt^2} = F\_gas\_back + F\_tension - F\_gas\_front - F\_oil \tag{4}$$

$$I\frac{d^2\alpha}{dt^2} = M_{gas} + M_{oil\_groove} + M_{oil\_cylinder} - M_{twist} \tag{5}$$

Regarding the consumption of lubricating oil, there are four ways of consumption, which are shown in Fig. 3; a) consumption through inertia, b) consumption due to the transport of blow-back gasses, c) consumption due to evaporation and consumption due to throw of the piston. In addition to this, a small part of the oil is carried with the blow-by gasses into the crankcase.

2.1 Simulation

The three-ring piston has been extensively analyzed so far, clearly specifying the oper-ation and function of each ring. The main sealing role is played by the first ring and the role of oil distribution is played by the third ring. The second ring performs a hybrid function, as it partly serves to stop the gases and partly to scratch the excess oil from the cylinder liner.

Also, each of these rings, but especially the first two, are energy consumers due to friction. Since the measures to reduce friction in internal combustion engines have

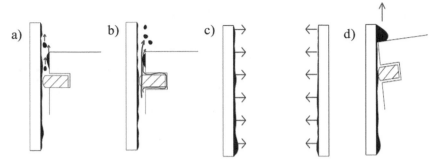

Fig. 3. Oil consumption

been intensive in recent years, we thought to analyze the behavior of the system by removing a ring. For this reason we will analyze the behavior of the system after we have completely removed the second ring. The simulations were performed on the parameters of a turbodiesel engine, operated at maximal load, and speed of 2000 rpm.

The simulations were done in three steps, first eliminating only the second ring, then eliminating its groove, subsequently reducing the height of the second crown and increasing it's diameter both with nearly 40%, as depicted in Fig. 4a, b and c.

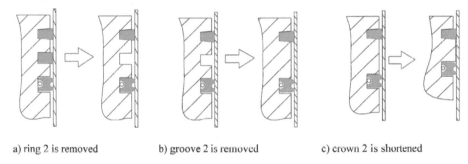

a) ring 2 is removed   b) groove 2 is removed   c) crown 2 is shortened

Fig. 4. Piston and inter-ring area in simulations

The original system together with the three cases have been implemented in ©Ricardo and solved in the RINGPAK solver.

3   Results and Discussion

The results have obtained for the original system and for each of the three sub-cases, will be presented through some tables. In each case, only the most relevant outcome of the simulations will be given. As a matter of comprehension, for the base case there will be given some graphs.

The first graph of Fig. 5 describes the axial location of the three rings in the respective grooves, and together with the third graph, which describes the radial locations of the rings in the grooves, are indicators of the ability of the rings to seal the gas route from

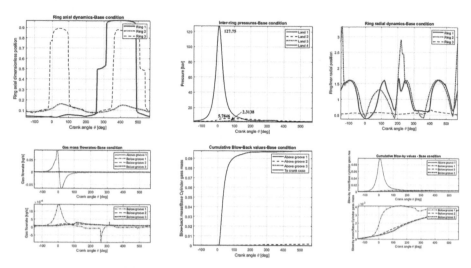

Fig. 5. Results of simulations for the base case

the combustion chamber to the crankcase. In this specific case, nor in the axial nor in the radial direction there are any irregularities, i.e. ring flutter or ring radial collapse. In the central graph are given the inter-ring pressures, while the other graphs of the cited figure shows the mass flowrates, the blow-by and the blow-back values of the gas for this case.

Table 1. Main results for the base case.

Base case		
Oil consumption		
Throw	0.00127	g/cycle
Evaporation	0.000112	g/cycle
Gas transport	$2*10^{-6}$	g/cycle
Total	0.001384	g/cycle
Direct blow-by	0.00632	L/cycle
Reverse blow-back	0.188	L/cycle
Mean Friction	110	N
Mean power loss	896	W

Table 1 shows the complete results of the simulations for the base case. In this table are given numerical values of the lubricating oil consumption, the gas flows and the mechanical losses for the base case condition at 2000 rpm and max load.

In Table 2, columns a), b) and c) are given the results of the simulations for cases explained in Fig. 4.

Table 2. Results for the modified cases.

Other cases	a) Ring 2 removed		b) Grove 2 removed		c) Crown 2 shortened	
Oil consumption						
Throw	0.00262	g/ cycle	0.00471	g/ cycle	0.000649	g/ cycle
Evaporation	0.000112	g/ cycle	0.000112	g/ cycle	0.000112	g/ cycle
Gas transport	0	g/ cycle	0	g/ cycle	$7.8*10^{-9}$	g/ cycle
Total	0.002732	g/ cycle	0.004822	g/ cycle	0.0007610	g/ cycle
Direct blow-by	0.0334	L/ cycle	0.0125	L/ cycle	0.00987	L/ cycle
Reverse blow-back	0.196	L/ cycle	0.184	L/ cycle	0.183	L/ cycle
Mean Friction	127	N	123	N	119	N
Mean power loss	1010	W	990	W	937	W

As it is possible to see from Table 2, passing from the removal of the second ring to the removal of the second groove, an increase of total oil consumption occurs. On the other hand, case c) shows an inversion of trend with respect to the previous two and with respect to the base case given in Table 1. In case c the second crown height is reduced with around 6 mm, and is increased in diameter, passing from 5 μm to 3 μm of tolerance of coupling to the cylinder.

The direct blow-by, leaked from the combustion chamber to the crankcase, results to be higher in the modified cases with respect to the base case. In cases a) and b) this parameter is increased with one order, while in case c) the increment is lower than in the previous two.

The reverse blow-by, also called the blow-back, seems to not vary too much between the base case and the modified cases. In particular, case a) shows an increase of this value while cases b) and c) are lower with respect to the base case.

At last observation, the mean friction and the power loss are higher for the modified cases with respect to the base case, with a steady decrease trend from case a) to case c).

This behaviour needs to be further analysed in order to understand the reasons behind those results. The removal of the second ring from a traditional 3-ring piston, affects the working conditions of the other two rings. The second ring, together with its upper and lower crowns, act as a pressure relief toward the third ring. When this ring is removed, the majority of the blow-by gasses from the first ring escapes in the crankcase through the third ring. For this reason there is an increase of blow-by values among the modified cases. This behaviour is verified in the third case where the reduction of the inter ring volume is accompanied with the reduction of the blow-by.

The second ring's main function is the scrape of excess oil from the liner. As long as the second ring is missing, the excess oil on the liner will be subject to various transport mechanisms and the total oil consumption will be higher.

In the third case, the lower value of oil consumption compared to the base case seems to be an exception. In fact we must clarify that the nominal oil film thickness for this model is 5 μm. As a consequence, the increase of the second land diameter till a 3 μmtolerance to the liner, transforms the entire second land into a ring itself because of the scrape of the oil from the liner. For this reason the oil consumption is one order lower but also the blow-by is lower.

The mean friction and power loss are also affected from this design change proposal. Although with one ring less, the friction should be lower, in fact the results showed the opposite. As long as it is expected to have a thicker oil film, there will also occur higher oil tangential stresses against dhe relative motion of the rings and the piston, which has also its secondary motion.

4 Conclusion

In this article the behaviour of a piston and ringpack was analysed through virtual simulations. Blow-by, oil consumption and mean friction were some of the parameters evaluated. Starting from a traditional 3-ring piston the article proposed the change of design toward a 2–ring piston. For this pourpose three cases were analysed: case a) where the second ring was removed, case b) where also the second groove was removed, case c) where the second land's height was reduced and its diameter increased. According to the results of the simulations, the removal of the second ring together with its groove brings to an increase of the blow-by, oil consumption and friction losses.

If, apart from the previous design change, the second land height is reduced and its diameter increased, with the aim to result in a smaller second land volume, then the oil consumption can be improved with respect to the previous cases, including the base case. The other two parameters, namely the blow-by and the mean friction, resulted in slightly higher-yet comparable values to the base case, but lower values in comparison to the previous cases. The obtained results are optimistic, in particular those of case c), where the second land volume was reduced, bringing closer the top and third rings and the piston to the liner.

These proposed changes must be further studied, in particular the 3 μm gap used in case c), which is the nominal piston-liner tolerance of some modern engines, hence this value must be deeply analysed for potential application.

Nomenclature

$M =$	Gas mass [kg]
$V =$	volume of inter-ring crevices [m^3]
$p =$	pressure of inter-ring crevices [Pa]
$T =$	temperature of gas in the crevices [K]
$mr =$	ring mass [Kg]
$x,y =$	positions of the ring center with respect of the groove center [K]

R =   characteristic gas constant [J/KgK]
Qin, Qout =   characteristic gas constant [Kg/s]

References

1. Keribar, R., Dursunkaya, Z., Flemming, M.F.: An integrated model of ring pack performance. ASME Trans. J. Eng. Gas Turb. Power **113**, 382–389 (1991)
2. Tian, T.: Dynamic behaviours of piston rings and their practical impact. Part 2: oil transport, friction and wear of ring/liner interface and the effects of piston and ring dynamics. Proc. Instit. Mech. Eng. Part J: J. Eng. Tribol. **216**(4), 229–248 (2002)
3. Selmani, E., Delprete, C., Bisha, A.: Simulation of the cylinder bore distortion and effect on the sealing capacity of the ringpack. SN Appl. Sci. 14 (2019). https://doi.org/10.1007/s42452-019-0303-0
4. Delprete, C., Selmani, E., Bisha, A.: Gas escape to crankcase: impact of system parameters on sealing behavior of a piston cylinder ring pack. Int. J. Energy Environ. Eng. **10**(2), 207–220 (2019). https://doi.org/10.1007/s40095-019-0296-x
5. Takiguchi, M., et al.: Characteristics of friction and lubrication of two-ring piston. JSAE Rev. **17**(1), 11–16 (1996)
6. Wannatong, K., Chanchaona, S., Sanitjai, S.: Simulation algorithm for piston ring dynamics. Simul. Model. Pract. Theory **16**(1), 127–146 (2008)

The Wider Picture of Value Creation in Life and Business Than Lean Production Concept

Ismar Alagić[1,2(✉)]

[1] TRA Tešanj Development Agency, Trg Alije Izetbegovića 1, 74260 Tešanj,
Bosnia and Herzegovina
ismar.alagic@gmail.com
[2] Faculty of Mechanical Engineering, University of Zenica, Fakultetska 1, 72000 Zenica,
Bosnia and Herzegovina

Abstract. Every system is intended to create a "value". Value is "all that contributes to the market form or function of a product or service in such a way those buyers are willing to pay for it". A very common term we encounter is waste. The Lean concept works by recognizing all activities that "do not add value" or what customers are not willing to pay. Lean production concept has recognized seven such losses, even more ones according to Author of article. The basic message of Lean Production concept is Lean Manufacturing = Common Sense. Think about the nuclear weapons race, recent wars, global warming, environmental pollution, GMO and fast food, various cancer causes, modern slavery, poverty, corrupt financial systems, the recent COVID- 19 outbreak, and so on. If we are smarter today why don't we use that knowledge to be happier and create positive changes both as a collective and as individuals? Isn't the most valuable form of knowledge that which can make us happier? Do we really know and understand what values are most important in our lives? Application of "custom design" value architecture through values type identification, time, value definition and value laws are shown in this article.

Keywords: Value · Value creation · Lean production · Time · Value laws

1 Introduction

As business professionals, we work passionately at our offices eight or more hours a day, committed to improving our corporate business performance and creating financial value for our shareholders. The moment we arrive home this professional attitude towards value creation is gone. Author of article has learned and implemented many businesses performance tools over the last 20 years of his business improvement and system control consulting career.

By doing so, Author has realized that most of these tools are directly applicable to our everyday lives. In this article, it will present you how to implement Value creation in order to create positive changes.

I. Karabegović et al. (Eds.): NT 2022, LNNS 472, pp. 360–367, 2022.
https://doi.org/10.1007/978-3-031-05230-9_42

2 Type of Values

The word "value" can be defined in several different ways [1–4]. Oxford dictionary defines value in two ways:

a) "the regard that something is held to deserve; the importance, worth, or usefulness of something."
b) "Principles or standards of behavior; one's judgement of what is important in life."

There are also numerous examples of synonyms for the word value because of its different meanings, such as: merit, worth, wealth, utility, usefulness, wellbeing, health, creativity,invention, reasoning, fairness, empathy, honesty, responsibility, ethics, morals, generosity, peace, love, etc. The term "value" can be applied to a number of academic and scientific disciplines, namely social sciences such as sociology, psychology, economics, and so on. Formal and natural sciences such as mathematics, physics, and chemistry are less likely to use the term "value".

We use the world "value" several times a day in different contexts. In the financial, business, corporate and economic sphere it is often related to the price of a product or service. Many scientists have researched the importance of different forms of value in relation to economy, finance and marketing. The knowledge that arose has been strongly incorporated into the discipline of management. During this survey survey Author has identifed following forms of value: financial value, economic value, added value, perceived value, emotional value, social value and sustainable value.

At some point during your career you have probably heard of PV (Present Value) or NPV (Net Present Value). Investopedia defines NPV as: *"Net present value (NPV) is the difference between the present value of cash inflows and the present value of cash outflows over a period of time."*

NPV can be expressed as an equation:

$$NPV = PV(benefits) - PV(costs) \tag{1}$$

The calculation is done in a specific currency, for example USD. A positive net present value indicates that the projected earnings generated by a project or investment – e.g. in present dollars - exceeds the anticipated costs, also in present currency. It is assumed that an investment with a positive NPV will be profitable, and an investment with a negative NPV will result in a net loss Fig. 1 provides a simplified cash flow model.

The cost represents a negative cash flow, a kind of an investment. If NPV calculation is applied to a project the cost would probably present cost of material, labor etc. In the initial period of time most of projects does not make any profit so "cash out flow" is accumulated in time. This is illustrated with the NPV straight line that initially goes downwards. From the moment that project generates profit, i.e. "cash in" is higher than cash out, this NPV line moves upwards. Basically we continuously sum up the "cash in flow" so our NPV gradually moves upwards. The moment we reach zero, the accumulated "cash out" is equal to accumulated "cash in". If this would be the end of a project there will be no profit but also no loss and NPV would be equal to zero. In financial terminology one can say that no "Value is Created", or "No Added Value is

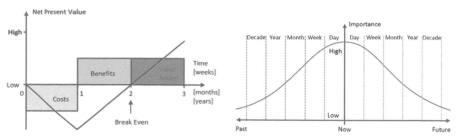

Fig. 1. Net present value.   **Fig. 2.** Time importance.

Created". If project continue accumulating "cash in" flow in period after break even NPV will be positive and then project "Creates Value". Projects with higher positive NPV are more attractive than projects with lower NPV. Projects with negative NPV will not be considered for investments as they "Destroy Value". The return is basically less than investment. In economics the focus is often on price formations, which is when the price determines the value of a product or service, in the marketplace. In the 19th century the pioneers of market economy, e.g. Karl Marx, proposed labor intensity as the most important measurement tool when estimating value. Others claimed that the value of a commodity (product or service) should depend on its utility and the cost of production. For example, Scottish economist John Law defined value using supply and demand; he explains his theory with the water-diamond paradox (although water is more useful than diamonds, it is sold for a lower price because the supply of water is immensely larger than that of diamonds). Likewise, Adam Smith, the father of political economics, introduced the "added value theory" [5].

The term "value added" describes the enhancement a producer incorporates into a product before putting it on the market. These extra features and enhancements added by the producer increase the value of a product. For example, If you think of raw materials as the input in a manufacturing process and the final product as the output, you will notice that the value of the final product has increased compared to the value of the raw material. Perceived value, a marketing concept, is a customer's evaluation of the benefits of a product and its ability to meet their needs and aspirations. Marketing professionals try to influence how customers determine the perceived value of a product by advertising the attributes of the product in a way that makes it sound superior to its competition. The equation below describes perceived product value works:

$$\text{(Perceived) Product Value} = \Sigma \text{ (Perceived) Benefits} - \Sigma \text{ (Perceived) Costs} \quad (2)$$

The symbol Σ represents the total sum. Perceived costs include costs such as procurement costs, installation, financing, servicing during ownership, and time. Reasonably, a higher amount of benefits and a lower cost will increase the perceived product value. It is important to remember that in this framework the purchasing price is just one variable involved in the customer purchasing decision process. In modern marketing, the value of a product is divided between functional and non-functional values. The functional value of a watch is that it tells time. However, despite having the same function, watch prices range from \$10 to \$1,000,000 or more. The marketing theory of value refers to the non-functional value of a product as emotional value. Big corporations define "social value" as Corporate Social Responsibility (CSR).

CSR is the amount of money (a budget) which a corporation spends on a yearly basis to demonstrate its social responsibility. Big corporations carefully consider how to advertise their CSR activities so that their brand value increases. Different communities are using different definitions and interpretations of sustainability. We will assume here that sustainability focuses on meeting the needs of the present without compromising the ability of future generations to meet their needs. In this approach sustainability consist of three main pillars: economic, environmental, and social—informally known as "profits, planet, and people". An other simplified definition of sustainability is to create "long-term value". Companies that focus exclusively on financial metrics (e.g. ROI for investors) often miss the longer-term issues.

3   Time

Let's look at a dictionary definition of time: "The indefinite continued progress of existence and events in the past, present, and future regarded as a whole."

Sounds great but to me sound as a "circular definition", something like: time is time. We all agree the clock does not stop and that everything has the beginning and the end but defining time is not a trivial task. Time is perceived with a kind of our inner biological clock. At some age of five years old we start intellectually perceiving the time and we learn clock concept as well as measuring concepts of second, minute and day. Later on gradually we learn to understand years, centuries or micro-second concepts. At an early age of our life perception is that time moves slowly and it will take ages until we get old.

Lucky ones who reach older age for example eighties, or nineties would comment how life and time moved incredibly quickly. Is this a shifted perception due to the fact that "our time" is coming the end or maybe a neurological phenomenon related to our inner biological clock? Generally if we are in the middle of a pleasant and happy event the perception is that time flies fast. On the other hand during difficult circumstances man has a perception that time flows slowly. In the following sections we will just touch upon the nature of time since the dimension of time is essential in defining the value in the next chapter. In different phases of our life our capability to absorb information from the environment, process it and store it in our brain is different. Our goals, objectives, interests change during life cycle and in some cases we may perceive the same event in a different way. In general the fact is that closer to the current time instant the higher is the relevance and importance of an event. This is the case for both past and future. The importance and relevance is higher for events closer to the "now". Also our capability to perceive events in a far future is limited as well as our capability to predict consequences of an events in far future. Think about weather forecast. Most of time we are interested in today and few days ahead. On longer periods the weather forecast are less predicable and less relevant for decisions we are about to make today. In that respect most of time we act and make decision based on our short terms: hours, days and weeks rather than years and decades. Figure 2 shows the time importance.

4   Value Definition Through Value vs. Happiness

Before we define Value, let's explore the category "happiness" in human life. Let's see what does google dictionary says about happiness [8]:

> "Happiness is that feeling that comes over you when you know life is good and you can't help but smile. It's the opposite of sadness. Happiness is a sense of well-being, joy, or contentment. When people are successful, or safe, or lucky, they feel happiness."

The above definition seems to look at the level of an individual, a kind of emotional feeling one has at some moment in time. A science which is mainly focused on happiness of an individual is "positive psychology science". There is a lot of research done recently on this subject mainly in USA [8]. There is so much literature about happiness out there also beside positive psychology. There are books with titles like "7 habits of happy people" etc. insinuating that it is possible to apply "quick and easy" solution or strategy in order to increase happiness quickly. It is important to emphasize that this book uses completely different methodology. We use notion of happiness, as one "performance indicator" in order to define value. Each individual and each business are unique and they need a unique value creation strategy. In this article we will continue to elaborate on the individual happiness. Recently I heard a funny but interesting definition of happiness: "free time multiplied by money you have". If we have money and free time seems we can go and to anything we want. We can do things which makes us happy. However, we know that not everything is about money and tangible stuff you can buy for money. You can purchase a house, a car, a suite or book a holiday in some remote destination but you cannot buy university diploma, knowledge, family, friends, trust, health and many other important values etc. In today's circumstances, especially in the urban areas, for sure, we cannot be happy without money. It is a "necessary" condition for happiness but not a "sufficient" one. For example, health is also an "necessary" condition too. Another similar definition of happiness is: "free time multiplied by health" Sometimes we do not appreciate our health enough unless we start losing it. In this context as long as we are healthy, the health is considered as "water": It is low price (value) as we have plenty of it. You might have heard of the water/diamond paradox. Finally, we come to the definition of individual value which will use in this article:

$$\text{Value} = \text{Happiness} \times \text{Time} \tag{3}$$

Individual Value is happiness, a positive and pleasant emotion which exists and lasts in time. That inner individual emotion can be more or less happy and it can take shorter or longer period of time.

5   Value Laws

The nature and universe do follow the laws of physics that we discovered, defined and systemized in what we call "science". Humans as part of the nature and the universe are just a specific manifestation of the nature as such. The idea that any set of laws is a subset of a higher sets of laws and dimensions is shown in Fig. 3.

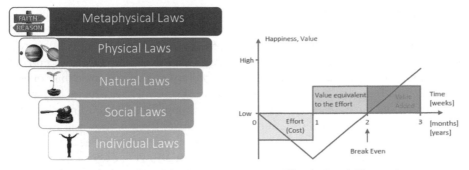

Fig. 3. Type of laws.   **Fig. 4.** Break Even point.

On top of the "intrinsic value" an adult person feels responsibility to create additional value, an additional value for our self, our family and often for our society. In that respect we define goals in our life: to finish university, to have employment, to buy a house in a nice neighborhood, to pursue our dream carrier, to support our kids in their effort to fulfill their dreams and aspirations just like we did ourselves. A very basic low in economics and corporate business is the investment low. In order to gain a benefit, we need to sacrifice some effort. In order to get some financial return, we need to invest some money first. This is illustrated in the Fig. 4.

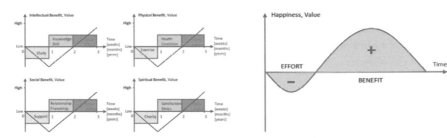

Fig. 5. Non-financial value creation.   **Fig. 6.** Value creation law.

We usually distinguish following three periods in time:

– Initial period of cost, effort and sacrifice;
– Period over we gain a value equivalent to the effort/cost we initially sacrificed/invested, and;
– Period over we get more value than what we initially sacrificed.

The end of the second period we call "break even". In business language we often use term "business opportunity", or "opportunity to create value". Frequent terminology is "opportunity realization process", or "value creation process". Corporates use so called "feasibility studies", "project proposals" and other methods and processes in order to prove that a project will create value over time. The non-financial value creation process is not different from the financial one. Some examples are provided in Fig. 5. From our life experience we know that an effort has to be made in order to gain benefits. There are following four examples: **Intellectual Benefit**: You need to study hard in order to gain knowledge and skills; **Physical Benefit**: You need to do be physical active in order to maintain physical condition and health; **Social Benefit**: When you support other people in office and home it is very likely they will try to give you something in return and; **Spiritual Benefits**: If you help someone in need you will have a moral and spiritual satisfaction. The same Benefit/Cost rule in finance is applicable for non-financial values.

A simplified graphical illustration in case of Break Even is shown in Fig. 4. The Fig. 6 illustrates the value creation law: in order to create Value, we need to put some effort first. In other words, CREATED VALUE = BENEFIT − COST = REWARD − EFFORT. A simplified illustration of a process which can be very complex, lengthy and non-linear in time is shown in Fig. 6. For example, you study very hard, over five years at your university. In that period had sleeping time, holiday time, maybe you were sick for couple of days or weeks etc. In that respect the Fig. 6, illustrates a time profile of a simplified, an average "VALUE CREATION CURVE" which again illustrates the law: No Effort − No Benefit (No Pain − No Gain). The another way to illustrate this law shown in Fig. 8. At this point I would like to make few conclusions: Effort (Cost) is required in order to create value; Time is required to create value, and; Created value (Benefit) is a non-intrinsic value. Although we will keep the terminology "value creation", we would like to reflect on the correctness of this terminology. The term "value extraction" might be more appropriate than value creation. The value was already out there. It is just a manifestation of the physical, natural, social laws and reality. In that respect the "value extraction", "value exploitation" or "value facilitation" would be more appropriate terminology. However, we will the term "created value" as it is well established terminology [6.7.8] (Fig. 7).

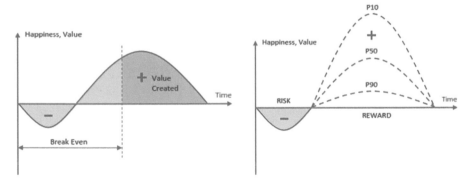

Fig. 7. Value creation and Break Even.   **Fig. 8.** Value creation by managing risk.

There is a saying: No Risk − No Profit. Let me briefly reflect on this law of "value creation" by undertaking a "risky" activity. This is a reality of the current complex financial systems. People invest their money in different way: saving, real-estate, stocks and stock options, state- securities, futures and all kind of so-called financial derivatives.

Some of these financial instruments are based on probability and some on value creation. The process where a financial value is put at risk with a probability to create or destroy value is shown in Fig. 8.

6 Conclusion

Rationally we should always do and plan an activity which create value (VA activities) and avoid any activity which does not (NVA activities) but this is not always possible. In Lean terminology the NVA activities are also called "Waste". The process of avoiding such activities is called the "waste elimination". It is not straightforward to identify NVA activities. They are part of our business as well as our everyday life. There are NVA activities which are an essential and necessary part of an entire process of value creation. An example from manufacturing is keeping inventories of the final products. In an ideal supply chain, a product would be delivered straight from the manufacturer to the customer without any storage in between. Keeping stocks of inventories destroys value but it cannot be avoided. In Lean terminology, an activity which is a non value added (NVA) but it is an unavoidable and essential part of the manufacturing process is called "Business Non Value Added" (BNVA).

The elimination and reduction of NVA activities is the main subject of Lean Production. In spirit of Lean Theory, the non-financial NVA activities will be defined as "Life Non Value Added" (LNVA). It is important to remember the Value Superposition Law, because it is a powerful tool when you manage your value improvement plan. The more activities you combine together the more value you create.

References

1. Alagić, I.: Industrial Engineering & Maintenance: Lean Production - Six Sigma with application of tools and methods in specific working conditions, pp. 391–404, ISBN 978–9958–074–09–7, COBISS.BH-ID 24232454, Štamparija-S, Tešanj (2017)
2. Alagić, I.: The application of DMAIC LSS methods in assembly technology design of filter W 1022/LE 19172, Proceeding, NT-2021, Sarajevo, pp. 401–409 (2021)
3. Alagić, I.: Application of lean six sigma tools in order to eliminate bottlenecks in working conditions firm from B&H, pp. 423–436, 19. međunarodni simpozij o kvaliteti, Plitvice, Hrvatska, 21.-23.3.2018. godine (2018)
4. Alagić, I., Božičković, R., Višekruna, V., Brkić, A.: "Primjena lean six sigma alata u konkretnim radnim uslovima firme iz automobilske industrije", pp. 151–160, Festival kvaliteta 2017, Jahorina, BiH (2017)
5. Milios, J.: Marx's value theory revisited, a value form approach. In: Proceeding of the Seventh International Conference in Economics, pp. 1–14, Ankara, Turkey (2003)
6. Deckers, L.: Motivation: Biological, Psyhological and Environmental, pp. 22–78, Routledge Press, Milton Park (2018)
7. Horn A.J.: Abstract labour & value in Marx's system. J. Soc. Theory 12–19 (2016)
8. Lyubomirsky, S.: The Myths of Happiness, pp. 34–47. Penguin Press, New York (2013)

The Value Management Through Value Architecture in Life and Business

Ismar Alagić[1,2](✉)

[1] TRA Tešanj Development Agency, Trg Alije Izetbegovića 1, 74260 Tešanj,
Bosnia and Herzegovina
ismar.alagic@gmail.com
[2] Faculty of Mechanical Engineering, University of Zenica, Fakultetska 1, 72000 Zenica,
Bosnia and Herzegovina

Abstract. Author of article has learned and implemented many business performance tools over the last 20 years of his business improvement and system control consulting career. In this article, it will present you how to implement Value management in order to create positive changes.

The objectives of this article are to familiarize readers with business concepts, methods, and tools, as well as to provide them with relevant examples on how to apply them in both their businesses and private lives.

It is important to realize that everything we plan and do on a daily basis should have one simple goal: to improve and maximize value creation for yourself and for the people around us. We are the CEOs of our lives. We know what we value and we make priorities and tough decisions every day. Some of us are more proactive and structured in our life-management process than others but after reading this article you will be equipped with essential managerial concepts and methods you will apply in both personal life and business environment. For those who are already familiar with business management concepts this article is a unique opportunity to explore it from a different angle.

Keywords: Value · Value management · Tools and methods · Change Management · Continuous improvement

1 Introduction

As humans we embrace and enjoy life. We exist in time and space yet most of the time we are unaware of all the factors that sustain human existence. We simply expect to enjoy it for as long as possible. However, without being consciously aware of it, any definition and perception of "value" starts and ends with "life". Something that we globally agree on, despite the religion or culture we belong to, is that the most valuable thing in existence is human life. Instinctively, we will do anything to protect, preserve, and prolong our life. Every conscious and educated adult knows that his or her life is limited and sooner or later will come to an end. The only physical continuation of our

© The Author(s), under exclusive license to Springer Nature Switzerland AG 2022
I. Karabegović et al. (Eds.): NT 2022, LNNS 472, pp. 368–375, 2022.
https://doi.org/10.1007/978-3-031-05230-9_43

lives are our successors and our children which is why the most important "value" after our own lives, is the lives of our loved ones including our children, family, and friends.

Regardless of the society or historical period we analyze, systems of "value" have always been based on both the quantity and quality of human lives [1]. An example of this would be the reason we consider the preservation and availability of essential natural resources as valuable; the reason it is perceived as a "value" is because it enables human life to continue existing. The need to protect and prolong our lives, as well as the lives of our loved ones, is instinctual and ingrained in our biology In the current global world at the beginning of the third millennium, most of us live a pretty comfortable life: we wake up early, go to school or office, we have our own business or work in a company, we are married or going to marry soon, we maybe enjoy our retirement and certainly enjoy our smart phones, google apps, Facebook, YouTube channels, summer holidays etc. In such circumstances the main imperative is to be a "happy" person. But what is relationship between our individual "happiness" and individual "value management"? We will try to provide answer on these and many other questions.

2 Value and Value Elements

The definition of individual value which is as follows [2]:

$$Value = Happiness \times Time \tag{1}$$

Individual Value is happiness, a positive and pleasant emotion which exists and lasts in time. That inner individual emotion can be more or less happy and it can take shorter or longer period of time. A combination of more happy feeling over longer period will create higher value than a less happy feeling over a short period of time.

The above definition is quite intuitive as we consider a "single individual" and his or her life over a period of time. In business we can use the same definition as we consider a group of people, i.e. stakeholders such as Customers, Employees, Investors, Suppliers, Partner, Owners, Creditors, Local Communities, Government, Public etc. Anyone who is impacted by our business activity in any way will be exposed to a emotional reaction of our business activity. If a stakeholder is happy about our business impact then we definitely create value to him, otherwise we destroy value to that stakeholder. For example, financial performance indicators such as salary or corporate profit ultimately result in employee or shareholder happiness.

Simple value architecture consisting of following five value elements: finance (Wealth); physical (Health); intellectual (Mind); social (People) and spiritual (Ethics). Figure 1 provides a high level analogy of individual five value elements applied to an enterprise.

VALUE ELEMENT	TOOLS, METHODS, CONCEPTS	MANAGMENT
Financial	Balance Sheet, Income Statement, Cash Flow, Dividend Policy etc.	Finance Department
Physical	Corporate physical infrastructure: Buildings, Offices, Production Facilities, IT infrastructure, etc.	Asset Management
Intellectual	Patents, Know How, Management System and Procedures, Controlled Documents, etc.	Intellectual Property Information Management
Social	All individual stakeholders: Customers, Employees, Suppliers treated as individuals with specific needs.	Customer Relationship Human Resources
Spiritual (Ethical)	Business Principles Compliance, Social & Environmental Responsibility, Anti-Trust, Anti-Corruption, Transparency, etc.	Legal Department

Fig. 1. The enterprise value elements.

Throughout an individual value creation, each value element exists simultaneously in some proportion. A simplified model is below:

1. No financial/natural resources (air, water, food, energy, shelter) - no life. The finance is represented by your surrounding natural and financial resources related to you. This is a kind of "external you".
2. No body - no life: The physical element is a kind of "internal you", your "hardware" and its physical components - organs.
3. No intellectual capacity (i.e., no intelligence) - no life: This is your "software", your operating system with all knowledge and experience. Even primitive living creatures need some primitive level of intelligence in order to survive.
4. No other people (no reproduction, no interconnection) - no life.
5. No spirit (conscience, ethics, morality) - no purpose of life: This is the fifth element connecting dots and making sense of it all.

In this regard we can express Value in a format of an equation:

$$\text{VALUE} = \text{time} \times (\text{WEALTH} + \text{HEALTH} + \text{PEOPLE} + \text{MIND} + \text{ETHICS})$$

$$(2)$$

There is usually one element dominating an event at the time. For example, if we are in a deep learning process, we "create intellectual value" however this will not be possible if we are hungry, sick, depressed, etc. In a way the value elements are enablers to each other and always present in a certain amount and ratio. If any of the five elements doesn't perform well, i.e., has a Low performance it will affect the overall individual value creation i.e., the overall happiness level. It is very natural that all value elements are present in our life simultaneously. During a sequence of events and activities these value elements are present in a certain proportion and create a combined feeling of happiness, see Fig. 2.

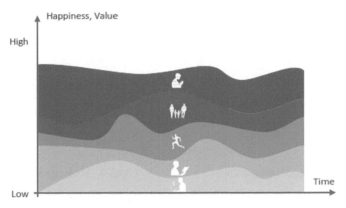

Fig. 2. Value composition.

3 Value Management

Corporate and company management is a complex discipline however its essence is very simple. The responsible manager needs to understand where is the company NOW (situation AS-IS), where it has TO-BE in future and finally HOW to get there. Performance Management System (PMS) is the most popular system in most of companies. A good PMS is focused on both financial and non-financial performance indicators. A company PMS objective is to increase value creation for the key stakeholders. The primary focus must be on customer [3–5].

here are two main ways to increase value creation:

1. Effectiveness (do the right thing): Create Maximum Value.
2. Efficiency (do it in a right way): Minimize Effort and Costs.

The purpose of management system is to be in control of the business activities and results. The same management system is applicable at the individual level. By applying the right management tools in your everyday life, you will be more proactive (you are in control) than reactive (the others are in control). A dictionary will tell us that management is: "the process of dealing with or controlling things or people". Nothing wrong with this definition but we need to modify it slightly for our purpose: "management is process of controlling things or people in order to create value". Figure 3 below depicts the individual management process of value creation.

The management process is really that simple. It starts at a point A and makes sure it arrives at a point B. During this process we create additional value. The starting point A, is the current time "now" and we need to control people and things (i.e., allocate resources) in order to arrive at desired point B in future. Before we manage a project, a company, an initiative or our individual lives we first need to define a set of goals that we want to achieve (point B). Then we need to design a plan of necessary activities and finally we will start to manage and implement this plan. Figure 4 provides more details about how to manage value creation.

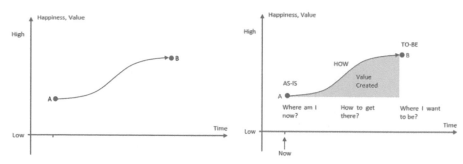

Fig. 3. Value management.   **Fig. 4.** Manage value creation.

The starting point we refer as "AS-IS", in order to describe a current reality. The arrival point we refer as "TO-BE" in order to describe the wanted situation at a specific time in future. The management process we refer as "HOW", which describes our activity plan. This plan is often called "STRATEGY". We should not be afraid to use the "plan" synonyms because they really mean exactly same thing:

HOW = PLAN = STRATEGY = STRATEGIC PLAN = MANAGEMENET PLAN = ROADMAP

The essence of the strategy design is very simple. You just need to answer three questions: - Where am I now? - Where I want to be? - How to get there?

Management of the non-financial value creation is equivalent to the management of the financial value creation. You need to have a solid plan or in other words "a strategy", which is nothing less or more than a structured list of activities which will take place at some specified time in the future. This structured analysis, design, planning and finally execution is basically a typical PROJECT approach. Implementation of project activities will change life or business daily routine and comfort zone. This is a critical moment when we need to accommodate for CHANGE. We are adopting new "ways of living and working" and new habits which will demand a high level of discipline and commitment. This is called "Change Management". Finally, once we went through a challenging CHANGE period there is a period of stability and sustainability with new habits and routines as "new normal" ways of living and working. This period we will consider as CONTINOUS IMPROVEMENT (CI) period. In business we use standard process performance management tools to ensure that we maintain a stable value creation process and also to make incremental improvements. An entire value creation process is depicted in the Fig. 5. The Fig. 5 illustrates entire individual value management process covering three distinct managing periods:

1. Project Management: Definition and Implementation of Critical Activities;
2. Change Management: Accommodating for new Ways of Living and Working;
3. Continuous improvement (CI): Sustaining the Change and Continuous Incremental Improvement.

This Value Creation Improvement process might look complex at the first glance but we will use a simple model for all three management models: Project, Change and CI.

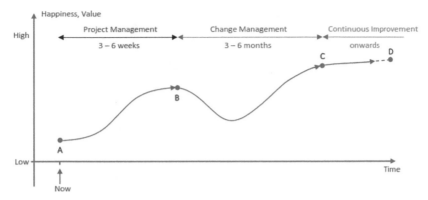

Fig. 5. Project management, change management and continuous improvement period.

You will be surprised how similar they are to each other and often it is just terminology which makes it look different. Figure 6 provides a simplified illustrations of these three value creation methods.

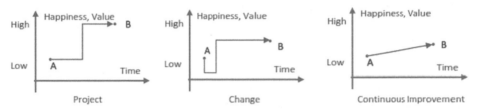

Fig. 6. Project management, change management and continuous improvement model.

Project Approach normally creates a large value. During project we define goals and execute activities. However changing habits and leaving comfort zone is about managing change. Change starts with a short value destruction period. This is the initial and critical "change resistance" period. Continuous Improvement is a slow but consistent value creation over long period of time. Before we provide a detailed Value Creation Model, we would like to provide a simple method applicable at the individual level. We will start with couple of very important questions. How to measure the value we, as individuals, create on daily, weekly or monthly basis? What kind of value do we create and how is it distributed between five value elements? Are we in a good balance? Is this the maximum value we can create or there is potential for additional improvement?

Naturally, in different phases of our lives, different activities and events have a different impact on our level of happiness. As a child we focus on learning through playing and socializing. The others take care of our finance, food, housing etc. As a student our intellectual growth is the main focus. Once retired finance is normally less important but health and socializing with our family are (Figs. 7 and 8).

There is a natural conflict between the value elements over time especially during "money making" period which is roughly from our 25th till 65th, which is in fact the longest life period. For some individuals, in this period, the financial value became

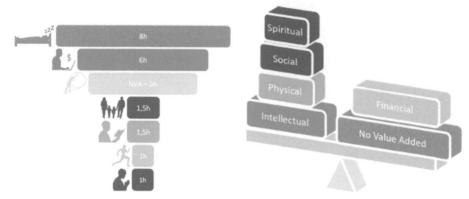

Fig. 7. AS-IS-DILO tornado diagram – 24 h value distribution.

Fig. 8. Time distribution per value element.

the most important one on expense of the others values. At this point I would like to introduce Mr. Ismar, a fictive character that is used as a role model in the article. He an IT professional, 40 years old, employed in a big international IT company. He is working on big data base design and maintenance for new customers. He is happily married and have two kids 10 and 6 years old. He and his wife Ms. Almina have very good salary and they almost completely paid off their big house and recently start thinking to buy another one as a kind of an investment. He has a wide network of friends. His retired parents live in another city about 2 h far away. His parents in law live in the same city. He is quite happy about his life. Like most of us Ismar is trying his best to be a good father, husband, son, friend, colleague and employee. He takes care of his health, personal development and has a strong ethical integrity. He struggles to synchronize all his obligations, hobbies and activities. He feels often he runs out of time. In office he is rushing from meeting to meeting, from one to another deadline. Situation at home seems often chaotic. He would like to spend more time with his kids, wife, parents etc. but there is no time left. He would also like to spend more time in nature or at least to do some more physical exercise as he used to do before. Often he and his family end up eating fast food just because there is no time for cooking or looking for healthy options. In short there is a lot of stress and he is convinced the life could be organized in a better way. Recently he familiarized himself about concepts of non- financial value management. He realizes that it is worth to do some analysis of his daily routines. He started with week days analysis. In the book he learned about method called DILO (a Day In Life Of). It is a tool which helps him to identify the non-value added (NVA) activities and time during one day. The final goal will be to eliminate NVA activities and replace them with VA activities. DILO is a method which maps individual activities, almost minute by minute, and record them in categories. So Mr. Ismar decided to use his mobile in order to record one typical day. He configured an alarm which sounds every 5 min, reminding him to record current activity in a very simple way: record current activity as VA or NVA; if it is VA record value element, and; sum up all activities and visualize. After looking at the data he collected on that particular day he created a tornado diagram showing all value elements of that particular day (Figs. 9 and 10).

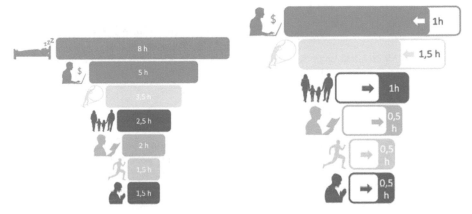

Fig. 9. Ideal value element DILO   Fig. 10. Ismar's day value gap.

4 Conclusion

We would like to emphasize that each and single individual has a specific "individual value element profile" which makes a perfect "value creation balance" for himself. We all need all value elements in a specific ratio, our individual ideal ratio. Having too much of one value element is often on expense of the other. Over-performing in one value element in short term may decrease the total value creation on long-term. The key is in the right "balance" i.e., a personalized, customized "individual value element profile". The DILO methodology can be used for business. This is usually applied on standard and routine jobs. We select a specific role in company and make full shift observation of that person. The management process is exactly same: 1) Identify AS- DILO 2) Define TO-BE DILO 3) Determine HOW action plan in order to achieve TO-BE goals. As already indicated the focus is "business efficiency": increase value creation for company by replacing NVA activities with VA activities.

References

1. Alagić, I.: Industrial Engineering & Maintenance: Lean Production - Six Sigma with application of tools and methods in specific working conditions, ISBN 978–9958–074–09–7, COBISS.BH-ID 24232454, Štamparija-S, Tešanj (2017)
2. Milios, J.: Marx's value theory revisited, a value form approach. In: Proceeding of the Seventh International Conference in Economics, Ankara, Turkey (2003)
3. Deckers, L.: Motivation: Biological, Psychological and Environmental. Routledge Press, Milton Park (2018)
4. Horn, A.J.: Abstract labour & value in Marx's system J. Soc. Theor. (2016)
5. Lyubomirsky, S.: The Myths of Happiness. Penguin Press, New York (2013)

Latest Trends and Possibilities in the Production of Stainless Steels Using WAAM

Nikola Knezović[✉], Angela Topić, and Franjo Gilja

Faculty of Mechanical Engineering, Computing and Electrical Engineering, University of Mostar, Matice hrvatske b.b, 88000 Mostar, Bosnia and Herzegovina
nikola.knezovic@fsre.sum.ba

Abstract. WAAM (Wire and Arc Additive Manufacturing) is a production technology that has been widely investigated in the last thirty years, although the first patent dates from almost one hundred years ago. It is becoming more and more popular to investigate since it enables the production of large near-net-shape metal products. It uses existing welding equipment to provide the heat source and the material feedstock, so the initial investment costs can be lower. Since it became interesting mostly thanks to its use in the aerospace industry, most used materials firstly included light metal alloys. In recent years, great development has been seen in the field of stainless steel products, enabling this technology to produce stainless steel parts for everyday use. This paper aims to present the newest trends and possibilities regarding the stainless steels parts produced using WAAM technology and to give suggestions on what can be done in the future.

Keywords: Welding · Additive manufacturing · Stainless steels

1 Introduction

1.1 WAAM

The evolution of the industry continuously pushes researchers to investigate and develop new production technologies. Since additive manufacturing technologies enable the production of more complex parts with higher productivity, that kind of technology became popular lately. However, the majority of the additive manufacturing technologies could not have been used to produce metal parts, until the extensive development of WAAM (Wire and Arc Additive Manufacturing). WAAM is basically the combination of welding and additive manufacturing, since it uses an electric arc as the heat source and welding wire as the material feedstock, while robotic hand or CNC machine can provide movement [1]. The first WAAM-related patent dates back to 1925 [2], but researchers' interest in it increased when it was found that it can save raw material waste in the aerospace industry [3]. Since most of the experiments and researches were focused to produce parts for aerospace applications, it defined the materials which were mostly used – titanium, aluminium and nickel-based alloys, etc. [4, 5].

I. Karabegović et al. (Eds.): NT 2022, LNNS 472, pp. 376–381, 2022.
https://doi.org/10.1007/978-3-031-05230-9_44

1.2 Stainless Steels

Stainless steels (known also as *rostfrei* or *inox* steels) are the family of steels known mostly for their corrosion resistance. Theoretically, they must fulfil two requirements: minimum of 10.5% chromium content by mass and homogeneous monophase microstructure. Its resistance to corrosion processes comes thanks to the presence of the chromium-rich oxide film which forms naturally on its surface. If there is enough oxygen available, that film is self-repairing. They are mostly divided into groups by their microstructure, which can be martensitic, ferritic and austenitic. There are also combinations of different microstructures(duplex) and in most cases, it is the combination of austenite and ferrite [6].

2 WAAM Researches Based on the Stainless Steels

Stainless steels were not popular amongst the WAAM-based researchers at the very beginning, mostly due to the problems which usually occurs because of the great heat input during welding. Nevertheless, significant progress has been made lately and numerous papers proved that stainless steel products for everyday use can be made using WAAM technology.

2.1 Austenitic and Martensitic Stainless Steels

In the paper [7], products were made using 308L stainless steel wire. The most important finding was the fact that parts were made without any major defects such as cracks. Voltage has been shown to greatly affect bead width, while bead height was significantly affected by welding current. Torch travel speed at the same time affected the whole bead, changing its width and height. This means that size and shape of a single bead and layer can be greatly influenced by varying welding parameters. The major drawback was the occurrence of the anisotropic properties of tensile strength, whose values were higher in deposition (horizontal or longitudinal) direction than in the building (vertical or transverse) direction.

Results in the paper [8] have shown great improvement, since anisotropy of the yield strength and the ultimate tensile strength were greatly reduced, mostly thanks to a stable austenite matrix and the work hardening effect caused by the higher nitrogen content in the wire. On the other side, nitrogen caused the formation of excessive micropores, so its amount has to be optimized in order to produce parts of better quality.

The authors of the paper [9] tried to form the desired microstructure in the ER420 martensitic stainless steel parts. They proved that it is possible to increase the austenite amount in martensitic matrix with the higher interpass temperatures, while delta-ferrite was observed in the microstructure of the parts produced with the lower interpass temperatures. This is an important finding since austenite should be more favourable due to its improved toughness, and having it when using higher interpass temperatures means that process can be done faster.

An idea to speed up the process was also introduced in the paper [10], where interpass air cooling was implemented during the production of some parts in order to cool down the layer as fast as it is possible, so the subsequent one could be deposited. However, no difference was found between the air-cooled parts and the ones produced without the cooling. All parts had similar problems since anisotropy of mechanical properties was found.

Problems with anisotropic properties were partially resolved and reduced in the paper [11], where additional heat treatment allowed the parts to reach the more homogenous microstructure which substantially enables the mechanical properties to distribute more uniformly. Martensitic stainless steel 410 welding wire was used as the feedstock and the final part exhibiteda high amount of uniformly distributed martensite phase, with only a negligible amount of unwanted delta-ferrite.

Another martensitic stainless steel (15–5 PH) was investigated in the paper [12]. Similar to the paper [11], the obtained microstructure was mostly martensitic, as expected. Cracks were not evidenced in the parts, but some minor inclusions were found (manganese and silicon oxides). The most important finding was the results of the microhardness testing – values were similar in both directions, which means it is possible to obtain isotropic microhardness properties (Fig. 1.)

Paper [13] is one of the few which deals with the testing of the different mechanical properties of the produced parts (stainless steel 316L) under dynamic conditions. The results obtained have shown that yield strength is 50% higher and the elastic modulus is twice higher for the dynamic conditions than for the quasi-static conditions.

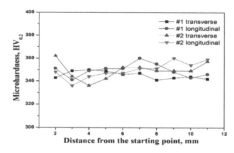

Fig. 1. Microhardness distribution of the 15–5 PH sample [12]

2.2 Duplex Stainless Steels (Austenitic/Ferritic)

Since most of the problems which occur when welding stainless steels are related to the ferrite, that microstructure was somehow neglected in the researches. However, since duplex stainless steels are widely used, it was important to prove that even parts with duplex microstructure containing higher amount of ferrite can be made with satisfying quality.

Paper [14] presented the parts with the 2209 stainless steel welding wire and it brought the initial conclusions which showed that the mechanical properties of the as-deposited parts are comparable to the wrought material in the longitudinal direction, while some properties were even better in the transverse direction. It has also shown that higher cooling rates prevented the formation of the detrimental intermetallic phases (sigma phase).

The same material was used in the research [15], where it was found that mechanical properties match the values exhibited in the cast parts of the same material. However, the austenite amount in these parts was higher than 70%, which is the upper recommended limit for duplex stainless steels.

The authors of the paper [16] also recorded a significantly higher austenite amount in the parts. Their main goal was to test pitting-corrosion resistance, which turned out to be excellent (probably because of the higher austenite amount).

Similar problems were found in the paper [17], where a lower ferrite amount caused a lower yield strength. Even if it means there were no detrimental intermetallic phases, deterioration of the mechanical properties is not desirable, which means the proper ferrite to austenite balance should be achieved in order to fulfil all the requirements.

A lower, but sufficient ferrite amount was recorded in the paper [18]. Substantially, the mechanical properties were higher and comparable to the values given in the welding wire data sheet (duplex stainless steel 2205 was used), which proves it is possible to achieve the desired properties.

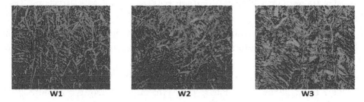

Fig. 2. Different austenite/ferrite microstructures (similar amounts of ferrite, red-coloured) [19]

The same material was used in the paper [19], where the authors tried to vary the interpass temperature in order to balance the austenite/ferrite ratio. It was shown that lower interpass temperatures lead to higher ferrite content, but even with the higher interpass temperatures ferrite amount was sufficient to achieve desirable mechanical properties. Figure 2 shows the microstructure of different specimens from that paper.

3   Conclusion

Although stainless steels were ignored initially, results in recent years have shown that these materials are too important and they have to be included in investigations related to the WAAM technology. Since their application spreads from specific high-level industries to everyday life, they can not be easily replaced with some other material. Stainless steels should be the choice for different applications in the future, and WAAM is a promising technology that will probably be used more and more. A combination of novel technology and material with superior characteristics is something that must catch the eye of researchers. There is still plenty of work to do and ideas to investigate, and some of the suggestions could be the following:

- different combinations of the shielding gases or feedstock components to obtain the desired microstructure;
- additional post-processing to reduce the drawbacks as residual stresses or distortions (for example, the approach suggested in the paper [20]);
- combinations of different types of steels, and even different alloys ([21–23]);
- *in-situ* non-destructive testing which could indicate the formation of defects during the process ([24]).

References

1. Busachi, A., Erkoyuncu, J., Colegrove, P., Martina, F., Ding, J.: Designing a WAAM based manufacturing system for defence applications. Procedia CIRP **37**, 48–53 (2015)
2. Williams, S.W., Martina, F., Addison, A.C., Ding, J., Pardal, G., Colegrove, P.: Wire + arc additive manufacturing. Mater. Sci. Technol. **32**(7), 641–647 (2016)
3. Mehnen, J., Ding, J., Lockett, H., Kazanas, P.: Design study for wire and arc additive manufacture. Int. J. Prod. Dev. **19**(1–3), 2–20 (2014)
4. Wang, F., Williams, S., Colegrove, P., Antonysamy, A.A.: Microstructure and mechanical properties of wire and arc additive manufactured Ti-6Al-4V. Metall. Mater. Trans. A **44**(2), 968–977 (2013)
5. Shen, C., Pan, Z., Cuiuri, D., Roberts, J., Li, H.: Fabrication of Fe-FeAl functionally graded material using the Wire-Arc additive manufacturing process. Metall. Mater. Trans. B. **47**(1), 763–772 (2016)
6. Karlsson, L.: Stainless steels: past, present and future. Svetsaren **1**, 1–6 (2004)
7. Le, V.T., Mai, D.S., Doan, T.K., Paris, H.: Wire and arc additive manufacturing of 308L stainless steel components: optimization of processing parameters and material properties. Eng. Sci. Technol. Int. J. **24**(4), 1015–1026 (2021)
8. Zhang, X., Zhou, Q., Wang, K., Peng, Y., Ding, J., Kong, J, et al.: Study on microstructure and tensile properties of high nitrogen Cr-Mn steel processed by CMT wire and arc additive manufacturing. Mater. Des. **166**, 107611 (2019)
9. Nemani, A.V., Ghaffari, M., Salahi, S., Lunde, J., Nasiri, A.: Effect of interpass temperature on the formation of retained austenite in a wire arc additive manufactured ER420 martensitic stainless steel. Mater. Chem. Phys. **266**, 124555 (2020)
10. Tonelli, L., Sola, R., Laghi, V., Palermo, M., Trombetti, T., Ceschini, L.: Influence of inter-layer forced air cooling on microstructure and mechanical properties of Wire Arc Additively Manufactured 304L austenitic stainless steel. Steel Res. Int. **92**(11), 202100175 (2021)
11. Zhu, B., Lin, J., Lei, Y., Zhang, Y., Sun, Q., Cheng, S.: Additively manufactured δ-ferrite-free 410 stainless steel with desirable performance. Mater. Lett. **293**(1), 129579 (2021)
12. Guo, C., Hu, R., Chen, F.: Microstructure and performances for 15–5 PH stainless steel fabricated through the wire-arc additive manufacturing technology. Mater. Technol. **36**(14), 831–842 (2021)
13. Chen, J., Wei, H., Zhang, X., Peng, Y., Kong, J., Wang, K.: Flow behavior and microstructure evolution during dynamic deformation of 316 L stainless steel fabricated by wire and arc additive manufacturing. Mater. Des. **198**, 109325 (2021)
14. Hejripour, F., Binesh, F., Hebel, M., Aidun, D.K.: Thermal modeling and characterization of wire arc additive manufactured duplex stainless steel. J. Mater. Process. Technol. **272**, 58–71 (2019)

15. Zhang, Y., Cheng, F., Wu, S.: The microstructure and mechanical properties of duplex stainless steel components fabricated via flux-cored wire arc-additive manufacturing. J. Manuf. Process. **69**, 204–214 (2021)
16. Kannan, A.R., Shanmugam, N.S., Rajkumar, V., Vishnukumar, M.: Insight into the microstructural features and corrosion properties of wire arc additive manufactured super duplex stainless steel (ER2594). Mater. Lett. **270**, 127680 (2020)
17. Lervåg, M., Sørensen, C., Robertstad, A., Brønstad, B.M., Nyhus, B., Eriksson, M., et al.: Additive manufacturing with superduplex stainless steel wire by CMT process. Metals **10**(2), 5–12 (2020)
18. Posch, G., Chladil, K., Chladil, H.: Material properties of CMT—metal additive manufactured duplex stainless steel blade-like geometries. Weld. World **61**(5), 873–882 (2017). https://doi.org/10.1007/s40194-017-0474-5
19. Knezović, N., Garašić, I., Jurić, I.: Influence of the interlayer temperature on structure and properties of wire and arc additive manufactured duplex stainless steel product. Materials **13**(24), 1–16 (2020)
20. Duarte, V.R., Rodrigues, T.A., Schell, N., Miranda, R.M., Oliveira, J.P., Santos, T.G.: Hot forging wire and arc additive manufacturing (HF-WAAM). Additive Manuf. **35**, 101913 (2020)
21. Mohan Kumar, S., Rajesh Kannan, A., Pravin Kumar, N., Pramod, R., Siva Shanmugam, N., Vishnu, A.S., et al.: Microstructural features and mechanical integrity of wire arc additive manufactured SS321/Inconel 625 functionally gradient material. J. Mater. Eng. Perform. **30**(8), 5692–5703 (2021)
22. Ozsoy, A., Tureyen, E.B., Baskan, M., Yasa, E.: Microstructure and mechanical properties of hybrid additive manufactured dissimilar 17–4 PH and 316L stainless steels. Mater. Today Commun. **28**, 102561 (2021)
23. Ahsan, M.R.U., Fan, X., Seo, G.J., Ji, C., Noakes, M., Nycz, A., et al.: Microstructures and mechanical behavior of the bimetallic additively-manufactured structure (BAMS) of austenitic stainless steel and Inconel 625. J. Mater. Sci. Technol. **74**, 176–188 (2021)
24. Knezović, N., Dolšak, B.: In-process non-destructive ultrasonic testing application during wire plus arc additive manufacturing. Adv. Prod. Eng. Manage. **13**(2), 158–168 (2018)

Information Measurement System for Determination of Cutting Force at Turning Technology

Nikola Šibalić[(✉)] and Marko Mumović

Faculty of Mechanical Engineering, University of Montenegro, Podgorica, Montenegro
nikola@ucg.ac.me

Abstract. As machining is one of the fundamental technological disciplines, knowledge of cutting forces is of great importance, so this paper presents an information measurement system, which was developed with the aim of measuring cutting forceat the machining process of longitudinal turning. The contribution of this paper is that it allows for a simple acquisition of experimental data using a dynamometer (piezo-plate) to measure the cutting force during cutting and thus form a procedure for calibration of measuring devices. Measurements of three components of force were conducted in three mutually perpendicular directions F_c, F_r and F_t for bronze alloy, using modern measuring equipment, piezo-plate, as well as an eight-channel universal amplifier. The obtained digital data were processed in the MX Assistant and MATLAB software.

Keywords: Information measuring system · Cutting force · Piezo-plate · Turning

1 Introduction

Machining due to its complexity stands out in the field of metal processing technology, which, above all, stands out in terms of the accuracy of the obtained geometry, as well as the quality of the finished surface from other technological disciplines. Today's modern industry of generation 4.0 has defined directions in terms of application and use of various cutting tools and accessories, as well as the application of modern, increasingly diverse materials, which have different mechanical characteristics, specific weight and of course machinability. Machinability is a broad term used to assess the suitability of a material for machining, i.e. it is a complex character of the material, which is primarily determined by the main factors of processing (cutting forces and technological cutting speeds) and the quality of the finishedsurface and the shape of the obtained chips. In this regard, it is necessary to use the developed procedure for determining the optimal cutting parameters through experimental research of cutting forces measurements (Fig. 1). Due to the stated reasons, the field of research of cutting resistance, i.e. measurement of cutting forces during turning, is very current and insufficiently researched for a large range of modern industrial materials.

The value of the cutting forces primarily depends on: the material of the workpiece and the material of the tool; processing parameter; tool geometries; coolant applications;

© The Author(s), under exclusive license to Springer Nature Switzerland AG 2022
I. Karabegović et al. (Eds.): NT 2022, LNNS 472, pp. 382–389, 2022.
https://doi.org/10.1007/978-3-031-05230-9_45

static and dynamic characteristics of the machine and other technological processing conditions [1, 2].

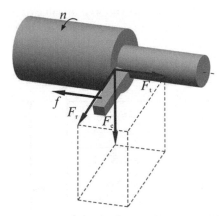

Fig. 1. Components of cutting force in longitudinal turning: F_c - main cutting force, F_r - radial force, F_t - thrust force, f - feed and n - rotational speed

Knowing the cutting forces allows you to: determine the energy balance of the machine tool; perform calculation and dimensioning of the elements of the kinematic system of the machine tool; perform calculation and dimensioning of cutting tools and accessories; perform a calculation of processing errors; determine the elements of the relevant processing parameters (feed, rotational speed) [1, 2].

Experimental investigations of the machining process were performed on a family of axisymmetric elements with a diameter of Ø20 mm, with a defined machining operation and shape of the processing contour, technological (machining) parameters, tolerance, and quality of the treated surface. A bronze alloy $CuSn14$ of mechanical characteristics are given in Table 1 were used as experimental material.

Table 1. Mechanical properties of bronze alloy $CuSn14$ [3]

Yield strength MPa	Tensile strength MPa	Elongation at break%	Brinell hardness HB	Density kg/dm^3
140	200	5	90	8.7

2   Development of Information Measuring System

The measurement of force components was performed using precise analog-digital measuring equipment, connected to the information measuring system. The measuring system consists of a multi-component dynamometer (piezo-plate), charge amplifier, analog indicator, and universal amplifier with AD/DA converter, which is presented in the block diagram (Fig. 2). The signal from the universal amplifier is obtained on a computer in the MX Assistant acquisition program, where it is later processed in the MATLAB program. The diagrams obtained in this way are a function of the electrical voltage (V) and time (s), which are then based on the calibration value of the measuring equipment convertedto the diagrams of forces (N) as a function of time (s). Figure 3 shows the research site.

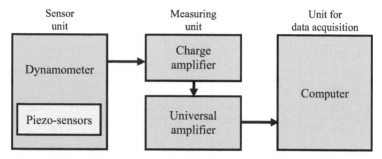

Fig. 2. Block diagram of the information measuring system

Fig. 3. Research site: 1 - universal lathe (processing machine), 2 - dynamometer (piezo plate), 3 - cutting tool (lathe tool with hard metal insert), 4 - charge amplifier (integrator), 5 - analog indicator, 6 - universal amplifier, 7 - a computer with software for data acquisition, 8 - calibration weight, 9 - calibration of measuring equipment and 10 - workpiece (bronze alloy $CuSn14$)

2.1 Sensor Unit

The Kistler 9257A multi-component dynamometer (piezo-plate) provides a dynamic and quasi-static measurement of the three components of the force (F_x, F_y, and F_z) acting from any direction on the top plate. The dynamometer has high stiffness and therefore a high natural frequency. Its high resolution allows even the smallest dynamic changes at high forces measured by the device. The measuring range is from −5kN to 5 kN, while at turning in case point A is the point of attack, the measuring range for the z-axis ranges from −7.5 kN to 15 kN. The design of the dynamometer is such that the forces acting on it cannot act by torque on the individual sensors in it, provided that they act in relation to the zero point of the dynamometer, which is located in the cross-section of the diagonal. The sensor can withstand a short overload of 50% of the nominal value. The operating temperature of the dynamometer is between 0 °C and 70 °C [4, 5] (Fig. 4).

Fig. 4. Multi-component dynamometer (piezo plate) Kistler 9257A with a mounted cutting tool (turning tool)

The dynamometer consists of four three-component force sensors placed under high preload between the base and top plates. Each sensor contains three pairs of quartz crystals, one sensitive to pressure in the z-axis direction, and the other two to shear in the x and y-axis directions [5]. These force components are measured practically without displacement, due to the high stiffness and preload of the sensor.

Depending on the direction of the force, positive or negative charges occur at the contacts. Negative charging gives positive voltages at the output of the charge amplifier, and vice versa. Four force sensors are mounted with the insulated ground. Loop current problems with the common ground are largely eliminated in this way. The dynamometer is protected against corrosion and moisture, meets the requirements of IP67 protection against water and solid particles. The lower surface of the device has a high quality of finish and requires a flat surface on which it is mounted, otherwise, internal stresses may occur which can damage the device. Also, if the dynamometer is not supported by the entire surface, there may be an increase in elasticity and a decrease in the resonant frequency [5].

The output with three channels of the dynamometer is the charge in pC. This signal needs to be integrated, i.e. it is treated with a charge amplifier, and in that way an electric voltage can be obtained, the acquisition of which can be done with the usual measuring amplifiers. As this signal is very low and sensitive to external influences, the cable that connects to the dynamometer must have a high degree of electrical insulation and a low value of the occurrence of triboelectricity [6].

2.2 Measuring Unit

The **charge amplifier** (integrator) Kistler 5806processes the data received from the sensor into a signal that can be processed by the Universal amplifier. The charge produced by a piezoelectric sensor is a variable that is difficult to measure due to its nature. For this reason, an electronic device known as an integrator or charge amplifier is connected after a sensor, to convert the charge into a voltage signal. The charge amplifier converts the negative charge produced by the piezoelectric sensor when it is subjected to a force load into a positive voltage that is proportional to the charge, and thus to the acting force. Due to their principle of operation, force sensors have a negative sensitivity and they produce a negative charge under load. Figure 5 gives a schematic representation of the used chargeamplifier, with its basic three components: a capacitor of the C_r range, a resistor of the time constant R_t, and a switch for reset, i.e. measurement. The output is V_{out} - output voltage, which is given by the expression:

$$V_{out} = -\frac{Q_{in}}{C_r} \tag{1}$$

where: Q_{in}- electric charge.

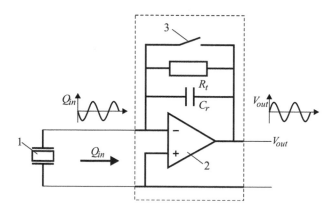

Fig. 5. Scheme of charge amplifier: 1 - sensor, 2 - integrator, 3 - switch

When using a charge amplifier, it is necessary to first define whether the signal we are measuring is quasi-static or dynamic. In case the signal is quasi-static, the value of the resistor of the time constant R_t should be equal to zero.

Due to its principle of operation, the measurement of the piezoelectric signal does not allow measurements with an absolute zero reference value. For quasi-static measurement, the zero value is defined at the beginning of the measurement, and the start is controlled by the Reset/Measure switch. For dynamic measurement, it is not possible to set a zero point because the measurements are performed without a reference zero point at the expense of the time constant [5, 7].

Universal amplifier HBM's QuantumX MX840B is an 8-channel universal amplifier (Fig. 6). Each channel allows acquisition with over 15 different types of sensors. Used as input connectors of the 15-pin DSUB15HD.

Fig. 6. HBM QuantumX MX840B

All channels are electrically insulated from each other and the power inputs. In order to unmistakably register the sensor at the entrance to the measuring amplifier, pins 4 and 9 on the connector are always short-circuited and when such a connector is connected to one of the channels, the measuring bridge gives feedback by changing the color of the lamp above that channel to green.

The connection between the output device and the measuring bridge itself is completely separated from the surroundings by the Faraday cage. This is achieved by using an adequate sensor cable with shielding and an electrically conductive fifteen-stage connector to which the shielding is connected and thus conducts electromagnetic interference to the housing of the measuring amplifier. In this way, the level of interference and noise is significantly reduced [8].

The MX840B universal amplifier can measure the signal from the piezo sensor directly in two ways.

The first method is piezo-resistant when piezo-sensors are connected in the formation of the Whiston bridge, which is not suitable due to the very nature of this sensor.

Another way is to supply the piezo sensor from the measuring bridge with a constant current and to measure its response during the process. In this way, it is possible to obtain only the dynamic response of the sensor, i.e. the alternating component of the signal. Since the purpose is to measure the force during cutting, which is a quasi-static load, we need a DC component of the signal and this method of measurement is therefore eliminated [8].

It was found that by using this device it is impossible to directly measure the signal from the piezo sensor and that is why the previously described charge amplifier was used, from which we take the electric voltage that this measuring amplifier can process.

3   Calibration and Measurement

The software used for data acquisition is *MX Assistant*, which allows the user to easily determine (calculate) the calibration curve of the sensor.

Since the output from the sensor is a quasi-static signal, it is possible to calibrate the sensor using a calibration weight whose mass is known. The calibration weights need to be placed so that their weight acts in the direction of the cutting force component that the piezo plate measures and passes through the point of attack at the top of the turning tool. As the axis markings given by Kisler on the piezo plate, we adopt equivalent designations in the paper ($F_c = F_z$, $F_r = F_y$, and $F_t = F_x$).

In the case of calibration of the force component F_c, the loading of the plate would be in the z-axis direction, while in the case of calibration of the force components F_r and F_t, auxiliary accessories would be made on which pulleys with bearings were placed (friction in bearings is neglected). A rope is placed over the pulley, which is collinear with the directions of force. The rope is Ø0.5 mm in diameter. Figure 7a shows the calibration procedure of the measuring equipment for component F_c - the main cutting force. The values of force F_c, F_r and F_t after calibration, showed a high degree of accuracy (100% ± 1%) for all three components.

By calibrating the sensors for measuring the force components F_c, F_r, and F_t the plate was loaded in the axial, longitudinal, and lateral directions with known calibration weights, where the dependence between load (N) and voltage value (V) from the universal measuring amplifier was obtained (Fig. 7b). When calibrating, the known masses were placed in the following order: 4.5 kg, 4 kg, 3 kg, and 2 kg. The diagram shows that the electric voltage is proportional to the load (known calibration weight).

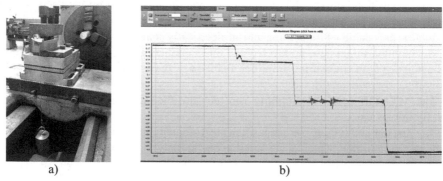

a)   b)

Fig. 7. Calibration of measurement equipment a) calibration of component F_c; b) sensor response for force component F_c (electrical voltage V - time s)

Longitudinal turning was performed on a machine tool with a universal lathe for the adopted contour, processing parameters, and according to the adopted experimental plan for the given material. The cutting force diagram is shown in Fig. 8.

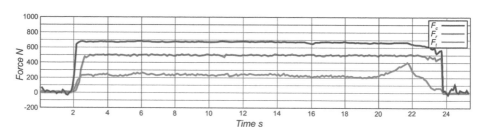

Fig. 8. Cutting force: F_c - main cutting force, F_r - radial force and F_t - thrust force

4   Conclusion

Machining is an exceptionally current technology characterized by complex processes that take place in the processing zone, which are conditioned by the action of numerous influential and interconnected factors, such as tool geometry, cutting parameters, workpiece material, etc.In order to investigate the cutting process, the paper presents the measurement of the force defined by the components in the x, y, and z directions. The measurement of forces was performed using precise analog-digital measuring equipment connected to the information measuring system.

In this way, a good experimental, measurement, and information basis has been created, which promises significant results in the research that follows.

The complexity of the turning process determines the performance of precise measurements during experimental research, through the developed information measurement system described in this paper. This information system can be used for a large number of different materials of the specified machining process.

References

1. Kalpakjian, S., Schmid, S.: Manufacturing Processes for Engineering Materials, Fifth Edition. Pearson Prentice Halle (2008). ISBN-13 978–981–06–7953–8
2. Stanić, J., Kalajdžić, M., Kovačević, R.: Merna tehnika u tehnologiji obrade metala rezanjem. IRO „GRAĐEVINSKA KNJIGA", Beograd (1983)
3. Bulatović, R.: Mašinski elementi I - Drugo izdanje. Univerzitet Crne Gore - Mašinski fakultet, Podgorica (2016). ISBN 978-9940-527-15-0
4. Kistler Group, Multicomponent Dynamometer, 9257B_000–151e-07.18 (2018). www.kistler. com
5. Kistler Group, Instruction Manual, 9257B_002–054e-03.15 (2015). www.kistler.com
6. Kistler Group, Cables, 1631C_000–346e-12.13 (2013). www.kistler.com
7. Kistler Instrument Corp, Piezo-Instrumentation, Nr. 11.5007, Ed. 8.82, P. 1...4
8. Hottinger Bruel & Kjaer GmbH, Operating Manual, HBM Test and Measurement, DVS: A03031_24_E00_01 (2021)

Genetic Modeling of Die Load in Hydroforming of Cross Tube

Mehmed Mahmić(✉), Edina Karabegović, and Ermin Husak

Technical Faculty Bihać, Univesity of Bihać, dr. Irfana Ljubijankića bb, 77000 Bihać, Bosnia and Herzegovina
mmahmic@gmail.com

ABSTRACT. In addition to the optimal process parameters, machines, tools and materials have a significant impact on the successful execution of the plastic forming process in the machining system. All these elements of the processing system are important for achieving maximum productivity, optimal quality, and minimum production costs. Previous research and theoretical analysis of dies as executive elements have shown that any deviation that may occur on the plastic forming die during operation reflects on the quality of the product, which does not ensure competitiveness in the market. The durability of the die depends on several parameters, depending on the plastic forming process, macrogeometry of the die, microgeometry of the die working zone, die load, tribological condition of contact surfaces, etc. The research provided in this paper relates to the modeling of die loads in the cross tube hydroforming process using the genetic algorithm method.

Keywords: Hydroforming · Die · Load · Experiment · GA modeling

1 Introduction

Market demands for continuous improvement of technical and technological solutions in production lead to the application of modern scientific methods in the field of stochastics, numerics, and genetics in modeling and optimization of production processes. In addition, the complexity of the shape and geometry of the product formed by the plastic forming process requires the development of new forming processes, and thus the research of new machining systems and dies as load carriers in the machining process. Hydroforming is a process of plastic forming whose characteristics are evidently superior to conventional plastic forming processes. Research has shown that the application of this process is significant in the automotive industry, and that one third of the process of forming with fluid under pressure relates to the hydroforming of tube elements. The tube hydroforming process is based on the synchronized action of the internal fluid pressure p_{uc} and axial puncher Fa on the tube installed in the die. Previous research on hydroforming processes has mostly been related to the analysis of processing parameters, tribological processing conditions, etc. In the recent years, evolutionary methods have been of special importance in processing research. Research on the die load in the process of plastic forming is mainly related to the development of software programs that

© The Author(s), under exclusive license to Springer Nature Switzerland AG 2022
I. Karabegović et al. (Eds.): NT 2022, LNNS 472, pp. 390–397, 2022.
https://doi.org/10.1007/978-3-031-05230-9_46

provide the ability to model and simulate these processes. It is known that the die load increases significantly with increasing contact friction, poor micro state of the working zone geometry, and the intensity of fatigue of the contact surfaces. In this paper, the method of genetic algorithm is used to model the die load the during the hydroforming of the cross tube [1–7].

2 Influence of Die Load on the Hydroforming Process

The success of the cross tube forming depends on the synchronized action of the working pressure of the fluid inside the pipe (p_{uc}) and axial punchers (Fa_1, Fa_2), so as to ensure smooth flow of specimen material (tube) into the radial space of the die designed to form a protrusion on both sides, as shown in Fig. 1.

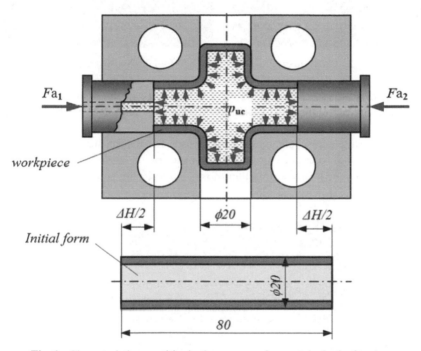

Fig. 1. Characteristic quantities in the process of cross tube hydroforming

The load force of the hydroforming die is a function of the geometry of the die working zone, the tribological state, the properties of the processing material, and the technological parameters of the process [1, 4, 6].

3 Modeling of Die Load in the Hydroforming Process

The outcome of the tube hydroforming process depends on the technological parameters of the process, tribological conditions, properties of the workpiece material, and die characteristics (working zone geometry, finishing, hardness, material, and die load). The analysis of the influence of the die load on the formation of the cross tube, Fig. 2, was performed in the laboratory research conditions of the hydroforming process of the cross tube, as shown in Fig. 3.

Fig. 2. Cross tube

Fig. 3. Hydroforming machine

The die closure system is essential for the stability of the hydroforming process. During the experiment, the prestressing force of the die clamping bolts had intensity F_s = 85 kN.

The measuring equipment consists of: sensor for measuring the pressure of the fluid in the tube, the force measuring dynamometer on the axial punch (2 pcs.), sensor for stroke/displacement measuring, measuring – amplifying equipment Spider and the computer (Fig. 4).

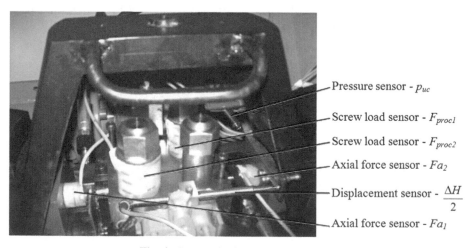

Pressure sensor - p_{uc}

Screw load sensor - F_{proc1}

Screw load sensor - F_{proc2}

Axial force sensor - Fa_2

Displacement sensor - $\dfrac{\Delta H}{2}$

Axial force sensor - Fa_1

Fig. 4. Sensors in the device workspace

By measuring the load on the bolts using dynamometers on two diagonal bolts, data were obtained that can be used for further analysis of the die load connected to this connection system. As modeling determines the validity of the interrelationships between input and output parameters, modeling method using a genetic algorithm (GA) was selected for modeling die loads in the cross tube hydroforming process [1–3, 6, 7] (Fig. 5).

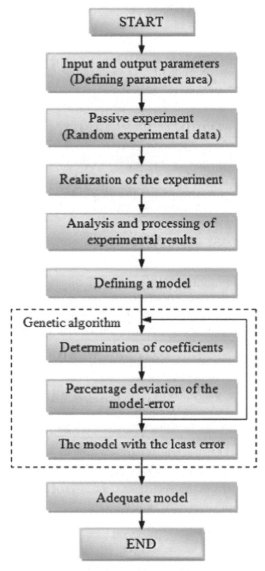

Fig. 5. *GA* model algorithm

Given that the application of the method of genetic algorithm for modeling and optimization of impact parameters on die load during the forming process can take a number of experiments and parameters, the experimental data given in Table 1 will be used for this analysis. Due to symmetry of the workpiece, mean value was taken on axial punchers (Fa_1, Fa_2) and bolt loads (F_{proc1}, F_{proc2}).

To apply the *GA* modeling method, interval values were used, as follows:

- axial force $X_1 = Fa = [86,53 \div 149,7]$ kN,
- internal fluid pressure in the tube $X_2 = p_{uc} = [755,6 \div 1478,5]$ bar,
- Tube shortening $X_3 = \Delta H/2 = [3,0 \div 14,0]$ mm.

As can be seen from the flow of *GA* modeling, the defined area of parameter selection is determined by the objective function, the die load model ($Y = F_p$), which is most influenced by random experimental data, as shown in Fig. 6.

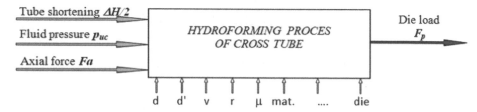

Fig. 6. Input-output quantities of the cross tube hydroforming process

In the process of hydroforming in this experimental study, the following constants were taken: diameter of tube and protrusion, shaping speed-motion of axial punchers, transition radius in the working zone of the die, friction, workpiece material, tools.

The processing of experimental data was carried out with the aim of obtaining a model for the load force on one die bolt related to the shaping process $F_p = f(Fa, p_{uc}, \Delta H/2)$, i.e., determining the coefficients of the predefined model (Ki).

Predefined model (GA) of the load of die closure bolt in polynomial form:

$$Y = K1 + K2 * X1 + K3 * X2 + K4 * X3 + K5 * X1 * X1 + K6 * X2 * X2 + K7 * X3 * X3$$
$$+ K8 * X1 * X2 + K9 * X1 * X3 + K10 * X2 * X3 + K11 * X1 * X2 * X3 \tag{1}$$

To determine the coefficients K1 to K11 of the assumed model, 10 independent starts of the GA system up to the generation of 10000 were performed. From these 10 independent starts, the three best solutions were selected and processed from the generation of 30000, where the best solution was obtained, i.e., the load model with the lowest percentage deviation.

After determining the coefficients, the shape of the obtained model is:

$$Y = 7.30719 - 0.000356546 * X1 - 0.0182451 * X2 + 0.623 * X3 - 0.00000771606 * X1 * X1$$
$$+ 0.00000873409 * X2 * X2 - 0.0294403 * X3 * X3 + 0.0000352117 * X1 * X2$$
$$- 0.00144261 * X1 * X3 - 0.000186122 * X2 * X3 + 0.00000424889 * X1 * X2 * X3 \tag{2}$$

Classic record of the obtained model of die load force, i.e., load of one screw for connecting the upper and lower die plate in the final form:

$$F_p = 7,30719 - 0,000356546 \cdot Fa - 0,0182451 \cdot p_{uc} + 0,623 \cdot \frac{\Delta H}{2} - 0,00000771606 \cdot Fa^2$$

$$+ 0,00000873409 \cdot p_{uc}^2 - 0,0294403 \cdot \frac{\Delta H^2}{4} + 0,0000352117 \cdot Fa \cdot p_{uc}$$

$$- 0,00144261 \cdot Fa \cdot \frac{\Delta H}{2} - 0,000186122 \cdot p_{uc} \cdot \frac{\Delta H}{2} + 0,00000424889 \cdot Fa \cdot p_{uc} \cdot \frac{\Delta H}{2} \qquad (3)$$

The deviation of the best solution or model is shown in Table 1.

Table 1. Experimental measurement data and calculated values according to the model

Ordinal number	Input parameters			Experimental measurement results	Calculated values according to the model (3)	Deviation of computational and experimental values	
	$X1 = Fa$ [kN]	$X2 = p_{uc}$ [bar]	$X3 = \Delta H/2$ [mm]	$Y = F_p$ (Exp) [kN]	F_p (GA) [kN]	Exp – GA	Difference [%]
1	92,7	1190,2	3	4,19	3,695	0,495	11,822
2	118,06	1478,5	8	10,89	10,890	0,000	0,000
3	136,11	1473,7	12	12,92	14,076	−1,156	8,944
4	96,85	1088,0	3	3,25	3,319	−0,069	2,138
5	130,35	1035,0	7	7,12	6,619	0,501	7,038
6	149,7	1053,0	14	9,815	9,665	0,150	1,527
7	104,51	999,4	3	3,28	3,278	0,002	0,062
8	128,25	989,4	7	6,555	6,209	0,346	5,277
9	141,45	1046,0	14	9,11	8,954	0,156	1,708
10	95,86	942,2	3	2,69	2,760	−0,070	2,597
11	126,75	964,9	7	5,93	5,990	−0,060	1,018
12	141,95	962,2	14	7,705	8,144	−0,439	5,692
13	101,87	841,8	3	2,95	2,827	0,123	4,167
14	123,75	867,2	7	5,37	5,401	−0,031	0,571
15	140,55	954,6	14	7,97	7,978	−0,008	0,099
16	86,53	755,6	3	2,36	2,362	−0,002	0,103
17	107,58	790,7	7	4,8	4,541	0,259	5,399
18	142,9	871,7	14	7,41	7,421	−0,011	0,152
19	87,67	774,8	3	2,395	2,373	0,022	0,910
20	119,8	799,2	7	5,005	5,037	−0,032	0,645
21	142,85	880,3	14	7,47	7,487	−0,017	0,226

The best model (model with the smallest deviation) was obtained in generation 26889, with a deviation of 2.86172% as shown in Fig. 7, and a graphical representation of the GA model is given in Fig. 8.

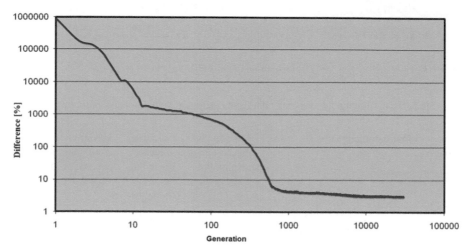

Fig. 7. Deviation of GA model from experimental values depending on model generation

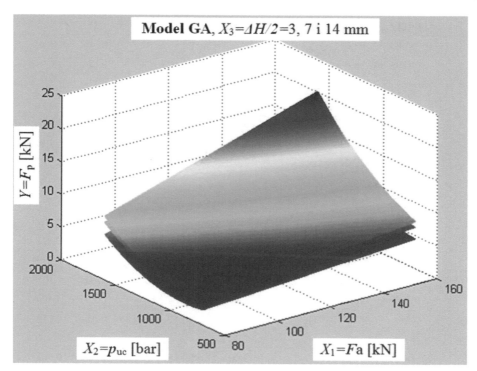

Fig. 8. Die load model depending on fluid pressure in the tube and axial force for tube shortening $\Delta H/2 = 3, 7$ i 14 mm

4 Conclusion

The application of the tube hydroforming process is justified by a number of advantages over conventional forming processes. Previous research and analysis of tube hydroforming systems have been mainly based on theoretical research, while experimental and other methods are evident for research into the parameters of the technological process of tube hydroforming. Existing resources were used to perform the experiment in this paper, and one of the evolutionary methods (GA) was used for modeling. The applied evolutionary GA method aimed to obtain a die load model depending on the stated input parameters. The obtained genetic model (expression 3) confirmed the positive influence of the defined parameters on the die load. By comparing the GA solution, i.e., the model, for the die load force (F_p) in relation to the solution obtained by experimental measurement (Table 1), an average deviation of 2,86% was determined. Solutions obtained by experimental and other methods in the research of the machining process, in this case the load of the die during operation, have a direct impact on its durability, as well as product quality.

References

1. Mahmić, M., Karabegović, E.: Mathematical modeling of die load in the process of cross tube hydroforming. J. Technol. Plasticity **40**(1), 47–55 (2015). Faculty of Technical Sciences, Novi Sad, Serbia. ISSN: 0354-3870-UDK 621.7
2. Jurković, M.: Matematičko modeliranje i optimizacija obradnih procesa, 1-411, Tehnički Fakultet Sveučilišta u Rijeci, Rijeka (1999). ISBN 953-6326-15-9
3. Brezočnik, M.: Uporaba genetskega programiranja v inteligentnih proizvodnih sistema, Fakultet za strojništvo, Maribor, Slovenija, Tiskarna Tehniških Fakultet, Maribor (2000)
4. Karabegović, E., Šemić, E., Poljak, J.: Control of hydroforming system of welded metal sheets. J. Technol. Plasticity **42**(2), 23–31 (2017). Faculty of Technical Sciences, Novi Sad, Serbia, ISSN: 0354-3870-UDK 621.7. https://doi.org/10.24867/jtp.2017.42-2.23-22
5. Karabegović, E., Poljak, J.: Experimental modeling of fluid pressure during hydroforming of welded plates. Adv. Prod. Eng. Manag. **11**(4), 345–354 (2016). Maribor, Slovenia, Journal home: apem-journal.org. ISSN 1854-6250. https://doi.org/10.14743/apem2016.4.232
6. Mahmić, M.: Modeliranje i optimizacija alata za plastično oblikovanje primjenom eksperimenta i baza znanja, Univerzitet u Bihaću, Tehnički fakultet Bihać, doktorska disertacija (2012)
7. Jurković, M., Brezočnik, M., Jurković, Z., Mahmić., M.: Stochastic and genetic modelling of tool life. In: Proceedings of the 11th TMT 2007 International Research/Expert Conference "Trends in the Development of Machinery and Associated Technology", University of Zenica, Universitat Politechnica de Catalunya-Spain, Bahcesehir University Istanbul-Turkey, 75–79, Hammamet, Tunisia (2007). ISBN 978-9958-617-34-8

Robotics Engineering Development Prospects for the Next Period

Branislav Dudić[1,2(✉)], Pavel Kovač[3], Borislav Savković[3], and Alexandra Mittelman[1]

[1] Faculty of Management, Comenius University in Bratislava, 820 05 Bratislava, Slovakia
branislav.dudic@fm.uniba.sk
[2] Faculty of Economics and Engineering Management, University Business Academy, 21000 Novi Sad, Serbia
[3] Faculty of Technical Sciences, University of Novi Sad, 21000 Novi Sad, Serbia

Abstract. Over the last 5 years, the global market has noticed a record in the number of robots delivered and manufactured, by different famous world producer. The total robot market is around $ 45 billion. As for the countries that are the largest customers of robots are China, South Korea, Japan, the USA and Germany. The goal of the paper is to show the very significant role in development of industrial robots, to show the situation on the world market and to show what are the innovations and investments in the field, as well to give guide on how the market will behave in the next period and, value of new robot installations in the automotive industry value of new robot installations in the electronics industry, the demand for robotics in the metal and manufacturing industry and sales value of new robot installations in the chemical and plastic items.

Keywords: Robots · Innovation · Technology · Production · Businesses · Sales · Demand · Value

1 Introduction

To succeed in the world market, companies must invest in the purchase of technology such as robots. Industrial robots have become very important tools that give manufacturers an edge over the competition in an challenge of global market. In the production process, industrial robots are able to perform a given activity better than humans in terms of quality of work. Robots are able to work continuously 24 h a day, where for comparison, human working time is on average 8–10 h. We also need to know that robots are flexible, so they are able to respond to a change in production faster than humans. Manufacturers who use this technology are gaining an effective economic tool sales value. Countries that are less technically advanced are unable to respond to industrial innovation and lag behind developed countries. Robotization is not only an innovation of global industrial production, but also to standardize outdated norms and designations of individual components of production for easier communication between foreign countries. The industrial robotics market worldwide allows them to lead an advantage on the world market which are the most important companies manufactured by industrial

I. Karabegović et al. (Eds.): NT 2022, LNNS 472, pp. 398–410, 2022.
https://doi.org/10.1007/978-3-031-05230-9_47

robots [1, 2]. Thanks to robots, almost all human problems can be solved, people's lives will be much better. Robots in industrial production are much more advanced than those of the 50's, they are fully automated and they handle a much wider range of tasks. Many activities can be robotic, such as logistics and transport, industry, public infrastructure maintenance, construction, demolition, medical and domestic assistance for dependents. Logistics represents the growth of the market for service robots. Areas of use are not limited to warehouses; they also include offices and hospitals. Thus, industrial robots took over certain work directly in production. Then there are the robots that perform dangerous tasks in various accident sites.

1.1 Industries Using Robotics

Help in hospitals. There is progress in the development of robots for everyday life, to assist in tasks such as carrying heavy objects or moving lying patients from one bed to another. At this time, a lot of attention is paid to robots that are able to respond to human expression and can be part of human life. Disabled assistance. Robotic devices can help a disabled person be a more independent in the life. These are mainly prosthetic technologies and support services for people without lower or upper limbs. Prostheses are active, allowing the user, for example, to grab objects or manoeuvre them.

Education. If we want to continue to support promising scientists and engineers, we should support children more and more in their programming skills, it is a fun and inspiring way through various activities. The reason to be focused on education and science and research in using robotics is the direct impact on a relatively wide section of the professional and lay public, as it deals with education from pre-primary through primary, secondary and tertiary to lifelong, with a direct impact on students and teachers, as well as other professional staff. Robotics in science and technology may improve infrastructure development programs, international scientific and technical cooperation [3].

Home help and maintenance. Because we live in a fast and busy time, people have less and less time to take care of the household. Robotic vacuum cleaners are already common in households, according to floor surfaces. The main manufacturers are Samsung, iRobot and Neato.

Software/applications. The user does not have to be a professional programmer to be able to use the robot fully - there are applications for this that can be easily downloaded from the Internet Security. Security robots help us keep our assets safe, but only a few are available.

Home Office during Covid 19. The only robotic technology available for this purpose is remote control. These are robots that know how to represent a person where he or she cannot appear. Today, Internet calls are commonly used in the world as corporate meetings, but this is still not the case, and it is such robots that enhance this experience.

Assistance to the elderly and infirm. As the population ages, the elderly, infirm people will soon have no one to care for. There are personal robots for this and it is one of the most sought-after areas of robotics.

Remote control. Robots that help you find where you can't be right now. They can be used to save on shipping costs and offer a more intense video conferencing experience. They differ from traditional video conferencing tools in that they allow the person at the other end of the video call to control what they see. It helps interaction, allows you to move around the office at eye level.

New generation robots have advantages over first-generation industrial robots such as: working together with workers, workers perform their tasks in a safe environment, robots take up less space, robots do not need to be separated by fences, robots are easy to manipulate and cheaper to implement. The paper shows the trend of application of industrial robots in the world, with reference to implementation in welding production processes. An analysis of the implementation of industrial robots in ten top countries in the world was conducted, as well as the implementation of industrial robots in ten top countries [4–7].

Significance of industrial robots in development of automobile industry [8]. We are in a time of change that affects the classic automotive industry in particular. Cities are struggling with strong growth and an ever-increasing traffic volume, which is accompanied by high emission levels [9]. The automotive industry can be described as one of the main industrial sectors of economic pillars in Central and Eastern Europe. Foreign direct investments come right from this source. Suppliers of the automotive industry are major contributors to the creation of total industrial production in the concerned countries and hence to the employment of the population. They employ qualified and educated workers in productive age as a relatively cheap workforce [10]. Safety Requirements for Collaborative Robots and Applications [11] Increased worker safety through the use of collaborative robots in the production processes. Innovative bionic systems in the context of sustainable development and environmental quality [12].

2 Literature Overview for Next Period

Technologies are constantly evolving, and every large amount of demand for certain jobs. Globalization has not limited itself to a single sector or sector, which has overwhelmed them, and has become an indispensable one in order to make the forerunner competitive in this dynamic marketplace. Their study, research, design and construction deals robotics. From the perspective of the robot subsystems - mechanical, electronic, control, propulsion and others talk about the robot as a mechatronic system [1].

2.1 Sales Value of the Industrial Robotics Market Worldwide from 2018 to 2022, by Application Area

2019 saw an unexpected dip in robotics sales, driven by a downturn in key manufacturing segments. This trend was expected to continue around the world in 2020, exacerbated by the COVID-19 recession. However, growth is expected to resume in all market segments, with China retaining over a third of the entire market, purchasing nearly 6.7 billion U.S. dollars in industrial robotics in 2022. Investment in robotics often increases when labor becomes more expensive. In the U.S. and Germany, where wages are largely stagnating, the incentive to invest in robots is less than in China, for example, where a rapidly urbanizing population is demanding higher pay. Similarly, East Asian countries tend to focus more on electronics manufacturing, which requires numerous and expensive robots. A more diverse manufacturing landscape, as seen in China and Germany, includes sectors in which robots have fewer advantages. Some of these applications are hard to automate, such as textiles, and some are simply lacking the specific robots to automate them, such as technical machinery [1].

2.2 Sales Value of the Industrial Robotics Market Worldwide from 2018 to 2022, by Main Country

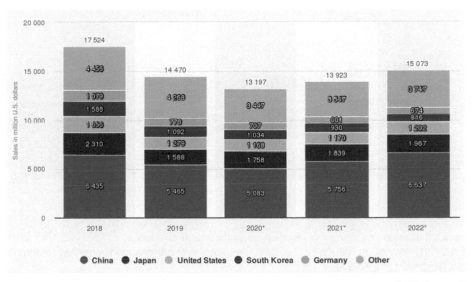

Fig. 1. Sales value of the industrial robotics market worldwide from 2018 to 2022, by main country (in million U.S. dollars) trend until 2020 and after that increasing for all main countries in millions US dollars [1]

2.3 Across the Automotive Industry in 2020, the Value of New Robot Installations Was Projected to Fall in Four of the Five Largest Markets

The automotive industry had one of its best years of robotics sales in 2018, but the market has become saturated, translating to a slowdown in new purchases by manufacturers. Even China, one of the fastest-growing auto markets in the world, saw a decline in new installations in 2019, and the forecast for 2020 was even lower. The trade war with the United States, combined with the COVID19 lockdown, has hit the country particularly hard. However, forecasts suggest a significant rebound. As seen previously, South Korea has a very high robot density, indicating that the marginal value of new installations may be lower. The expected lifespan of an industrial robot is 12 years, suggesting that many of these new installations are simply replacing outdated machinery. Germany and the United States also have established automotive sectors and exhibit similar signs of saturation. Japan was the only country forecast to grow in 2020. This could be driven in part by the industry's transition to electric cars. Such a fundamental change in the product would require entirely new manufacturing robots.

2.4   Sales Value of New Robot Installations in the Automotive Industry Worldwide from 2018 to 2022, by Country

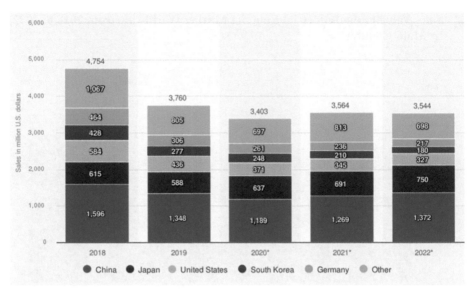

Fig. 2. Sales value of new robot installations in the automotive industry worldwide from 2018 to 2022, by country (in million U.S. dollars), the automotive robotics market hss incresing trend after depression in 2020 [1]

Worldwide, the sales value of new robotic installations in the automotive sector peaked in the 2018, when industrial robots worth approximately USD 4.7 billion US Dollars were installed worldwide. Most of the new robots will be installed in Asia (China, South Korea, Japan), a large part in the USA and European countries in Germany. China accounted for about 33% of global sales, making it a world leader in industrial robotics in the automotive industry, followed by Japan (16%), the United States (14%), South Korea (13.5%) and Germany (10%). Other European countries, Italy, Spain and France, are relatively far behind.

2.5   In the Next Few Years, the Sales Volume of Robots in the Electronics Industry Is Not Expected to Hit the Heights Reached in 2018

Electronics robots were a casualty of the China-United States trade war, which had spillover effects on other major players, particularly South Korea. The beneficiaries of this conflict seem to have been the smaller manufacturing nations. However, the other emerging manufacturers were also exposed to the COVID-19 pandemic. Fortunately, this depressed growth is expected to rebound in 2021, allowing for a year of catch-up growth before returning to the 2019 rate. This sector still does not have continuous stability, changes are expected in terms of the representation of certain countries, some

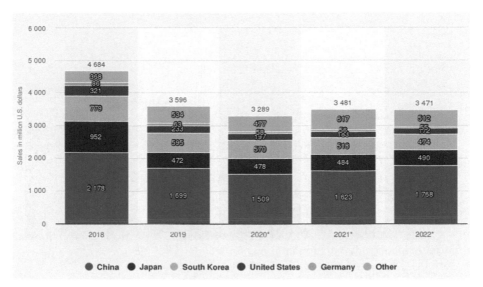

Fig. 3. Sales value of new robot installations in the electronics industry worldwide from 2018 to 2022, by country (in million U.S. dollars) [1]

will develop stronger and some less by 2022. The assumption is that China and Japan will be dominant in this field of activity.

The electronics market is the second largest market for industrial robotics. In 2018, the world's new robotic installations sold a value of more than 4.7 billion US dollars. In 2019, Kim's sales value dropped significantly to about $ 3.6 billion. By the end of 2022, 80% of the robots will be sold in China, Korea and Japan. Other European countries, Italy, Spain and France, are relatively far behind.

2.6 The Demand for Robotics in the Metal and Machinery Industry Is Projected to Favour Incumbent Countries from 2020 to 2022

Only China and Japan are expected to record more demand for metal and machinery robotics in 2022 than in 2018. However, sales in Germany and the United States are expected to be the most consistent over that period. Notably, the rest of the world is forecast to suffer a major loss in market share between 2018 and 2022, falling by ten percentage points. Smaller countries tend to have lessdeveloped business apparatus, and they recover more slowly from a recession. These countries show large expected growth in 2022, but the gain only amounts to 20 million U.S. dollars in sales. For comparison, Japan's forecast gain in the same period is 37 million U.S. dollars.

2.7 Sales Value of New Robot Installations in the Metal and Machinery Industry Worldwide from 2018 to 2022, by Country

The sales value of new robot installations in the global metal and engineering industries will be around $ 1.9 billion between 2018 and 2022. However, China and Japan are

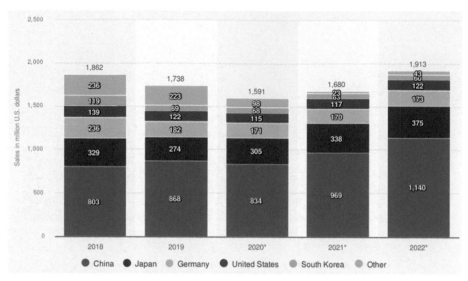

Fig. 4. Sales value of new robot installations in the metal and machinery industry worldwide from 2018 to 2022, by country (in million U.S. dollars) [1]

expected to grow in demand for industrial robots in the metal and engineering industries and drive sales value growth. China currently has the highest sales of industrial robots, and the Chinese market accounts for almost about 60% of the global sales value.

2.8 Sales Value of New Robot Installations in the Plastic and Chemical Products Industry Worldwide from 2018 to 2022, by Country

The COVID-19 pandemic increased demand for single-use plastics as businesses shifted away from reusable items for hygienic purposes. It also emphasized the importance of biochemical manufacturing, particularly pharmaceuticals and personal protective equipment, such as makes and gloves. However, falling demand for new installations is expected to continue in most markets. For example, South Korea is facing a saturated market, resulting in a decline in new robotics purchases. The COVID-19 pandemic has exacerbated this trend by disrupting supply chains, though 2021 is expected to reap the benefits of pent-up demands that were not realized in 2020. The less-developed robotics markets are set to gain market share, highlighted by the positive growth seen in 2021. They could use the pandemic as an opportunity to take market share from their incumbent competitors. However, in a complicated and delicate industry like this, such action also carries a high risk of failure.

2.9 Sales Value of New Robot Installations in the Plastic and Chemical Products Industry Worldwide from 2018 to 2022, by Country

The highest sales value of new robots in the global plastics and chemicals industry in 2018 is approximately $ 847 million US Dollars. Between 2019 and 2022, the value is

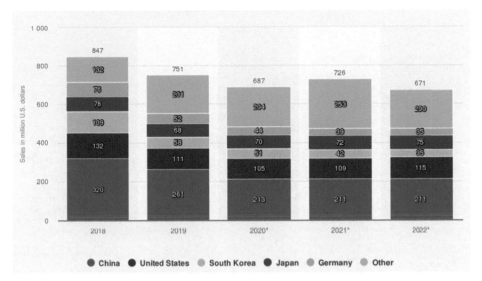

Fig. 5. Sales value of new robot installations in the plastic and chemical products industry worldwide from 2018 to 2022, by country (in million U.S. dollars) [1]

expected to fluctuate to approximately $ 671 million, which is lower than the recorded period. The market consists of five countries, namely China, the USA, Korea, Japan and Germany. China will continue to dominate the robotics market for plastic and chemical products, with sales of US $ 211 million in 2022.

2.10 Sales Value of New Robot Installations in the Food and Beverages Industry Worldwide from 2018 to 2022, by Country

The food and beverage industry is expected to have roughly the same volume of new robot installations in 2022 as it did in 2018, but this cannot be attributed to stagnation in the market. Instead, a few major players win, while others see their market share plummet. The United States is likely to increase its purchases by nearly a third in that time period, recovering quickly from the trade war that hit the agricultural sector particularly hard.

China's growth forecast is smaller in percentage terms but still amounts to 47 million U.S. dollars. This growth is balanced by contractions in Germany and the rest of the world, both of which suffer a streak of declining sales. As a share of food industry revenue, Germany's new robot installations appear to have plateaued. Elsewhere, there is the capacity to expand in China, Japan, and the United States.

2.11 Sales Value of New Robot Installations in the Food and Beverages Industry Worldwide from 2018 to 2022, by Country

The sales value of new robot installations in the global food and beverage industry will fall to approximately $ 426 million in 2020. The value of sales is expected to improve in 2022 and will be around $ 523 million. Demand for industrial robots in this sector will

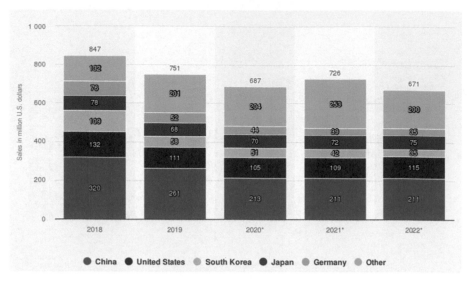

Fig. 6. Sales value of new robot installations in the food and beverages industry worldwide from 2018 to 2022, by country (in million U.S. dollars) [1]

increase in China, Japan and the United States, while in the rest of the world, demand will stagnate and decline over the forecast period.

2.12 Global Automotive Robot Market Size by Region 2020–2028

Automotive robotics covered a $ 5.7 market in 2020, more than half of which was covered in the Asia-Pacific region ($ 3.5). This market is projected to grow in the coming period, where it is projected that turnover in this field will reach 14 million in 2028. Of that, over $ 9 million is projected for the Asia-Pacific region.

2.13 Global Industrial Robotics Market Revenue 2018–2028

Overall, the global industrial robot market has surpassed $ 43 billion in 2021. With this growth rate (CAGR) of around 10%, it is assumed that in 2028 it will exceed the level of $ 70 billion.

2.14 Size of the Global Market for Automotive Robotics in 2020, with a Forecast Through 2028, by Region

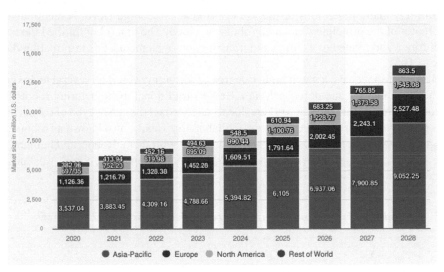

Fig. 7. Size of the global market for automotive robotics in 2020, with a forecast through 2028, by region (in million U.S. dollars) [1]

2.15 Size of the Global Market for Industrial Robots from 2018 to 2020, with a Forecast for 2021 Through 2028

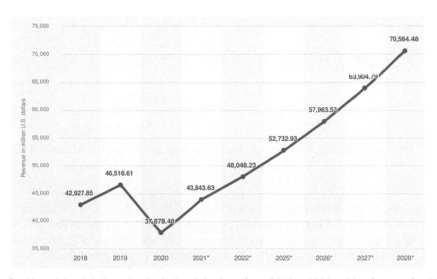

Fig. 8. Size of the global market for industrial robots from 2018 to 2020, with a forecast for 2021 through 2028 (in million U.S. dollars) [1]

2.16   Widening Applications of Industrial Robots

The characteristics of the robot are very impressive, they are designed to move in all three orthogonal axes. Their spectrum of action is very wide, from lifting very heavy elements to infinitely many repetitions of the same movements. George Devol is mentioned as the creator of the principles on which robots are based. The first robot fulfilled the tasks set before it and related to welding and transmission of elements, this information dates back to 1961.

Today's robots have a much higher degree of autonomy based on several technological advancements made in recent years. Besides traditional manufacturing robots that operate separated from human workers due to safety concerns, a new type of robots have been gaining popularity in recent years. Collaborative robots, also known as robots, are designed to work alongside humans and collaborate with them. Robots and humans are expected to complement each other in a workspace, making up for the other's weaknesses and leveraging their strengths.

2.17   Non-industrial Robotics Market Size Worldwide from 2019 to 2025

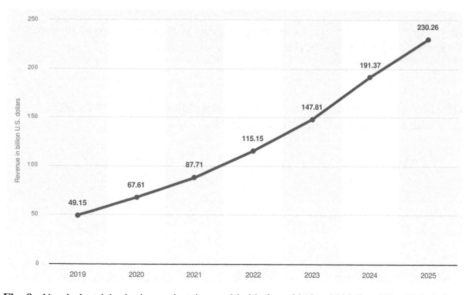

Fig. 9. Non-industrial robotics market size worldwide from 2019 to 2025 (in million U.S. dollars) [1]

Speaking of non-industrial robots, it is predicted that by 2025, this market will have a turnover of $ 230 billion. In general, the main market for robots is focused on the industrial sector, mostly in Japan and European countries, but gradually non-industrial robots are finding their application, such as personal assistance robots, some customer support, various types of vehicles and aviation in the form of drones.

3 Conclusion

The current trend is that all want to robotize, the visions for the future are very positive for robotics. The new world age is in a hurry and companies want to produce more and faster, it happens that human power is replaced by robotic. It is usually considered that robotics will keep moving forward and IT companies will start developing programs for robots in the same amount. In addition, the rate of robotization will have an increasing impact on the change in demand for labor, not only in terms of the number of jobs created, but also in terms of the structure of demand for specific activities. Production robotization will increasingly influence the development of the supply network, as well as the wider involvement of global companies in global value chains. From the point of view of industrial policy, in addition to a proactive approach to reducing the inefficient allocation of labor in the world, there is a growing challenge to support promising areas of business based on use of robotic production. In the future, we can expect not only the replacement of labor by demand for robots, but especially the changed structure of labor demand. While several routine activities are replaced by robots in automation, skilled work activities complement robotic production.

References

1. In Statista - The Statistics Portal. https://www.statista.com/statistics/. Accessed 07 Dec 2021
2. Dudić, B., Kovač, P., Savković, B.: Industrial robots' application. In: Proceedings of the 14th International Scientific Conference MMA 2021-Flexible Technologies, Faculty of Technical Sciences, Novi Sad, pp. 211–216 (2021). ISBN 978-86-6022-364-9
3. Stoličná, Z., Barjaková, J.: The possibilities of the optimization of managing organizations of education sector. In: Bilgin, M.H., Danis, H., Demir, E., Karabulut, G. (eds.) Eurasian Business and Economics Perspectives. ESBE, vol. 20, pp. 97–114. Springer, Cham (2021). https://doi.org/10.1007/978-3-030-85304-4_6
4. Karabegovic, I., Husak, E., Karabegovic, E., Mahmic, M.: Implementation of industry 4.0 - robotic technology in production processes: a review of welding processes in the world. In: Proceeding of International Conference ETIKUM 2021
5. Grushko, S., Vysocky, A., Suder, J., Glogar, L., Bobovsky, Z.: Analysis of increasing the friction force of the robot jaws by adding 3D printed flexible inserts. https://doi.org/10.17973/MMSJ.2021_12_2021127
6. Karabegovic, I., Husak, E.: Significance of industrial robots in development of automobile industry in Europe and the World. J. Mob. Veh. 40(1), 7–16 (2014). University of Kragujevac, Fakulty of Engineering, Kragujevac, Serbia
7. Karabegovic, I., Husak, E.: Industry 4.0 based on industrial and service robots with application in China. J. Mobility Veh. Mech. 44(4), 59–71 (2018)
8. Karabegovic, I., Karabegovic E.: Increased worker safety through the use of collaborative robots in the production processes of Industry 4.0. J. Saf. J. Saf. Work Living Environ. 62(1), 11–18 (2020)
9. Groos, A., Stoličná, Z.: Technology trends and their impact on the German automotive industry. In: Science and Innovation: Collection of Scientific Articles, Steyr, pp. 10–27 (2020). ISBN 978-3-953794-29-5
10. Stoličná, Z., Grožák, A.: The impact of the automotive industry on unemployment in Slovakia. In: Vision 2020: Sustainable Economic Development and Application of Innovation Management From Regional Expansion to Global Growth. Norristown: International Business Information Management Association, pp. 5854–5863 (2018). ISBN 978-0-9998551-1-9

11. Matthias, B.: Industrial safety requirements for collaborative robots and applications. In: Workplace Safety in Industrial Robotics: Trends, integration and Standard ERF, Columbia, 1 October 2014

12. Stevovic, S., Mirjanic, S., Golubovic, D.: Innovative bionic systems in the context of sustainable development and environmental quality. In: 4th International Scientific Conference, COMETa 2018 "Conference on Mechanical Engineering Technologies and Applications", Faculty of Mechanical Engineering East Sarajevo, East Sarajevo-Jahorina, RS, B&H, 27th–30th November 2018, pp. 611–620 (2018). ISBN 978-99976-719-4-3

Local Dynamics and Global Stability of Certain Second Order Rational Difference Equation in the First Quadrant with Quadratic Terms

Jasmin Bektešević, Vahidin Hadžiabdić$^{(\boxtimes)}$, Midhat Mehuljić, and Adnan Mašić

Faculty of Mechanical Engineering, Department of Mathematics and Physics, University of Sarajevo, St. Vilsonovo šetalište 9, 71 000 Sarajevo, Bosnia and Herzegovina
hadziabdic@mef.unsa.ba

Abstract. In this paper we present a local dynamics and investigate the global behavior of certain second order rational difference equation of type

$$x_{n+1} = \frac{ax_n x_{n-1} + ax_{n-1}^2 + bx_{n-1}}{cx_n + cx_{n-1} + d},$$

with positive parameters a, b, c, d ($a \neq c$) and initial conditions. We establish the relations for local stability of equilibriums and necessary conditions of existence of period-two solution. We then use this result to give global behavior results and determine the part of the basins of attraction of all equilibrium points.

Keywords: Local stability · Global stability · Equilibrium

1 Introduction

In this paper we study the local and global stability character, the periodic nature and the boundedness of solutions of rational second order difference equation of type

$$x_{n+1} = \frac{ax_n x_{n-1} + ax_{n-1}^2 + bx_{n-1}}{cx_n + cx_{n-1} + d}, \quad n = 0, 1, 2, \ldots \tag{1}$$

We restrict our attention to positive parameters a, b, c, d with condition $a \neq c$ and initial conditions x_1 and x_0 are arbitrary nonnegative numbers which will make our results more special but also more precise and applicable. Some special cases of Eq. (1), more precisely, difference equation of type

$$x_{n+1} = \frac{bx_{n-1}}{cx_n + cx_{n-1} + d}, \quad n = 0, 1, 2, \ldots \tag{2}$$

were investigated in [1]. It is an amazing fact that the Eq. (2) contains a large nuber of equations whose dynamics have not been thoroughly understood yet and remain a great challenge for further investigation. As it was shown in [1] such equation can exhibit

© The Author(s), under exclusive license to Springer Nature Switzerland AG 2022
I. Karabegović et al. (Eds.): NT 2022, LNNS 472, pp. 411–419, 2022.
https://doi.org/10.1007/978-3-031-05230-9_48

the whole range of different global behaviors such as global asymptotic stability of the equilibrium, global periodicity (i.e. all solutions are periodic with the same period) and chaos. The first systematic study of global dynamics of a special rational difference equations with quadratic terms was performed in [2, 3]. Dynamics of some related quadratic fractional difference equations was considered in the papers [4–12] and [13–15]. In this paper we will perform the local stability analysis of all equilibrium points of Eq. (1) and we will give the necessary and suffcient conditions for the equilibrium to be locally asymptotically stable, a saddle point or a non-hyperbolic equilibrium. Some of our results are based on number of theorems which hold for monotone difference equations, which will be described in the next section. To investigate global behavior of Eq. (1) we are forced to avoid well known theory of monotone maps, and in particular cooperative maps, which guarantee the existence and uniqueness of the stable and unstable manifolds for the fixed points and periodic points. Also, Eq. (1) has infinitely many period-two solutions. The first time, difference equation with infinitely many period-two solutions have been studied in [16], and corresponding difference is

$$x_{n+1} = ax_n x_{n-1} + ax_{n-1}^2 + bx_{n-1}, \quad n = 0, 1, 2, \ldots$$

The following result can be proved by using the techniques of proof of Theorem 11 in [17].

Theorem 1: Consider the equations $x_{n+1} = f(x_n, x_{n-1})$ where f is increasing function in its arguments and assume that there is no minimal period-two solution. Assume that $E_1(x_1, y_1)$ i $E_2(x_2, y_2)$ are two consecutive equilibrium points in North-East ordering that satisfy $(x_1, y_1) \preccurlyeq_{ne} (x_2, y_2)$ and that E_1 is a local attractor and E_2 is a saddle point or a non-hyperbolic point with second characteristic root in interval $(-1, 1)$, with the neighborhoods where f is strictly increasing. Then the basin of attraction $B(E_1)$ of E_1 is the region below the global stable manifold $W^s(E_2)$. More precisely,

$$B(E_1) = \{(x, y) : \exists y_u : y < y_u, (x, y_u) \in W^s(E_2)\}.$$

The basin of attraction $B(E_2) = W^s(E_2)$ is exactly the global stable manifold of E_2. The global stable manifold extend to the boundary of the domain of equations $x_{n+1} = f(x_n, x_{n-1})$. If there exists a period-two solution, then the end points of the global stable manifold are exactly the period two solution.

Now, the theorems that are applied in [17] provided the two continuous curves $W^s(E_2)$ (stable manifold) and $W^u(E_2)$ (unstable manifold), both passing through the point E_2 from Theorem 1, such that $W^s(E_2)$ is a graph of decreasing function and $W^u(E_2)$ is a graph of an increasing function. The curve $W^s(E_2)$ splits the first quadrant of initial conditions into two disjoint regions, but we do not know the explicit form of the curve $W^s(E_2)$. In this paper we expose the explicit form of the curve that separates the first quadrant into two basins of attraction of a locally stable equilibrium point and of the point at infinity of Eq. (1). In general, In complex domain, if $f(z) = \frac{P(z)}{Q(z)}$, where $z \in \mathbb{C} \cup \{\infty\}$ and P and Q are polynomials without common divisors, then Julia set J_f is the set of points z which do not approach infinity after $f(z)$ is repeatedly applied (corresponding to a attractor). At the same way, in real domain, corresponding Julia set J_f is connected and it is boundary of set of initial conditions for which the orbit of $f(n)$ does not tend to infinity. One of the major problems in the dynamics of polynomial maps in real domain is determining the basin of attractions of the point at infinity and in particular the boundary of the that basin known as the Julia set. We precisely determined the Julia set of Eq. (1) (boundary of set of initial conditions in the first quadrant for which the solutions of Eq. (1) does not tend to infinity) and we obtained the global dynamics in the interior of the Julia set, which includes all the points for which solutions are not asymptotic to the point at infinity. It turned out that the Julia set for Eq. (1) is the union of the stable manifolds of some saddle equilibrium points, nonhyperbolic equilibrium points or period-two points. In general, it is very important to mention that there is no explicit form of stable and unstable manifolds for the fixed points and periodic points of any difference equation (or system of difference equations), so the disadvantage of all results is that these manifolds are continuous decreasing (increasing) functions of which the parametrization is uncomfortable and we can only obtain their asymptotic formulas by using the method of undetermined coefficients. So the advantage of our results is that we obtain the exact formula of our Julia set of Eq. (1). We first list some results needed for the proofs of our theorems. The main result for studying local stability of equilibria is linearized stability theorem (see Theorem 1.1 in [1]):

Theorem 2: (linearized stability): Consider the difference equations

$$x_{n+1} = f(x_n, x_{n-1}) \tag{3}$$

And let $\bar{x}$ be an equilibrium of Eq. (3). Let $p = \frac{\partial f(\bar{x}, \bar{x})}{\partial u}$ and $q = \frac{\partial f(\bar{x}, \bar{x})}{\partial v}$ denote the partial derivatives of $f(u, v)$ evaluated at the equilibrium $\bar{x}$. Let λ_1 and λ_2 are the roots of the quadratic equation $\lambda^2 - p\lambda - q = 0$.

a) If $|\lambda_1| < 1$ and $|\lambda_2| < 1$, then $\bar{x}$ is locally asymptotically stable (sink).
b) If $|\lambda_1| > 1$ or $|\lambda_2| > 1$, then $\bar{x}$ is unstable.
c) If $|\lambda_1| < 1$ and $|\lambda_2| < 1 \Leftrightarrow |p| < 1 - q < 2$.
d) If $|\lambda_1| > 1$ and $|\lambda_2| > 1 \Leftrightarrow |q| > 1$ and $|p| < |1 - q|$, then $\bar{x}$ is repelier.
e) If $|\lambda_1| > 1$ and $|\lambda_2| < 1 \Leftrightarrow |p|) |1 - q|$, then $\bar{x}$ is saddle point.
f) If $|\lambda_1| = 1$ or $|\lambda_2| = 1 \Leftrightarrow |p| = |1 \quad q|$ or $q = 1$ and $|p| \leq 2$, then $\bar{x}$ is non-hyperbolic point.

The next theorem (Theorem 1.4.1 in [18]) is a very useful tool in establishing bounds for the solutions of nonlinear equations in terms of the solutions of equations with known behaviour.

Theorem 3: Let I be an interval of real numbers, let k be a positive integer, and let $F : I^{k+1} \rightarrow I$ be a function which is increasing in all its arguments. Assume that $\{x_n\}_{n=-k}^{\infty}, \{y_n\}_{n=-k}^{\infty}, \{z\}_{n=-k}^{\infty}$ are sequences of real numbers such that

$$x_{n+1} \leq F(x_n, \ldots, x_{n-k}), \quad n = 0, 1, \ldots$$

$$y_{n+1} = F(y_n, \ldots, y_{n-k}), \quad n = 0, 1, \ldots$$

$$z_{n+1} \geq F(z_n, \ldots, z_{n-k}), \ n = 0, 1, \ldots$$

and

$$x_n \leq y_n \leq z_n \text{ for all } -k \leq n \leq 0.$$

Then

$$x_n \leq y_n \leq z_n \text{ for all } n > 0.$$

2 Main Results

The equilibrium solutions $\bar{x}$ of Eq. (1) satisfy the equation

$$\bar{x} = \frac{\bar{x}(2a\bar{x} + b)}{2c\bar{x} + d}.$$

This immediately leads to the following cases:

a) There always exists zero equilibrium $\bar{x} = 0$.
b) There exist one positive equilibrium solution $\bar{x} = \frac{d-b}{2(a-c)}$ if and only if $\frac{d-b}{a-c} > 0$.

Set $f(x, y) = \frac{axy + ay^2 + by}{cx + cy + d}$ and let

$$p = \frac{\partial f(\bar{x}, \bar{x})}{\partial x} = \frac{(ad - bc)\bar{x}}{(2c\bar{x} + d)^2} \text{ and } q = \frac{\partial f(\bar{x}, \bar{x})}{\partial y} = \frac{4ac\bar{x} + (bc + 3ad)\bar{x} + bd}{(2c\bar{x} + d)^2}$$

denote the partial derivatives of $f(x, y)$ evaluated at the equilibrium $\bar{x}$. Clearly, if $\bar{x} = 0$, then $p = 0$ and $q = \frac{b}{d}$. Now, by using the Theorem 2, we obtained the following result on local stability of the zero equilibrium of Eq. (1):

Proposition 1: The zero equilibrium of Eq. (1) is one of the following:

a) locally asymptotically stable if $0 < 1 - \frac{b}{d} < 2 \Leftrightarrow b < d$;
b) non-hyperbolic and locally stable if $b = d$;
c) unstable if $b > d$:

Similarly, if $\bar{x} = \frac{d-b}{2(a-c)}$, then $p = -\frac{(a-c)(d-b)}{2(bc-ad)}$ and $q = \frac{a(b-3d)+c(b+d)}{2(bc-ad)}$. Now, in view of Theorem 2, the precise description of local stability of positive equilibrium solution $\bar{x}$ is given with the following result:

Proposition 2: If $\frac{d-b}{a-c} > 0$, the positive equilibrium $\bar{x}$ of Eq. (1) is non-hyperbolic point.

Proof. After straightforward calculation we get $1 - q = \frac{(a-c)(d-b)}{2(bc-ad)}$. Therefore the conditions (c), (d) and (e) of Theorem 2 are impossible to hold. Since $|p| = |1 - q|$ that implies the positive equilibrium $\bar{x}$ of Eq. (1) is non-hyperbolic point.

Theorem 4: If $\frac{min\{a,b\}}{max\{c,d\}} > 1$, then every solutions $\{x_n\}$ of Eq. (1) satisfies $\lim_{n\to\infty} x_n = \infty$.

Proof. If $\{x_n\}$ is solution of Eq. (1), then $\{x_n\}$ satisfies inequality

$$x_{n+1} = x_{n-1}\frac{ax_n + ax_{n-1} + b}{cx_n + cx_{n-1} + d} \geq x_{n-1}\frac{min\{a,b\}}{max\{c,d\}}\frac{x_n + x_{n-1} + 1}{x_n + x_{n-1} + 1} = x_{n-1}\frac{min\{a,b\}}{max\{c,d\}}.$$

which in view of the result on difference inequalities, see Theorem 3, implies that

$$x_n \geq y_n, n \geq 1$$

where $\{y_n\}$ is a solution of the initial value problem

$$y_{n+1} = y_{n-1}\frac{min\{a,b\}}{max\{c,d\}}, y_{-1} = x_{-1} \text{ and } y_0 = x_0, \quad n = 0, 1, \ldots$$

Consequently, $x_{-1}, x_0 > 0$ then $y_{-1}, y_0 > 0$ for all n and

$$x_n \geq y_n = \lambda_1 \sqrt{\frac{min\{a,b\}}{max\{c,d\}}}^{\,n} + \lambda_2 \left(-\sqrt{\frac{min\{a,b\}}{max\{c,d\}}}\right)^n, n = 1, 2, \ldots$$

where $\lambda_1, \lambda_2 \in \mathbb{R}$ such that $y_n \geq 0$ for all n, which implies $\lim_{n\to\infty} x_n = \infty$.

Theorem 5: Consider the difference equation Eq. (1) in the first quadrant of initial conditions, where $0 < b < d$ and $0 < c < a$. Then Eq. (1) has a zero equilibrium and unique positive equilibrium $\bar{x} = \frac{d-b}{2(a-c)}$. The line $y = -x + \frac{d-b}{a-c}$ is the Julia set and separates the first quadrant into two regions: the region below the given line is the basin of attraction of zero equilibrium point and the region above the line is the basin of attraction of the point at infinity and every point on the line except $E(\bar{x}, \bar{x})$ is a period-two solution of Eq. (1).

Proof. Since $0 < b < d$, then by applying Proposition 1 the zero equilibrium is locally asymptotically stable. In a view of Proposition 2 the positive equilibrium $\bar{x} = \frac{d-b}{2(a-c)}$ is non-hyperbolic point. Period-two solution ϕ, ψ satisfies the system

$$\psi = \frac{a\psi\phi + a\phi^2 + b\phi}{c\psi + c\phi + d}, \psi = \frac{a\phi\psi + a\psi^2 + b\psi}{c\phi + c\psi + d}.$$

Obviously, the point $(0, 0)$ is solution of the system above, but it is not period-two solution. Hence, this yields $\frac{a\psi + a\phi + b}{c\psi + c\phi + d} = 1$. Therefore every point of the set $\left\{(\phi, \psi) : \frac{a\psi + a\phi + b}{c\psi + c\phi + d} = 1\right\}$ is a period-two solution of Eq. (1) except point $E(\bar{x}, \bar{x})$. It is clear that the line $y = -x + \frac{d-b}{a-c}$ is the graph of decreasing function in the first quadrant. Let us note that $y = -x + \frac{d-b}{a-c} \Leftrightarrow \frac{a(x+y)+b}{c(x+y)+d} = 1$ and set $h(x, y) = \frac{a(x+y)+b}{c(x+y)+d}$. Since $0 < b < d$ and $0 < c < a$, then $0 < bc < ad$ which implies $ad - bc > 0$. Further, $\frac{\partial h(x,y)}{\partial x} = \frac{(ad-bc)x}{(cx+cy+d)^2} > 0$ and $\frac{\partial h(x,y)}{\partial y} = \frac{ad-bc}{(cx+cy+d)^2} > 0$ and all this leads that the function $h(x, y)$ is increasing function in both variables in the first quadrant. Assume that $\{x_n\}$ is solution of Eq. (1) for initial condition (x_0, x_{-1}) which lies below the line $y = -x + \frac{d-b}{a-c}$. Then

$$h(x_0, x_{-1}) = \frac{a(x_0 + x_{-1}) + b}{c(x_0 + x_{-1}) + d} < 1 \text{ and } x_{n+1} = x_{n-1}h(x_n, x_{n-1}).$$

Now,

$$x_1 = x_{-1}h(x_0, x_{-1}) < x_{-1},$$

and

$$x_2 = x_0 h(x_1, x_0) < x_0 h(x_{-1}, x_0) = x_0 h(x_0, x_{-1}) < x_0.$$

Thus (x_2, x_1) and (x_0, x_{-1}) are two points in North-East ordering $(x_2, x_1) \preceq_{ne} (x_0, x_{-1})$ which implies that the point (x_2, x_1) is also below the line $y = -x + \frac{d-b}{a-c}$ and holds $h(x_2, x_1) < 1$. Similarly we obtain the following:

$$x_3 = x_1 h(x_2, x_1) < x_1,$$

and

$$x_4 = x_2 h(x_3, x_2) < x_2 h(x_1, x_2) = x_2 h(x_2, x_1) < x_2.$$

Continuing on this way we get

$$(0, 0) \preceq_{ne} \dots \preceq_{ne} (x_{2n}, x_{2n-1}) \preceq_{ne} \dots \preceq_{ne} (x_4, x_3) \preceq_{ne} (x_2, x_1) \preceq_{ne} (x_0, x_{-1})$$

which implies that both subsequences $\{x_{2n}\}$ and $\{x_{2n-1}\}$ are monotonically decreasing and bounded below by 0. Since below the line $y = -x + \frac{d-b}{a-c}$ there are no period-two solutions it has to be $x_{2n} \to 0$ and $x_{2n-1} \to 0$. On the other hand, if we consider solution $\{x_n\}$ of Eq. (1) for initial condition (x_0, x_{-1}) which lies above the line $y = -x + \frac{d-b}{a-c}$, then $h(x_0, x_{-1}) > 1$. By using the method shown above we obtain the following condition:

$$(x_{2n}, x_{2n-1}) \succ \succeq_{ne} \dots \succeq_{ne} (x_4, x_3) \succeq_{ne} (x_2, x_1) \succeq_{ne} (x_0, x_{-1}).$$

Hence both subsequences $\{x_{2n}\}$ and $\{x_{2n-1}\}$ are monotonically increasing. Above the line there are no equilibrium points or period-two solutions thus $x_{2n} \to \infty$ and $x_{2n-1} \to \infty$.

The next Fig. 1 is visual illustration of Theorem 5 obtained by using Mathematica 9.0, with the boundaries of the basins of attraction obtained by using the software package Dynamica 3.0 for $a = 4, b = 1, c = 2, d = 3$.

Proposition 3: Consider the difference equation of type

$$x_{n+1} = \frac{a x_n x_{n-1} + A x_{n-1}^2 + b x_{n-1}}{c x_n + C x_{n-1} + d}, \tag{4}$$

in the first quadrant of initial conditions. Assume that the given positive parameters a, A, b, c, C, d satisfy $b < d, a + A > c + C$ and $min\{a, A\} > max\{c, C\}$. Then the global stable manifold of the positive equilibrium is between two lines

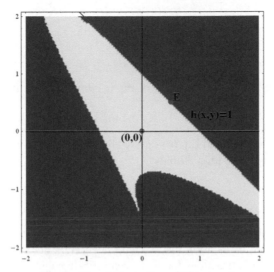

Fig. 1. Visual illustration of Theorem 5

$$p_1 : \frac{max\{a, A\}(x + y) + b}{min\{c, C\}(x + y) + d} = 1$$

And

$$p_2 : \frac{min\{a, A\}(x + y) + b}{max\{c, C\}(x + y) + d} = 1.$$

Proof. Set $g(x, y) = \frac{axy + Ay^2 + by}{cx + Cy + d}$ and let $p = \frac{\partial g(\bar{x}, \bar{x})}{\partial x}$ and $q = \frac{\partial g(\bar{x}, \bar{x})}{\partial y}$ denote the partial derivatives of $g(x, y)$ evaluated at the equilibrium $\bar{x}$. It easy to that Eq. (4) has two equilibrium points: zero equilibrium and unique positive equilibrium $\bar{x} = \frac{d - b}{a + A - c - C}$. If $\bar{x} = 0$, then $p = 0$ and $q = \frac{b}{d}$. Since $b < d$, after straightforward calculation by applying Theorem 2 the zero equilibrium is always locally asymptotically stable, thus the positive equilibrium must be unstable equilibrium point. The theorems applied in [17] provided the following global behavior. More precisely, if the positive equilibrium is a saddle point or a non-hyperbolic point then there exists a global stable manifold which contains point $E(\bar{x}, \bar{x})$; where $\bar{x}$ is the positive equilibrium. In this case global behavior of Eq. (4) is described by Theorem 9 in [19]. If the positive equilibrium is a repeller then there exists a period-two solution and we obtained that the period-two solution is a saddle point and there are two global stable manifolds which contain points $P_1(\phi, \psi)$ and $P_2(\psi, \phi)$ where (ϕ, ψ) unique period-two solution of Eq. (4). In this case the global behavior of Eq. (4) is described by Theorem 10 in [19]. So, the global dynamics of Eq. (4) is exactly the same as the global dynamics of equations described by Theorems 9 and 10 in [19]. Let

$$z_{n+1} = z_{n-1} \frac{min\{a, A\}(z_n + z_{n-1}) + b}{max\{c, C\}(z_n + z_{n-1}) + d} \tag{5}$$

and

$$z_{n+1} = z_{n-1} \frac{max\{a, A\}(z_n + z_{n-1}) + b}{min\{c, C\}(z_n + z_{n-1}) + d} \tag{6}$$

By applying Theorem 3 for solution $\{x_n\}$ of Eq. (4) the following inequality holds

$$y_n \leq x_n \leq z_n$$

for all n. Obviously Eq. (5) and Eq. (6) satisfy all conditions of Theorem 5 this implies that the statement of Proposition 3 holds.

3 Conclusion

In this paper we restrict our attention to certain rational second order difference equation Eq. (1). It is important to mention that we have accurately determined the Julia set of Eq. (1) and the basins of attractions for the zero equilibrium and the positive equilibrium point. In general, all theoretical concepts which are very useful in proving the results of global attractivity of equilibrium points and period-two solutions only give us existence of global stable manifold(s) whose computation leads to very uncomfortable calculus. Also, in a view of Proposition 3, if distance between the lines p_1 and p_2 small enough real number then lines p_1 and p_2 we can use as approximations of the global stable manifold which contains the point $E(\bar{x}, \bar{x})$.

References

1. Kulenović, M.R.S., Ladas, G.: Dynamics of Second Order Rational Difference Equations with Open Problems and Conjectures. Chapman and Hall/CRC, Boca Raton (2001)
2. Amleh, A.M., Camouzis, E., Ladas, G.: On the dynamics of a rational difference equation, part I. Int. J. Difference Equ. **3**, 1–35 (2008)
3. Amleh, A.M., Camouzis, E., Ladas, G.: On the dynamics of a rational difference equation, Part II. Int. J. Difference Equ. **3**, 195–225 (2008)
4. Chan, D.M., Kent, C.M., Ortiz-Robinson, N.L.: Convergence results on a second-order rational difference equation with quadratic terms. Adv. Difference Equ. **7** (2009). https://doi.org/10.1155/2009/985161. Article no. 985161
5. Dehghan, M., Kent, C.M., Mazrooei-Sebdani, R., Ortiz, N.L., Sedaghat, H.: Monotone and oscillatory solutions of a rational difference equation containing quadratic terms. J. Difference Equ. Appl. **14**, 1045–1058 (2008)
6. Dehghan, M., Mazrooei-Sebdani, R., Sedaghat, H.: Global behaviour of the Riccati difference equation of order two. J. Difference Equ. Appl. **17**, 467–477 (2011)
7. DiPippo, M., Janowski, E.J., Kulenović, M.R.S.: Global stability and attractivity of second order quadratic fractional difference equation. Adv. Difference Eq. **179** (2015). 13 p.
8. Kalabušić, S., Kulenović, M.R.S., Mehuljić, M.: Global period-doubling bifurcation of quadratic fractional second order difference equation. Discrete Dyn. Nat. Soc. (2014). 13p.
9. Kent, C.M., Sedaghat, H.: Global attractivity in a quadratic-linear rational difference equation with delay. J. Difference Equ. Appl. **15**, 913–925 (2009)
10. Kent, C.M., Sedaghat, H.: Global attractivity in a rational delay difference equation with quadratic terms. J. Difference Equ. Appl. **17**, 457–466 (2011)

11. Kulenović, M.R.S., Pilav, E., Silić, E.: Local dynamics and global attractivity of a certain second order quadratic fractional difference equation. Adv. Difference Equ. **2014**(68), 32p (2014)
12. Mazrooei-Sebdani, R.: Convergence in homogeneous difference equations of degree 1. J. Difference Equ. Appl. **19**, 13–26 (2013)
13. Mazrooei-Sebdani, R.: Homogeneous rational systems of difference equations in the plane. J. Difference Equ. Appl. **19**, 273–303 (2013)
14. Sedaghat, H.: Nonlinear Difference Equations. Theory with Applications to Social Science Models. Mathematical Modelling: Theory and Applications, vol. 15. Kluwer Academic Publishers, Dordrecht (2003)
15. Sedaghat, H.: Global behaviours of rational difference equations of orders two and three with quadratic terms. J. Difference Equ. Appl. **15**, 215–224 (2009)
16. Bektešević, J., Hadžiabdić, V., Mehuljić, M., Mujić, N.: The global behavior of a quadratic difference equation. Filomat **32**(18), 6203–6210 (2018)
17. Brett, A., Kulenović, M.R.S.: Basins of attraction of equilibrium points of monotone difference equations. Sarajevo J. Math. **5**(18), 211–233 (2009)
18. Camouzis, E., Ladas, G.: Dynamics of Third Order Rational Difference Equations with Open Problems and Conjectures. Chapman and Hall/CRC, Boca Raton (2008)
19. Bektešević, J., Kulenović, M.R.S., Pilav, E.: Global dynamics of quadratic second order difference equation in the first quadrant. Appl. Math. Comput. **227**, 50–65 (2014)

Analysis and Behavior of a Competitive System of Differential Equations

Vahidin Hadžiabdić[1]([✉]), Midhat Mehuljić[1], Jasmin Bektešević[1], and Mirsad Trobradović[2]

[1] Faculty of Mechanical Engineering, Department of Mathematics and Physics, University of Sarajevo, St. Vilsonovo šetalište 9, 71 000 Sarajevo, Bosnia and Herzegovina
hadziabdic@mef.unsa.ba
[2] Faculty of Mechanical Engineering, Department of Engines and Vehicles, University of Sarajevo, Sarajevo, Bosnia and Herzegovina

Abstract. The aim of this study was to investigate a certain competitive system of ordinary differential equations is observed. The analysis of local stability was studied in detail. Using nullclines, an insight into the global behavior of this system of equations was obtained. From the biological point of view, it is necessary to emphasize the persistence of the given system for certain values of parameters a and b. At the end, a graphical representation of the obtained results is shown.

Keywords: Equilibrium point · Sadlle point · Source · Sink · Comb · Eigenvalues · Stability · Jacobian matrix

1 Introduction

Differential equations are very important from several aspects, such as equations of motion in physics, models of prey and predators in ecology, models of neurons, models of dynamical systems, which relate to understanding and classifying solutions. They are important and are used in geodesy in constructing diffeomorphisms from vector fields, to construct the exponential map of a Riemannian manifold, to relate curvature and volume.

We want to investigate the behavior of the system of differential equations given by

$$\begin{cases} \dot{x} = x - x^2 - \frac{xy}{x+a} \\ \dot{y} = by - by^2 - xy \end{cases} \tag{1}$$

where a and b are real positive parameters. Here we will limit ourselves to examining the behavior of the positive equilibrium points alone. We consider populations of two species that interact, whose sizes are denoted by $x(t)$ and $y(t)$, modeled by a system of first-order differential equations. Our system is a form.

$$\begin{aligned} \dot{x} &= F(x, y) \\ \dot{y} &= G(x, y) \end{aligned}.$$

© The Author(s), under exclusive license to Springer Nature Switzerland AG 2022
I. Karabegović et al. (Eds.): NT 2022, LNNS 472, pp. 420–426, 2022.
https://doi.org/10.1007/978-3-031-05230-9_49

Such systems have been studied [1–3].

There are different types of biological interactions that are mathematically represented, different signs for the rates of change of growth rate $F(x, y)$ and $G(x, y)$ with respect to the size of the population $x(t)$ and $y(t)$. We start with the case of a species in competition, with two species perhaps competing for the same nutrient resource, they can compete against each other. Competition means increasing the size of any population tends to reduce the growth rate of the rest of the population. Research that studies the relationships of the two or more species has been dealt with in their work [4–6].

2 Existence of Equilibrium Points and Their Analysis

We will now determine the equilibrium points and the conditions when they exist.

Theorem 1: System (1) has a maximum of five equilibrium points: $E_1(0, 0)$, $E_2(0, 1)$, $E_3(1, 0)$, $E_4\left(\frac{1+b-ab-\sqrt{1+(-1+a)b(-2+(3+a)b)}}{2b}, \frac{-1+b(-1+a+2b)+\sqrt{1+(-1+a)b(-2+(3+a)b)}}{2b^2}\right)$

and $E_5\left(\frac{1+b-ab+\sqrt{1+(-1+a)b(-2+(3+a)b)}}{2b}, -\frac{1-b(-1+a+2b)+\sqrt{1+(-1+a)b(-2+(3+a)b)}}{2b^2}\right)$.

Where the equilibrium point E_4 exists, if the following conditions are fulfilled $0 <$ $b \leq 1$ and $0 < a < 1$ or $b > 1$ and $2\sqrt{\frac{-1+b}{b}} + \frac{1-b}{b} \leq a < 1$. The equilibrium point E_5 exists if the condition is fulfilled $b > 1$ and $2\sqrt{\frac{-1+b}{b}} + \frac{1-b}{b} \leq a <$ 1.Especially if it is $b > 1$ and $2\sqrt{\frac{-1+b}{b}} + \frac{1-b}{b} = a$, equilibrium point $E_4 = E_5 =$ $E\left(1 - \sqrt{\frac{-1+b}{b}}, \frac{-1+\sqrt{\frac{-1+b}{b}}+b}{b}\right)$.

Proof: Equilibrium points are solutions of the system

$$x - x^2 - \frac{xy}{x+a} = 0,$$

$$by - by^2 - xy = 0.$$

From the second equation we find that it is $y = 0$, $y = 1 - \frac{x}{b}$. By including $y = 0$ in the first equation, we get that $x - x^2 = 0$. Based on this, we get points $E_1(0, 0)$ and $E_3(1, 0)$. Similarly, including $y = 1 - \frac{x}{b}$, we come to a relation $x\left(1 - x + \frac{-b+x}{b(a+x)}\right) = 0$. From here we find the remaining equilibrium points. The conditions under which there are equilibrium points are obtained by direct calculations.

Numerical methods cannot give the clearest idea of the behavior of the system at equilibrium points. Therefore, it is necessary to perform a linear approximation of the system, around these points of known coordinates, i.e. we will linearize both functions that appear in a given system. For this purpose, a matrix of partial derivatives is used, which is called the Jacobian matrix. We will use the technique as in the works [7–9].

For this purpose, we write our system in matrix form $\dot{x} = f(x)$, where we have that

$$f(x) = \begin{pmatrix} x - x^2 - \frac{xy}{x+a} \\ by - by^2 - xy \end{pmatrix}.$$

The Jacobi matrix in this case is in the following form:

$$D_f = \begin{pmatrix} 1 - 2x - \frac{ay}{(a+x)^2} & -\frac{x}{a+x} \\ -y & b - x - 2by \end{pmatrix}.$$

Theorem 2: The equilibrium point E_1 is real (nodal) source.

Proof: The corresponding Jacobian matrix for E_1 has the following form $D_f(E_1) = \begin{pmatrix} 1 & 0 \\ 0 & b \end{pmatrix}$, it follows that eigenvalues are $\lambda_1 = 1$ and $\lambda_2 = b$, which are both positive. For trace and determinant we obtain that $trD_f(E_1) = 1 + b > 0$ and $detD_f(E_1) = b > 0$, based on the theory from [1] and [10–12], the accuracy of the above theorem follows.

Let's examine the character of the point E_2. We have that $D_f(E_2) = \begin{pmatrix} \frac{-1+a}{a} & 0 \\ -1 & -b \end{pmatrix}$, it follows that it is $\lambda_1 = \frac{-1+a}{a}$ and $\lambda_2 = -b$. By direct calculation we have that $trD_f(E_2) = \frac{-1+a}{a} - b$ and $detD_f(E_2) = -b + \frac{b}{a}$. We can see that $\lambda_2 < 0$ always and $\lambda_1 > 0$ if $a > 1$, while for $0 < a < 1$ we have that $\lambda_1 < 0$. From here we conclude that equilibrium point E_2 is a saddle point for $a > 1$ and for $0 < a < 1$ equilibrium point E_2 is nodal sink. Therefore we can say that the following theorem holds.

Theorem 3: If $0 < a < 1$ equilibrium point E_2 is a nodal sink, if $a > 1$ this point E_2 is saddle point.

Let us discuss the behavior of the equilibrium point E_3. In this case we have that $D_f(E_3) = \begin{pmatrix} -1 & -\frac{1}{1+a} \\ 0 & -1+b \end{pmatrix}$. The corresponding eigenvalues are given by $\lambda_1 = \frac{-1-a}{1+a}$ and $\lambda_2 = -1 + b$. Besides we have that it is $trD_f(E_3) = -2 + b$ and $detD_f(E_3) = 1 - b$. Here we have that $\lambda_1 < 0$ always, while $\lambda_2 < 0$ if $0 < b < 1$ and $\lambda_2 > 0$ if $b > 1$. From this we can formulate the following theorem.

Theorem 4: The equilibrium point E_3 is a nodal sink for $0 < b < 1$ and E_3 is a saddle point for $b > 1$.

For the equilibrium point E_4 we get a very large expression for $D_f(E_4)$, so we won't list it, as well as expressions for eigenvalues. The expressions for $detD_f(E_4)$ and $trD_f(E_4)$ are a little simpler and are given in the form

$$detD_f(E_4) = \frac{1}{2b^2}(-1 + (-1 + a)a(3 + a)b^3 + \sqrt{1 + (-1 + a)b(-2 + (3 + a)b)}$$
$$+ b(-2 + \sqrt{1 + (-1 + a)b(-2 + (3 + a)b)} + a(3$$
$$- 2\sqrt{1 + (-1 + a)b(-2 + (3 + a)b)})) + (-1 + a)b^2(-3$$
$$+ 2\sqrt{1 + (-1 + a)b(-2 + (3 + a)b)} + a(-3 + \sqrt{1 + (-1 + a)b(-2 + (3 + a)b)}))),$$

$$trD_f(E_4) = \frac{1}{2}(-\frac{b(a^2(2 + a) + 2b)}{a + b} + \frac{-1 + \sqrt{1 + (-1 + a)b(-2 + (3 + a)b)}}{b}$$
$$+ \frac{b - a(\sqrt{1 + (-1 + a)b(-2 + (3 + a)b)} + a(-2 + \sqrt{1 + (-1 + a)b(-2 + (3 + a)b)}))}{a + b}).$$

Theorem 5: The equilibrium point E_4 is saddle point if $b > 1$ *and* $2\sqrt{\frac{-1+b}{b}} + \frac{1-b}{b} <$ $a < 1$. For $b > 1$ equilibrium point E_4 is sink, nodal sink for

$$-\frac{1}{b^2}2(-1+(-1+a)a(3+a)b^3 + \sqrt{1+(-1+a)b(-2+(3+a)b)}$$

$$+ b(-2 + \sqrt{1+(-1+a)b(-2+(3+a)b)} + a(3$$

$$- 2\sqrt{1+(-1+a)b(-2+(3+a)b)})) + (-1+a)b^2(-3$$

$$+ 2\sqrt{1+(-1+a)b(-2+(3+a)b)} + a(-3$$

$$+ \sqrt{1+(-1+a)b(-2+(3+a)b)}))) + \frac{1}{4}(-\frac{b(a^2(2+a)+2b)}{a+b}$$

$$+ \frac{-1+\sqrt{1+(-1+a)b(-2+(3+a)b)}}{b}$$

$$+ \frac{b-a(\sqrt{1+(-1+a)b(-2+(3+a)b)} + a(-2+\sqrt{1+(-1+a)b(-2+(3+a)b)}))}{a+b}) > 0.$$

Nodal source for

$$-\frac{1}{b^2}2(-1+(-1+a)a(3+a)b^3 + \sqrt{1+(-1+a)b(-2+(3+a)b)}$$

$$+ b\left(-2 + \sqrt{1+(-1+a)b(-2+(3+a)b)} + a(3\right.$$

$$- 2\sqrt{1+(-1+a)b(-2+(3+a)b)}\right)) + (-1+a)b^2(-3$$

$$+ 2\sqrt{1+(-1+a)b(-2+(3+a)b)} + a(-3$$

$$+ \left.\sqrt{1+(-1+a)b(-2+(3+a)b)}\right))) + \frac{1}{4}\left(-\frac{b(a^2(2+a)+2b)}{a+b}\right.$$

$$+ \frac{-1+\sqrt{1+(-1+a)b(-2+(3+a)b)}}{b}$$

$$+ \frac{b-a(\sqrt{1+(-1+a)b(-2+(3+a)b)} + a(-2+\sqrt{1+(-1+a)b(-2+(3+a)b)}))}{a+b}) < 0.$$

Proof: In the case where it is $b > 1$ *and* $2\sqrt{\frac{-1+b}{b}} + \frac{1-b}{b} < a < 1$ worth it $detD_f(E_4) <$ 0, so the given point is a saddle point. In the second case, when only $b > 1$, $trD_f(E_4) < 0$, always. So it is necessary to examine the relationship $\left(trD_f(E_4)\right)^2 - 4detD_f(E_4)$. If this ratio is greater than zero we have nodal source, and if it is less than zero we have nodal sink. The expression given in the theorem setup just represents a given relation.

Determine the character of the remaining equilibrium point E_5. We find that the expression for $D_f(E_5)$, as in the previous case very large and inconspicuous so we will not list it, nor the expressions of the corresponding eigenvalues. The determinant and trace for this point are

$$detD_f(E_5) = \frac{1}{2b^2}(-1 - 2b + 3ab + 3b^2 - 3a^2b^2 - 3ab^3$$

$$+ 2a^2b^3 + a^3b^3 - \sqrt{1+(-1+a)b(-2+(3+a)b)}$$

$$- b\sqrt{1+(-1+a)b(-2+(3+a)b)} + 2ab\sqrt{1+(-1+a)b(-2+(3+a)b)}$$

$$+ 2b^2\sqrt{1+(-1+a)b(-2+(3+a)b)} - ab^2\sqrt{1+(-1+a)b(-2+(3+a)b)}$$

$$- a^2b^2\sqrt{1+(-1+a)b(-2+(3+a)b)}),$$

$$trD_f(E_5) = \frac{1}{2b(a+b)}(-a - b + 2a^2b + b^2 - 2a^2b^2 - a^3b^2 - 2b^3$$

$$- a\sqrt{1 + (-1 + a)b(-2 + (3 + a)b)} - b\sqrt{1 + (-1 + a)b(-2 + (3 + a)b)}$$
$$+ ab\sqrt{1 + (-1 + a)b(-2 + (3 + a)b)} + a^2b\sqrt{1 + (-1 + a)b(-2 + (3 + a)b)}).$$

Theorem 6: The equilibrium point E_5 is sink for $0 < a < 1$ and $1 < b < 2\sqrt{-\frac{1}{(-1+a)(3+a)^2} + \frac{1}{3+a}})$ or for $a \geq 1$ and $b > 1$.

Proof: For this point it is always true that it is $detD_f(E_5) > 0$ and $trD_f(E_5) < 0$ under the given assumptions of the theorem. In addition, under the same conditions, direct verification is valid and $\left(trD_f(E_5)\right)^2 - 4detD_f(E_5) > 0$. Which means it's our point E_5 is nodal sink (Fig. 1).

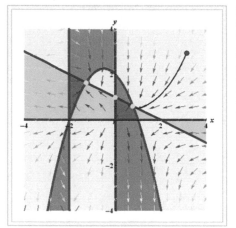
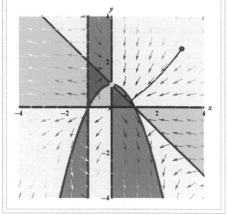
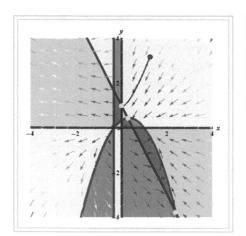
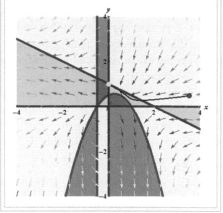

Fig. 1. Visual representation for different parameter values

Let it be now $b > 1$ and $2\sqrt{\frac{-1+b}{b}} + \frac{1-b}{b} = a$, in this case we have that equilibrium point $E_4 = E_5 = E$. For this point we find

$$\text{that } D_f(E) = \begin{pmatrix} \dfrac{-2+\sqrt{\frac{-1+b}{b}}-\sqrt{-1+b}\sqrt{b}+b}{1+2\sqrt{-1+b}\sqrt{b}-b+b^2} & \dfrac{b\left(-2+\sqrt{\frac{-1+b}{b}}-\sqrt{-1+b}\sqrt{b}+b\right)}{1+2\sqrt{-1+b}\sqrt{b}-b+b^2} \\ -1-\dfrac{\sqrt{-1+b}}{b^{3/2}}+\dfrac{1}{b} & 1-\sqrt{\dfrac{-1+b}{b}}-b \end{pmatrix}, \; trD_f(E) =$$

$\dfrac{1+b\left(-1+\sqrt{-1+b}\sqrt{b}-2b+b^2\right)}{1+b-b^2}$, $detD_f(E) = 0$, we know that the relation is known, $trD_f(E) = \lambda_1 + \lambda_2$ and $detD_f(E) = \lambda_1 \cdot \lambda_2$. Now from this we can conclude that one eigenvalue is zero, and the other is equal to the trace. By direct verification we get that it is $trD_f(E) < 0$ for $1 \le b\left(\frac{1}{2}\left(1+\sqrt{5}\right)orb\right)\frac{1}{2}\left(1+\sqrt{5}\right)$. So we conclude that the point E is comb.

3 Conclusion

Given the limitations of the given system of equations, the obtained results give a detailed overview of the behavior of the system for various values of parameters a and b. The analysis of local stability provides an answer to the possibility of coexistence for two species. This, in addition to persistence for certain values of parameters a and b, has a very great significance from the biological point of view. In addition, the extinction n of one of the populations does not necessarily lead to the extinction of the other population, which is also important from a biological point of view. The study of such models is very topical as we can see from [13–15]. Note that this was not the case with the predator-prey model. At that time, we had a situation where predators became extinct in the event of prey extinction. Of particular importance is the existence of two equilibriums inside the first quadrant, both of which, for certain parameter values, represent locally stable equilibrium points. Of course, these are not two equilibriums that are stable at the same time, but the possibility that they are alternately locally stable for certain parameter values.

References

1. Verhulst, F.: Nonlinear Differential Equations and Dynamical Systems. UTX. Springer, Heidelberg (1990). https://doi.org/10.1007/978-3-642-97149-5
2. Kant, S., Kumar, V.: Dynamics of a preypredator system with infection in prey. Election. J. Diff. Equ. **209**, 1–27 (2017)
3. Fulda, J.S.: The logistic equation and population decline. J. Theor. Biol. **91**(2), 255–259 (1981)
4. Seifert, G.: A Lotka-Volterra predator-prey system involving two predators. Methods Appl. Anal. **2**(2), 248–255 (1995)
5. Hadžiabdić, V., Mehuljić, M., Bektešević, J., Šarić, I.: Application of the nullcline method to a certain model of competitive species. TEM J. **8**(1), 73–77 (2019)
6. Hadžiabdić, V., Mehuljić, M., Bektešević, J., Mujić, N.: Coexistence between predator and prey in the modified Lotka - Volterra model. TEM J. **7**(2), 330–334 (2018)
7. Hadžiabdić, V., Mehuljić, M., Bektešević, J.: Lotka-Volterra model with two predators and their prey. TEM J. **6**(1), 132–136 (2017)
8. Imamović, Z., Hadžiabdić, V., Mehuljić, M., Bektešević, J., Burgić, D.: Modified gauss-type competitive system. In: Karabegović, I. (ed.) NT 2021. LNNS, vol. 233, pp. 1183–1187. Springer, Cham (2021). https://doi.org/10.1007/978-3-030-75275-0_131

9. Hadžiabdić, V., Mehuljić, M., Bektešević, J., Mašić, A.: Dynamics and stability of Hopf bifurcation for one non-linear system. TEM J. **10**(2), 820–824 (2021). https://doi.org/10. 18421/TEM102-40

10. Perko, L.: Differential Equation and Dynamical Systems. TAM, vol. 7. Springer, New York (2001). https://doi.org/10.1007/978-1-4613-0003-8

11. Hirsch, M.W., Smale, S., Devaney, R.L.: Differential Equations, Dynamical Systems, and An Introduction to Chaos, 2nd edn. Elsevier, San Diego (1974)

12. Wiggins, S.: Introduction to Applied Nonlinear Dynamical Systems and Chaos. TAM, vol. 2. Springer, New York (2003). https://doi.org/10.1007/b97481

13. Farhan, A.: Lotka-Volterra model with prey-predators food chain. Iraqi J. Sci. 56–63 (2020)

14. Jia, L.: Analysis for a delayed three-species predator-prey model with feedback controls and prey diffusion. J. Math. **2020** (2020). Article no. 5703859, 26 pages. https://doi.org/10.1155/ 2020/5703859

15. Nagy, I., Romanovski, V.G., Tóth, J.: Two nested limit cycles in two-species reactions. Mathematics **8**, 1658 (2020). https://doi.org/10.3390/math8101658

Global Period-Doubling Bifurcation of a Certain Second-Order Quadratic Rational Difference Equations

Midhat Mehuljić[1]([⊠]), Jasmin Bektešević[1], Vahidin Hadžiabdić[1], and Naida Mujić[2]

[1] Faculty of Mechanical Engineering, Department of Mathematics and Physics, University of Sarajevo, St. Vilsonovo šetalište 9, 71 000 Sarajevo, Bosnia and Herzegovina
mehuljic@mef.unsa.ba

[2] Faculty of Electrical Engineering, University of Sarajevo, Sarajevo, Bosnia and Herzegovina

Abstract. In this paper we considered local dynamics of the following difference equation $x_{n+1} = \frac{\alpha x_n^2 + \beta x_n + \gamma x_{n-1}}{A x_n^2 + B x_n + C x_{n-1}}$, where α, β, γ, A, B, $C \geq 0$ and $\alpha + \beta + \gamma > 0$, $A + B + C > 0$ and where the initial conditions x_{-1} and x_0 are arbitrary nonnegative real numbers such that $x_{-1} + x_0 > 0$. The local stability of the equilibrium was fully examined for all values of parameters α, β, γ, A, B, $C \geq 0$. It has been shown that a single equilibrium point can be a locally asymptotically stable, saddle or non-hyperbolic point, but it cannot be a repeller for any of the parameter values. Also, for certain values of the coefficients, the global asymptotic stability of the equilibrium is shown.

Keywords: Boundedness · Difference equation · Equilibrium · Global stability · Local dynamics

1 Introduction

In this paper we investigate the global dynamics of the following difference equation

$$x_{n+1} = \frac{\alpha x_n^2 + \beta x_n + \gamma x_{n-1}}{A x_n^2 + B x_n + C x_{n-1}}, \quad n = 0, 1, \ldots \tag{1}$$

where α, β, γ, A, B, $C \geq 0$ and $\alpha + \beta + \gamma > 0$, $A + B + C > 0$ and where the initial conditions x_{-1} and x_0 are arbitrary nonnegative real numbers such that $x_{-1} + x_0 > 0$. This equation is a special case of the general rational difference equation of the second order with quadratic terms in the numerator and denominator

$$x_{n+1} = \frac{A x_n^2 + B x_n x_{n-1} + C x_{n-1}^2 + D x_n + E x_{n-1} + F}{a x_n^2 + b x_n x_{n-1} + c x_{n-1}^2 + d x_n + e x_{n-1} + f}, \quad n = 0, 1, \ldots \tag{2}$$

Many special cases of Eq. (1) have been studied before. Special cases:

$$x_{n+1} = \frac{\gamma x_{n-1}}{B x_n + C x_{n-1}}, \quad n = 0, 1, \ldots \tag{3}$$

© The Author(s), under exclusive license to Springer Nature Switzerland AG 2022
I. Karabegović et al. (Eds.): NT 2022, LNNS 472, pp. 427–435, 2022.
https://doi.org/10.1007/978-3-031-05230-9_50

$$x_{n+1} = \frac{\beta x_n + \gamma x_{n-1}}{C x_{n-1}}, \quad n = 0, 1, \ldots \tag{4}$$

$$x_{n+1} = \frac{\beta x_n}{B x_n + C x_{n-1}}, \quad n = 0, 1, \ldots \tag{5}$$

$$x_{n+1} = \frac{\beta x_n + \gamma x_{n-1}}{B x_n}, \quad n = 0, 1, \ldots \tag{6}$$

$$x_{n+1} = \frac{\beta x_n + \gamma x_{n-1}}{B x_n + C x_{n-1}}, \quad n = 0, 1, \ldots \tag{7}$$

are observed in [1]. Some other special cases of the Eq. (2) which have quadratic terms have been observed in [2, 3], where $A = C = D = a = c = d = 0$. Also, in [4–6], the equation obtained for $C = D = F = c = d = f = 0$, have been studied in much more detail. There are several results obtained in [7–10] on the global stability of equations which are special cases of Eq. (2), and which have quadratic terms. In this paper, the local stability of the unique equilibrium will be done in detail, as well as the existence and local stability of the prime period-two solution. Necessary and sufficient conditions for the existence of equilibrium point as locally asymptotically stable point, saddle point and non-hyperbolic point and also for period two solutions. The analysis of the stability of these points will give us an insight into possible dynamic scenarios involving period-doubling bifurcation, as well as the global attractivity of the equilibrium. The following results of global asymptotic stability for quadratic fractional difference equation were obtained in [11].

Theorem 1: Suppose that Eq. (2) has the unique equilibrium $\bar{x}$. If the following condition holds

$$\frac{(|A - a\bar{x}| + |B - b\bar{x}| + |C - c\bar{x}|)(U + \bar{x}) + |D - d\bar{x}| + |E - e\bar{x}|}{(a + b + c)L^2 + (d + e)L + f} < 1$$

where L and U are lower and upper bounds of all solutions of Eq. (2) and $L + f > 0$, then $\bar{x}$ is globally asymptotically stable.

Theorem 2: Assume that Eq. (2) has the unique equilibrium $\bar{x}$ in the interval $[m, M]$, where $m = min\{\bar{x}, x_{-1}, x_0\}$ and $M = max\{\bar{x}, x_{-1}, x_0\}$ are lower and upper bounds of a specific solution of Eq. (2) and $m + f > 0$. If the following condition holds

$$(|A - a\bar{x}| + |B - b\bar{x}| + |C - c\bar{x}|)(M + \bar{x}) + |D - d\bar{x}| + |E - e\bar{x}|$$
$$< (a + b + c)m^2 + (d + e)m + f$$

then $\bar{x}$ is globally asymptotically stable on the interval $[m, M]$.

For Eq. (1), Theorems 1 and 2 give the special results.

Corollary 1: If the following condition holds

$$\frac{(|\alpha - A\bar{x}|)(U + \bar{x}) + |\beta - B\bar{x}| + |\gamma - C\bar{x}|}{AL^2 + (B + C)L} < 1$$

where $L > 0$ and U are lower and upper bounds of all solutions of Eq. (1), then $\bar{x}$ is globally asymptotically stable.

Corollary 2: If the following condition holds

$$(|\alpha - A\bar{x}|)(M + \bar{x}) + |\beta - B\bar{x}| + |\gamma - C\bar{x}| < Am^2 + (B + C)m$$

where $m = min\{\bar{x}, x_{-1}, x_0\}$ and $M = max\{\bar{x}, x_{-1}, x_0\}$ are lower and upper bounds of a specific solution of Eq. (1), then unique equilibrium $\bar{x}$ is globally asymptotically stable on the interval $[m, M]$.

2 Local Stability of the Positive Equilibrium

In this section we investigate the equilibrium points of Eq. (1), where

$$\alpha, \beta, \gamma, A, B, C \in [0, \infty) \text{ with } \alpha + \beta + \gamma, A + B + C \in (0, \infty)$$

and where the initial conditions x_{-1} and x_0 are arbitrary nonnegative real numbers such that $x_{-1} + x_0 > 0$. In view of the above restriction on the initial conditions of Eq. (1), the equilibrium points of Eq. (1) are the positive solutions of the equation

$$\bar{x} - \frac{\alpha\bar{x}^2 + \beta\bar{x} + \gamma\bar{x}}{A\bar{x}^2 + B\bar{x} + C\bar{x}}$$

which is equivalently

$$Ax^2 + (B + C - \alpha)x - (\beta + \gamma) = 0.$$

If

$$\beta + \gamma = 0, A > 0 \text{ and } \alpha > B + C$$

then the unique positive equilibrium of Eq. (1) is given by

$$\bar{x} = \frac{\alpha - B - C}{A}.$$

If

$$A = 0, B + C > \alpha \text{ and } \beta + \gamma > 0$$

then the unique positive equilibrium of Eq. (1) is given by

$$\bar{x} = \frac{\beta + \gamma}{B + C - \alpha}.$$

If

$$A > 0 \text{ and } \beta + \gamma > 0,$$

then the only equilibrium point of Eq. (1) is the solution

$$\bar{x} = \frac{\alpha - B - C + \sqrt{(\alpha - B - C)^2 + 4A(\beta + \gamma)}}{2A}.$$

It is important to note that if Eq. (1) has a positive equilibrium that it is unique, which greatly facilitates the analysis of the local stability of the positive equilibrium.

Let

$$f(u, v) = \frac{\alpha u^2 + \beta u + \gamma v}{Au^2 + Bu + Cv},$$

Then

$$\frac{\partial f}{\partial u} = \frac{Cv(2\alpha u + \beta) + B(\alpha u^2 - \gamma v) - Au(\beta u + 2\gamma v)}{(Au^2 + Bu + Cv)^2}$$

and

$$\frac{\partial f}{\partial v} = \frac{u(-C(\alpha u + \beta) + \gamma(B + Au))}{(Au^2 + Bu + Cv)^2}.$$

Let's mark now with $\bar{x}$ equilibrium point of Eq. (1), then the linearized equation of Eq. (1) at the point of equilibrium $\bar{x}$ has the form

$$z_{n+1} = pz_n + qz_{n-1},$$

where

$$p = \frac{\partial f}{\partial u}(\bar{x}, \bar{x})$$

and

$$q = \frac{\partial f}{\partial v}(\bar{x}, \bar{x}).$$

Theorem 3: Suppose it is

$$\beta + \gamma = 0, A > 0 \text{ and } \alpha > B + C,$$

then the only positive equilibrium point

$$\bar{x} = \frac{\alpha - B - C}{A}$$

of Eq. (1) is locally asymptotically stable.

Proof: It's not hard to see that

$$p = \frac{\partial f}{\partial u}(\bar{x}, \bar{x}) = \frac{B + 2C}{\alpha}$$

and

$$q = \frac{\partial f}{\partial v}(\bar{x}, \bar{x}) = -\frac{C}{\alpha}.$$

Then, from the fact that it is

$$1 - p - q = 1 - \frac{B + C}{\alpha}$$

$$1 + p - q = \frac{\alpha + B + 3C}{\alpha} > 0$$

and

$$q + 1 = 1 - \frac{C}{\alpha}$$

the proof of the above theorem follows from Theorem 1 in [1].

Theorem 4: Suppose it is

$$A = 0,\, B \mid C > \alpha \text{ and } \beta + \gamma > 0,$$

then the only positive equilibrium point

$$\bar{x} = \frac{\beta + \gamma}{B + C - \alpha}$$

of Eq. (1) is

a) locally asymptotically stable if $B\beta + C(3\beta + \gamma) + \alpha(\beta + 3\gamma) > B\gamma$;
b) saddle point if $B\beta + C(3\beta + \gamma) + \alpha(\beta + 3\gamma) < B\gamma$;
c) nonhyperbolic point if $B\beta + C(3\beta + \gamma) + \alpha(\beta \mid 3\gamma) = B\gamma$.

Proof: We have that

$$p = \frac{C\beta - B\gamma + \alpha(\beta + 2\gamma)}{(B + C)(\beta + \gamma)}$$

and

$$q = \frac{-C\beta + (B - \alpha)\gamma}{(B + C)(\beta + \gamma)}.$$

Then it is

$$p - q + 1 = \frac{B(\beta - \gamma) + C(3\beta + \gamma) + \alpha(\beta + 3\gamma)}{(B + C)(\beta + \gamma)},$$

$$1 - p - q = 1 - \frac{\alpha}{B + C},$$

$$q + 1 = \frac{B\beta + (2B + C - \alpha)\gamma}{(B + C)(\beta + \gamma)},$$

$$q - 1 = -\frac{(B + 2C)\beta + (C + \alpha)\gamma}{(B + C)(\beta + \gamma)}.$$

We conclude that based on the conditions the theorem is always valid $1 - p - q > 0, q + 1 > 0, q - 1 < 0$, and

$$p - q + 1 = \frac{B(\beta - \gamma) + C(3\beta + \gamma) + \alpha(\beta + 3\gamma)}{(B + C)(\beta + \gamma)} > 0$$

if

$$B\beta + C(3\beta + \gamma) + \alpha(\beta + 3\gamma) > B\gamma.$$

Theorem 5: Suppose that $A > 0$ and $\beta + \gamma > 0$, then the only positive point is the equilibrium

$$\bar{x} = \frac{\alpha - B - C + \sqrt{((\alpha - B - C)^2 + 4A(\beta + \gamma))}}{2A}$$

of Eq. (1) is

a) locally asymptotically stable if $\bar{x} > \frac{2\gamma}{(B+3C+\alpha)}$;
b) saddle point if $\bar{x} < \frac{2\gamma}{(B+3C+\alpha)}$;
c) nonhyperbolic point if $\bar{x} = \frac{2\gamma}{(B+3C+\alpha)}$.

Proof: We've seen before that if it is $A > 0$ and $\beta + \gamma > 0$, then the only positive equilibrium point of Eq. (1) is

$$\bar{x} = \frac{\alpha - B - C + \sqrt{((\alpha - B - C)^2 + 4A(\beta + \gamma))}}{2A}.$$

Based on identity

$$A\bar{x}^2 + (B + C - \alpha)\bar{x} = \beta + \gamma$$

We see that it is

$$p = \frac{\partial f}{\partial u}(\bar{x}, \bar{x}) = \frac{(C(2\alpha\bar{x} + \beta) + B(\alpha\bar{x} - \gamma) - A\bar{x}(\beta + 2\gamma))}{(\bar{x}(A\bar{x} + B + C)^2)}$$

$$= \frac{C(2\alpha\bar{x} + \beta) + B(\alpha\bar{x} - \gamma) - A\bar{x}(\beta + 2\gamma)}{(A\bar{x} + B + C)(\alpha\bar{x} + \beta + \gamma)},$$

$$q = \frac{\partial f}{\partial v}(\bar{x}, \bar{x}) = \frac{-C(\alpha\bar{x} + \beta) + (A\bar{x} + B)\gamma}{\bar{x}(A\bar{x} + B + C)^2}$$

$$= \frac{-C(\alpha\bar{x} + \beta) + (A\bar{x} + B)\gamma}{(A\bar{x} + B + C)(\alpha\bar{x} + \beta + \gamma)}.$$

Now is

$$p - q + 1 = \frac{B + 3C}{(A\bar{x} + B + C)} + \frac{\alpha\bar{x} - 2\gamma}{(\alpha\bar{x} + \beta + \gamma)} = \frac{(B + 3C)\bar{x}}{(\alpha\bar{x} + \beta + \gamma)} + \frac{\alpha\bar{x} - 2\gamma}{(\alpha\bar{x} + \beta + \gamma)}$$

$$= \frac{(B + 3C + \alpha)\bar{x} - 2\gamma}{(\alpha\bar{x} + \beta + \gamma)},$$

$$1 - p - q = \frac{2A\bar{x} + B + C}{(A\bar{x} + B + C)} - \frac{\alpha\bar{x}}{(\alpha\bar{x} + \beta + \gamma)} > 0,$$

$$q + 1 = 1 + \frac{-C(\alpha\bar{x} + \beta) + (A\bar{x} + B)\gamma}{(A\bar{x} + B + C)(\alpha\bar{x} + \beta + \gamma)} > 0,$$

$$q - 1 = -\frac{C}{(A\bar{x} + B + C)} - \frac{\alpha\bar{x} + \beta}{(\alpha\bar{x} + \beta + \gamma)} < 0.$$

Based on the last four relations and Theorem 1 in [1] the proof of the above theorem follows.

3   Global Asymptotic Stability Results

Theorem 6: Consider Eq. (1) where all coefficients are positive. If the following condition holds

$$\frac{(|\alpha - A\bar{x}|)(U + \bar{x}) + |\beta - B\bar{x}| + |\gamma - C\bar{x}|}{AL^2 + (B + C)L} < 1$$

where $L = \frac{\min\{\alpha,\beta,\gamma\}}{\max\{A,B,C\}}$ and $U = \frac{\max\{\alpha,\beta,\gamma\}}{\min\{A,B,C\}}$ are lower and upper bounds of all solutions of Eq. (1), then $\bar{x}$ is globally asymptotically stable.

Proof: In this case the lower and upper bounds for all solutions of Eq. (1) for $n \geq 1$ were obtained as follows

$$x_{n+1} = \frac{\alpha x_n^2 + \beta x_n + \gamma x_{n-1}}{A x_n^2 + B x_n + C x_{n-1}} \geq \frac{\min\{\alpha, \beta, \gamma\}}{\max\{A, B, C\}} \frac{x_n^2 + x_n + x_{n-1}}{x_n^2 + x_n + x_{n-1}} = \frac{\min\{\alpha, \beta, \gamma\}}{\max\{A, B, C\}} = L > 0$$

and

$$x_{n+1} = \frac{\alpha x_n^2 + \beta x_n + \gamma x_{n-1}}{A x_n^2 + B x_n + C x_{n-1}} \leq \frac{\max\{\alpha, \beta, \gamma\}}{\min\{A, B, C\}} \frac{x_n^2 + x_n + x_{n-1}}{x_n^2 + x_n + x_{n-1}} = \frac{\max\{\alpha, \beta, \gamma\}}{\min\{A, B, C\}} = U.$$

Based on Corollary 1, the proof of the above theorem follows.

Theorem 7: Consider Eq. (1) where $B = 0$ and all other coefitients are positive. If the following condition holds

$$\frac{(|\alpha - A\bar{x}|)(U + \bar{x}) + \beta + |\gamma - C\bar{x}|}{AL^2 + CL} < 1$$

where $L = \frac{\min\{\alpha,\beta,\gamma\}}{\max\{A,C\}}$ and $U = \frac{\max\{\alpha,\gamma\}}{\min\{A,C\}} + \frac{\beta}{AL}$ are lower and upper bounds of all solutions of Eq. (1), then $\bar{x}$ is globally asymptotically stable.

Proof: In this case the lower and upper bounds for all solutions of Eq. (1) for $n \geq 1$ were obtained as follows

$$x_{n+1} = \frac{\alpha x_n^2 + \beta x_n + \gamma x_{n-1}}{A x_n^2 + C x_{n-1}} \geq \frac{\min\{\alpha, \beta, \gamma\}}{\max\{A, C\}} \frac{x_n^2 + x_n + x_{n-1}}{x_n^2 + x_{n-1}} \geq \frac{\min\{\alpha, \beta, \gamma\}}{\max\{A, C\}} = L > 0$$

and

$$x_{n+1} = \frac{\alpha x_n^2 + \beta x_n + \gamma x_{n-1}}{A x_n^2 + C x_{n-1}} = \frac{\alpha x_n^2 + \gamma x_{n-1}}{A x_n^2 + C x_{n-1}} + \frac{\beta x_n}{A x_n^2 + C x_{n-1}}$$

$$\leq \frac{\max\{\alpha, \gamma\}}{\min\{A, C\}} \frac{x_n^2 + x_{n-1}}{x_n^2 + x_{n-1}} + \frac{\beta}{A x_n} = \frac{\max\{\alpha, \gamma\}}{\min\{A, C\}} + \frac{\beta}{A x_n}$$

$$\leq \frac{\max\{\alpha, \gamma\}}{\min\{A, C\}} + \frac{1}{AL} = U.$$

Based on Theorem 1 and Corollary 1, the proof of the above theorem follows.

Remark 1: The global asymptotic stability of the equilibrium follows from Theorem 2 and Corollary 2, in the interval $[min\{\overline{x}, x_{-1}, x_0\}, max\{\overline{x}, x_{-1}, x_0\}]$, where $min\{\overline{x}, x_{-1}, x_0\} > 0$, i.e. when $x_{-1} x_0 > 0$. In this case $max\{\overline{x}, x_{-1}, x_0\}$ can be replaced with $U = \frac{\max\{\alpha, \beta, \gamma\}}{\min\{A, B, C\}}$.

4 Conclusion

Examining the behavior of the general quadratic rational Eq. (2) is quite complicated. The behavior of its special cases is currently being examined and in this way an attempt is made to infer the behavior of Eq. (2). Examining the stability of Eq. (1), as well as some other special cases, helps us in this. The existence of a unique positive equilibrium for a given equation greatly facilitates the analysis of its stability. By analyzing the local stability, we see that there is only a bifurcation of the doubling of the period. The absence of a repeller for any parameter value does not allow the existence of Neimark-Sacker bifurcation.

References

1. Kulenović, M.R.S., Ladas, G.: Dynamics of Second Order Rational Difference Equations with Open Problems and Conjectures. Chapman and Hall/CRC, Boca Raton (2001)
2. Amleh, A.M., Camouzis, E., Ladas, G.: On the dynamics of a rational difference equation, part I. Int. J. Differ. Equ. 3, 1–35 (2008)
3. Amleh, A.M., Camouzis, E., Ladas, G.: On the dynamics of a rational difference equation, part II. Int. J. Differ. Equ. 3, 195–225 (2008)
4. Kalabušić, S., Kulenović, M.R.S., Mehuljić, M.: Global period-doubling bifurcation of quadratic fractional second order difference equation. Discrete Dyn. Nat. Soc. **2014** (2014). Article no. 920410, 13 pages

5. Kalabušić, S., Kulenović, M.R.S., Mehuljić, M.: Global dynamics and bifurcations of two quadratic fractional second order difference equations. J. Comput. Anal. Appl. **21**(1), 172–184 (2016)

6. Bektešević, J., Mehuljić, M., Hadžiabdić, V., Kalabušić, S.: Global asymptotic behavior of some quadratic rational second-order difference equations. Int. J. Differ. Equ. **12**(2), 169–183 (2017)

7. Janowski, E.A., Kulenovic, M.R.S.: Attractivity and global stability for linearizable difference equations. Comput. Math. Appl. **57**, 1592–1607 (2009)

8. Kent, C.M., Sedaghat, H.: Global attractivity in a quadratic-linear rational difference equation with delay. J. Differ. Equ. Appl. **15**, 913–925 (2009)

9. Kent, C.M., Sedaghat, H.: Global attractivity in a rational delay difference equation with quadratic terms. J. Differ. Equ. Appl. **17**, 457–466 (2011)

10. Sedaghat, H.: Global behaviours of rational difference equations of orders two and three with quadratic terms. J. Differ. Equ. Appl. **15**, 215–224 (2009)

11. DiPippo, M., Janowski, E., Kulenović, M.: Global asymptotic stability for quadratic fractional difference equation. Adv. Differ. Equ. **2015**(1), 1–13 (2015). https://doi.org/10.1186/s13662-015-0525-4

Electrical Engineering, Computer Science, Information and Communication Technologies, Control Systems

Energy-Efficient AI Systems Based on Memristive Technology

Adnan Mehonic[(⊠)]

Department of Electronic and Electrical Engineering, UCL, Torrington Place,
London WC1E 7JE, UK
adnan.mehonic.09@ucl.ac.uk

Abstract. The future progress of AI is dependent on the capability of hardware systems. Intense AI-driven data-centric applications lead to unsustainably large compute power demands. Current state-of-the-art hardware consumes far too much energy and is restrictive to many applications where energy resources are limited (e.g. IoT devices). This is becoming unsustainable and might halt the future rapid progress of AI, which is becoming even more critical in a wide range of rapidly growing data-centric technologies spanning Big Data, IoT, transport, medicine, security, entertainment. This challenge led to the exploration of post-CMOS technologies, from innovations in materials and novel nanoelectronics to better-suited computer architectures (e.g. non-Von Neumann systems that solve the data movement bottleneck). This paper presents memristor technology as a potential solution for implementing energy-efficient AI systems.

Keywords: Memristors · Energy-efficient AI · Neuromorphic

1 Introduction

Among three factors (accessible data, algorithmic solutions and powerful hardware) needed for continuous development in artificial intelligence (AI), hardware is becoming the main bottleneck. The increasing compute power demands vastly outpace the efficiency improvements obtained through Moore scaling or innovative computer architecture solutions. Although Graphical Processing Units (GPUs) improved performance by more than >300x from 2012 until 2020, this is not enough to meet the demand [1]. The compute power demands now double every two months, which results in the cost of AI increasing dramatically. More specifically, the cost of training has grown from a few $ in 2012 to ~$10m in 2020 [1]. Fundamental limitations to improving the energy efficiency of computing are the excessive growth of data transfer costs when upscaling the Von Neumann architecture and limits of CMOS technologies scaling. Von Neumann architecture is not the best suited for data-intense tasks, as data transfer from memory to compute units leads to inefficiencies in most cases. This article introduces the concept of memristors, novel beyond-complementary metal–oxide–semiconductor (CMOS) technology. For the last two decades, memristors have been studied by different research

© The Author(s), under exclusive license to Springer Nature Switzerland AG 2022
I. Karabegović et al. (Eds.): NT 2022, LNNS 472, pp. 439–442, 2022.
https://doi.org/10.1007/978-3-031-05230-9_51

communities - mathematicians, experimental material engineers, nanoelectronics, circuit designer and computer scientists. The article focuses on the use of memristors for AI applications.

1.1 Memristive Technology

Memristor technology is a strong candidate for beyond CMOS and Von Neumann computing solutions. Memristor technology includes a broad family of devices, such as redox-based resistive random-access memory (ReRAM), phase-change memories (PCMs), and magneto resistive random-access memory (MRAM), to name a few, all based on different physical mechanisms and materials systems.

Memristor operation is based on a resistance switching process, typically obtained from simple metal-insulator-metal structures. An insulating thin film can change its electrical resistance between at least two stable states (commonly called the high resistance state – HRS, and the low resistance state - LRS). Switching devices from HRS to LRS is called SET, and the reverse process, transition from the LRS to the HRS, is called RESET. The typical device behaviour (SiOx memristor) is shown in Fig. 1 [2]. Figure 1a shows I/V voltage sweep curves where devices start in the pristine (highly insulated state) and go through the electroforming process when sufficiently high voltage is applied. The electroforming is a one-time process that prepares the device for switching operation. Switching the device back to the HRS is commonly achieved by reversing the voltage polarity, while for switching to the LRS the same polarity used for electroforming is applied. Current compliance is typically used for electroforming and SET processes to prevent extensive materials damage (e.g. hard breakdown). The physical process that underpins the resistance switching depends on a specific materials system; in the case of metal oxides, the switching process is associated with creating and dissolution of conductive filaments. Apart from two resistance states, it is possible to operate devices in such a way as to obtain multiple resistance states, which is more relevant in the context of this paper (Fig. 1b).

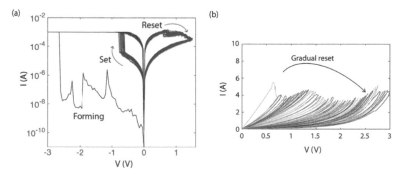

Fig. 1. (a) I/V curve showing resistance switching in SiOx memristor devices. (b) Analogue resistance modulation in SiOx memristor devices. Reproduced from [2, 5].

1.2 Deep Learning Accelerators Based on Memristive Crossbar Arrays

In the Von Neumann architecture, memory and processing units are physically separated, and data needs to be shuttled back and forth between them during the execution
of computational tasks. This leads to energy cost and latency, especially challenging for
mobile and edge applications, where energy resources are restricted. Although there are
many ways how memristive technology can be utilised for energy-efficient computing,
this paper focuses on deep learning accelerators based on memristive crossbar arrays.
Memristive crossbar arrays use the concept of in-memory computing to solve the Von
Neumann bottleneck on the most fundamental level. The main physical computational
primitive is the matrix-vector multiplication (MVM) or multiply-and-accumulate (MAC)
accelerator, critical for most deep learning algorithms (Fig. 2). To perform the multiplication, $Ma = b$, where M is a matrix, and a and b are vectors, the elements of M are
mapped and represented as conductances of memristors. Vector a is mapped to amplitudes of applied voltages. The resulting vector b can be obtained as the resulting current.
Therefore, it is possible to implement MVM operation utilising adjustable conductances
of memristive devices, Ohm's law, and Kirchhoff's current summation law. This approach leads to massive energy-efficiency improvements (>100x) over state-of-the-art
CMOS systems [3].

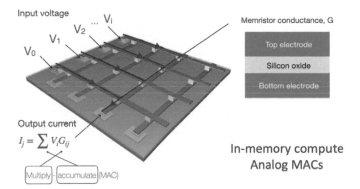

Fig. 2. **MAC** Multiple-and-accumulate (MAC) accelerator based on memristive crossbar array.

The memristor technology promises to bring significant efficiency improvements;
however, several technological challenges still need to be addressed. These challenges
are mostly related to reliability at scale and overcoming device- and system-level non-
idealities associated with the analogue nature of information processing. Apart from
materials engineering solutions, there is an emphasis on hardware-algorithm co-design
that could significantly mitigate the nonidealities [4].

2 Conclusion

Memristive technologies are versatile both in terms of their applications (e.g. non-volatile
digital memory, analogue memory, neuromorphic computational primitives) and physical implementations, but they are still in the development phase, and it is to be seen

if they are going to achieve their full potential.. Memristor technology will likely augment the current, very successful CMOS systems, and provide improvements for specific computational tasks and applications.

References

1. Mehonic, A., Kenyon, A.J.: Brain-inspired computing needs a master plan. Nature **604**, 255–260 (2022). https://doi.org/10.1038/s41586-021-04362-w
2. Mehonic, A., et al.: Intrinsic resistance switching in amorphous silicon oxide for high performance SiOx ReRAM devices. Microelectron. Eng. **178**, 98–103 (2017)
3. Chang, H.-Y., et al.: AI hardware acceleration with analog memory: microarchitectures for low energy at high speed. IBM J. Res. Dev. **63**(6), 8:1–8:14 (2019)
4. Joksas, D., et al.: Committee machines—a universal method to deal with non-idealities in memristor-based neural networks. Nat. Commun. **11**(1), 4273 (2020)
5. Mehonic, A., Joksas, D., Ng, W.H., Buckwell, M., Kenyon, A.J.: Simulation of inference accuracy using realistic RRAM devices. Front. Neurosci. **13**, 593 (2019)

K-means Clustering and Hadamard Metric for Graphs Modelling

Piercarlo Cattani[1](✉) and Francesco Villecco[2]

[1] Department of Computer, Control and Management Engineering,
University of Rome "La Sapienza", via Ariosto 25, 00185 Rome, Italy
`cattani.1642354@studenti.uniroma1.it`
[2] Department of Industrial Engineering, University of Salerno, via Giovanni Paolo II 132,
84084 Fisciano, Italy

Abstract. Modelling a geometric surface or contour surface is a modern field involving a wide range of topics such as mathematics, computer science, engineering design. The main task in modelling a geometric surface is to approximation the shape of the contour surface of a 3D solid object the most efficient way. In this paper we will give a method to optimize the geometric model, by defining a local sampling of the mesh, based on k-means method and minimization of the Hausdorff measure with an homologous model.

Keywords: Geometric modelling · Hausdorff distance · k-means · Clustering

1 Introduction

Geometric modelling of a surface is based on a polygonal tessellation, but the most efficient models are using triangular meshes. However, by increasing the density of triangles in the mesh would reduce the approximation error, thus giving a better rendering. Nevertheless also the computational cost increases thus reducing the efficiency of the geometric modeling. Moreover, for a given surface there not exists a unique polygonal shape so that is still unsolved the problem of chosing the optimal mesh. For a given solid we can have two geometric models which are characterized by several parameters like e.g., the polygons of the tesselation, the discrete curvature in the meshes, computed at the sampling points, the minimal circuits, the color intensity. In order to compare two models it has been recently proposed [1] a measure of the difference between two shapes by using the so called Hadamard shape. However, in order to measure all distances between polygons on two shapes, has some computational costs that can be reduced by measuring the distances between some suitably chosen points on the mesh. This subset of points is selected by using the k-means algorithm. In the following we will propose a comparison between homologous geometric models which is based on the k-means method, thus reducing the computational costs. Moreover, it will be shown that the Hadamard distance decreases by increasing the number of k-means points, so that a 1/r dependence holds true.

© The Author(s), under exclusive license to Springer Nature Switzerland AG 2022
I. Karabegović et al. (Eds.): NT 2022, LNNS 472, pp. 443–448, 2022.
https://doi.org/10.1007/978-3-031-05230-9_52

2 Preliminary Remarks

2.1 Hausdorff Distance

Let S be convex solid object, and $\{V_i\}$, $i = 1, ...n_V$ a finite set of sampling points (vertices) on the boundary $\partial \overline{S}$; a mesh on S is the surface $\partial \overline{\Pi}$ of a convex polyhedron Π or convex polytope, approximating S, which is obtained by suitably connecting the set $\{V_i\}$ with non intersecting edges $\{E_k\}$, $k = 1, ...n_E$ and faces $\{F_h\}$, $h = 1, ...n_F$. For regular polyhedron the Euler's formula holds

$$n_V - n_E + n_F = 2$$

Let us denote with F_{ik} and E_{ik} faces and edges through the vertex V_i, the Gauss-Curvature at each vertex V_i is defined as

$$C(V_i) = 2\pi - \sum_{\alpha_k \in F_{ik}} \alpha_k$$

The 3D mesh can be plot into 2D by an homology so that we can have a tessellation in the plane where each vertex is characterized by the coordinates and a finite set of N additional parameters, such as curvature $C(V_i)$, intensity of light $I(V_i)$, so that a vertex V_i on the boundary of the 3D solid corresponds to a point X_i in the N-dimensional space $\mathbb{R}^N$

$$X_i = (V_i, C(V_i), I(V_i)......) \tag{1}$$

Let A_1 and A_2 two subset of the set of vertices S:

$$A_1 = \{V_i\}_1 \quad i = 1, ..., m_1, \quad A_2 = \{V_j\}_2 \quad j = 1, ..., m_2,$$
$$A_1 \cap A_2 = \emptyset, m_1 + m_2 \leq n_V$$

We define a metric on the partition of S, i.e. the set of subsets of S, by

$$\overline{A_1 A_2} = \min \overline{V_i V_j}, \quad \forall V_i \in A_1 \wedge \forall V_j \in A_2$$
$$\overline{V_i V_j} = \sqrt{(x_i - x_j)^2 + (y_i - y_j)^2 + (z_i - z_j)^2}$$

The Hausdorff metric is defined as (see e.g. [1])

$$e(A_1 A_2) = \max_{V_i \in A_1} \overline{A_1 A_2} = \max_{V_i \in A_1} \min_{V_j \in A_2} \overline{V_i V_j}, \forall V_i \in A_1 \wedge \forall V_j \in A_2 \tag{2}$$

There immediately follows from the definition that

$$e(A_1 A_2) \neq e(A_2 A_1)$$

In any case, this function enables to compute the proximity of sets and then to classify the subsets of the given mesh. This approach can be extended to generalized mesh $S_X = \{X_i\}$ in the space of parameters $\mathbb{R}^N$, so that the proximity in $\mathbb{R}^N$, is intended as a close distance not only geometrically but also qualitatively, in curvature, light, colors, and of the other parameters of the N-dimensional space.

2.2   *K*-means Clustering

Let S_k be a subset of $S_X = \{X_i\}$, with cardinality $c_k = |S_k|$, that is the number of points $X_j \in S_k$, in the N-dimensional space of parameters $\mathbb{R}^N$

$$S_k \subseteq S_X, k = 1, ...p \le n_V, \quad \bigcup_{k=1,...p} S_k = S_X, \quad \sum_{k=1,...p} k = n_V \tag{3}$$

and μ_k the centroid of S_k that is

$$\mu_k = \frac{1}{c_k} \sum_{j=1,...,c_k} X_j, \quad (X_j \in S_k, c_k = |S_k|, k = 1, ...p) \tag{4}$$

There follows that the centroid it is also the Hausdorff distance for each subset, in fact it can be easily shown, directly from Eq. (4), that

$$e(\mu_k X_i) = min, \quad \forall X_i \in S_k$$

In other words, the centroids of the subsets (3) of S_X can be used to define some clusters around the centroids, and each cluster can be represented by its centroid. So that a comparison of geometrical objects, and meshes, can be done by comparing centroids of clusters, thus reducing the computational costs of the geometrical modelling. Indeed, the centroids depend on a-priori partition of the set S_X. A simple algorithm to define these cluster (subsets) is the Lloyd algorithm, or Voronoi iteration [2], which consists in chosing arbitrarely an initial number, then a preliminary clustering (definition of subsets) is done by using the Hausdorff measure, so that the closest points are assigned to a each cluster. Then the centroids of each each cluster are computed, and as a second step, the original set S_X is partitioned by considering the Hausdorff distance from the centroids, and repeated again until we have the optimal partition (see Fig. 1).

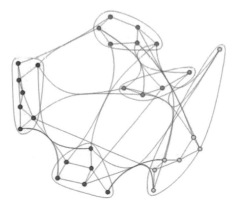

Fig. 1. K-means clustering

2.3   Optimal Path in the Mesh

In order to made a comparison between partitions, subsets and geometric solids, we have defined a suitable mesh $S_X = \{X_i\}$ in an N-parameters space $\mathbb{R}^N$. By the k-means method the mesh S_X can be partitioned into some suitable subsets and as a consequence of the partition we obtain a set $S_\mu = \{\mu_k\}, k = 1, ..., p$, of centroids μ_k where each centroids represents a cluster. There follows that the initial mesh S_X is replaced by the lower cardinality set S_μ which, however, is made of more significant points (centroids) defined by the Hausdorff distance. Each centroid can represent its cluster and a minimal path S_μ can be defined as

$$\min \sum_{k,h=1,...,p} \overline{\mu_k\mu_h}, \quad \forall \mu_k, \mu_h \in S_\mu$$

The minimum path (see e.g. Fig. 2) is a fundamental parameter that can be used both in solid modeling, for instance in order to search for the minimal values of lights, or colors, and it migt have also some interesting application in other fields, such as R&D networks.

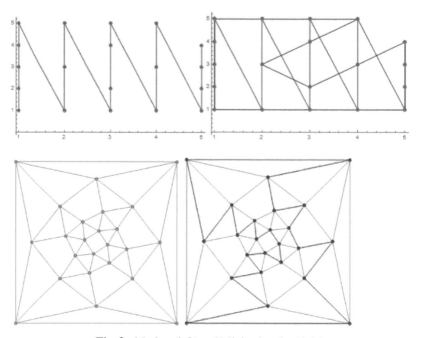

Fig. 2. Meshes (left) and Minimal paths (right)

3   Applications - R&D Networks

An R&D network is a fundamental model for economics networks optimization [3–6]. The main idea is that the R&D organization can be represented by a planar graph

(Figs. 2 and 3) where the nodes are players (such as customers, firms, brokers,..) and the edges represents the partnership, (cooperation, agreement, strategy,..). A linear-quadratic function of consumers [5] is

$$U_X = a \sum_{i=1,..,nv} X_i - \frac{1}{2} \left(b \sum_{i=1,..,nv} X_i^2 + c \sum_{i,j=1,..,nv} \overline{X_i X_j} \right)$$

This function can be easily minimized by using the k-means and the Hausdorff distance between centroids.

$$U_X \geq U_\mu = a \sum_{i=1,..,nv} \mu_i - \frac{1}{2} \left(b \sum_{i=1,..,nv} \mu_i^2 + c \sum_{i,j=1,..,nv} \overline{\mu_i \mu_j} \right)$$

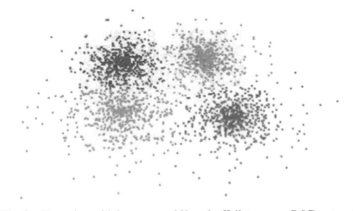

Fig. 3. Clustering with k-means and Hausdorff distance on R&D network

4 Conclusion

In this paper, the k-means algorithm and the Hausdorff distance have been used in the N-dimensional parameter space to improve the solid geometric modeling and the network analysis.

References

1. di Filippo, A., Villecco, F., Cappetti, N., Barba, S.: A methodological proposal for the comparison of 3D photogrammetric models. In: Rizzi, C., Campana, F., Bici, M., Gherardini, F., Ingrassia, T., Cicconi, P. (eds.) ADM 2021. LNME, pp. 930–937. Springer, Cham (2022). https://doi.org/10.1007/978-3-030-91234-5_94
2. Lloyd, S.P.: Least squares quantization in PCM. IEEE Trans. Inf. Theory **28**(2), 129–137 (1982)

3. Cantner, U., Graf, H.: The network of innovators in Jena: an application of social network analysis. Res. Policy **35**, 463–480 (2006)
4. Goyal, S., Moraga-Gonzalez, J.L.: R&D networks. RAND J. Econ. **32**, 686–707 (2001)
5. Konig, M.D., Battiston, S., Napoletano, M., Schweitzer, F.: The efficiency and stability of R&D networks. Games Econ. Behav. **75**, 694–713 (2012)
6. Westbrock, B.: Natural concentration in industrial research collaboration. RAND J. Econ. **41**(2), 351–371 (2010)

Realization of Single Image Super-Resolution Reconstruction Based on Wavelet Transform and Coupled Dictionary

Wei Qin[1], Min Zhao[1], Shuli Mei[1], Piercarlo Cattani[2], Vincenzo Guercio[3], and Francesco Villecco[4(✉)]

[1] College of Information and Electrical Engineering, China Agricultural University, Beijing 100083, China
[2] Department of Computer, Control and Management Engineering, University of Rome "La Sapienza", via Ariosto 25, 00185 Roma, Italy
[3] Engineering School, Deim, University of Tuscia, Largo dell'Università, 01100 Viterbo, Italy
`guerciovincenzo2005@libero.it`
[4] Department of Industrial Engineering, University of Salerno, via Giovanni Paolo II 132, 84084 Fisciano, Italy
`fvillecco@unisa.it`

Abstract. In various tasks of machine vision, image resolution is one of the important factors that affect the performance of the model. Generally, crop images with low resolution and lack of detail informa-tion may be collected. The picture is not good for the accuracy of yield prediction and crop pest identification. In this paper, tomato leaves are used as the target image, and the super-resolution reconstruction process takes advantage of the sharpness of the image, which has the characteristic of Scale invariance. Firstly, each image block is classified by using clustering algorithm accord-ing to the sharpness value of the image, and then wavelet transform is used to extract image features from each class of image blocks to get wavelet subbands respectively, subbands of each class not only train a union dictionary, but also learn a separate mapping function. Joint dictionary training and separate mapping matrix learning are helpful to optimize the high resolution and low resolution sparse coefficients. In the Reconstruction Stage: in order to reduce the image reconstruction time, the wavelet transform is only applied to the image blocks with a certain sharpness value, while the image reconstruction performance is basically unchanged, then the high-resolution image blocks are reconstructed by using the mapping function, coupled dictionary and the sparse representation coefficients of the image blocks. When the sharpness of the image block is lower than a certain sharpness value, the high and middle resolution image blocks will be superimposed to finally get the high resolution image.In the various tasks of machine vision, image resolution is one of the important factors that affect average PSNR value of the algorithm in this paper is 3.94 dB, 3.54 dB, 3.36 dB, 3.23 dB, 3.01 dB and 1.51 dB higher respectively.

Keywords: Sharpness value · Clustering · Sparse representation · Coupled dictionary

I. Karabegović et al. (Eds.): NT 2022, LNNS 472, pp. 449–456, 2022.
https://doi.org/10.1007/978-3-031-05230-9_53

1   Introduction

Image is one of the important media for visual information transmission and plays animportant role in the field of agriculture. This paper selects tomato leaves as the research object to study the problem of super-resolution reconstruction [1–4].

Super-resolution algorithms can be divided into traditional algorithms [5] and deep learning algorithms [6] in terms of time and effect. References [7–9] mainly use a single overcomplete dictionary for learning. However, it is shown in the literature [10] that excessive redundancy will lead to the degradation of the signal recovery and lead to instability. Therefore, on this basis, the literatures [11–13] all use clustering data into different templates. In the process of dictionary learning, there is no coupling between the sparse representation coefficients of HR and LR. References [14–18] show that the sparse coefficients of high and low resolutions have a certain room for improvement. Combining the above advantages and disadvantages, this paper uses the scale-invariant sharpness value to build a dictionary, jointly trains the low-resolution dictionary and the high-resolution dictionary, and further optimizes the sparse parameters of both [19–23].

2   The Proposed Dictionary Training Algorithm

2.1   Wavelet-Based Feature Extraction

Firstly, use the high-resolution training images to downsample to form low-resolution images, and then cluster the low-resolution images according to the clustering criteria. The high-resolution image blocks corresponding to each type of low-resolution image blocks are also transformed into four word bands through wavelet transformation, namely their high-resolution training data [24–27].

2.2   Clustering

The sharpness value feature is used as the clustering criterion. The sharpness value of point a in the image block is calculated as follows:

$$S(a) = \|I(a) - \mu_{A(a)}\| \tag{1}$$

where S(a) is the sharpness of point a, I(a) is the pixel value of point a, and $\mu A(a)$ is the average value of 8 neighboring pixels of point a.The sharpness value of the entire image block is the average of the sharpness values of all points. The sharpness of an image is scale invariant. The method of clustering refers to the method of literature.

2.3   Dictionary Training

In the low-resolution image block, first specify the number of clusters c, and then use the above algorithm to indicate the size of each category. We need to minimize the following energy function to solve the required coupling dictionary and mapping function, W_H and W_L represent HR and LR training data, subbands of high-resolution image patches and corresponding subbands of low-resolution image patches, respectively, and C represents the corresponding clusters.

$$
\begin{aligned}
min_{D_H^C, D_L^C, f(\cdot)} & E_{data}\left(D_H^C, W_H^C\right) + E_{data}\left(D_L^C, W_L^C\right) + \lambda_H^C\left\|a_H^C\right\|_1 + \lambda_L^C\left\|a_L^C\right\|_1 \\
& + \gamma E_{map}\left(f\left(a_H^C\right), a_L^C\right) + \lambda E_{reg}\left(a_L^C, a_H^C, D_L^C, D_H^C, f(\cdot)\right)
\end{aligned}
\tag{2}
$$

In the formula, $E_{data}\left(D_H^C, W_H^C\right)$ is the data fidelity item of the data description error, and $E_{map}\left(f(a_H^C), a_L^C\right)$ is the mapping fidelity item, indicating the difference between the two sparse coefficients. E_{reg} is the regularization term used to regularize the sparse coefficients and mapping, and $f(\cdot)$ is the coupling relationship between HR and LR data sparse coefficients.

When considering the mapping function as a linear function M, the problem can be transformed into a ridge regression and dictionary learning problem as follows.

$$
\begin{aligned}
min_{D_H^C, D_L^C, M^C} & \left\|W_H^C - D_H^C a_H^C\right\|_F^2 + \left\|W_L^C - D_L^C a_L^C\right\|_F^2 + \gamma\left\|a_L^C - M^C a_H^C\right\|_F^2 \\
& + \lambda_M^C\left\|M^C\right\|_F^2 + \lambda_H^C\left\|a_H^C\right\|_1 + \lambda_L^C\left\|a_L^C\right\|_1 \\
s.t. & \left\|D_{H,i}^C\right\|_2 \leq 1 \cap \left\|D_{L,i}^C\right\|_2 \leq 1, for\ all\ i
\end{aligned}
\tag{3}
$$

where $\gamma, \lambda_H, \lambda_L, \lambda_L$ are the regularization coefficients of the optimized performance, $D_{H,i}$ and $D_{L,i}$ are the D_H and D_L atoms respectively, and M is the mapping function.

Equation (3) is not jointly convex for D_H, D_L and M. When one of the variables is fixed, Eq. (3) becomes convex. Therefore, we design an iterative algorithm to optimize the variables alternately. To solve the energy minimization problem in Eq. (3), In this paper, we choose LARS [28] because Its efficiency and strong stability.

$$
\left\{
\begin{aligned}
\min_{\alpha_H^C} & \left\|W_H^C - D_H^C \alpha_H^C\right\|_F^2 + \gamma\left\|\alpha_L^C - M^C \alpha_H^C\right\|_F^2 + \lambda_H^C\left\|\alpha_H^C\right\|_1 \\
\min_{\alpha_L^C} & \left\|W_L^C - D_L^C \alpha_L^C\right\|_F^2 + \gamma\left\|\alpha_H^C - M^C \alpha_L^C\right\|_F^2 + \lambda_L^C\left\|\alpha_L^C\right\|_1
\end{aligned}
\right.
\tag{4}
$$

when α_H and α_L are fixed, the dictionary pair can be updated as follows:

$$
\min_{D_H^C, D_L^C} \left\|W_H^C - D_H^C \alpha_H^C\right\|_F^2 + \left\|W_L^C - D_L^C \alpha_L^C\right\|_F^2 s.t. \left\|D_{H,i}^C\right\|_2 \leq 1 \cap \left\|D_{L,i}^C\right\|_2 \leq 1
\tag{5}
$$

Equation (5) is a quadratic constrained quadratic programming problem (QCQP). We address this problem by adopting a continuously updated strategy [13]. Finally, with the dictionary pairs and sparse coefficients unchanged, M can be updated as.

$$
\min_{M^C} \left\|\alpha_L^C - M^C \alpha_H^C\right\|_F^2 + \left(\frac{\lambda_M^C}{\lambda_M^C \gamma \gamma}\right)\left\|M^C\right\|_F^2
\tag{6}
$$

The problem is a ridge regression problem and its solution can be analytically deduced as:

$$
M^C = \alpha_L^C(\alpha_H^C)^T\left(\alpha_H^C(\alpha_H^C)^T + \left(\frac{\lambda_M^C}{\lambda_M^C \gamma \gamma}\right) \cdot I\right)^{-1}
\tag{7}
$$

where I is the identity matrix. The sparse coefficients, dictionary pairs, and mapping functions are updated in each iteration until the error satisfies the requirement and stops the class loop.

3   The Proposed Image Reconstruction Algorithm

Input a low-resolution image, obtain its (MR) medium-resolution image through bicubic interpolation, divide the LR image into image blocks, and classify according to the size range of each category above [29–33]. When the sharpness of the image block is higher than a certain value, use the normalization dictionary and feature block of this type of subband to learn the sparse coefficient α_L, use the mapping function M to obtain the corresponding sparse coefficient α_H, use the normalization dictionary D_H and the sparse coefficient α_H to reconstruct the imageblock. Corresponding sub-bands, and then obtain a high-resolution image through inverse wavelet transform, denoted as SR2, when the image block sharpness is lower than a certain value, use the corresponding MR image block as a reconstruction block, denoted as SR1, reshape and merge overlapping patches SR1 and SR2, the HR image estimates are obtained, the unreconstructed regions use the same loxel values as the MR images, and the overlapping regions of the images are replaced by their average values.

4   Experimental Results and Analysis

In this paper, the same training set as Timofte [12] is used, and the images are divided into five categories based on the sharpness value of the image during initial training. The classification results of each category are: [0–1], [1–4], [4–8], [8–10], each image block uses haar wavelet to extract features in the SWT process, the high and low resolution image block size is 6×6, the number of iterations is set to 20. The Our1 in this paper processes each class interval. The Our2 in this paper does not process the sharpness interval of the first category, and only performs SR reconstruction on the blocks of the remaining categories. If the sharpness of the first category belongs to the interval, the block corresponding to the MR image position is directly used as the reconstructed block (Fig. 1 and Tables 1, 2 and 3).

From the quantitative results, the PSNR value of the Our2 proposed in this paper is 3.94 dB, 3.54 dB, 3.36 dB, 3.23 dB, 3.01 dB higher than Bicubic, Yang, ANR, Yeganli [15], NCSR [14], Yang Q [16] respectively. The average SSIM value of the Our2 proposed in this paper is 0.08 dB, 0.06 dB, 0.04 dB, 0.04 dB, 0.02 dB, 0.01 dB higher than Bicubic, Yang, ANR, Yeganli, NCSR, and Yang Q, respectively. the Our2 proposed in this paper has similar performance in PSNR and SSIM than Our1, but the image reconstruction time is greatly reduced [34–40].

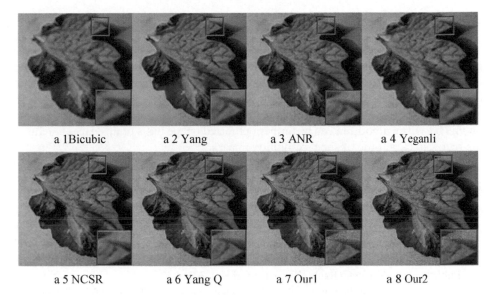

| a 1Bicubic | a 2 Yang | a 3 ANR | a 4 Yeganli |
| a 5 NCSR | a 6 Yang Q | a 7 Our1 | a 8 Our2 |

Fig. 1. SR comparison results of tomato slices

Table 1. Comparison of PSNR values under different SR algorithms

	Bicubic	Yang	ANR	Yeganli	NCSR	Yang Q	Our1	Our2
P1	28.39	28.86	29.04	29.16	29.32	30.82	31.97	31.96
P2	29.19	29.43	29.64	29.68	29.98	31.56	32.65	32.65
P3	25.07	25.60	25.71	25.94	26.15	27.56	29.86	29.86
Aver	27.55	27.96	28.13	28.26	28.48	29.98	31.49	31.49

Table 2. Comparison of SSIM values under different SR algorithms

	Bicubic	Yang	ANR	Yeganli	NCSR	Yang Q	Our1	Our2
P1	0.83	0.86	0.87	0.88	0.89	0.90	0.91	0.91
P2	0.80	0.82	0.84	0.84	0.85	0.86	0.89	0.89
P3	0.79	0.80	0.81	0.82	0.83	0.84	0.85	0.859
Aver	0.80	0.82	0.84	0.84	0.86	0.87	0.89	0.88

Table 3. Comparison of Time under different SR algorithms

	Our1	Our2
P1	725	654
P2	737	676
P3	746	680

5   Conclusion

The algorithm proposed in this paper is classified based on the sharpness value of the image, and the fuzzy C-means clustering method is used for clustering, which is further improved compared to Yeganli's algorithm. By using an overcomplete dictionary will lead to the degradation of signal recovery and lead to instability, so through some work, create a cluster dictionary instead of a single dictionary, and future work can continue on this basis.The algorithm can be further optimized on the basis of this paper to reduce the reconstruction time of super-resolution images.

References

1. Celenta, G., De Simone, M.C.: Retrofitting techniques for agricultural machines. In: Karabegović, I. (ed.) NT 2020. LNNS, vol. 128, pp. 388–396. Springer, Cham (2020). https://doi.org/10.1007/978-3-030-46817-0_44
2. Dašić, P.: Scientific and technological trends: Selected scientific-professional papers. SaTCIP Publisher Ltd., Vrnjačka Banja (2020). (in Serbian). 305 str
3. Pappalardo, C.M., Lombardi, N., Guida, D.: A model-based system engineering approach for the virtual prototyping of an electric vehicle of class l7. Eng. Lett. **28**(1), 215–234 (2020)
4. Sicilia, M., De Simone, M.C.: Development of an energy recovery device based on the dynamics of a semi-trailer. In: Ivanov, V., Pavlenko, I., Liaposhchenko, O., Machado, J., Edl, M. (eds.) DSMIE 2020. LNME, pp. 74–84. Springer, Cham (2020). https://doi.org/10.1007/978-3-030-50491-5_8
5. Pandey, G., Ghanekar, U.: Improved single image super-resolution based on compact dictionary formation and neighbor embedding reconstruction. In: Verma, G.K., Soni, B., Bourennane, S., Ramos, A.C.B. (eds.) Data Science. TCSN, pp. 89–97. Springer, Singapore (2021). https://doi.org/10.1007/978-981-16-1681-5_7
6. Datta, R., Mandal, S., Umer, S., et al.: Single Image Reconstruction using Novel Super Resolution Technique for Large Scaled Images (2021)
7. Fang, F., Li, J., Zeng, T.: Soft-edge assisted network for single image super-resolution. IEEE Trans. Image Process. **29**, 4656–4668 (2020)
8. Colucci, F., De Simone, M.C., Guida, D.: TLD design and development for vibration mitigation in structures. In: Karabegović, I. (ed.) NT 2019. LNNS, vol. 76, pp. 59–72. Springer, Cham (2020). https://doi.org/10.1007/978-3-030-18072-0_7
9. Pappalardo, C.M., Lettieri, A., Guida, D.: A general multibody approach for the linear and nonlinear stability analysis of bicycle systems. Part II: application to the Whipple-Carvallo bicycle model. J. Appl. Comput. Mech. **7**(2), 671–700 (2021)
10. Zuo, J., Wang, Z., Zhang, Y., et al.: Research on image super-resolution algorithm based on mixed deep convolutional networks. Comput. Electr. Eng. **95**, 107422 (2021)
11. Yang, J., Wright, J., Huang, T.S., et al.: Image super-resolution via sparse representation. IEEE Trans. Image Process. **19**(11), 2861–2873 (2010)
12. Timofte, R., De Smet, V., Van Gool, L.: Anchored neighborhood regression for fast example-based super-resolution. In: Proceedings of the IEEE International Conference on Computer Vision, pp. 1920–1927 (2013)
13. Gorodnitsky, I.F., Rao, B.D.: Sparse signal reconstruction from limited data using FOCUSS: a re-weighted minimum norm algorithm. IEEE Trans. Signal Process. **45**(3), 600–616 (1997)
14. Dong, W., Zhang, L., Shi, G., et al.: Image deblurring and super-resolution by adaptive sparse domain selection and adaptive regularization. IEEE Trans. Image Process. **20**(7), 1838–1857 (2011)

15. Yeganli, F., Nazzal, M., Unal, M., et al.: Image super-resolution via sparse representation over coupled dictionary learning based on patch sharpness. In: 2014 European Modelling Symposium, pp. 203–208. IEEE (2014)
16. Yang, Q., Wang, H.: Super-resolution reconstruction for a single image based on self-similarity and compressed sensing. J. Algorithms Comput. Technol. **12**(3), 234–244 (2018)
17. De Simone, M.C., Guida, D.: Experimental investigation on structural vibrations by a new shaking table. In: Carcaterra, A., Paolone, A., Graziani, G. (eds.) AIMETA 2019. LNME, pp. 819–831. Springer, Cham (2020). https://doi.org/10.1007/978-3-030-41057-5_66
18. Dašić, P.: Response surface methodology: Selected scientific-professional papers. SaTCIP Publisher Ltd., Vrnjačka Banja (2019). (in Serbian, Slovenian and Russian). 301 str. ISBN 978-86-6075-054-1
19. Manrique-Escobar, C.A., Pappalardo, C.M., Guida, D.: A multibody system approach for the systematic development of a closed-chain kinematic model for two-wheeled vehicles. Machines **9**(11), 245 (2021)
20. De Simone, M.C., Ventura, G., Lorusso, A., Guida, D.: Attitude controller design for microsatellites. In: Karabegović, I. (ed.) NT 2021. LNNS, vol. 233, pp. 21–31. Springer, Cham (2021). https://doi.org/10.1007/978-3-030-75275-0_2
21. Pappalardo, C.M., Lettieri, A., Guida, D.: Identification of a dynamical model of the latching mechanism of an aircraft hatch door using the numerical algorithms for subspace state-space system identification. IAENG Int. J. Appl. Math. **51**(2), 346–359 (2021)
22. Rivera, Z.B., De Simone, M.C., Guida, D.: Unmanned ground vehicle modelling in Gazebo/ROS-based environments. Machines **7**(2) (2019). Article no. 42
23. Pappalardo, C.M., Lettieri, A., Guida, D.: A general multibody approach for the linear and nonlinear stability analysis of bicycle systems. Part I: methods of constrained dynamics. J. Appl. Comput. Mech. **7**(2), 655–670 (2021)
24. De Simone, M.C., Guida, D.: Control design for an under-actuated UAV model. FME Trans. **46**(4), 443–452 (2018)
25. Dašić, P., Dašić, J., Antanasković, D., Pavićević, N.: Statistical analysis and modeling of global innovation index (GII) of Serbia. In: Karabegović, I. (ed.) NT 2020. LNNS, vol. 128, pp. 515–521. Springer, Cham (2020). https://doi.org/10.1007/978-3-030-46817-0_59
26. Pappalardo, C.M., Manca, A., Guida, D.: A combined use of the multibody system approach and the finite element analysis for the structural redesign and the topology optimization of the latching component of an aircraft hatch door. IAENG Int. J. Appl. Math. **51**(1), 175–191 (2021)
27. De Simone, M.C., Guida, D.: Modal coupling in presence of dry friction, Machines **6**(1) (2018). Article no. 8
28. Efron, B., Hastie, T., Johnstone, I.: Robert Tibshirani "least angle regression,." Ann. Stat. **32**(2), 407–499 (2004)
29. Yang, M., Zhang, L., Yang, J., et al.: Metaface learning for sparse representation based face recognition. In: 2010 IEEE International Conference on Image Processing, pp. 1601–1604. IEEE (2010)
30. Tošović, R., Dašić, P., Ristović, I.: Sustainable use of metallic mineral resources of Serbia from an environmental perspective. Environ. Eng. Manag. J. (EEMJ) **15**(9), 2075–2084 (2016). https://doi.org/10.30638/eemj.2016.224. ISSN 1582-9596
31. Manrique-Escobar, C.A., Pappalardo, C.M., Guida, D.: On the analytical and computational methodologies for modelling two-wheeled vehicles within the multibody dynamics framework: a systematic literature review. J. Appl. Comput. Mech. **8**(1), 153–181 (2022)
32. De Simone, M.C., Rivera, Z.B., Guida, D.: Obstacle avoidance system for unmanned ground vehicles by using ultrasonic sensors. Machines **6**(2) (2018). Article no. 18
33. Dašić, P., Dašić, J., Crvenković, B.: Applications of access control as a service for software security. Int. J. Ind. Eng. Manag. (IJIEM) **7**(3), 111–116 (2016). ISSN 2217-2661

34. Manrique Escobar, C.A., Pappalardo, C.M., Guida, D.: A parametric study of a deep reinforcement learning control system applied to the swing-up problem of the cart-pole. Appl. Sci. **10**(24), 9013 (2020)
35. Pappalardo, C.M., Guida, D.: Dynamic analysis and control design of kinematically-driven multibody mechanical systems. Eng. Lett. **28**(4), 1125–1144 (2020)
36. Formato, A., Ianniello, D., Romano, R., Pellegrino, A., Villecco, F.: Design and development of a new press for grape marc. Machines **7**(3), 51 (2019)
37. Guida, R., De Simone, M.C., Dašić, P., Guida, D.: Modeling techniques for kinematic analysis of a six-axis robotic arm. IOP Conf. Ser. Mater. Sci. Eng. **568**(1) (2019). Article no. 012115
38. Dašić, P.: Reliability of technical systems: Selected scientific-professional papers. SaTCIP Publisher Ltd., Vrnjačka Banja (2019). (in Serbian and Russian). 308 str. ISBN 978-86-6075-051-0
39. Naviglio, D., et al.: Study of the grape cryo-maceration process at different temperatures. Foods **7**(7) (2018). https://doi.org/10.3390/foods7070107. Article no. 107
40. Ardila-Parra, S.A., Pappalardo, C., Estrada, O.A.G., Guida, D.: Finite element based redesign and optimization of aircraft structural components using composite materials. IAENG Int. J. Appl. Math. **50**(4), 1–18 (2020)

Shearlet and Patch Reordering Based Texture Preserving Denoising Method for Locust Slice Images

Shuli Mei[1], Leiping Zhu[1], Matteo d'Amore[2], Andrea Formato[3],
and Francesco Villecco[4]([✉])

[1] College of Information and Electrical Engineering, China Agricultural University, Beijing 100083, People's Republic of China
[2] Department of Pharmacy, University of Salerno, via Giovanni Paolo II 132, 84084 Fisciano, Italy
[3] Department of Agricultural Science, University of Naples "Federico II", via Università 100, 80055 Portici, Naples, Italy
[4] Department of Industrial Engineering, University of Salerno, via Giovanni Paolo II 132, 84084 Fisciano, Italy
fvillecco@unisa.it

ABSTRACT. Locust slice image is a kind of cartoon-like images, in which the texture possesses the property of self-similarity. Both of the texture and the noises belong to the high-frequency signals, and so it is difficult to tell the difference between them for most denoising methods. Aim to the problem, we propose a novel denoising method by combining the patch reordering with the shearlet transform. In the reordering process, the patches are divided into smooth and texture components. The filters obtained from the training set are employed to process the patches in smooth regions and the shearlet transform are employed to process the texture regions. The experiments show that the values of PSNR and SSIM of the processed images obtained by the proposed method are better than the common methods.

Keywords: Texture preserving · Image denoising · Patch reordering · Shearlet transform

1 Introduction

The study on the interaction mechanism of biological pesticides and organisms plays an important role in the development of the new pesticide. The research on locust microscopic section images is an important tool in observing the cell structure of locusts. The slice images of the locust have rich texture structures, some of which have self-similar fractal structures [1–5]. Both of the noise in the images and texture belong to the high-frequency signals, so it is difficult to tell the difference between them by mean of the common image processing methods. As the common methods are employed to remove the high-frequency signals in order to denoise, while the texture and contour of the image

© The Author(s), under exclusive license to Springer Nature Switzerland AG 2022
I. Karabegović et al. (Eds.): NT 2022, LNNS 472, pp. 457–463, 2022.
https://doi.org/10.1007/978-3-031-05230-9_54

is blurred. Time-fractional gas dynamics equation [6–10] and fractal logic equation [4] are actively used to explore the convergence point with the field of image processing. Considering the multi-scale characteristics, some researchers use wavelet transform to denoise the image. Wavelet, as one of the mathematical tools, plays an important role in multi-resolution image analysis. However, wavelet is only sensitive to a limited number of directions [11–15]. Due to the lack of directional characteristics, it is impossible to achieve the maximum sparse description of image texture by wavelet transform. The shearlet transform proposed based on affine transform has anisotropic characteristics, whose can describe multiple directions [16–18]. Furthermore, the images texture suitable for described using shearlet transform sparse representative, which has directive and self-similarity [19–22]. However, texture in high noise condition, artifacts were prone to appear in image reconstruction using the shearlet transform. In order to effectively eliminate the boundary effect generated during shearlet transform, our method is proposing a shearlet transform based on patch reordering for locust slice images with texture preserved [23–26].

2 Denoising Algorithm in Locust Images

We take the slice images of the interaction between locusts and microbial pesticides at the tissue level as the research object. The infected locusts were cut perpendicular to the long axis with LKB-3 ultramicrotome to prepare 0.05 um series ultrathin sections [27–30]. Then, in order to facilitate observing the slices, the slices were stained with uranyl acetate and lead citrate, and then were collected through the H-7500 microscope. In the process of preparing micro slices, it is easy to be affected by the collection environment, operation process, material properties, which produce the noise in images.

2.1 Patchreordering

The image denoising model is defined as

$$z = y + V \tag{1}$$

where Y is an image of $N \times N$ pixels and Z is a noisy image. Let z and y be the column stacked operations of Z and Y, respectively. The matrix V denotes the environmental noise with zero mean and variance σ^2 (Gaussian noise). The denoising method is applied to the noisy image Z to reconstruct the image Y. It is based on the following steps:

Step 1: Two-dimensional image converted into one-dimensional signal.
Step 2: Reordering and smoothing.
Use a sorted matrix to cluster similar patches together to form a smooth 1D signal.
Step 3: Patch classification.

Let $std(x_i)$ denote the standard deviation of the patch x_i and let C be a scalar design parameter. If $std(x_i) < C\sigma$ then $x_i \in$ smooth patches, otherwise $x_i \in$ texture. The patch classification image are shown in Fig. 1.

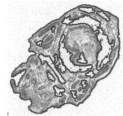

a.Locust slice image with   b.Noisy image   c. patch classification
texture structure

Fig. 1. Noisy image of locust slice and patch classification

2.2 Shearlet Transform

The accurate description of the image texture required the basis function to have multi-scales characteristics, such as anisotropic scaling, translation invariance, and directionality. Continuous shearlet transform is a direction-sensitive wavelet transform, The mathematical expression of shearlet transform of function $f \in L^2(R^2)$ is defined as

$$SH_f(a, s, t) = \langle f, \psi_{a,s,t} \rangle \tag{2}$$

$$\psi_{a,s,t}(x) = a^{-3/4} \psi \left(A^{-1} S^{-1} (x - t) \right) \tag{3}$$

where the $SH_f(a, s, t)$ describes shearlet system. $\langle \bullet \rangle$ denotes inner product. x denotes the signal to be processed. The scalar a denotes the scale. The vector s denotes the direction, the vector t denotes the translation. The matrix $A = \begin{pmatrix} a & 0 \\ 0 & a^{1/2} \end{pmatrix}$ is used to control the scale, the matrix $S = \begin{pmatrix} 1 & s \\ 0 & 1 \end{pmatrix}$ is used to change the direction, so that Eq. (3) can be rewritten as

$$\psi_{a,s,t} = a^{-3/4} \psi \left(\begin{bmatrix} \frac{1}{a} & -\frac{s}{a} \\ 0 & \frac{1}{\sqrt{a}} \end{bmatrix} (x - t) \right) \tag{4}$$

2.3 Application of Denoising Algorithm in Locust Slice Images

The specific steps are as follows:

Step 1: Add Gaussian noise with different variances to the prepared locust slice images.
Step 2: Classify the locust slice image into a smooth region and a texture region by patch classification algorithm.
Step 3: Learning filters were used in the smooth region. Shearlet transform was performed in the texture region.
Step 4: Finally, the patch reverse ordering was performed to obtain the denoised locust slice image.

As can be seen in Fig. 2, take Gaussian noise with added variance σ = 30 as an example. In order to verify the effectiveness of the algorithm in this paper, we calculated the expected signal-to-noise ratio PSNR and structural similarity (SSIM) indicators to objectively measure the effectiveness of the algorithm.

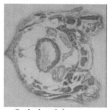   

a.Original image   b. Nosiy image

Fig. 2. Prepared work

Table 1 and Table 2 respectively show the PSNR and SSIM of the proposed method and other methods under different variances.

As shown in Table 1, as the noise variance increases, the noise becomes more and more obvious and the PSNR decreases.

Table 1. PSNR comparison between the proposed method and other methods

	σ = 10	σ = 20	σ = 30	σ = 40	σ = 50	σ = 60
WT	37.1429	32.9814	30.4991	28.6997	27.4121	26.3524
WF	35.5207	29.7209	26.0881	23.4834	21.5350	19.9273
MF	34.4705	29.4151	26.1061	23.6925	21.7824	20.2198
MNF	32.2743	29.5638	27.0798	25.0021	23.3079	21.8474
OMP	37.8122	33.5815	31.0733	29.4485	28.0475	26.8882
KSVD	38.3714	34.1393	31.5161	29.6918	28.3149	27.3208
NLM	35.7690	31.8769	29.5585	28.0599	26.6715	25.6640
Proposed method	**39.2712**	**35.9246**	**34.1269**	**33.3188**	**32.6558**	**32.9212**

As shown in Table 2, SSIM also decreases with the increase of the noise variance. Compared with Wavelet Transform (WT), Wiener Filter (WF), Median Filter (MF), Mean Filter (MNF), Patch-based OMP denoising with a fixed dictionary (OMP), K-SVD Method, and Non-Local Mean (NLM), the PSNR has increased by an average of 14.44% (WT), 36.82% (WF), 36.68% (MF), 32.20% (MNF), 12.10% (OMP), 10.64% (K-SVD) and 17.89% (NLM), respectively. The SSIM has increased by an average of 99.47% (WT), 170.05% (WF), 171.75% (MF), 131.96% (MNF), 86.55% (OMP), 85.51% (K-SVD) and 146.11% (NLM) or more. Various methods were considered in processing the locust slice image of Fig. 3.

Table 2. SSIM comparison between the proposed method and other methods

	$\sigma = 10$	$\sigma = 20$	$\sigma = 30$	$\sigma = 40$	$\sigma = 50$	$\sigma = 60$
WT	0.6166	0.5066	0.4248	0.3622	0.3172	0.2788
WF	0.5891	0.4356	0.3340	0.2636	0.2146	0.1762
MF	0.5529	0.4205	0.3290	0.2638	0.2141	0.1762
MNF	0.5914	0.4652	0.3766	0.3116	0.2619	0.2212
OMP	0.6487	0.5308	0.4469	0.3884	0.3419	0.3056
KSVD	0.6611	0.5430	0.4562	0.3902	0.3393	0.3020
NLM	0.5827	0.4473	0.3544	0.2950	0.2459	0.2090
Proposed method	**0.9171**	**0.8525**	**0.7862**	**0.7425**	**0.7037**	**0.7471**

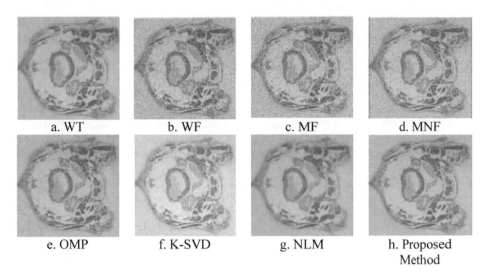

a. WT   b. WF   c. MF   d. MNF

e. OMP   f. K-SVD   g. NLM   h. Proposed
                                                       Method

Fig. 3. Denoising images with different methods

3 Conclusion

The proposed method combines the shearlet transform with the patch reordering, which not only preserves the directionality of the texture, but also reduces the artificial artifacts caused by the shearlet in the smooth region [31–34]. Compared with Wavelet Transform, Wiener Filter, Median Filter, Mean Filter, Patch-based OMP denoising with a fixed dictionary, K-SVD method, and Non-Local Mean, the proposed method has better denoising effects. The PSNR has increased by an average of 14.44%, 36.82%, 36.68%, 32.20%, 12.10%, 10.64% and 17.89% or more, respectively, and the SSIM has increased an average of 99.47%, 170.05%, 171.75%, 131.96%, 86.55%, 85.51% and 146.11% or more.

References

1. Li, L., Guo, S., Mei, S., Zhang, N.: Image restoration of locust slices based on nearest unit matching. Trans. Chin. Soc. Agric. Mach. **46**(8), 15–19 (2015)
2. Li, L., Guo, S., Mei, S., Zhang, N.: Mosaic and repair method of locust slices based on feature extraction and matching. Trans. Chin. Soc. Agric. Eng. **31**(7), 157–165 (2015)
3. Rivera, Z.B., De Simone, M.C., Guida, D.: Unmanned ground vehicle modelling in Gazebo/ROS-based environments. Machines **7**(2) (2019). Article no. 42
4. Pappalardo, C.M., Lettieri, A., Guida, D.: A general multibody approach for the linear and nonlinear stability analysis of bicycle systems. Part I: methods of constrained dynamics. J. Appl. Comput. Mech. **7**(2), 655–670 (2021)
5. Colucci, F., De Simone, M.C., Guida, D.: TLD design and development for vibration mitigation in structures. In: Karabegović, I. (ed.) NT 2019. LNNS, vol. 76, pp. 59–72. Springer, Cham (2020). https://doi.org/10.1007/978-3-030-18072-0_7
6. Srivastava, H.M., Saad, K.M.: Some new models of the time-fractional gas dynamics equation. Adv. Math. Models Appl. **3**(1), 5–17 (2018)
7. Celenta, G., De Simone, M.C.: Retrofitting techniques for agricultural machines. In: Karabegović, I. (ed.) NT 2020. LNNS, vol. 128, pp. 388–396. Springer, Cham (2020). https://doi.org/10.1007/978-3-030-46817-0_44
8. Dašić, P.: Scientific and technological trends: Selected scientific-professional papers. SaTCIP Publisher Ltd., Vrnjačka Banja (2020). (in Serbian). 305 str
9. Pappalardo, C.M., Lombardi, N., Guida, D.: A model-based system engineering approach for the virtual prototyping of an electric vehicle of class l7. Eng. Lett. **28**(1), 215–234 (2020)
10. Sicilia, M., De Simone, M.C.: Development of an energy recovery device based on the dynamics of a semi-trailer. In: Ivanov, V., Pavlenko, I., Liaposhchenko, O., Machado, J., Edl, M. (eds.) DSMIE 2020. LNME, pp. 74–84. Springer, Cham (2020). https://doi.org/10.1007/978-3-030-50491-5_8
11. Manrique Escobar, C.A., Pappalardo, C.M., Guida, D.: A parametric study of a deep reinforcement learning control system applied to the swing-up problem of the cart-pole. Appl. Sci. **10**(24), 9013 (2020)
12. Pappalardo, C.M., Guida, D.: Dynamic analysis and control design of kinematically-driven multibody mechanical systems. Eng. Lett. **28**(4), 1125–1144 (2020)
13. Guida, R., De Simone, M.C., Dašić, P., Guida, D.: Modeling techniques for kinematic analysis of a six-axis robotic arm. IOP Conf. Ser. Mater. Sci. Eng. **568**(1) (2019). Article no. 012115
14. Dašić, P.: Reliability of technical systems: Selected scientific-professional papers. SaTCIP Publisher Ltd., Vrnjačka Banja (2019). (in Serbian and Russian). 308 str. ISBN 978-86-6075-051-0
15. Ardila-Parra, S.A., Pappalardo, C., Estrada, O.A.G., Guida, D.: Finite element based redesign and optimization of aircraft structural components using composite materials. IAENG Int. J. Appl. Math. **50**(4), 1–18 (2020)
16. Khalili Golmankhaneh, A., Cattani, C.: Fractal logistic equation. Fractal Fractional **3**(3), 41–48 (2019)
17. Manrique-Escobar, C.A., Pappalardo, C.M., Guida, D.: A multibody system approach for the systematic development of a closed-chain kinematic model for two-wheeled vehicles. Machines **9**(11), 245 (2021)
18. De Simone, M., Ventura, G., Lorusso, A., Guida, D.: Attitude controller design for microsatellites. In: Karabegović, I. (ed.) NT 2021. LNNS, vol. 233, pp. 21–31. Springer, Cham (2021). https://doi.org/10.1007/978-3-030-75275-0_2
19. Guo, K., Labate, D.: Optimally sparse multidimensional representation using shearlets. SIAM J. Math. Anal **39**(1), 298–318 (2007)

20. Pappalardo, C.M., Lettieri, A., Guida, D.: Identification of a dynamical model of the latching mechanism of an aircraft hatch door using the numerical algorithms for subspace state-space system identification. IAENG Int. J. Appl. Math. **51**(2), 346–359 (2021)
21. De Simone, M.C., Guida, D.: Experimental investigation on structural vibrations by a new shaking table. In: Carcaterra, A., Paolone, A., Graziani, G. (eds.) AIMETA 2019. LNME, pp. 819–831. Springer, Cham (2020). https://doi.org/10.1007/978-3-030-41057-5_66
22. Dašić, P.: Response surface methodology: Selected scientific-professional papers. SaTCIP Publisher Ltd., Vrnjačka Banja (2019). (in Serbian, Slovenian and Russian). 301 str. ISBN 978-86-6075-054-1
23. Li, X., Mei, S., Li, L.: Image denoising of locust slices based on meyer window function shearlet transform. Trans. Chin. Soc. Agric. Mach. **47**(z1), 449–456 (2016)
24. Pappalardo, C.M., Lettieri, A., Guida, D.: A general multibody approach for the linear and nonlinear stability analysis of bicycle systems. Part II: application to the Whipple-Carvallo bicycle model. J. Appl. Comput. Mech. **7**(2), 671–700 (2021)
25. De Simone, M.C., Guida, D.: Control design for an under-actuated UAV model. FME Trans. **46**(4), 443–452 (2018)
26. Dašić, P., Dašić, J., Antanasković, D., Pavićević, N.: Statistical analysis and modeling of global innovation index (GII) of Serbia. In: Karabegović, I. (ed.) NT 2020. LNNS, vol. 128, pp. 515–521. Springer, Cham (2020). https://doi.org/10.1007/978-3-030-46817-0_59
27. Pappalardo, C.M., Manca, A., Guida, D.: A combined use of the multibody system approach and the finite element analysis for the structural redesign and the topology optimization of the latching component of an aircraft hatch door. IAENG Int. J. Appl. Math. **51**(1), 175–191 (2021)
28. De Simone, M.C., Guida, D.: Modal coupling in presence of dry friction. Machines **6**(1) (2018). Article no. 8
29. Tošović, R., Dašić, P., Ristović, I.: Sustainable use of metallic mineral resources of Serbia from an environmental perspective. Environ. Eng. Manag. J. (EEMJ) **15**(9), 2075–2084 (2016). https://doi.org/10.30638/eemj.2016.224. ISSN 1582-9596
30. Manrique-Escobar, C.A., Pappalardo, C.M., Guida, D.: On the analytical and computational methodologies for modelling two-wheeled vehicles within the multibody dynamics framework: a systematic literature review. J. Appl. Comput. Mech. **8**(1), 153–181 (2022)
31. Formato, A., Ianniello, D., Romano, R., Pellegrino, A., Villecco, F.: Design and development of a new press for grape marc. Machines **7**(3), 51 (2019)
32. De Simone, M.C., Rivera, Z.B., Guida, D.: Obstacle avoidance system for unmanned ground vehicles by using ultrasonic sensors. Machines **6**(2) (2018). Article no. 18
33. Naviglio, D., et al.: Study of the grape cryo-maceration process at different temperatures. Foods **7**(7) (2018). https://doi.org/10.3390/foods7070107. Article no. 107
34. Dašić, P., Dašić, J., Crvenković, B.: Applications of access control as a service for software security. Int. J. Ind. Eng. Manag. (IJIEM) **7**(3), 111–116 (2016). ISSN 2217-2661

Shearlet Transform and the Application in Image Processing

Hu Haitao[1]([✉]), Piercarlo Cattani[2], Vincenzo Guercio[3]([✉]), and Francesco Villecco[4]

[1] School of Information and Electrical Engineering, China Agricultural University, Beijing, China
sy20213081590@cau.edu.cn
[2] Department of Computer, Control and Management Engineering, University of Rome "La Sapienza", via Ariosto 25, 00185 Roma, Italy
[3] Engineering School, University of Tuscia, Largo dell'Università, 01100 Viterbo, Italy
fvillecco@unisa.it
[4] Department of Industrial Engineering, University of Salerno, via Giovanni Paolo II 132, 84084 Fisciano, Italy

ABSTRACT. Shearlet is a multi-dimensional function used for sparse representation, which has many excellent characteristics such as multi-resolution and multi-direction. It can detect the position of singular points and the direction of singular curves, and is more sensitive to the geometric structure of the image. Therefore, this paper introduces the shearlet transform and its application in image processing, and introduces the bendlet transform proposed on this basis.

Keywords: Shearlet · Bendlet · Transform · Image processing

1 Introduction

Fourier theory is widely used in signal analysis, but it cannot provide frequency information in the corresponding time period, and there is no time-frequency analysis. As a result, the wavelet analysis tool was born. It can obtain better time resolution in the high-frequency part of the signal by replacing the infinite-length trigonometric function basis with the finite-length decaying wavelet basis. And obtaining better frequency resolution in the low-frequency part of the signal.

The continuous wavelet transform of the signal x(t) is defined as

$$Wx(a,b) = \int_{-\infty}^{+\infty} x(t)\psi_{a,b}^*(t)dt \tag{1}$$

In which

$$\psi_{a,b}(t) = \left(\frac{1}{\sqrt{a}}\right)\psi[(t-b)/a] \tag{2}$$

is the wavelet basis function, a is the scale factor, and b is the translation factor.

I. Karabegović et al. (Eds.): NT 2022, LNNS 472, pp. 464–470, 2022.
https://doi.org/10.1007/978-3-031-05230-9_55

Wavelets have the characteristics of multi-scale, which can flexibly and accurately analyze the relationship between time and frequency, and can flexibly select different basis functions for different problems. Wavelet can also realize the clustering of important information in the image. These characteristics of wavelet make it particularly suitable for the processing of non-stationary signals such as image signals. Using wavelet can realize the processing of image denoising, compression, fusion, etc.

The two-dimensional separable wavelet transform has only a limited direction, which cannot express the curve singular features in the image explictly. In order to overcome the shortcomings of wavelet transform, some scholars have proposed Curvelet transform [1], Contourlet transform [2], etc. In 2007, Guo et al. [3] studied the use of affine systems with synthetic expansion to construct a Shearlet function close to the optimal representation. Compared with other multi-scale set analysis tools, it has a simpler mathematical structure, and also has excellent multi-resolution analysis characteristics and nonlinear approximation performance.

2   A Brief Introduction of Shearlet Transform

Shearlet transform is good at detecting singular points with geometric features through its anisotropy of odd functions. At the same time, the shearlet has the directionality. When the shearlet direction is consistent with the image texture direction, the shearlet coefficientsare the largest, and when the intersection of the shearlet and the image texture is empty, the shearlet coefficientsare 0.

The definition [3] of shearlet transform is as follows:

$$SH_\psi f(a,s,t) = f, \psi_{a,s,t} \tag{3}$$

where f represents the original function, which belongs to the function in the two-dimensional square integrable space $L^2(R^2)$, and a, s, t respectively correspond to the scale, direction and conversion parameters.

$$\psi_{a,s,t}(x) = a^{-\frac{3}{4}} \psi(A^{-1}B^{-1}(x-t)) \tag{4}$$

In the formula, A is an anisotropic expansion matrix, B is a shear matrix, $t \in R^2$ is a translation parameter, x = x, y is the position vector of pixels. Matrix A and B can be expressed as $A = \begin{pmatrix} a & 0 \\ 0 & \sqrt{a} \end{pmatrix}, B = \begin{pmatrix} 1 & s \\ 0 & 1 \end{pmatrix}$.

As shown in Fig. 1, it is the subdivision diagram of shearlet in frequency domain, indicating that it can form vertebral bodies in different directions, and detect any direction according to needs, and the direction books at different scales are multiplied layer by layer.

Fig. 1. Frequency domain subdivision of shearlet

3   Application of Shearlet Transform

3.1   Image Denoising

Shearlet transform is the problem of feature extraction and low-pass filtering. And the actual signal in the shearlet coefficients after the shearlet transformation is often large in absolute value, while the noise still has strong randomness after transformation, so the absolute value of the shearlet coefficients of the noise is small.

According to this characteristic, the threshold shrinkage method is often used to realize the noise reduction, the threshold and threshold function need to be determined by the threshold shrinkage method by global threshold or hierarchical adaptive threshold. The global threshold is that each layer of shearlet coefficient matrix is processed by the same threshold. Layered adaptive threshold is defined for each layer of coefficient matrix.

There are three kinds of threshold functions: hard threshold function, soft threshold function and semi-soft threshold function. The hard threshold function is $\varphi(X) = \begin{cases} X, x > T \\ 0, x \leq T \end{cases}$; the soft threshold function is $\varphi(w) = sgn(w)(|w| - T)$, sgn (x) is a symbolic function, 1 when $x > 0$, -1 when $x < 0$. The hard threshold can retain the features such as image edges, but there will be ringing effects. Although the soft threshold can make the processed image smoother, its edge and other features will become blurred. Bruce [4] et al. improved this and proposed a semi-soft threshold method, which uses two upper and lower thresholds.

On the Method of Estimating Threshold, Hu et al. [5] proposed to use the Monte-Carlo method to estimate the high-frequency coefficient, and then perform the shrinkage noise reduction. This algorithm has achieved good visual effect and higher PSNR value while suppressing noise and maintaining the edge [6–10].

Shearlet is sensitive to the texture direction. When noises occur in the smooth region of the image, these noises organized according to certain rules are understood to be distributed over a texture, and thus artifacts appear on the reconstructed image. To solve this problem, Mei et al. [11] proposed the method of shearlet and total variation coupling to remove artifacts while denoising, which improved the effect of image denoising.

3.2   Image Overlaying

Image fusion algorithm can synthesize a high-quality image after extracting useful information from multiple images of different sources or different focuses [12–15]. And there are two types of algorithms based on multi-scale transformation or not. The methods based on non-multi-scale changes include average method and PCA method [16, 17].

The multi-focus image fusion can obtain more detailed information of the image. When the multi-focus image is fused, the two images are transformed, and the low frequency and high frequency components of the two images are fused according to the corresponding fusion rules. The panchromatic image can reflect the spatial structure information, while the multispectral image has abundant spectral information. The fusion methods of the two are as follows. The multispectral and panchromatic images often convert the image to HSV space first, and only perform shearlet transform on the V component, and then fuse them accordingly to obtain the new V'.

In the process of image fusion, after transforming the image into the shearlet domain, appropriate image fusion rules should be processing.Quanet al. [18] use genetic algorithm to optimise the weighted factors in the fusion rule. Biswaset al. [19] used link distortion graph (LTM) to realize the fusion of multi-focus images.

3.3 Texture Recognition

Shearlet transform has directional detection ability because of adding shear matrix. When the shearlet direction is consistent with the texture direction, the shearlet coefficient is the largest [20–24]; when the shearlet direction is parallel to the texture direction, the shearlet coefficient is 0; otherwise, when the two intersect, the shearlet coefficient is very small [25] This feature enables the shearlet transform to identify the texture information in the image, identify the direction of the texture, and also distinguish texture and noise accordingly.

Wang et al. [26] used shearlet transform to realize interval weed identification, which was better than the traditional identification method, because the texture of wheat seedlings was regular and the direction was relatively single, while the texture of weeds was complex.

3.4 Edge Detection

Shearlet has a strong sensitivity to the curve structure characteristics in the image, which is used in edge detection. The candidate edge points are obtained by applying the modulus maximum principle to detect the image, and then the non-maximum points are suppressed along the edge direction. Finally, the threshold is used to correct the edge points, which has a good edge detection effect [27–30].

Kinget al [31] presented a novel edge and ridge (line) detection algorithm based on complex-valued wavelet-like analyzing functions-so-called complex shearlets-displaying several traits useful for the extraction of flame fronts. Yanget al. [32] performed shearlet transform on magnetic tile images to obtain low-frequency and high-frequency subbands, and denoising and enhancement were carried out. Then, the threshold segmentation method was used to extract the crack defects [33–36].

4 Conclusion

Shearlet has excellent sparse representation ability and sensitivity to direction, so it is applied to many aspects of image processing and has achieved certain results. Shearlet transform is particularly suitable for images with texture characteristics. For image fusion and noise reduction, shearlet transform has unique advantages. However, there are detailed textures everywhere in the default image of the shearlet transform, and the noise is often mistakenly identified as texture features. At the same time, the shearlet transform fits the curve in a straight line, which is not suitable for images with curvature. These problems need to be studied and improved [37–41].

In this regard, A second-order shearlet transform proposed by Lessing [42] et al. studied those problems, which can precisely character the curvature of discontinuities in piecewise constant images. Kushol et al. [43] used the sensitivity of bendlet to curvature direction to extract the features of retinal blood vessels, combined with ensemble classifier to achieve the segmentation of blood vessels, the accuracy rate was 95%. Bendlet transform is a powerful tool when aimming at the image with curvature.

References

1. Emmanuel, J.C., Donoho, D.L.: Curvelets - A Surprisingly Effective Nonadaptive Representation For Objects with Edges. Curve & Surface Fitting (2000)
2. Do, M.N., Vetterli, M.: The contourlet transform: an efficient directional multiresolution image representation. IEEE Trans. Image Process. **14**(12), 2091–2106 (2005)
3. Guo, K., Labate, D.: Optimally sparse multidimensional representation using shearlets. SIAM J. Math. Anal. **39**(1), 298–318 (2007)
4. Bruce, A.G., Gao, H.Y.: Understanding WaveShrink: Variance and bias estimation. Biometrika **83**(4), 727–745 (1996)
5. Hu, H., Sun, H., Deng, C., et al.: Image-denoising algorithm based on Shearlet transform. J. Comput. Appl. **30**(06), 1562–1564 (2010)
6. Manrique-Escobar, C.A., Pappalardo, C.M., Guida, D.: A multibody system approach for the systematic development of a closed-chain kinematic model for two-wheeled vehicles. Machines **9**(11), 245 (2021)
7. De Simone, M.C., Ventura, G., Lorusso, A., Guida, D.: Attitude controller design for micro-satellites. In: Karabegović, I. (ed.) NT 2021. LNNS, vol. 233, pp. 21–31. Springer, Cham (2021). https://doi.org/10.1007/978-3-030-75275-0_2
8. Pappalardo, C.M., Lettieri, A., Guida, D.: Identification of a dynamical model of the latching mechanism of an aircraft hatch door using the numerical algorithms for subspace state-space system identification. IAENG Int. J. Appl. Math. **51**(2), 346–359 (2021)
9. De Simone, M.C., Guida, D.: Experimental investigation on structural vibrations by a new shaking table. In: Carcaterra, A., Paolone, A., Graziani, G. (eds.) AIMETA 2019. LNME, pp. 819–831. Springer, Cham (2020). https://doi.org/10.1007/978-3-030-41057-5_66
10. Dašić, P.: Response surface methodology: Selected scientific-professional papers. SaTCIP Publisher Ltd., Vrnjačka Banja (2019). (in Serbian, Slovenian and Russian). 301 str. ISBN 978-86-6075-054-1
11. Mei, S., Li, X., Zhao, H., et al.: Method of denoising and removing artifacts for farm remote sensing image based on shearlet and total variation. Trans. Chin. Soc. Agric. Eng. **33**(S1), 274–280 (2017). (in Chinese)
12. Rivera, Z.B., De Simone, M.C., Guida, D.: Unmanned ground vehicle modelling in Gazebo/ROS-based environments. Machines **7**(2) (2019). Article no. 42
13. Pappalardo, C.M., Lettieri, A., Guida, D.: A general multibody approach for the linear and nonlinear stability analysis of bicycle systems. Part I: methods of constrained dynamics. J. Appl. Comput. Mech. **7**(2), 655–670 (2021)

14. Colucci, F., De Simone, M.C., Guida, D.: TLD design and development for vibration mitigation in structures. In: Karabegović, I. (ed.) NT 2019. LNNS, vol. 76, pp. 59–72. Springer, Cham (2020). https://doi.org/10.1007/978-3-030-18072-0_7

15. Pappalardo, C.M., Lettieri, A., Guida, D.: A general multibody approach for the linear and nonlinear stability analysis of bicycle systems. Part II: application to the Whipple-Carvallo bicycle model. J. Appl. Comput. Mech. 7(2), 671–700 (2021)

16. Hu, G., Liu, Z., Xu, X., et al.: Research and recent development of image fusion at pixel level. Appl. Res. Comput. 03, 650–655 (2008)

17. De Simone, M.C., Guida, D.: Control design for an under-actuated UAV model. FME Trans. 46(4), 443–452 (2018)

18. Quan, Yining, Song, et al.: Remote sensing image fusion based on shearlet and genetic algorithm. Int. J. Bio-Inspired Comput. 9(4), 240–250 (2017)

19. Biswas, B., Chatterjee, P., Ghoshal, S., et al.: Multi-focus image fusion method based on linked twist map (LTM) in shearlet domain. In: 2nd International Conference on Signal Processing and Integrated Networks (SPIN) 2015, pp. 526–531 (2015)

20. Dašić, P., Dašić, J., Antanasković, D., Pavićević, N.: Statistical analysis and modeling of global innovation index (GII) of Serbia. In: Karabegović, I. (ed.) NT 2020. LNNS, vol. 128, pp. 515–521. Springer, Cham (2020). https://doi.org/10.1007/978-3-030-46817-0_59

21. Pappalardo, C.M., Manca, A., Guida, D.: A combined use of the multibody system approach and the finite element analysis for the structural redesign and the topology optimization of the latching component of an aircraft hatch door. IAENG Int. J. Appl. Math. 51(1), 175–191 (2021)

22. De Simone, M.C., Guida, D.: Modal coupling in presence of dry friction. Machines 6(1) (2018). Article no. 8

23. Tošović, R., Dašić, P., Ristović, I.: Sustainable use of metallic mineral resources of Serbia from an environmental perspective. Environ. Eng. Manag. J. (EEMJ) 15(9), 2075–2084 (2016). https://doi.org/10.30638/eemj.2016.224. ISSN 1582-9596

24. Manrique-Escobar, C.A., Pappalardo, C.M., Guida, D.: On the analytical and computational methodologies for modelling two-wheeled vehicles within the multibody dynamics framework: a systematic literature review. J. Appl. Comput. Mech. 8(1), 153–181 (2022)

25. Mei, S.: Denoising for locust slice image with texture preserving based on coupling technology of variational method and shearlet transform. Trans. Chin. Soc. Agric. Eng. 32(17), 152–159 (2016)

26. Wang, H., Zhu, M., Li, L., et al.: Regional weed identification method from wheat field based on unmanned aerial vehicle image and shearlets. Trans. Chin. Soc. Agric. Eng. 33(S1), 99–106 (2017)

27. De Simone, M.C., Rivera, Z.B., Guida, D.: Obstacle avoidance system for unmanned ground vehicles by using ultrasonic sensors. Machines 6(2) (2018). Article no. 18

28. Alinsaif, S., Lang, J.: Texture features in the Shearlet domain for histopathological image classification. BMC Med. Inform. Decis. Mak. 20, 312 (2020). https://doi.org/10.1186/s12911-020-01327-3

29. Dašić, P., Dašić, J., Crvenković, B.: Applications of access control as a service for software security. Int. J. Ind. Eng. Manag. (IJIEM) 7(3), 111–116 (2016). ISSN 2217-2661

30. Manrique Escobar, C.A., Pappalardo, C.M., Guida, D.: A parametric study of a deep reinforcement learning control system applied to the swing-up problem of the cart-pole. Appl. Sci. 10(24), 9013 (2020)

31. Reisenhofer, R., Kiefer, J., King, E.J.: Shearlet-based detection of flame fronts. Exp. Fluids 57(3), 1–14 (2016). https://doi.org/10.1007/s00348-016-2128-6

32. Yang, C., Yin, M., Jiang, H., et al.: Detection of surface crack defects in magnetic tile images based on nonsubsampled shearlet transform. Trans. Chin. Soc. Agric. Mach. 48(03), 405–412 (2017)

33. Pappalardo, C.M., Guida, D.: Dynamic analysis and control design of kinematically-driven multibody mechanical systems. Eng. Lett. **28**(4), 1125–1144 (2020)
34. Guida, R., De Simone, M.C., Dašić, P., Guida, D.: Modeling techniques for kinematic analysis of a six-axis robotic arm. IOP Conf. Ser. Mater. Sci. Eng. **568**(1) (2019). Article no. 012115
35. Dašić, P.: Reliability of technical systems: Selected scientific-professional papers. SaTCIP Publisher Ltd., Vrnjačka Banja (2019). (in Serbian and Russian). 308 str. ISBN 978-86-6075-051-0
36. Ardila-Parra, S.A., Pappalardo, C., Estrada, O.A.G., Guida, D.: Finite element based redesign and optimization of aircraft structural components using composite materials. IAENG Int. J. Appl. Math. **50**(4), 1–18 (2020)
37. Celenta, G., De Simone, M.C.: Retrofitting techniques for agricultural machines. In: Karabegović, I. (ed.) NT 2020. LNNS, vol. 128, pp. 388–396. Springer, Cham (2020). https://doi.org/10.1007/978-3-030-46817-0_44
38. Dašić, P.: Scientific and technological trends: Selected scientific-professional papers. SaTCIP Publisher Ltd., Vrnjačka Banja (2020). (in Serbian). 305 str
39. Pappalardo, C.M., Lombardi, N., Guida, D.: A model-based system engineering approach for the virtual prototyping of an electric vehicle of class l7. Eng. Lett. **28**(1), 215–234 (2020)
40. Sicilia, M., De Simone, M.C.: Development of an energy recovery device based on the dynamics of a semi-trailer. In: Ivanov, V., Pavlenko, I., Liaposhchenko, O., Machado, J., Edl, M. (eds.) DSMIE 2020. LNME, pp. 74–84. Springer, Cham (2020). https://doi.org/10.1007/978-3-030-50491-5_8
41. Formato, A., Ianniello, D., Romano, R., Pellegrino, A., Villecco, F.: Design and development of a new press for grape marc. Machines **7**(3), 51 (2019)
42. Lessig, C., Petersen, P., Schaefer, M.: Bendlets: a second-order shearlet transform with bent elements. Appl. Comput. Harmonic Anal. **46**(2), 384–399 (2019)
43. Kushol, R., Kabir, M.H., Abdullah-Al-Wadud, M., et al.: Retinal blood vessel segmentation from fundus image using an efficient multiscale directional representation technique Bendlets. Math. Biosci. Eng. **17**(6), 7751–7771 (2020)

Backup Technologies Applicable for Securing the Databases in Intellectual Property Organizations

Radu Costin Moisescu[1,2(✉)]

[1] Polytechnic University of Bucharest, Splaiul Independenţei nr. 313, 6th District, Bucharest, Romania
radu_moisescu@yahoo.com
[2] State Office for Inventions and Trademarks, 5 Ion Ghica Street, Bucharest, Romania

Abstract. The proposed paper argues from a scientific point of view the need to implement the international standard ISO 27001:2013 by analyzing security measures to be integrated in organizations whose object of activity is the protection of intellectual property in order to protect data and databases. At the same time, the need to analyze and implement adequate technologies for centralized backup systems in intellectual property organizations is taken into account using dynamic simulations through which systems performance can be determined and evaluated by analyzing experimental results. The research refers to an analysis of database security technologies in the information systems within these organizations. The issue of analysis of architectures and technologies that can be applied in the field of database security by implementing modern centralized backup mechanisms from existing information systems in these organizations is an extremely relevant aspect analyzed and explained in this research. The proposed research can be considered a reasoned and documented point of view in a topical field.

Keywords: Information system · Computer system · Data architectures · Data · Backup

1 Introduction

Modern economies are driven by digital information and new communication technologies, giving organizations and individuals easy access to a wide range of information, regardless of their forms of existence, storage or geographic location. All these mechanisms for the transmission, storage and processing of information encourage the development of new products and services.

In this context, developed countries have implemented both government programs for the development and implementation of new technologies and the consolidation and security of information and communication infrastructures.

I. Karabegović et al. (Eds.): NT 2022, LNNS 472, pp. 471–479, 2022.
https://doi.org/10.1007/978-3-031-05230-9_56

The development of the use of electronic means for communications in all areas of activity and fundamental transformations in organizations and throughout society has led to a sustained increase in the volume of data to be processed and stored, thus raising the issue of securing them.

Changes in technology at all times and the complexity of the new type of company using electronic communications devices have increased the need to secure electronic data. Electronic communication has determined the confluence of three evolutions - changing the nature of publishing in the digital world, growth, the use of licenses despite their sale and the use of technical protection services creates unprecedented opportunities for individuals to access information using enhanced technologies. This could also have a negative impact on public access to information, creating the need to closely monitor developments over time.

From the information security management point of view, an Intellectual Property organisation may be able to define, implement, maintain, and improve based on International Standards ISO/IEC 27001:2018 and ISO/IEC 27002:2013:

a) an Information Security Management System (hereafter "ISMS") based on International Standard ISO/IEC 27001:2018, incorporating backup and recovery processes [1].
b) the information security objectives and measures based on International Standard ISO/IEC 27002:2013, specific to backup and recovery processes [2].

2 Experimental Research

Experimental plans are used in various industrial sectors to develop and optimize technological processes. In this sense, the use of the term optimize expresses the need to streamline or improve a certain process, the experimenter being responsible in establishing the analysis strategy that leads to the desired results.

The analysis of the results of the experimental determinations performed is currently facilitated by the existence of computer programs and applications that provide in addition to an immediate mathematical calculation and appropriate graphical representations that facilitate the understanding and interpretation of these results [3].

Intelligent planning of experimental determinations as well as a correct statistical analysis in order to obtain a mathematical model with which to evaluate the evolution of the answers in the field studied are the essence of an experimental plan [4].

In this research report, the experimental determination was used to study the behavior of data backup systems in real operating conditions. In this context, factor plans and factor analysis are essential both for the investigation of the effects of various factors on the systems and for the determination of possible interactions between these factors.

The objective of the experimental plan is to assess the impact of increasing the volume of data to be saved/restored in intellectual property organizations on the recently implemented centralized rescue/restoration systems, as well as the analysis of technological improvements and changes that will be proposed in order to increase the quality of the analyzed processes [5].

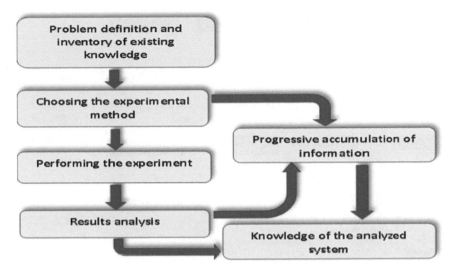

Fig. 1. The process of knowledge accumulation

The scope of the experimental research is to determine the factors that can be modified and updated in order to reduce the backup processes execution times, decrease the percentage of failed sessions and ensure data recovery in case of disaster.

Thus, in the factor analysis, there are three types of decisions that must be considered:

a) the factors to be studied will be selected according to the objectives of the experimental plan. In most cases, it is necessary to establish the relative importance of each factor and its influence on the phenomenon studied;
b) the degree of variation of the influencing factors, measured in percentages in which the researcher's experience can play a key role in their evaluation;
c) the answers by which the studied phenomena are calculated and evaluated.

The process of knowledge accumulation is presented schematically in Fig. 1 and takes place in four stages [6]:

– Defining the problem and inventorying existing knowledge;
– Choosing the experimental method;
– Results analysis;
– Progressive accumulation of information.

The equipment used in this experimental research is:

– SAN Storage system model Dell-EMC Unity D400 with a capacity of 80TB (a mixt of SAS and SSD volumes of disks);
– DD Boost disks appliance system model Dell-EMC Data Domain DD6300 for backup to disks;

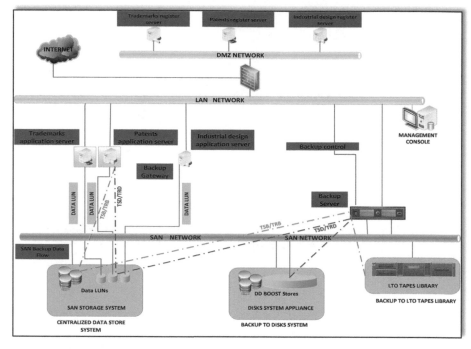

Fig. 2. The IT architecture of the experimental centralized backup system

- LTO tape library model Overland Neo for backup to tapes;
- Microfocus data protector application running on a Lenovo server technology.

In this experiment, the independent variables represent the formulation or process factors that have a level set at the beginning of the experimental plan [7]. For conducting the experimental study, input parameters processed were: the patent database size, the monthly increase of the database (that is estimated to be achieved by submitting protection applications within the intellectual property organization for a period of 12 months), monthly estimated number of protection applications and data transfer rates on magnetic tape media and disk volumes [8].

Input parameters were noted with:

X_1 = the size of the database measured in TB where $X_1 \in$ (1TB... .10 TB);

X_2 = estimated monthly increase of the studied database measured in GB where $X_2 \in$ (10 GB... .30 GB);

X_3 = estimated monthly number of protection applications filed where $X_3 \in$ (100... .2000);

X_4 = data transfer rate measured in Mbps on magnetic tape where $X_4 \in$ (320 Mbps... .400 Mbps);

X_5 = data transfer rate measured in Mbps on disks where $X_5 \in$ (520 Mbps... .600 Mbps);

Table 1. Experimental data recorded

No.	Tsd vs X_1, X_3										
	X1 (TB)	X2 (TB)	X3 (No.)	X4 (Mbps)	X_5 (Mbps)	Tsd (minutes)	Tsb (minutes)	Trb (minutes)	Trd (minutes)	FITS1 (minutes)	RESI1 (minutes)
1	1.2034	0.8	876	381	562	385.5041	541.2132	589.63	402.41	385.0330	0.471
2	1.2054	1	1040	379	574	380.0001	534.6412	611.24	398.25	392.2893	−12.28
3	1.2074	0.7	763	352	553	391.1151	544.2131	605.78	387.23	383.7151	7.399
4	1.2094	0.6	626	356	586	375.1763	540.1145	617.76	401.12	380.1664	−4.990
5	1.2114	0.9	968	348	534	402.4719	565.5241	623.73	403.36	393.8123	8.659
6	1.2134	0.9	988	332	547	395.7708	572.4457	592.76	398.34	395.8995	−0.128
7	1.2154	1	1030	335	522	410.4469	581.8532	612.64	423.74	398.7763	11.670
8	1.2174	0.9	863	348	561	389.3404	567.2134	613.89	435.43	394.1507	−4.810
9	1.2194	0.9	930	364	550	395.6121	588.8716	632.76	424.23	397.9250	−2.312
10	1.2214	1	998	354	553	394.4906	596.1762	653.12	411.45	401.7352	−7.244
11	1.2334	0.8	881	369	528	411.4646	621.8172	662.65	428.63	405.7504	5.714
12	1.2594	0.9	903	349	521	422.2008	627.7881	640.21	435.34	424.3397	−2.138

Output sizes (also called output parameters) were noted with:

Tsd = execution time of a disk save session measured in minutes – represented by red dotted line in the Fig. 2;

Tsb = execution time of a tape save session measured in minutes - represented by green dotted line in the Fig. 2;

Trd = runtime of a disk restores information session measured in minutes - represented by red dotted line in the Fig. 2;

Trb = execution time of a tape restore session measured in minutes - represented by green dotted line in the Fig. 2;

The experimental records are presented in Table 1. A number of 12 tests were performed representing simulations of the growth of the patents database in the 12 months of a calendar year based on monthly estimates of the number of applications for protection that can be submitted to a national intellectual property organization.

The Fig. 3 present the 3D Surface Plot, to examine the relationship between Tsd (minutes) response variable and two predictor variables X_1 (TB) and X_3 (the number of application protection), by viewing a three-dimensional surface of the predicted response.

The following regression Eq. (1) was identified during the experimental research:

$$Tsd = -470 + 685 * X_1 + 0.0359 * X_3 \qquad (1)$$

Fitted Line represented in Fig. 4 is a statistical model based on the data collected in the experiment and which determines how one factor influences the other factor. The points in the graph in Fig. 4 represent the individual values and the line represents the trajectory that the model should follow.The red line represents how appropriate the model is for the data entered.

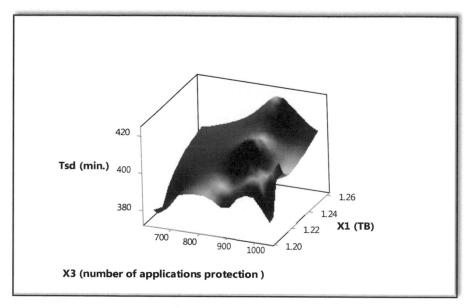

Fig. 3. Surface plot Tsd vs. X1, X3

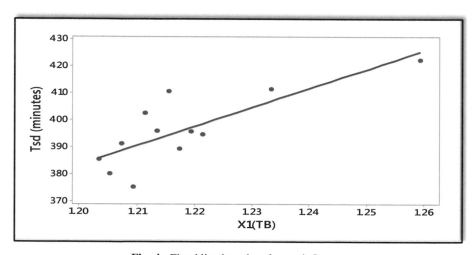

Fig. 4. Fitted line based on factors influences

The following regression Eq. (2) was identified during the experimental research:

$$\text{Tsd} = -461.8 + 704.4 * X_1 \qquad (2)$$

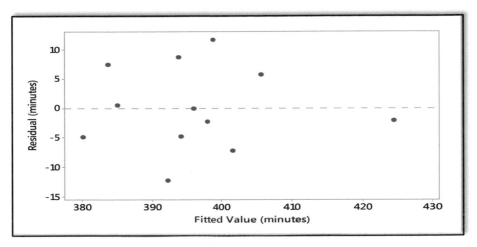

Fig. 5. Residuals versus fits

In the Fig. 5 are represented the residuals versus fits graph plots (the residuals on the y-axis and the fitted values on the x-axis). In the experiment, the graph of residues versus fits shows that this verifies the hypothesis that the residues are randomly distributed and have constant variability. Ideally, the dots should fall randomly on both sides of 0, with no patterns recognized in the dots.

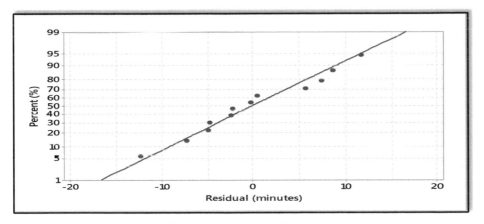

Fig. 6. Probability plot of RESI1

The graph in Fig. 6 represents the probability and evaluates how closely the data points follow the adjusted distribution line. As can be seen, the specific theoretical distribution fits well and the points fall almost along the straight line. We can also see that the points in the normal probability graph follow the line well and the normal distribution matches the data well.

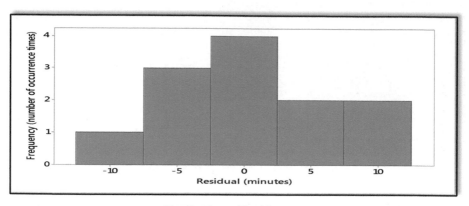

Fig. 7. The residue histogram

The residue histogram represented in Fig. 7 is used to determine if the data is distorted or if there are incorrect values in the data.

3 Conclusion

The main purpose of the use of new communication technologies and IT applications to the central and local public administration is to increase the quality of public services and to streamline the activity of the administrative apparatus. Therefore, intellectual property organizations have implemented online services for patents, trademark or industrial design.In this context, to ensure the proper methods to secure and protect the electronic data, theintellectual property organizations must implement the new technologies in the area of backups and restore of electronic data.

In order to increase the quality of backup processes and implicitly those of data restoration, intellectual property organizations need to consider what measures can be implemented to reduce process execution times, decrease the percentage of failed sessions and ensure data recovery in case of disasters.

References

1. ISO/IEC 27001:2018: Information Technology – Security Techniques – Information Security Management Systems – Requirements, pp. 4–22 (2018)
2. ISO/IEC 27002:2013: Information Technology – Security Techniques – Code of Practice for Information Security Management, pp. 10–22 (2013)
3. Țîțu, M., Oprean, C., Boroiu, A.: Cercetarea experimentală aplicată în creșterea calității produselor și serviciilor. Editura AGIR, pp. 60–92 (2011)
4. Oprean, C., Țîțu, M.: Managementul calității în economia și organizația bazate pe cunoștințe. AGIR, pp. 90–124 (2008)
5. Chang, G., Shams, K., Callas, J., Kern, A.: A novel approach to automated, secure, reliable, & distributed backup, pp. 52–87 (2012)
6. Chomsiri, T.: Sniffing packets on LAN without ARP spoofing. In: Third International Conference on ICCIT 2008, vol. 2, pp. 62–98. IEEE (2008)

7. Mellado, D., Rosado, D.G.: An overview of current information systems security challenges and innovations. J. Univ. Comput. Sci. **18**(12), 33–47 (2012)
8. Troppens, U., Erkens, R., Muller-Friedt, W., Wolafka, R., Haustein, N.: Storage networks explained: basics and application of fibre channel SAN, NAS, iSCSI, InfiniBand and FcoE, pp. 169–218 (2009)

Deduplication Technologies Applicable for the Backup Systems Inintellectual Property Organizations

Radu Costin Moisescu[1,2] and Aurel Mihail Titu[3(✉)]

[1] Polytechnic University of Bucharest,
Splaiul Independenței nr. 313, 6th District, Bucharest, Romania
radu_moisescu@yahoo.com
[2] State Office for Inventions and Trademarks, 5 Ion Ghica Street, Bucharest, Romania
[3] Lucian Blaga University of Sibiu, 10 Victoriei Street, Sibiu, Romania
mihail.titu@ulbsibiu.ro

Abstract. The sustained pace of data growth as well as the attempt to ensure that the management of this data can be achieved successfully is a growing challenge nowadays. Increasing the volume of data and their complexity, concerns about performance in trying to save/restore or move this data. In this context, in intellectual property organizations, the centralized storage technologies must be closely monitored to ensure that the storage mechanism implemented is appropriate to these types of data. Thus, new data deduplication technologies are an essential tool to help address these concerns. The research refers to an analysis of the impact that the implementation of data deduplication mechanisms in the information systems within these organizations has. The impact of the development of data deduplication mechanisms is presented by analyzing the results of dynamic simulations that can determine and evaluate the performance of centralized backup systems in these organizations and which represents an extremely relevant aspect analyzed and explained in this research. The proposed research can be considered a reasoned and documented point of view in a topical field.

Keywords: Information system · Deduplication · Data architectures · Management · Backup

1 Introduction

Intellectual property will certainly develop, adapting to the digital age. But, in order to achieve a balance between the public interest in digital information and the private rights to information, additional efforts are needed, but also an adaptation over time. In the context of the development of IT infrastructures and applications that facilitate access to digital information, the protection of intellectual property is an important mission in which decision makers and stakeholders will have to work together to ensure the protection of these rights. Increasing and improving access to digital information for society as a whole is thus ensured by the development of information infrastructure and

I. Karabegović et al. (Eds.): NT 2022, LNNS 472, pp. 480–487, 2022.
https://doi.org/10.1007/978-3-031-05230-9_57

the implementation of new technologies for storing, saving and restoring data from these systems [1].

The development of information systems in terms of hardware infrastructure but also of specific applications have facilitated the access of both people and companies to online filing systems, publishing in national registers or in official bulletins in national and international intellectual property organizations [2].

The main advantages of using these methods of filing protection claims are primarily speed and accessibility. But this development of electronic services brings with it the problem of increasing the volumes of electronic data that need to be processed, stored and saved. In this context, the research of technologies that can improve the parameters of storage and saving of data in intellectual property organizations is a topic of interest that needs to be studied and explained in detail.

2 Experimental Research

The technology that ensures the compression of data that is saved by removing redundant or repeating data is known as the deduplication technique. This technology can be implemented in backup to disks systems and allows saving and storing a single instance of a data, when it is repeated, in a database or information system.This technology is also known as single instance storage, smart compression or data reduction.

Data deduplication technology involves an analysis and comparison of received data segments to be saved and stored later in disk volumes. Thus, a reference is created by deduplication algorithms if the data segment is already saved. In this case the data is compared and if there is no difference between the already saved data segment and the one to be saved, the new segment is discarded creating a reference of that segment.

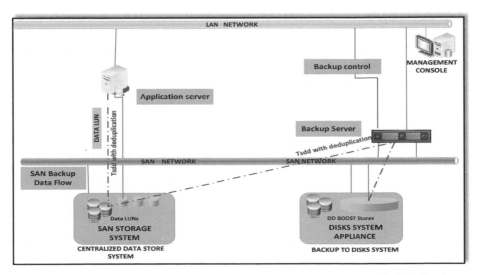

Fig. 1. The IT architecture of the experimental backup to disks system with deduplication

In the same way, a deduplication algorithm analyzes the output data through a dedicated network connection eliminating redundant data, reducing the volume of data to be saved and thus increasing their transfer speed [3].

A simple analysis of the market shows that the technologies for data deduplication are very different and the choice of a manufacturer can be more or less inspired, especially for backing up databases, applications more sensitive to the process of data deduplication. In this context, the idea of data deduplication is not seen by all experts in the field as an effective solution to reduce the volume of data to be saved.

The experimental plan's main objective is to assess the impact of increasing the volume of data to be saved to disks using the deduplication technology, in intellectual property organizations on the recently implemented centralized backup systems [4].

A data volume of 1.1TB was chosen for this experiment and a number of 10 tests were performed, simulating an increase in data volume with an approximate rate of 100GB.The scope of the experimental research is to determine the influence of the deduplication technology on the time needed to execute a backup to disks session [5]. In this context, the application server in the Fig. 1 has a volume of disks with 1,1TB data attached from the SAN. The backup process involves the transfer of data through the SAN Network, processed and deduplicated at the backup application level, running on the backup server and finally, saved deduplicated to the backup to disks system appliance.

When using deduplication technology, the following should be considered:

a. Deduplication backs up only to disk-based devices;
b. Deduplication backs up cannot be used on backup tapes;
c. Because the application uses only deduplication algorithms, there is no need for dedicated and specific hardware infrastructure to store the backup data;
d. Duplicated data is dropped and the backup applicationtransfer only a copy of the data to be stored on disks systems with reference to the singlecopy;
e. In the deduplication process only, unique data is stored. This contributes to the reduction of the storage capacity needed;
f. Software deduplication uses fragmentation technology to split data flow into segments of data [6];
g. A device with volume of disks must be shared to the backup system.

The equipment used in this experimental research and represented graphically in the Fig. 1 is composed by:

− A Dell-EMC Unity D400 SAN Storage system having the capacity of 80TB (a mixt of Solid-StateDisks and Serial Attached SCSI (SCSI Stands for Small Computer System Interface), volumes of disks);
− One Data Domain DD6300 for backup to disks,DD Boost disks appliance system model Dell-EMC;
− A backup server and Microfocus data protector application running on this server.

In this experiment, the independent variables represent the formulation or process factors that have a level set at the beginning of the experimental plan. For conducting the experimental study, input parameters processed were: a volume of 1TB data size with the predicted monthly increase of the data volume and data transfer rates on disk volumes [6].

In this experiment, input parameters were noted with:

X_1 = the volume of data measured in TB where X1 $\in$ (1TB... .10TB);

X_2 = data transfer rate measured in Mbps on disks where X5 $\in$ (520 Mbps... .600 Mbps);

Output size (also called output parameters) was noted with:

Tsdd = execution time of the backup to disks with deduplication measured in minutes – represented by red dotted line in the Fig. 1.

Table 1. Experimental data recorded

Tsdd vs X1, X2					
No.	X1 (TB)	X2 (Mbps)	Tsdd (minutes)	FITS1 (minutes)	RESI1 (minutes)
1	1.1034	532	325.3441	324.418	0.92565
2	1.2054	554	331.0201	332.116	−1.09580
3	1.3024	523	341.1751	345.159	−3.98386
4	1.4124	506	355.1213	357.949	−2.82770
5	1.5114	514	367.4789	366.892	0.58704
6	1.6134	549	378.7105	373.156	5.55408
7	1.7154	531	388.4869	385.263	3.22412
8	1.8174	544	399.1604	393.952	5.20816
9	1.9194	558	402.6921	402.531	0.16063
10	2.0214	562	404.4606	412.213	−7.75233

The experimental records are presented in Table 1. A number of 10 tests were performed representing simulations of the growth of the volumes of data in the application server for patents (online filing) based on monthly estimates of the number of applications for protection that can be submitted to a national intellectual property organization [7].

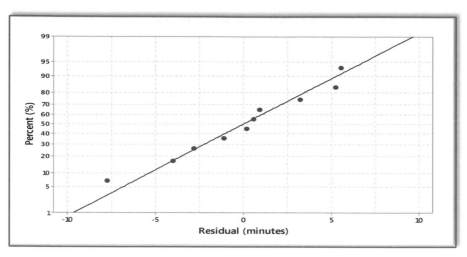

Fig. 2. Probability plot of RESI1

The graph in Fig. 2 represents the probability plot of the calculated RESI1 and evaluates how closely the data points follow the adjusted distribution line [8].

It can be observed that the points in the normal probability graph follow the line well and the normal distribution matches the data well.As can be seen, the specific theoretical distribution fits well and the points fall almost along the straight line.

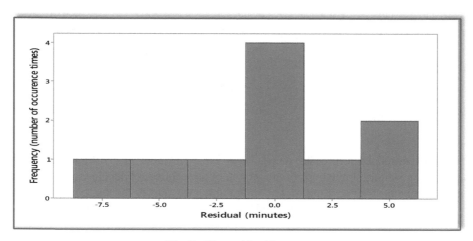

Fig. 3. The residue histogram

The residue histogram represented in Fig. 3 is used to determine if the data is distorted or if there are incorrect values in the data.

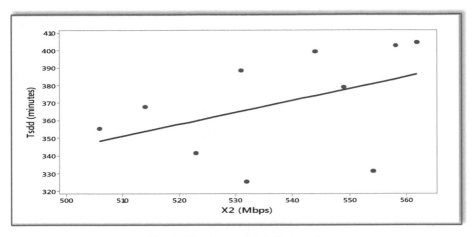

Fig. 4. Fitted line in statistical model

In the Fig. 4, thered line represents how appropriate the model is for the data entered.Fitted Line represented in Fig. 4 is a statistical model based on the data collected in the experiment and which determines how one factor influences the other factor.

The points in the graph in Fig. 4 represent the individual values and the line represents the trajectory that the model should follow [9].

The following regression Eq. (1) was identified during the experimental research:

$$Tsdd = 273.6 + 99.24 * X_1 - 0.1102 * X_2 \tag{1}$$

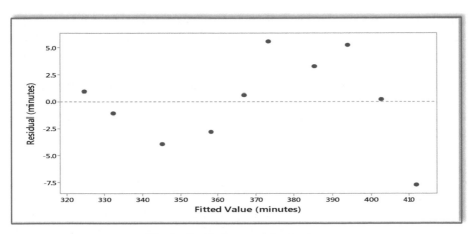

Fig. 5. Residuals versus fitted values

In the Fig. 5, on the Y axis are represented the residuals and on the X axis the fitted values.In this experiment, the hypothesis that residues have constant variability and are

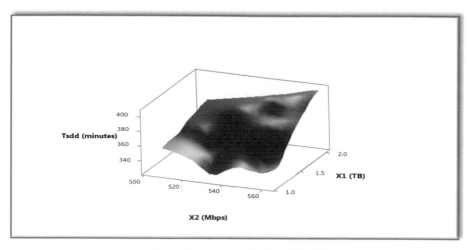

Fig. 6. Surface plot Tsdd vs. X1, X2

randomly distributed is confirmed by the graph of residues versus fits [10]. In an ideal model, the dots should be randomly represented on both sides of 0.

The Fig. 6 present the 3D Surface Plot, to examine the relationship between Tsdd (minutes) response variable and two predictor variables X_1 (TB) and X_2 (the speed rate of data written on disks at the SAN level), by viewing a three-dimensional surface of the predicted response [11].

3   Conclusion

Deduplication of data or as many experts in the field say, intelligent compression, is a technology used to reduce the need for storage by eliminating redundant data. Different files may share single instances, but are stored separately. Data deduplication solutions keep a single instance for all files.

The demand for such solutions has grown in the interest of more and more companies to find a solution to reduce the costs of doubling the annual storage requirements. This direction is determined by increasing the use of business applications and expanding their diversity.

There are several different levels at which the data deduplication process can take place. The first level is the source that produces the files, which can be scanned, indexed and fingerprinted. Subsequently, each file with an identical fingerprint is deleted, creating a link to the source file. The deduplication process using a secondary storage (SAN, NAS) represent another level, these solutions being offered even by the manufacturers of these systems. In this case, the files may be redundant on the hard disk of the application servers or databases of the intellectual property organizations, but at the time of data storage, scanning and fingerprinting processes are applied and, subsequently, the removal of multiple instances. Depending on the manufacturer, data deduplication may occur at the same time as copying to storage equipment or later, based on a programmable routine.

In the experiments performed, we can conclude in the first phase that data deduplication is a technology that consumes a lot of processing power. This makes deduplication on SAN/NAS systems quite expensive, requiring a large volume of internal memory, as well as CPU-level performance for file analysis, indexing, and fingerprinting. Thus, the process of data deduplication implemented in data backup systems in intellectual property organizations can bring important benefits to e-mail, file servers. However, the situation is changing in the case of transactional databases and their back-up, especially if they already contain data compression features.

Moreover, when applied incorrectly, data deduplication technologies can adversely affect back-up and recovery time and create network traffic jams. The general conclusion is that data deduplication is indicated to be implemented for the storage of existing data at the level of application and data servers, email or web servers, but less recommended for database backups, for which solutions as data compression can be found with most important benefits.

References

1. Blokdyk, G.: Data deduplication: Beginner's Guide, pp. 20–66 (2017)
2. Savage, A.: The data deduplication handbook - everything you need to know about data deduplication, pp. 12–34 (2016)
3. Kim, D., Song, S., Choi, B.-Y.: Data deduplication for data optimization for storage and network systems, pp. 132–150 (2018)
4. Ipsen, A.: BackupAssist 10 Feature Series, pp. 16–38 (2017)
5. Thwel, T.T., Sinha, G.R.: Data deduplication approaches concepts, strategies, and challenges, pp. 106–155 (2015)
6. Thornburgh, R.H., Schoenborn, B.J.: Storage Area Networks: Designing and Implementing a Mass Storage System, 1st edn., pp.156–187 (2000)
7. Vissers, C.A., Pires, L.F., Quartel, D.A.C., van Sinderen, M.: Architectural Design Conception and Specification of Interactive Systems, pp. 217–254. Springer, Cham (2016). https://doi.org/10.1007/978-3-319-43298-4
8. Brooks, C., Dachuan, H., Jackson, D., Miller, M.A., Rosichini, M.: IBM TotalStorage SAN File System, 4th edn., pp. 451–489 (2006)
9. Oprean, C., Țîțu, M.: Managementul calității în economia și organizația bazate pe cunoștință. AGIR, pp. 70–89 (2008)
10. NAS Guide: DIY NAS Guide: NAS Configuration Guide with Open-Source Software on Raspberry Pi or PC for Network Hard Disk Drive, Backup and Data Share, pp. 20–46 (2019)
11. Nelson, S.: Pro Data Backup and Recovery, pp. 152–178 (2011)

A Strategy for Applying Open Data Initiatives

Lucija Brezočnik(✉), Gregor Polančič, Sašo Karakatič, Grega Vrbančič, Špela Pečnik, and Vili Podgorelec

Intelligent Systems Laboratory, Faculty of Electrical Engineering and Computer Science, University of Maribor, Koroška cesta 46, 2000 Maribor, Slovenia
{lucija.brezocnik,gregor.polancic,saso.karakatic,grega.vrbancic, spela.pecnik}@um.si

Abstract. The advances of information technology have brought us to a period of increasingly rapid creation, sharing and exchange of information. In general, we do not perceive just how much information we are creating by using modern information technology devices. Still, the reality here is that we are creating data on a business and personal level while it continues to grow in both importance and volume. Therefore, this paper provides guidelines for a successful open data initiative implementation. For the latter, we present key stakeholders with their engagement reasons, critical success factors, and strategic themes for the designated area.

Keywords: Open data · Open government · Freedom of information · Transformation · Guidelines

1 Introduction

Nowadays, most individuals and organisations generate a broad range of data when performing their operations; however, the data is often left inaccessible in a local database or even forgotten after the initial use. Because of that, properly prepared and presented data is a largely untapped resource. However, the question that arises is why would such data be of public interest? There are many areas where we can expect open data (OD) to be of value and where examples of its use already exist [1]. Many different groups of people and organisations can benefit from the availability of OD, including the government itself [2]. At the same time, it is impossible to precisely predict how and where value will be created in the future because the use of data often creates new possibilities for further uses. The nature of innovation is that developments often come from unlikely places.

OD is data that anyone can access, use and share. It should be available online, open-licensed, machine-readable, available in bulk, and free of charge [1]. Opening up data is not a complex process, though it must be implemented wisely. Foremost, one must choose a dataset(s) and apply desired open license. There are several OD licenses from which one could choose [3]: the creative content is usually licensed using a Creative Commons License, and datasets using an Open Data Commons license. However, lately, the Creative Commons 4.0 license has been recommended for data as well as creative content.

I. Karabegović et al. (Eds.): NT 2022, LNNS 472, pp. 488–497, 2022.
https://doi.org/10.1007/978-3-031-05230-9_58

The demand for OD is rising, and currently, we are speaking about the emergent third wave, which concept is to reuse the public and private data [4]. The philosophy behind the third wave is that the attention to the demand is at least as important as the supply side of the data equation. As a selected region always defines the OD specifics, this paper provides guidance on carrying out this specialization.

The main missions of the paper are:

- to present the key stakeholders of open data with their benefits and challenges,
- to present the current status of the open data, and
- to provide recommendations and guidelines for utilising the open data initiative in the selected region.

2   Stakeholders of Open Data

When defining key stakeholders [5], the government and, more precisely, local governments, i.e., municipalities, have precedence in the OD area. In those, vital role plays geospatial data, which is a key for smart, sustainable, and prosperous communities, revealing crime rates in neighbourhoods, road closures or nearest health facilities. Another key stakeholder is civil society, which has been a motivity in progressing OD plans, primarily via raising awareness, setting standards, and specifying public expectations. The list also includes journalists and media covering various practices from data science to finding stories, all the way to creating interactive content and visualizations for articles. Researchers strive to produce knowledge on OD, either if they are part of academia or part of the international development sector. Through data collaboratives, private sector actors can be data users, data intermediaries, and/or data providers. Lastly, we must mention multilateral organizations formed between three or more countries that work on their joint issues. An example of that could be development banks, which are crucial in promoting development outcomes in low- and middle-income nations.

Table 1 presents multiple benefits for stakeholders of OD and some challenges.

Table 1. Benefits and challenges of open data stakeholders.

Stakeholder	Benefits	Challenges
Civil society	The activities which publish OD, have positive impacts on the public awareness about the availability of OD, commonly building online platforms for hosting and sharing OD. They leverage interaction between government institutions and community groups. Besides, they train other organisations as potential users and educate them on how OD can help to fulfil their missions	The major challenge of smaller civil society organisations is the lack of resources which are required to develop the technical capacities needed to take the advantage of OD

(continued)

Table 1. (*continued*)

Stakeholder	Benefits	Challenges
Government	Government progress on OD is being monitored with OECD studies and the UN E-Government Reports. In general, institutionalisation requires clear and transparent frameworks for the governance of OD. This allows OD being made an integral part of government business and enablesengagement across traditional programming silos and well across stakeholder groups	A growing awareness of the need to upgrade policies, institutional structures, programmes, and practices may be spotted. It includes production, and management, and well it ensures an effective reuse of government data to secure long-term sustainability and continuity of OD initiatives
Private sector	Business sector applies OD in its procedures to provide new products and services. Incubators and accelerators have managed large amounts of businesses worldwide in applying and utilising OD	OD demonstrates considerable potential to SMEs (small and medium-sized enterprises) to discover its benefits to innovations
Journalists and media	Journalists tend to acquire, extract, analyse, report on, and also produce OD. Journalists can alter the raw data into new insights, which informs customers, boosts public engagement in the democratic process, and hold powerful organisations responsible	If more media houses focus on making OD an essential source, more examples of automation tools may be seen in the future. However, the promise of "automated journalism" based on OD still remains unfulfilled
Researchers	A significant increase in research volume on OD is evident.This could lead to presumptions that OD research created a space for emerging areas of enquiry, e.g. privacy, data justice, and rights	OD should become more collaborative across disciplines and regions
Multilateral organizations	Multilateral organizations invest in many dimensions of the data ecosystem. They could obtain a considerable impact by mainstreaming OD work in their practices and increasing the ROI (return on investments) in operations with a strong data component	Multilateral organizations should put "open by default" and "open by design" ideas into application. Both initiatives will require distinct perceptiveness to security and privacy as well in-depth research and learning to utilize continuous advancement

It is vital to emphasize, that each stakeholder could be either supplier of OD, consumer of OD, or even both.

3   Current Status of Open Data

Few websites provide a brief overview of the OD sites worldwide (see Fig. 1). Nonetheless, two of the most known are Data.gov [6], comprising the USA government's open data, and data.europa.eu [7], including the European countries. The latter arose from the initiative of the European Commission around five years ago. Annually, they provide an OD maturity report, where they assess the status of OD maturity in the 27 EU member states and three EFTA countries, Liechtenstein, Norway, and Switzerland. Assessment is conducted with four main topics [7]: Policy, Portals, Impact, and Quality. The Policy topic covers policy framework, governance of open data, and open data implementation; Portals topic covers portal features, portal usage, data provision, and portal sustainability; Impact topic covers strategic awareness, political impact, social impact, environmental impact, and economic impact; and Quality topic covers monitoring and measures, currency and completeness, DCAT-AP compliance, and deployment quality. Based on those topics or dimensions, each country gets a maturity score. Figure 2 presents Slovene maturity level rating.

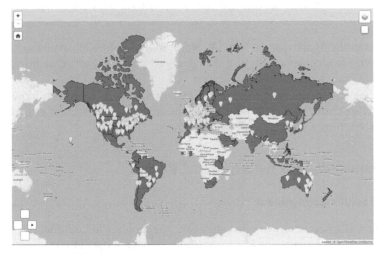

Fig. 1. Map representation of open data sites [6].

Even if we consider the whole world or specifically aselected region, the stakeholders providing regional data are usually local governments. However, all those local government stakeholders are always united in the central government OD websites. For example, the Italian government manages the Italian National Open Data portal, the Austrian government manages the Open Data Austria, and the French government manages the Open Data Portal France. Besides, it is also standard practice for other organizations to present data on their websites which consist of multiple data providers. One great example of that is Geoportal.de [8], which explores over a few hundred data providers who contributed their data, all grouped in different categories, i.e., Federation, Countries, Municipalities, and Science and Research. Furthermore, stakeholders could

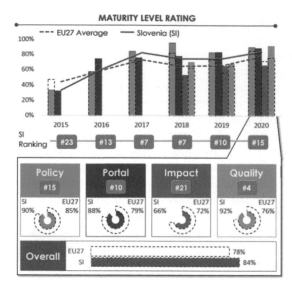

Fig. 2. Slovene open data maturity level rating [7].

also be non-profit organizations like 52° North Initiative for Geospatial Open Source Software [9], whose primary mission is to develop spatial research data infrastructures to foster information derivation from data.

In general, non-regarding the stakeholder type, OD usually comprises topics of Agriculture, Environment, Transport, and Government, followed by Economy, Education, Society, Technology, Culture, Tourism, and others.

4   Recommendations for a Selected Region

4.1   Stakeholder Engagement

To enhance OD awareness, we must stimulate stakeholders to contribute data. One of the biggest reasons for opening data is visibility. Opening data allows the stakeholder to become an established partner in the field of the provided data topic. This does not include purely providing the data but also publishing the results obtained while analyzing it. Such an approach even increases the recognition of the stakeholder. Another reason is promotion. Link to published data can be easily shared on social media, newsletters, local news stations, or others. Social media posts can be shared on the company's social media profiles, which employees can reshare. Newsletters are also standard practice in organizations to spread the voice of their novelties, and news stations would spread the news about it if adequately presented. Opening up the data also shows professionalism. When stakeholders think of opening data, they must think about the sharing format, e.g., Excel file, pure CSV, XML, JSON, and service, e.g. database, API, interactive website. Non-regarding the way, a stakeholder must first decide on the delivery type, prepare data and then provide it to others. If this is done correctly, the professionalism of the stakeholder increases.

When people, organizations or consortiums recognize that some stakeholders provided data that could benefit them, they can offer a potential partnership. Such partnership could not be purely on the data level but could also be expanded to a profound research project(s). Defined OD requirements increasingly limit the latter and force partners to share data. Lastly, stakeholders already use some OD in their business. That could be Google Scholar if we are talking about Academic data; World Health Organization or HealthData.gov for organizations working with health and scientific data; World Bank Open Data if an organization is doing some business in the financial or economic field; or Climate Data Online if the environmental topic is relevant to the organization. Consequently, it is reasonable for a stakeholder to give something back to the community.

4.2 Critical Success Factors

To perform effective OD initiatives in the selected region, several factors should be addressed. Table 2 presents six generic success factors, which have been defined by the specialization of more generic OD success factors [10] tailored to the selected region specifics. These factors are applicable at a macro-level, e.g. when reviewing a government OD initiative, and at a micro-level, e.g. when an organization or municipality wants to start publishing OD.

Table 2. Success factors tailored to the selected region.

CSF of OD initiatives	Generic	Selected RegionSpecifics
Policy	A clear OD policy enables a shared vision for all stakeholders with defined overall objectives and a strategy with a defined action plan, timeline and corresponding responsibilities. Such policy should embed in a broader digital, innovation, or reform policy package of a government or organization	The selected region can comprise territories with contrasted demographic, social, and economic trends and a significant cultural and linguistic diversity. For example, Alpine Space specifics are identified and addressed in the EUSALP strategy [11]
Governance	A clear Governance structure with a dedicated owner who drives the policy's implementation, reviews its progress, and measures its success is vital for a successful OD policy. There also needs to be responsible parties for each of the actions within the roadmap	Potential diversity in the selected region accompanies a great variety of governance systems and traditions, implying a need to establish a multilateral organization for monitoring and controlling an OD initiative

(continued)

Table 2. (*continued*)

CSF of OD initiatives	Generic	Selected RegionSpecifics
Capacity	An OD policy with an adequate governance structure forms a solid foundation, whereas the capacity or resources are needed to publish and manage OD on the operational level. Ongoing communication and training are vital to ensure people from technical (e.g., IT, geospatial, data scientists) and non-technical backgrounds (e.g., business owners, project managers, domain experts) have the necessary skillsets to engage with OD projects actively	OD initiatives in the selected region should include staff training, focusing on developing new skillsets. For example, action group 3 of the EUSALP strategy [11] aims to improve the competence of the job market, education and training with employment prospects in strategic sectors in the region and increase the region's employment levels through shared macro-regional activities
Technical	A set of Technical Publishing Guidelines with simple examples can support a streamlined approach to making valuable OD available	The technical aspect of OD publishing should be part of the innovation aspect of the selected region strategy
Engagement	The OD initiatives should analyze and address the needs of stakeholders who want to discover, understand, and access information. User engagement and following a user-centric approach are paramount to understanding user needs, fostering collaboration, and building trust	A think tank of public interest on the availability of services could be implemented in the selected region. Sharing awareness of OD should also be part of a multilateral strategy
Demand	The actual demand for publishing OD should be identified. Working directly with stakeholders to identify challenges and projects that could benefit from OD is the most effective way to achieve substantive impact	A multilateral organization responsible for OD should also have the responsibility to promote it and share best practices and cases

4.3 Strategic Themes Guiding the Open Data Initiative

Based on generic or regional OD initiatives [12], the strategic themes used for guiding the OD initiative are then tailored to the specifics of the selected region and the underlying strategy [11] (see Table 3).

Table 3. Strategic themes guiding the open data initiative in the selected region.

Strategic theme	Description
1	Broaden the range of public bodies (governments, local governments, multilateral organizations, HEI) actively engaged in the OD initiative
2	Widen the scope and improve the quality, quantity and range of OD and associated meta-data. Improve the quality and range of services provided in the selected region
3	Continue to engage with all stakeholders and encourage the use of OD, which includes civil society, governments, municipalities, private sector, journalists and media, researchers, and multilateral organizations
4	Support and encourage various OD users in the selected region, including civil society, governments, local governments, private sector, journalists and media, researchers, and multilateral organizations
5	Provision of a framework tailored to the selected region, to support and train all Data Providers and build capacity in the management and use of OD
6	Monitor and evaluate the impact, benefits, and risks of the OD initiative and benchmark against other jurisdictions
7	Ensure that effective governance structures and multilateral organizations are in place to implement the OD strategy

4.4 Possible Specializations for a Selected Region

It is essential to identify possible specializations for a selected region. They could be divided to feature ideas familiar to the countries and regions in the selected region and those that arise due to differences in the area's features.

Firstly, a group of people preparing such an initiative must define the most distinctive and unique common features of all the regions in the selected region. If such features comprise shared natural resources and natural biodiversity, a shared OD platform covering them could be implemented. For example, tracking the biodiversity of flora and fauna across the region can boost the pattern extraction about possible adverse effects which may arise as the consequence of any economic decisions. This can be especially useful as the flora, fauna, rivers, lakes and other natural habitats stretch across multiple regions and countries. Furthermore, the common weather and soil patterns manifest in similar agriculture and livestock farming.

Another shared topic could be a way of transportation across the selected region. Even though some international automotive vehicle and train systems exist, there is a big area for improvement as there are no common regional goals in this field. For example, traffic and freight control across the countries enable quicker responses to the changing patterns.

Since some countries and regions have excellent working OD platforms, they could share them as examples or guidance for other countries. For example, Switzerland has an excellent OD platform that provides data on public transport. Therefore, it would be helpful to create an OD portal that would include data, e.g. on transport from all countries around Switzerland with regard to the selected region. A similar portal (Research Alps Dataset) already exists and covers data on private and public research entities and business companies located in these countries.

One essential feature is also cultural diversity. The latter could be an advantage for a shared OD platform, where different cultures could display different work procedures and goals. Working on a shared platform would further connect these regions, boosting the cooperation in academic, economic, and other sectors. Nonetheless, even the tourism sector can benefit from mentioned differences. There are strives of tourists that visit aregion because of its cultural and ecological diversity, so providing a shared data platform about this topic can lead to potential new connections and synchronizations among different regions.

5   Conclusion

This paper demonstrates how OD plays an essential role in all sectors globally. We presented the key stakeholders and provided their main benefits and challenges of using OD. Then, we displayed the current status of the OD and pointed out websites where people can see and search publicly available datasets. Later, we focused on providing guidance and recommendations for users applying OD initiatives to the selected region. We highlighted reasons why stakeholders would contribute data, emphasized critical success factors to perform an effective OD initiative, and presented possible specializations.

In the future, we would like to conduct a specific case study for a selected area. Such a case study could result in concrete recommendations and a framework that would serve as a manual for OD initiatives in the designated area.

Acknowledgements. The authors acknowledge the financial support from the Slovenian Research Agency (Research Core Funding No. P2-0057).

References

1. Verhulst, S., Young, A.: Open Data Impact When Demand and Supply Meet Key Findings of the Open Data Impact Case Studies, pp. 1–56. SSRN (2016)
2. Janssen, M., Charalabidis, Y., Zuiderwijk, A.: Benefits, adoption barriers and myths of open data and open government. Inf. Syst. Manag. **29**(4), 258–268 (2012). https://doi.org/10.1080/10580530.2012.716740
3. Open Data Institute. Publisher's Guide to Open Data Licensing. https://theodi.org/article/publishers-guide-to-open-data-licensing/. Accessed 25 Dec 2021
4. Verhulst, S.G., Young, A., Zahuranec, A.J., Ariel Aaronson, S., Calderon, A., Gee, M.: The Emergence of a Third Wave of Open Data. https://opendatapolicylab.org/images/odpl/third-wave-of-opendata.pdf. Accessed 20 Dec 2021
5. State of Open Data. Open Data Stakeholders. https://www.stateofopendata.od4d.net/chapters/stakeholders/introduction.html. Accessed 20 Dec 2021
6. Data.gov. The home of the U.S. Government's open data. https://www.data.gov/. Accessed 10 Dec 2021

7. Data.europa.eu. The official portal for European data. https://data.europa.eu/en. Accessed 10 Dec 2021

8. Geoportal.de. The Spatial Data Infrastructure Germany. https://www.geoportal.de/. Accessed 10 Dec 2021

9. north. Spatial Information Research. https://52north.org/. Accessed 10 Dec 2021

10. D. Lee. Success Factors for Open Data Initiatives. https://www.linkedin.com/pulse/success-factors-open-data-initiatives-deirdre-lee/. Accessed 20 Dec 2021

11. European Commission. EUSALP, Alpine Space Programme and Alpine Convention. https://ec.europa.eu/regional_policy/sources/cooperate/alpine/eusalp_alpine_space_alpine_convention.pdf. Accessed 15 Dec 2021

12. Data.gov.ie. Open Data Strategy 2017–2022. https://data.gov.ie/pages/open-data-strategy-2017-2022. Accessed 10 Dec 2021

Digital Transformation Using Artificial Intelligence and Machine Learning: An Electrical Energy Consumption Case

Vili Podgorelec[1,2](✉), Sašo Karakatič[1], Iztok Fister Jr.[1], Lucija Brezočnik[1], Špela Pečnik[1], and Grega Vrbančič[1]

[1] Intelligent Systems Laboratory, Faculty of Electrical Engineering and Computer Science, University of Maribor, Maribor, Slovenia
vili.podgorelec@um.si

[2] Faculty of Electrical Engineering and Computer Science, University of Maribor, Koroška cesta 46, 2000 Maribor, Slovenia

Abstract. Companies nowadays eagerly compete in providing their customers with the best possible services, where the companies in the electrical energy market are no exception. As artificial intelligence and machine learning are considered the fundamental multi-purpose technologies and the innovation entity with the most significant potential for disruption, the companies strive to adopt these technologies and integrate them into their business processes. To test the possibilities for the introduction of AI and ML methods in their information system and business processes, we established a pilot project with a company operating in the electrical energy domain. An electrical energy consumption forecasting model has been developed alongside with some additional components. The obtained results show that a proper use of AI and ML methods can offer means for providing new and advanced services to different kinds of company's customers.

Keywords: Digital transformation · Artificial intelligence · Machine learning · Power engineering · Electrical energy consumption · Prediction

1 Introduction

Efficient processing of business data has always been at the heart of the electrical power engineering industry and its related activities, enabling it to provide secure, reliable and high-quality services and transparent operations [1]. In times of constant growth of both the volume and details of data collected on the one hand, and the desire and need for their best use on the other, we are witnessing extremely rapid development of information technology, software solutions and services. As the overall amount of generated data grows daily, so does the ability of computer systems and approaches to process this mass of data and discover new, more accurate insights. In doing so, modern methods of artificial intelligence (AI) and machine learning (ML) are the ones that enable the designers of information systems and software engineers to develop new services based on in-depth automated processing of the captured data. When properly implemented,

I. Karabegović et al. (Eds.): NT 2022, LNNS 472, pp. 498–504, 2022.
https://doi.org/10.1007/978-3-031-05230-9_59

software solutions based on AI methods and ML algorithms can thus offer a range of functionalities as well as new and improved business solutions which provide customers with a new level of user experience, while offering to provider a competitive advantage.

However, advanced ML methods and the theory behind such increasingly capable intelligent solutions are becoming more and more sophisticated and complex. While bringing ever new functionalities to users, software engineers, IT designers and developers of new systems and solutions on the other hand are faced with a serious challenge from such a development, as they do not have enough time and opportunity to keep up with the technological progress alongside with their everyday operational tasks. In fact, the challenges are all the greater because, as technology evolves, the concepts of their use often change as well. For this purpose, we established a pilot project with the company Informatikad.d., in order to test the possibilities for the introduction and integration of AI & ML methods in their information system and business processes.

The main contributions of this paper are:

- to present a brief overview of the digital transformation using AI and ML,
- to elaborate the digital transformation process with AI and ML on a real-world electrical energy consumption case, and
- to present some general results of a pilot implementation of the electrical energy consumption forecasting model, that has been developed, alongside with possible benefits of such a model for the company adopting it.

2   AI & ML for Digital Transformation

AI is a technology that is transforming businesses as well as our everyday life. It is a wide-ranging tool that enables people to rethink how we integrate information, analyse data, and use the resulting insights to improve decisionmaking. AI is considered the innovation entity with the most significant potential for disruption [2] and the fundamental multi-purpose technology, especially in relation to machine learning [3].

With the proper introduction of intelligent software and services companies can improve product capabilities and service quality, communicate better with customers, streamline operations, and create predictive and accurate business strategies. However, not many companies nowadays have the skills and capacity to adequately address and tackle the whole digital transformation process using AI & ML. For this purpose, they need help from a competent knowledge provider. Such help commonly encompasses three main phases:

- identification phase, with the search for business cases where it makes the most sense to implement AI solutions,
- design phase, where the research is performed on how to design and develop AI & ML methods and solutions, and how to integrate such solution into the company's business model, and
- implementation phase, encompassing a pilot implementation and its evaluation, as well as the transfer of knowledge from a knowledge provider to the company.

If the company decides to fully adopt the tested technologies on the selected, and possibly other, business cases, it will need to build the proper capacity and skills. Thus, it is of vital importance, that counselling and education accompany all these phases.

3   Digital Transformation with AI & ML in the Electrical Power Engineering

The process of digital transformation with the use of AI & ML methods have been established between Informatikad.d. (the company seeking to leverage its business processes with the help of AI & ML) and the Intelligent Systems Laboratory from the University of Maribor (the knowledge provider). The basic purpose of the established cooperation was to implement a digital transformation in a part of the company's business model. AI &ML are certainly tools that can be used to implement digital transformation that helps in gaining a competitive edge with customers at present and in the future. As the company's core business is providing services for electricity distribution companies, the focus of the transformation was analysing and predicting electrical energy consumption. For this purpose, weperformed a pilot project to introduce AI &ML methods in the company's business solutions.

Informatikad.d. is a company that performs custom information services for most of the Slovenian electricity distribution companies. Their major services include:

– the billing system for electricity distribution (for about 800,000 customers),
– the billing system for selected electricity suppliers (with over 140,000 customers),
– the provision of the electricity market data exchange platform (single entry point), which is used by all Slovenian electricity distributors and all major electricity suppliers (over 30),
– an integration platform based on a service bus, through which over 20 different information systems are already integrated, and
– applications for the management of all consumer-related processes on the side of the distributor (issuing of opinions and guidelines, project consents, connection consents, system use agreements, etc.)

3.1   The Development and Introduction of AI & ML Methods

The basis of the performed pilot project was the development of a prototype back-end software solution for an in-depth analysis of consumer electrical energy consumption data and the use of the results of such an analysis for the implementation of various functionalities. The solution was based on the use of various state-of-the-art AI & ML approaches, methods and techniques. The primary source of data was the captured historical data on measured consumption at individual metering points.

The basic developed functionalities include:

– machine learning models for predicting electrical energy consumption for each consumer (daily, hourly, and 15-min forecasts),

– detection of unusual consumption patterns and identification of deviations in consumption, and
– clustering of users (i.e. consumers) based on their consumption behaviour patterns in groups of similar users regarding the electrical energy consumption behaviour.

The fundamental part of the system is the electrical energy consumption forecasting model. Consumption forecasting is of prime importance for the restructured energy management environment in the electricity market [4]. Despite the increase of smart grids technologies and energy conservation research, many challenges remain for accurate forecasting of electricity production and/or consumption using big data or large-scale datasets [5]. Different approaches have been used recently for electrical energy consumption forecasting, from various regression methods [6], different probabilistic forecasting techniques [7], all the way to the most advanced ML methods such as artificial neural networks [8] and especially deep learning methods [4].

We designed a hybrid forecasting algorithm, using polynomial regression with regularization for fast generalized long-term forecasting and deep Long Short-Term Memory (LSTM) recurrent neural network for accurate short-term forecasting, which turned out to provide accurate predictions (see an example in Fig. 1). The multivariate LSTM neural network comprised a series of LSTM and Dropout layer pairs with decreasing dropout probability rates, the final Dense layer and the RMSprop optimizer. The network was fed with daily measurements of at least two years to be able to capture seasonal effects. The number of neurons of the LSTM layers was set as 5 times the degrees of freedom in the used data (number of samples multiplied by the dimension of each sample). The LSTM has been commonly trained until the stability of loss rate has been achieved.

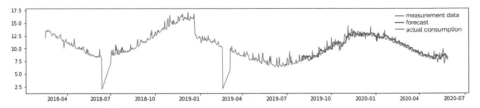

Fig. 1. Electrical energy consumption forecast for one selected household.

The second part of the system is the anomaly detection component, developed for automatic identification of unusual electrical energy consumption and detecting characteristic deviations from the estimated consumption for an individual user. To this end, methods and models for detecting both point (e.g., today's consumption deviates from everyday) and contextual anomalies (e.g., consumption did not increase/decrease as much as would be expected given the circumstances) have been developed. While the point anomalies might represent possible failures, outages, etc., the contextual anomalies might indicate a changed regime of a specific consumer's electrical energy consumption. Figure 2 represents the consumption for a selected household, where both types of anomalies have been detected – the first identified anomaly from the left-hand side represents a point anomaly (a possible measurement error or failure), while the small group of several anomaly points represents a possible concept drift (changed mode of operation, see how the overall consumption has dropped since that point).

The anomaly detection was based on the forecasting model. The exponentially smoothed forecasted values have been used to determine the anticipated "normal" value, with the calculated error (plus standard deviation) on the training set representing the threshold for the "region of normality". If a real measurement exceeded the anticipated region of normality, it was identified as possible point anomaly. If several consecutive measurements were identified as anomalies, the group was identified as possible contextual anomaly.

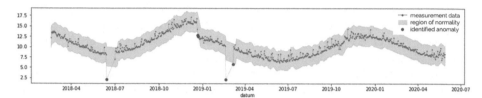

Fig. 2. Identification of anomalies in electrical energy consumption for a selected household.

The third major part of the system is the clustering component, aimed at the analysis and segmentation of consumers into groups of similar users based on electrical energy consumption patterns (Fig. 3 shows two very different profiles, each quite typical for a group of consumers with similar monthly consumption behaviour). Such clustering may provide means for better user management, advanced consumption statistics, and more accurate forecasting, to name a few.

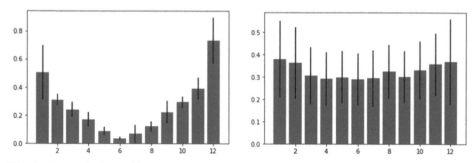

Fig. 3. Two typical monthly user consumption profiles, indicating two different groups of users.

3.2 Electrical Energy Consumption Forecasting Results

The developed forecasting models have been tested on a set of electrical energy consumption measurements. The results of the short-term forecasting using recurrent neural networks (RNN) turned out to be very good (see Fig. 4), achieving less than 10% mean-absolute error (MAE) on average, and around 12% rootmeansquare error (RMSE) – both metrics are calculated on normalized scales and thus expressed as percentages. As the consumption may be very dispersed on some occasions, the standard deviation of predictions is unfortunately quite large (around ± 10% for MAE, and ± 12% for RMSE).

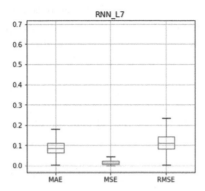

Fig. 4. The predictive performance results of electrical energy consumption forecasting.

3.3 Possible Benefits for the Company

By providing an efficient and accurate forecasting model for the electrical energy consumption (together with other components), the company may be able to offer a set of advanced new services to its various customers:

– for distributors and suppliers: optimization of the amount of electricity, optimization of alternative sources, planning of the electricity distribution network development and investments in the electricity network, …
– for external stakeholders (e.g., regulators): visualization and analysis of consumption by regions, municipalities,…, analysis of consumption behaviour habits, determination of the tariff system and new billing rules,…
– for end-users (e.g., consumers): better monitoring and planning of consumption, detection of faults on devices, more optimal consumption of electricity,…

In this manner, we helped the company to identify, design, and implement (currently in a pilot setting) advanced AI & ML based solutions, which can enable new and advanced services directly to their business users as well as end consumers.

4 Conclusion

In this paper, we provided an overview of the digital transformation based on AI and ML methods, algorithms and tools. We briefly explained how a knowledge provider could help a company, operating within the electrical energy market, when engaging in the adoption of AI and ML methods for their business operations and services. A case of such a pilot project, aimed at providing new advanced services based on an efficient and accurate electrical energy consumption forecasting model, has been provided.

Although the developed forecasting model shows very promising results, the challenging task of providing a production-ready electrical energy consumption forecasting technology is by far not finished. No specific learning model outperforms other learning models for every forecasting problem. Thus, the best algorithm choice depends on specific forecasting tasks and challenges. Even then, there is an enormous space of possibilities for fine-tuning the chosen algorithms' parameters and settings, as they can significantly influence the results, both from the predictive performance point of view, as well as regarding their computational complexity.

Finally, the development of accurate and efficient ML models alone is not even remotely enough to achieve all the benefits the proper digital transformation can offer. However, it is definitely a very good and promising start in this direction, which we will certainly pursue in the future.

Acknowledgements. The authors acknowledge the financial support from the Slovenian Research Agency (Research Core Funding No. P2–0057).

References

1. Zhang, Y., Huang, T., Bompard, E.F.: Big data analytics in smart grids: a review. Energy Inform. **1**(1), 1–24 (2018). https://doi.org/10.1186/s42162-018-0007-5
2. Duan, Y., Edwards, J.S., Dwivedi, Y.K.: Artificial intelligence for decision making in the era of Big Data–evolution, challenges and research agenda. Int. J. Inf. Manage. **48**, 63–71 (2019)
3. Lichtenthaler, U.: Building blocks of successful digital transformation: Complementing technology and market issues. International Journal of Innovation and Technology Management **17**(1), 2050004(2020)
4. Hafeez, G., et al.: A novel accurate and fast converging deep learning-based model for electrical energy consumption forecasting in a smart grid. Energies **13**(2244), 1–25 (2020). https://doi.org/10.3390/en1309224
5. Almalaq, A., Zhang, J.J.: Deep learning application: load forecasting in big data of smart grids. In: Pedrycz, W., Chen, S.-M. (eds.) Deep Learning: Algorithms and Applications. SCI, vol. 865, pp. 103–128. Springer, Cham (2020). https://doi.org/10.1007/978-3-030-31760-7_4
6. Liu, H., Wang, Y., Wei, C., Li, J., Lin, Y.: Two-stage short-term load forecasting for power transformers under different substation operating conditions. IEEE Access **7**, 161424–161436 (2019)
7. Nespoli, L., Medici, V., Lopatichki, K., Sossan, F.: Hierarchical demand forecasting benchmark for the distribution grid. Electric Power Syst. Res. **189**, 106755 (2020)
8. Veeramsetty, V., Deshmukh, R.: Electric power load forecasting on a 33/11 kV substation using artificial neural networks. SN Appl. Sci. **2**(5), 1–10 (2020)

Petri Net-Based Model of Master/Slave Dataset Replication in Big Data

Ilija Hristoski[1]($\boxtimes$) and Tome Dimovski[2]

[1] Faculty of Economics - Prilep, "St Kliment Ohridski" University - Bitola, Prilepski Braniteli St. 133, 7500 Prilep, North Macedonia
ilija.hristoski@uklo.edu.mk
[2] Faculty of ICTs, "St Kliment Ohridski" University - Bitola, 7000 Bitola, North Macedonia

Abstract. In today's corporate world, the Big Data paradigm assists organizations in harnessing their plentiful data in their never-ending pursuit for new business possibilities. The main problem that such businesses have is keeping Big Data infrastructure performant, scalable, dependable, and accessible. All of these issues are often solved by implementing fault-tolerant systems. Dataset replication is one of the critical procedures that must be performed regularly to increase system resilience. The purpose of this work is to propose a performance model regarding the execution of reading and writing operations found in one of the most used Big Data replication architectures, the Master/Slave architecture, which is based on the usage of the class of Generalized Stochastic Petri Nets. Such a model may be used for performance analysis to acquire various performance metrics concerning different working scenarios with varying input parameters.

Keywords: Big Data infrastructure · Master/Slave dataset replication · Read/write operations · Performance modeling · Generalized Stochastic Petri Nets (GSPNs)

1 Introduction

The total quantity of data generated, shared, recorded, copied, and consumed by organizations, businesses, industry, and individuals are expected to increase dramatically throughout the world in the forthcoming years. It has already reached 64.2 zettabytes in 2020 (1 zettabyte equals 1 trillion gigabytes) [1]. In addition, global data generation is expected to exceed 180 zettabytes during the next five years, up to 2025. Despite this intensely rising trend, only a portion of this freshly created data is stored, as only 2% of the data produced and consumed in 2020 was saved and held until 2021. Following such a significant rise in data volume, the installed base of storage capacity is expected to grow at a compound annual growth rate (CAGR) of 19.2% from 2020 to 2025. In 2020, the installed base of storage capacity already surpassed 6.7 zettabytes on a global scale [1]. The Big Data paradigm reflects the unstoppable process of 'datafication' where virtually anything and everything is being documented, measured, as well as digitally captured, and transformed into data [2]. Faced with the so-called 'data deluge', inherent to the Big Data era, where an overwhelming amount of complex and heterogeneous data is pouring from anywhere, anytime, and any device, organizations require effective solutions to cope successfully with the challenge of turning massive amounts of widely divergent data into meaningful comprehensions to support evidence-based decision-making, relying on insight extraction using Big Data platforms. In such a turbulent high-tech landscape, some of the most prominent challenges that organizations face regularly vis-à-vis the Big Data infrastructure, are attaining adequate high levels of data dependability (availability, reliability, maintainability, durability, safety, and security), as well as exhibiting high performance, i.e. timely responses to READ and WRITE requests. All of these issues are addressed effectively through the concept of dataset replication. As a practice of storing the same data in various places (at least two), dataset replication provides enhanced data availability and accessibility, improved scalability and flexibility, lower data access costs, and higher fault tolerance. In addition, the application of this approach also increases the robustness and dependability of Big Data systems. As such, it has become a must-have for efficient and effective digital transformation.

Recognizing the significance and multiple benefits of dataset replication in today's highly competitive business environment, we develop and propose a performance model encompassing the execution of READ and WRITE requests in a Master/Slave dataset replicated architecture, comprised of a single master node and two slave nodes, based on the utilization of the class of Generalized Stochastic Petri Nets (GSPNs). It is a model designed to define and capture all of the key characteristics of how a real system runs in terms of resources utilized, resource contention, and delays caused by processing and/or physical constraints (e.g. arrival rates of READ and WRITE requests, the processing speed of nodes, etc.), having minded the general idea and logic found within the Master/Slave dataset replication. The resulting performance model can be used "to gain greater understanding of the performance phenomena", i.e. "to estimate the runtime of a job when the input parameters for the job are changed, or when a different number of processors is used in a parallel computer system" [3].

The paper is structured as follows. Section 2 briefly overviews the recent research made on this topic. Section 3 elaborates the Master/Slave dataset replication. Sections 4 proposes a corresponding GSPN-based model of the processing of READ and WRITE data requests in a Master/Slave dataset replicated architecture, comprised of a single master and two slave nodes. The related discussion and conclusions are given in the last section.

2   Brief Overview of the Recent Research

Since their introduction in the late 1990s, many kinds of stochastic Petri Nets (i.e. Petri Nets that include a temporal specification as a characteristic of firing delays associated with some types of firing transitions) have been routinely used for performance modeling and evaluation of distributed computer systems. Lately, these types of Petri Nets are increasingly being utilized for modeling and evaluating the performance and dependability of the infrastructural components of Big Data systems and architectures, including replicated systems. The following is a synopsis of the relevant research made in this area during recent years.

The work of Gianniti et al. (2017) provides a novel approach to modeling and analysis of MapReduce job execution times using the class of Fluid Stochastic Petri Nets (FSPNs) [4]. Their work confirmed that FSPN-based models can be used successfully to study not only the MapReduce framework but also even more sophisticated ones, such as Spark [5]. In their research, Chattaraj et al. (2019) propose and evaluate a failure-repair mathematical model of the Hadoop Distributed File System (HDFS) cluster's Master/Slave architecture, based on the class of Generalized Stochastic Petri Nets (GSPNs), to analyze the dependability of HDFS data storage system [6]. Chiang et al. (2020) have used WoPeD, an open-source software system dedicated to the analysis of Petri Nets, to perform reachability analysis of a Resource-Oriented Petri Net (ROPN) model of Hadoop MapReduce Framework, comprised of two nodes for parallel processing of the sub-jobs after splitting [7].

3 Master/Slave Dataset Replication

During the process of dataset replication, the system saves several copies of a dataset, known as replicas, on multiple nodes (at least two). Because the same data is duplicated on several nodes, replication automatically delivers scalability and availability. Fault tolerance is also accomplished because data redundancy ensures that data is not lost when a single node fails. One of the most prominent methods to implement replication is the Master/Slave approach, depicted in Fig. 1. In such architecture, there is a single master node, and at least one slave node. Data are written to the master node and afterward replicated over to slave nodes. External WRITE data requests, such as insert, update, and delete operations, are processed by the master node, whereas READ data requests can be processed by any slave node. In Fig. 1, the master node manages all WRITE requests, whilst data can be READ from either Slave Node #1 (Replica A) or Slave Node #2 (Replica B) [8]. As opposed to reading operations, all writing operations are consistent, since they are coordinated solely by the Master Node.

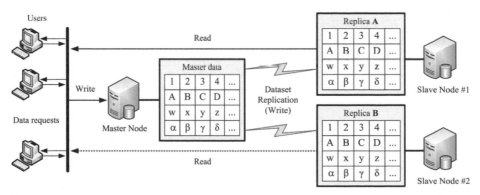

Fig. 1. Schematic representation of a Master/Slave dataset replication using a single master node and two slave nodes (Source: Authors' representation)

The exclusive inclusion of the slave nodes in performing read operations points out the fact that the Master/Slave dataset replication is ideal for managing READ-intensive workloads, rather than WRITE-intensive ones, since growing READ data requests can be successfully addressed by adding extra slave nodes, i.e. by horizontal scaling of the system infrastructure. If the Master Node fails, then it is either recovered from its backup, or one of the slave nodes has to be configured as a new master node.

4   GSPN-Based Model of Read/Write Operations

For modeling purposes, the class of generalized Stochastic Petri Nets (GSPNs) has been utilized [9, 10], in conjunction with TimeNET 4.5, a special-purpose software tool intended for modeling and analysis of various types of stochastic Petri Nets, including both exponentially and non-exponentially distributed firing times [11]. GSPNs have been extensively used for modeling and evaluating the performance of complex distributed systems exhibiting non-deterministic behavior. They represent a graphical tool, used for formalizing stochastic discrete-event dynamic systems including features such as concurrency, synchronization, parallelism, mutual exclusion, blocking, and conflict, and a mathematical tool for performing advanced formal analysis, as well [12].

The proposed simulation model is portrayed in Fig. 2, depicting a system with a single master node and two slave nodes. It is an integral model showing the execution of a single request, which can be either a READ or a WRITE request. The arrival of requests in the system is a Poisson process, which occurs at a rate of $\lambda_{arrival}$. The incidence of a READ request happens with a probability of p_{READ}, and therefore, the one of a WRITE request is $1 - p_{READ}$. These are the firing probabilities of the immediate transitions *T_read_request* and *T_write_request*, respectively.

The processing of the WRITE request starts in the Master Node (a token in the place *P_master_start_write*) and lasts, on average, $1/\mu_{master\_write}$ time units. The firing of the exponential transition *T_master_write*, which puts two tokens in the place *P_master_end_write*, is conditioned by the existence of a token in the place *P_master_idle*, i.e. only when the Master Node is idle. Those two tokens are immediately removed from the place *P_master_end_write* by firings of the immediate transitions *T_slave1_replication* and *T_slave2_replication*, which denote the beginning of the simultaneous replication of the master data (i.e. previously written data) to Slave Node #1 and Slave Node #2 (as a result, a single token is put in the places *P_slave1_start_replication* and *P_slave1_start_replication*). The writing operation is performed by each slave node only when they are idle, i.e. when there is a token in the places *P_slave1_idle* and *P_slave2_idle*. The process of replication lasts for an exponentially distributed time and occurs when exponential transitions *T_slave1_write* and *T_slave2_write* fire with rates of μ_{slave1_write} and μ_{slave1_write}, correspondingly. The successful replication is denoted by placing single tokens in places *P_slave1_end_replication* and *P_slave2_end_replication*, which are both terminal places for the writing operation.

When it comes to the modeling of the READ requests' processing, there are several possible solutions, which are conceded by the combination of at least three main aspects: (1) the determinism/indeterminism in the order of placing the READ request to slave nodes; (2) the usage/misusage of a load-balancing scheme; (3) the assumption whether the datasets, residing on slave nodes, are up-to-date, or not. In this regard, the hereby proposed GSPN model implements the case when there is no determinism in the order of placing the READ request to slave nodes, a load balancing scheme is used, and the datasets, residing on slave nodes, may not be up-to-date/synchronized.

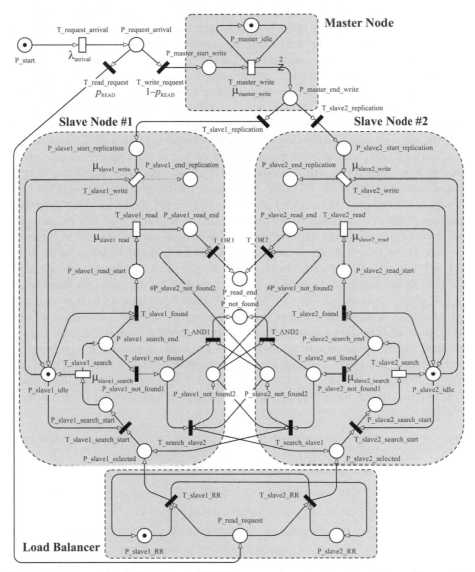

Fig. 2. GSPN model of READ and WRITE requests' processing in a Master/Slave replicated architecture comprised of a single master node and two slave nodes (Source: Authors' representation)

This way, the GSPN model addresses one of the biggest concerns of the Master/Slave dataset replication: the problem of inconsistent reads, which occurs in a case when either of the slave nodes is read before being updated during the replication process.

The processing of the READ request begins with the firing of the immediate transition $T\_read\_request$ with a probability of p_{READ}, which puts a token in the place $P\_read\_request$. The load balancer, implementing the Round Robin scheme, selects one of the slave nodes to be read from. No matter which slave node is chosen (a token in the place $P\_slave1\_selected$ or $P\_slave2\_selected$), the reading operation is preceded by a searching operation, which completes, on average, in $1/\mu_{slave1\_search}$ and $1/\mu_{slave2\_search}$ time units, respectively, after firing of the exponential transition T_slave1_search or T_slave2_search. As a result, the searched data is either located or not. If it is located, the reading operation starts (a token in the place $P\_slave1\_read\_start$ or $P\_slave2\_read\_start$). After $1/\mu_{slave1\_read}$ time units, the firing of the exponential transition T_slave1_read denotes the successful ending of the reading operation from Slave Node #1, which puts a token in the place $P\_slave1\_read\_end$. No matter whether a token resides in the place $P\_slave2\_not\_found2$ (if the reading operation has been already carried out unsuccessfully by Slave Node #2) or not, the firing of the immediate transition T_OR1 puts a token in the place P_read_end, which is one of the two possible terminal places for the READ request processing (the other one is P_not_found). However, if the searched data is not located on Slave node #1 (a token in the place $P\_slave1\_not\_found1$), then the system tries to read the searched data from Slave Node #2, by firing the immediate transition T_search_slave2 and by putting a token in the place $P\_slave2\_selected$, and also in the place $P\_slave1\_not\_found2$, to denote that the reading operation has been already carried out unsuccessfully by Slave Node #1. The operation of reading the searched data from Slave Node #2 follows the same steps as described earlier. If the searched data is found and read, then a token is put into the place $P\_slave2\_read\_end$. This enables the immediate transition T_OR2, due to a token already residing in the place $P\_slave1\_not\_found2$. The firing of this transition puts a token in the terminal place P_read_end. If the searched data was not found on Slave

Table 1. Parameters of the Master/Slave replicated architecture used in the proposed GSPN model

Parameter	Meaning
$\lambda_{arrival}$	Incoming requests' arrival rate
p_{READ}	Incoming request's probability of being a READ request
$1 - p_{READ}$	Incoming request's probability of being a WRITE request
μ_{master_write}	The processing rate of the WRITE request in the Master Node
$1/\mu_{slave1\_search}$	Mean searching time for data to be read by the Slave Node #1
$1/\mu_{slave2\_search}$	Mean searching time for data to be read by the Slave Node #2
μ_{slave1_read}	The processing rate of the READ request in the Slave Node #1
μ_{slave2_read}	The processing rate of the READ request in the Slave Node #2

Node #2 after being not found on Slave Node #1 previously, the immediate transition *T_AND2* becomes enabled and fires to put a token in the terminal place *P_not_found*. The parameters utilized in the GSPN model are summarized in Table 1.

5 Conclusion

The need of providing performance models that can be a valuable support for Big Data systems designers and developers comes out of the fact that predicting the performance of contemporary Big Data architectures can be a costly, time-consuming, and daunting task. In general, performance models, such as the hereby proposed one, allow for carrying out various "what-if" analyses while improving the observed system in various dimensions, regardless of whether it is still in the design phase or has already been deployed. Such analyses are beneficial to system designers, assisting them in optimizing newly designed system architectures for the best long-term performance through the process of fine system tuning. They have the potential to be very effective in assisting both computing centers and data centers in selecting the optimal system infrastructure during the procurement phase. Moreover, application developers can use performance modeling as a method to significantly improve the performance of their programming code by better understanding which application tuning measures yield the most improvements towards consistent performance. In this context, the proposed GSPN model can provide significant insights into the performance of the Master/Slave replicated architecture by obtaining the following performance metrics for various operating scenarios: average response time, average queue lengths, the average number of READ and WRITE requests waiting in queues to be processed by nodes, node throughput as a function of request arrival rates, node utilization, and so on. Future research will focus on a performance evaluation of the proposed GSPN model concerning various input parameters and conveying a comparative assessment of the obtained results with those yielded by the performance analysis of a GSPN model representing the execution of READ and WRITE requests within the alternative Peer-to-Peer (P2P) replicated architecture.

References

1. Holst, A.: Volume of data/information created, captured, copied, and consumed worldwide from 2010 to 2025, Statista.com (2021). https://www.statista.com/statistics/871513/worldwide-data-created/
2. Mayer-Schönberger, V., Cukier, K.: Big Data: A Revolution That Will Transform How We Live, Work, and Think. An Eamon Dolan/Houghton Mifflin Harcourt, Boston, MA, USA (2013)
3. Bailey, D.H., Snavely, A.: Performance modeling: understanding the past and predicting the future. In: Cunha, J.C., Medeiros, P.D. (eds.) Euro-Par 2005. LNCS, vol. 3648, pp. 185–195. Springer, Heidelberg (2005). https://doi.org/10.1007/11549468_23
4. Gianniti, E., Rizzi, A.M., Barbierato, E., Gribaudo, M., Ardagna, D.: Fluid Petri Nets for the performance evaluation of MapReduce applications. In: Proceedings of the 10th EAI International Conference on Performance Evaluation Methodologies and Tools (VALUETOOLS 2016), Taormina, Italy, pp. 243–250 (2017). https://doi.org/10.4108/eai.25-10-2016.2267025

5. Gianniti, E., Rizzi, A.M., Barbierato, E., Gribaudo, M., Ardagna, D.: Fluid petri nets for the performance evaluation of MapReduce and spark applications. Perform. Eval. Rev. **44**(4), 23–36 (2017). https://doi.org/10.1145/3092819.3092824

6. Chattaraj, D., Sarma, M., Samanta, D.: Stochastic Petri Net based modeling for analyzing dependability of big data storage system. In: Abraham, A., Dutta, P., Mandal, J.K., Bhattacharya, A., Dutta, S. (eds.), Emerging Technologies in Data Mining and Information Security: Proceedings of IEMIS 2018, Volume 2, Advances in Intelligent Systems and Computing, vol. 813, 473−484, Springer, Singapore (2019). https://doi.org/10.1007/978-981-13-1498-8_42

7. Chiang, D.-L., et al.: Modeling and analysis of Hadoop MapReduce systems for big data using Petri Nets. Appl. Artif. Intell. 1−25 (2020). https://doi.org/10.1080/08839514.2020.1842111

8. Erl, T., Khattak, W., Buhler, P.: Big Data Fundamentals: Concepts. Drivers and Techniques. Prentice-Hall/Arcitura Education Inc, North Vancouver, Canada (2015)

9. Ajmone Marsan, M., Balbo, G., Conte, G., Donatelli, S., Franceschinis, G.: Modelling with Generalised Stochastic Petri Nets. Wiley, Hoboken (1995)

10. Balbo, G.: Introduction to generalized stochastic Petri Nets. In: Bernardo, M., Hillston, J. (eds.) SFM 2007. LNCS, vol. 4486, pp. 83–131. Springer, Heidelberg (2007). https://doi.org/10.1007/978-3-540-72522-0_3

11. TimeNET 4.5 Official Website (2021). https://timenet.tu-ilmenau.de/

12. Zimmermann, A.: Stochastic Discrete Event Systems: Modeling, Evaluation, Applications. Springer, Heidelberg (2008). https://doi.org/10.1007/978-3-540-74173-2

Cybercrimes Ways to Pick up Electronic Devices

Petrică Tertereanu[1,2](✉)

[1] Polytechnic University of Bucharest, Bucharest, Romania
tertereanupetrica@yahoo.com
[2] County Police Inspectorate Vâlcea, Calea lui Traian Street, no. 95, Rm Vâlcea, Vâlcea County, Romania

Abstract. Currently, almost every crime has an electronic component, such as computers or an electronic technology that facilitates crime. The more time we spend in the digital world, the more opportunities we offer cybercriminals. The investigation refers to the way of singing up electronic devices containing evidence related to the crime being investigated, whether it is a regular one or terrorist action. In today's society, people use different computers, electronic devices and other electronic media in many activities in their lives. Criminals use the same devices to facilitate their illegal activities. Technologies enable criminals to commit crimes internationally, remotely, with complete anonymity. Cyber attacks are a complex category of threats, through the accentuated dynamics, the global character, the difficulty of identifying the source of the attack and establishing effective countermeasures. The likely targets of these attacks can be both civilian critical infrastructure objectives and communication systems and information technologies used by military structures.

Keywords: Information system · Electronic devices · Cyber attack · Cybercrime · Electronic technology

1 Introduction

In recent decades, cyberspace has had a major impact on all components of society. The uninterrupted functioning of information and communication technology has made our daily lives, our fundamental rights, our social interactions and our economies dependent on them. Fundamental rights, democracy and the rule of law need protection in cyberspace. But online freedom requires both safety and security [1].

The more we spend in the digital world, the more opportunities we offer cybercriminals. Cybercrime is a form of crime with one of the fastest rates of development, which causes worldwide damage to at least one million people a day.

In today's society, in the constituent elements of the crime most often there is an electronic part such as the computer or an electronic component that contributes to its commission.The computing technique used contains e-evidence that relates to the crime committed, whether we are referring to a common crime or a terrorist offence [2]. Information and Communication Technology (ICT) is the backbone of economic growth and is a source of strategic importance underpinning the sectors of the economy.

I. Karabegović et al. (Eds.): NT 2022, LNNS 472, pp. 513–521, 2022.
https://doi.org/10.1007/978-3-031-05230-9_61

Information technology is part of complex systems that ensure the functioning of economies, in strategic areas such as:energy, health, transport and finance, while many business models are made upon the basis of the idea of the continuous functioning of the internet and the good use of IT systems [2].

Cybercrime means non-compliance with the legislation by using computing to transfer or store data that allows a criminal attack (accessing computer systems by overcoming technical security difficulties). Cybercrime is represented by 4 categories respectively: crimes againstconfidentiality, integrity, the availability of data and information systems, electronic communications offences, content-related offences (child pornography) and offences against intellectual property [3].

2 Ways of Lifting Electronic Devices

The digital world offers a wide range of advantages, but it has come to the confluence that it is very unstable.Basic services such as: electricity supply, water supply, medical or mobile communications services may be affected by security incidents, whether intentionally or negligently. The EU economy is already targeted by cybercrime targeting the private sector and individuals. Cybercriminals are taking advantage of innovative methods in information technology to illegally access information systems to access critical data or request ransoms from businesses. A new form of cyber-attacks is the development of cyber espionage in the economic sector and the activities of the state commander [2]. The European Union and the Member States need to adopt robust legislation to effectively fight cybercrime [4]. In order to combat illegal operations carried out online by organised crime groups, especially those targeting electronic banking, the sexual exploitation of children and attacks targeting critical infrastructure, the European Cybercrime Centre was established in 2013. This Center provides specialized support to criminal investigations and promotes solutions throughout the European Union. The Centre also provides support and provides high-level technical, analytical and forensic expertise in the framework of joint investigations in the EU [2]. In everyday life, people use different computers, electronic devices and other electronic media. Criminals use the same devices to facilitate their illicit activities. The advantages of the use of technology by criminals have the effect of committing criminal acts on a planetary scale regardless of distance and with unknown identity.It is worth noting that electronic devices and computers have the ability to be used in committing criminal acts and in this way they store data that are considered the evidence and that describe the relationship between the victim and the perpetrator [5]. In the event that a breach of legal norms is found, in which computers or any electronic technology have been involved, rules must be followed as set out in the following figure (Figs. 1 and 2):

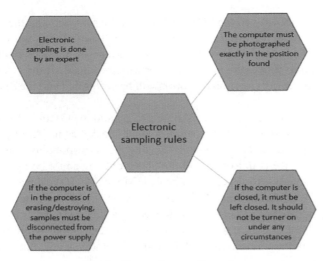

Fig. 1. Electronic sampling rules

For the preservation of samples, the procedure must be followed in the orderof thefollowingfigue:

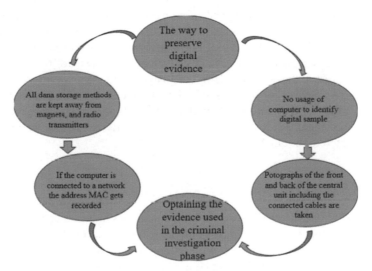

Fig. 2. The way to preserve digital evidence

In the case of taking samples and preserving them from mobile phones, tablets, GPS units that can store information directly on an internal memory or on an external storage device, the following general rules must be observed:

- interrupting the device to communicate with any network or receive wireless signal;
- it will hold the device charged and rise with all the cables [4].

If the device is turned off, the SIM card slot will be searched and the card removed to prevent contact with the telephone network, the battery will be removed if possible and the device will be stored radio waves (fraday or Ramesey Shielded Enclosure) to cut off the connection to Bluetooth, Wi-Fi or other wireless signals.If the electronic device is working, the sim card extraction operation will be performed and the device is placed in "Airplane Mode".

In the event that the device has an Apple iOS operating system (IPhone, iPad, etc.) and the screen is locked with a PIN code, it will put the device in "Airplane Mode" as a result of vertically sliding the screen and selecting the corresponding icon.

If the Apple device is unlocked and the above option cannot be applied, it will be searched in the phone's settings option "Airplane Mode" [4]. If the device is unlocked, it is positioned in "USB Debugging Mode" and "Stay Awake Mode" before opening. This activity is only performed if the device is locked by a PIN code. The raised device will be putin a special housing that turns off radio signals (Faraday mesh or Ramsey Shielded Enclosure) to block the connection to GPS, Wi-Fi, Bluetooth or any other wireless signal [5].

Smart mobile phones are mostly used for socialization and information and less for calls.In this way, smart devices store sensitive data, about the behavior of its owner.

Thus, the forensic evolution of mobile devices has developed, which has the role of recovering information from a mobile device.

The development of large-scale technologieshas led to the rise of cybercrime. The data stored in a mobile device, containsdetails on how the owner behaves, which become the target of traditional cyber attacks as well as those who perfect malicious software in order to manipulate them. Mobile devices contain data collection in 3 different locations relative to respective hardware: SIM, internal memory and SD card [6].

The architecture of the operational system for extracting and archiving the data stored in these electronic devices must consist of a digital camera, an external memory support (HDD, SSD, NAS, etc.), a write-blocking device for SD cards, a SIM card reader, a data collection device (hardware/software) and a set of adapters related to the multitude of connectors related to mobile hardware structures [6].

3   Digital Evidence

Digital devices have the role of ensuring communications between people, both locally and globally. Most people believe that the only sources of digital evidence are only mobile phones, computers and the internet, without taking into account that any electronic product that processes information can be used to commit a crime.

An example would be portable games that have the ability to carry encoded messages between suspects [7]. It is worth noting that specialists in the field participating in judicial activities must have the capacity to properly identify and collect this digital evidence. Digital samples can be defined as valuable information and data for an investigation, which is stored, primated or transmitted by an electronic device.

From a forensic perspective, the devices from which digital samples can be taken are structured into three categories: computers, samples on the Internet or other devices that are not related to the Internet or other mobile devices [8]. The data stored by the elctrtonic devices was given the designation "digital" due to the fact that it was divided into digits, binary units of digits one(1) and zero (0), recovered and saved using a set of instructions called software or code.

Using these instructions any type of information: photos, words, or spreadsheets could be saved [8, 12]. The exploitation of this type of evidence saved in this way is a branch in the process of developing forensics.

The Internet facilitated access to information but at the same time offered the opportunity for illegal trafficking of images, data and espionage.

Cybercriminals exploit this possibility and can access financial and communication systems in government networks and large corporations, in order to steal money, information identities or to sabotage systems. Ensuring computer security is a vast topic that will become more and more important given that the planet is becoming extremely interconnected, and networks must withstand critical transactions [9, 10].

The biggest challenge for internet crime is that technical staff and investigators understand how processes work and are in the cross-up to the upskilling of software innovations and tracking technologies [11].

Cybercrime continues to be an expanding problem in both the public and private sectors. A single computer can contain evidence of criminal activity on the web. Digital samples can be identified on the hard drive and in computer peripheral equipment, including removable media such as memory drives and CD-ROM disks [12].

Digital evidence plays a very important role in any criminal investigation. Data relating to aspects existing prior to the commission of the offence and information stored after the offence has been committed are relevant during the investigation [13].

An example may be that a suspect can use an online program such as Google Maps or Street View to determine how to position the home to determine where they can break into.

The suspect has the opportunity to list the stolen goodies through commercial platforms such as Craigslist or E-Bay. Some of the crimes can be committed entirely by digital means. In all the situations presented there is an electronic "trace" of information. Instant messages, pictures, stored documents, emails, text messages, and browsing history are considered proof.

As a rule, electronic devices use online managed safety systems that are known as the "cloud" [8].

Such a system stores an average of 1000 and 1500, perhaps even more and more text messages sent and received from that phone. As a rule, mobile devices store information about the locations where the device was located as well as the time period.

During research, the last 200 locations of GSM cells accessed by a mobile device can be accessed on average. Even photos posted on social networks such as Facebook may contain information that relates to locations.

Images uploaded to a social network such as Facebook may indicate data indicating where they were taken.Images taken with a mobile device that has a GPS function may contain data indicating the place where they were taken [8].

Among those facts are those shown in the following figure (Fig. 3):

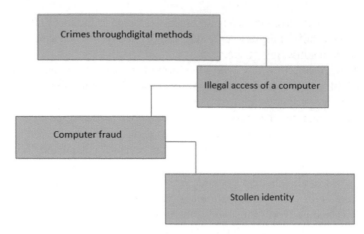

Fig. 3. Crimes through digital methods

The most important aspect about digital evidence is that which refers to its authenticity because in most trials the lawyer invokes the authenticity,respectively, the method of collecting and preserving digital evidence.

The technical objective of proving the authenticity of a digital object is to have checks in place to be able to demonstrate stories of how the data have been managed, which leads to the idea that, the data has not been modified, replaced or corrupted [14].

During the month of October 2021, a network of 12 people was dismantled who were acting around the world through ransomware attacks that targeted critical infrastructure.

The judicial and law enforcement operation involved 8 countries. It is estimated that cyber attackers have claimed at least 1800 victims in 71 countries.

The attackers chose large corporations as their targets and aimed to block the activity.

The cyber attackers were well organized each having different roles within the professional criminal organization.

A group of members of the criminal network were tasked with network penetration,using methods to corrupt IT networks, including brute force attacks, SQL injections, theft of personal data, and phishing emails with malicious attachments. After this operation, cyber attackers focused on lateral displacement,distributing in the system malware such as Trickbot or subsequent exploit instructions such as Cobalt Strike or PowerShell Empire,so that they cannot be identified and to collect additional information [15, 16].

Attackers can stay on the compromised network for months by checking the weak bridges of the IT network, prior to making the profit from the infection, by deploying a ransomware. It is known that these cyber attackers have implemented the LocherGoga, MegaCortex, and Dharma ransom tips, among others. Given that the attackers were not identified in the residence, they had the opportunity to get extraordinary benefits from ransomware attacks.

The victims were asked for such ransoms and were asked to pay in virtual currency Bitcoin in exchange for decryption keys.

Cooperation between the investigative teams at international level was coordinated by Europol and Eurojust as a result of the fact that the victims were located in different regions of the globe.

A key role in this operation was played by Europol's European Cybercrime Centre (EC3), which was intended to make the operational links and to present technical data on digital evidence and malware [15, 16].

At the time of submission of the notification document or of the finding of the commission of the crime, the investigating bodies of the judicial police must identify the violated legal norms, the manner of committing the deed, the establishment of the injured person and the civil party, these aspects being important in order to establish the way of investigation.

In order to carry out the the activity of criminal investigation in a process according to the provisions of the Code of Criminal Procedure (art.305 al.1), the criminal investigation bodies order the initiation of criminal prosecution "in rem" (regarding the deed).

This procedural stage of the criminal investigation involves the administration of evidence that establishes the existence or non-existence of a crime, when identifying the perpetrator(s) and establishing all the circumstances committed including the determination of the damage caused [17]. With regard to criminal activity carried out online, a website considers that it can constitute, for example, screenshots, documents submitted by the injured person, etc. as evidence.

Our legislation defines evidence as "any element of fact that serves to establish the existence or non-existence of a crime, to identify the person who committed it and to know the circumstances necessary for the just settlement of the case and which contributes to finding out the truth in criminal proceedings".

Inother news, through computer systems, crimes are committed by infringing copyright and related rights, thus processing data and information that may constitute evidence in the criminal trial. In order to obtain this evidence for materialization, a computer search is carried out as provided for in the Code of Criminal Procedure.

Computer perquisition may be ordered 'where the discovery and collection of evidence requires the search for a computer system or a computer data storage medium.

The computer search can be carried out only on the basis of obtaining a computer search warrant that from a procedural point of view can be obtained either at the criminal investigation stage, which implies only the initiation of criminal prosecution regarding the deed or at the trial stage.

In the first stage, the computer search warrant is issued by the judge of rights and freedoms of the court competent to hear the case at first instance at the motivated request of the prosecutor.

The search in the computer system or of a computer data storage medium is carried out by a specialist from the judicial bodies or from outside them or by a specialized policeman.

The computer search is necessary to be attended by the suspect or defendant and the prosecutor or the investigating body of the judicial police [17].

4 Conclusions

Given that there are many recommendations for lifting electronic devices containing evidence, it is necessary to bear in mind that an essential condition is to interrupt communications between the device and the network.

This activity is done by removing the battery from the device if possible or by shielding it from radio/bluetooth/wireless signals).

This operation is vital because there are several ways in which a suspect or accomplice has the possibility of destroying or altering evidence on a device that is not in his physical possession.

Also, at the time of connecting the device to the network, several elements can be changed, including the option for the device's memory to be deleted.

Compact electronic devices are in a continuous hardware and software development ensuring both communication and personal data storage or remote operation of other mobile devices.

The means of proof are represented by the files stored in the internal memory of mobile devices, which is why for the most correct authentication of the downloaded data depends on the use of specialized software that has the ability to select and transfer the information to a space outside the device.

It should be noted that, in practice, the computer search is carried out in order to discover evidence consisting of computer data, stored either on a computer system, when the means of storing computer data is an integral part of the computer system (for example, telephone/TV/smart watch or tablet that has an internal memory), or on a storage means independent of the computer system (memory card, memory stick, SIM card, CD, DVD, external drive).

In that regard, it has been shown that the logical interaction with a medium (means) of storing computer data, is the only one that can result in the collection of digital evidence relevant to the case.

Artificial intelligence can be used by cyber attackers because it has the ability to detect network vulnerabilities by analyzing every aspect in seconds.

Similarly, artificial intelligence can be used to scan and identify all vulnerabilities and if it is intended to attack that vulnerability, it can automatically be locked before being exploited by the attacker.

References

1. https://european-union.europa.eu/
2. Securitate Cibernetică: Amenințări și politici de răspuns, Sinteză Documentară nr.1(64), ISSN1841–2998 (2016)
3. Soltan, V.: Frauda informatică, Revista Procuraturii Republicii Moldova (2019). https://ibn. idsi.md/sites/
4. U.S. Departament of Homlend Security, Best practices for Seizing Electronic Evidence v.4.2 (2021). www.crime-scene-investigator.net
5. Meilă,A.E.,: Infracțiuni Cibernetice-modalități de ridicare a dispozitivelor electronice. Biroul Informare Documentare, nr.1(162), ISSN 2065–9318 (2021)

6. Moloșag, N.: Examinarea criminalistică a dispozitivelor mobile. Teză de master (2020). http:// repository.utm.md/xmlui/bitstream/handle/
7. http://www.forensicsciencesimplified.org/digital/principles.html
8. National Forensic Science Technology Center: Digital evidence. www.forensicsciencesimpli fied.org
9. Chudasama, D., Patel, D., Shah, A., Shaikh, N.: Research on cybercrime and its policing. Am. J. Comput. Sci. Eng. Surv. **8**(3), 14 (2020)
10. https://www.consilium.europa.eu/ro/policies/e-evidence/
11. Reedy, P.: The risks for digital evidence, Strategic Leadership in Digital Evidence, pp. 7174, ISBN9780128196182 (2021). https://www.sciencedirect.com/science/article/
12. Weir, G.R.S., Mason, S.: The foundation of evidence in electronic from, pp. 36–39 (2017). https://www.jstor.org/stable/j.ctv512x65.10
13. Nickson, M.K., Victor, R.K., Venter, H.S.: Diverging deep learning cognitive computing techniques into cyber forensics. Forensic Science International: Synergy **1**, 61–67, ISSN 2589–871X (2019). www.sciencedirect.com/science/article/
14. Mason, S., Stanfield, A.: Authenticating electronic evidence, Electronic Evidence Book Editor(s): Stephen Mason and Daniel Seng Published by: University of London Press; Institute of Advanced Legal Studies Stable pp.193–260 (2017). http://www.jstor.org/stable/j.ctv512 x65.14
15. Homeland Security Today. 12 Targeted for Involvement in Ransomware Attacks Against Critical Infrastructure (2021). www.hstoday.us
16. https://www.hstoday.us/subject-matter-areas/cybersecurity/12-targeted-for-involvement-in-ransomware-attacks-against-critical-infrastructure/
17. Udroiu, M., Zlati, G.: Sinteze de Procedură penală Partea Generală, p. 585. Editura C.H.Beck-București (2020)

Recursive Method of Forming a Technical Object Description and Design Process Organization

Viktoriia Antypenko(✉), Viktor Nenia, Anna Marchenko, and Bohdan Antypenko

Faculty of Electronics and Information Technologies,
Sumy State University, Sumy 40007, Ukraine
`victoriait@ukr.net`

Abstract. This article presents the recursive method of forming a description of any mechanical engineering object, which provides its practical computer support, regardless its type. Creative activity on new technical objects development is considered as manufacturingactivity. It is a subject to control and management, organizational and administrative regulation. Such activity has three features of its substantiation. These are designing, projection and aesthetic design. The last two ones are quite simple according to control and management of the process of the technical object project development as they are based on a previously accepted and approved concept. However, in the first case, where the designed object is considered as the one containing many complex processes, the "decomposition" approach is chosen. Its development process is algorithmized on the basis of formal Terms of References (ToR) meaning the selection of components that would be specially combined into one unit and ensure compliance with the ToR requirements. Such object description using ICOM approach of the IDEF0 method, form the interfaces that are necessary for the object functioning. It provides the clarification of the component description if it is absent or formation the last ToR in the design chain, if there is an understanding of the component practical implementation or choosing an appropriate item according to the catalog, if there is a real component description. These three cases cover all possible variants of the output decomposition results and guarantee the completion of an organized recursive process of detailing the designed object description.

Keywords: Recursive method · Technical object description · Computer support · Decomposition · Component description · Design process organization · Manufacturing

1 Introduction

Nowadays there is a regularity: the law of human society on growing consumption, is that people's activities in everyday life, in manufacturing and in the public field require new types of objects which should be more functional, more convenient and easier to use. In general, they must be of higher quality. Furthermore, business as the driving force of modern progress demands to produce these new types of objects as soon as possible. Mainly, these requirements are laid down and partially implemented at the designing stage. Exactly this feature determines the design requirement for fast and reliable development of the future object project while guaranteeing the set quality level. According to these conditions, design should be considered not as art, science or a product of mental activity, but as a manufacturing process. At the same time, the previously mentioned aspects should either be taken into account, or it is needed to provide the designers with an opportunity to implement them. Designing, like other manufacturing processes, requires its own tool and machine support, as well as automation of the control process. The desire to develop them drives this research, some results of which are published for discussion and improvement.

2 Related Work

The importance of the objects designing subject has led to a huge number of publications. It has become almost impossible to study and comprehend them, especially recently with a significant increase in the variety of things created for use and expanding the range of specialists involved in their development. This determines the subjectivity of both the choice of works for analysis and the points of view on certain designing aspects. Conceptually, authors' position is that designing is a technology, but in some situations (insufficient development of theory and technology as something new always is designed) it has the character of a craft. In this case, authors support Suzanne Rivard [1], especially the compliance with established rules and discipline. Although the process of formation the designing theory is considered along with the designing practice, but proposed in [1] spiral process of formation the designing theory cannot be transferred to the process of creating the objects project. Iterations, which are inevitable, must be minimized and local. In this case, global cycles of repetition of the whole process or its individual phases or stages are not allowed. Everything must be solved within one project operation.

These requirements can only be met by organizing the designing process according to a «top-down» approach. This is not news, it is normal practice. It is thoroughly elaborated and makes it possible to present the designing results quite clearly [2]. The Structured Analysis & Design Technique (SADT) methodology has received comprehensive support and further development and has gained the status of Integrated Computer-Aided Manufacturing (Integrated DEFinition IDEF) standards. Once developed several years ago, these standards still play their role, and there is no alternative to the main standard of functional modeling. However, one thing is missing. It is the details needed to formalize the modeling and obtain specific values of indicators on this basis. The central idea of the SADT methodology about the focusing on the main aspects of the modelled objects has worked and still is topical. Nevertheless, it left in the shadow the question of ensuring the concreteness of the performed analysis. Until now, there are several ways to translate the compilation of diagrams (IDEF0, UML, etc.) into other forms of models and finally into machine code. This is due to the need for syntactic and semantic control of the model and the opportunity to objectively verify possible options for the model implementing.

In [3], it is proposed to use an inhibitory Petri net for models with unidirectional links. Colored and regular Petri nets are also used [4–6]. Some progress has been made, but the authors of all these works introduce many rules, restrictions and exceptions on model transformation. However, there is no mutually unambiguous correspondence between models in different forms, especially semantic. Confirmation of the fact that the reserves of the SADT methodology have not yet been exhausted is the work [7], in which this methodology is applied to the functional modeling of technical energy facilities.

3 Research Method

The performed analysis of the conceptual method of structural analysis and designing [1] and examples of its application [2, 3, 7] has showed us the range of opportunities and advantages. It will be used by the authors as the main tool for design procedures. The method is known, therefore, there is no need to present it.

Organization of the design procedures implementation is a problematic issue in general. This is due to the need to apply a set of automation tools of design works to designing the various technical objects that have different composition and schemes of combining their components into one whole, which has its own unique properties and purpose to perform original functions. This means that it is impossible to compose a program code for monitoring, control and management of the design process in advance. At the same time, the development of designing methods and workable schemes for their implementation, taking into account the focus on the computers use always rests on the questions whether the algorithm will be implemented or whether the number of transactions will be over. The last ones are fundamental. These questions are the focus ofmathematicians. Here the authors have been helped by the theory of algorithms and its section of recursive functions [8]. This section of the encyclopedia is written at the modern formalized level of mathematical logic. Thus, the authors provide a meaningful basis that allows understanding, developing and controlling the offered results.

The basis is made up of three provisions (axioms) that for sets of arguments and sets of functions there are the following:

- a function that returns zero for any argument;
- a function that returns an argument by number;
- a function that returns the function at this position by number.

As well as next two operations:

- compositions of functions, i.e. functions from a function: one function for another is an argument;
- operation of simple recursion.

This recursion operation is performed according to the following algorithm. There are three functions $f(x_1, x_2, x_3,..., x_n)$, $g(x_1, x_2, x_3,..., x_n, x_{n+1}, x_{n+2})$, $h(x_1, x_2, x_3,..., x_n, x_{n+1})$ respectively for n, $n + 2$ and $n + 1$ arguments.

Let's start with the step.

$$h(x_1, x_2, x_3, ..., x_n, 0) = f(x_1, x_2, x_3, ..., x_n).$$

And it continues to repeat until the result or the fulfillment of a predetermined condition is obtained.

$$h(x_1, x_2, x_3, \ldots, x_n, i + 1) = g(x_1, x_2, x_3, \ldots, x_n, i, h(x_1, x_2, x_3, \ldots, x_n,)).$$

In the theory of algorithms [8] (based on the works of Church, Post, Turing and Markov) it is argued that a problem can be solved if the process of its solution can be reduced to a recursive process, i.e. a finite set of identical actions that are repeated. This is the second method that the authors use. Of course, a systems approach is also used. Where necessary, during the material presentation, either the features of the methods use, or authors' interpretation of them, or authors' improvement of methods and what it consists of, are going to be given.

4 Results

4.1 Researched Processes

The concept of designing in terms of ontology characterizes two processes associated with artificial objects. Firstly, designing is one of the stages in the life cycle of any artificial object. At this stage, the object does not actually exist. The object is represented by a project. It is a collection of documents in paper or electronic form, presented as data files, layouts, etc. The project contains a description of the future object. The essence of this life cycle stage is filling the design documents with information meaning only data that describe different aspects of the object must appear in the documents. The last one is intended for manufacturingthe object in the conditions of the selected enterprise. This means that the description of the object is focused on a specific manufacturing technology: specific materials, sets of technological operations, tools, machines and equipment, methods of processing the materials and/or energy and information. Secondly, designing is the activity name of specialists in creating a description of the future object. This is a purely organizational aspect of the matter containing information when, what and who performs.It is clear that these are two completely different processes and, accordingly, approaches to their description and implementation should be different, although they have the same end result. However, first of all, it is necessary to understand how the project of technical object behaves and only then organize its filling according to it. After that it is time to arrange the work of experts on implementation the technique of formation the technical object description. Unfortunately, the first two processes have completely fallen out of researchers' sight and are not found in publications. This justifies the need to return to basics and adjust approaches to to research directions.

The authors consider creative activity on development the new technical objects as manufacturing activity. It is the subject of control and management, organizational and administrative regulation. This activity has three features of its rationale. They are designing, projection and aesthetic design. The last two ones are quite simple according to control and management of the process of the technical object project development as they are based on a previously accepted and approved concept, designing solutions of which are ready and already tested prototypes or proposals of aesthetic designers. However, in the first case, where the designed object is considered as the one that contains many components with complex processes, it is necessary to use modeling and simulation and to examine several options. Exactly these processes are the subject of the authors'consideration.

The world experience summarized in [9] shows that the most appropriate approach to designing is «top-to-bottom» approach (from general to detailed), as it allows step-by-step verification of design decisions and design process management.

The choice of design status as amanufacturing process involves its rationalization and pragmatism. That is why it has been decided to algorithmize its development process on the basis of a formalized Terms of References (ToR). The presence of a clear TOR ensures the formation of a reasonable management order to perform tasks along with guarantees of responsibility for the execution of the assigned work. In this case, it becomes possible to carry out objective control. The presence of a TOR formalized form allows to establish monitoring of the step-by-step increasing the information description of the designed object in the design documents and operational control over the progress of works performance by machine.

4.2 Description of the Object as a Whole Unit. The Beginning of the Formation the Conceptual Part of the Project

In the first step, the object is designed as a whole unit, which will be used in the intended environment. In this case, the relationships of the object are specified with other objects. The description of these relationships (interface) is performed by forming the sets of parameters of substances and artefacts in accordance with the accepted arrangements in the SADT methodology and shown by the Eqs. (1):

$$Interface = I \cup C \cup O \cup M \cup Pe \cup Pi \tag{1}$$

A description of the sets of external and internal parameters has been added to the generally accepted notations in the SADT and IDEF0 methodologies on the Input (I), Command, Output and Mechanism diagrams, as it was done in [10]. In addition to the traditional meaningful description of substances and artefacts in the diagrams, the authors have introduced their characteristics.

REMARK. The authors follow traditional guidelines not to provide redundant information. However, the authors insist on strict control of all necessary but sufficient data for a complete description of the design objects.

The functioning of an object as a whole unit is described in general case by a system of Eqs. (2):

$$F\ (I, C, O, M, Pe, Pi, D) = 0 \tag{2}$$

Here, a set D is additionally introduced, which includes those of the design object parameters that determine its functioning.

In principle, the system of Eqs. (2) can be used as any description, but the rigor of the performance of the object analysis and its functioning as well as further computer realization certainly presupposes a formalized mathematical description.

The interface description sets of the object (1) and its generalized model (2) form the basis and initial description of the object in its project as a system S_0.

4.3 Description of the Object as a Component. The Continuation of the Formation the Conceptual Part of the Project

Now the decomposition of the function F is performed. The last one is going to be considered as $F_{i,j}$ for the j-th object of the i-th level. For decomposition, a set of $K_{Fi,j}$ functions is introduced and described by the Eq. (3):

$$f_{i,j,k}\left(i_{i,j,k}, c_{i,j,k}, o_{i,j,k}, m_{i,j,k}, pe_{i,j,k}, pi_{i,j,k}, d_{i,j,k}\right) = 0, \quad k = 1, 2, 3, \ldots, K_F \quad (3)$$

The sets $i_{i,j}, c_{i,j}, o_{i,j}, m_{i,j}, pe_{i,j}, pi_{i,j}, d_{i,j}$ have, respectively, the same content as the sets I, C, O, M, Pe, Pi, D in the case of the first decomposition level or similar to ones of the previous level.

Relationships between functions, as well as relationships between any other sets, are introduced based on the Cartesian product of the considered sets. Since the result of the Cartesian product has a large cardinality (an analogue of the product of the matrices sizes), therefore, the set of connections $R_{i,j}$ is formed as such that contains tuples for all K_R actual relationships and represented by the Eq. (4):

$$R_{i,j} = \{k, a, i, b, j, type\}, k = 1, 2, 3, \ldots, K_R \quad (4)$$

Here k identifies the relationship between the i-thelement of the set a and the j-th element of the set b of type $type$. Moreover, a and b are the only possibly coinciding sets of $i_{i,j}, c_{i,j}, o_{i,j}, m_{i,j}, pe_{i,j}, pi_{i,j}, d_{i,j}$. On the set of relationships for each individually and for their aggregates, K_{Cm} dependences and constraints of functions are introduced and described by the Eq. (5):

$$C_{i,j,k}\left(R_{i,j}, i_{i,j,k}, c_{i,j,k}, o_{i,j,k}, m_{i,j,k}, pe_{i,j,k}, pi_{i,j,k}, d_{i,j,k}\right) = 0, k = 1, 2, 3, \ldots, K_{Cm} \quad (5)$$

For substances such as matter or energy, such dependences will be conservation laws. For other substances and entities, different balances are used according to the nature of the considered objects and processes.

The aggregate system (3), (4) and (5) describes the object functioning as a component.

If the solution of system (3), (4) and (5) provides that the corresponding parameters of $i_{i,j,k}, c_{i,j,k}, m_{i,j,k}, pe_{i,j,k}, pi_{i,j,k}, d_{i,j,k}$ coincide with their values in I, C, M, Pe, Pi, D or similar to ones of the previous level, and the corresponding values of $o_{i,j,k}$ coincide with those in the set O, this is a sign of the correct decomposition for the subsystem S_{ij}. S_0 is not an exception.

The results of the performed decompositions for all objects of all levels, which are grouped accordingly, form a complete model of the complete function decomposition and, respectively, the content of the first part of the object project being developed.

4.4 Organization of Recursive Design Process

Now that the formalization of the decomposition process is available, let's consider how the process of content filling the project of the created object is organized. The authors adhere to the chosen approach and the task for each work is going to be supported by the presence of a formalized Terms of Reference on designing. For the first step, there is a ToR for the object design as a whole unit. The designing process at the initial stage consists of functions decomposition. When performing decomposition, it should be taken into account the experience gained by the company and use such functions that have already been practically implemented and can be directly applied after adjustment or regulation. In this case, a ToR is formed to organize the purchase of the required component or to include in the project the relevant componentsfrom other project. Thesecond option involves the use of a function that does not require further decomposition.

In this case the ToR on constructive formalization of such functionality is formed. After full implementation of these two options, any ToR is not created, and the last function is assigned the status «Implemented».If further decomposition is required (this is the third and last option), theToR is formed according to such scheme as to propose a system that would perform the function (2) within the interface (1).

Since functions become simpler and simpler with each level of decomposition, for each last one of them in the decomposition tree there will eventually be either a ready-made constructive solution or even a ready-made object. This is a guarantee of completion the recursive process, as new ToR will no longer be generated.

At the same time, it must be noted that the control over the compliance with the ToR requirements is a separate process. Requirements can form separate tree structures. There are possible options for incorrect construction of requirements structures and some their parts may not be implemented in the main process. Under these conditions, the opportunity of machine or manual placement the initial ToR for the implementation of certain or missed requirements is provided.

4.5 Organization of the Design Works

Initialization of design works begins at some moment of time $T0$, at which the ToR presence for design with the status «Not processed» (Fig. 1) will be revealed. From the moment of time $T1$ to the moment of time $T2$ planning will be carried out. As a result, for the ToR implementation the deadline for performance, executor and responsible person will be appointed. In the period from $T3$ to $T4$ this position of the plan will be accepted for implementation, and during the period from $T5$ to $T6$ the task will be brought to the executor. The design itself is carried out from time $T7$ to $T8$ and this is the longest period of time. The design results go through the stage of control, agreement and approval ($T9\ldots T10$).

After that, the materials of this design operation will be available to other participants in the design process from time $T12$. For all design solutions that require any amends, KToR pieces will be created and they will be the basis for launching more such K processes. During computer realization, green gaps should be tried to be minimized or completely removed.

Moment of time	Project artefact	Process	Artefact of operational work
T0	Terms_of_References_ToR$_{ij}$		
T1 T2		Planning	Plan$_{ij}$
T3 T4		Dispatch	Operational_plan$_{ij}$
T5 T6		Document workflow	Task_for_work$_{ij}$
T7 T8		Designing	Design_solution_DS$_{ij,1}$ ... Design_solution _DS$_{ij,N}$
T9 T10		Control and approval	
T11	Design_solution_DS$_{ij}$ Terms_of_References _ToR$_{ij,1}$	Document workflow	Report$_{ij}$
T12	Terms_of_References _ToR$_{ij,K}$		

Time

Fig. 1. Scheme of the current design procedure implementation

5   Conclusion

This paper presents a recursive method of the technical object designing. It is suitable for objects of any type. The method is based on a separate consideration of changes in the project state as a description of the technical object and the processes performed by the design team. The proposed method is algorithmic and formalized. The information content of the project filling is given.

Thus, a recursive method of forming a description of any technical object has been developed providing its practical computer support, regardless of its type.

References

1. Rivard, S.: Theory building: neither an art nor a science, but a craft. In: Advancing Information Systems Theories: Rationale and Processes, pp. 131–159 (2021)
2. DeMarco, T.: Structured Analysis and System Specification. NY: Yourdon (1978)
3. Vojt, N.N., Kanev, D.S., Kirillov, S.: Metod translyacii diagrammy s odnim tipom napravlennoj svyazi v ingibitornuyu set' petri. Avtomatizaciya processov upravleniya **3**, 38–45 (2019)
4. Godlevskyi, M.D., Orlovskyi, D.L., Kopp, A.M.: Structural analysis and optimization of IDEF0 functional business process models. Radioelektronika, informatika, upravlinnya (3 (46)) (2018)
5. Fang, K., Lin, S.: An integrated approach for modeling ontology-based task knowledge on an incident command system. Sustainability **11**(12), 3484 (2019)
6. Lakhoua, M.N.: The need for systemic analysis and design methodology of medical equipments. Int. J. Appl. Syst. Stud **8**(1), 76–88 (2018)
7. Fedorova, N., Shcheglov, Y., Kobylyackiy, P.: Application of IDEF0 functional modeling methodology at the initial stage of design the modernization of TPP in ETC. InE3S Web of Conferences(Vol. 209, p. 03013). EDP Sciences (2020)

8. Dean, W.: Recursive Functions (Stanford Encyclopedia of Philosophy) (2021)
9. Cloutier, R.J., Hutchison, N.: Guide to the Systems Engineering Body of Knowledge (SEBoK), version 2.2. INCOSE, Stevens Institute of Technology (2020)
10. Antypenko, V., Nenia, V., Marchenko, A., Antypenko, B., Kovpak, A.: Information technology for implementation the functional modeling of a technical object. In: Karabegović, I. (ed.) NT 2021. LNNS, vol. 233, pp. 504–512. Springer, Cham (2021). https://doi.org/10.1007/978-3-030-75275-0_55

Using SWOT Analysis for More Efficient Communication and Performance at the Level of Public Administration

Iuliana Moisescu[1,2]($\boxtimes$)

[1] Polytechnic University of Bucharest,
Splaiul Independenței nr. 313, 6th District, Bucharest, Romania
iuliana_moise@yahoo.com
[2] Ministry of Culture, 22 Unirii Boulevard, 3rd District, Bucharest, Romania

Abstract. Various strategic management and planning tools and techniques are available for organizations to analyze their current situation, and one powerful yet simple technique used in strategic planning is the SWOT analysis. This type of analytical tool is also used at the level of public administration, as it offers the opportunity to identify the key elements necessary to fulfill the purpose for which those public authorities were created. Knowing the strengths, weaknesses, opportunities, and threats represents essential information that must underpin certain decisions necessary in a particular economic and social context to converge with the objectives assumed by government programs. Also, identifying the weaknesses through a SWOT analysis can lead to implementing the most suitable and significant strategy, which could optimize the activities carried out at the level of a public institution. Public institutions need improvements so a SWOT analysis could create the premises for more efficient communication and performance concerning public services.

Keywords: Public administration · SWOT · Improve · Management · Communication

1 The Use of a SWOT Analysis

The first step in any effective communication is to carefully and accurately identify the situation facing an organization. A situation may identify an opportunity to be embraced because it offers a potential advantage to the organization [1]. Strategic management is a set of managerial decisions and actions that determines the long-run performance of a corporation. The study of strategic management emphasizes the monitoring and evaluating of external opportunities and threats in light of acorporation's strengths and weaknesses to generate and implement a new strategic direction for an organization [1]. Different strategic factors like Strengths, Weaknesses, Opportunities, and Threats, known by their acronym, SWOT, should be analyzed by every organization, on various

I. Karabegović et al. (Eds.): NT 2022, LNNS 472, pp. 531–538, 2022.
https://doi.org/10.1007/978-3-031-05230-9_63

occasions, for better strategic decisions. SWOT analysis on the communication component was never developed, in the last five years, at least, at the level of the ministry. Usually, the management attention is held by the economic and budgetary situations, which decrease annually. Without the budget, the leadership can't establish new cultural projects or support successful ones. Every new minister wants to identify himself with a new project, but time doesn't allow them to finish new projects due to permanent governmental changes. The numbers speak for themselves, in the last five years, seven ministers have succeeded the ministry. The leaders neglect the fact that it is not necessary to innovate something, and the permanent changes make it even impossible. Sometimes it's better to optimize the process that doesn't excel or to improve what already exists. All the ministers are politically enrolled, so they look first for a good image and notoriety, probably for future electoral success. Thus, all the ministers want the best communication possible, not necessarily for the organization but to emphasize how good they are as leaders. From my point of view, for successful communication matters what kind of information you disseminate. A press release informing the public about creating a new platform for endorsement should be a real success both for a ministry and minister. The need for digitalization for less bureaucracy and easy connectivity of citizens with the public institution will produce good public reactions, could enhance the good image for both parties: the minister and the ministry. But to understand what subjects may bring and stimulate the positive public perception of any organization, it can use the tools that science has made available, respectively tools used in the field of management.

2   SWOT Analysis on Communication Component

"Strategical planning it is about the management of the total organisation, in order to create the future" [3]. A manager needs to analyze the internal and external environment of the organization to have a vision about the factors that can effectively build the strategic approach. To improve communication and the activities carried out by an institution, the experts recommend the most common tool used in strategic planning: the SWOT analysis. Today, citizens want quick access to information, and public services provided through the central public authorities are often inefficient and nonconforming to the current connectivity and efficiency needs. The efforts made by the government to streamline intra-governmental activity are constant. Thus, every year, the government develops various programs, strategies, projects, and strategic plans to optimize the work of ministries. Such an effort is the Institutional Strategic Panel 2021–2024 [4], carried out by the Ministry of Culture, the central public authority responsible for developing and implementing strategies and policies in the culture field. The analysis, carried out by the Public Policy Unit, started with the interpretation of the questionnaire, applied at the level of the minsitry, in February 2021, to which 48% of the total number of employees in the ministry responded.

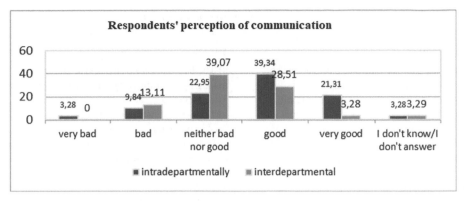

Fig. 1. Respondent's perception of communication [3]

As shown in Fig. 1, intradepartmental communication is considered good with a value of 39.34. At the interdepartmental level, communication is considered neither good nor bad, with a value of 39.07%. These values reveal the need for a considerable improvement in the inter-and intra-departmental communication process.

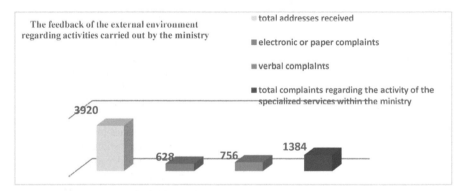

Fig. 2. The feedback of the external environment regarding activities carried out by the ministry

Analyzing data from the registers with addresses at the level of the communication office within the ministry indicates that 1384 citizens contacted the institution to notice that it has submitted a file for approval and has not received it. The analysis was carried out between January-December this year. Also, the citizens complained about not receiving information regarding the status of the file they submitted to the specialized committees. Thus, more than 35% of addresses received by the communication office are complaints related to the lack of communication of specialized services within the ministry. Using the experience gained in the communication office and the analysis of numerous strategies, plans, and internal documents, I carried out a SWOT analysis, presented in Table 1, focused mainly on the communication component.

Table 1. The SWOT analysis form MC

Strengths	Weakness
a) A prestigious public institution in the cultural field at the national level;	a) The lack of a centralized system for managing petitions/requests at the level of the institution;
b) Unchanged organizational structure;	
c)Existence of a core of civil servants with considerable experience in their field of activity;	b) Lack of an institutionally functional procedure regarding petitions;
	c) Lack of document approval management at the minister's cabinet-level;
d) An institution with representation at the level of each county;	d) Insufficient budget;
e) Existence of a system procedure for communication of information of public interest;	e) Lack of investment in informatic systems;
	f) Lack of investment in purchasing special software to streamline database management;
f) Management of EU funding & grants;	g) Lack of essential office equipment to streamline the activity;
g) Openness to collaboration at national and international level;	h) Lack of an IT department
h) The existence of institutions subordinated to the ministry with a high level of quality of the services provided;	i) Lack of adaptation to current connectivity needs;
	j) Weak and outdatedinformatic systems;
i) It offers, at the national level, diversified cultural services and offers, through the subordinated institutions;	k) Lack of a system for informing citizens about the status of documents submitted to specialized commissions;
j) Communicates verbal public interest information, through a specialized department;	l) Human Resources with the lowest salary from the budget system;
k) Permanent updating of the institution's website;	m) Lack of human resources in the most complex and essential structures;
	n) Legal barriers for promoting new approaches;
	o) Institutional priorities incompatible with the real needs of the institution;
	p) Difficulties in intradepartmental and interdepartmental communication;
	q) Lack of a crisis communication strategy at the level of the institution;
	r) The inefficiency of Dignitaries offices;

(continued)

Table 1. (*continued*)

Opportunities	Threads
a) Digitization of the institution; b) Creating a platform for submitting documentation and requests for obtaining approvals; c) Implementation of an integrated petition/request registration system; d) Working efficiently; e) Concluding partnerships with the private sector to finance cultural projects; f) Implementation of policies promoted by the EU; g) Adoption of European legislation on branches of activity in the field of culture; h) Human resources management adapted to the volume of activity carried out;	a) Lack of administrative capacity to implement digitization; b) Impossibility of adapting to the needs of public service beneficiaries regarding connectivity and efficiency; c) Austere budgets or budget cuts; d) Impossibility to cover losses during the Pandemic suffered by artists or organizations working in the cultural-creative sector; e) Unstable political system; f) Frequent changes in the management of the institution; g) Frequent legislative changes; h) COVID-19 Pandemic

Analyzing each strength, weakness, opportunity, and threat.

Strenghs

a) The Ministry of Culture is one of the most prestigious national institutions, being the national central public authority responsible for developing and implementing strategy and policies in the field of culture [5].
b) The organizational structure of the ministry hasn't been changed in the last 6 years.
c) More than 23% of the civil servants that work in the ministry have experience of more than 20 years.
d) According to the law, the MC has decentralized services at the level of each county. The ministry also has theaters, operas, or museums of national importance that operate in different development regions of the country.
e) The MC has a system procedure for the communication of information of public interest, which is approved and adapted to current legislation.
f) The MC's Project Management Unit is managing EU funding & grants.
g) The MC collaborated with local authorities for cultural programs and participate annually at cultural events like Biennale di Venezia or international book fairs.
h) The ministry has subordinate institutions internationally appreciated like the Madrigal Choir, George Enescu Philharmonic, National Theatre of Bucharest, Romanian Operas from Bucharest, Iasi, Timisoara, Cluj.

Weekness

a)–c) The documents enter the ministry through several channels, so often the addresses are duplicated or tripled, and without an integrated centralization of the documentation, the data analysis processes become inefficient. Documents approved by the minister are released but not according to their submission date.

d)–m) Due to the insufficient funds, the informatic system is outdated, with seven-year-old processors which operate currently. The Windows operating system is obsolete, and there is no specialized IT structure. The experts say that" the technology and project management are often identified as key catalysts and drivers enabling the public sector to morph, modernise and mimic the private sector in order to be more effective and efficient" [6]. Also, the low salaries are not attractive for the very well trained young people from different fields such as legal, economic, architecture, or archeology.

n) Modifying the legislation is time-consuming, as the normative acts have to be approved by different ministries, and blockages may occur.

o) The public administration must adapt to the government decision, and sometimes this means suspending your projects and getting involved with other projects.

p) The poor communication at the inter and intradepartmental level can originate in the personality of civil servants. Another cause may be the lack of authority of the heads of different specialized structures shown towards their subordinates.

s) Dignitaries offices are, sometimes, occupied by councilors who do not know the specifics of the ministry and lows and are constantly seeking advice from civil servants, although they are working as advisers. This aspect can lead to the inefficiency of the work of a department, which should process documents instead of providing specialized advice to dignitaries.

Opportunities:

a) The organization of the future will be the organization that will best adapt to the trends imposed by the new generations. Specialists who study human behavior say that the children of the Alpha Generation are familiar, even addicted to smartphones and tablets, and therefore the technology will be part of their communication and networking process [7]. In this context, organizations will be constrained to implement digitization systems to meet these needs.

b)–d) The covid-19 pandemic has strengthened the need to build a variant for the submission of documents to obtain approvals or approvals. There were times when public institutions were closed, and online became the only way citizens could contact an institution. An endorsement may have three or four bookshelves filled with documents that can also get done in electronic format. Debureaucratization may get accomplished by implementing electronic platforms that take over existing documents electronically. The need for connectivity is vital computer document processing systems are the most suitable options for the present and the near future, especially for public institutions.The electronic process system may bring efficiency to activities carried out by the public institution.

e)–g) As a European state, there is an obligation to adopt various policies developed by the EU, including those in the culture domain.

h) Because human resources are significant, along with other types of resources, improved management of human resources can generate a more efficient activity of a specialized structure and implicitly may boost the image of an institution.

Threats:

a)–b) As the budgets are reduced, especially now with the Corona 19 Pandemic situ-
       ations, it might get difficult to implement a digitized system that will optimize
       the activities carried out by a minister. By streamlining the processes carried out,
       chances are to improve the institutional communication as well.
c)–d) As the public budget is reduced year by year, a solution may be to identify
       external funding, such as UE funding, or to conclude public-private partnerships
       to implement various cultural projects.
e)–g) The alternation of political parties in government also brings changes at the level
       of ministries, as ministers are appointed politically. Depending on the governing
       programs of the political parties, different laws are changed.

3 Conclusion

Successful management must rely on excellent public programs and strategies, which
should always be doubled by effective managerial communication. Today, citizens
express whenever they have the opportunity discontent with the public system in general, the quality of public services provided by civil servants, the quality of professional
training for civil servants, and how public institutions in Romania are managed. Citizens,
who are the beneficiaries of the public services provided by public institutions, are, in
the end, the evaluators of the managerial activity of a public institution.

By analyzing the internal environment of the central public authority, various
conclusions have been obtained:

- the current human resources, with an experience of more than 20 years within the
  institution, are a strong point on which very few organizations can rely.
- according to the normative acts in force, each public institution is obliged to invest
  in the professional training of it's employees, a fact that may positively influence
  the development of the activity at the level of the specialized structure, in which the
  trained employees will operate.
- as there is no financial incentive for employees, there is a shortage of staff in various
  specialized structures. The shortage of staff determines, usually, a blockage in the
  current activity of the institution, and the overload of the employees with an increased
  volume of documents will end up with the request of the civil cervant to end his
  activity or be transferred to another institution.
- as the ministry is responsible for issuing opinions and approvals, a platform for submitting documents to obtain endorsements is, in my view, the best investment that
  a public institution can make. Only at the level of the Cultural Heritage Directorate,
  there are 19 commissions.

The efficiency of the processes carried out at the level of a central public authority
depends on the informatic system. As computer systems become obsolete and the speed
of the internet runs slowly, the duration for searching information necessary for the
activity, such as the legislative program, electronic registers, or data access, increases.

One way to streamline business and improve intradepartmental communication is to use free tools such as Google Drive or Dropbox, where you can upload documents that can be viewed and edited by all employees in a compartment or office. The advantage of this tool is that it can be accessible to employees who are in telework. Another way to improve the ministry communication is to promote the work of subordinate institutions, which bring to the public's attention large-scale cultural services, such as the George Enescu Festival or the concerts of the Madrigal Choir, well known at the international level. With the pandemic, many cultural activities and projects were closed. Most cultural organizations have suffered as a result of having to close down. The economic losses were very high, which led the government to provide financial support to organizations working in the creative cultural sector. However, the financial support was granted only for companies or authorized persons who had documented the economic losses suffered from the pandemic lockdown. Freelance artists and non-profit associations in the cultural sector did not receive financial support. Due to the lack of budget funds, the financial aid schemes proposed by the ministry for artists were postponed. Unfortunately, the artistic and cultural environment was left without financial support. The image of the institution was affected, although the citizens did not want to understand that a financial effort cannot be supported by a ministry but by the government who allocates funds to all central public authorities. The allocation of funds by a public institution to citizens will never be enough, regardless of the budgetary effort. For a better image of a public institution, it is important not to have an infinite number of press releases, but the most important is what you communicate. Citizens can change their attitude their perception about an institution not for the photos of leaders in festive backgrounds, but for efficient services. Citizens will not remember how many meetings a minister attended, but will not forget how long they waited for an answer or an endorsement from a ministry. So streamlining the services provided to citizens is the best way to communicate.

References

1. Smith, R.D.: Strategic Planning for Public Relations. Lawrence Erlbaum Associates, Mahwah (2002)
2. David Hunger, J., Wheelen, T.L.: Essentials of Strategic Management, 5th edn. Pearson Education, Inc., New Jersey (2011)
3. Hussey, D.: Strategic Management from Theory to Implementation, 4th edn. Butterworth Heinemann, Oxford (1998)
4. The Institutional Strategic Panel 2021–2024 for Ministry of Culture, December, 2021
5. Decision no. 90 of February 10, 2010, on the organization and functioning of the Ministry of Culture, December 2021. http://legislatie.just.ro/Public/DetaliiDocument/116380
6. Breen, L., Hannibal, C., Huatuco, L.H., Dehe, B., Xie, Y.: Service improvement in public sector operations – a European comparative analysis. Eur. Manage. J. **38**(3), 489–491 (2020)
7. Williams, A.: Meet Alpha: The Next 'Next Generation'. The New York Times newspaper, 19 September 2015. https://www.nytimes.com/2015/09/19/fashion/meet-alpha-the-next-next-generation.html. Accessed 14 Feb 2021

Design of Experiments, A Tool for Streamlining an Approval Process, Which Takes Place at the Level of Central Public Administration

Iuliana Moisescu[1,2(✉)] and Aurel Mihail Titu[3(✉)]

[1] Polytechnic University of Bucharest, Sp.Independenţei nr. 313, 6th District, Bucharest, Romania
iuliana_moise@yahoo.com
[2] Ministry of Culture, 22, UniriiBlvd., 3rd District, Bucharest, Romania
[3] Lucian Blaga University of Sibiu, 10 Victoriei Street, Sibiu, Romania
mihail.titu@ulbsibiu.ro

Abstract. The world of science offers numerous possibilities through development, modernization, testing, and applying knowledge to create quality, more efficient products or services. The public institution offers various services, but how they have carried the process needs improvement, as citizens are constantly expressing their dissatisfaction with the inefficiency of their activities. This paper aims to use tools used in the statistics field and apply them in public administration institutions. The role of this approach is to identify significant variables that can lead to the optimization of an intricate approval methodology, which is carried out at the level of a ministry. In our quest to improve a process, we will use the DOE statistical tool, used in areas like computer science, pharmacology, biochemistry and genetics, engineering, or medicine for the past 20 years, but hasn't been implemented in administration at its potential. This tool provides a possibility to run various determinations by a software program, which provides experimental data that can support the decision-making for process optimization.

Keywords: Public administration · DOE · Optimization · Management · Experimental data

1 Introduction

The starting point of DOE is considered the One Factor At a Time (OFAT), a scientific method used by Fisher at the beginning of 1920 [1], to determine the best condition for increasing the growing crops using different fertilizers, running experiments on different blocks of land. The importance of his experiments runs at the Rothamsted Experimental Station, in Hertfordshire, consisted of the approach by examining various factors simultaneously, and that statisticians should be consulted first in the design of experiments rather than at the end of testing.By using this tool, which has origins in mathematical modeling, an experimenter can obtain results that lead to the improvement of a system, a process, or a product [2]. Currently, the field of statistics is supported by

I. Karabegović et al. (Eds.): NT 2022, LNNS 472, pp. 539–547, 2022.
https://doi.org/10.1007/978-3-031-05230-9_64

companies that produce software programs such as MINITAB or Statgrafics [3], which provides support to plan, analyze, and interpret experimental data.The processes that take place at the level of a ministry have a high degree of complexity, which involves the correlation of normative acts, instructions, regulations, implementing rules, in the context of a large volume of documents, which must be processed in a limited time by a restricted number of civil servants, we considered implementing a methodology that has proven useful.

2   The Steps for Planning and Conducting an Experiment

Various processes that are carried out at the level of the central public authorities are not efficient, which generates numerous complaints formulated by the beneficiaries of the public services. The inefficiency of the processes is due to the implementation of a considerable number of normative acts, processing a large volume of documents in a limited time by a restricted number of civil servants. Thus, some of the processes executed at the level of the pubic institutions are not efficient enough, therefore need improvements. The design of experiments, often abbreviated as DOE, is an important statistical discipline and the most frequently used factorial designs are 2^k designs, that is designs involving k factors that are varied on two levels [4]. DOE is a tool offered by science that, so far, has not been used at the level of the Ministry of Culture.

- *Defining the objective*

Different factors can influence system performance [5], like a human, technological, material, and financial resources, impacting the processes carried out by central public authorities. Because these resources are limited, we used a statistical tool, which is a mathematical modeling component [6], called full factorial design.Under the control of the examiner, experimental determinations or experiments are used to define an investigation of a system.The investigation process starts with the elaboration of an experimental plan, in which the examiner should consider specific aspects, as the selection of factors, the determination of the levels of variation for each factor and the the answers, obtained from the examination of the process, must be expressed numerically. The role of experimental determinations is to optimize a phenomenon or process by generating and collecting as much information as possible, subsequently analyzing the results obtained from experimental determinations. Thus, the experimentation, respectively the investigation of the analyzed system, supposes a process that includes several stages.

- *Defining the measurements of the response variable*

The objectives functions established for the experimental design presented in Table 1. are to evaluate the value of the response functions. For the experiment, we selected two objective functions.

Table 1. The objective functions established for the experimental design

Name	Units	Analyze	Goal
Z1 - No. endorsements issued by the NCMC	No	Mean	Max
Z2 - The endorsement transmission time	Days	Mean	Min

Through the first objective function, named Z1, we want to identify the factors that can maximize the number of endorsements issued by the ministry.Through the second objective function, called Z2, we want to identify the factors that can minimize the endorsement transmission time to the beneficiaries.

● *Determine factors and levels*

A process of analysis was carried out with the specialists working within the Cultural Heritage Directorate, public managers, and the advisers within the public policy structure.

As a result, for the experiment was identified 36 variables that could have an impact on the process.

Table 2. The response variables

Name	Units	Low	High
A: The number of files submitted per month to the commission - Ndd	No	−1	+1
B: The number of people processing the commission's files - Nppd	No	−1	+1
C: Number of meetings organized monthly - Nso	No	−1	+1
D: Number of complaints for non-issuance of the endorsement- Nrn	No	−1	+1
E: Non-compliant or incomplete documentation - Di	No	−1	+1
F: The number of types of approvals in a single committee - Naoz	No	−1	+1
G: The period in which the endorsements are signed - Tsa	Days	−1	+1

Based on the information obtained both a priori and empirically regarding the approval process within the National Commission of Museums and Collections, seven influencing variables were selected, which are summarized in Table 2.

● *Selection of the experimental design type*

The experimental design type selected was the central factorial experiment composed of the second order. According to the literature, the central composed experiment is considered a complete factorial experiment of type 2^k, with 2-factor interactions. To carry out this experiment we used a specialized computer program such as Statgraphics. We selected two response variables and seven experimental factors, as is presented in the Tabel 3. The selected design had 128 runs, with 1 sample taken during each run.

- *Performing the experiment using the design matrix*

To carry out this experiment we used Statgraphics, a specialized computer program for analyzing data.

Tabel 3. The design matrix and signs for effects in 2^7 factorial designs

No.	Ndd	Nppd	Nso	Nrn	Di	Naoz	Tsa
1	1	0,5	1	1	1	1	1
2	84	0,5	1	1	1	1	1
...	...	...	...	...	...	...	...
13	84	0,5	1	1	1	1	122
14	84	0,5	3	5	1	1	1
...	...	...	...	...	...	...	...
127	1	0,5	1	1	1	1	122
128	84	3	3	5	6	9	122

Hence, the data, presented in Table 3, were introduced. The two-level factors model design was fit to the experimental results. Thus, the factorial design contains 128 determinations.

- *Analysing the experimental results*

After entering the variables, calculations were performed on the values of the objective functions Z1 and Z2, which were entered in the program, resulting in the following data.How the statistical models fitted to the response variables is presented in the Table 4. Models with P-values below 0,05, of which was 1, demonstrate that the experiment is appropriate, from a statistical point of view, at the 5,0% importance level.

Table 4. Statistical models fitted to the response variables.

Model	Z1	Z2
Model d.f.	28	28
P-value	**0,0081**	0,0671
Error d.f.	99	99
Stnd. error	1,0365309	7,064167475
R-squared	**35,71**	**30,14**
Adj. R-squared	17,53	10,38

Additionally, of importance is the R-squared results, which reveals, as a value measured in percent, the fluctuations of the response in our experimental model. Our R-squared results ranged between 30,14% and 35,71%.

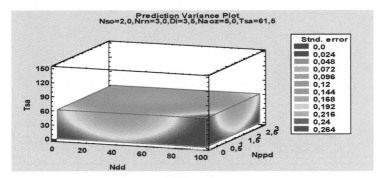

Fig. 1. The prediction Variance Plot

The plot shows in Fig. 1 the standard error of the anticipated reaction relative to the variables Ndd, Nppd, and Tsa, when the rest of the variables are kept constant. For our experiment, the lower values of the standard error are considerable because the smaller they are, the more accurately the average evaluated response indicates.

- *Analizing the Experiment for the Response Function Z1*

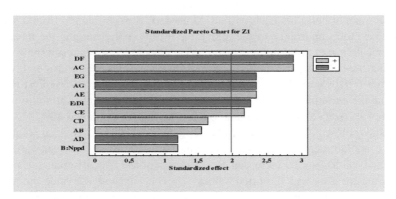

Fig. 2. The standardized Pareto Chart for the reponse function Z1

Figure 2 illustrates each calculated result in decreasing level of importance. For our model, respectively for the Z1 objective function response, are important the longest bars because they reflect the standardized effect, representing the calculated impact diverged by its standard error. Only the bars from the right side of the vertical line are relevant to our experiment, and they are, according to Fig. 2, seven variables interaction: DF, AC, EG, AG, AE, E: Di and CE.

Table 5. Analysis of variance for Z1

Source	Sum of Squares	Df	Mean Square	F-Ratio	P-Value
A:Ndd	0,048828125	1	0,048828125	0,05	0,8316
B:Nppd	1,423828125	1	1,423828125	1,33	0,2524
E:Di	5,080078125	1	5,080078125	4,73	**0,0321**
...	...	...	...	...	......
AC	8,251953125	1	8,251953125	7,68	**0,0067**
AE	5,486328125	1	5,486328125	5,11	**0,0260**
AG	5,486328125	1	5,486328125	5,11	**0,0260**
CE	4,689453125	1	4,689453125	4,36	**0,0393**
DF	8,251953125	1	8,251953125	7,68	**0,0067**
EG	5,486328125	1	5,486328125	5,11	**0,0260**
Total error	106,3652344	99	1,074396307		
Total (corr.)	165,4511719	127			

R-squared = **35,71200907%**
R-squared (adjusted for d.f.) = **17,52954698%**
R-squared (predicted) = 0,0%
Standard Error of Est. = **1,0365309**
Mean absolute error = **0,7313842773**

Table 5 presents the ANOVA table separations of the oscillations in Z1 into distinct components, separately, for each output. Afterward, it experiments, from a statistical point of view, the importance of the individual result, approximating the mean square with the calculation of the experimental error. For our model, seven results have P-values smaller than the value of 0,05, implying the fact that variables have considerably distinct values from zero, when the importance level is situated at the 95,0%. The R^2 datum reveals that the example as fitted represents 35,71200907% of the variability in Z1. R^2 also called the coefficient of determination, is always between 0 and 100%. Even if the value of R2 is below 50%, but each variable is significant from a statistical point of view, we can extrapolate pertinent deductions about the relationships between variables. For the specified values, more increased R^2 results denote fewer discrepancies between the experimental data and the fitted values [7]. The adjusted R2 value, known to be appropriate for analogizing examples with miscellaneous data set of different variables, represents 17,52954698%. The standard error of the calculation reveals that the standard deviation of the residuals is 1,0365309. The mean absolute error (MAE) at the value of 0,7313842773 suggest the moderate significance of the residuals. Considering the P-value is less than 5,0% represents a suggestion of a possible correlation at the 5,0% significance level. Figure 3 shows the estimated values of the objective function Z1 response to thevariables used in the experiment. In each plot, the selected variables alternate to their minimum and maximum values, while the other variables are kept stable at their central level.

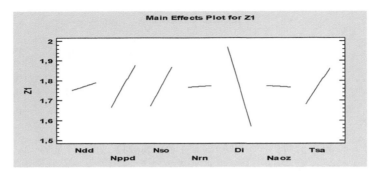

Fig. 3. The main effects plot for Z1

Figure 3 shows that the most important factors are the line. The larger the line, the more significant is the factor. In our case, for the Z1 objective function and according to the line length, the following factors are important: Di, Nppd, Nso and Tsa.

- *Analizing the experiment for the response function Z2*

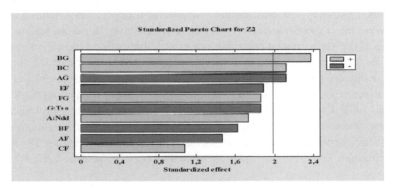

Fig. 4. The standardized Pareto Chart for the reponse function Z2

Figure 4 represents each calculated result in decreasing level of importance. For the Z2 objective function response are important the longest bars because they reflect the standardized effect, representing the calculated impact diverged by its standard error. Only the bars from the right side of the vertical line are relevant to our experiment, and they are three variables interaction: BG, BC, and AG.

Table 6. Analysis of variance for Z2

Source	Sum of squares	Df	Mean square	F-ratio	P-value
A:Ndd	140,28125	1	140,28125	2,81	0,0968
…	…	…	…	…	…

(*continued*)

Table 6. (*continued*)

Source	Sum of squares	Df	Mean square	F-ratio	P-value
AG	210,125	1	210,125	4,21	**0,0428**
BC	210,125	1	210,125	4,21	**0,0428**
BG	264,5	1	264,5	5,30	**0,0234**
...	...	...	...	...	...
Total error	4940,34375	99	49,90246212		
Total (corr.)	7071,96875	127			

R-squared = **30,14188941%**
R-squared (adjusted for d.f.) = **10,38403996%**
PRESS = 8258,605448
R-squared (predicted) = 0,0%
Standard Error of Est. = **7,064167475**
Mean absolute error = **4,832763672**

Table 6 presents the ANOVA table [8] separations of the oscillations in Z2 into distinct components, separately, for each output. Afterward, it experiments, from a statistical point of view, the importance of every result, approximating the mean square with the calculation of the experimental error. For our model, three effects have P-values smaller implying the fact that they are considerably distinct from zero at the 95,0% importance level. The R-Squared datum reveals that the example as fitted represents 30,14188941% of the oscillation in Z2. The adjusted R-squared datum, proper for analogizing examples with varied variables, is 10,38403996%. The standard error calculation indicates that the standard deviation of the residuals is 7,064167475. The mean absolute error (MAE) of 4,832763672 represents the medium value of the residuals. Since the P-value has a smaller value than 5,0%, it might represent a potential correlation at the 5,0% significance level.

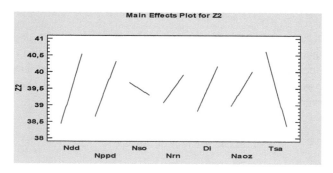

Fig. 5. The main effects plot for Z2

Figure 5 illustrates the estimated values of the objective function Z2 response to the variables used in the experiment. In each plot, the selected variables alternate to their minimum and maximum values, while the other variables are kept stable at their central level.The most semnificant factor is reprezented by the largest line, which, in our case, is the Tsa factor, followed by the Ndd and Nppd factors.

3 Conclusion

By applying mathematical modeling, supported by software specially created for statistics, we were able to identify the significant factors that influence the objective functions to optimize the approval process carried out byy a ministry. Following the experiment, the subsequent factors are important for the objective function Z1: Di, Nppd, Nso, and Tsa, and for the objective function Z2 Tsa, Ndd, and Nppd are significant. Experimenting provided us with information about the most statistically significant interactions between factors. Thus, to achieve a higher number of endorsements, the interactions between the Nrn-Naoz factors must be minimized, and the interaction between the Ndd-Nso factors must be maximized. The goal set for the Z2 objective function can be achieved only by maximizing the interaction between Nppd-Tsa and Nppd-Nso factors and minimizing the interaction between Ndd-Tsa factors. We must specify that mathematical modeling was never used to optimize a process at the level of a ministry. Process optimization should be the first goal of a manager when he is appointed. For a citizen, obtaining an endorsement in a short period remains a desideratum. This mathematical modeling can be applied to all 19 commissions carried out within the Cultural Heritage Directorate, one of the most significant components of the ministry and whose mission is to protect and save the quintessential cultural heritage values of a state.

References

1. Fisher, R.A.: The Design of Experiments, Oliver and Boyd, Edinburgh (1935)
2. Durakovic, B.: Design of experiments application, concepts, examples: state of the art. Periodicals Eng. Nat. **5**(3), 421–439 (2017)
3. Polhemus, N.W.: Process Capability Analysis: Estimating Quality. Taylor & Francis Group (2018)
4. Velten, K.: Mathematical Modeling and Simulation. Wiley-VCH (2009)
5. Oprean, C., Țițu, A.M.: Managementul calității în economia și organizația bazate pe cunoștințe, București, AGIR (2008)
6. Jankovic, A., Chaudhary, G., Goia, F.: Designing the design of experiments (DOE) (2021). https://www.sciencedirect.com/science/article/pii/S037877882100582X
7. Dean, A., Voss, D.: Design and Analysis of Experiments. Springer, New York (1999). https://doi.org/10.1007/b97673
8. Everitt, B.S.: The Cambridge dictionary of statistics. Cambridge University Press, Cambridge (2002)

Process Management in a Public Organization Providing Services to Citizens

Constantin- Dorin Olteanu[1,2(✉)] and Iuliana Moisescu[1,3]

[1] Faculty of Industrial Engineering and Robotics, Polytechnic University of Bucharest, 313, Splaiul Independenţei, Bucharest, Romania
`ocosti@gmail.com`
[2] Directorate for the Registration of Persons, 14 Câmpului Street, Sibiu, Romania
[3] Ministry of Culture, 22, UniriiBlv., 3rd District, Bucharest, Romania

Abstract. The scientific paper proposes a pragmatic perspective of presenting some theoretical and practical aspects regarding a new approach within the studied organization, a process-based approach. The point of view submitted is a small part of the possibilities to implement the process-based approach. For an overview, were presented general aspects of the process types. An important aspect was a realization of the personal analysis by categories of the identified processes and their classification in basic processes, central processes, and management processes. We determined the connections that appear between them. Based on these connections, we made a current map of the processes. The next step is to make some personal contributions regarding the improvement of the current situation regarding the map of the institutional processes within the studied organization, creating a new map of the processes with the brought changes. In the last part, some modeling methods used in modeling and optimizing organizational processes were presented. One direction of research will be the modeling of the processes within the studied public organization.

Keywords: Process · Process map · Management · Organization

1 Introduction

The developments of managing in public administration evolve in small steps, and resistance to change is high. Public organizations need to change and function more performing. This aspect appears in the context of the evolution of society, of alignment with the international context and needs, to work efficiently.

Implementing a quality management system in an organization requires a process-based approach. This requires the management of processes, their inputs-outputs as well as interactions between processes. The process-based approach appears in ISO 9001:2015 where it is a requirement for an efficient quality management system. The process-based approach can also be used, in the management system of a public organization providing services to citizens.

A process can be defined, according to ISO 9001: 2015, as "a set of interrelated or interacting activities, which transforms a set of input elements to achieve output

I. Karabegović et al. (Eds.): NT 2022, LNNS 472, pp. 548–556, 2022.
https://doi.org/10.1007/978-3-031-05230-9_65

elements that an internal or external client needs" [1]. Inputs and outputs can be tangible (equipment, materials, or components) or intangible (information or knowledge).

We can consider a process, a sequence of predefined activities, which by applying them achieve desired results, which combine the operations of the organization with the requirements of customers. In general, processes are cross-functional, horizontally, attached to the vertical hierarchical structure of the organization, and not just a single person is responsible for the entire process [2].

All processes must be correlated with the objectives of the organization and must design to add value, depending on the field of activity and the complexity of the organization. The process-based approach involves establishing the organization's processes to operate as an integrated and complete system. The management system integrates processes and measures to achieve the objectives, and the processes define the interdependent activities and verifications to provide the desired outputs.

2   Process Category Analysis and Proposal of a Process Map

The studied public organization has a hierarchical organizational form, with well-defined responsibilities at the level of each structure divided into services and compartments. The activities within the organization are periodically analyzed, and new opportunities are sought to improve the quality of services. These improvements can be found in the rules of organization and operation, which are reviewed annually or as often as necessary. Currently, the quality management system within the studied organization does not take a process-based approach. Our proposal for a process-based approach respects the principles of quality management presented in ISO 9000: 2015 and ISO 9004: 2018. The process-based approach has the advantage of controlling the link between individual processes as well as control and interaction between processes. The processes that we have identified within the organization are based on the organization and functioning regulation in which the activities corresponding to the services and compartments within the organization are pointed out. Within the studied organization there are and interact three types of processes: management processes, central processes, basic processes or support processes.

Management processes are institutional, strategic processes that coordinate the development of central processes and basic processes. These processes involve actions on setting the goals of the organization, setting responsibilities, establishing the possibilities and means of communication, identifying the need for human and material resources, infrastructure and working environment, processes related to the evaluation, analysis, and improvement of results. The management processes were divided into:

- General administration processes;
- Executive administration processes;
- Civil status administration processes by the leader of the civil status service;
- Processes of administration of the service record of persons by the leader of the service record of persons.

The central processes are the processes of direct realization of the services following some requirements of the citizens and of the institutions with which they collaborate. Central processes create value and determine the quality of services. The central processes were divided into:

– Coordination and Control processes within the civil status process;
– The processes of processing and entries the mentions in the Civil Status Registers within the civil status process;
– The processes of records of persons within the process records of persons the general.

The base or support processes are the processes that, through punctual, discontinuous, and limited interventions, support the main processes and the managerial processes, ensuring their normal and efficient functioning. Support processes are professional activities that serve the base activity. The support processes are divided into:

– Contentious-Legal Processes;
– Human resources management processes;
– Public relations processes;
– Information systems management processes;
– Financial processes - accounting;
– Material technical assurance processes;
– Registration processes.

After identifying and classifying the processes, we make a map of the processes.

The process map makes, in a visual way, a presentation of the existing processes in the organization. The processes map illustrates the process flows that influence quality. This tool allows the development of vision and understanding of the dynamics of the processes and their correlations in a global way within the studied organization. In this way, we can highlight ways to improve within a process and identify the constraints to which it is subject. A process map is used when changes occur in the company and when you want to explain how the organization works. We made the map of the processes starting from the base processes, then the central ones, to the management processes realizing the connections that appear between them. A representation of the process map is shown in Fig. 1.

Process-based approach:

– Allows focusing on integration, alignment, and effective connection of processes with planned objectives and targets;
– It also allows the organization to focus on improving the effectiveness and efficiency of processes;
– Facilitates performance which, in turn, provides customer confidence in the organization's ability to deliver quality;
– Promotes a transparent and constant flow of operations within the organization;
– Contributes to reducing costs and shortening operational cycles through the efficient use of resources;

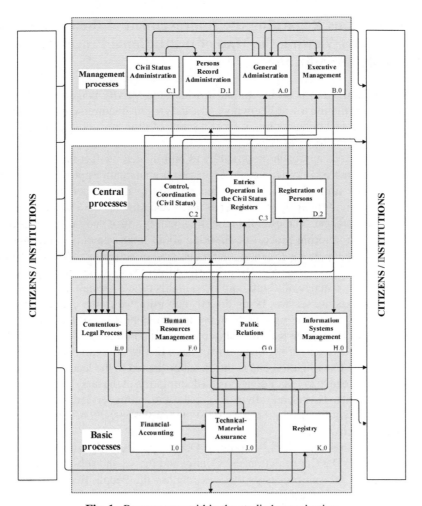

Fig. 1. Process map within the studied organization

- Allows interest focused on improving process results and generating consistently and predictably improved results;
- Facilitates involvement, clear definition of responsibilities, and authorities of employees.

After identifying and classifying the processes, we make a map of the existing process. The processes map illustrates, visually, the process in the organization. The processes map shows the process flows.

3   Contributions Regarding the Improvement of the Current Situation by Viewing the Map of Institutional Processes Within the Study Organization

The studied organization is an institution in which the activity is carried out in good conditions the orientation towards solving and satisfying the requirements of the citizens is a priority. As in any organization, it faces a series of problems, obstacles, challenges that influence the quality of activities.

The process-based approach proposed by us highlighted the fact that at the level of the organization, there is no dedicated office with human staff employees specialized in quality. At this time, the control, verification, and quality of activities within the organization are monitored by the leader of each department, the executive director, and the general manager. The establishment of a quality office and the emergence of a separate quality monitoring process could bring about improvements in the quality as a whole, following the interaction and interconnection of activities in the organization. The beneficiaries of this office would be primarily citizens who come with requests to the organization. The existence of a staff employed in this office, specialized in quality, not belonging to an existing service, being directly subordinated to the executive director, would bring objectivity and impartiality in the specific quality control activity.

The activity of the People Records service is carried out in optimal conditions, but at this moment this service has only one compartment. Its division into two compartments Unique Office and Coordination and Control, subordinated to the leader of the service, we believe would lead to a clearer division and delimitation of the tasks of each employee in these departments. This clearer division of tasks within this service, we consider that would bring an extra quality in the activity to the employees, the beneficiaries of this quality being the citizens.

Following this restructuring, sub-process D.2 People Records will be divided into two sub-processes, D.2 Coordination and Control and the second sub-process D.3 One-stop-shop. Part of the sub-processes that were within the People Records department will return to the new department, One-stop-shop.

We also consider that the Financial, Accounting, and Technical - Material Insurance department can be divided into two departments subordinated to the Deputy Executive Director. This division would lead to a clearer division and delimitation of the tasks of each employee in these departments.

We consider that the Contentious-Legal, Human Resources, and Public Relations Department can be divided into two, a Contentious-Legal department and another Human Resources and Public Relations department. And these two compartments would remain subordinate to the deputy executive director.

Of course, these divisions of compartments will not affect the existing interactions and collaborations within the old compartments, instead, this division comes with a clearer delimitation of tasks, the management of the new compartments will be more efficient, the quality of services will increase.

In line with the proposed improvements, we redraw the process map, presented in Fig. 2. The new map shows the reorganization of the processes and the connections between them.

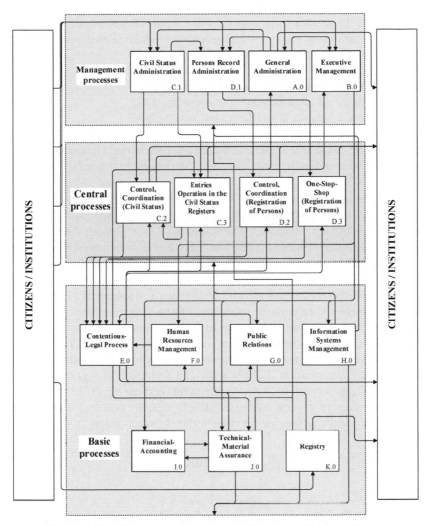

Fig. 2. Map of the proposed processes within the studied organization

The process-based approach proposed by us highlights the need to form a dedicated office with quality employees. Within the Persons Records service, we consider that this service can be divided into two compartments Unique Office and Coordination and Control, subordinated to the leader of the service this would lead to a clearer division and delimitation of the tasks of each employee in these departments.

4 Modeling Methods Used in Modeling and Optimizing Organizational Processes

Pointing and highlighting processes is an important step in implementing a quality management system in an organization. Implementing the process-based approach within an organization puts the customer in the spotlight his requirements and needs will become a priority in the organization's policies. Using process modeling comes with the following advantages:

- The operation of the system will be better understood;
- The relevant data will be separated by using the models;
- A model allows testing the sensitive points of the system;
- Changes can be applied very quickly;
- Changes can be tested more easily on the model than on real system testing;
- Changes can be tested more easily on the model than on real system testing;

For the subsequent realization of modeling the processes within the researched organization, we will focus on one of the modeling languages presented below.

CIMOSA (Computer Integrated Manufacturing Open System Architecture) is a modeling language that covers organizational, functional, informational, and resources aspects at different levels within the organization with a detailed model [4].

The IDEFO language was derived from the SADT (Structured Analysis Design Technique) modeling language, developed by Douglas T. Ross and SofTech, Inc. IDEF0 is a graphical modeling method for modeling actions, decisions, and activities within an organization or system. Its purpose was to describe, through graphical visualization, the functional aspect of a system. Being a visual way, modeling with IDEF0 allows an easier understanding of the interactions between processes and functional aspects of any system within the organization. IDEF0 includes a definition of a graphical modeling language and a description of a model development methodology [5].

The graphical representation used in the IDEF0 language is in the form of a rectangle that represents a particular specific function within the studied system (Fig. 3).

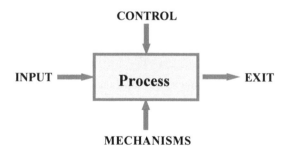

Fig. 3. Process representation

IDEF1X is an information modeling technique that can be used as a tool in the design of relational databases. The IDEF1X modeling method uses data as a resource [6].

IDEF3 is a graphical modeling method that can be considered a way of describing processes. The method simulates the sequence of actions and the connections between them in the processes. IDEF3 complements IDEF0 and can be used to build process models and as a process creation method [7].

GRAI and GRAI/GIM. The GRAI method with the GRAI grid and GRAI networks focuses on the decision-making aspects in the organization. This language was later extended to cover the functional, informational, and organizational perspectives of the organization in the form of the GIM method. The models produced are static descriptive models [8].

PETRI Networks. PETRI networks are powerful mathematical graphical tools for representing and analyzing the events of dynamic systems characterized by parallelism and synchronization. Various extensions of PETRI networks, in particular top-level and object-oriented networks, can be used to model the organization model functional, informational, and resource aspects of the organization. The proposed models are analytical models that support quantitative and qualitative systemic analysis methods and are executable [9].

00A (Object-Oriented Modeling) and OMT. Object-oriented methods are increasingly, used to model the dynamic and static aspects of information. They can be used at the requirements and design levels, but only cover informational perspectives [10].

IEM (Integrated Enterprise Modeling). It is a method based on object-oriented modeling. It uses an IDEFO block, in which inputs and outputs are objects such as commands, products, and resources [11].

ARIS (Architecture for Information System). It is first and foremost a reference architecture for the analysis of information systems consisting of a functional perspective, a command, an informational, and an organizational perspective. The modeling language has many elements in common with the CIMOSA language [8].

In the above presentations, an attempt was made to make a comparative analysis of the modeling languages, to later choose a method of modeling the processes within the studied organization.

5   Conclusion

This paper aims to identify the processes and make the process map in a public organization subordinated to a county council. We approached the topic of process management within the studied organization by identifying and classifying processes.

By performing process analysis, we drew up a map of the existing processes with the current situation within the organization. The process map thus made illustrates visually existing processes in the organization. In the next step, we have contributed to improving the current state of the process map. In the last part, some methods used in modeling and optimizing organizational processes were briefly presented.

The process-based approach proposed by us highlights the need to form a dedicated office with quality staff. Within the Personnel Records service, we consider that this service can be divided into two compartments, One-Stop-Shop and Coordination and Control, subordinated to the leader of a service. This reorganization leads to a clearer division and delimitation of the tasks of each employee in these departments.

In this paper, following the documentation and collaboration with the management and colleagues from the research institution, we managed, as a personal contribution, to identify and describe the existing processes. I mention that at the moment, within the organization, there is no process-oriented approach. Also, as a personal contribution is making the process map that presents a current situation and following the proposals made to obtain improvements, the map of the processes has been redone with the corresponding modifications. As a later direction, we will continue to model the processes with a method that we will establish later.

References

1. Tricker, R., Bruce, S.L.: ISO 9001:2000 in Brief. Butterworth-Heinemann, Oxford (2001)
2. Lee, R., Dale, B.G.: Business process management: a review and evaluation. Bus. Process Manage. J. **4**(3), 214–225 (1998)
3. Rooda, J.E.: Modelling Industrial Systems: Systems Engineering Group. Eindhoven University of Technology, Eindhoven (2000)
4. Cimosa, A.: Cimosa - a primer on key concepts, purpose and business value (1996)
5. IDEF0, F.I.: Integration Definition for Function Modelling. National Institute of Standards and Technology (1993)
6. IDEF1X, F.I.: Integration Definition for Information Modelling. National Institute of Standards and Technology (1995)
7. IDEF3, F.I.: Process Description Capture. Method Report. National Institute of Standards and Technology (1995)
8. Vernand, F.B.: Enterprise modelling languages. In: Proceedings of ICEIMT 1997. Italia: International Conference on Enterprise Integration and Modelling Technology (1997)
9. Călin, S., Popescu, T., Sima, V., Jora, B.: Conducerea adaptivă și flexibilă a proceselor industriale. București: Editura Tehnică (1988)
10. Oprean, C., Kifor, C.: Arhitecturi și modele generice în sisteme CIM. Tehno 98, Timișoara (1998)
11. Williams, T., et al.: Architectures for Integrating Manufacturing Activities and Enterprises. Computers in Industry (1994)

The Influence of Covid-19 Pandemic on Digitalization of Medical Services in Montenegro

Ivana Katnić[1](✉) and Anđela Jakšić-Stojanović[2]

[1] Faculty of International Economics, Finance and Business, University of Donja Gorica, Oktoih 1, Donja Gorica, 81 000 Podgorica, Montenegro
ivana.katnic@udg.edu.me
[2] University of Donja Gorica, Podgorica, Montenegro

Abstract. Despite the significant number of negative consequences caused by COVID 19 pandemic on the global level, the fact is that it has simultaneously stimulated the process of digitalization not only in various business segments, but also in everyday life. Due to numerous lock downs, restrictive measures and fear of pandemic, people have started to use modern technologies much more intensively realizing their numerous advantages, especially in terms of saving time and money. The most intensive changes were particularly noticeable in developing countries, such as Montenegro, especially in the sectors of health and education. This paper deals with the influence of COVID-19 pandemic on digitalization of medical services in Montenegro.

Keywords: COVID-19 · Digitalization · ICT · Medical services · Montenegro

1 Introduction

COVID-19 pandemic has had an extremely negative impact on the economy on the global level. There has been an extremely decline in economic activity in Montenegro as well, especially due to the fact that Montenegrin economy is dominantly based on the tourism industry and this industry has faced one of the greatest crises in history. But in the same time, the pandemic has had a major impact on businesses, society and the economy in general and has significantly caused the need for digital transformation of companies.

The fact is that many companies in Montenegro did not recognize the importance of digitalization until they faced new circumstances as a result of the pandemic. But, due to a pandemic, it became clear that digitalization has become a necessity which enables companies to adapt to new circumstances such as: working from home, online education, online shopping, using technologies in everyday life etc.

The most intensive changes have taken place in the health sector, which faced many challenges, but, despite many unsuccessful attempts of digitalization in the last decade, has digitized the most important segments of its business in a very short period and at quite satisfactory level.

I. Karabegović et al. (Eds.): NT 2022, LNNS 472, pp. 557–561, 2022.
https://doi.org/10.1007/978-3-031-05230-9_66

2 Digitalization of Medical Services in European Countries

In many European countries like the UK and the Netherlands, primary care has been 'paperless' for many years and a lot of medical services is already digitalized [1]. Electronic health records, e-prescribing, e-referral, etc. are also widely used in many medical systems. It should mention widely used SMS appointment reminders, phone check-in, etc. which have significantly increased not only efficiency of the medical system, but also the safety of patients, that was extremely important during a COVID-19 pandemic [2–4]. Also, there is a significant increase of e-mail consultations [5], although some concerns regarding legal liability, confidentiality and some other issues appeared [6]. In order to overcome these challenges, some countries have already introduced new legal requirements in order to define and solve all possible problematic issues [7].

Also, numerous online patient portals and online applications are being widely used especially regarding appointment booking, ordering repeat medications and viewing test results [8]. Thanks to these portals, patients may download relevant medical documents needed for work, travel, etc. without going to medical institution which significantly safes their time and money. Remote, virtual or tele-consulting has become reality as well, bringing a lot of advantages to medical staff and patients as well [9]. In some areas, such as mental health, the provision of medical services has completely moved to the virtual sphere, and all trends suggest that this type of service provision will be dominant in the future, while there are some areas in which the movement to virtual sphere is very limited, due to the nature of medical services.

Digitalization of health services has already both many advantages for patients- by saving their time and money (through lowering direct and indirect costs of medical services) and enabling them to prevent lifestyle changes and facilitate disease self-management [10] (Fig. 1).

Fig. 1. Digitalization of health services: advantages, source: authors

On the other side, there are some challenges that should be faced in the future such as possible workload of medical staff, legal liability which represent a very complex issue, as well as the issues of confidentiality and patients' safety (Fig. 2).

Fig. 2. Digitalization of health services: challenges, source: authors

3 Digitalization of Medical Services in Montenegro

Covid-19 pandemic has significantly encouraged the process of digitalization in Montenegro. The most intensive changes were particularly noticeable in the health sector.

During the pandemic, the most online services were created by the Health Insurance Fund.

These are: e-Ordering (prescription extension), e- Insurance (health insurance) and e-Pharmacy (availability of medicines in all pharmacies that have a contract with the Health Insurance Fund). Additionally, e-Finding service has been created, which enables patients to see the results of biochemical laboratory analyzes via the Internet, as well as e-Recipe service, which enables patients to see the prescribed and realized recipes. Although the service e-Schedule for online scheduling of visits to doctors in health centers existed before, this service actually came to life at the time of the pandemic. Also, the e-serviceMedical Commission was created, which enables patients to have an overview of medical commissions by cities in Montenegro, as well as e-serviceMedicines that displays a list of prescription medicines and medicines used in healthcare facilities.

The process of digitalization of medical services in Montenegro is realizing very fast, so at the moment there are intensive works on the further digitalization of primary and secondary health care. It is also important to mention that the digitalization of the Clinical Center, which represents the only tertiary health care institution in Montenegro, is in progress. Additionally, the digitalization of emergency medical care is also in the final phase, so it is quite expected to have a comprehensive e-platform that will optimize and facilitate the flow and management of processes at all three levels.

Regarding the Covid-19 pandemic and the process of vaccination, in the first phase the national digital COVID certificate was introduced, while its connection withthe European Digital Certificate is in the final phase and it is expected to be finished until the end of the year.

4   Conclusion

Pandemic COVID-19 has significantly improved the digitization of business in Montenegro. The biggest strides have been made in the field of medical services. Many new e-services have been introduced. Additionally, patients began to use those e-services that were previously available to them, but have not been used before the pandemic. There are still intensive works on the further digitalization of medical services and creation of a comprehensive e-platform that includes primary, secondary and tertiary health care institution is in the final stage which may bring a lot of benefits to different stakeholders, primary patients.

References

1. European Commission: eHealth adoption in primary healthcare in the EU is on the rise (2021). https://digital-strategy.ec.europa.eu/en/library/ehealth-adoption-primary-healthcare-eu-rise. Accessed Dec 2021
2. Schwebel, F.J., Larimer, M.E: Using text message reminders in health care services: a narrative literature review. Internet Interv. **13**, 82–104 (2018). https://doi.org/10.1016/j.invent.2018.06.002
3. InTouch: New check-in app helps social distancing measures and infection prevention at Poole Hospital. https://www.intouchwithhealth.co.uk/patient-mobile-check-in-poole-hospital/. Accessed Dec 2021
4. Greenhalgh, T., Jimenez, J.J., Prather, P.A., Tufekci, Z., Fisman, D., Schooley, R.: Ten scientific reasons in support of airborne transmission of SARS-CoV-2. Lancet **397**, 1603–1605 (2021). https://doi.org/10.1016/S0140-6736(21)00869-2
5. Newhouse, N., LupiáñezVillanueva, F., Codagnone, C., Atherton, H.: Patient use of email for health care communication purposes across 14 European countries: an analysis of users according to demographic and health-related factors. J. Med. Internet Res. **17**, e58 (2015). https://doi.org/10.2196/jmir.3700
6. Antonio, M.G., Petrovskaya, O., Lau, F.: The state of evidence in patient portals: umbrella review. J. Med. Internet Res. **22**, e23851 (2020). https://doi.org/10.2196/23851
7. Huygens, M.W.J., Swinkels, I.C.S., Verheij, R.A., Friele, R.D., Van Schayck, O.C.P., De Witte, L.P.: Understanding the use of email consultation in primary care using a retrospective observational study with data of Dutch electronic health records. BMJ Open **8**, e019233 (2018). https://doi.org/10.1136/bmjopen-2017-019233
8. Essen, A., Scandurra, I., Gerrits, R., Humphrey, G., Johansen, M.A., Kierkegaard, P.: Patient access to electronic health records: differences across ten countries. Health Policy Technol. **7**, 44–56 (2018). https://doi.org/10.1016/j.hlpt.2017.11.003

9. Walker, R.C., Tong, A., Howard, K., Palmer, S.C: Patient expectations and experiences of remote monitoring for chronic diseases: Systematic review and thematic synthesis of qualitative studies. Int. J. Med. Inform. **124**, 78–85 (2019). https://doi.org/10.1016/j.ijmedinf.2019.01.013
10. Pagiliari, C.: Digital health and primary care: Past, pandemic and prospects. J. Global Health. **11**, 01005 (2021)

Digitalization and Implementation of Modern ICT in Tourism and Creative Industries on the Example of the Cultural Routes

Nenad Vujadinović[1]([✉]) and Andjela Jaksic-Stojanovic[2]

[1] Faculty of Arts, University of Donja Gorica, Oktoih 1, DonjaGorica, 81 000 Podgorica, Montenegro
nenad.vujadinovic@udg.edu.me

[2] University of Donja Gorica, Podgorica, Montenegro

Abstract. Creative industries have extremely important role in the global economy and they represent one of the key drivers of development of the world's most developed countries. They have an increasing share in GDP and significant participation in employmenton a global level. Additionally, they significantly contribute to the sustainability and inclusive growth. The creative industries represent the heart ofthe tourism industry, especially in the last decade and, according to trends and perspectives on the global tourism market, this influence will be even greater in future.The process of digitalization and implementation of modern ICT plays an extremely important role in tourism and creative industries and this paper analyzes their influence on the quality of tourist offer and level of tourists' satisfaction on the example of the cultural routes, especially through the prism of experiences.

Keywords: Creative industries · Cultural routes · Tourism · Digitalization · ICT

1 Introduction

International tourism is undergoing a serious transformation- on the supply and demand side. Regarding the demanding side, the tourists' expectations have been significantly changed. Tourists are seeking for authentic experiences, trying to make a strong connection with particular destination itself, its history, tradition, culture, customs, local people etc. They are looking for products and services which will completely meet their needs and expectations. Additionally, they are much better informed - they collect the information about the destination from different sources before and during the trip, and they have the possibility to share their experiences and impressions with other tourists after the trip [1].

On the supply side, there are significant changes as well. There are new innovative tourist products specially designed according to needs and expectation of tourists. The rapid development of the Internet and the popularity of social networks influenced many stakeholders in tourism and hospitality industries and changed the way they function [1]. Tourist companies recognize the importance of using digital marketing in order not only to attract, but to keep customers as well.

I. Karabegović et al. (Eds.): NT 2022, LNNS 472, pp. 562–567, 2022.
https://doi.org/10.1007/978-3-031-05230-9_67

In conditions of increased competition, stakeholders in tourism industry make great efforts to identify and implement strategies that will differentiate them from the competition and enable them to create an attractive image on the global tourist market. The continuous improvement of quality of tourist offer and increasing the level of tourists' satisfactionbecamethe imperative in tourism industry.

The creative industries represent the heart of tourism development, especially in the last decade and, according to trends and perspectives on the global tourism market, this influence will be even greater in future.

2 Tourism and Creative Industries

The creative industries have an increasing share in GDP, as well as extremely important role in the global economy [2]. They have a significant participation in employment and represent a key driver of development of the world's most developed countries [3]. This concept was promoted in 1990s as a result of policies introduced by developed countries governments which tended to promote culture, the technology-intensive sectors, entertainment etc. [4, 5].

According to some relevant statistics, the value of the global market for creative goods doubled from $208 billion in 2002 to $509 billion in 2015 which represents an impressive growth [6]. Additionally, developed economies such as US or UK significantly increased the values and shares of exports of goods related to the creative industries with a significant year average growth rate in the same period [7]. In 2018, the global market value of the creative industries was estimated at $1.3 trillion, and according to some relevant predictions, the global annual growth rates will be between 5 and 20% in OECD countries [8].

One important issue regarding the creative industries is the fact that they significantly contribute to the sustainability and inclusive growth dominantly because of the wide range of its diverse activities [9].

In the last decade, the creative industries have an extremely important role in tourism development. They significantly contribute to the improvement of quality of tourist offer – at the national, regional and international level. They lead to creation on new innovative, specially design products and services as well as creation of tourist attractions. They contribute to more balanced regional development and extension of tourist season. Also, they significantly influence the level of satisfaction of tourists through creation of authentic unique experiences.

3 Digitalization and Implementation of Ict in Tourism and Creative Industries

The process of digitalization and implementation of modern ICT plays an extremely important role in tourism and creative industries. The increasing emphasis on the global tourist market is being placed on "theexperiences', so the concept of "3S" (*sun, sea, sand*) is replaced with the concept of "3E" (*Excitement, Entertainment, Education*) [10]. In order to base tourism and creative industries development on principles of sustainability,

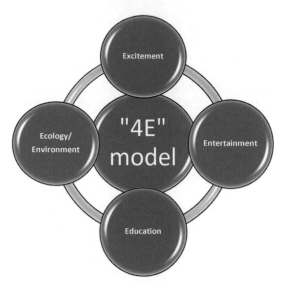

Fig. 1. "4E" model of tourism and creative industries development, source: authors

additional "E" should be added (*Ecology/Environment*). In that sense, the new model "4E" is proposed (Fig. 1).

Experiences represent the heart of tourism and creative industries. The process of creating experiences is quite complex and multidimensional. In fact, it comprises four main elements which are: tourism products, customer services, stories and interpretation (Fig. 2).

Fig. 2. The process of creating experiences in tourism and creative industries

The last two are extremely important from the point of view of tourism and creative industries. One of the main reasons is the fact that the story and its interpretation may transform tourist products and services in authentic, unique, lasting and memorable experiences. They unable tourists not to "consume" the tourist product or service, but to be a part of it i.e. to be an integral part of the experience.

The importance of these elements will be analyzed on the example of The Routes of the Olive Tree and the Old Olive Tree in Bar which is the integral part of this route.

The traditional story about the Old olive tree presented by the tourist guides and described on the websites, portals, social networks etc. would sound like this:

"The Old Olive tree in Bar often called Old Lady is the most famous tree in Montenegro and a great tourist attraction at the same time. It is located in Mirovica, just next to the Bar-Ucinj highway. The tree is one of the oldest in Europe, as well as in the whole world, and has been under protection of the law since 1957, when it was declared a

natural monument. In its more than 2200 years history, the olive tree suffered fires twice, managing to recover each time, and it still bears fruit today.According to a legend, this tree was used as a means of calming down of blood feuds when this phenomenon was plentiful in Montenegro, as well as the place where families, brotherhoods and armies came to settle their differences[11].*"*

But, the modern tourists would rather hear this story from The Old lady. It may sound like this:

"Dear Children, Come. Come Closer. It's Cold Outside, and My Branches, Though Old, Can Still Keep You Warm.

You know, they call me the old lady. I don't not remember the exact date or even year of my birth! It was so long ago! More than 2,200 years ago. Or even 2700. God knows! But I remember that it was winter, just like now. On the night of my birth, the wind was whispering, and the waves were dancing gracefully.

Oh, look out, put those candles away. I almost lost my life twice in the fire. There were some people here playing cards, they were careless, they threw a match and a fire broke out. Even now I shudder at the force of the fire. The flames were everywhere, raging like a dragon. Have a look! The scars are still there. I thought I wouldn't survive. But it seems that fate wanted otherwise.

Anyway, we won't talk about it anymore. Let us return to our story. They call me the old lady, and you know that old ladies like to tell beautiful stories. And I am really excited to tell you so many stories. Some of them may sound like fairytales, but believe me, they are all true. I will tell you magical stories about couples in love whispering the most beautiful and sweetest words, just there where you stand, stories about revenge, anger, reconciliation and love that took place under my branches, stories about the past that shape our present and future as well [11].*"*

This kind of presentation would be definitely more interesting for tourists because of the fact that it involves personal experience. Because of this personal element, it is more credible and, as such, it may provoke much more intense emotions among tourists. It is informative and educative, but in the same time dynamic, dramatic, exciting, located somewhere between reality and imagination. And the most important thing- it creates authentic experience.

But, the stories does not refer only to the written or spoken spoken word. Indeed, in order to achieve full effect and complete the tourists' experiences, visual stimuli should be introduced. In this segment, the implementation of modern ICT technologies may play a crucial role.

So, the Old Lady could lead the tourists virtually through different periods of Montenegrin history, they could enjoy in legends and beautiful fairytales from its past, in whispering of lovers under the olive tree and their promises about the eternal love, they could learn more about the phenomenon blood feuds, terrifying law of vendetta among South Slavic peoples. Also, tourists would be given the opportunity to participate virtually in the process of olive cultivation and growing, the production of olive oil and olive-based cosmetics products etc.

Interactive maps that consist of all important information regarding particular cultural route (history, interesting tourist spots, fairytales and legends from the past, food and accommodation services etc.) may improve the quality of tourist offer before, during and after the visit, while virtual tours, the elements of virtual and augmented reality etc. could enrich the tourists' experiences and increase the level of their satisfaction.

4 Conclusion

In conditions of increased competition as well as dynamic changes on the global tourist market, the concept "4E" (*Excitement, Entertainment, Education, Ecology/Environment*) in tourism and creative industries becomes especially important in order to create high quality innovative tourist products and services on the global tourist market.

The process of digitalization and implementation of modern ICT plays an extremely important role in tourism and creative industries because it enables creation of intensive, authentic, unique, lasting and memorable experiences which represent the heart of tourism and creative industries. Such experiences may significantly improve the quality of tourist offer and increase the level of tourists' satisfaction.

Additionally, the process of digitalization and implementation of modern ICT in tourism and creative industries may differentiate destinations from their competition and enable them to create an attractive image on the global tourist market.

References

1. Jakšić-Stojanović, A., Drakić, M., Buhalis, D.: Social networks strategy. In: Buhalis, D. (ed.) EncyclopediaofTourism Management and Marketing. Edward Elgar Publishing (2022, forthcoming)
2. Landoni, P., Dell'era, C., Frattini, F., MesseniPetruzzelli, A., Verganti, R., Manelli, L.: Business model innovation in cultural and creative industries: insights from three leading mobile gaming firms. Technovation 92–93 (2020)
3. Cooke, P., De Propris, L.: A policy agenda for EU smart growth: the role of creative and cultural industries. Pol. Stud. **32**(4), 365–375 (2011)
4. Caves, R.E.: Creative Industries: Contracts Between Art and Commerce. Harvard University Press (2000)
5. Lampel, J., Germain, O.: Creative industries as hubsofneworganizationalandbusinesspractices. J. Bus. Res. **69**(7), 2327–2333 (2016)
6. UNCTAD: Creative economy outlook: trends in internationaltrade in creativeindustries (2018). https://unctad.org/system/files/official-document/ditcted2018d3_en.pdf. Accessed Nov 2021
7. World Bank: TCdata 360 (2020). https://tcdata360.worldbank.org/indicators. Accessed Nov 2021
8. UNCTAD: Creative economy outlook: trends in internationaltrade in creativeindustries (2018). https://unctad.org/system/files/official-document/ditcted2018d3_en.pdf. Accessed Dec 2021
9. UNESCO: Culturalandcreativeindustries in the face of COVID-19: aneconomicimpactoutlook (2021). https://unesdoc.unesco.org/ark:/48223/pf0000377863?posInSet=1&queryId=18d8b725-72cd-4018-ad79-bfdd0ee274e4. Accessed Dec 2021

10. Orlandic, M., JaksicStojanovic, A.: Implementation of new technologies in the promotion of the cultural routes - practices and challenges. In: Karabegovic, I. (ed.) NT 2021. LNNS, vol. 233, pp. 607–614. Springer, Cham (2021). https://doi.org/10.1007/978-3-030-75275-0_66
11. Orlandic, M., Jakšić-Stojanović, A.: The importance of digitalisation and implementation of modern ICT in the valorisation of the culturalroutes -trends and perspectives, 1st Cultural Routes Academic workshop, online, 9th December, 2021

Nonlinear Transformations with Fourier Series as Applied to Electrotechnical Problems

Volodymyr Chenchevoi[1], Serhii Serhiienko[1], Vira Shendryk[2(✉)], Andrii Nekrasov[1], and Maksim Fed[1]

[1] Kremenchuk Mykhailo Ostrohradskyi National University, Kremenchuk, Ukraine
[2] Sumy State University, Sumy, Ukraine
v.shendryk@cs.sumdu.edu.ua

Abstract. In recent years, the mathematical apparatus for studying the energy modes of electrical circuits, analyzing the processes of energy conversion in non-linearities, identifying the parameters of electrical equipment using the theory of instantaneous power, which is based on methods for converting signals represented in the form of approximation dependences, in particular, Fourier series, has been developing. As a rule, processes are analyzed in the description of which nonlinear operations with Fourier series are used – their multiplication, division, exponentiation, etc. The complexity of these operations virtually eliminates analytical expressions to describe the final dependencies. When performing these mathematical operations, as a rule, the convolution operation is used, which has found wide application when working with signals in radio engineering. The execution of the above operations is available to the researcher using applied mathematical packages. In electrical problems of the above nature, it is possible to use solutions obtained without the full–fledged use of computing facilities. At the same time, another problem is solved – the study of the convolution mechanism and the receipt of intermediate results as a source material for determining the specific data of the analyzed processes. The work focuses on the issues of obtaining results using series with a finite number of members.

Keywords: Fourier series · Series multiplication · Instantaneous power of electrical signals · Identification of nonlinearity parameters

1 Introduction

In studies in the analysis of energy processes in electrical systems and complexes, operations are carried out on functional dependencies in the form of exact mathematical or approximation expressions [1, 2]. Despite the currently available possibility of obtaining acceptable results using modern technologies in an acceptable time, it becomes necessary to determine intermediate results in analytical form as the basis for calculating approximate solutions for creating effective information devices and systems. [3]. Intermediate information in a fairly complete volume allows to consider the energy process taking into account the features and characteristics of generation, energy exchange, dissipation, energy storage [4].

© The Author(s), under exclusive license to Springer Nature Switzerland AG 2022
I. Karabegović et al. (Eds.): NT 2022, LNNS 472, pp. 568–578, 2022.
https://doi.org/10.1007/978-3-031-05230-9_68

Considering the problem from a slightly different position – the position of using energy methods for the tasks of direct analysis of the energy process, identification of the parameters of electrical equipment using modern methods, we have to state that nonlinear operations with functional dependencies describing energy processes are the only tool for obtaining the expected results. Since in these problems trigonometric Fourier series are mainly used as approximating functions, it becomes obvious the need for multiplication, raising to a power and dividing these series. Such tasks are implemented using special mathematical techniques, in particular, by using the well–known convolution operation when processing continuous discrete signals [5]. Nonlinear transformations during multiplication, division, raising to the power of the series consist, first of all, in the transformation of frequencies, as a result of which the resulting signals do not contain harmonic ones from the composition of the original dependencies. This property is provided quite important, since it allows the separation of the resulting signals by frequency, which makes it possible to draw up equations for the energy balance of instantaneous power at the corresponding frequency of its measurement or conversion [6].

This study presents an example of determining the parameters of time–varying nonlinearity using approximating functions in the form of Fourier series as the initial ones for multiplication and division.

Purpose of the study: to show the capabilities of the instantaneous power method in solving problems of assessing energy modes, identifying the parameters of devices and systems by using the results of nonlinear transformations of Fourier series, namely, the operations of their multiplication, division, exponentiation.

2   Analysis of Previous Studies

In problems of electromechanics and power engineering, a method for studying power engineering is used, which consists in using trigonometric functions such as: approximation dependences of currents, voltages, moments, rotation speed in the form of trigonometric series, their products, details, etc. In this case, the results of these operations are the source material for solving problems of an identification nature, for assessing the indicators of energy processes, etc. [7–11].

The studies indicated that the use of the energy method involves obtaining functions of instantaneous power at each of the elements of the equivalent circuit in the form of products of voltage and current, torque and speed, etc. The next step is to represent the instantaneous power in the form of Fourier series over the repetition interval of the processes. The balance equations for the instantaneous power components for the power supply and the elements of the equivalent circuit are formed independently of each other:

- Equations of the balance of constant components of instantaneous power for the power source and elements of the equivalent circuit;
- Equations of the balance of the cosine components of the instantaneous power for the power supply and the elements of the equivalent circuit;
- Equations of balance of sinus components of instantaneous power for the power supply and elements of the equivalent circuit.

The solution of identification problems by any of the methods, as a rule, comes across an insufficient number of equations in the identification system, and this is associated with a significant number of harmonic power components. It is necessary to point out the distinctive feature of the product of trigonometric series. This is manifested in the fact that from the number of harmonic functions in the product, the power is expressed, the number of resulting components of different frequencies rapidly increases with growth. So, for example, when multiplying cosines with two frequencies, the number of resulting components is equal to two with frequencies $\Omega_1 = \omega_1 + \omega_2$ and $\Omega_2 = \omega_1 - \omega_2 \ldots$ When multiplying three components, the resulting number of components is four a, in general for m factors, the number of components is determined by the expression

$$K_{\Sigma} = 2^{m-1},\tag{1}$$

where K_{Σ} – the number of components with different frequencies, as a result of multiplication m harmonic functions of different frequencies. Fast growth K_{Σ} confirms the earlier concluded conclusion regarding the advisability of using trigonometric series when approximating the instantaneous power curve to create a mathematical basis for the energy method. An increase in the number of components is a prerequisite for using the energy method in solving identification problems in relation to devices with rather complex circuit structures, as well as containing nonlinearities.

3 Materials and Results of the Study

The elements of harmonic analysis are widely used in converting technology, in the analysis of indicators of energy processes. Less often, the tasks of analyzing dependencies or processes are solved, which are not presented in the form of sums of individual functions, but in the form of their products, particulars, degrees, etc. under the assumption that the sought functional dependencies are also trigonometric series. In practical application, the multiplication and division of series is most often encountered in determining the parameters of the energy regime: determining the components of the instantaneous power in networks with non–sinusoidal voltages and currents in converter systems of various functional purposes, in the study of operating modes of nonlinearities, identification of the parameters of electrical devices. The definition of the components of instantaneous power is described in sufficient detail in the scientific literature [12–16]. Characteristically, the main task is to multiply functional dependencies $U(t)$ and $I(t)$, which, as a rule, are approximated either by polynomials of the corresponding degrees, or by trigonometric series with a finite number of harmonic ones. Due to the fact that when analyzing energy issues most often one has to deal with sinusoidal or non–sinusoidal periodic dependences, then, as a rule, approximation expressions are used for $U(t)$ and $I(t)$ in the form of trigonometric series.

In the problems of research and identification of nonlinearities, the mathematical aspects can be much more complicated than when determining energy indicators in non–sinusoidal modes. For illustration, we indicate the established classification of elements of electrical circuits into linear, nonlinear, parametric, nonlinear–parametric. In this case, nonlinear are those characteristics whose parameters depend on one or more parameters characterizing the mode (current, voltage, temperature, etc.); parametric non-linearities have time–dependent characteristics, i.e. the parameters are changed accordingly according to the specified algorithm; nonlinear–parametric – those that depend both on the parameters of the operating mode and on time. The general property of non-linear elements in the complexity of processes, flowing in them under periodic influences from a power source or non–linear element. This complexity is primarily associated with frequency transformations, inevitably arising from the interaction of time–varying signals and parameters of nonlinear elements. In this case, the need to multiply the series arises both when determining the dependences of an energy nature, and when analyzing the simplest equations of electrical equilibrium.

To illustrate what has been said, consider the simplest electrical circuit consisting of a constant voltage source U and changing resistance $R = R_0 + \sum\limits_{i=1}^{i=\infty} R_i \sin(\Omega t)$ – active resistance. For illustration, the dependences of the resistance, changing over time, of the current passing through the circuit are presented at: $U = 100$ V, $R(t) = 10+5\sin(\Omega t)+ 4\sin(2\Omega t)$, $\Omega = 314\text{s}^{-1}$. As a result of a change in resistance over time, a current flows through the circuit

$$I(t) = I_0 + \sum_{m-1}^{m=\infty} I_m \sin(m\Omega t - \phi_m)$$

$$= I_0 + \sum_{m=1}^{m=\infty} I_{ma} \cos(m\Omega t) + \sum_{m=1}^{m=\infty} I_{mb} \sin(m\Omega t) \qquad (2)$$

where I_{ma}, I_{mb} – quadrature components of the current component with a relative frequency m; ϕ_m v – phase shift of m harmonic of the current.

Figure 1 it can be seen that even with the indication of the simplest law of resistance, the current flowing through the circuit has a complex harmonic composition. As a result, we can write the equality:

$$U(t) = R(t)I(t) = \left(R_0 + \sum_0^{i=2} R_i \sin(\Omega t) \right)$$

$$\times \left(I_0 + \sum_{m=1}^{m-\infty} I_m \sin(m\Omega t - \phi_m) \right) \qquad (3)$$

in which the operation of series multiplication is obvious in the same way as in the case of determining the cardinality. The result becomes much more complicated if the supply voltage also has a periodic form; in particular, the complexity increases if the periods of variation of the alternating voltage and current components are not the same.

When defining specific targets, such as current $I(t)$ with a known $R(t)$ there is a dependence:

$$I(t) = \frac{U(t)}{R(t)} = \frac{U}{R_0 + \sum\limits_0^{i=2} R_i}, \qquad (4)$$

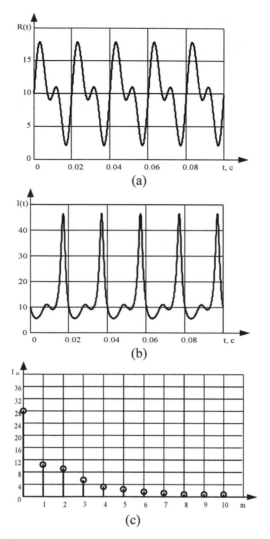

Fig. 1. Curves of changes in resistance, current over time (a, b), current spectrum (c).

In determining $R(t)$ – its identification – will have a different result

$$R(t) = \frac{U(t)}{I(t)} = \frac{U}{I_0 + \sum\limits_{m=1}^{m=\infty} I_m \cos(m\Omega t - \phi_m)} \tag{5}$$

As follows from the example, the face of a more complicated operation than multiplication, the operation on the rows is their division. In this case, the quotient:

$$\frac{1}{I_0 + \sum\limits_{m=1}^{m=\infty} I_m \cos(m\Omega t - \phi_m)} = G(t) \tag{6}$$

represents the instantaneous value of the conductivity of the circuit – a new series with the same harmonic order as the series approximating $I(t)$.

Moving on to energy parameters, for example, determining the instantaneous power on an active resistance, we get:

$$P(t)_R = R(t) \cdot I^2(t)$$

$$= \left(R_0 + \sum_0^{i=2} R_i \right) \left(I_0 + \sum_{m=1}^{m=\infty} I_m \cos(m\Omega t - \phi_m) \right)^2. \tag{7}$$

In Eq. 7, the most interesting is the variable component of the resistance times the square of the variable component of the current. Obviously, this is a multiplication of three strings.

An illustrative example is the case of nonlinearity with saturable inductance. Suppose that the dependence of inductance on current [4] has the form:

$$L(I) = a_1 I + a_3 I^3 + a_5 I^5 + \ldots + (a_n) I^n \tag{8}$$

and the current is presented in the general form of a Fourier series: $I(t) = \sum_{m=1}^{m=\infty} I_m \cos(m\Omega t - \phi_m)$ Obviously, to obtain the dependence $L(t) I(t)$ substitute in expression for $L(t)$, i.e. a series describing $I(t)$ it is necessary to consistently build into the third, fifth, etc. power:

$$L(t) = a_1 I(t) + a_3 I^3(t) + a_5 I^5(t) + \ldots + (a_n) I(t)^n \tag{9}$$

For other elements of the electrical circuit, you should use the general dependence for power $P(t) = U(t) I(t)$.

For the energy balance, the power of the inductor should be presented in the form:

$$U_L(t) = \frac{d}{dt}(L(t) I(t))$$

$$P_L(t) = U_L(t) I(t) = \frac{d}{dt}(L(t) \cdot I(t)) \cdot I(t)$$

$$= \frac{dL(t)}{dt} I^2(t) + \frac{dI(t)}{dt} L(t) \cdot I(t). \tag{10}$$

The following conclusion follows from the expression: the representation of energy dependences in the form of expressions of instantaneous power, equations of electrical equilibrium in circuits with nonlinearities using Fourier series is complicated due to frequency transformations when performing operations (their multiplication, division, raising to a power).

Obviously, both when multiplying the series, and when dividing them, when identifying nonlinearities, a new resulting series with a certain harmonic composition is obtained as a result. This concerns the series that determines the harmonic composition of the instantaneous power $\sum_0^{\infty} P_k(t)$ (when multiplying $U(t)$ on the $I(t)$); multiplying the series determining the current $I(t) = I_0 + \sum_{0}^{\infty} I_m(t)$ per row approximating the dependence of resistance in time $R(t)$ (in the case of identifying the parameters of the analyzed nonlinearity).

The considered example can be appropriately complicated by representing the supply voltage in the form of a series $U(t) = \sum_{0}^{\infty} U_n(t)$. In this case the equation of electrical equilibrium can be written as follows: $\sum_{0}^{\infty} U_n(t) = \sum_{i=0}^{\infty} R_i(t) \sum_{0}^{\infty} I_m(t)$, which in the conditions of the above example also means the product of the series. As a result, different approaches can be considered in the formation of problems with nonlinear transformations of series:

– by determining the product of two series with a known harmonic composition and harmonics, but unknown coefficients of the series in the result;

– by determining the product of two series, and one of the factors is a series with a known harmonic composition, but with unknown coefficients, the other factor with known coefficients and harmonic composition, with the result being a known series.

When solving specific electrical and electromechanical problems, the determination of certain indicators and modes is carried out through the implementation of these mathematical operations on, dependencies, which are a mathematical description of physical processes.

Checking the effectiveness of the mathematical apparatus is carried out in accordance with the given example.

When the resistance changes, a current flows along the specified circuit, which is represented in the form of a harmonic series:

$$I(t) = I_0 + \sum_{m=1}^{m=M} \left(I_{a_m} \cos(m\Omega t) + I_{b_m} \sin(m\Omega t) \right) \tag{11}$$

The series coefficients are determined according to the dependencies:

$$I_0 = (1/T) \int_0^T I(t)dt$$

$$I_{a_m} = (2/T) \int_0^T I(t) \cos(m\Omega t)dt \tag{12}$$

$$I_{b_m} = (2/T) \int_0^T I(t) \sin(m\Omega t)dt$$

where T is the recurrence period of the processes.

Restricting to a finite number of harmonic components, for example, $m = 14$, we obtain the dependence for $I(t)$ in the form

$$\begin{aligned}
I(t) = &I_o + I_{a_1} \cos(\Omega t) + I_{b_1} \sin(\Omega t) + I_{a_2} \cos(2\Omega t) + I_{b_2} \sin(2\Omega t) \\
&+ I_{a_3} \cos(3\Omega t) + I_{b_3} \sin(3\Omega t) + I_{a_4} \cos(4\Omega t) + I_{b_4} \sin(4\Omega t) \\
&+ I_{a_5} \cos(5\Omega t) + I_{b_5} \sin(5\Omega t) + I_{a_6} \cos(6\Omega t) + I_{b_6} \sin(6\Omega t)
\end{aligned}$$

$$+ I_{a_6} \cos(6\Omega t) + I_{b_7} \sin(7\Omega t) + I_{a_8} \cos(8\Omega t) + I_{b_8} \sin(8\Omega t)$$
$$+ I_{a_9} \cos(9\Omega t) + I_{b_9} \sin(9\Omega t) + I_{a_{10}} \cos(10\Omega t) + I_{b_{10}} \sin(10\Omega t)$$
$$+ I_{a_{11}} \cos(11\Omega t) + I_{b_{11}} \sin(11\Omega t) + I_{a_{12}} \cos(12\Omega t) + I_{b_{12}} \sin(12\Omega t)$$
$$+ I_{a_{13}} \cos(13\Omega t) + I_{b_{13}} \sin(13\Omega t) + I_{a_{14}} \cos(14\Omega t) + I_{b_{14}} \sin(14\Omega t)$$

The values of the coefficients for the sine and cosine components are given in Table 1. Constant component $I_0 = 14,045$ A.

Table 1. Value of harmonics in the current curve

m	I_{am}, A	I_{bm}, A	m	I_{am}, A	I_{bm}, A
1	4.141	−9.479	8	0.09	− 0.501
2	−2.588	−8.379	9	−0.227	− 0.232
3	−4.904	−0.752	10	−0.201	0.032
4	−2.497	2.031	11	−0.058	0.115
5	0.269	2.028	12	0.038	0.072
6	1.101	0.634	13	0.051	6.63–10–3
7	0.736	−0.352	14	0.022	− 0.024

After multiplying the series describing the current and the series describing, the voltage balance equation is obtained in the form:

$$U = \sum_{i=0}^{i=3} R_i(t) \cdot \sum_{m=0}^{N} I_m(t) = \sum_{k=0}^{N+3} P_k(t) \tag{13}$$

The result of multiplication is a new series containing sine and cosine components, as well as a constant – the average value of the resulting harmonic series. Obtaining the coefficients for sine and cosine components manually is a rather difficult task, even with a relatively small number of analyzed current harmonics, which takes a long time and at the same time the final result obtained has low accuracy. For this, it is advisable to use specialized software products, such as Maple, Mathematica or Mathcad.

Then the problem is to compose a system of equations from the result of multiplying the series, i.e. essentially out of the line $\sum_0^{2M} p_k(t) \ldots$ The basis for drawing up dependencies as a basis for a system of equations for determining three unknowns (R_0, α, β), are the following conditions:

– the average value of the product of two series on the process repeatability interval, equal to the source voltage

$$\frac{1}{T} \int_0^T R(t)I(t)dt = I_0 R_0 + \frac{I_{1a}\alpha}{2} + \frac{I_{2b}\beta}{2} = U;$$

- supply voltage value at time $t = 0$ ($\Omega t = 0$)

$$R_0 \left(I_0 + \sum_1^M I_{am} \right) = U;$$

- supply voltage value at time $t = 0,005$ from $\left(\Omega t = \frac{\pi}{2} \right)$

$$(R_0 + \alpha) \left(I_0 + \sum_1^M I_{bm} \right) = U.$$

Taking these provisions into account, we obtain a system of equations for determining the required parameters:

$$I_0 R_0 + \frac{I_{1b}\alpha}{2} + \frac{I_{2b}\beta}{2} = U$$

$$R_0 \left(I_0 + \sum_1^M I_{am} \right) = U$$

$$(R_0 + \alpha) \left(I_0 + \sum_1^M I_{bm} \right) = U$$

When solving the system of equations, the software product Maple was used. The method of Cramer and Gauss was used as methods for solving the system of equations.

Table 2. Calculated values of parameters for different numbers of significant current harmonics

Tumber of harmonics k	Coefficients		
	α	β	R_0
5	1.634328406	13.88065166	11.81195370
7	5.24624221	2.73420543	9.70591090
10	4.78627918	4.35821253	10.0351229
14	5.04509914	3.88743211	9.98203234

Table 2 shows the values of the parameters for different numbers of current harmonics in the system of equations.

As follows from the data obtained, acceptable results of identification of resistance parameters take place when the number of harmonic is equal to a sufficiently large (in our case $k \geq 10$).

The solution of the inverse problem – determining the dependence of the current $I(t)$ with known dependences of voltage and resistance in time is the case of dividing voltage by resistance $U(t)R(t)^{-1}$, which corresponds to the previously presented equality $U(t) = I(t) R(t)$, from which it follows that if $U(t)$ – a series and resistance– a series, then $I(t)$– the desired series, the harmonic composition of which must meet the following conditions. First, in accordance with the equations of electrical equilibrium, which means the equality $_i$ – that voltage harmonic of the left side of the equation to the value $_i$ – that harmonic – right. In its simplest form, this is the equality of a constant value $U(t)$ – a constant component of the work $I(t) R(t)$. A similar correspondence must be observed for each of the harmonics. This means that as a result of the multiplication of the series $\sum_{n=0}^{n=\infty} I_n(t) \sum_{r=0}^{r=\infty} R_r(t)$ we obtain a number of vector equations. On the left side of each of them – we have a vector – a harmonic of the corresponding order, and on the right – a vector sum formed by the harmonics of current and resistance in the course of their frequency transformations.

4 Conclusion

In this work, using the example of a test problem, the possibilities of the instantaneous power method are revealed in solving problems of assessing energy modes, identifying the parameters of devices and systems by using the results of nonlinear transformations of Fourier series, operations of their multiplication, division, exponentiation.

The results presented in the work can be used to determine the energy indicators by the values of the instantaneous power with the allocation of: constant (active), alternating active and reactive components in the space of canonical components, pseudo–active and pseudo–reactive components in the space of non–canonical components, alternating non–canonical components. Taking into account the named components, clarifying the nature of some of them and, first of all, pseudo–active and pseudo–reactive components, will make it possible to complete the protracted discussion on the problem of inactive components of the power of non–sinusoidal signals.

References

1. Voldek, I., Popov, V.V.: Electrical Machines. Machines of alternating current, SPb, Piter Publ (2010). 356 p.
2. Zagirnyak, M., Kalinov, A., Romashykhina, Z.: Decomposition of electromotive force signal of stator winding in induction motor at diagnostics of the rotor broken bars. Sci. Bull. Natl. Min. Univ. 4(154), 54–61 (2016)
3. Rodkin, D.I., Romashihin, Y.V.: Rationale for settlement circuit for induction motors. Tech. Electrodyn. 2, 89–90 (2012)
4. Zagirnyak, M., Rodkin, D., Romashykhin, Iu., Rudenko, N., Chenchevoi, V.: Identification of nonlinearities of induction motor equivalent circuits with the use of the instantaneous power method. In: Proceedings of 17th International Conference Computational Problems of Electrical Engineering, CPEE (2016)
5. Hasegawa, M., Ogawa, D., Matsui, K.: Parameter identification scheme for induction motors using output inter-sampling approach. Asian Power Electron. J. 2(1), 15–23 (2008)
6. Lee, S.H., Yoo, A., Lee, H.J., Yoon, Y.D., Han, B.M.: Identification of induction motor parameters at standstill based on integral calculation. IEEE Trans. Ind. Appl. (2017). 99 p.
7. Zagirnyak, M., Romashykhina, Zh., Kalinov, A.: Diagnostic signs of induction motor broken rotor bars in electromotive force signal. In: Proceedings of 17th International Conference Computational Problems of Electrical Engineering, CPEE (2016)

8. Mosyundz, D.: Energy method of nonlinear inductance parameters identification. In: XIV International PhD Workshop, OWD 2012, 20–23 October, 2012, pp. 456–460

9. Zagirnyak, M., Rodkin, D., Korenkova, T.: Estimation of energy conversion processes in an electromechanical complex with the use of instataneous power method. In: 16th International Power Electronics and Motion Control Conference and Exposition, Antalya, Turkey, pp. 238–245 (2014)

10. Rod'kin, D.I.: Decomposition of polyharmonic signal power components. J. Electr. Eng., Iss. **6**, 34–37 (2003)

11. Zagirnyak, M., Rod'kin, D., Korenkova, T.: Enhachement of instantaneous power method in the problems of estimation of electromechanical complexes power controllability. Przeglad Elektrotechniczny (Electrical Rewer), No 12b, pp. 208–212 (2011)

12. Maga, D., Zagirnyak, M., Miljavec, D.: Additional losses in permanent magnet brushless machines. In: Proceedings of EPE–PEMC – 14th International Power Electronics and Motion Control Conference, pp. 12–13 (2010)

13. Zagirnyak, M.V., Prus, V.V., Nikitina, A.V.: Grounds for efficiency and prospect of the use of instantaneous power components in electric systems diagnostics. Przeglad Elektrotechniczny (Electrical Rewer) **82**(12), 123–125 (2006)

14. Shimoni, K.: Teoreticheskaya elektrotekhnika [Theoretical electrical engineering], Ripol Klassik, Moscow (2013). 778 p.

15. Zagirnyak, M., Rod'kin, D., Romashykhin, Iu., Romashykhina, Zh., Nikolenko, A., Kuznetsov, V.: Refined calculation of induction motor equivalent circuit nonlinear parameters by an energy method. Eastern–Eur. J. Enterpr. Technol. **3**(5(87)), 4–10 (2017)

16. Akagi, H., Watanabe, M.: Instantaneous Power Theory and Applications to Power Conditioning. Wiley, New York (2007). 379 p.

Technosphere – Alchemy of Technicque

Halima Sofradžija[✉], Fahira Fejzić Čengić, Abdel Alibegović, and Enita Kapo

Faculty of Political Sciences, University of Sarajevo, Skenderija 72, Sarajevo, Bosnia and Herzegovina
mima8@hotmail.com

ABSTRACT. Technology has reshaped the world. In the digital age, technology creates a new reality, a technosphere that brings radical changes to the world and life, at all levels of human experience. There is no end in sight to the technological modification of the world, nor are the consequences known - we are talking only about the possibilities of technology. The hyperreality of the technosphere is fascinating precisely because of its seductiveness. Changing our experiences in technoculture invites us to rethink new social rationality.

Keywords: Technoculture · Technosphere · Technology · Network society · Society digitalization

1 Introduction

Modern society is marked with accelerated historical transformations as well as with recognizable changes of existing paradigms, in the light of progress of new media, information technology, and communication technologies. In digital era technology makes a new reality, some authors named "the culture of real virtuality", here we think first of Manuel Castells [1]. So, in the era of new technologies, there is no doubt, the world is more and more submerged in the technosphere. In technosphere, technology is the most recognizable media. Technological civilization speaks and it directs us to recognition and understanding of a new technogenetic construction of the world. Certainly, this situation requires a new way of thinking about life and the world, in the technological age in which we live.

2 Technosphere - Technology as a Medium

The challenges of modern society in technological civilization are numerous and extremely complex; therefore, the authors from various fields are engaged in the analysis of the whole society in that context, precisely contemplating the contemporary moment and the situation where "technology becomes nature", as many thinkers have pointed out. As we already mentioned in the introduction of this text, submergence in the world of technosphere, where the society as the whole is placed in the tight hug of technic and technology, speaks to the fact that contemporary man/human being has found himself in a different experience of the world such as previous epochs do not know. Some

I. Karabegović et al. (Eds.): NT 2022, LNNS 472, pp. 579–584, 2022.
https://doi.org/10.1007/978-3-031-05230-9_69

mediologists and sociologists remind us: "The most recognizable media in hypermedia society is – technology. New technologies and irreversible digitalization of society lead to greater mediation within all social relations. In culture of real virtuality it is technology that made possible the existence of virtual classrooms (or classrooms without walls, as A. Giddens notices), intermediation through technology in knowledge transfer, digital platforms as meeting points, and finally, existence of electronic universities. Innovative technologies have already immensely influenced education processes, which imply different trials of knowledge transfer and the education itself" [2]. Technology is a medium in the literal sense, the most important medium, in the complex phenomenon which is an essential determinant of our time, and we call that technosphere. It was technology that made the internet possible, as the most recognizable communication medium/media, what Manuel Castells [1, 3] was talking about, emphasizing that the influence of the internet to the society is immeasurable and unquestionably creates a "new communication environment" or as Castells [3] would put it even more clearly: The internet is a communication medium ("The internet is a communication medium with its own logic and language, but it is not limited to a specific area of expression - it extends to all areas") [3, p. 222]. The modern man is submerged in the world of technology, we know today that submergence exists on a daily basis. Of course, it changes social practices and makes one completely new quality of life. And the way of the communication itself is changing [4] and 21st century society is in deep historical transformations [5] which indicates the complexity of the overall social processes, when the current world is changing in all directions and levels. When we talk about technosphere, we can say that we are talking about multiverse of technology. In this multiverse of technology all areas of sociability and social life change in a specific way with the pervasive application of technology. The technosphere opens up the possibility of countless new worlds created by innovative technologies. The alchemy of technology is revealed/visible in the multiverse of technology, in the very seductiveness of technology, a kind of technomagic/techno-magic. The magical moment of innovative technologies undeniably has its charming: "Immersed and dragged into their technologized created spaces, seduced by the siren's call and the dazzle of technology, millions of people move into parallel virtual worlds every day. In this immersion in digitalized spaces, they exercise their sociability through technology; therefore, we talk about techno-sociality. Technosociality reveals, identifies and realizes a completely new experience of sociability - the realization of sociability and remote connectivity through technology, where technology serves as the medium" [6]. The media multiverse speaks to the incredible mediation of society. M. Galović [7] reminds us that today there is hardly anything significant that has not been mediated, affected by them, processed by them, broadcast through them, interfered in reality. Of course, it is not an accident Manuel Castells speaks about situation of rising of networked society. As we read in a perfect/great text, which also deals with a topic that preoccupies us here, under the title: "The Process of Technicizing The World, Technoculture and Technology as a Medium" [8] - "The alchemy of technology has created a technosphere, technospaces, telepresentation, telepresence, possibility of network systems existence. The adjustment of society to new technological discoveries is continuing, there is no end to innovations coming from that field. Integration of biotechnologies, nanotechnologies, informatics, multimedia, changes the existing world as we know it in many ways. Unmanned aerial vehicles, drones, remote wireless control, crypt-currencies, floating capital and all features of fluid life mentioned by Zygmunt Bauman [9], artificial intelligence, robotics, holograms, have become our reality long time ago" [8]. Some modern authors, in serious, studious pursuits of gaining understanding this issue of the technosphere, consider that the very concept of the technosphere is a completely new paradigm of thought. Paić [10] speaks in the introductory pages of his book: that the very concept of the technosphere is also a "new" paradigm of thought, a logic of the technogenetic structure of worlds, where a continual transformation of the condition takes place. Actually, everything that happens during the rule of the technosphere, according to this author, takes place through five overlapping fields of action: anthropology, cybernetics, art, architecture, up to design, because everything is distinct and different in the synthetic "other nature" [10]. And it is important, in the frame of this observation: the driving force of the technosphere is in information. Cybernetic space is different from any other experience of communication up to now.

3   The World as a Technoconstruction

Gianni Vattimo is the author of a well-known and much-quoted sentence: "And technique is a fairy tale, a message sent" [11]. Technique and technology have capillary permeated our world, transforming everything into "technical sets". The technical approach to the nature and technicalization of many areas of life have instructed and brought modern theorists of technology to give it the appropriate expression/word – *pantechnicism*. M. Galović [12] says that many theories of technology have emerged in parallel, considering different fields of technical relations: man and technology (anthropologists), history and technology (historians of technology, and trials in philosophy and sociology of technical interpretation of history), society and technology (sociologists trying to interpret social changes by the progress of science and technology: post-industrial society, innovation society, information society), economics and technology, science and technology, future and technology (futurists) [12]. But it has to be added here, mediologists as well as communicologists because the modern media are the result of high technology and obviously redefine the field of communication. Since communications take place in cyberspace, *tele-space*, "at a distance", they are called *telecommunications*, and as the content is about information, the word was created   *telematics* (telecommunications + informatics) which, according to some mediologists, means that the society we enter is not only informational, but *telematic* what Galović [12] is talking about, but not only him; beside, it has to be emphasized that moving into cybernetic space is just one sort of technicizing, which leads to further domination and spreading of the technical, technogenic [12, p. 316]. Previously, we mentioned that the driving force of technosphere is in information. According to Galović [12], communicologists are stick to the term of information as a message. Information is promoted and clearly observed as a resource. Actually, information became universal currency which is profitably traded, as an incentive, instruction, modern theorists draw special attention to that as J. Ridderstale/K. Nordstorm [13]; one who rules the information - rules the world. There is no doubt, that what is rapidly transforming society, are technics and technology, they are modeling the world, we say so patternmakers of the postmodern world. Many authors recall to Lewis Mumford [14], to his understanding of technology: "One part of our civilization-one dedicated to technology - usurped power over other components, geographic, biological and anthropological: in fact, the craziest proponents of the process are publicly announcing that today the entire biological world is being replaced by technology" [14, p. 301.]. In the dialectic of technology, as well as in the dialectical game of man and technology, the capture of society in *networked electronic digital systems*, the irreversible digitalization of society, is visible; Janus' character of technology is a challenge of our time, which must be the subject of serious reflection and understanding of a new culture determined by technology, technoculture. Vertovšek and Gregurić [15] leads us to knowledge: "The technological and biological reality of the upcoming era cannot be viewed as separated from all the sociological, psychological, and media aspects of the society and the individual. What will these irreversible consequences in the networked reality of the media and humans imply in bioethical terms? How should we already now think of man, the human community, and the emerging, globalized world? We are approaching a great "tipping point" in which human life and the possible coexistence of artificial and natural intelligence will be tested and one can only speculate on the media picture of such a reality (...) Thinkers and scientists, philosophers and technologists try to describe the future as a time that will open up new perspectives and human opportunities" [15]. Redefining of the world, mediation in a *hypermedia environment in the networked society*, also bears numerous reductions. Creating real virtuality, suffocated by mediations, we cancelled the world present before. Everything is in transformation, started up with technology from the technical transformation of the media and communication, emerging interactivity, technosociality (McLuhan speaks of technical devices as human prostheses, that parable is well known) up to the complete media globalization of the world [16] certainly in understanding the ambiguity of the term of media i.e. what is - a medium. In the modern construction of the world, only the transformation of that world should be mentioned in a specific way, in a set of *character/Greek:* σημεῖον, semeion [10, p. 13, 17–19]. Under subtitle *Jean Baudrillard's Media Theory as Intertext*, K. Purger [20, p. 263] mentions that Baurillard in his first book: "The System of Objects", describes the functioning of modern societies and the way they operate through ubiquitous labeling systems: "It is a system in which all objects, messages and products, all history, culture, meaning, relationships and experience are transformed into signs that replace experience and mediate it. For Baudrillard, therefore, consumption is not merely the acquisition of objects, but the appropriation of the meaning that a product possesses or is believed to possess. The goods market is a *market of signs*, where each item has its inscribed symbolic value" [20]. Jean Baudrillard (Irony of Technology) [21], he concludes that our illusion of the purposefulness of technology has reached its peak, technique and technology in the techno-construction of the world have become a way that enables every other relationship to the world, where there is always a hidden and evident combination of **possibilities** and **risks**, a symbolic amalgam of this time we live in.

4 Conclusion

In technoculture, innovative technologies are a source of options, on several levels, and mean a constant possibility of producing something new. Technology is driving changes, in fact it is the main initiator of the 4th industrial revolution, and every industrial revolution in history has profoundly and irreversibly changed the world and social patterns. In the digital age, technology creates a new reality, a technosphere that brings radical changes to the world and life, at all levels of human experience. As mentioned earlier in the text, the technosphere opens up the possibility of creating countless worlds with innovative technologies. The media influence of technical images, sublime messages and technical sets, "alternative worlds", forms a completely new image of reality. Alchemy of the technics is visible in multiversum of technology, and technology itself is a medium. In multiversum of technology, a kind of special magic in the cybernetic space, actually - techno-magic, an everyday electronic magic is visible to all of us well known in the era of accelerated digitalization of the whole society. In the technosphere, technosociality, telepresence, technorelation, technoculture, the way of human action and understanding of life is changing.

References

1. Castells, M.: Uspon umreženog društva, Golden marketing, Zagreb (2000)
2. Fejzić Čengić, F., Sofradžija, H.: Hypermedia Sphere and Virtual Classrooms In medias res, časopis filozofije medija, Zagreb **10**(18), 2871–2882 (2021). https://doi.org/10.46640/imr.10.18.7
3. Castells, M.: Internet galaksija - Razmišljanja o Internetu, poslovanju i društvu, Naklada Jesenski, Turk, Zagreb (2003)
4. Čustović, E.: Communication behavior in public and media discourse. DHS **3**(9), 361–370 (2019). Filozofski fakultet Tuzla; dhs.ff.untz.ba/index.php/home
5. Alibegović, A., et al.: 21st century society – key features and issues. DHS **1**(1), 151–163 (2016). Filozofski fakultet Tuzla. dhs.ff.untz.ba/index.php/home
6. Sofradžija, H.: Technosociality and the rise of the network society. In: Karabegović, I. (ed.) NT 2020. LNNS, vol. 128, pp. 432–436. Springer, Cham (2020). https://doi.org/10.1007/978-3-030-46817-0_49
7. Galović, M.: Uvod u filozofiju znanosti i tehnike – znanost i tehnika u razdoblju nagovještaja povijesnog obrata, Bibl")teka filozofska istraživanja, Zgb (1997)
8. Sofradžija, H., Alibegović, A., Bakić, S., Arifhodžić, M.: The process of technicizing the world, technoculture and technology as a medium. In: Karabegović, I. (ed.) New Technologies, Development and Application IV, NT 2021., Lecture Notes in Networks and Systems, vol 233, pp 1158–1165. Springer, Cham (2021). https://doi.org/10.1007/978-3-030-75275-0_128, ISBN978-3-030-75274-3; https://www.springer.com/gp/book/9783030752743 (2021)
9. Bauman, Z.: Fluidni život. Mediteran Publisching, Novi Sad (2009)
10. Paić, Z.: Tehnosfera I, Žrtvovanje i dosada: Životinja – Čovjek – Stroj, Sandorf & Mizantrop, Zagreb (2018)
11. Vattimo, G.: Kraj moderne, Biblioteka Parnas, MH, Zagreb (2000)
12. Galović, M.: Rastanak od čovjeka – mizantropija znanosti i pad u tehnički bezdan, Demetra, Filozofska biblioteka, Zagreb (2017)
13. Ridderstale, J., Nordstorm, K.: Kapital igra (Funky Business) Plato, Bgd (2004)
14. Mumford, M.: Mit o mašini, GZH, Zagreb (1986)
15. Vertovšek, N., Gregurić, I.: The scientific-technical and media-related future of man and the world. In medias res, časopis filozofije medija **8**(14), 2263–2279 (2019)
16. Fejzić, F.: Medijska globalizacija svijeta, Promocult, Oko, Sarajevo, (2004) and see more Fejzić Čengić F: Medijski zapis o zemlji - Medijska kultura u Bosni i Hercegovini u tradiciji i modernom vremenu i prostoru, FPN, Dobra knjiga, Sarajevo (2021)

17. Lyotard, J.F.: Postmoderno stanje, Svetovi, Novi Sad (1988)
18. Alić, S: Masmediji zatvor bez zidova/ Tekstovi filozofije medija Zagreb (2012)
19. Sofradžija, H., Šehić, S., Alibegović, A, Bakić, S., Camo, M.: Education as a Process and result. Int. J. Contemporary Educ. Redfame Publishing, 4(1), 56–64 (2021). http://redfame.com/journal/index.php/ijce
20. Purger, K.: Slike u tekstu – Talijanska i američka književnost u perspektivi vizualnih studija, Durieux, Zagreb (2013)
21. Baudrillard, J.: Savršeni zločin, Circulus, Beograd (1998)

Era of (Bio)technicized Corporeality – from the Culture of Real Virtuality to the New Sphere of the Transhuman Reality

Abdel Alibegović[✉], Sanjin Mahmutović, Halima Sofradžija, and Sanja Raljević Jandrić

Faculty of Political Sciences, University of Sarajevo, Skenderija 72, 71000 Sarajevo, Bosnia and Herzegovina
abdel.alibegovic@fpn.unsa.ba

Abstract. The authors are trying to contemplate still opened possibilities for the future of a mankind considering various processes of technicization in our lifeworld. Deep understanding of present-day patterns of the mankind culture, different social systems, as well as general and individual human action/behavior – challenges us with difficult questions regarding future of a mankind like the one that is already happening before us in the shapes of (bio)technicization of different aspects of human existence. Thus, the authors are presuming that mankind will come in a new phase of transhuman reality from a position of technosociety no matter how distant or unreal it may seem.

Keywords: Biotechnology · Technicization of the lifeworld · Transhumanism · Culture of real virtuality · Transhuman reality

1 Introduction

The results of technical-technological evolution and revolution should be observed in a manner of critical ethos, and such a vision of the order of things is not possible if uncritically taken for granted the assumption how technical-technological innovations are destined as absolutely certain, undoubtedly destined for betterment and success of a humankind, as accomplishments which should not be questioned and think about different aspects of their impact on our lifeworld. The concept of a human development, relation between technique and society should be considered and analyzed in the context of the spirit of the times, in a way that all-time technization of lifeworld is approached from humanistic, as a societal side, and from the other side from new-instrumental, new-positivistical or indeed techno-utopian comprehension of that spirit of the times. Thanks to the reference authors, everything what we have viewed as an upcoming future and what was related to technization of our overall reality and panhuman existences, now it is becoming a matter of our reality and present in a quicker and more dynamic way. Technization of lifeworld can be analyzed in multiple ways, because only then we have the

I. Karabegović et al. (Eds.): NT 2022, LNNS 472, pp. 585–592, 2022.
https://doi.org/10.1007/978-3-031-05230-9_70

possibility to a holistic perspective, which won't be limited with neither exclusive techno-optimism nor techno-pessimism. Techno-sociality is one of the phases in technization of human existence and this phase manifests in technization of our communication forms. Gradually but surely and even guided, humanity is switching from a phase of a culture of real virtuality to a new phase of transhuman reality, which in its mildest form, is a question of a universal human meaning, because it will either announce the end of a concept of human or it will contribute to the new forms of post-human existence.

2 Reshaped Galaxy – Mcluhan's Galaxy

Communication is a main characteristic of society. Thanks to that kind of ability, human is a social being. Human speech has perfected through the process of evolution. Society has been inventing new ways of communication throughout the time. Appearance of writing system represents first significant change in human communication, since the written word has left mark from the past and it allowed us to get to know life of our ancestors, but it has also been a connection with ancient theories and philosophies. In fact, that was crucial for human cultural development until nowadays. In last decades, modern technologies have significantly modified our way of communication inside the society, which has led to changes in social order and habitus of individuals on a global level. Manuel Castells [1] had analyzed new networked society and he had also foreseen creation of a new global culture which is isolated from physical space. By analyzing *the rise of networked society*, Castell starts off with McLuhan's theory about development of mass media. Marshall McLuhan [2] had believed that strengthening the influence of television after Second World War had created a new communication galaxy, meaning it was the moment of "overcoming Gutenberg's technology" what McLuhan [2] himself marks as *a reshaped galaxy*. Actualization of television represents the end of Gutenberg's galaxy or the end of a communication system in which are prevailing „typographic mind and phonetical alphabetical order" [1, p. 360]. Although the television has taken a primary spot in communication system with wider audience, other media are still present, but they have transformed in order to survive in a new galaxy. Their content is determined by television content. Castells [1] emphasizes that the question why television has become primary media was long-discussed and considers Russell Neumann's [3] thesis which says it is "a consequence of primary instinct of lazy audience" [1, p. 359] as a most acceptable hypothesis. In other words, people are accepting the form of communication that requires the least effort and television fulfils that demand completely. We can identify this new form of culture as a screening culture [4, p. 2873]. In communication through television people are just making a decision to watch and decision about program is being made later and spontaneously: "To the audience of millions of viewers similar messages are being sent at the same time by centralized senders" [1, p. 360]. The main goal of sender in this case is homogenization of audience. McLuhan's galaxy is a productive area for manipulation and spreading propaganda content, considering that viewers are not thinking critically about program they are following. The process of changing human's habits and perfecting techniques of propaganda and manipulation has begun, but because of its non-critical relation towards media content, human being is not aware of them. McLuhan [2] considered that viewers directly participate in television program, since television is a live image in which a viewer is emotionally involved. Mass media is a sort of one-sided communication system what is different than actual communication system. Thereby, television represents stage for all social events that need to be transmitted to society and it is also what makes a television a powerful tool. Relation between media and society Castells [1] sees as feedback between distorted mirrors: "Media is an expression of our culture and our culture acts primarily through material given by media" [1, p. 365.]. Since television has possibility to connect with viewers on emotionally more intense level, the space for manipulation and indoctrination has been created through mass media. Affected by new technologies, beginning of 1980's has brought changes in media world. During that time radio went through specialization, newspapers are being written and edited from far away and using videorecorder becomes a popular thing. Possibility of recording television program and subsequent reproduction of content was exactly what changed audience's habits which led to watching program selectively. By the end of the previous century the number of TV stations has been increased and cable and satellite televisions have been developed. This led to segmentation of previously mass audience and segmented audience is related to sender of information to a larger extent. It should be taken into consideration that construction of global village itself [5, p. 25–28] has become useless and redundant, because Castells [1] would say how media "has become indeed globally connected, programs and messages are being spread through world network and we no longer live in global village but in adjusted cottages which are produced globally and distributed locally" [1, p. 370.]. However, diversification of media hadn't changed their one-sided character, since the audience was sending feedbacks only through market reaction. If we consider this, it would become clear that diversification of media and segmentation of audience hasn't decreased a possibility of manipulation. In fact, manipulation in these conditions becomes stronger, because people will identify with certain media depending on their character and thoughts. Therefore, people are still unaware of manipulation and impossibility of critical relation towards content.

3 Internet Galaxy – Software-Mediated Communication

As communication via internet becomes more popular, new communication galaxy is being developed. Castells [1] emphasizes that internet network is the spine of "world computer-mediated communication" [1, p. 374.] and it began to spread rapidly in the 90's of the last century. New communication galaxy allowed free flow of information in internet network. Thanks to digital technology development, information began to travel across network in other forms like sound or image, what created conditions for flow of information of any kind. In computer-mediated communication, it first seemed like the principle of one-sided communication between network actors has weakened: "opposite of mass media of McLuhan's galaxy, CMC networkshave technologically and culturally ingrained traits of interactivity and individualization" [1, p. 386]. A new galaxy has opened possibility towards greater interaction, but experience form last decades had shown that new forms of manipulation have been born. Castells [1] thought that by using new media, the line between separated parts of society – family and business, free time and fun is fading. In that way incompatible social fragments are overlapping in new *real virtuality*. If we could follow development of modern media in an appropriate manner, it would first of all require analysis of transformation of media content as well as their influence on social actors. Therefore, we could say that new media contributed to the creation of virtual communities and we had transformation in understanding the concept of space itself, but social actors are also facing new forms of *social isolation* [6]. By merging globalized and transformed mass media and computer-mediated communication by the end of 20th century, a new system of electronic communication has been born – a multimedia. The multimedia emphasizes interaction between different media and electronic communication is being extended to all segments of life. One of the characteristics of social pattern, which is supported by multimedia system, is communication of all kinds of messages in the same system, even though system is selective, meaning we have "integration of messages in one common cognitive pattern" [1, p. 398] what leads to mixing different meanings. In the new communication galaxy different contents are overlapping through one multimedia. People have a possibility to experience contents through one electronic device, when usually they would be experiencing this in separated physical environments like school activity, sport activity, trial, sex. This has significantly shaken cultural norms and the system of symbols. Now we have a chance for these different cultural contents in *digital universe* to connect in one *supertext,* regardless if these contents are the most elite ones or the most popular. Thereby, they are creating completely new environment, a new form of reality. Castells [1] observes how "all kinds of cultural expressions, from the worst to the best, from the most elite to the most popular are standing together in this digital universe and connecting to the previous, present and future manifestations of communication mind in a huge historical supertext. That way, they are forming a new symbolic environment. These expressions are making virtuality become our reality" [1, p. 399].

4 Culture of Real Virtuality

Key elements of culture are communication processes based on production and consummation of symbols, therefore societies act through world of symbols. Inside a new communication system, it begins with a creation of *real virtuality*. This definition Castells [1] derives from meaning of terms: virtual (how it actually is, maybe not strictly or by name) and *reality* (actually existing) and Milardović [5] writes about nine definitions of virtual reality [5, p. 77–80]. Reality as we know has always been virtual considering different perception based on cultural symbols. Consequently, reality as we know has always been virtual, because we are seeing it through symbols which "frame practice with some meaning that is slipping out of their rigorous semantic definition" [1, p. 399–400]. Communication system which produces real virtuality Castells [1] defines as a special system in which "reality itself (i.e. human material/symbolic existence) is completely embraced, entirely immersed in virtual setup of images, in imaginary world where phenomena exist not only on screen, through which experience is communicated, but they become experience themselves" [1, p. 400]. In the internet galaxy a new culture has been born characterized by significantly different norms of behavior and cultural symbols from the ones that we already knew. New electronic communication integrates cultural differences and weakens the power of traditional message senders. Traditional message senders through history, like religion needs to integrate in order to ensure its own place in culture of real virtuality. New communication galaxy transforms time and space. Thus, parts of space are being relocated from physical environment to internet galaxy. On the other hand, time in internet galaxy cannot be programmed what erases boundaries between past, present and future. Modern technology has fundamentally changed the nature of communication what directly creates one new culture. Today's development of communication in newly born culture has shown that new galaxy does not mean the end of mass media, manipulation, and non-critical audience. Has development of new technologies only created an illusion of major interaction and omnidirectional communication in internet galaxy? Should we ask ourselves if manipulation, propaganda, and indoctrination are just better disguised? Vertovšek [7] emphasizes how there is no clear transition between old and new media and also highlights that under the influence of new media human perception of reality has changed and we are not aware of that. By globalization of mass media, important social topics like economic inequality or poverty are being replaced with superficial and less important topics, since they require less effort of audience: "In the nature of mass media is not to deal particularly with actual changes, but to change image of reality which, behold, produces itself new images of reality of new media" [7, p. 24]. Based on that, we could conclude how media in modern age has sacrificed its baselines, informing of public, freedom and professionalism for the sake of corporation's interests and power centers. Distracting from relevant contents by placing socially irrelevant topics is one of the ways to manipulate through media, especially in the era of digitalization. Manipulation is also done by placing stereotypes and prejudices, especially in case of other cultures because that is the easiest way to satisfy an average member of society. It is becoming clearer how media and their methods are more violent and more aggressive what reflects in creating new stereotypes, mainly "untrue and unverifiable, but very convincing" stereotypes [7, p. 37], what results in destabilization of human relations generally and concentration and contemplation especially. For a reason Nicholas Carr [8] wrote:"... it seems to me that internet corrodes ability of concentration and contemplation. Whether I am plugged in or not, my mind now expects that it will receive information in the same way internet sends them in uninterrupted flow of small particles. I used to be a diver in the sea of words. Now I sail on the surface like those guys on water skis" [8, p. 18]. Vertovšek [7] opens the question of practice for a reason: whether the internet era is the human peak of global information exchange or it is a "dark age of mediocrity, narcissism and dehumanization" [7, p. 113]. Existing experience in the new communication galaxy didn't show tendency to strengthen people's critical thinking. It can be said that completely opposite is happening. As in McLuhan's galaxy, people are not thinking critically about content that is being sent to them. Access to all kinds of information in any moment has caused additional distancing from thoroughly analyzing the content. As Vertovšek [7] mentions, books are being read less and often interactions with books are limited to the superficial analysis. It is possible to find all the answers in the network and the way how we perceive social reality and media content has been significantly changed. Under these circumstances, the whole population is being dehumanized. Gradual facilitation of man's life by developing new technologies, basic activities are disappearing from human lives and everyday life. At first, in the computer-mediated communication it was possible to exchange text messages and later the content of other formats like sound and image. Today we are witnessing a transition of the entire social systems from well know space to a new and invisible, but very much real space. Pupils and students are attending classes online [4, p. 2876.], companies are working only in cyberspace, all necessary components for daily activities are incorporated in mobile phones. Social systems are integrated into the new communication galaxy what has made the culture of real virtuality deeply embedded. Transformation that already happened and it is continuously happening in front of our eyes, defines our future in a specific way. We talk about new codes, symbols and meanings, about new perception of overall reality and it is justified to consider how much mankind will be prepared for upcoming future [9, p. 13–15].

5   Era of (Bio)technicized Corporeality – New Spheres of Transhuman Reality

It is important to bear in mind that we are facing real difficulties in adequate understanding the role of technical-technological innovations, new forms of communication, new forms of technization of human's mental and physical existence. In previous titles we strived to show the new forms of understanding and perceptions of our reality, in a way of digital-virtual image of the world. It is inevitable to emphasize how with technization of our communication, the processes of technization of our perception of reality are also occurring and based on that we could talk about concepts of techno space and techno time. We are noting more certainly that we overcame the era in which we announced the world of cyborg, drones, non-humanoid and humanoid robots, superhuman/post-human identities [10, p. 284] and we are already in phase where we have reports about murders via drones, shopping and delivering goods by these exact drones, about humanoid and non-humanoid robots protecting us, serving or doing as we say. From the culture of real virtuality, humankind is now going to a phase of transhuman reality [11]. Therefore, Anne Balsamo [9] in her efforts to read the body in technological culture will write:" In our hyper-mediated techno-culture consciousness about body is enhanced, so with a help of technology we could witness, if not rule, to performance of body processes on molecular level" [9, p. 306]. (Bio)technization of our corporeality is one of the specific elements and characteristics of biotechnological century [12], which basically does not necessarily mean catastrophic consequences, but warnings coming from modern scientists (Rifkin [12], Žižek [13], Tofler [14, 15], Touraine [16, 17], Sloterdijk [18], Jonas [19], Beck [20, 21], Baumann [22, 23]…), considering different forms of bio(technicized) discrimination, digital control, surveillance and manipulation, are demanding a special form of precaution and in-depth analysis for indicated process of transformation. Humankind is, truth to be told one circle of worried individuals, in the phase when we are still trying to name properly the overturn which happened, epistemological turn that we are witnessing, twist to a certain transhuman era. Shall we name this post-human or after-human state or transhuman turn defined by worldview when it comes to sense and position of a man in lifeworld. It is significant how transhuman reality is marked by striving towards self-evolution of humankind, in a way of techno-scientific rationality and all of that to surpass what an average person is, as well as to have attributions of a transcendent being, provided/transmitted onto transhuman. This way, traditional understanding of human identity has been questioned [24]. With intention for self-evolution into transhumancreation, but also with intention to strengthen the potentials of all kinds of artificial intelligence, with constant modification of different forms of our lifeworld, it is more and more realistic to have a state where only intelligence makes sense, which with its own capacities and still unreached ambitions surpasses and unreservedly cancels man's world in a sense of corporeality, mental constitution and all kinds of attributions that we consider human's from its creation/evolution.

Discussion about how digital and robotic reality [7, p. 107], transhuman reality opens even ethical questions is so weak and late that every attempt of its revitalization, systematically within conventional knowledge, dominantly belittles. Unfortunately, human's action primarily realizes in a way what can be done and must be done and afterwards returning and referring to ethical questions, in the state when the whole image of world is covered by techno-scientific and transhuman domination, seems like dead-end situation. We strongly believe that, in spite of all current parameters which go in favor of the domination of all unnatural forms of perceptions of life, just temporarily, or so we hope at least, humankind will come to the state of post domination and real responsibility towards itself, its essence and responsibility towards the lifeworld in general. Our hopes are justified if the volume of cognitions and perspectives for our future enlarges in favor of holiness of Life and its understanding. However, what lies ahead for us in the near future is deeper "immersion in cybernetic reality" [11, p. 100], which was already named as digital reality or robotic reality [7].

6   Conclusion

The authors tended to make a sort of demystification of techno-optimistic and techno-pessimistic conceptions in understanding technical-technological evolution and revolution, as well as the influence of (bio)technization of our lifeworld on social actors and society as a whole. We still have open possibilities for more pleasant future of mankind, but also for other species. However, the authors have shown fear that by uncontrollable processes of technization of our lifeworld, the future of mankind will probably transform into something unfavorable for human nature, environment, freedom, creativity. The main assumption in this paper was how mankind from digital and robotic reality and techno-sociality will come to a new phase of transhuman reality, no matter how far or unreal it may seem. Deep understanding of present patterns of virtual culture of mankind and era of transhuman reality confronts us with complex questions of mankind's future, the one that is already happening in front of our eyes, as well as the one on which we should responsibly and profoundly influence.

References

1. Castells, M.: Uspon umreženog društva. Golden marketing, Zagreb (2000)
2. McLuhan, M.: Gutenbergova galaksija (Nastajanje tipografskog čoveka). Nolit, Beograd (1973)
3. Neuman, W.R.: The Future of Mass Audience. Cambridge University Press, New York (1991)
4. Fejzić-Čengić, F., Sofradžija, H.: Hipermedijska sfera i virtualne učionice. In Medias Res, Časopis filozofije medija **10**(18), 2869–2880 (2021). Centar za filozofiju medija i mediološka istraživanja, Zagreb

5. Milardović, A.: Globalno selo: sociologija informacijskog društva i cyber culture. Centar za politološka istraživanja, Zagreb (2010)
6. Alić, S.: Masmediji, zatvor bez zidova. Centar za filozofiju medija i mediološka istraživanja, Zagreb (2012)
7. Vertovšek, N.: Drveno željezo medija. Medijska kultura, Nikšić (2020)
8. Carr, N.: Plitko. Jesenski i Turk, Zagreb (2011)
9. Featherstone, M., Burrows, R. (ur.): Kiberprostor, kibertijela, cyberpunk: kulture tehnološke tjelesnosti. Naklada Jesenski i Turk, Zagreb (2001)
10. Hayles, K.N.: How We Became Posthuman (Virtual Bodies in Cybernetics, Literature, and Informatics. The University of Chicago Press, Chicago-London (1999)
11. Vertovšek, N., Gregurić, I.: Filozofija budućih kiberprostora i transhumanistička stvarnost. Filozofska istraživanja **38**(1), 99–116 (2018). Zagreb
12. Rifkin, J.: Biotehnološko stoljeće: trgovina genima u osvit vrlog novog svijeta. Naklada Jesenski i Turk, Hrvatsko sociološko društvo, Zagreb (1999)
13. Žižek, S.: *Disparities*, Bloomsbury Academic. An Imprint of Bloomsbury Publishing Plc., London (2016)
14. Toffler, A.: Treći talas. Prosveta, Beograd (1983)
15. Toffler, A.: Šok budućnosti. Grmeč, Beograd (1997)
16. Touraine, A.: Kritika modernosti. Politička kultura, Zagreb (2007)
17. Touraine, A.: Nova paradigma: za bolje razumevanje savremenog društva. Službeni glasnik, Beograd (2011)
18. Sloterdijk, P.: U istom čamcu: ogled o hiperpolitici. Časopis Beogradski krug, Beograd (2001)
19. Jonas, H.: Princip odgovornost: pokušaj jedne etike za tehnološku civilizaciju. Veselin Masleša, Sarajevo (1990)
20. Bek, U.: Rizično društvo: u susret novoj moderni. Filip Višnjic, Beograd (2001)
21. Bek, U.: Svetsko rizično društvo: u potrazi za izgubljenom sigurnošću. Akademska knjiga, Novi Sad (2011)
22. Bauman, Z.: Fluidni život. Mediterran Publishing, Novi Sad (2009)
23. Bauman, Z.: Tekuća modernost. Naklada Pelago, Zagreb (2011)
24. Fukujama, F.: Naša posthumana budućnost: posledice biotehnološke revolucije. CID, Podgorica (2003)

Experiences of Application of Zoom Application in Higher Education on Examples of Teaching Mother Tongue Spelling

Milena Burić[1,2(✉)]

[1] Univerzitet Crne Gore, Filološki fakultet, Podgorica, Montenegro
`milenab@ucg.ac.me`
[2] Faculty of Philology, University of Montenegro, Danila Bojovica bb,
81 400 Niksic, Montenegro

Abstract. The paper analyzes the advantages of the usage of the Zoom application in teaching and learning process during the COVID-19 pandemic, as well as the examples of good practice in teaching mother tongue spelling.It also reviews the current perception of the disadvantages of online teaching, citing the challenges posed by the sudden transition from traditional to online work format, due to the unpreparedness of the educational system and reference legislation. It is stated and exemplified that the mentioned type of teaching provided the necessary tools for combining appropriate teaching methods, thus achieving the primary outcomes which are reflected in the acquisition of knowledge of spelling and the improvement of positive orthoepic and orthographic habits. It is concluded that it is necessary to include the Distance Learning System in the forthcoming reaccreditation process, as well as to define the ways of its incorporation into the "classical" form of teaching, which will have to be supported by adequate legal norms. It is also concluded that teaching through the Zoom application intensified the two-way teaching process, encouraged the development of digital competencies of its participants and raising them to a more advanced level.

Keywords: Zoom application · Spelling · Spelling teaching methodology · Method (dialogical monologue text) · Experiences

1 Introduction

The planet-known virus is the cause of sudden changes in the forms of work at all levels of education, including higher education. In the conditions of general isolation, lecturers in higher education, depending on the epidemiological measures of the state, as well as on the policy of the institution, were offered several possibilities of conducting lectures at a distance, popularly called "online". While at the primary and secondary level of education, as shown by research in Western Balkan countries, some other models of online teaching were more popular (such as Google classrooms, Viber, and even Microsoft Teams) [1], the Zoom application experienced the widest affirmation in the higher education system, primarily because of easy and fast connection, but also because

I. Karabegović et al. (Eds.): NT 2022, LNNS 472, pp. 593–598, 2022.
https://doi.org/10.1007/978-3-031-05230-9_71

of a number of other advantages that will be discussed in the paper. Of course, this application, as well as other online forms of work, is marked by certain shortcomings, but in conditions of general isolation, the educational goals achieved through its application are worthy of respect, which we will hopefully show by examples of teaching native language spelling.

2   Limits and Challenges

In the literature, the most common limits of online teaching are: lack of direct communication (live), questionable objectivity of assessment and difficult examination, lower educational efficiency compared to traditional teaching, technical problems [1]. We would like to add that we should not forget that the most striking shortcomings and challenges in teaching through the Zoom application are the result of the lack of adequate legislation and the absence of methodological-professional and digital competencies of certain actors in the online teaching process.

Current law on higher education, as well as other sub-laws that regulate educational activities in Montenegro, mention online teaching only as a possibility of "distance learning" without precise and detailed elaboration of the application of this teaching model [2, 3], so the question of the regularity of its implementation arises. For example: in order for the recording of participants in Zoom classes to be legitimate, it would be necessary to obtain the consent of each student (as an adult citizen) individually. Since the study contracts, which students usually conclude after enrolling in college in Montenegro, do not include any articles regarding distance learning, the question arises whether students are required to attend lectures in this format, whether the professor has the right to categorically require their visual (by turning on the camera) and not only auditory and verbal participation, whether the participants in the teaching process are obliged to provide themselves computer, mobile phone and internet connection, as well as adequate work space etc. So far, the university has not been in any way responsible for solving mentioned challenges, and experience shows that there were a lot of challenges during previous period. Namely, poor internet connection, interruptions of classes due to technical problems (loss of electricity, and thus home internet), as well as impaired audibility, disrupt the continuity of teaching and weaken the concentration and patience of its actors. Misunderstanding between the professor and the student also arises because it is difficult to interpret the signs of non-verbal communication, gestures, facial expressions on the screen, which are sometimes quite enough in live teaching to establish good cooperation and resolve possible misunderstandings.

Insufficient IT awareness and lack of digital competence have been leading professors to some unpleasant situations. There were many cases of participation of uninvited persons in Zoom classes, where the lecturer was not able to protect himself or other invited participants. The lecture was also unprotected in the sense that students sometimes followed classes via mobile phone while traveling by bus or some other public place, so the teacher could never be sure who was able to listen and evaluate him, which certainly creates a feeling of discomfort and indirectly negatively affects the quality of his exposure.

Wide legal regulations defines that a student completes studies according to the law and the rules of studies that were valid on the day of enrollment. The scoring of attendance at classes and the preparation of seminar papers and reports provided by the program was changed althought it is not obliged to do any changes after the official beginning of the academic year.

It seems that the digitalization of society, the development of information and communication technology and its application in higher education is not accompanied by the timely adoption of legal regulations that would regulate the conditions and ways of implementing the most modern online forms of work, including teaching using the Zoom application. Due to all the above, we agree that "distance learning will be improved in both practical and legislative terms" [4].

The credibility of oral answers has been constantly questioned, and for successful cooperation between professors and students, as well as for any functional interpersonal relationship, trust is necessary. Due to doubts about the authenticity of the assignments, answers to the requests and questions, the lecturers avoided entrusting the students with work assignments that they would carry out in a monologue, by reading papers, seminar papers etc. This violated the objectivity of assessment, and the academic ethics of all participants in the teaching process was often tempted.

Insufficient methodological information, as well as the lack of digital training, used to limit professors so much that they resorted to reading materials and dictation, which is the most unpopular form of work not only when it comes to presenting spelling materials but in teaching methodology in general. According to the testimony of students, a classic academic lecture via the Zoom application, ie. reproductive monologue communication lasting longer than 40 min passivized them and, in extreme cases, put them to sleep. The mentioned methodical approach necessarily implies antipathies towards the professor and, more importantly, aversion or indifference towards the subject.

It has already been noticed that one of the shortcomings of online teaching is the lack of developing students' communication competencies [1]. This shortcoming, when it comes to teaching mother tongue spellingthrough the Zoom platform, can be manifested only in cases of absence of methical preparation of the lecturer, ie only when the lecturer excludes the dialogical method. Thus, during the teaching through the Zoom application, the statement that methodology cannot be a normative discipline was confirmed more clearly than ever, because "valid teaching practice is based on inventive behavior" [5].

3   Advantages and Examples of Good Practice

The advantages of teaching mother tongue spelling through the Zoom application may be divided into general and specific. In general, they refer to the advantages that this application has in relation to conventional or traditional, but also to other types of online teaching, while specific ones would refer exclusively to positive experiences in teaching materials on linguistic or spelling topics.

The most important value of teaching "at a distance" is certainly the prevention of the spread of the current infectious disease. The possibility of illness of participants in the teaching process has been reduced to zero, which is impossible to achieve in the conditions of "normal" lectures and exercises, given that the epidemic is spreading due to physical contact.

Zoom application gives the lecturer the opportunity to record a complete lecture, subsequently view the recording and thus eliminate or correct any methodological and other errors and shortcomings that are detected by careful and objective analysis of the class and self-evaluation. It must be acknowledged that this is a very valuable opportunity to improve one's own work, and indirectly to improve the teaching process.

The advantage that this application has over other online forms (eg. Moodle platform) is the possibility of very simple, fast and successful connection of students and the control that the organizer of the Zoom lecture has over this process. Also important is the fact that a large number of students have already participated in the "Zoom Rallies" at the high school level, so it was not necessary to spend time on registration and mastering the tools offered by some other platforms.

It has already been noticed that online teaching enables mass participation in it, twenty-four-hour availability, asynchronous communication, development of digital skills, access to various sources on the Internet [1]. Thus, during the work through the Zoom application, simultaneous individual and group work is enabled. The lecturer can address a larger number of people orally and at the same time answer the questions of individual students through a "chat" without disturbing the attention of the team. Such situations require from the professor increased intellectual engagement, quick reactions, the ability to synchronize mental, motor and speech activities. In this way, his skill of multioperativeness is encouraged and perfected.

When it comes to specific advantages, the following should be emphasized: since the orthographic topic belongs to the synchronic linguistic perspective, there were no barriers in Zoom classes that were present in the classes of diachronic (historical) language disciplines (such as Old Slavonic, language history, dialectology), for whose thematic processing requires high audibility, absence of noise and interference with connections, in order to adopt the correct pronunciation of former voices (eg. semivowels, nasal vowels).

Zoom teaching enables the implementation of methodical actions of immanent so-called traditional teaching, with the need to choose a combination of methods adapted to the online format. Thus, the inductive method proved to be effective in teaching the mother tongue through the Zoom application, especially when processing topics that are somewhat familiar or close to the student population. Such topics are capital letter writing, compound and disassembled word writing, spelling signs, punctuation, writing abbreviations. For example: when processing the teaching unit of capital letter writing, students, after giving selected and grouped examples (names of geographical terms, institutions and organizations, holidays, historical events), drew their own conclusions, ie the orthographic norm that is used in cases of the same type. By combining the inductive with the dialogical method, the attention of the lecture participants was permanently at a very high level, liveliness, interestingness and healthy competitiveness were provided. Students are continuously "awake", active and motivated to exemplify the rule they have come to, to help each other by correcting and analyzing the mistakes made.

When processing spelling and punctuation marks, as well as abbreviations, the text method proved to be successful, which is also supported by the Zoom application tools, since students can be immediately presented with the text in which they need to make spelling corrections.

Accustomed to a dialogical way of working, encouraged to talk and free to ask questions, when processing composed and disassembled word writing, students showed an incredible interest in contemporary examples that are treated differently by spelling authors [6–10]. They themselves searched for answers in their recommended spellings, some of which are also available online.In this case, the professor is obliged to direct them to the literature that offers systemic, not exemplary (individual) solutions [11], because only in this way will students be fully informed and able to detect errors very easily, and thus protected from the negative impact of spelling errors that abound in the media and social networks.

Lesser known topics, such as vocal alternations, were more successfully treated by applying deduction and monologue method in the introductory part of the class. When processing the mentioned teaching units, students should first be acquainted with the rules and conditions under which changes are made. Then, in order to break the monotony of the monologue, by giving research tasks and questions, students would be motivated to explain the examples by applying the above rules. After that, again in the lecturer's monologue, students would be introduced to deviations from the rules, ie exceptions, followed again by examples for the exercise.

By harmonizing the logical with the methods of communication, e.g. inductive methods with dialogic, known material from orthography is successfully placed in the service of knowing the unknown, students are allowed to express and defend their own, and refute other people's opinions, build a culture of speech and feel joy for their own affirmation and results of the heuristic method [5]. On the other hand, integrated methods of deduction and monologue enable saving time and systematic processing of the teaching unit, but their combination with other methods, primarily textual and dialogical, is necessary in order to achieve all other educational goals.

4 Conclusion

The future task of the methodology of teaching spelling and mother tongue in general is to form guidelines and suggestions for improving teaching and improving curricula based on proven positive experiences of online lectures and through their inclusion in the so-called traditional format or live teaching. The re-accreditation of study programs at Montenegrin universities in the near future already envisages the mandatory intro duction of online teaching in a certain percentage, which will determine the conditions and methods of its implementation. In this regard, the support of laws and sub-laws is necessary.

Although there was no social interaction with all its educational and psychological benefits during the Zoom lecture, this type of teaching resulted in a special kind of closeness, occasionally familiarity in communication, the use of youth jargon - which can be justified by the fact that the workspace was usually a home environment and that it was difficult to get rid of the habits and freer behavior that is present in the home and family.

The Zoom application has gained enormous popularity and applicability in higher education by offering quick and easy connectivity, tools that enable a successful and timely combination of logical and communication methods, while encouraging the improvement of multi-interoperability of all participants in the teaching process. It enabled the realization of the primary goals of spelling teaching - the acquisition and development of knowledge, as well as positive orthoepic and orthographic habits.

It may be said that the younger generations are already determined by their birth as members of a technological, not to say technocratic society, and their digital competencies are at a more advanced level than the representatives of the more mature, ie older population who are now forced to learn from those they teach. The Zoom application has intensified and clarified the two-way teaching process: teachers and students learn from each other, and as long as the professor is willing to learn rather than just teach, the education system and all its participants benefit.

References

1. Matijašević, J., Carić, M., Škorić, S.: Online nastava u visokom obrazovanju – prednosti, nedostaci i izazovi, XXVII Skup Trendovi razvoja: "On-line nastava na univerzitetima" (zbornik radova), Novi Sad, pp. 165–168 (2021). www.trend.uns.ac.rs
2. Zakon o visokom obrazovanju. https://www.paragraf.me/propisi-crnegore/zakon-o-visokom-obrazovanju.html
3. Statut Univerziteta Crne Gore. https://www.ucg.ac.me/objava/blog/6/objava/3708-statut-uni verziteta-crne-gore
4. Bingulac, N., Matijašević, J., Škorić, S.: Aspekti učenja na daljinu posmatrano iz pravne perspektive, XXVII Skup Trendovi razvoja: "On-line nastava na univerzitetima" (zbornik radova), Novi Sad, pp. 23–26 (2021). www.trend.uns.ac.rs
5. Nikolić, M.: Metodika nastave srpskog jezika i književnosti, Beograd (1992)
6. Perović, M., Silić, J., Vasiljeva, LJ., Čirgić, A., Šušanj, J.: Pravopis crnogorskoga jezika, Ministarstvo prosvjete i nauke, Podgorica (2010)
7. Ivić, P., Klajn, I., Pešikan, M., Brborić, B.: Jezički priručnik, Beograd (1991)
8. Halilović, S.: Pravopis bosanskoga jezika, Sarajevo (1996)
9. Pešikan, M., Jerković, J., Pižurica, M.: Pravopis srpskoga jezika, Novi Sad – Beograd (2002)
10. Pravopis srpskohrvatskoga književnog jezika, Novi Sad – Zagreb (1960)
11. Burić, M.: Precizna kodifikacija – neophodan uslov za kultivaciju norme, in: Jezička situacija u Crnoj Gori – norma i standardizacija (Zbornik radova br. 87), Crnogorska akademija nauka i umjetnosti, Podgorica, pp. 185–193 (2008)

IoT System for Structural Monitoring

Angelo Lorusso$^{(\boxtimes)}$ and Domenico Guida

Department of Industrial Engineering, University of Salerno, Via Giovanni Paolo II, 132, 84084 Fisciano, SA, Italy
{alorusso,guida}@unisa.it

Abstract. Structural monitoring of civil buildings is extremely important and topical, especially in monitoring the life cycle of reinforced concrete construction. It is possible to assess the impact of weather and external factors that can affect the safety of buildings by monitoring their natural vibrations. However, conventional vibration monitoring systems are complicated to install and operate and costly. Furthermore, not all monitoring systems integrate real-time data processing and remote data storage. Taking advantage of discoveries in the field of information technology, a monitoring system based on the paradigms of the Internet of Things (IoT) has been developed to provide a viable alternative to tried-and-tested vibration monitoring systems. A prototype system consisting of a microprocessor (Raspberry Pi) and a low-cost accelerometer for microelectromechanical systems (MEMS) was used to minimize the cost and size of the system. The architecture of the IoT vibration monitoring and display system and its working mechanism are explained in all the steps. The performance of the developed IoT vibration monitoring system was visualized through a cloud application that gives the possibility to view the collected data in real-time. The future of this study is the application of Machine Learning techniques about the data acquired.

Keywords: IoT · Sensor · Digital Twin · Structural Health Monitoring

1 Introduction

In the last decade, the problem of mechanical and physical monitoring structures for civil use has begun to undergo a radical change from the approach adopted in previous years. In fact, structures, especially those made of reinforced concrete and steel, were designed for an indefinite period of use and duration and were only checked occasionally and, usually, and usually only after unexpected events in which damage was presumed, such as earthquakes, landslides or other [1]. Thus, in addition to monitoring, the concept of maintenance was added, in the sense that proper maintenance is of fundamental importance to significantly increase the useful life of a structure or infrastructure, while also having a minimal environmental impact due to a large amount of waste materials demolished and the materials needed to renew the existing construction. So, especially for large public heritage structures, the idea was to intervene periodically by monitoring and controlling them through a supervision program [2, 3]. But because of both the lack of investment of public funds and the mismanagement of these funds, this planning is not

I. Karabegović et al. (Eds.): NT 2022, LNNS 472, pp. 599–606, 2022.
https://doi.org/10.1007/978-3-031-05230-9_72

done, and most of the time, there is no effective control, causing terrible collapses that have claimed many lives [4]. On the other hand, in the case of small buildings or those that do not form part of the public heritage, maintenance planning becomes even more difficult and complex to carry out, especially in private buildings, because the owners show that they do not have the capacity to perceive the danger that a damaged structure may entail [5]. All this has led to questions about new paradigms of action and thus to the birth of a completely innovative and modern concept of "structural health" and the strict necessity of its monitoring through systems linked to new technologies coming from the concept of IoT and Digital Twin [6].

1.1 State of Art

Thus, the concept of Structural Health Monitoring (SHM) [7] is introduced, a new subject involving several scientific disciplines by incorporating knowledge and technologies from different backgrounds such as civil technologies, mechanical engineering, and computer science. SHM provides a new approach to the detection of structural damage in buildings, giving advance notice to maintenance engineers and thus the ability to intervene through inspections, repairs, reinforcement, or otherwise [8]. Very often, these systems are limited to this type of application, but there is an increasing need to be able to predict the location and quantification of any damage. This goal is not trivial, and to obtain satisfactory results it is necessary to introduce artificial intelligence techniques, such as neural networks or genetic algorithms, whose "training" is obtained thanks to computational mechanics techniques of the structure by predicting the presence of various damages within the structure. Thus, the SHM discipline comprises several steps:detection, localization, quantification, and prediction.

Given the multidisciplinary nature of SHM, many new methodological applications and materials allow a practical assessment of the health condition of a structure or infrastructure: from materials used for diagnostics [9], or composites, to modeling of structures with finite element methods or Digital Twin, and SHM connected with IoT [10]. They represent a technology based on the Internet of Things (IoT), in which physical objects can live and interact virtually with other machines and people. One of the areas in which the Digital Twin could potentially provide significant contributions, and in part is already doing so, is related to monitoring the conditions of buildings and structures [11]. The ability to continuously monitor the main parameters of a building structure allows the identification of possible structural problems and the implementation of actions to prevent and manage risky events [12]. The growing evolution of BIM has given a solid development for this approach: Building Information Modeling indicates a method for optimizing the design, construction and management of buildings with the help of software. All relevant data of a building can be collected, combined and linked digitally. The virtual building can also be visualized as a three-dimensional geometric model.

According to Gartner, among the five emerging trends that will drive technology innovation for the next decade, one place must be given to the Digital Twin. It's not hard to imagine how digital twins incorporating big data, artificial intelligence, machine learning, and the Internet of Things will become central to Industry 4.0. A Digital Twin is a virtual representation of a physical entity, living or non-living, a person or a system, even a complex one[13]. The digital component is somehow connected with the physical part, with which it can exchange data and information, both synchronously (in real-time) and asynchronously (at later times). The digital twin can develop into a true digital copy of real, physical elements (physical twin), represent processes, replicate people, reproduce places and infrastructures, systems, and devices used for various purposes [14, 15]. In general terms, the digital twin is characterized by the set of data and information that compose it in any way related to the element to be represented as the digital twin. Another characteristic is how the elements of the physical component are connected with the corresponding virtual part [16]. Furthermore, how to access data and information resources through web platforms, with the possibility of searching and analyzing information (machine learning, deep learning, big data, artificial intelligence). Finally, how to exchange data and information between the virtual and physical components, using sensors and actuators [17]. The data representing the digital twin can be derived from a wide variety of sources: from sensors that transmit various aspects of its operating conditions; historical data relating to past conditions; data provided by human experts, such as engineers, technicians, doctors, with specific and relevant knowledge; data collected from other similar machines, or from the systems and environment of which it may be part [18]. Information is retrieved from any database accessible through the Internet [19]. The digital twin may also use machine learning and artificial intelligence systems, to process data and produce new knowledge or predict operational scenarios by analyzing the collected data. The term digital twin is particularly popular in the industry, where it is specifically used to refer to the virtual reproduction of a real process or service made through data collected by sensors. In many industries, digital twins are already widely used to optimize the operation and maintenance of both physical assets and production systems and processes [20].(Fig. 1)

Fig. 1. Prototype

A possible vibration monitoring system can solve several problems related to vibrations that arise during the operation of civil structures and reinforced concrete and steel infrastructures. Consequently, the development of vibration monitoring and analysis systems is widely effective.

2 Development

2.1 The Aim

The objective of the study was to develop a system for monitoring civil structures in reinforced concrete, using a device characterized by its low cost and suitable to adapt easily to different environmental conditions that are proposed, and the analysis of the vibration acceleration spectrum [21]. For this, it was necessary to study different approach issues, for example, set up a workflow based on the modularity of events, then develop a basic helpful prototype for initial tests and different environmental tests, finally integrate data from the prototype to the cloud system to visualize and use the results obtained through a Python code.

2.2 Prototype

The prototype assembled in the laboratory consists of an advanced microprocessor such as a Raspberry Pi4 board used for the acquisition and transformation of data from the MEMS accelerometer, a sensor chosen for this type of acquisition is a triaxial MPU digital accelerometer, an additional board suitable for the Raspberry where to allocate a data card to send data via the Internet to the cloud system; a cloud system, Thinksboard for storing, displaying and managing the collected data [22].

The prototype developed is based on the principle of modularity. It can be easily customised according to needs and environmental conditions and allows it to be quickly developed and monitored for further improvements. This type of prototype ensures a low cost of the project proposal thanks to the use of hardware elements that are readily available on the market and easy to use by free, i.e. non-proprietary software. The accelerometer chosen for testing is the ADXL345, a miniature triaxial digital sensor, which ensures low power consumption, high data resolution (13 bits) and an acceleration measurement range of up to ± 16 g. The measurement range can be selected from the row: ± 2 g, ± 4 g, ± 8 g and ± 16 g. The measurement result can be read byte by byte via the SPI or I2C digital interface in a 16-bit data form. The chosen type allows the measurement of low-frequency vibrations, dynamic accelerations, static gravity accelerations, tilt angle and movements. The sensor can be safely placed both indoors and outdoors, and can be used at temperatures from $-40\,^{\circ}$C to $+85\,^{\circ}$C. It also has a high impact resistance of up to 10,000 g.

The prototype system provides constant data for real-time monitoring of natural building vibrations, e.g. building vibrations as a function of external influences, and analysis of vibration parameters. The system can issue warning alerts and specialized alert operators, thus preventing possible malfunctions of the analyzed structure. The system provides the user with sufficient information to identify potential problems and choose preventive measures based on the analysis of the vibration spectrum for a concrete building, comparing the vibration spectrum under the same conditions at different times [23].

This prototype proves to be very advantageous for the prevention and monitoring of reinforced concrete buildings as a whole and the protection of buildings and infrastructures in civil use, since the system can prevent significant long-term damage due to its rapid response to previously recorded changes in natural vibrations [24].

Natural vibrations must be monitored both in time and in frequency bands.

For a building, vibration sampling can be carried out continuously 24 h a day, but this would be excessive and extremely heavy for storing and processing data. Therefore, given the ease of setting the sampling through commands applied to the Raspberry, it is more useful to study case by case the type of helpful sampling to have comparable data from one sampling to another.

In our case, only for study purposes, a sampling of 15 minutes every 6 h was foreseen, equal to 4 daily samples. During the sampling phase, the sensor's raw data is transformed into more useful data for reading and analysis. Then an algorithm was used to develop the acquired data locally in real-time; in fact, a Fast Fourier Transformation (FFT) and a bandpass filter are applied to the raw acceleration signals to remove noise and static components. Sufficient flash memory size should be selected to avoid data loss due to lack of space. In general, the Raspberry Pi 4 will transfer all this data simultaneously to the cloud for permanent data storage and visualization. Due to the increased computational potential, the Raspberry Pi 4 provides a greater capacity for real-time processing of the stored data; in fact, a local data processing algorithm is developed within the prototype sensing node. Sufficient size of the flash memory should be selected to avoid data loss due to lack of space. In general, the Raspberry Pi 4 will transfer all these data simultaneously to the cloud for permanent data storage and display [25].

2.3 Cloud - ThingsBoard

ThingsBoard is used for data visualization and storage. ThingsBoard is an open-source cloud platform dedicated to the management of IoT devices. Thanks to the MQTT protocol, it can interact with the devices in real time; in fact, MQTT is used to connect the sensors for structural monitoring to the platform. In addition, ThingsBoard allows the easy visualization of data through dashboards created by the operators, who can modify and generate them according to specific needs. They were then exploited to implement the visualization layer of the architecture. The platform enables the secure prediction, monitoring, and control of Internet of Things data elements using external APIs, thus defining the relationship between devices, resources, clients, or any other entity; it also collects and stores telemetry data. The platform also collects and stores telemetry data. Thanks to the presence of integrated widgets, the platform allows intuitive data visualization and customization of the interface display mode. Dashboards can be shared with other users and thus expand the functionality, allowing devices to be controlled remotely and device data to be sent to other systems [6] (Fig. 2).

For data storage, ThingsBoard uses the database to store the entities. The chosen platform supports several database options common in the computing environment, such as SQL, which stores all entities and telemetry in a SQL database. However, the Things-Boardwebsite recommends the use of Postgre SQL, which is the main SQL database ThingsBoard uses.

Fig. 2. Workflow

The authors of ThingsBoard recommend using a Hybrid system (PostgreSQL+Cassandra) that stores and stores all the entities in the PostgreSQL database and the time series data in the Cassandra database [26].

3   Conclusion

This research presents the improvement and design for prototyping IoT monitoring hardware for real-time control and monitoring of natural vibrations in a reinforced concrete building and evaluation of the data. The developed IoT sensing system consists of a Raspberry Pi board, a MEMS accelerometer, a 4G Internet dongle and a cloud-based data storage system. The Raspberry Pi is used to collect and process the signals sent by the MEMS accelerometer. Signal processing is done locally, with data being filtered, integrated and evaluated in real-time against previously stored data. The cloud system's data storage system and warning functions are directly linked and developed. In addition, a dashboard-based user interface has been developed for real-time data visualization, which can be easily interrogated, for real-time measurements and warning signals, which can be accessed directly from the computer through dedicated permissions or from mobile device applications. Notable features of the IoT vibration monitoring system include ease of installation due to the small size of the prototype, limited hardware costs and low power consumption, simplicity of sensor nodes connected directly to the microcomputer, reliable lossless data transmission due to the protocols used, real-time data processing, vibration assessment and alerting capability. The Raspberry Pi's data processing and storage capabilities of the Thingboard make it possible to calculate various vibration indicators in real-time and to assess the impact of vibration by comparing various pre-set vibration limits [27].

Finally, the Raspberry Pi allows connections to many types of sensors thanks to its open-source features. Thus, the developed IoT sensing prototype can be safely used for other monitoring functions, such as temperature and humidity sensors for measuring and monitoring the environment and the operating conditions of the device and how other external factors may affect its characteristics. Furthermore, as possible future developments, structural monitoring through real-time vibration analysis can support the reconstruction of Digital Twin where conditions different from the natural conditions of a reinforced concrete structure can be prevented and simulated. Furthermore, the amount of data collected through the prototype will be used as an initial dataset to train Machine Learning and Deep Learning systems to prevent possible risks and problems to the structure [9]. Finally, this system could be used to validate renovation and consolidation interventions on reinforced concrete structures by comparing the vibration spectrum before and after the intervention.

References

1. Pappalardo, C.M., Lettieri, A., Guida, D.: A general multibody approach for the linear and nonlinear stability analysis of bicycle systems. Part II: Application to the Whipple-Carvallo bicycle model. J. Appl. Comput. Mech. **7**(2), 671–700 (2021)
2. Dašić, P., Dašić, J., Antanasković, D., Pavićević, N.: Statistical analysis and modeling of global innovation index (GII) of Serbia. In: Karabegović, I. (ed.) NT 2020. LNNS, vol. 128, pp. 515–521. Springer, Cham (2020). https://doi.org/10.1007/978-3-030-46817-0_59
3. Dašić, P., Dašić, J., Crvenković, B.: Applications of access control as a service for software security. Int. J. Ind. Eng. Manage. (IJIEM) **7**(3), pp. 111–116 (2016). ISSN 2217-2661
4. Sun, X., Liu, H., Song, W., Villecco, F.: Modeling of eddy current welding of rail: three-dimensional simulation. Entropy **22** (2020). Art. no. 947
5. Dašić, P.: Response surface methodology: selected scientific-professional papers (in Serbian, Slovenian and Russian). Vrnjačka Banja: SaTCIP Publisher Ltd. (2019). 301 str. ISBN 978-86-6075-054-1
6. Guida, C.G., Gupta, B.B., Lorusso, A., Marongiu, F., Santaniello, D., Troiano, A.: An Integrated BIM-IoT approach to support energy monitoring. In: International Conference on Smart Systems and Advanced Computing (Syscom-2021) (2021)
7. Sarah, J., Hejazi, F., Rashid, R.S.M., Ostovar, N.: A review of dynamic analysis in frequency domain for structural health monitoring. IOP Conf. Ser. Earth Environ. Sci. **357** (2019). https://doi.org/10.1088/1755-1315/357/1/012007
8. Sicilia, M., De Simone, M.C.: Development of an energy recovery device based on the dynamics of a semi-trailer. In: Lecture Notes in Mechanical Engineering, pp. 74–84 (2020)
9. Deivasigamani, A., Daliri, A., Wang, C.H., John, S.: A review of passive wireless sensors for structural health monitoring. Modern Appl. Sci. **7**(2) (2013). https://doi.org/10.5539/mas.v7n2p57
10. Pappalardo, C.M., Guida, D.: Experimental identification and control of a frame structure using an actively controlled inertial-based vibration absorber. In: Proceedings - 2017 International Conference on Control, Artificial Intelligence, Robotics and Optimization, ICCAIRO 2017, January 2018, pp. 99–104 (2018)
11. Dašić, P.: Reliability of technical systems: selected scientific-professional papers (in Serbian and Russian). Vrnjačka Banja: SaTCIP Publisher Ltd. (2019). 308 str. ISBN 978-86-6075-051-0.
12. De Simone, M., Guida, D.: Experimental investigation on structural vibrations by a new shaking table. In: Carcaterra, A., Paolone, A., Graziani, G. (eds.) AIMETA 2019. LNME, pp. 819–831. Springer, Cham (2020). https://doi.org/10.1007/978-3-030-41057-5_66
13. Colace, F., Elia, C., Guida, C.G., Lorusso, A., Marongiu, F., Santaniello, D.: An IoT-based Framework to Protect Cultural Heritage Buildings (2021). https://doi.org/10.1109/smartcomp52413.2021.00076.
14. Tošović, R., Dašić, P., Ristović, I.: Sustainable use of metallic mineral resources of Serbia from an environmental perspective. Environ. Eng. Manage. J. (EEMJ) **15**(9), 2075–2084 (2016). ISSN 1582-9596. https://doi.org/10.30638/eemj.2016.224
15. Dašić, P.: Scientific and technological trends: selected scientific-professional papers (in Serbian). Vrnjačka Banja: SaTCIP Publesher Ltd. (2020). 305 str. ISBN 978-86-6075-072-5
16. Manrique-Escobar, C.A., Pappalardo, C.M., Guida, D.: A multibody system approach for the systematic development of a closed-chain kinematic model for two-wheeled vehicles. Machines **9**(11), 245 (2021)
17. Formato, A., Ianniello, D., Pellegrino, A., Villecco, F.: Vibration-based experimental identification of the elastic moduli using plate specimens of the olive tree. Mach. **7**(2) (2019). Art. no. 46, https://doi.org/10.3390/machines7020046.

18. Pappalardo, C.M., Lettieri, A., Guida, D.: Identification of a dynamical model of the latching mechanism of an aircraft hatch door using the numerical algorithms for subspace state-space system identification. IAENG Int. J. Appl. Math. **51**(2), 346–359 (2021)

19. Castiglione, A., Colace, F., Moscato, V., Palmieri, F.: CHIS: a big data infrastructure to manage digital cultural items. Future Gener. Comput. Syst. **86** (2018). https://doi.org/10.1016/j.future.2017.04.006

20. Pappalardo, C.M., Lettieri, A., Guida, D.: A general multibody approach for the linear and nonlinear stability analysis of bicycle systems Part I: methods of constrained dynamics. J. Appl. Comput. Mech. **7**(2), 655–670 (2021)

21. Salvati, L., d'Amore, M., Fiorentino, A., Pellegrino, A., Sena, P., Villecco, F.: Development and testing of a methodology for the assessment of acceptability systems. Machines **8**(47) (2020)

22. Formato, A., Ianniello, D., Romano, R., Pellegrino, A., Villecco, F.: Design and development of a new press for grape marc. Machines **7**(3), 51 (2019)

23. De Simone, M., Ventura, G., Lorusso, A., Guida, D.: Attitude controller design for micro-satellites. In: Karabegović, I. (ed.) NT 2021. LNNS, vol. 233, pp. 21–31. Springer, Cham (2021). https://doi.org/10.1007/978-3-030-75275-0_2

24. Celenta, G., De Simone, M.: Retrofitting techniques for agricultural machines. In: Karabegović, I. (ed.) NT 2020. LNNS, vol. 128, pp. 388–396. Springer, Cham (2020). https://doi.org/10.1007/978-3-030-46817-0_44

25. Bezas, K., Komianos, V., Koufoudakis, G., Tsoumanis, G., Kabassi, K., Oikonomou, K.: Structural health monitoring in historical buildings: a network approach. Heritage **3**(3) (2020). https://doi.org/10.3390/heritage3030044

26. Arcadius Tokognon, C., Gao, B., Tian, G.Y., Yan, Y.: Structural health monitoring framework based on internet of things: a survey. IEEE Internet Things J. **4**(3), (2017). https://doi.org/10.1109/JIOT.2017.2664072

27. di Filippo, A., Lombardi, M., Marongiu, F., Lorusso, A., Santaniello, D.: Generative design for project optimization (2021). https://doi.org/10.18293/dmsviva21-014

Mobile Application mTransporters

Suad Sućeska[✉]

Brčanska 7, 71000 Sarajevo, Bosnia and Herzegovina
suad_s@bih.net.ba

Abstract. mTransporteri (mTransporters) is mobile application. It is intended for getting contact data of transport firms from remote database on smartphone. Obtained dataset is displayed in table with fields as follows: Firma (Firm), telefon (telephone), SMS and email. It also enables direct contact of transport firms by tap on appropriate contact data: phone, SMS or email. The application is made in JAVA programming language and works on smartphones with operating system Android. The version of the Android operating system for the application must be at least 25.

Keywords: Transport firms · Contact data · Remote database · Android · Mobile application

1 Introduction

Web applications can get data from remote database and present them in appropriate form on the display. Mobile applications running on smartphone in the case of getting contact data add to these applications more functionalities. Smartphone enables making many sorts of communication with another phone, smartphone and computer. Today's smartphones necessarily enable three kinds of communication: phone, SMS and email. The mobile application mTransporteri uses this opportunity. It enables getting contact data of transport firms and their direct contact from the application.

2 About the Application

Mobile application mTransporteri enables getting contact data of transport firms of specified type from some city in Bosnia and Herzegovina (B&H). It also enables to directly contact got firms from the application using some of three kinds of communication: phone, SMS or email. Selected number of the data rows is obtained for selected type of transport and city of firm headquarter. The data are obtained from remote database through the Internet [1].

Basic dataset is displayed in table with columns as follows: name of a firm, phone number, mobile phone number and email address.

Shema of dataflow of the application is shown on the Fig. 1.

I. Karabegović et al. (Eds.): NT 2022, LNNS 472, pp. 607–612, 2022.
https://doi.org/10.1007/978-3-031-05230-9_73

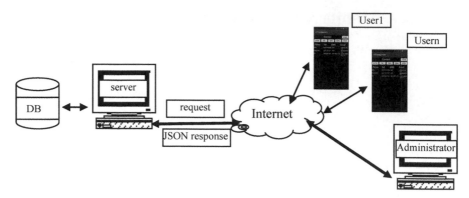

Fig. 1. Schema of dataflow of mobile application mTransporteri

The Web address for download of the application is as follows: http://aplikacije. suads.com.ba/mTransporteri/mTransporteri.apk.

The application is made in JAVA programming language for Android mobile operating system minimum version 25. Download of the application has to be made with smartphone with turned on Internet connection (Wi-Fi,...). After download of the application it is installed by tap on the button Install.

2.1 Instructions to Work with the Application

Mobile application mTransporteri connects to remote database on mySQL server, where it takes selected data from. Therefore it requires turned on Internet connection (Wi-Fi,...).

Fig. 2. Main activity of mobile application mTransporteri

Purpose of the application is to give selected number of rows with data on transport firms from B&H for selected: type of transport and city. Starting activity of the application is the Main activity, and it is got after running the application. It is shown on the Fig. 2 [3]. On the top of this activity is action bar containing name of the application and activity, as well as a menu icon with three vertical dots. In the row below them are textboxes with names of the fields for selection the parameters and button List. Row below them contains 3 fields for selection of parameters: type of transport, size (number of displayed rows of data), and city. Next to them is button Send. TextView below them is intended for display selected parameters of the query.

After getting the data, it is necessary to tap on the button List in order to get activity DisplayListView containing table with selected number of rows with the data [2].

Fig. 3. Tabular presentation of the data of firms of mobile application mTransporteri

The table contains following columns: Firma (name of the firm), Tel. (phone number), SMS (mobile phone number) and Email (email address). Activity DisplayListView above the table contains field for display of name of selected transport firm or its contact data: Contact. Next to it is a button Clear for deleting the content of textbox Contact. Below them is row with five buttons: Detalji (details), Tel, SMS, Email, and Back. The button Detalji serves for display more data of selected firm which name is at the moment displayed in textbox Contact. The button Tel serves for direct calling of a number in textbox Contact through the application Phone. The button SMS serves for direct starting of application for sending SMS messages with SMS number displayed in textbox Contact. The button Email serves for direct starting email application with filled in contact data from the textbox Contact. The button Back serves for returning the display back on the main activity [4].

Fig. 4. Activity DisplayListView with selected mobile phone number

Action bar in main activity contains Menu with following options: Uputstvo, Info and Close. Tap on the option Uputstvo gives litle user manual in the form of dialogue. The option Info gives general information about the application, also in the form of dialogue. Tap on the option Close to close the application. The menu is showed on the Fig. 5 [3].

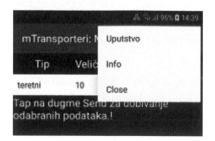

Fig. 5. The menu of application mTransporteri

To introduce with the application before usage read short manual. It opens by tap on the option Uputstvo in the menu shown on the Fig. 5 [3].

Fig. 6. Dialogue Uputstvo of mobile application mTransporteri

The text in the manual can be scrolled by moving finger up or down. Tap on the button Close in the manual to close it.

Tap on the option Info in the menu to get short information about the application in form of dialogue. This dialogue is also closed with it's button Close.

3 Conclusion

Mobile application mTransporteri enables getting contact data of transport firms using smartphone. Selected parameters of the query: type of a transport, size (number of data rows) and city; are sent to remote Web site, which returns back requested data. Got data are displayed in the table with columns: Firm, Tel., SMS and Email. The application also enables making contact with transport firms direct from the application through communication services: phone, SMS, and email, which have to be active on the smartphone. To run the application it is neccessary to have mobile operating system Android minimum version 25.

References

1. McWherter, J., Gowell, S.: Professional Mobile Application Development. Wiley, Indianapolis (2015). (Kindle Edition. Wrox)

2. Fain, Y.: JAVA 24-Hour Trainer –2nd edition. Wiley Publishing Inc., Indianapolis (2015). (Kindle Edition. Wrox)
3. Introduction to Android Application Development, LINK group 1998–2017
4. Advanced Android Application Development, LINK group 1998–2017

Intelligent Transport Systems, Logistics, Traffic Control

Application of Big Data Sets and Data Science in Transportation Engineering

Krešimir Vidović[1,2(✉)]

[1] Ericsson Nikola Tesla, Krapinska 45, Zagreb, Croatia
kresimir.vidovic@ericsson.com

[2] Faculty of Transport and Traffic Sciences, University of Zagreb, Vukelićeva 4, Zagreb, Croatia

Abstract. The development and application of information and communication technologies (ICT) have started an era of big data. Big data sets originating from mobile telecommunication networks, besides their primary purpose within the telecom environment, are becoming more popular in other application areas such as transportation engineering. Existing research have proven that analytics of such data, by using data science methodology and tools can be used to draw a complete and representative picture of urban migrations and mobility assessment. The application of this analytics is used in transportation planning, transportation management and for improvements in traffic safety. Further developments in both communication technology and data science indicates that real application of mobile network data, especially as a substitute for traditional measurements have great potential that needs to be exploited.

Keywords: Big data · Data science · Intelligent transport systems

1 Introduction

In recent years telecommunication network has been recognized by scientific community as a valuable source of data, that can be utilized for geospatial analysis in application in different business verticals (marketing, transport, health…). The basic principles of operation of mobile communication network requires that mobile terminal is in constant (when used) or periodic (when idle) communication with mobile network base station antennas. Since that base station antenna is characterized, among others, with its location and coverage parameters (including radius, azimuth etc.…), fact that the user's mobile terminal is connected to particular antenna enables its approximate positioning in space. The accuracy of this type of positioning may vary, and is dependent on technology (2G, 3G, 4G, 5G), signal strengths, environment, obstacles etc.…The positioning accuracy isn't comparable to GNSS systems since this type of mobile positioning may result with significant location uncertainty (position error of 100–500 m in urban environment) [1, 2]. Therefore, use cases for this type of data should be resistant to this position error. Besides this flaw, this type of data is characterized and with important advantages, including significant sample (mobile phone penetration is more than 100% in development countries [3], that covers almost entire population, data is created instantly and

I. Karabegović et al. (Eds.): NT 2022, LNNS 472, pp. 615–623, 2022.
https://doi.org/10.1007/978-3-031-05230-9_74

collected continuously, data is collected without user intervention and is immune to subjectiveness. This type of data is huge in scale and may be referred to as big data, and it requires a large computer infrastructure, as well as application of data science in order to processed data effectively end to extract valuable knowledge. Application of data science principles overtelecom originated big data sets can be used to draw a complete and representative picture of urban migrations and mobility assessment, and therefore are applicable in transportation engineering. In general, pillars are transportation planning [4], transportation safety [5] and mobility assessment [6]. This paper will elaborate on the possibilities of application of big data sets in aforementioned pillars and will present several use cases of successful application with results and lessons learnt.

2 State of Art Analysis

The current era is characterized by availability of structured and unstructured information, which might be collected from traditional data sources (sensors) and nontraditional sources, like mobile communication network. Availability of different data sources, availability of powerful data processing possibilities and development of new methods have initiated a boost in inspiration for new developments. This trend is resulting in new data-driven analytical paradigm that provides a new analytical possibilities and new insights in transportation related themes.

Two trends are key enablers of this phenomena: Big data and Data sciences. The main themes of Big Data are information, technology, methods and impact and therefore, it can be defined as the information asset characterized by such a High Volume, Velocity and Variety to require specific Technology and Analytical Methods for its transformation into Value [7]. Data science is an interdisciplinary field that uses scientific methods, processes, algorithms and systems to extract knowledge and insights from noisy, structured and unstructured data, and apply knowledge and actionable insights from data across a broad range of application domains [8]. Advanced data analysis and Data Science concept is identified as anvaluable tool for scientists and transport engineers.

Transport and urban mobility related stakeholders can in general, be divided in three categories regarding the big data utilization in operations. Some have not leveraged big data yet but have data to process (f.e. data from ticketing system, data from road use charge system…), some have started their activities in implementing big data but have yet to apply (testing big data analytical possibilities and its application in comparison to traditional data sources), and third, have implemented a full-spectrum solution and have established a robust big data environment (small number of stakehoders, in general use cases are based on crowdsourcing data and its utilization in traffic simulations).

Data collected from the telecommunications environment is characterized by comprehensiveness (one data collection covers an arbitrarily large space and a significant sample), reliability (analysis results are objective because they do not contain the effects of self-perception that are characteristic of other types of data collection, such as surveys), repeatability recording allows analysis of trends) and adaptability in terms of geographical and temporal and configuration as needed.Data originating from telecom environment can be divided in two categories. First categories consists of data originating from mobile telecommunication network, which includes Call Data Record data and signaling data. Second categories is data which is passively or actively collected from mobile terminals, and this data source can be identified as crowdsourcing data. This paper will utilize data from mobile telecommunication network, therefore Call Data Records and signaling data. Telecom originated big data may be considered as sensitive, so all organizational and technical measures should be performed when this data is used outside of telecom environment and data have to be handled in accordance with the General Data Protection Regulation (EU) 2016/679 (GDPR), which regulates data protection and privacy in the European Union (EU) and the European Economic Area (EEA).

Analysis of literature have shown that scientific community have already recognized telecom big data as a valuable tool for transportation related scenarios of application. Majority of papers are dealing with utilization of big data for transport demand identification as a part of transportation strategies planning process, since the general conclusion is that mobile data can be used to create a comprehensive and characteristicimage of mobility flows over ananalyzed area. The most common use case is determination of origin destination matrix (ODM) as an indicator of population movement in general, and more recent evenenriched by transportation mode [9–13]. Several papers are elaborating of usability of this type of data for real time traffic management, since this type of information can be used for real time road traffic analytics and connected applications [14–17]. Recently a significant effort have been made to utilize this type of data in traffic safety, in particular on identification of driver behavior (speeding, distraction as a result of usage of mobile phone while driving etc....) [5, 18–21]. Analysis of literature have shown that the authors have identified the benefits and potentials of telecom big data usage in transport engineering and that mobile phone data can be used to augment or complement the traditional sources data. As with any novel data sources, several shortcomings have been recognized as well (in particular in regarding to position accuracy, size of the sampled data, validity of traffic probes, and privacy issues) and in general this is still rather an un-exploited domain with a significant space for improvements.

3 Methological Principles

The role of this chapter is to present a general methodological concept that is used for telecom big data analysis and includes key steps, from data collection to analytical reports. Methodology overview is presented on Fig. 1.

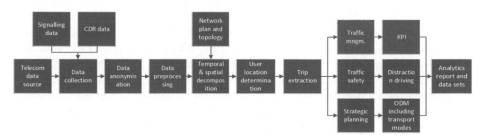

Fig. 1. Overview of methodology

First step include identification of telecom operator which will act as a data source. Telecom operator should have operational and technical capabilities, as well as business argumentation to extract the required data, and should have a significant market share in order to ensure statistical representativeness of the population sample. Then, based on the use case of the analysis, data collection should be performed. Data is collected for defined project area, and for defined time frame. Call Record Data is collected all the time by the MNO since this data is used for billing purposes. Signaling data may not be collected all the time, or might not be collected at all, so technical prerequisites have to be fulfilled in order for data collection to be performed. Following the data collection, appropriate steps have to be performed regarding anonymization. The data have to be properly and irreversibly anonymized, so the usage of this data should not reveal the identity of individual, but shall contain minimum viable information in order to remain useful. Following the data anonymization the data can be transferred from telecom environment for analytical processing. Analytical processing starts with data preprocessing. It includes preparation of data and my include detection and removal of outliers based on predefined spatial and technical criteria, data smoothing and similar techniques in order to increase the data quality. Data preprocessing shall end with enhanced information on positions of mobile terminal in observed period and in observed area. In order to translate telecommunication information on location based on antenna location, it is necessary to utilize network plan and topology data in order to determine user position in space.. The result of this step is approximate user location in space based on telecom data only. The user location is determined from telecommunication logs either from signaling or from CDR data. Both data sets might have significant number of data in short period of time (f.e. several hundreds thousand sevents in one second). Therefore, this data has to be clustered (f.e. in one second) in order to reduce the data size and make it more useful for analysis. Following this step, information on user location is used to determine its locations of retention. Location of retention, or sometimes called "stay location" is location where the user has spentenough time in order to presume that it is stationary on this location. Those locations may be home, office, hospital, cinema, shopping mall, or any other location where people tend to stay longer, and which might be considered as a goal (or destination) of its movement (trip). Locations that are not stay locations can be considered as moving locations, or locations where the user presence was logged, but only for shortperiod of time (example. staying for one minute during the red light on crossroad). The movement among location of retention can be considered as trip. This step ends with extraction of trip data and depending of the use case and the goal of the analysis, following steps might differ. For strategic planning purposes, trips are usually summed in origin destination matrix. Also, different techniques (heuristic model, application of machine learning) might be utilized to identify the most probable trip mode, or identify the purpose of the trip. Additional data might be used for proper identification, like the list of points of interest in observed area, locations of public transport lines and stations, land use information etc. For traffic safety, both CDR and mobile traffic volume data can be used to identify the usage of mobile phone during the trip, and to identify whether the user have actively or passively used mobile phone while on move. Additional information might be needed as well, like the traffic counting data and data on the vehicle occupancy in order to identify the drivers against passengers and to identify the location of the road network where this trend occurs, which represents the safety risk. For previously mentioned use cases, the analytics usually doesn't require usage of real time data, and the analytics can be performed on historical data. For traffic management purposes near real time analytics is required, and therefore, due to the complexity of big data sets, analytical process has to be optimized in order to enable fast processing. The common use case might include calculation of various key performance indicator (like identification of the number of people in observed area, represented in the form of heat maps etc.) Since this data source is one of many that are used in traffic management it might be needed to consider its integration within the traffic management center using data aggregation platform (IoT data platform) or national access point as described in EU Directive 2010/40. [22, 23] If data aggregation platform integration is envisaged, standardized data protocols should be used (like Datex, Netex, Siri or 3PP protocols) suited for particular KPI generated from MNO data.

4 Application of the Methodology

This chapter will provide insights in several projects that have been completed using the proposed methodology in Sect. 3. First two projects have been performed using telecom big data sets originating from mobile network operator in Croatia, where both CDR and signaling data was used, and the third research was tested using CDR data from China.

Project Sustainable Urban Mobility ToolBox [4], financed by European Institute of Technology had a goal to use a anonymized telecom big data sets in order to identify migration patterns of population in three Croatian cities (Zagreb, Rijeka and Dubrovnik). The goal of the big data analysis was to identify trips made by particular transport mode and to identify the zone pairs where patterns are under or above specific treshold. For example, transport sector pairs were to be identified where the majority of trip is performed by car, since this might implicate that the sustainable transport mode offer is not sufficient, does not exist or have some shortcomings. During the project, a conventional research's have been performed in parallel to validate the results of big data analysis, and it was performed by surveying, traffic counting and license plate recordings on selected sections. The modal split comparison differs for passenger vehicles for 2.49%, public

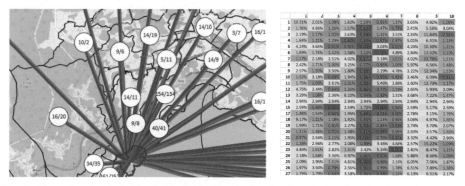

Fig. 2. Example of origin–destination matrix visualized using desire lines, and matrix for one day relativized on the base of one zone, based on [4]

transport 2.56% and active modes of transport of about 0.70%.Comparative analysis of daily distribution revealed the existence of a strong connection between the two sets of mobility patterns (R2 = 96%).Example of origin–destination matrix as a result of EIT project is presented on Fig. 2.

Project "Development of Methodology for Defining a Pattern of Drivers Mobile Phone Usage While Driving", [5] had a goal to identify a pattern of usage of the mobile phone (both voice calls, SMS and applications) by drivers while driving. This findings is is used to identify segments of the road infrastructure on which mobile phone usage is more intense. For identification of voice calls while driving CDR data is used (as it contains information on calls and messages), and for data on application usage traffic volume data was used. The research results indicate that the share of drivers who use their mobile phone for voice calls or for mobile app usage is high and presents a traffic safety risk (31.5% of drivers are actively using mobile phone on some sections). In average every fifth driver is using its phone to make a voice call or is using some of his mobile apps. If the maximal established values are observed, the mobile phone is used by every third driver on certain sections (Fig. 3).

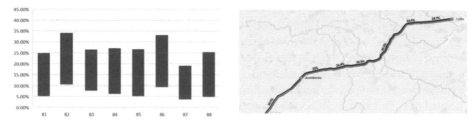

Fig. 3. Share of drivers who actively use their mobile phone while driving on the section of D1 road in Croatia, and its visualisation on map based on [5]

Research Estimating Urban Mobility Based on the Records of User Activities in Public Mobile Networks had a goal to establish a process of the passenger urban mobility estimate as a quantitative measure of the process of urban migrations caused by socioeconomic activities, in the function of determining a new urban mobility estimation index. Indicators are generated from a database of mobile phone users call data records and are integrated into the urban mobility index of the population based on the model defined through the adaptive neuro-fuzzy inference system (ANFIS). The validation of the model hows that deviations are possible in the range from 3% to 11%, relative to the value obtained through using the basic dataset (Fig. 4).

The validation of data in all mentioned project have proven that proposed analytical possibilities driven by data science principles over anonymized telecom originated big data sets can either complement or substitute traditional transport measurements and established ways of obtaining data.

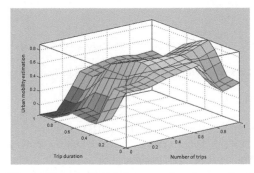
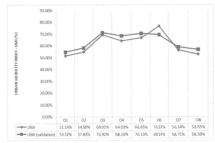

Fig. 4. The relationship between the trip duration indicators and the number of trips with respect to the mobility estimate (a) and the comparison of the urban mobility index from the basic dataset with the validation data, based on [6]

5 Conclusion

Approach that combines big data science methods with traditional methodologies is appropriate and applicable for transport planning and that the symbiose of those methods can significantly improve conventional research. Appropriately anonymized and processed mobile network data has proven to be an important data set that can be utilized for mobility planning and optimization. Different lessons have been learnt, in particular in relation to operational aspects regarding the utilization of big data in transportation planning.

During the application of the methodology, it was found that the big data set sometimes provides a more detailed database than required for understanding of transport system. The big data detects all the trips that user makes during the day. It records all type of movement, short trips within the zone or even short recreative walking. That kind of movement are usually not the transport research objective. Also, since the mobile phone penetration is continuously rising and since mobile telecom users are spending more time actively using their devices, the data available for research is becoming more massive, resulting with the fact that mobile network user footprint is becoming more complete since there is more data available to reconstruct potential movement with potential data gaps decreasing. Also, the evolution of mobile network generations (majority of users in urban environment is using 4G network, 5G is on the rise) supports intentions to decrease position error and enhance positioning accuracy, since the position accuracy is key input parameter that directly influence the overall analysis results. Therefore, this methodology can be considered as futureproof since every new generation will bring enhancement in terms of position determination, which will result in more accurate analysis.

Acknowledgment. The author wish to thank "Laboratory for Data Science in traffic and logistics" at University of Zagreb, Faculty of Traffic and Transportation sciences for support with data and software during the creation of this article.

References

1. Campos, R.S.: Evolution of positioning techniques in cellular networks, from 2G to 4G. Wirel. Commun. Mob. Comput. **2017** (2017)
2. Zandbergen, P.A.: A comparison of assisted GPS, WiFi and cellular positioning. Trans. GIS **13**(SUPPL. 1), 5–25 (2009)
3. GSM Association: Mobile economy (2021)
4. Šoštarić, M., Vidović, K., Jakovljević, M., Lale, O.: Data-driven methodology for sustainable urban mobility assessment and improvement. Sustainability **13**, 7176 (2021)
5. Čolić, P., Jakovljević, M., Vidović, K., Šoštarić, M.: Development of methodology for defining a pattern of drivers mobile phone usage while driving. Sustainability **14**(3), 1681 (2022)
6. Vidović, K., Šoštarić, M., Mandžuka, S., Kos, G.: Model for estimating urban mobility based on the records of user activities in public mobile networks. Sustainability **12**(3), 838 (2020)
7. De Mauro, A., Greco, M., Grimaldi, M.: A formal definition of Big Data based on its essential features. Libr. Rev. **65**(3), 122–135 (2016)
8. V. Dhar, "Data Science and Prediction," *SSRN Electron. J.*, 2012
9. Lee, S.: The use of mobile phone data in transport planning. Int. J. Technol. Policy Manag. **20**(1), 54–69 (2020)
10. Bera, S., Rao, K.V.K.: Estimation of origin-destination matrix from traffic counts: the state of the art. Eur. Transp. - Trasp. Eur. **49**, 3–23 (2011)
11. Caceres, N., Romero, L.M., Benitez, F.G.: Exploring strengths and weaknesses of mobility inference from mobile phone data vs. travel surveys. Transp. A Transp. Sci. **16**(3), 574–601 (2020)
12. Çolak, S., Alexander, L.P., Alvim, B.G., Mehndiretta, S.R., Gonzalez, M.C.: Analyzing cell phone location data for urban travel: current methods, limitations and opportunities. In: TRB 2015 Annual Meeting, pp. 1–17 (2015)
13. Alexander, L., Jiang, S., Murga, M., González, M.C.: Origin-destination trips by purpose and time of day inferred from mobile phone data. Transp. Res. Part C Emerg. Technol. **58**, 240–250 (2015)
14. Costa, C., Chatzimilioudis, G., Zeinalipour-Yazti, D., Mokbel, M.F.: Towards real-time road traffic analytics using Telco Big Data. In: ACM Interbational Conference on Proceeding Series, vol. Part F130527 (2017)
15. Liu, L., Biderman, A., Ratti, C.: Urban mobility landscape: real time monitoring of urban mobility patterns. In: Proceedings of the 11th International Conference on Computing Urban Plann ing Urban Managagement, pp. 1–16 (2009)
16. Laboshin, L.U., Lukashin, A.A., Zaborovsky, V.S.: The big data approach to collecting and analyzing traffic data in large scale networks. Procedia Comput. Sci. **103**, 536–542 (2017)
17. Galloni, A., Horváth, B., Horváth, T.: Real-time monitoring of hungarian highway traffic from cell phone network data. In: ITAT 2018 Proceedings, 2018, vol. 2203, pp. 108–115 (2018)
18. Zhang, L., Cui, B., Yang, M., Guo, F., Wang, J.: Effect of using mobile phones on driver's control behavior based on naturalistic driving data. Int. J. Environ. Res. Public Health **16**(8), 1464 (2019)
19. Jovović, I., Peraković, D., Husnjak, S.: The impact of using modern information and communication equipment and services on driving safety. PROMET - Traffic Transp. **30**(5), 635–645 (2018)
20. Oviedo-Trespalacios, O., Haque, M.M., King, M., Washington, S.: Understanding the impacts of mobile phone distraction on driving performance: a systematic review. Transp. Res. Part C Emerg. Technol. **72**, 360–380 (2016)
21. Khan, I., Rizvi, S.S., Khusro, S., Ali, S., Chung, T.S. Analyzing drivers' distractions due to smartphone usage: evidence from autolog dataset. In: Mobile Information System, vol. 2021 (2021)

22. European Parliament and of the Council EU: Directive 2010/40/EU Of The European Parliament and of the Council of 7 July 2010 on the framework for the deployment of Intelligent Transport Systems in the field of road transport and for interfaces with other modes of transport. Off. J. Eur. Union, pp. 1–13 (2010)
23. Mandžuka, S., Žura, M., Horvat, B., Bićanić, D., Mitsakis, E.: Directives of the European Union on intelligent transport systems and their impact on the Republic of Croatia. PROMET - Traffic Transp. **25**(3), 273–283 (2013)

Multicriteria Decision Support System for Motorways Safety Management

Sadko Mandžuka[1,5(✉)], Luka Dedić[2], Goran Kos[3], and Marko Šoštarić[4]

[1] Faculty of Traffic and Transport Sciences, Vukelićeva 4, 10000 Zagreb, Croatia
sadko.mandzuka@fpz.hr
[2] Promettis Ltd., CvijeteZuzorić 5, Zagreb, Croatia
[3] Institut for Tourism, Vrhovec 5, 10000 Zagreb, Croatia
[4] Faculty of Traffic and Transport Sciences, Vukelićeva 4, 10000 Zagreb, Croatia
[5] Department of ITS, University of Zagreb, Vukeliceva 4, 10000 Zagreb, Croatia
smandzuka@fpz.unizg.hr

Abstract. The paper describes the decision support system for safety management on the motorway section. The system is based on the estimation of the probability of traffic accidents on the motorway - Crash potential. Based on this assessment, the system recommends active measures to reduce the likelihood of their actual occurrence. The paper presents a model based on the ANFIS methodology and is based on measuring the values of selected traffic flow parameters and external factors that affect the flow of traffic. This approach is important for improving existing algorithms for managing variable traffic signs on highways.

Keywords: Road safety · Crash potential · Intelligent transport system · ANFIS system

1 Introduction

Road traffic safety is big challenge of today's society. The World Health Organization (WHO) in its annual publication stated that approximately 1.3 million people die each year as a result of road traffic crashes that road traffic crashes cost on average 3% of country's gross domestic product, more than half of all road traffic deaths are among vulnerable road users: pedestrians, cyclists, and motorcyclists, 93% of the world's fatalities on the roads occur in low- and middle-income countries, even though these countries have approximately 60% of the world's vehicles, and road traffic injuries are the leading cause of death for children and young adults aged 5–29 years[1].

Ministry of Interior, Department for strategic planning, analytics and development [2] stated that the Republic of Croatia has a direct loss of 2,4% of GDP, while the indirect losses are much bigger. In the last 10 years, approximately 33,554 traffic accidents have occurred on Croatian roads. In 32.02% of accidents, people were injured. In other words, 14,803 people are hurt in traffic every year. Of that number, 78,95% of injured suffered from minor injuries, 18,81% suffered from serious injuries, and 2,24% died. The three-year statistic (2018–2020.) shows that on average 5.8% of total accidents occur on

I. Karabegović et al. (Eds.): NT 2022, LNNS 472, pp. 624–630, 2022.
https://doi.org/10.1007/978-3-031-05230-9_75

Croatian motorways, making up 8,7% of total deaths caused by traffic accidents [2–4], which speaks to the severity of such accidents.

Road safety represents the interaction between a man, a road, and a vehicle. By affecting each factor, the safety status on the roads can be improved. The paper studied and defined the multicriteria decision support system for the prevention of accidents on the motorway. Based on the estimation of the probability of traffic accidents on the motorway - Crash potential, the Decision support system recommends active measures to reduce the likelihood of their occurrence.

The paper consists of six chapters, of which the introduction and conclusion are the first and the last chapter, respectively. A description of the model is presented in the second chapter. The third chapter presents the factors that can influence the decision to assess the risk of collision. In the fourth chapter, the decision support system based on the ANFIS methodology is presented.

2 Crash Potential Model

Calculation of risk assessment is conducted through combination of two models. One of the models uses the basic traffic flow input like speed, density, velocity, etc., while other model uses the external factors that influence the traffic flow. Communication between a driver or a user and the Crash potential system is conducted through the Variable Message signs (VMS). Once the risk is calculated the massage on action is send to a driver via VMS. In other words, the VMS will help to inform the drivers to take percussion and recommended actions in order to minimize a possibility for crash occurrence.

The crash potential model is based on the research made by Lee et al. (2002) [5] which is further modified in 2004. by Lee et al. [6], and research [7] The basic linear logarithmic model of crash developed by Lee et al. is as follows:

$$
\begin{aligned}
\ln(F) = \Theta + \lambda_{CVS(i)} + \lambda_{Q(j)} + \lambda_{COVV(k)} + \ldots \\
\ldots + \lambda_{R(l)} + \lambda_{P(m)} + \beta \ln(EXP)
\end{aligned}
\tag{1}
$$

where:

F - expected number of accidents in the observed period,

Θ - constant,

$\lambda_{CVS(i)}$ - coefficient of variation of speed (CVS),

$\lambda_{Q(j)}$ - difference of average speed between the beginning and end of an observed section (km/h),

$\lambda_{COVV(k)}$ - influence of traffic lane change (COVariance of Volume difference between the upstream and downstream of a specific location),

$\lambda_{R(l)}$ - effect of road geometry (control factor),

$\lambda_{P(m)}$ - effect of peak/off-peak traffic volume (control factor),

β - parameter for exposure,

EXP - exposure to accident occurrence based on vehicle/kilometre.

Even dough the mentioned model is derived from empirical data for a six-lane motorway the model that will be used in research that follows will be used for four-line motorways (two in each direction), with pull-up lane and dividing fences between directions.

The calibration of the model will be based on additional factors mentioned in Huzjan et al. [7]:

- Variety of vehicle speed in traffic flow
- Change in average traffic flow speed
- Overtaking frequency
- Degree of saturation
- Impact of structures
- Meteorological conditions
- Time of day

Using these two combined mathematical models the developed model will be easier to adapt to the conditions in the Republic of Croatia and it will also facilitate the determination of the test corridor.

3   Other Influencing Factors

Besides factors that are consisted and mentioned in the previous chapter, other (external) factors can influence decisions for minimizing the risk of accident occurrence. Based on research [5, 6, 8] the additional five factors can be derived. The mentioned factors are formulated in the mathematical form of model:

$$n = \frac{1}{4}[n(CVS) + n(Q) = n(p) + n(V)]M \tag{2}$$

where:
  n - total crash potential,
  n(CVS) - crash potential due to speed variance on a fixed location,
  n(Q) - crash potential due to speed variance between two points,
  n(P) - crash potential due to overtaking,
  n(V) - crash potential due to traffic volume,
  M. - crash potential due to night.
  The equal weighting factor is determined to each of these five elements in the model in the total crash potential. In that way, the most unfavourable values of all elements will give the highest value of the crash potential. the presented model will be used as the base for creating a more valid model that can be used in real-time. Even dough capturing and preciselydetermine the weighting factor of each component can be exhausting manual work, percentage of accuracy can be enhanced with empirical data. With more empirical data on traffic accidents each component in the model can be better calibrated and accurately presented. To accomplish that, traffic data for the accurate representation of the characteristics of traffic flow and traffic data on accidents for the accurate model calibration in the past several years should be taken into account. Here is important to say that the more data collected and incorporated the more robust and accurate model will be.

4   Decision Support System

ANFIS is a multilayer adaptive fuzzy inference system based on a neural network, defined by Yang, [9]. It contains a neural network as a structured network consisting of nodes (neurons) interconnected by directional connections. Each node is characterized by a specific function with fixed or variable parameters. Changes in parameter values are caused by adaptive (meaning that their outputs depend on parameters that change through the network passing through links and other nodes) and non-adaptive (fixed) network nodes, but in conjunction with learning rules that specify how these parameters will change to minimize the prescribed error measure. ANFIS combines the advantages of two machine learning methods (fuzzy logic and neural network). ANFIS works on the principle of applying inference methods characteristic of neural networks in order to determine the parameters of the Fuzzy Inference System - FIS. The result is a system that allows the use of well-known algorithms for learning artificial neural networks, which cannot be used directly in fuzzy inference systems, while providing the ability to use fuzzy logic and fuzzy inference.

ANFIS enables the design and fine-tuning of fuzzy causal IF-THEN rules that describe the behaviour of a complex decision-making system. The ANFIS method enables work with large data sets, supports a different number of membership functions, enables great possibilities of generalization, enables the creation of linguistic and numerical fuzzy rules for solving problems [10]. The advantage of designing ANFIS is that it combines fuzzy decision-making ability with the learning ability provided by the neural network to model the dynamics of different nonlinear systems. ANFIS can be created as a five-layer network of multilayer perceptron (MLP), Fig. 1. In this figure, the circle indicates a fixed node, and the rectangle indicates an adaptive node.

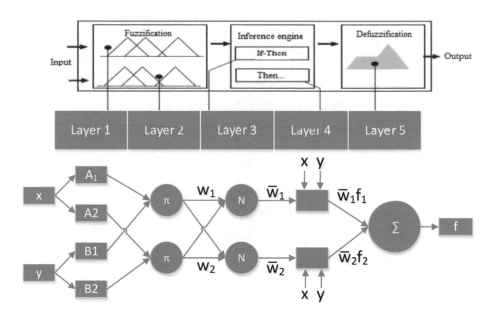

Fig. 1. ANFIS network architecture, Source: [9]

In the first layer (Fuzzification layer), two typical input values X and Y are entered into two input nodes, which then transform these values through the corresponding membership functions. In the second layer (Rule layer), each node multiplies the input signals, and the output (*firing strengths* for the rules) is calculated using fuzzy operators. They have the symbol π which indicates that it is a simple multiplier. The third layer contains fixed nodes N, and they normalize the *firing strengths*. Each i-th node calculates the share of the activation value of the i-th rule in the sum of all activation values of the ANFIS algorithm rule. The outputs of this layer are called *normalized firing strengths*. In the fourth layer, the nodes are adaptive, where their output value is the product of normalized firing strengths and corresponding node function with consequent parameters. There is only one node in the fifth layer, which calculates the overall weight as output from the system.

The proposal of the architecture of the decision support system for motorway safety management is shown in Fig. 2. The core of the system is ANFIS as machine learning technology. Based on predefined algorithms and learned rules that use the crash potential model, recommendations are given for the operation of road signalisations. In principle, there are two types of impact on traffic. These are primarily variable messages signs (VMS) and setting appropriate variable speed limits (VSL) signs. The system has the ability to learn, which allows the best solutions to be used based on lessons learned in the past.

Rigorous testing on simulation models and selected tests in reality is necessary for the future application of this technologies. In this regard, it is of particular importance

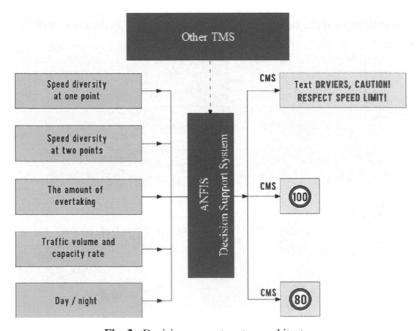

Fig. 2. Decision support system architecture

to analyse driver behaviour with respect to compliance with speed limits. This is a key prerequisite for the success of the use this motorway safety management approach.

The real advantages of the approach presented in this paper will be possible through cooperative traffic management systems [11, 12]. Through the cooperative intelligent vehicle speed adaptation system, problems related to compliance with speed limits will be eliminated.

5 Conclusion

In the Republic of Croatia, a smaller share of traffic accidents occurs on motorways and expressways, but this smaller percentage still comes with a higher mortality rate. In order to reduce number of accidents,in this research theoretical mathematical crash potential model is proposed. Proposed Crash potential model with combined external factors that are formulated in mathematical models is part of decision support system based on ANFIS methodology. ANFIS is a multilayer adaptive fuzzy inference system based on a neural network. ANFIS enables the design and fine-tuning of fuzzy causal IF-THEN rules that describe the behaviour of a complex decision-making system.The proposed methodology aims to reduce the crash potential and increase the level of traffic safety on Croatian motorways. In Republic of Croatia there is a network of interconnected motorways which are mostly in full profile with two lanes and a stop lane in each direction and to a lesser extent in a semi-profile with one lane and one stop lane in each direction. Having that in mind, the further research will be directed in the development and testing of the proposed model in a highly calibrated traffic microsimulation environment that uses roads and traffic flow characteristic for Croatia.

Acknowledgements. The paper was realized within the project *"Proposal of a model for the management of variable traffic signals on the motorway section based on multicriteria optimization"*, approved within the National Road Safety Program of Republic of Croatia for 2022.

References

1. World Health organization. https://www.who.int/news-room/fact-sheets/detail/road-traffic-injuries. Accessed 02 Feb 2022
2. Bulletin on road traffic safety 2020. Ministry of Interior, Department for strategic planning, analytics and development, Zagreb (2021)
3. Bulletin on road traffic safety 2019. Ministry of Interior, Department for strategic planning, analytics and development, Zagreb (2020)
4. Bulletin on road traffic safety 2018. Ministry of Interior, Department for strategic planning, analytics and development, Zagreb (2019)
5. Lee, C., Saccomanno, F., Hellinga, B.: Analysis of Crash Precursors on Instrumented Freeways. Transportation Research Record 1748, TRB, National Research Council, Washington, D.C., pp. 1–8 (2002)
6. Lee, C., Hellinga, B., Saccomanno, F.: Assessing Safety Benefits of Variable Speed Limits. Transportation Research Record No. 1897, pp. 183–190 (2004)
7. Huzjan, B., Mandžuka, S., Kos, G.: Real-time traffic safety management model on motorways. Tehnicki Vjesnik - Technical Gazette **24**(5) (2017)

8. Lee, C., Hellinga, B., Saccomanno, F.: Real-time crash prediction model for application to crash prevention in freeway traffic. Transp. Res. Record **1840**, 67–77 (2003)
9. Jang, J.S.R.: ANFIS: adaptive-network-based fuzzy inference system. IEEE Trans. Syst. Man Cybern. **23**(3), 665–685 (1993)
10. Jang, J.-S. R., Sun, C.-T., Mizutani, E.: Neuro-fuzzy and soft computing: a computational approach to learning and machine intelligence. PHI Learning (2010)
11. Mandzuka, S., Skorput, P., Vujic, M.: Architecture of cooperative systems in traffic and Transportation. In: 23rd Telecommunications Forum (TELFOR), Belgrade, pp. 25–28 (2015)
12. Gregurić, M., Ivanjko, E., Mandžuka, S.: The use of cooperative approach in ramp metering. PROMET Traffic Transp. **28**(1), 11–22 (2016)

Overview of Resilience Processes in Transport Management Systems

Lucija Bukvić, Jasmina Pašagić Škrinjar, Pero Škorput[✉], and Maja Tonec Vrančić

Faculty of Transport and Traffic Science, University of Zagreb, Zagreb, Croatia
pskorput@fpz.unizg.hr

Abstract. This paper proposes an overview of transport system vulnerability assessment models that allow identifying critical links for the development of future high quality transport management systems (TMS). The challenges of increasing congestion and negative environmental impacts, shifting trips from personal vehicles to other transport options is generally seen as one of the most important actions. In terms of resilience and business continuity, transport systems need to be efficient as well as robust, as their vulnerability may cause various negative impacts. The methodological approach is particularly useful for planning resilient response in the preparedness stage, prioritizing investment for mitigation and adaptation, and prioritizing the rehabilitation (access restoration) of the disrupted links in the response and recovery stages. Resilience accounts for not only the ability of the system to absorb externally induced changes, but also cost-effective and efficient, adaptive actions that can be taken to preserve or restore performance post-event.

Keywords: Transport management system · Resilience · Transport network · Vulnerability · Attack

1 Introduction

Malicious interventions are assumed to have a greater impact if they act on the transport network because a single attack affects multiple individual road users, and this assumption can greatly increase the risk of cyber incidents. To illustrate the effect of estimated cyber-attacks, it is necessary to observe the traffic system in an urban environment. Some of the interventions are not effective in terms of measuring the increase in capacity (congestion) and travel time on the network, while some of them significantly increased the travel time and costs of the entire network. Theoretical framework that describes the effect of the analyzed cyber-attacks on the urban transport network, and on certain elements of the network, enables the assessment of the risks and impact that certain cyber-attacks have on transport management systems. New methodologies allow operators to identify vulnerabilities in terms of online vulnerabilities, especially regarding remotely controlled malicious interventions.

© The Author(s), under exclusive license to Springer Nature Switzerland AG 2022
I. Karabegović et al. (Eds.): NT 2022, LNNS 472, pp. 631–638, 2022.
https://doi.org/10.1007/978-3-031-05230-9_76

2   Resilience in Public Transport

Papers focusing on the Public Transport Network (PTN) which, using simulation tools and data analyses, conduct research on vulnerabilities of the system and the limited significance of resilience, determine the levels of flexibility and ability of the system to ensure the persistence of key functions, even in cascading incidents. System vulnerability is a key concept used to analyze network structure and is defined by various forms [1]. In recent years, studies on the vulnerability of the Public Transport Network (PTN) have attracted increasing attention due to the possible negative consequences they may have on the day-to-day functioning of the city. Recent studies [2] examine the vulnerability of the public transport network. The proposed methods integrate stochastic supply and demand models, dynamic route choice and limited operational capacity [3]. This dynamic agent-based modelling of network performance enables to capture cascading network effects as well as the adaptive redistribution of passenger flows [4]. Topological attributes of a transportation system significantly affect its resilience to disaster events in terms of throughput, connectivity or compactness [5]. Resilience is considered with and without the benefits of preparedness and recovery actions [6].

Resilience is one of the key factors contributing to preserving the functionality of critical infrastructure elements. This represents the ability of the elements to mitigate the intensity of the impact caused by the interventions and reduce the duration of their harmful effects. The required level of resilience can be achieved by continuously improving the system through five basic areas: (1) readiness, (2) absorption, (3) response, (4) ability to recover, and (5) adaptability. These areas and their criteria, shown in Fig. 1, determine the level of resilience of critical infrastructure elements, which significantly reduces their vulnerability [7].

There are several ways to assess and model cascading effects between critical infrastructures that can be found in the literature, [8], which aim to provide an overview of methods for assessing critical infrastructure interdependence as part of risk and vulnerability analysis [9]. One of the methods of quantitative risk assessment is defined as the product of frequency and consequence ($R = FC$), based on the parameters: frequency (F), probability (P), range (E), i.e., the number of people affected by the unavailability of the service and, duration (D), as in the time period (h) when the service is not available.

Based on all the mentioned studies of the negative impact on the transport network, the vulnerability of the public transport network by targeted attacks and cascading effects can be observed through graph theory applied to model PTN and to analyze possible network service vulnerability to relevant disruption, such as closure and unavailability of one or more nodes [10].

The presented concept for the establishment of such a defined methodology could result in resistance in the acceptance of automated technology among road network users, and thus lower penetration of automated vehicles. On the other hand, with a larger number of vehicles, there is a greater number of security threats for the entire system and transport network, which requires additional human and material resources. Clear and decisive long-term commitment to the development strategy is needed [11]. This research provides a recommendation for work and future research, although the problems of defining a unique way of resisting cyber-attacks and improving the effectiveness of security systems that have not yet been resolved.

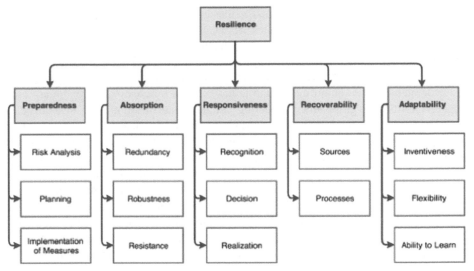

Fig. 1. Areas and criteria that determine the resilience of critical infrastructure elements. Source: [9]

2.1 Security Risk Analysis Techniques and Modeling of Cyber-Attacks

A framework for assessing safety risk in attempt to establish an effective measure for the safety of connected vehicles was built through the study [12]. The proposed framework configured an assessment process based on the conventional standardized security model of risk analysis GMITS (Guidelines for the Management of IT Security) ISO13335 and the attack tree analysis used to assess system threats and vulnerabilities. The security risk assessment framework uses the results of an analysis of assets, threats/vulnerabilities and risks to finally determine the risk assessment. The proposed framework has been applied to identify risks in case of increased speed of attacked vehicles and leakage of personal data, which are the main threats faced by autonomous vehicles. The risk analysis model is based on GMITS ISO13335 developed by NIST (National Institute of Standards and Technology), currently the leading standard model of safety risk analysis. The GMITS guidelines are an integral part of the standard report issued by the Technical Committee. The report consists of five parts: (1) IT Security Concept and Model; (2) Management and Planning; (3) Management Technique; (4) Countermeasures; (5) Network Connection Guideline. Security risk management is described by the proportional relationship of assets, threats and vulnerabilities. The defined proportional relationships in the equation change relatively according to the risk management criteria of each organization or system. These patterns are used in conducting security risk analysis. Therefore, when defining the threats, assets and vulnerabilities of connected systems, it is important to follow the guidelines based on the definition provided by GMITS ISO13335 and the EVITA project which configures the importance of evaluation items [12].

The analysis of cyber-attack modeling techniques provides an overview of three models for analyzing the same cyber-attack [13]. Each of the methods is unique and represents the same attack in different ways thus gaining insight using three attack modeling techniques. These three techniques are: Diamond Model, Kill Chain Method, and Attack Diagram. The attack graph shows whether an attacker is successfully attacking, and presents several alternative ways to create remote attacks, such as Distributed Denial of Service (DDoS). Remotely denying the service of navigation systems can be observed with a diamond model and attack method by creating an imaginary load (capacity). Although graph attack modeling techniques are different, they all share common attributes such as attacker, victim, network, attack plan, cargo, and transportation [14]. One of the important parts of any cyber-attack is that an attacker does not attack any high-profile network without first properly studying all the limitations of the network. The plan is made using collected data on targeted infrastructure and resistance capabilities. To carry out an attack, an attacker must create an imaginary load, in this case a program file (.exe), which is a very common way to attack any workstation of a particular operating system. Also, the social engineering method is used to assess the success of the delivery of cargo to the victim of the attack. So, the chance of success is 50%, because once the file is delivered, it is up to the victim whether to execute it or not [13].

Macroscopic control strategies are introduced for simulation and modelling, which rely on real-time observation of relevant spatially aggregated measures of traffic performance [15]. Observation of chaotic scatter-plots of speed versus density from individual fixed detectors was aggregated the scatter nearly disappeared and points grouped neatly along a smoothly declining curve. This evidence suggests that an MFD (macroscopic fundamental diagram) exists for the complete network because the fixed detectors only measure conditions in their proximity, which may not represent the whole network [16, 17]. A two-dimensional generalization of the MFD, the so-called Generalized Macroscopic Fundamental Diagram (GMFD) is also proposed [18], which relates the average flow to both the average density and the (spatial) inhomogeneity of density. Quantization of the macroscopic turning flows into units of single vehicles is necessary to obtain realistic fluctuations in the traffic variables, and how this can be implemented in a fluid-dynamic model [19]. New method to simulate destination flows without the requirement of individual route assignments is proposed by combining both methods that allow different simulation scenarios [20]. Incorporating user equity considerations into a road link importance measure, as a complement to measuring the total increase in vehicle travel time, it is also measured the disparity in the distribution among individual users [21, 22]. The vulnerability assessment is considered in terms of planning systems processes in which the performance of network components is tested against established performance criteria. The risks and consequences associated with failures at different locations need to be accounted for [23].

2.2 Attack Scenarios in Urban Environment

The methodology for modelling attacks on transport systems contains the parameters which can be implemented in the PTV Visum simulation tool, using a general model developed based on the description of the expected penetration of the CAV vehicles. The task assignment process on the transport network is performed by the equilibrium method as an optimization problem because the function aims to minimize travel times if input data are available [13]. In order to define the expected distance reduction factor, the average distance travelled is calculated based on the average transport performance and weighted speed parameters in the case of the road network where the penetration rate is equal to or greater than 80%. The value of the factor of optimizing the capacity of CAVs was estimated by the authors at 40%, and for verification they compare the distances travelled by previous and future transport systems. These data are obtained by conducting periodic surveys and research in this area, the so-called BPR patterns which are behavioural patterns of travellers.

The analyzed locations of the attack need to be described in detail and with the graphical representation of the transport network. Also, this way of classifying CAV penetration and levels of autonomy appears in other studies [24, 25].

The efficiency drop caused by the operational work of the network is not critical at first. Even in the event of a single attack, the average travel time of the affected vehicles increases by 20–30 min and thus has a significant psychological impact on road users. On the other hand, even given the only monthly repeated single attack and assuming an hour of travel is € 15, the annual loss caused by incidents exceeds the level of € 200,000. This impact is of greater importance in environments of poorer socio-economic status, especially if the action is long-term. In the case of simultaneous interventions of the multiplier effect of infection at the network level, the results are multiple losses [13].

Furthermore, considering congestion dynamics in large urban areas, another study presented a new method for analyzing the relationship between city size and network structure [26]. Such a new approach suggests the application of a macroscopic baseline diagram to analyze the relationship between land use and traffic congestion [27]. The authors [28] made a thorough comparison of transport systems in European cities, especially taking into account aspects of sustainability. The results confirm that the classification of the urban transport system leads to a division into groups characterized by different cultural, structural and patterns of size. Similar findings are provided by other researchers [29, 30], who strongly support such an approach to specification. One of the most important results of the research was to support the application of simple management concepts in the case of a certain city size, for certain network structures and traffic density characteristics. These results were conducted by the authors in their research [31] because they strongly support the assumption that cities with similar network structure and patterns of behaviour of residents can be characterized by similar transport processes, and thus similar consequences of network attacks.

3   Conclusion

With the development of technology of connected and automated transport systems, it is expected to move in the direction of continuous improvement if we observe the significant impacts of today, such as the elimination of vehicle ownership which puts emphasis on the resilience of future transport management systems, also as the ecological footprint of the population. However, clear assessments of the negative impact of new systems are needed. The aim of further research is to analyze the vulnerabilities, threats and incurred costs of the impact of disruptions and attacks in the case of connected and automated systems. Interventions are targeted at the transport network, rather than at individual road users, and it is assumed that in this way the risk can be multiplied when an incident occurs. The priority of a conservative assessment of the approach is that only a relative increase in capacity and travel time is taken into account as a saving factor. A limitation of the current studies is that those allow the assessment of only a limited number of types of attacks. Accordingly, a new cyber-attack can primarily be well described for modeling purposes and, if possible, its effect on capacity, congestion, distance travelled, speed, or traffic volume can be reliably estimated.

Effective transport management in these conditions should ensure the loading of the transport network to the point of its capacity and maintain a continuous uniform movement, at relatively low speeds. In addition, to meet the needs of the modern economy and society in reducing transport costs and increasing the speed of passenger travelling and delivery of goods, transport operators have to use the means of modern technology and modernize the organization of their business. Devotion to the formation of a knowledge-based system analysis, methods of mathematical modelling and simulation, the enlarged model for evaluating the performance of transport interchanges, which is supposed to solve the problem of high load of the road network, taking into account not only quantitative parameters, which characterize the system under consideration, but also the legal basis for the formation of relations between all subjects of the transport management systems. [32].

Generally, the consideration of real-time and stochastic data is very limited in current TMS. The future developments in such platforms should enable aggregation of input data from several sources that is shared between included information providers. Using advanced models and algorithms can help improve the modal split and reduce transport times and slack, as well as response times to unexpected events during transport. Future research directions include: 1) More effective hybrid algorithms that can support very large-scale network simulations; 2) Incorporating complex time-space service network design problems in simulations; 3) Focusing on social impacts of intermodal transport policies at local, regional and international levels.

References

1. Mattsson, L.G., Jenelius, E.: Vulnerability and resilience of transport systems—a discussion of recent research. Transp. Res. Part A Policy Pract. **81**, 16–34 (2015)
2. Rodríguez-Núñez, E., García-Palomares, J.C.: Measuring the vulnerability of public transport networks. J. Transp. Geogr. **35**, 50–63 (2014)
3. Cats, O., Jenelius, E.: Planning for the unexpected: the value of reserve capacity for public transport network robustness. Transp. Res. Part A Policy Pract. **81**, 47–61 (2015)
4. Cats, O., Jenelius, E.: Beyond a complete failure: the impact of partial capacity degradation on public transport network vulnerability. Transportmet. B **6**, 77–96 (2018)
5. Zhang, X., Miller-Hooks, E., Denny, K.: Assessing the role of network topology in transportation network resilience. J. Transp. Geogr. **46**, 35–45 (2015)
6. Cats, O., Yap, M., van Oort, N.: Exposing the role of exposure: public transport network risk analysis. Transp. Res. Part A Policy Pract. **88**, 1–14 (2016)
7. Rehak, D., Slivkova, S., Brabcova, V.: Indication of critical infrastructure resilience failure. Cepin, M., Bris, R. (eds.) Safety and Reliability – Theory and Application (ESREL), pp. 955–962. CRC Press, London (2017)

8. Bukvić, L., Smolković, M., Pašagić Škrinjar, J.: Implementation of blockchain in logistics and digital communication-short overview. In: Proceedings of 11th Supply Chain Management for Efficient Consumer Response Conference, p. 12 (2020)

9. Utne, I.B., Hokstad, P., Vatn, J.: A method for risk modeling of interdependencies in critical infrastructures. Reliab. Eng. Syst. Saf. **96**(6), 671–678 (2011)

10. McGee, S., Frittman, J., Ahn, S.J., Murray, S.: Risk relationships and cascading effects in critical infrastructure: implication for the hyogo framework. In: Applied Systems Thinking Institute (ASysT) Global Assessment Report on Disaster Risk Reduction 2015. 36 p. (2014)

11. Candelieri, A., Galuzzi, B.G., Giordani, I., Archetti, F.: Vulnerability of public transportation networks against directed attacks and cascading failures. Publ. Transp. **11**, 27–49 (2019)

12. Kong, H.-K., Hong, M.K., Kim, T.-S.: Security risk assessment framework for smart car using the attack tree analysis. J. Ambient. Intell. Humaniz. Comput. **9**, 1–21 (2017). https://doi.org/10.1007/s12652-016-0442-8

13. Török, A., Szalay, Z., Uti, G., Verebélyi, B.: Modelling the effects of certain cyber-attack methods on urban autonomous transport systems, case study of Budapest. J. Ambient. Intell. Humaniz. Comput. **11**, 1629–1643 (2020)

14. Al-Mohannadi, H., Mirza, Q., Namanya, A., Awan, I., Cullen, A., Disso, J.: Cyber-attack modeling analysis techniques: an overview. In: 2016 IEEE 4th International Conference on Future Internet of Things and Cloud Workshops (FiCloudW), pp. 69–76 (2016). https://doi.org/10.1109/W-FiCloud.2016.29

15. Daganzo, C.F.: Urban gridlock: macroscopic modeling and mitigation approaches. Transp. Res. Part B **41**(1), 49–62 (2007)

16. Geroliminis, N., Daganzo, C.F.: Macroscopic modeling of traffic in cities. In: The 86th Transportation Research Board Annual Meeting. Paper No. 07-0413, Washington, DC (2007)

17. Geroliminis, N., Daganzo, C.F.: Existence of urban-scale macroscopic fundamental diagrams: some experimental findings. Transp. Res. Part B **42**(9), 759–770 (2008)

18. Geroliminis, N., Sun, J.: Properties of a well-defined macroscopic fundamental diagram for urban traffic. Transp. Res. Part B **45**, 605–617 (2011)

19. Knoop, V.L., Hoogendoorn, S.P., Van Lint, J.W.C.: Impact of traffic dynamics on the macroscopic fundamental diagram. In: Transportation Research Board 92nd Annual Meeting (2013)

20. Mazloumian, A., Geroliminis, N., Helbing, D.: The spatial variability of vehicle densities as determinant of urban network capacity. Phil. Trans. R. Soc. A **368**, 4627–4647 (2010)

21. Jenelius, E.: Geography and road network vulnerability: regional equity vs. economic efficiency. In: 9th Network on European Communications and Transport Activities Research (NECTAR) Conference, Porto, Portugal (2007)

22. Jenelius, E.: User inequity implications of road network vulnerability. J. Transp. Land Use **2**(3–4), 57–73 (2010)

23. Jenelius, E., Mattsson, L.G.: Road network vulnerability analysis of area-covering disruptions: a grid-based approach with case study. Transp. Res. Part A **46**, 746–760 (2012)

24. Taylor, M.A.P., D'Este, G.M.: Critical infrastructure and transport network vulnerability: developing a method for diagnosis and assessment. In: Proceedings of the Second International Symposium on Transportation Network Reliability (INSTR 2004). Department of Civil Engineering, University of Canterbury, Christchurch, pp. 96–102 (2004)

25. Takács, Á., Drexler, D.A., Galambos, P., Rudas, I.J., Haidegger, T.: Assessment and standardization of autonomous vehicles. In: IEEE 22nd International Conference on Intelligent Engineering Systems (INES), pp. 185–192 (2018)

26. Debernard, S., Chauvin, C., Pokam, R., Langlois, S.: Designing human-machine interface for autonomous vehicles. IFAC, Elsevier **49**(19), 609–614 (2016)

27. Tsekeris, T., Geroliminis, N.: City size, network structure and traffic congestion. J. Urban Econ. **76**, 1–14 (2013)

28. Saeedmanesh, M., Geroliminis, N.: Clustering of heterogeneous networks with directional flows based on "Snake" similarities. Transp. Res. Part B Methodol. **91**, 250–269 (2016)
29. Alonso, A., Monzón, A., Cascajo, R.: Comparative analysis of passenger transport sustainability in European cities. Ecol. Ind. **48**, 578–592 (2015)
30. Klinger, T., Kenworthy, J.R., Lanzendorf, M.: Dimensions of urban mobility cultures - a comparison of German cities. J. Transp. Geogr. **31**, 18–29 (2013)
31. Berki, Z., Monigl, J.: Trip generation and distribution modelling in Budapest. Transp. Res. Procedia **27**, 172–179 (2017)
32. Panteleeva, M., Borozdina, S.: Model development of expert estimation of quality of transport interchange functioning. In: MATEC Web of Conferences, vol. 193, p. 01011. EDP Sciences (2018)

Concept of Road Traffic Noise Monitoring in the Function of Environmental and Health Protection

Osman Lindov$^{(\boxtimes)}$ and Adnan Omerhodžić

Faculty of Traffic and Communications, University of Sarajevo, 71000 Sarajevo,
Bosnia and Herzegovina
osman.lindov@fsk.unsa.ba

Abstract. Traffic noise is one of the most significant negative impacts on the environment. Road traffic is the largest noise source in urban areas. The negative consequences of the impact of road traffic noise in urban areas due to the high concentration of the population are particularly pronounced in the segment of the impact on human health. Therefore, an adequate treatment of road traffic noise requires a comprehensive noise monitoring concept. The comprehensive concept of noise monitoring from road traffic is the basis for the management and implementation of noise reduction measures within the permitted and acceptable levels. This paper will present the basics of the concept of noise monitoring from road traffic with special reference to the impact of road traffic noise on human health. Possibilities of application of innovative tools and procedures in the monitoring process in the function of reducing road traffic noise and protection of human health will also be presented.

Keywords: Road traffic noise · Noise monitoring · Environmental and health protection

1 Introduction

Urbanization and technological progress are providing more comfortable way of life and realization of everyday activities of the population. However, due to the increasing concentration of population in urban centers and the need for mobility, certain environmental problems arise. Observed from the aspect of environmental requirements and protection of public health, road traffic is one of the main sources of air and noise pollution in cities. Road traffic noise is an inevitable phenomenon that causes serious consequences for the health of the population exposed to excessive noise levels.

Noise pollution from road traffic is one of the main causes of degradation of the quality of life and as a serious growing problem requires more intensive engagement and treatment of the full range of competent and interested entities. Viewed from the aspect of responsibility towards one's own health, the problem of road traffic noise requires the engagement and proactive action of the population through the available elements of the noise management system. This means involving the population in the

© The Author(s), under exclusive license to Springer Nature Switzerland AG 2022
I. Karabegović et al. (Eds.): NT 2022, LNNS 472, pp. 639–650, 2022.
https://doi.org/10.1007/978-3-031-05230-9_77

processes of problem identification and the process of creating action plans to reduce road noise pollution. Noise pollution in urban areas is one of the priority tasks of experts dealing with issues of improving the quality of life. Noise problems should be solved multidisciplinary, and measures to solve noise pollution should be based on technical knowledge and achievements of modern science.

This paper emphasizes the importance of adequate treatment of road traffic noise as one of the biggest environmental problems in urban areas. Also, the paper emphasizes the importance of noise monitoring as a key mechanism in the noise management system of road traffic. The aim of this paper is to emphasize the importance of noise pollution problem from road traffic and to encourage further theoretical and practical work within this issue. This especially refers to the integration of innovative techniques and tools which are providing more efficient implementation of road traffic noise monitoring in relation to existing systems that are not able to fully meet the requirements such as technical-technological, personnel and financial.

2   Concept of Noise Monitoring from Road Traffic

Noise monitoring from road traffic is a comprehensive and continuous process of recording and monitoring noise levels from road traffic. That is an integral approach to treating the impact of noise on environmental segments with an appropriate methodology defined by regulations. The basic mechanisms of road traffic noise monitoring are methods and procedures for measuring and recording noise levels in the vicinity of the road.

Adequate and efficient monitoring of road traffic noise requires several key elements, namely:

– Methodological procedure and procedures for monitoring implementation.
– Information system and database on recorded noise parameters.
– Equipment and measuring instruments for measuring and recording noise parameters.
– Software tools for creating noise maps.
– Personnel with appropriate competencies and skills for the implementation of monitoring.

The main purpose of noise monitoring in road traffic is reflected in the creation of a documentation basis for measured and modeled values of noise levels that will enable efficient and adequate management of noise impacts on the environment. Observed from point of view of the results and benefits of monitoring road traffic noise process, two key groups of stakeholders can be identified:

– Direct stakeholders (road infrastructure managers).
– Indirect stakeholders (population).

The key stakeholders in road traffic noise monitoring are road infrastructure and road managers, as they represent the owners of noise sources. Road infrastructure managers, as owners of noise sources, have an obligation to manage noise pollution and environmental impacts. This includes taking measures and activities aimed at preventing, reducing or mitigating noise levels at the source, on transmission routes and at the receiver's location. The efficiency of the mentioned process of road traffic noise pollution control directly depends on the quality and functionality of road traffic noise monitoring.

Unlike direct stakeholders in noise monitoring as owners of noise sources whose priority is the obligation to manage noise pollution, the priority and needs of indirect stakeholders in noise monitoring are information based on the results of noise monitoring. Therefore, the population, as indirect stakeholders in noise monitoring, needs reliable, accurate and timely information on noise pollution monitoring. The main purpose of this information is reflected in the support for planning and decision-making such as choosing the location of housing or choosing a location for leisure activities, all in the function of protecting health from the negative effects of road noise pollution.

3   Noise Monitoring from Road Traffic in EU Countries

Commitment to European Union policy is the pursuit for highly ranked protection of health and environment, as well as noise protection, as environmental noise has been identified as one of the biggest environmental problems in European Union. A very wide range of different national approaches to noise issues in the countries of the European Union have initiated the need for joint action in order to more effectively manage the problems of noise impact on health and the environment. Based on the above and in accordance with the provisions of the Green Paper, Directive 2002/49/EC [1] to the management and assessment of environmental noise was adopted. This Directive is the basic legal regulation of the countries of the European Union which, among other things, treats road traffic noise.

3.1   Noise Monitoring Mechanisms in the Function of Road Traffic Noise Management

From a monitoring point of view, Directive 2002/49/EC encourages a common understanding of road traffic noise problems and coordinated action. This includes the use of harmonized noise indicators, noise measurement and modeling methods and procedures, noise limit values and permitted values, noise mapping procedures, public information mechanisms, and the creation of action plans and measures to reduce noise.

The main indicators in the monitoring process are L_{den}, which assesses noise disturbances, and L_{night}, which assesses sleep disorders. In addition to the mentioned countries, they can use additional indicators, but during the joint reporting, it is necessary to transform and adjust the values to the stated basic indicators. The values of the L_{den} and L_{night} indicators are determined by assessment methods, i.e. by calculation or noise measurement according to national regulations and standards. If countries do not have national methods and standards for the calculation of noise from road traffic, it is recommended to use the "NMPBRoutes 96" [1].

One of the most important mechanisms for monitoring noise from road traffic, which needed for development of action plans for noise pollution management, but also a highly adapted tool for informing the population, is the strategic noise map. The Strategic Noise Map is intended for the integrated assessment of population exposure to noise from several sources or for the purpose of comprehensive forecasts for a given area. Each state is obliged in accordance with regulation of Directive 2002/49/EC to draw up a strategic noise map for all areas with population over 250000 people and also for all roads over 6000000 passing vehicles on yearly basis. In addition, states is obligated to develop action plans to monitor noise and impact within their territories, including noise abatement, if necessary, for: sites near roads over 6000000 passing vehicles on yearly basis as well as populated places, populated over 250000 people within its territories [1].

3.2 Institutional Structures of Road Traffic Noise Monitoring

States, through the competent institutional structures and on the basis of national regulations, implement the process of monitoring and management of road traffic noise pollution. National regulations should comply with regulation of Directive 2002/49/EC. In accordance with the needs of joint action and noise monitoring, states shall designate competent authorities responsible for enforcement of mentioned Directive, and for the coordination of reporting and experience exchange of activities in order to define appropriate noise protection action plans.

According to the above, the umbrella institution is The European Environment Agency (EEA), which enables the institutional integration of designated national stakeholders and provides a wide range of information necessary for adequate noise pollution management. EEA, through its European Environment Information and Observation Network (EIONET) service, collects environmental information from individual countries. Among other things provided data are used for road noise pollution management processes.

EEA manages EIONET and coordinates its activities in cooperation with the National Focal Points (NFP). NFP are national environmental agencies or ministries of the environment. NFP are in charge for coordination of National Reference Centers (NRCs) networks, which are organized groups of which consist of different experts on environmental issues from several state institutions. There are currently 24 National Reference Centers, of which the 17th National Reference Center deals exclusively with noise issues.

The National Noise Reference Center implements the Directive 2002/49/EC by controlling noise measurement and modeling procedures, controlling noise maps, controlling action plans and reporting procedures. Within the EIONET, seven European Topic Centers (ETCs) have been integrated, representing consortia of organizations with an expert approach in specific areas of environmental protection, including noise impact.

4   Impact of Noise from Road Traffic on Population Health

Noise pollution in urban areas is very significant environmental problem. The specifics of urbanization, the concentration of a large number of inhabitants in relatively small areas in the immediate vicinity of the most important traffic arteries, and the continuous trend of increasing the number of vehicles, have influenced road traffic pollution to be the biggest environmental problem in urban areas. The consequences of road noise in urban areas and the impact on the population are considered a serious health problem.

Research studies have identified an association with noise exposure and threats to health and quality of life through consequences like heart disease, anxiety, and hearing loss [2, 3]. For this reason, activities to improve regulations that enable more efficient monitoring and reporting of noise pollution have recently been intensified. A very important mechanism for noise monitoring of road traffic is reporting on the exposure of inhabitants to noise from road traffic. This enables quality, verified and accurate information based on which is possible to more efficiently and responsibly create measures and action plans to reduce noise pollution. The results of reporting on the exposure of inhabitants to noise from road traffic should also serve as a support for the population to make decisions when planning their own activities and a more responsible attitude towards the protection of health potentially endangered by road noise pollution.

Noise exposure data and related action plans in line with Directive 2002/49/EC are used to assess noise from environment in European Union. This data is available from The Noise Observation & Information Service for Europe, which displays noise contour maps for major noise sources in Europe. Some countries publish data on noise as well as an assessment of the likelihood of health consequences due to noise. EEA uses data to assess the likelihood of covering areas affected by noise pollution in a report named as Quiet areas report [4].

Table 1 shows indicators of the exposed number of people to harmful levels of road noise with more than 3000000 passing vehicles on yearly basis, which are above the levels of noise indicators set by Directive 2002/49/EC:

- L_{den}: Indicator is used to assess discomfort and represents the annual average day, evening and night exposure periods with an evening weight of 5 dB (A) and a night weight of 10 dB (A).
- L_{night}: Indicator is used to assess sleep disorders, and represents the annual average night period of exposure.

Table 1. Number of people exposed to $L_{den} \geq 55$ dB and $L_{night} \geq 50$ dB in Europe [5]

Noise source	$L_{den} \geq 55$ dB	$L_{night} \geq 50$ dB	Aglomerations
Roads	78236200	55088700	In urban areas
Main roads	30574800	20705700	Outside urban areas
Roads	81668800	57479600	In urban areas
Main roads	31142900	21096500	Outside urban areas

Noise from road traffic is one of the main health risks, including discomfort, sleep disturbance and ischemic heart disease. Table 2 shows more detailed data on the approximative number of people with different health outcomes due to environmental noise.

Table 2. Approximative number of persons with different health outcomes due to noise from road traffic [6]

Areas	High annoyance	High sleep disturbance	Ischemic heart disease	Premature mortality
In urban areas	12525000	3242400	29500	7600
Outside urban areas	4625500	1201000	10900	2500

The recommendations of the World Health Organization (WHO) are based on the reliability that noise reduction below the levels of $L_{den} = 50$ dB and $L_{night} = 45$ dB, will outweigh the potential harmful consequences. These values represent exposure levels above which there is an increase in adverse health effects.

The results of scientific evidence recommend a set of health outcomes that can be quantified by assessing the impact of road traffic noise on health (Health Impact Assessment - HIA): incidence of ischemic heart disease [7, 8], anxiety [9], sleep disorders [10], brain incidence stroke [11] and the incidence of diabetes [3].

The World Health Organization has introduced methods for quantifying the Burden of Disease (BoD) [12] from environmental noise using DALYs (Disability-Adjusted Life-Year) [13], which combines years of life lost due to premature mortality and years of life lost due to time spent in conditions weaker than the state with complete health. This methodology was used to calculate the burden of disease as a result of disturbance, sleep disturbance and reading impairment, using exposures and reactions as well as the proportion attributed to the population for Ischemic Heart Disease is (IHD).

Table 3 shows the burden of disease outcomes estimated on the basis of noise data resulting from reports under Directive 2002/49/EC, and the results roughly show that around 1000000 healthy years of life are lost each year due to the impact of environmental noise on effects of health.

Table 3. Burden of disease outcomes estimated on the basis of noise data resulting from reports under Directive 2002/49/EC [6]

Health effect	Public health impact (DALYs/year)	Public health impact (DALYs/year) per million
High annoyance	453000	900 per million people
High sleep disturbance	437000	800 per million people
Ischemic heart disease	156000	300 per million people

Noise from environment, in addition to the negative health consequences of the impact on the population, also has economic consequences. One of the most important methodologies for quantifying the economic costs of noise due to health, involves the allocation of monetary costs according to "DALY" [14]. As the monetary cost of DALY is around € 78500, the resulting economic impact of noise is estimated at € 35 billion for distress, € 34 billion for sleep disturbances, € 12 billion for Ischemic Heart Disease is and € 5 million for cognitive impairment in children.

5 Possibilities of Application of New Techniques and Tools in the Process of Noise Monitoring from Road Traffic

The most important mechanism of the road traffic noise monitoring process is the measurement and modeling of noise from road traffic, processing and adjustment of these data into adequate forms necessary for reporting. Roads are linear noise sources, so given the size of road networks, leaders of the process of monitoring and measuring noise have difficult circumstances to provide measuring equipment that will enable the collection of necessary data within acceptable time limits throughout the jurisdiction. For this reason there is objective need to create, develop and use new techniques and tools that will enable more efficient collection of data on road noise levels. In this process, it is extremely important to choose a proven technique that guarantees accurate and quality data.

One such technique is "Mobile crowdsensing" [15], which is based on the collection and recording of data by a big group of individuals using smartphone, tablet, etc. The data is then shared, processed, and filtered together to measure, map, analyze, evaluate, and predict processes of common interest. The basis of the "Mobile crowdsensing" technique is mobile devices with an integrated set of different sensors (light, temperature, sound, movement, location, touch, etc.) that can collect huge amounts of useful data for different needs. Therefore, the technique of "Mobile crowdsensing" implies "Crowdsourcing" of sensor data from mobile devices.

"Mobile crowdsensing" is a segment of sensing which enables regular citizens to provide information given from their mobile devices to aggregate and emerge this datas into the cloud for crowd intelligence extraction and humancentric service delivery [16]. Depending on user involvement, there are two types of "Mobile crowdsensing":

- Participatory crowdsensing, where users personally and voluntarily participate in the collection and creation of sensory data [17].
- Opportunistic crowdsensing, where sensory data is recorded, collected and exchanged automatically without explicit user action, and often without user knowledge [16, 18].

The main phases of the "Mobile crowdsensing" process are: data collection based on sensors available via the IoT [19], data storage on user devices and data upload by stakeholders to undertake certain activities. The problems of implementing "Mobile crowdsensing" are selection of users for data collection, i.e. designing incentive mechanisms for the selection of appropriate participants [20]. Problems can also be in resource constraints such as energy required, bandwidth, and computing power.

"Mobile crowdsensing" has significant potential to use sensory information to record road noise pollution data [21]. "Crowdmapping" is a technique that together with "Mobile crowdsensing" can be implemented as a support in the process of monitoring road traffic noise. "Crowdmapping" is a technique that aggregates recorded data generated by users (e.g. noise measurement) and combines it with geographic data to create a digital map (e.g. noise map). "Crowdmapping" is an effective way to visually show the geographical distribution of a phenomenon.

GIS tools enable a significant role and support for the implementation of "Mobile crowdsensing" techniques in noise monitoring from road traffic. Originally [22], GIS tools in the function of noise monitoring from road traffic were used as a platform for the implementation of developed models that established a correlation between the spatial distribution of noise impacts and relevant activities in space. Improving and developing the model required integration of additional input data, which led to public involvement in data collection and the development of PGIS (Participatory GIS) tools.

Participatory GIS (PGIS) or Geographic Information System with Public Participation (PPGIS) is a participatory approach to spatial planning and management of spatial information and communications. Although the application of PGIS is very sensitive due to the possible bias of the involved public, as well as the need to ensure a certain level of experience and knowledge of the involved public, the application of PGIS in the process of road traffic noise monitoring has great potential benefits [23].

The development of sensor components of smartphones has enabled significant support for "Mobile crowdsensing" in the process of monitoring road traffic noise. The ability to use smartphone microphones to measure ambient noise levels has led to the development of several significant and proven applications to support and improve road noise monitoring: NoiseTube[24], Ear-Phone [21] and SoundOfTheCity [25].

The "NoiseTube" is a low-cost tool which involving population to record noise levels using their mobile phones as noise sensors. The approach is developed by experience of population and it's defined as Web-based ICT resources that enable the population to record and archive information relevant to noise pollution management. As a result of the recorded data and their processing is the production of maps of noise and noise exposure of the population. "NoiseTube" is a system consisting of a mobile phone, an application and a web portal. The application allows you to measure noise and record on a website, and view on Google Earth. Application provides users with a noise exposure dosimeter that informs the user about his noise pollution. Figure 1 shows an example of a noise map made in the Sound Plan software, based on data on noise levels recorded by the "NoiseTube" application. Tests were performed to visualize the exposure of objects in the vicinity of the road on noise maps and showed a relatively good match for different ways of collecting input data: measuring noise with Bruel and Kjaer 2260, modeling noise with NMPB model and measuring noise with "NoiseTube" application.

"Ear-Phone" is an open platform for the use of mobile phones in the process of recording ambient noise levels. "EarPhone" leverages context-aware sensing and is very successful in solving the challenge of using the phone as a sensor and measurement variations depending on the phone's orientation and user context. Classifiers are used to accurately determine the context, after which the sensor with the handset decides whether to sense and record noise. Experimental research and simulations have shown that "EarPhone" is a feasible platform for the assessment of noise pollution that allows high accuracy in the process of making noise maps [21].

Fig. 1. Example of a noise map created in Sound Plan software

"SoundOfTheCity" is a mobile application that enables participatory participation of the population in recording noise from the environment and submitting recorded data to a central server that performs visualization and creation of noise maps. The application allows the user to monitor their own exposure to noise from their environment. This application allows you to identify, classify and filter different measurement locations. Thus, data recorded indoors or in a vehicle is not downloaded to a central server. Noise maps, which are generated on the basis of measured values, enable the public to be actively involved in the activities of creating action plans and implementing measures to reduce noise in their environment.

Research [26] is explaining the details of "Mobile crowdsensing" technique and comparison of results which are related to practical examples and results of usual techniques which are used as an instrument to measure noise levels. Research has shown that with "Mobile crowdsensing" is possible to develop maps of noise on simply way, using less resources and saving time. The aim is to present that efficiency and accuracy for measured data provided from calibrated smartphones is acceptable.

The results of the conducted research and the indicators of success and acceptability of the "Mobile crowdsensing" technique are the basis for future research in order to improve the road traffic noise monitoring system. This primarily refers to the possibilities of implementing monitoring based on the "Mobile crowdsensing" technique with recording, storage and visualization of road noise pollution in real time. Research [27] refers to the possibilities of developing a methodology for monitoring road traffic noise in real time with the identification of locations and daily periods of extreme noise. This is the basis for the management of noise pollution from road traffic, and the adoption of measures and actions to reduce noise in real time and real traffic conditions.

6   Conclusion

Current indicators in the noise pollution management system indicate that road traffic is the largest source of noise pollution in urban areas. Due to the tendency of accelerated urbanization and increase in the number of vehicles, the problem of noise in urban areas from road traffic requires special treatment in order to take preventive and corrective measures.

The results of scientific research indicate significant cause-and-effect relationships between the impact of noise from road traffic and the health consequences of residents. Therefore, the results of road traffic noise monitoring and population exposure data should be available to the public. These data should enable more responsible access of people to their own health when making decisions and choosing locations for various activities, and more active participation of people in creating measures and action plans to reduce road traffic noise in their environment. Developed methodologies for monetary quantification of noise impact on population health should be used when creating measures and action plans to reduce noise pollution from road traffic.

The success and efficiency of noise pollution management from road traffic requires harmonized regulations, technical-technological and human resources. Directive 2002/49/EC specifies clear guidelines for the management of road traffic noise pollution, but the greatest contribution should be made by states by creating national regulations and management systems in line with current scientific knowledge. Management of road traffic noise pollution requires an appropriate system of monitoring, measurement and modeling of noise, production of noise maps, reporting on population exposure to noise, creation and implementation of action plans and measures to reduce noise. The most important segment in the process of road traffic noise management is noise monitoring, which enables recording of necessary data on noise levels on the basis of which activities in other segments are implemented, such as noise mapping, population reporting and creating action plans.

On the road as a linear source of noise, large variations are possible in different locations at daily basis. Therefore, a responsible road noise pollution management system requires continuous monitoring. The implementation of continuous monitoring is a very demanding process that requires significant technical, technological, human and financial resources. Existing systems of measuring equipment and sensor stations are not able to fully meet the requirements of road noise monitoring when it comes to the requirements of space coverage, accuracy of necessary data and large variations in the value of noise on roads.

"Mobile Crowdsensing" techniques and platforms have shown significant progress in their ability to integrate into road traffic noise monitoring processes. For this reason, it is necessary to develop sensibility and flexibility to use the public as support in the implementation of road traffic noise monitoring systems. It is certainly necessary to take into account when selecting participants with regard to expertise, knowledge and skills. In addition, a key step in the process of integration and improvement of road traffic noise monitoring system is the creation of more flexible regulations that will enable the integration of proven "Mobile Crowdsensing" platforms into certain bylaws such as regulations, procedures and procedures for road traffic noise monitoring.

References

1. EUR-Lex: Directive 2002/49/EC. EUR-Lex (2002). https://eur-lex.europa.eu/legal-content/EN/TXT/?uri=celex%3A32002L0049. Accessed 28 Dec 2021
2. Stansfeld, S.A., Matheson, M.P.: Noise pollution: non-auditory effects on health. Br. Med. Bull. **68**, 243–257 (2003)

3. Thacher, J.D., et al.: Long-term exposure to transportation noise and risk for type 2 diabetes in a nationwide cohort study from Denmark. Environ. Health Perspect. **129**(12) (2021)
4. Nugent, C., Blanes, N., Fons, J., Sáinz de la Maza, M.: Quiet areas in Europe—the environment unaffected by noise pollution, Copenhagen (2016)
5. European Environment Agency: Exposure of Europe's population to environmental noise, Copenhagen (2019)
6. E. Peris and European Environment Agency: Environmental noise in Europe – 2020, Copenhagen (2019)
7. Lim, Y.H., et al.: Long-term exposure to road traffic noise and incident myocardial infarction: a Danish nurse cohort study. Environ. Epidemiol. (Philadelphia, PA) **5**(3) (2021)
8. Ndrepepa, A., Twardella, D.: Relationship between noise annoyance from road traffic noise and cardiovascular diseases: a meta-analysis. Noise Health **13**(52), 251–259 (2011)
9. Stansfeld, S., Clark, C., Smuk, M., Gallacher, J., Babisch, W.: Road traffic noise, noise sensitivity, noise annoyance, psychological and physical health and mortality. Environ. Health Glob. Access Sci. Source **20**(1) (2021)
10. Kim, M., et al.: Road traffic noise: annoyance, sleep disturbance, and public health implications. Am. J. Prevent. Med. **43**(4), 5-1–5-32 (2012)
11. Sørensen, M., et al.: Transportation noise and risk of stroke: a nationwide prospective cohort study covering Denmark. Int. J. Epidemiol. **50**(4), 1147–1156 (2021)
12. Haneef, R., et al.: Recommendations to plan a national burden of disease study. Arch. Publ. Health **79**(1), 1–8 (2021)
13. Grosse, S.D., Lollar, D.J., Campbell, V.A., Chamie, M.: Disability and disability-adjusted life years: not the same. Publ. Health Rep. **124**(2), 197 (2009)
14. Department for Environment Food & Rural Affairs UK: Environmental noise: valuing impacts on: sleep disturbance, annoyance, hypertension, productivity and quiet, London (2014)
15. Ganti, R.K., Ye, F., Lei, H.: Mobile crowdsensing: current state and future challenges. IEEE Commun. Mag. **49**(11), 32–39 (2011)
16. Guo, B., et al.: Mobile crowd sensing and computing. ACM Comput. Surv. (CSUR) **48**(1) (2015)
17. Burke, J., et al.: Participatory sensing. In: Workshop on World-Sensor-Web (WSW 2006): Mobile Device Centric Sensor Networks and Applications, pp. 117–134 (2006)
18. Trivedi, A., Bovornkeeratiroj, P., Breda, J., Shenoy, P., Taneja, J., Irwin, D.: Phone-based ambient temperature sensing using opportunistic crowdsensing and machine learning. Sustain. Comput. Inform. Syst. **29**, 100479 (2021)
19. Kumar, S., Tiwari, P., Zymbler, M.: Internet of things is a revolutionary approach for future technology enhancement: a review. J. Big Data **6**(1), 1–21 (2019). https://doi.org/10.1186/s40537-019-0268-2
20. Li, S., Shen, W., Bilal, M., Xu, X., Dou, W., Moustafa, N.: Fair and size-scalable participant selection framework for large-scale mobile crowdsensing. J. Syst. Architect. **119**, 102273 (2021)
21. Rana, R., Chou, C.T., Bulusu, N., Kanhere, S., Hu, W.: Ear-phone: a context-aware noise mapping using smart phones. Pervasive Mob. Comput. **17**, 1–22 (2015)
22. Oliveira, M.P.G., Medeiros, E.B., Davis, C.A.: Planning the acoustic urban environment. In: Proceedings of the 7th International Symposium on Advances in Geographic Information Systems, pp. 128–133 (1999)
23. Cinderby, S., Snell, C., Forrester, J.: Participatory GIS and its application in governance: the example of air quality and the implications for noise pollution. **13**(4), 309–320 (2008). https://doi.org/10.1080/13549830701803265
24. Maisonneuve, N., Stevens, M., Ochab, B.: Participatory noise pollution monitoring using mobile phones. Inf. Polity **15**(1–2), 51–71 (2010)

25. Ruge, L., Altakrouri, B., Schrader, A.: SoundOfTheCity-continuous noise monitoring for a healthy city. In: Proceedings - IEEE International Conference on Pervasive Computing and Communications Workshops, pp. 670–675 (2013)
26. Grubeša, S., Petošić, A., Suhanek, M., Đurek, I.: Mobile crowdsensing accuracy for noise mapping in smart cities. **59**(3–4), 286–293 (2018). https://doi.org/10.1080/00051144.2018.1534927
27. Jezdović, I., Popović, S., Radenković, M., Labus, A., Bogdanović, Z.: A crowdsensing platform for real-time monitoring and analysis of noise pollution in smart cities. Sustain. Comput. Inform. Syst. **31**, 100588 (2021)

Digitalization of the Railway Transport System - The Applicability of Blockchain Technology in the Context of the Industrial Revolution 4.0

Cătălin-Laurențiu Bulgariu[1,2]([⊠])

[1] Faculty of Industrial Engineering and Robotics, Polytechnic University of Bucharest, 313 Splaiul Independenței, 6 District, Bucharest, Romania
catabulgariu@gmail.com
[2] National Railway Company C.F.R. S.A., 38 Dinicu Golescu Avenue, 1 District, Bucharest, Romania

Abstract. This work aims to expose a detailed analysis of the main characteristics of blockchain technology, its benefits and applicability in the rail transport system. The need for research in this regard lies in the system's need to adapt to new challenges, to increase the quality of services and attractiveness. Since this technology is still in an early stage, this paper opens new horizons of knowledge, opens new directions of research and helps to understand the phenomenon.

Keywords: Digitalization · Railway system · Blockchain · Decentralization · Industry 4.0

1 Introduction

The digitalization of the railway system means a leap of quality and a prominent answer to several problems of operation and attractiveness. Still in an early stage, the new blockchain technology that seems to revolutionize several aspects of our social life, has a vast applicability in several areas and its usefulness in the field of transport can revolutionize this system in many ways. In this paper is presented blockchain technology in the context of the industrial revolution 4.0, applicabilities and utility in the field of railway transport. Among the main aspects of blockchain technology are safety, transparency, data security and speed of operations [1]. This revolutionary technology, designed in the first phase with a financial purpose having applicability in the creation of a cryptocurrency register has been extended and offers a generalized framework in several industries such as the transport system, the health system, the information system, etc.. Engineering and technology are the basic pillars that will determine the success of the new industrial revolution (Industry 4.0). Industrial sectors such as electric/electronic, automotive, aerospace, rail and telecommunications are the main potential adopters of revolutionary and digital technologies in the context of Industry 4.0, this being the only way to meet the needs of users, to be competitive and to increase the quality of the services offered. The current transport system, created to provide and ensure efficient

I. Karabegović et al. (Eds.): NT 2022, LNNS 472, pp. 651–658, 2022.
https://doi.org/10.1007/978-3-031-05230-9_78

connectivity between the different modes of transport (air, naval, road, rail), is surpassed by the mobility of users and hardly responds to the growing needs of globalization. In order to facilitate the travel needs of passengers, intermodal transfer points for different modes of transport should be correlated from several technical and organizational considerations, which with the help of new technologies will allow a quick and punctual transfer of passengers. In its current state, transport management and correlation between different modes of transport are still limited, due to barriers related to the lack of interconnection of networks, schedules, etc. For example, a delay in the arrival of the bus or train at an airport can lead to the loss of the flight. In the future, this could be adjusted if the schedules of buses, flights and passengers are interconnected by the new technologies of the 4th industrial revolution, integrating physical and cyber technologies. Industry 4.0 includes several technologies that find their use in several fields of activity and that would contribute to a large extent in the development of rail transport as a whole [2] (Figs. 1 and 2).

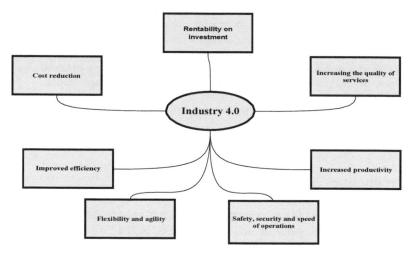

Fig. 1. The advantages of industry 4.0 in the railway sector

The main technologies incorporated in industry 4.0 are [3]:

Fig. 2. Technologies in Industry 4.0.

2   About and How Blockchain Technology Works

In order to be able to determine the main advantages and the potential of the applicability of blockchain technology in the rail transport system, we will have to understand what this technology is and what are its most important characteristics. Being the foundational technology in the world of cryptocurrencies, the birth of blockchain technology is generally considered to have taken place with the launch of bitcoin [4] in 2008, however many concepts that we are now using and underlying blockchain technology are the result of much older research, the idea behind the technology being described since 1991, when researchers Stuart Haber and W. Scott Stornetta [5] introduced a cryptographic computing system to mark digital documents so that they could not be updated or modified, later updated by other researchers over time. Blockchain is essentially a decentralized database that comprehensively uses the distributed storage of data shared through the nodes of a digital network, the information being stored electronically in a computer system. Blockchain provides decentralized, transparent and secure systems. It is a technology having as a structure a series (chaining) of decentralized data (distributed), under the name of blocks, connected and secured through the use of cryptography [6]. The connection between the distributed databases, called blocks, is made by hash [7]. Each storage space for information, transactions, etc. (block) is connected and has a link (a hash) of a previous block. When the storage space of a block is exhausted, it closes and will remain tied to the previous block, also full, thus forming a data chain that has the name of blockchain. The succession of the blocks will continue as the next newly formed block will be completed and joined to the chain and the previous blocks. The successive block will contain information about the previous block and therefore form a chain of data that cannot be changed retroactively. Each block in the data chain contains a unique hash, the transaction data and the hash of the previous block. The initial block is called a genesis block. A genesis block does not contain a previous hash. The participants of the blockchain network can be legal entities, companies, institutions or individuals that share a copy of the accounting ledger containing the transactions and information valid in a sequential manner [7, 8]. Blockchain technology is characterized by the way data and transactions are encrypted using specific mathematical formulas (algorithms). The data is written in structures called "transaction blocks", where no changes can be made to the old data but only new ones are added, forming new data blocks, functioning as a distributed data ledger [9]. Since 2009, blockchain technology has evolved, if initially used in finance or the field of cryptocurrencies, thoroughly improving the quality of different applications in terms of transaction speed, data security, privacy and ease of use, then this technology has gained its usefulness over time in several important areas, becoming an integral part of the implementation of Industry 4.0. To explore the application possibilities and utilities of blockchain technology in various industries, many companies have established their research centers for its development [10, 11]. Blockchain technology has gone through several stages of defining its utility, starting from an initial phase of exploring possible applicability to the point where it can be implemented and put into practice, also increasing its initial utility area [12] (Fig. 3).

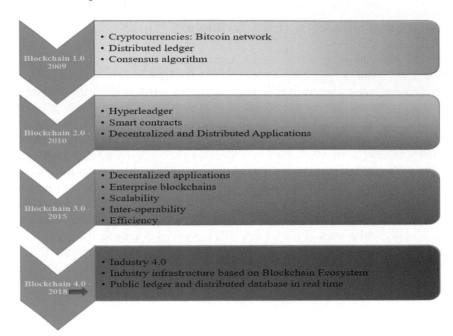

Fig. 3. Stages of the evolution of blockchain technology

To understand how blockchain technology is operated, in Fig. 4 the general structure of a blockchain is presented. Each chain block is identified by its cryptographic hash and each block refers to the hash of the block before it. The hash establishes a link between blocks, thus creating a chain of blocks [13]. The mode of operation on which blockchain technology is based is characterized by the participation of several actors in a network, where each actor can store, share and perform operations through network blocks, each being a node of the network, thus forming a distributed network without a central control entity or other intermediaries, the prototype being more solid.

The data of a blockchain is usually considered safe and immutable [12].

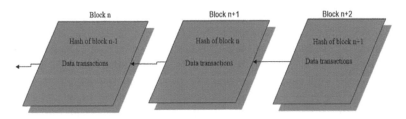

Fig. 4. The simplified base structure of the blockchain

In conclusion, in Fig. 5, there are concentrated and delimited the main key features of the blockchain that make it an attractive technology in the context of industry 4.0 [14, 15].

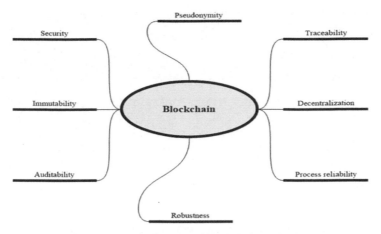

Fig. 5. The main features of blockchain technology

3 Application of Blockchain - Based Technology in the Digitalization of the Rail Transport System - Approach Methodology

Railways are an integral part of the critical infrastructure of any country. Freight services as well as passanger services have strict reliability, availability, maintenance, safety and security requirements. The consistency of rail transport in terms of capacity, time, schedules and, consequently, regularity is affected by various factors related to the operation and maintenance of infrastructure by the factors involved in these operations. The growing requirements of the railways in terms of data collection, transmission, storage, processing and general management have increased the need for digitalization. However, the digital world is affected by cyber security issues and cyber threats [16]. With their expansion, railways have an increasing need for communication, computing and dependence on digitalization creates exposure to cyber threats. In this respect, new technologies embedded in industry 4.0, such as blockchain technology, can respond to these issues and increase network safety simplifying execution processes. Digitalization acts as one of the main trends in the development of economy and society as a whole and is integrated into the economic development strategies of many states. An efficient transport system must be an integral part of society's progress by adopting innovative technologies that ensure intelligent, safe and environmentally friendly mobility. Digitalization can help create a new generation of rail systems that increase competitiveness over other modes of transport. The research, adoption and implementation of new digital technologies and solutions will accelerate the transformation and transition to the rail 4.0 [17]. Proposed transition steps to a digitalization rail system are presented in Fig. 6.

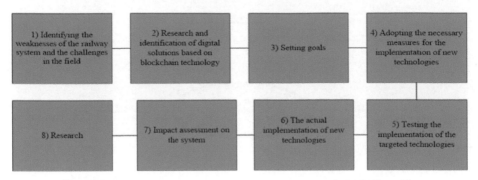

Fig. 6. Steps proposed for the adoption of blockchain-based technology in rail transport

Digital transformation of transport systems is a priority in strategic development of a state, requires a common approach and interaction between all system factors [18]. The main objectives of the digitalization of the rail system are [19]:

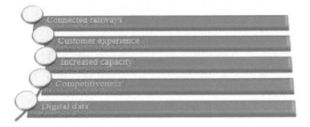

The use of digital technologies in the railway industry can lead to:

- Improving the safety of freight traffic and passengers through the exchange of real-time data on the current traffic situation;
- Creating a digital space for the interaction between carriers, shippers and passengers;
- Software optimization for determination routes;
- Using blockchain technologies to increase the level of transparency of the activities of all participants in the transport process, the logistics chain;
- Software development for data collection and analysis of information about the location of the goods, road conditions, etc. [20].

The main potential applications of blockchain technology in the railway industry are found in various areas of major interest including: decentralization of traffic management and control of rail traffic [21], supply chain [22, 23], the communication between all the factors of the transport system and the distribution of information, the management of passenger flows and goods.

4   Conclusion

In conclusion, some key points on the potential of blockchain technology in rail transport were identified and traced:

- Sharing information: Each actor of the rail system would have access to all shared information in the register, resulting in direct coordination between all those involved, the speed of operations, better data quality, safety storage, time reduction and effort;
- reduction of hardware components;
- Automatic processes between infrastructure and means of transport;
- Simplification of system processes and infrastructure;
- Automatic payments between the different parts of the system when consuming the transport relationship using smart contracts.

References

1. Javaid, M., Haleem, A., Singh, R.P., Khan, S., Suman, R.: Blockchain technology applications for Industry 4.0: a literature-based review. Blockchain Res. Appl. (2021). https://doi.org/10.1016/j.bcra.2021.100027
2. Ahamat, M.A.: Transportation in the 4th industrial revolution. (02), 10–11 (2017). https://www.mbot.org.my/MBOT/files/a9/a9b3dd97-dfb4-4320-ad41-6076c93db4fb.pdf
3. Pieriegud, J.: Digital transformation of railways. Department of Transport, SGH Warsaw School of Economics, Siemens Sp. z o.o., Poland (2018). ISBN 978-83-950826 0-3. www.sgh.waw.pl
4. Nakamoto, S.: A Peer-to-Peer Electronic Cash System - Bitcoin (n.d.). https://bitcoin.org/bitcoin.pdf
5. Haber, S., Stornetta, W.S.: How to time-stamp a digital document. J. Cryptol. (3), 99–111 (1991). https://doi.org/10.1007/BF00196791
6. Narayanan, A., Bonneau, J., Felten, E., Miller, A., Goldfeder, S.: Bitcoin and Cryptocurrency Technologies: A Comprehensive Introduction. Princeton University Press, Princeton (2016)
7. Johar, S., Ahmad, N., Asher, W., Cruickshank, H., Durrani, A.: Research and applied perspective to blockchain technology: a comprehensive survey. Appl. Sci. 11(14) (2021). https://doi.org/10.3390/app11146252
8. Deepak, P., Nisha, M., Saraju, M.P., Elias, K., Chi, Y.: The blockchain as a decentralized security framework. IEEE Consum. Electron. Mag. 7(2), 18–21 (2018). https://doi.org/10.1109/MCE.2017.2776459
9. Harish, N., Solvej, K., Helen, G.: Distributed ledger technology (DLT) and blockchain. FinTech Note (1) (2017). http://hdl.handle.net/10986/29053
10. Shrimali, B., Patel, H.B.: Blockchain state-of-the-art: architecture, use cases, consensus, challenges and opportunities. J. King Saud Univ. Comput. Inf. Sci. (2021). https://doi.org/10.1016/j.jksuci.2021.08.005
11. Umesh, B., et al.: Blockchain for industry 4.0: a comprehensive review. IEEE Access 8 (2020). https://doi.org/10.1109/ACCESS.2020.2988579
12. Díaz, R.M., Figueroa, L.V., Pérez, G.: Blockchain implementation opportunities and challenges in the Latin American and Caribbean logistics sector. FAL Bull. (3) (2021). https://hdl.handle.net/11362/47165

13. Konstantinos, C., Michael, D.: Blockchains and smart contracts for the internet of things. IEEE Access **4**, 2292–2303 (2016). https://doi.org/10.1109/ACCESS.2016.2566339

14. Asma, L.: Distributed management framework based on the blockchain technology for industry 4.0 environments – Ph.D. thesis, Telecom SudParis, École doctorale n◦ ED 626 Ecole doctorale IP Paris. Institut Polytechnique de Paris, Paris (2020)

15. Astarita, V., Giofrè, V.P., Mirabelli, G., Solina, V.: A review of blockchain- based systems in transportation. Information **11**(1)(21) (2020). https://doi.org/10.3390/info11010021

16. Patwardhan, A., Thaduri, A., Karim, R.: Distributed ledger for cybersecurity: issues and challenges in railways. Sustainability **13**(18) (2021). https://doi.org/10.3390/su131810176

17. Naser, F.: Review: the potential use of blockchain technology in railway applications: an introduction of a mobility and speech recognition prototype. In: 2018 IEEE International Conference on Big Data (Big Data), pp. 4516–4524. IEEE, Seattle (2018). https://doi.org/10.1109/BigData.2018.8622234

18. Vasilenko, M., et al.: Digital technologies in quality and efficiency management of transport service. In: E3S Web Conference, vol. 244 (2021). https://doi.org/10.1051/e3sconf/202124411046

19. CER The Voice of European Railways: A Roadmap for Digital Railways (2016). https://www.cer.be/sites/default/files/publication/A%20Roadmap%20for%20Digital%20Railways.pdf

20. Bobrova, V.V., Berezhnaya, L.Y.: Digitization of the transport industry in Russia: advances in economics. Bus. Manag. Res. (2019). https://doi.org/10.2991/mtde-19.2019.33

21. Kuperberg, M., Kindler, D., Jeschke, S.: Are smart contracts and blockchains suitable for decentralized railway control? Ledger 5 (2020). https://doi.org/10.5195/ledger.2020.158

22. Koh, L., Dolgui, A., Sarkis, J.: Blockchain in transport and logistics – paradigms and transitions. Int. J. Prod. Res. **58**(7), 2054–2062 (n.d.). https://doi.org/10.1080/00207543.2020.1736428

23. Aritua, B., Wagener, C., Wagener, N., Adamczak, M.: Blockchain solutions for international logistics networks along the new silk road between Europe and Asia. Logistics **5**(3) (2021). https://doi.org/10.3390/logistics5030055

Object Detection and Reinforcement Learning Approach for Intelligent Control of UAV

Zoran Miljković(✉) and Đorđe Jevtić(✉)

Faculty of Mechanical Engineering, Department of Production Engineering,
University of Belgrade, Kraljice Marije 16, 11120, Belgrade 35, Serbia
{zmiljkovic,drjevtic}@mas.bg.ac.rs

Abstract. In recent years, the development of deep learning models that can generate more accurate predictions and operate in real-time has brought both opportunities and challenges across the various domains of robotic vision. This breakthrough enables researchers to design and deploy more challenging tasks on intelligent mobile robots, which require emphasized abilities of learning and reasoning. In this paper, a new method for intelligent robot control, based on deep learning and reinforcement learning is proposed. The fundamental idea of this work is how the UAV equipped with a monocular camera can learn significant information about the object of interest in the context of its localization and navigation. For such purpose, the object detection system based on Tiny YOLOv2 architecture is employed. Furthermore, bounding box data generated by a convolution neural network is utilized for depth estimation and determining object boundaries. This information has shown how the state-space dimensions can be significantly reduced, which was essential for further implementation of the Q-learning algorithm. In order to test the proposed framework, a model is developed in MATLAB Simulink. The simulation, which covered different scenarios, was carried out on the UAV within the 3D scene rendered by Unreal Engine. The obtained results have demonstrated the applicability of the proposed methodology for depth estimation, gathering information about the object, object-driven navigation, and autonomous localization and navigation.

Keywords: Unmanned aerial vehicles · Autonomous localization and navigation · Q-learning · Convolution neural networks · Deep learning · Intelligent control

1 Introduction

Affordability and availability of small UAVs and sensory equipment, as well as rapid development in deep learning-based approaches, make suitable robotic platforms for performing a variety of complex tasks. To develop intelligent control for such a UAV, it is necessary to consider several different research fields: robotics, aerospace, computer vision, deep learning, reinforcement learning, electronics, etc.

The overview of the potentials, challenges, and limits of deep learning in robotics are comprehensively presented in the scientific paper [1]. Some of the significant research

© The Author(s), under exclusive license to Springer Nature Switzerland AG 2022
I. Karabegović et al. (Eds.): NT 2022, LNNS 472, pp. 659–669, 2022.
https://doi.org/10.1007/978-3-031-05230-9_79

directions pointed out within this journal paper, are prior knowledge about the environment available to the vision system and employing active vision should lead to better scene understanding and improvement of the perception system's performance. The Convolutional Neural Network (CNN) implementation for detecting and tracking the object by UAV is presented in the paper [2]. The authors utilized UAV's dynamics and model based on SSD architecture to develop a tracking algorithm. The authors of [3] integrated CNN and a line-of-sight guidance algorithm to fly a racing drone through the gates as quickly as possible. Furthermore, they developed a deep-learning model called ADRNet, which improves inference speed and enables precise flight control.

The contribution of this paper is given through the implementation of the original methodology for depth estimation and the novel reinforcement learning (RL) algorithm based on the Q-learning method, which should enable performing intelligent tasks such as localization and navigation. Furthermore, it is shown that the state space dimensions can be significantly reduced by utilizing CNN, so that the Q-learning algorithm can converge to the optimal solution successfully. To test the proposed framework, the 3D scene rendered by Unreal Engine was utilized within a model developed in MATLAB Simulink. Finally, the simulation was conducted using the UAV as a robotic system.

This paper has the following structure. Section 2 describes the object detection system, i.e. the chosen CNN model and the training process. The methodology for depth estimation based on object detection is presented in Sect. 3. In Sect. 4, the novel reinforcement learning algorithm based on Q-learning is given. In Sect. 5, the architecture of the intelligent control is explained. Section 6 provides the simulation settings and obtained results, while the discussion about the proposed framework takes place in Sect. 7. Finally, Sect. 8 gives the concluding remarks.

2   Object Detection

The first process in every object detection system acquires an image by the external sensor. Then, the obtained image represents the input to the CNN model. If the object is detected, the system returns the following information: label, bounding box (BB), and scores for every class. BB is formulated as a vector row consisting of four elements (x_1, y_1, x_2, y_2), where each dimension represents the boundary location in such a way that (x_1, y_1) defines the upper-left corner while the (x_2, y_2) defines the lower-right corner. Based on this information, a system for estimating the depth and object dimensions and robot motion control can be developed.

Our own custom-made dataset is obtained within the US City Block Scene (3D environment), which is rendered using the Unreal Engine® from Epic Games®. The dataset contains 679 images which are resized to 416×416 pixels and then divided into training data (581 images), validation data (49 images), and test data (49 images). Three different classes were considered, and Building_2 is the one in central focus within this paper. Since the last YOLO [4] architecture implemented in MATLAB R2021a is YOLOv2, the version of this CNN model called Tiny YOLOv2 is chosen for the training process. Because of its size, this model is suitable for real-time object detection tasks, but it also has some drawbacks, such as multiple detections of the same object within the image. Therefore, non-max suppression is utilized to keep the detections with the highest score. Finally, the output from the object detection block is the image with one BB for every detected class.

The initial learning rate, the learn rate drop period and learn rate drop factor were set to 0.001, 10, and 0.3, respectively. For the training process, the Adam optimizer is used, and the model is trained for 80 epochs with a mini-batch size set to 16. The training process is performed by using MATLAB software package running on the computer with the following specifications: AMD Ryzen 7 4800H processor, Nvidia GeForce RTX 3050 Laptop GPU, and 16 GB of RAM.

3   Depth Estimation Using Object Detection

Within this section, a method for depth estimation will be presented by using the object detection system. The given task is the determination of the depth considering the left and right boundaries of the cuboid-shaped building (Fig. 1).

As it shown in Fig. 1, the first step that needs to be done is finding the angles $\alpha_{L,1}$ and $\alpha_{R,1}$. Angle $\alpha_{L,1}$ is determined by rotating the flying mobile robot around its local z-axis to the right in increments equal to 0.01 rad until the right boundary of the building is on the left side of the image. As shown in Fig. 1, this rotation is carried out in the pose in which the left boundary of the building is obtained (pose A). Angle $\alpha_{R,1}$ is determined in the same way, i.e. the mobile robot rotates to the left until the left boundary of the building is on the right side of the image. Similarly, as for the left side, this rotation is carried out in the pose in which the right boundary of the building is obtained (pose B). By determining w, $\alpha_{L,1}$ and $\alpha_{R,1}$ it is possible to calculate $b_{L,1}$, $b_{R,1}$, as well as approaching distance h per the following Eq. (1) (the sine rule is used):

$$b_{L,1} = w \cdot \frac{\sin(90° - \alpha_{R,1})}{\sin(\alpha_{L,1} + \alpha_{R,1})}; \quad b_{R,1} = w \cdot \frac{\sin(90° - \alpha_{L,1})}{\sin(\alpha_{L,1} + \alpha_{R,1})};$$
$$h = b_{L,1} \cdot \sin(90° - \alpha_{L,1}) \tag{1}$$

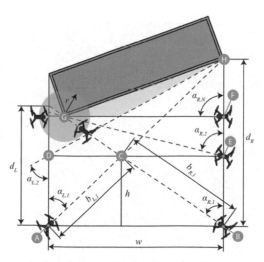

Fig. 1. Method for depth estimation

If $b_{R,1} > b_{L,1}$, this means that the right boundary of the building is at distance d_R. The next poses pursued by the robot are represented by points D and E. Furthermore, angles $\alpha_{L,2}$ and $\alpha_{R,2}$ are determined in the same way as $\alpha_{L,1}$ and $\alpha_{R,1}$. This procedure is repeated until the angles $\alpha_{L,N}$ and $\alpha_{R,N}$ are equal to or greater than 90°. However, as Fig. 1 shows, the yellow triangles represent the space where the edges of the building may be located. Therefore, in the last step, the flying mobile robot is orbiting along the circular path with radius r so that the front side of the robot is always oriented towards the center (point G in Fig. 1). The initial pose in the final step is determined by the intersection of the circle and dashed line containing points C and G, while the end pose is determined by the intersection of the circle and line containing points F and G.

4   Q-Learning Algorithm

Q-learning is a model-free, off-policy temporal difference control algorithm introduced by Watkins in 1989. This RL algorithm is a tabular method that iteratively updates every visited state-action pair into the Q-table. For more information about this method, the authors refer to [5].

The flowchart of the developed RL algorithm is shown in Fig. 2. Firstly, the start and target pose, set of all possible actions, and the number of learning episodes should be set. Then, the Q-table is initialized as a zero matrix with the following dimensions: number of states × number of actions. Every state has its unique ID in Q-table, represented by Eq. (2):

$$State\_ID = a \cdot LENGTH \cdot HEIGHT + b \cdot HEIGHT + 494 + c \tag{2}$$

where a, b, and c are the components of the vector which represents the pose of the UAV's center of gravity at time step t, while the $LENGTH$ and $HEIGHT$ represent state space

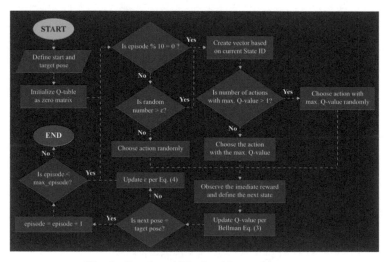

Fig. 2. Developed RL algorithm architecture

dimensions. If the random number is greater than ε or the current learning episode is divisible by 10, the agent chooses the action with the maximum Q-value from the vector constructed for the current *State_ID*. The algorithm checks if the vector contains more than one maximum Q-value. If this is the case, the agent randomly chooses one of the actions with a maximum Q-value. Otherwise, it chooses the action with the maximum Q-value. On the other side, if the current learning episode is not divisible by ten and if the random number is less than ε, the agent chooses a random action from the finite set of actions (up, down, forward, backward, left, right). In the following step, the reward and the next state are calculated based on the current state information and selected action.Then, the Q-value is updated by using the Bellman Eq. (3):

$$Q_{t+1}(s_t, a_t) = (1 - \alpha) \cdot Q_t(s_t, a_t) + \alpha \left[r_{t+1} + \gamma \max_a Q_t(s_{t+1}, a_t) \right] \qquad (3)$$

where: $Q_{t+1}(s_t, a_t)$ denotes state-value function at time step $t + 1$; r_{t+1} represents the reward; a_t is the action taken at the state s_t; γ and α are the discount factor, and learning rate respectively. Furthermore, it is noticeable that this algorithm contains two loops. The first one is completed after the maximum number of learning episodes is reached, while the inner loop breaks once the agent reaches the target pose. Before the new episode starts, the ε parameter is updated by the Eq. (4):

$$\varepsilon_{t+1} = 1 - e^{0.005 \cdot current\_episode - 12} \qquad (4)$$

5   Intelligent Control System

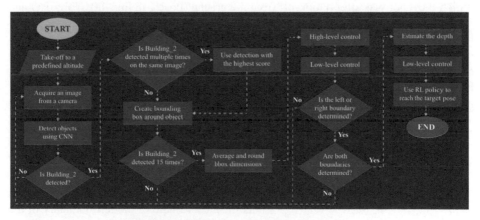

Fig. 3. Flowchart of the intelligent controller

In this section, the intelligent controller presented in Fig. 3 will be explained. As we can see, it consists of the object detection system, depth estimation block, and the RL policy. At the beginning, the flying mobile robot is placed in an arbitrary position with orientation so that the image plane is parallel or almost parallel to the front surface of the building that needs to be cleaned. This paper assumes that the building exterior is cuboid-shaped, where the robot is able to localize itself within the environment, and that the system is fully observable. The overall aim of the intelligent agent is to reach the goal pose, which is defined at the upper right corner on the front side of the building. In the beginning, the robot is set to take-off to a predefined altitude. After the altitude is reached, the object detection system is activated. As mentioned before, this system outputs images with BBs for every detected class. It is noted that the object detection system is activated only once the robot starts to hover in the new state. This is considered reasonable since the UAV is the underactuated system. In other words, in order to move forward, backward, left, or right, it has to rotate towards the direction of movement in the horizontal plane [6]. Therefore, the roll and pitch angles are minimal during the hovering regime.

In order to achieve the system reliability, BBs on 15 consecutive detections of the object of interest (Building_2) are summed, averaged, and rounded. The obtained values represent the input information of the high-level control system which determines the next pose. Furthermore, three different cases are considered in this paper (Eq. (5)):

$$1. \ x_1 > m_L \ \&\& \ x_2 > (416 - m_R); \quad 2. \ x_1 < m_L \ \&\& \ x_2 < (416 - m_R);$$
$$3. \ x_1 < m_L \ \&\& \ x_2 > (416 - m_R); \tag{5}$$

where m_L and m_R denote the left and right margin width, respectively. If the first or the third case is obtained, the flying mobile robot moves 1 m to the left. This process repeats until $m_L < (208 - m_C/2)$ where m_C represents the center margin width (pose 1 in Fig. 4a). After the previous step is terminated, the flying mobile robot starts to rotate until the $\alpha_{L,1}$ is determined as described in the previous section. Furthermore, since the robot has to move towards the opposite side, the next pose it should visit is the pose in which the robot was after it took-off from the ground (pose 1 in Fig. 4a). Now, if x_1 and x_2 satisfy the cases one or three given by the Eq. (5), the mobile robot moves 1 m to the right. This is considered reasonable since the left boundary of the building has been obtained already. Then, the robot continues to move to the right until the right boundary of the building is obtained (pose 4 in Fig. 4a). After that, it starts to rotate once again, but this time, on the opposite side. If case two takes place, the opposite of the previous mentioned situations occurs (Fig. 4b). When $\alpha_{R,1}$ is determined (pose 5 in Fig. 4a), the estimated depth can be calculated per the following Eq. (6):

$$d = round\left(\frac{(d_L + d_R)}{2}\right) = round\left(\frac{w}{2} \cdot \left(\frac{\sin(90° - \alpha_{L,1})}{\sin\alpha_{L,1}} + \frac{\sin(90° - \alpha_{R,1})}{\sin\alpha_{R,1}}\right)\right) \tag{6}$$

The obtained value is decreased by 4, so the next pose (pose 6 in Fig. 4a and b) is defined as $s_t = [x_L, d - 4, 0]$, where x_L represents the x coordinate of the left boundary of the building, while the last element represents the altitude. From this pose, the intelligent agent navigates to the pose depicted with 7 (Fig. 4) by using the learned RL policy.

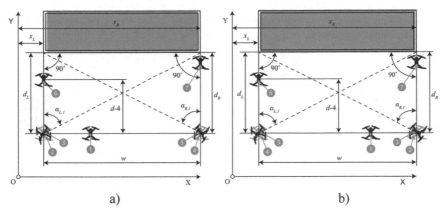

Fig. 4. Characteristic poses during the localization and navigation tasks

6 Simulation Results

Within this section, the simulation setup is described, and the obtained results are shown. In order to test the proposed approach, the model is created in MATLAB Simulink. It is noted that low-level control, the sensor block, and the 3D scene (US City Block) are implemented from Simulink Library, while the object detection system and high-level control are originally developed by the authors.

Inputs of simulation model are the initial pose and the take-off altitude, while the following parameters were tracked during every algorithm run: poses in which the building's boundaries are obtained, estimated depth, as well as the overall functionality of the algorithm, such as the activation of the object detection system and actions chosen when the robot was in hovering regime.

Before the evaluation process started, the left and right margins widths were set to 30 pixels, while the center margin width was set to 10 pixels. During the evaluation process, the algorithm is run for 15 different scenarios.

One of the experimental runs is depicted in Fig. 5. The poses in which the left and right boundaries of the building are obtained are shown in Fig. 5a and Fig. 5c, respectively. Also, one can notice that the object detection system is activated, which means that the robot is in a hover flight regime. Figure 5b shows the situation in which α_L is determined, followed by returning the robot to the pose 4 depicted in Fig. 4a. Finally, the depth is estimated, and the robot starts the approaching maneuver (Fig. 5d).

Building_2 boundaries are accurately found in all scenarios. The depth is estimated based on d_L and d_R values which are shown in Table 1. Based on these results and true depth information d_T extracted from the Simulink, the relative error δd is calculated per the following Eq. (7):

$$\delta d = \frac{d_T - \left(\frac{(d_L + d_R)}{2}\right)}{d_T} \cdot 100\% \tag{7}$$

Table 1 displays that even though d_L and d_R were not equal in every experimental run, the relative error was less than 5% in all of the cases, which can be considered a good estimation.

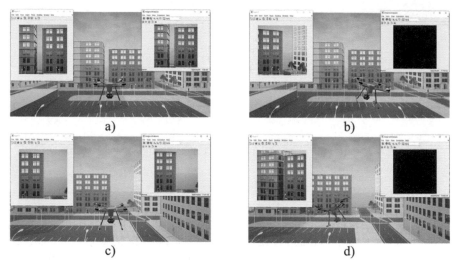

Fig. 5. Object (Building_2) detected by CNN running in the MATLAB Simulink. Snapshots represent the UAV during its localization and navigation tasks.

Table 1. Simulation setup and obtained results presented for eight different scenarios

Initial pose [m]	Take-off altitude [m]	x_L	x_R	d_L	d_R	d_T	δd	d
(36, 48, 0)	14	29	62	68.32	70.1	70.5	1.83	113
(38, 54, 0)	16	29	62	63.39	61.87	64.5	2.90	113
(39, 52, 0)	14	29	62	63.39	66.61	66.5	2.26	113
(33, 60, 0)	17	29	62	56.32	56.32	58.5	3.73	112
(35, 64, 0)	12	29	62	52.64	52.64	54.5	3.41	113
(54, 48, 0)	16	29	62	70.1	68.32	70.5	1.83	113
(57, 54, 0)	17	29	62	61.87	61.87	64.5	4.08	112
(59, 58, 0)	9	29	62	59.0	59.0	60.5	2.48	113

By determining x_L and x_R, the simulation environment in which the intelligent agent operates has the following dimensions: $2 \times 33 \times 28$ m. Since the number of possible states (2958) is greater than the number of states presented in [7], in this paper, the maximal number of training episodes is set to 2400. The actions that an agent can take at every time step, the reward function, the learning rate, and the discount factor are defined as in the paper [7]. Within this RL problem, the agent has to navigate from the start pose at (0, 17, 1) to the target pose at (2, −16, 28) by using the minimal number of learning steps. Obtained results are presented in Fig. 6. It is noticeable that the algorithm converged to the optimal solution successfully, i.e. that the agent learned the optimal number of steps after reaching 460th episode (Fig. 6 down).

Some of the generated paths are presented in Fig. 7. The initial poses are represented with green markers, while the target pose is depicted in red.

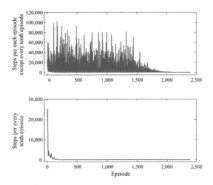 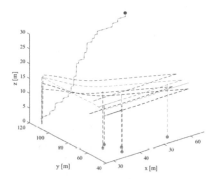

Fig. 6. Number of steps taken by the agent during the learning process

Fig. 7. Paths generated by the intelligent agent for five different scenarios

7 Discussion

Even though the proposed algorithm has demonstrated that it can potentially be utilized as a part of the commercial solution regarding the addressed problem, it makes sense to use it only as a turn-key solution, i.e. if the cleaning system is custom-made for a specific building. Otherwise, the authors' opinion is that the proposed approach is not a profitable solution. For example, if the cleaning company owns such a system, a new dataset must be created for every building, which is more expensive than hiring a robot operator. In such cases, it would make sense to develop an object detection system based on specific characteristics of the building, such as texture or color. Based on this fact, the authors tried to develop the object detection algorithm by using the most apparent feature of Building_2 - the red color. For such a purpose, MATLAB Image Processing Toolbox is utilized. The obtained results have shown that the algorithm could not distinguish something evident to our eyes. The answer to this is related to the color context, i.e. human eye seeing that the Building_2 is distinctly red due to the environment around the red pixels.

Another important fact should be pointed out. As explained in Sect. 3, the angles α_L and α_R are determined by the robot's rotation in increments equal to 0.01 rad until the left or right boundary of the building is located on the right or left side of the image. For example, in the case where the angle is determined one increment after the increment in which the boundary of the building should be obtained, the error in estimating the depth would be minor if the robot is closer to the building than if it is further away from it, which is undoubtedly favorable from the aspect of its navigation.

In addition, the challenge of developing an object detection system arises if the object should be recognized within a scene crowded with objects with the same or similar characteristics. The possible solution to this problem is introducing relations between objects which is recently done by MIT researchers [8]. If object relations present in the scene can be provided to the vision system as prior knowledge, more promising object detection performance can be expected. In other words, context can help in understanding the scene better, in improving predictions and refining detections [1]. Furthermore, one significant advantage of intelligent flying mobile robots compared to ground mobile robots is the possibility of operating in a higher altitude range, which with the methodology presented in [9], can undoubtedly be utilized to improve the robot's perception system.

8   Conclusion

In this paper, the novel intelligent control architecture consisting of the system for object detection, block for depth estimation, and the Q-learning policy is presented. Furthermore, all proposed methodologies are explained in detail, and the following is concluded: (i) the object detection system can be used for estimating the object dimensions and depth, (ii) forcing the agent to exploit the obtained experience every ten episodes has proved to be reasonable, (iii) proposed approach has shown how state space dimensions can be firstly reduced, so that the methods such as Q-learning can be successfully applied for localization and navigation tasks. The performance of the developed framework is tested for fifteen different scenarios, and results have shown that the intelligent flying mobile robot successfully found the building boundaries, and that the depth is estimated with the relative error of less than 5% in all cases. Also, this intelligent agent learned the optimal path to the goal pose. Therefore, the obtained simulation results have demonstrated that the proposed approach can be utilized for applications that require the intelligent behavior of the flying mobile robot. Future research in this field should include implementation of the presented methodology for various relative poses of the building with respect to an intelligent flying mobile robot. Moreover, the implementation of the system for obstacle avoidance based on the state-of-the-art reinforcement learning algorithms, as well as the determination of the remaining boundaries of the building with the ultimate aim of creating the map of glass surfaces should be considered. Besides mentioned, changes in lighting conditions, using the building with complex geometry, differences in the presence of non-building objects (cars, trees, advertising, etc.) around the building, fusing the network's predictions with information about realized robot movement, and accumulated data of previous robot's behavior, need to be investigated in detail in order to improve the reasoning ability of the intelligent mobile robot.

Acknowledgment. This work has been financially supported by the Ministry of Education, Science and Technological Development of the Serbian Government, through the project "Integrated research in macro, micro, and nano mechanical engineering – Deep learning of intelligent manufacturing systems in production engineering", under the contract number 451-03-9/2021-14/200105, and by the Science Fund of the Republic of Serbia, grant No. 6523109, AI - MISSION4.0, 2020–2022.

References

1. Sünderhauf, N., et al.: The limits and potentials of deep learning for robotics. Int. J. Robot. Res. **37**(4–5), 405–420 (2018)
2. Rohan, A., Rabah, M., Kim, S.H.: Convolutional neural network-based real-time object detection and tracking for parrot AR drone 2. IEEE Access **7**, 69575–69584 (2019)
3. Jung, S., Hwang, S., Shin, H., Shim, D.H.: Perception, guidance, and navigation for indoor autonomous drone racing using deep learning. IEEE Robot. Autom. Lett. **3**(3), 2539–2544 (2018)
4. Redmon, J., Farhadi, A.: YOLO9000: better, faster, stronger. In: Proceedings of the IEEE Conference on Computer Vision and Pattern Recognition, USA, HI, Honolulu, pp. 6517–6525 (2017)
5. Sutton, R.S., Barto, A.G.: Introduction to Reinforcement Learning. MIT Press, Cambridge (1998)
6. Jevtić, Đ., Svorcan, J., Radulović, R.: Flight mechanics, aerodynamics and modelling of quadrotor. In: Karabegović, I. (ed.) NT 2021. LNNS, vol. 233, pp. 681–689. Springer, Cham (2021). https://doi.org/10.1007/978-3-030-75275-0_75
7. Miljković, Z., Jevtić, Đ., Svorcan, J.: Reinforcement learning approach for autonomous UAV navigation in 3D space. In: 14th International Scientific Conference MMA 2021 – Flexible Technologies, Serbia, Novi Sad, pp. 189–192 (2021)
8. Liu, N., Li, S., Du, Y., Tenenbaum, J., Torralba, A.: Learning to compose visual relations. In: 35th Conference on Neural Information Processing Systems (2021)
9. Choi, S., Kim, J.T., Choo, J.: Cars can't fly up in the sky: improving urban-scene segmentation via height-driven attention networks. In: IEEE/CVF, pp. 9373–9383 (2020)

Real-Time Mobile Robot Perception Based on Deep Learning Detection Model

Aleksandar Jokić[(✉)], Milica Petrović, and Zoran Miljković

Faculty of Mechanical Engineering, Department of Production Engineering, University of Belgrade, Kraljice Marije 16, 11120 Belgrade, Serbia
{ajokic,mmpetrovic,zmiljkovic}@mas.bg.ac.rs

Abstract. The recent advances in deep learning models have enabled the robotics community to utilize their potential. The mobile robot domain on which deep learning has the most influence is scene understanding. Scene understanding enables mobile robots to exist and execute their tasks through processes such as object detection, semantic segmentation, or instance segmentation. A perception system that can recognize and locate objects in the scene is of the highest importance for achieving autonomous behavior of robotic systems. Having that in mind, we develop the mobile robot perception system based on deep learning. More precisely, we utilize an accurate and fast Convolution Neural Network (CNN) model to enable a mobile robot to detect objects in its scene in a real-time manner. The integration of two CNN models (SSD and MobileNet) is performed and implemented on mobile robot RAICO (Robot with Artificial Intelligence based COgnition). The experimental results show that the proposed perception system enables a high degree of object recognition with satisfying inference speed, even with limited processing power provided by Nvidia Jetson Nano integrated within RACIO.

Keywords: Perception system · Mobile robots · Convolutional neural networks · Object detection

1 Introduction

Research conducted in the field of robotics in the second decade of the XXI century is increasingly focused on achieving autonomous behavior of robotic systems. One of the basic requirements needed to achieve autonomous behavior is the ability of a robotic system to understand the environment in which it operates. A deep learning-based perception system, which would enable mobile robotic systems to achieve this cognitive function, is developed within the proposed research paper. This problem implies the development of a perception system that can recognize objects in its environment in real-time and pass this useful information to the decision-making system to adequately reason about future actions that need to be taken to achieve the proposed task. The advantage of Deep Learning (DL) algorithms is shown in the ability to learn various objects with high recognition accuracy and the possibility of real-time implementation. In this paper, a mobile robot perception system based on DL detection model is developed and implemented on the intelligent mobile robot RAICO.

I. Karabegović et al. (Eds.): NT 2022, LNNS 472, pp. 670–677, 2022.
https://doi.org/10.1007/978-3-031-05230-9_80

The potential application of deep machine learning in the field of robotics, with particular reference to the application of these algorithms within the perception system, was analyzed in [1]. The authors emphasize the possibility of applying DL for the tasks of active observation and semantic recognition of objects of interest, which are a direct result of the work of the perception system. The authors of [2] propose a perception system based on ROS (Robot Operating System). Numerous CNN models have been developed to detect people and/or objects, estimate the position of people, localize and build an environment map, and generate a semantic description of a scene. The main disadvantage of the proposed concept is that the application of the perception system cannot be realized in real-time, even though the processing has been relocated to a centralized computer platform with high processing performance. The perception system of a mobile robot that continuously needs to learn new detected classes of objects was analyzed in [3]. Based on the state of objects in the current scene, their geometric, and semantic dependencies, the system can generate a list of potential classes to which the detected object may belong. By applying this methodology, with minimal interaction with operators, the continuous operation of mobile robots can be ensured, even in the scenes and environments where they have not been trained. Various CNN models have shown high accuracy of object detection, with satisfactory processing speed, of which the most efficient and popular are Yolo [4], Faster-RCNN [5], and SSD [6]. Within this work, the proposed perception system is based on CNN architecture that integrates MobileNet architecture [7] and SSD models [6].

2   Object Detection Based on Deep Machine Learning

The perception system based on deep machine learning provides the high-precision detection of objects, and with the application of efficient CNN models, the detection can be performed in real-time. One of the most efficient CNN models used for mobile and embedded vision systems is MobileNet [7]. The characteristic of the MobileNet model is reflected in the decoupling of the traditional convolutional layer into two separate operations. The first operation is a convolution performed separately for each feature map (Depthwise convolution). In this way, a new multidimensional structure is formed with appropriate dimensions in width and height. However, it is necessary to perform another convolution process (Pointwise convolution) so that the output structure has dimensions according to the required number of used filters. Also, within the MobileNet model, normalization of current inputs (Batch normalization) and/or positive linear activation function (Rectified Linear Unit - ReLU) is applied after each convolution layer. By comparing the number of parameters and the algorithmic complexity of the standard and decoupled convolution, it is concluded that decoupled convolution is 8 to 9 times more efficient [7], without significant loss in the model's accuracy.

The proposed CNN architecture consists of a backbone network (MobileNet) and a detection network (SSD). By integrating these two CNN models, excellent results have already been achieved in identifying objects of interest [6] and the detecting construction machinery [8]. Applying the SSD model makes it possible to detect several objects in the image simultaneously. In order to perform the required detection, the network's outputs are the coordinates of the center, the width and height of the bounding box, and the network's confidence in the classification of a particular object. Having this in mind, the SSD model performs a functional approximation of the position of the bounding box in the image and the classification of the object located in it. Before starting the training of SSD models, it is necessary to define the initial set of bounding boxes of different dimensions and aspect ratios used for localization of objects. In Fig. 1 it can be seen that within the gird box with indexes (2, 2), there are four assigned bounding boxes (shown by dashed red lines) of different dimensions.

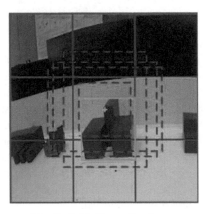

Fig. 1. Initial bounding boxes

For each layer whose feature maps are used for object detection, r results are generated - four for the functional approximation of the bounding boxes, and c for each class (Eq. 1):

$$r = kmn \cdot (c + 4), \tag{1}$$

where c is the number of different classes, k is the number of initial bounding boxes, and m and n are the dimensions of the feature map. Objects of different dimensions can be detected within different layers of the SSD model, where smaller objects are detected in the initial layers (because feature maps are larger), and larger ones in the final layers. As there are a large number of bounding boxes created based on Eq. (1), not all of them are used in each iteration of SSD model training, but only those bounding boxes that have an Intersection Over Union (IOU) value greater than 0.5. IOU is calculated according to initial bounding boxes (Fig. 1 - red rectangles) and the entire region of the object (Fig. 1 - green rectangle). Several bounding boxes can adequately include one object; therefore, it is necessary to define a variable $x_{ij}^p = \{1, 0\}$ with a value of 1 if the i-th initial bounding box has IOU > 0.5 for the j-th actual bounding box and the p-th class, otherwise has a value of 0. The hybrid cost function used to train SSD models is shown by Eq. (2) [6]:

$$L(x, c, l, g) = \frac{1}{N} \left(L_{conf}(x, c) + \alpha L_{loc}(x, l, g) \right) \tag{2}$$

where N represents the number of adequately paired bounding boxes in the image, the classification cost function is denoted $L_{conf}(x, c)$, and the localization cost function is

$L_{loc}(x, l, g)$. The classification cost function (Eq. 3) is a standard Softmax activation function for multi-class problems, which also takes into account negative examples:

$$L_{conf}(x, c) = - \sum_{i \in Pos}^{N} x_{ij}^{p} \log(\hat{c}_i^{p}) - \sum_{i \in Neg} \log(\hat{c}_i^{0}), \tag{3}$$

$$\hat{c}_i^{p} = \frac{\exp(c_i^{p})}{\sum_{p} \exp(c_i^{p})}, \tag{4}$$

where c_i^{p} (Eq. 4) represents the network's output related to the confidence of the classification of the i-th initial bounding box for the p-th class. The cost function used to learn the location of the bounding boxes is represented by Eqs. (5), (6), and (7):

$$L_{loc} = \sum_{i \in Pos}^{N} \sum_{m \in \{cx, cy, w, h\}} x_{ij}^{k} smooth_{L1}\left(l_i^{m} - \hat{g}_j^{m}\right), \tag{5}$$

$$smooth_{L1}(x) = \begin{cases} 0.5x^2 & \text{if } |x| < 1 \\ |x| - 0.5 & \text{otherwise} \end{cases} \tag{6}$$

$$\hat{g}_j^{cx} = \frac{g_j^{cx} - d_i^{cx}}{d_i^{w}}, \hat{g}_j^{cy} = \frac{g_j^{cy} - d_i^{cy}}{d_i^{h}}, \hat{g}_j^{w} = \log\left(\frac{g_j^{w}}{d_i^{w}}\right), \hat{g}_j^{h} = \log\left(\frac{g_j^{h}}{d_i^{h}}\right). \tag{7}$$

where the network outputs are denoted by l, the desired output by g, and the initial bounding box by d, the bounding box center is defined by the coordinates cx and cy, while the width and height of the region are represented by w and h.

3   Training of Convolutional Neural Network

The MobileNet-SSD artificial neural network training is performed on our data set gathered in the laboratory model of the manufacturing environment, using the intelligent mobile robot RAICO with the stereo vision subsystem. RAICO is placed in different positions within the manufacturing environment, and both cameras generate the images. A total of 258 images are collected. The collected images are then manually labeled by defining the classes and locations of the objects.

Information regarding bounding boxes and classes of objects is saved in .xml file for each image, defined in VOC format [9]. Out of 258 images (sample of the dataset is given in Fig. 2), 80% are used for training and 20% for validation. The mobile robotic system is trained to recognize five classes of different machine tools (M#1–M#5) in the laboratory model of the manufacturing environment. For the mobile robot to learn to recognize machines that are not fully visible, images in which additional elements are set to obscure the machines are generated. The initial parameters used for convolutional neural network training are defined as follows. The utilized training algorithm is a stochastic gradient with momentum, the minibatch size is set to 2, the rest of the parameters are varied, and their value will be defined in Sect. 4.

Fig. 2. A sample of dataset. Images generated by both camers at the same pose.

4 Experimental Evaluation

In order to determine the optimal parameters for CNN training, eight experiments were performed. The experiment plan is shown in Table 1. The training was conducted for 50 epochs, with image resolution of 800 × 600 × 3 pixels. Also, the techniques for data augmentation are used, and the weights for the considered model are initialized based on the CNN model trained on a COCO data set [9].

Table 1. Experimental plan

Exp. no.	Learning rate	Momentum	Schedule for LR
1.	0.01	0.9	Cos
2.	0.001	0.9	Cos
3.	0.01	0.8	Cos
4.	0.001	0.8	Cos
5.	0.01	0.9	MultiStep
6.	0.001	0.9	MultiStep
7.	0.01	0.8	MultiStep
8.	0.001	0.8	MultiStep

Two scheduling methods for Learning Rate (LR) are utilized (i) Cos – Cosine annealing schedule and (ii) MultiStep schedule. Network testing is performed based on mean Average Precision (mAP) metric. The results for all experiments are shown in Table 2. The convolutional network selected for implementation on the mobile robot perception system is the network from Experiment no. 3 due to the highest level of mAP achieved on the validation data set.

Table 2. Average precision per class and mean Average Precision (mAP)

Exp. no.	M #1	M #2	M #3	M #4	M #5	mAP
1.	0.8811	0.9341	0.7487	0.9300	0.9948	0.8977
2.	0.8521	0.9641	0.7478	0.8578	0.9948	0.8833
3.	0.8868	0.9855	0.8128	0.9203	0.9924	**0.9195**
4.	0.8972	0.9659	0.7294	0.8998	0.9973	0.8979
5.	0.8751	0.8035	0.4886	0.7692	1	0.7873
6.	0.9218	0.9726	0.7247	0.8176	0.9948	0.8863
7.	0.8902	0.9628	0.6969	0.9051	0.9901	0.8890
8.	0.8700	0.9763	0.8212	0.8948	1	0.9125

Objects which are detected by using the intelligent mobile robotic system RAICO with integrated perception system after implementing the selected CNN model are shown in Fig. 3. During image generation, the possibility of applying a real-time perception system is tested. The entire perception system achieved the maximum processing speed of 46 FPS (Frame Per Second). For each generated image, the most CPU time is spent on evaluating the CNN model – an average of 25 ms.

Fig. 3. Object detected by mobile robot RAICO's perception system.

By analyzing Fig. 3, it can be concluded that not all machine tools are detected in all positions of the mobile robot. Detection of machine #3 is the biggest challenge for the perception system, which is the expected conclusion based on results from Table 2. The authors believe that it is more challenging to detect machine #3 due to its size difference compared to other machines. Moreover, the achieved accuracy seen in Fig. 3 is entirely consistent with the results shown in Table 2, based on which it can be concluded that the process of training and testing of CNN models is adequately performed.

5 Conclusion

In this work, we have shown the development of a novel mobile robot perception system. The deep learning approach for object detection is analyzed, and SSD and MobileNet convolutional neural network architectures are integrated. Afterward, the integrated CNN model is trained on the custom-developed dataset. The experiment that included the different parameter settings for training is executed, and the CNN with the best mean average precision is selected to be implemented on the mobile robot's perception system. Finally, the entire perception system is tested in a real-world setting on mobile robot RAICO within the experimental manufacturing environment. The real-world evaluation confirmed the high accuracy (over 90%) of the proposed model and real-time capabilities by reaching a maximum speed of around 46 FPS.

Acknowledgment. This research has been financially supported by the Ministry of Education, Science and Technological Development of the Serbian Government, through the project "Integrated research in macro, micro, and nano mechanical engineering – Deep learning of intelligent manufacturing systems in production engineering" (contract No. 451-03-9/2021-14/200105), and by the Science Fund of the Republic of Serbia, Grant No. 6523109, AI – MISSION 4.0, 2020–2022.

References

1. Sünderhauf, N., et al.: The limits and potentials of deep learning for robotics. Int. J. Robot. Res. **37**(4–5), 405–420 (2018)
2. Lee, C.Y., Lee, H., Hwang, I., Zhang, B.T.: Visual perception framework for an intelligent mobile robot. In: 17th International Conference on Ubiquitous Robots (UR), pp. 612–616 (2020)
3. Young, J., Basile, V., Kunze, L., Cabrio, E., Hawes, N.: Towards lifelong object learning by integrating situated robot perception and semantic web mining. In: 22nd European Conference on Artificial Intelligence (ECAI), vol. 285, pp. 1458–1466. (2016)
4. Redmon, J., Divvala, S., Girshick, R., Farhadi, A.: You only look once: unified, real-time object detection. In: IEEE Conference on Computer Vision and Pattern Recognition, pp. 779–788 (2016)
5. Ren, S., He, K., Girshick, R., Sun, J.: Faster R-CNN: towards real-time object detection with region proposal networks. IEEE Trans. Pattern Anal. Mach. Intell. **39**(6), 1137–1149 (2016)
6. Liu, W., et al.: SSD: single shot multibox detector. In: Leibe, B., Matas, J., Sebe, N., Welling, M. (eds.) ECCV 2016. LNCS, vol. 9905, pp. 21–37. Springer, Cham (2016). https://doi.org/10.1007/978-3-319-46448-0_2

7. Howard, A.G., et al.: MobileNets: efficient convolutional neural networks for mobile vision applications. ArXiv Preprint arXiv:1704.04861 (2017)
8. Arabi, S., Haghighat, A., Sharma, A: A deep learning based solution for construction equipment detection: from development to deployment. ArXiv Preprint arXiv:1904.09021 (2019)
9. Lin, T.-Y., et al.: Microsoft COCO: common objects in context. In: Fleet, D., Pajdla, T., Schiele, B., Tuytelaars, T. (eds.) ECCV 2014. LNCS, vol. 8693, pp. 740–755. Springer, Cham (2014). https://doi.org/10.1007/978-3-319-10602-1_48

Indicators that Model the Quality of Electric Vehicles and Services Provided

Aurel Mihail Titu[1](✉) and Gheorghe Neamțu[2]

[1] "Lucian Blaga" University of Sibiu, 10 Victoriei Street, Sibiu, Romania
mihail.titu@ulbsibiu.ro
[2] Faculty of Industrial Engineering and Robotics, University Politehnica of Bucharest, Splaiul Independenței nr. 313, 6th District, Bucharest, Romania

Abstract. The scientific paper shows our own research and an original point of view on how to accept the electric vehicle from a technical and economic point of view. It started from the idea of making a SWOT analysis regarding the electric vehicle and the classic one. The main qualitative and quantitative indicators that characterize the quality of electric vehicles have been defined here and analyzed, then the formulation of corrective measures is reached. Towards the end of the scientific paper, a statistic on the sale and registration of ecological vehicles at European Union level is proposed. Authors offer their own view on the role of ecological vehicles in the sustainable development of the road transport system.

Keywords: Sustainable development · Electric vehicle · Ecological road transport · Quality indicators · Non-conformities

1 Introduction

Sustainable development in transport derives from a comprehensive structure that aims to meet the travel needs of today's generation without compromising the possibility for future generations to meet their own needs [13]. A rechargeable (PEV) electric vehicle [1] is any vehicle that can be recharged from an external source of electrical energy, such as wall outlets, and the energy stored in rechargeable batteries acts or contributes to the driving of wheels [2]. As in the case of the renewable energy over the last decade, the automotive industry is currently investing heavily in the emergence of low- and zero-emission vehicle technologies such as electric vehicles [9]. In road transport, the quality of service is an indicator of the success of all operators [3]. It has direct implications for travellers, setting the level of satisfaction [3]. The differentiation of transport services from that of material goods is highlighted by certain characteristics that define quality [3]. The particular characteristics of a technical nature, specific to the electric means of transport, may lead to a decrease in the quality of road transport services. They create and cause discomfort, delays, failure to comply with transport plans and programs on time, create stress, uncertainty and distrust among users, but alsonon-conformities in ensuring quality services. Travelling with electric vehicles in Europe should be simple: recharging with electricity should be as easy as filling the tank [4].

© The Author(s), under exclusive license to Springer Nature Switzerland AG 2022
I. Karabegović et al. (Eds.): NT 2022, LNNS 472, pp. 678–686, 2022.
https://doi.org/10.1007/978-3-031-05230-9_81

2 SWOT Analysis of the Electric Vehicles. Proposals

SWOT analysis of electric vehicles is carried out with the aim of presenting the advantages and the disadvantages of the electric vehicle, in comparison with the vehicle with heat engine, its characteristics, stage of development, place and role in the sustainable development of road transport (Tables 1, 2, 3 and 4).

Table 1. SWOT analysis of electric vehicles – strengths (**S**)

Strengths	Proposals for operationalization
S_1. They are environmentally friendly	Better electric accumulators and their recharging only from renewable sources [8]; using modern, clean manufacturing technologies; using environmentally friendly materials and raw materials; full information/advertising about e-mobility [7]
S_2. Simple power supply	Power supply anywhere, at any charging station [7]
S_3. Low operating costs	Development of infrastructure/batteries and intelligent charging [7]
S_4. High reliability	Reduced fallrate; extension of the warranty period
S_5. Recover energy	Increasing the storage capacity of electric accumulators; more powerful and more efficient electric generators/engines
S_6. Attractive incentives when purchasing a new vehicle	Assignment of routes or specific areas and reserved parking spaces [8]; maintaining eco-bonuses and informing/educating the population; rental agreements for 48–60 months with 10,000 km/year included in electric vehicles or accumulators [7]
S_7. Power	Improving the technical performance of electric engines
S_8. Simplified maintenance procedures	Development of the network of service units; the training of specialists in the field of maintenance; the continuous instruction of the specialists in the field
S_9. They are extremely silent	Require mounting sound generators [8]; Training/educating the population
S_{10}. Fast amortization of the purchase	Informing, stimulating and educating the population; maintaining or lowering the price of electricity
S_{11}. Simplified transmission	Increasing the capacity, power and autonomy of the batteries; increasing the power and decreasing the dimensions of the electric engine

The SWOT analysis was developed on the basis of information from the specialized literature and the online environment.

Table 2. SWOT analysis of electric vehicles – weaknesses (**W**)

Weaknesses	Removal measures
W_1. Low autonomy	Design and development of cheaper electric batteries with a high capacity, with low weight and dimensions
W_2. Are dependent on power infrastructure	Allocation and development of electricity charging infrastructure in all parking lots, on all motorways and public roads [8]; development of innovative mobility programmes [7]
W_3. Slow charging	Endowment of all charging points with powerful filling stations (100 kw; 400–500 Vcc) [7]
W_4. High purchase price	Maintenance of incentives by the authorities [8]; lower prices for the electric vehicle/accumulators [8]
W_5. Infrastructure and poor service	Programs for the extension of renewable energies [8]; conversion of classical filling stations in electrical stations; developmentof dedicated service units
W_6. High-cost price of batteries	Decrease in the price per 1/kwh for electric accumulators; use of cheaper and high-quality materials in the manufacture of electric accumulators; development of new technologies [8]
W_7. Lowercapacityof electric batteries in the frost	Manufacture of heated accumulator batteries with improved energy storage capacity and charging [6]; avoidance of excessive vehicle load, strong accelerations and aggressive driving mode (sporty)

Table 3. SWOT analysis of electric vehicles – opportunities (**O**)

Opportunities	Development/valorization measures
A_1. Global warming	Elimination of any cost advantage for petrol vehicles [8]
A_2. The evolution of technology	Development of new technologies in the field of electromobility, safety, ergonomics and comfort of electric vehicles
A_3. It is the product of research	Continue research in the development of electromobility technologies; allocate the necessary resources for research in this field
A_4. Environment	Strengthening legislation on environmental protection
A_5. Taxes, duties and facilities	Exemption from the payment of annual taxes andduties; eco-bonuses; free and dedicated parking places, equipped with power stations
A_6. Easy access to cities	Facilities for movement in the central areas of major cities

Table 4. SWOT analysis of electric vehicles – threats (**T**)

Threats	Mitigation/reduction/counteraction measures
T_1. Energy crisis	Maintaining or decreasing the price of electricity [8]; developing renewable energies [8]
T_2. Financial crisis	Growth of GDP, living standards and purchasing power; maintaining the leu-foreign currency ratio at the closest possible value

3 Indicators for Assessing the Quality of Electric Vehicles

Particular characteristics, specific to the means of electric transport, can lead to the decrease of the quality of road transport services. The characteristics are determined by attributes that set the indicators of appreciation of the quality of electric vehicles. We propose this definition in order to be able to make an assessment below, according to the data in Tables 5, 6, 7 and 8.

Table 5. Assessment indicators defining technical qualities

Quality indicator	Requirement to which the quality indicator for the motor vehicle or road transport service relates	How the indicator influences the expected quality of the user/client		Corrective measures
		Positive	Negative	
Autonomy and capacity of the electric battery	Regularity, rhythmicity, punctuality, comfort	If it is high	If it is low	Being directly influenced by the capacity, weight, size and temperature of the environment, investments and research are proposed in development of electric batteries
Electric battery charging time and station power	Regularity, rhythmicity, punctuality, comfort	If minimal	If maximum	Equipping the supply points with stations of 50–100 kw/400–500 Vcc/100–400 A Charging times decrease from 15 h to 29 min [5]
Power of electric traction engines	Regularity, rhythmicity, punctuality, comfort	If they have high power	If they have low power	Equipping electric vehicles with engines that ensure an appropriate weight-to-traction ratio

(*continued*)

Table 5. (*continued*)

Quality indicator	Requirement to which the quality indicator for the motor vehicle or road transport service relates	How the indicator influences the expected quality of the user/client		Corrective measures
		Positive	Negative	
Weight of vehicle	Regularity rhythmicity punctuality comfort, safety	If it is low	If it is high	They have a higher weight due to additional equipment. The use of light materials (e.g. aluminum, titanium, carbon, plastic, etc.)

Table 6. Assessment indicators defining ergonomics and comfort

Quality indicator	Requirement to which the quality indicator for the motor vehicle or road transport service relates	How the indicator influences the expected quality of the user/client		Corrective measures
		Positive	Negative	
Operating noise level	Vehicle safety	If they emit noise	If they do not emit noise	Equipping vehicles with sound (noise) generators
Sound level emitted by audible warnings	Comfort of the driver, vehicle safety	If it is low and pleasant to hearing	If it is loud and it bothers hearing	Equipping means of transport with audible warning devices with a pleasant sound, withan acoustic level corresponding to the perception of the human ear in any situation [10]
Intuitive positioning of the controls	Comfort of the driver, vehicle safety	If ergonomically arranged	If not ergonomically arranged	Mounting elements in a visible place, operated intuitively, effortlessly, without deflecting the attention or gaze of the user [10]

Table 7. Assessment indicators defining active safety

Quality indicator	Requirement to which the quality indicator for the motor vehicle or road transport service relates	How the indicator influences the expected quality of the user/client		Corrective measures
		Positive	Negative	
Curtain airbag	Safety and comfort	With amenities	No amenities	Their introduction in standard versions [10]
Other active safety systems provided by regulation (EU) 858/30.05. 2018	Safety and comfort of passengers, pedestrians and other vulnerable persons	If provided	If not provided	Equipping the standard variants with: rear cross traffic alert – RCTA; panoramic view monitor; panoramic view parking assist; intelligent front and rear parking sensors with automatic braking – ICS [10, 11]

Table 8. Assessment indicators defining reliability and maintenance

Quality indicator	Requirement to which the quality indicator for the motor vehicle or road transport service relates	How the indicator influences the expected quality of the user/client		Corrective measures
		Positive	Negative	
Probability of good functioning R(t)	Regularity, rhythmicity, punctuality, vehicle safety	$R(t) = p(t) = Prob$ $(t > T)$ [12]	$R(t) = p(t) = Prob$ $(t < T)$ [12]	Qualitative materials and parts, defensive driving, timely performance of maintenance work
Probability of failure F(t)	Regularity, punctuality rhythmicity, comfort	$F(t)\ Prob\ (t < T)$ [12]	$F(t)\ Prob\ (t > T)$ [12]	Qualitative materials and parts, defensive driving, timely performance of maintenance work
Rate (intensity) of falls (defects) z(t)	Regularity, rhythmicity, punctuality, comfort, vehicle safety	Low fall rate $z(t) <$	High fall rate $z(t) >$	Using qualitative materials, defensive driving, timely performance of maintenance works, oualitative materials and parts

(continued)

Table 8. (*continued*)

Quality indicator	Requirement to which the quality indicator for the motor vehicle or road transport service relates	How the indicator influences the expected quality of the user/client		Corrective measures
		Positive	Negative	
Average operating times MTBF = m	Regularity, rhythmicity, punctuality, comfort, vehicle safety	If it is high MTBF = m >	If it is low MTBF = m <	Defensive driving, timely performance of maintenance works, qualitative materials and parts

4   Research on the Evolution of Sales and Registration of Clean Vehicles in Europe

Figure 1 shows the situation of salesand registration of electric and hybrid rechargeable vehicles at EU level, during the period 2014–2020 [14, 15], after the development of the Project *Electric Vehicles Network in Urban Europe - URBACT II, (EVUE)*, carried out during the period November 24, 2009–May 24, 2010 [5, 7]. According to the data presented in Fig. 1, the electric vehicle market has matured strongly, especially in the case of light electric vehicles and buses (both battery-powered electric vehicles and plug-in hybrid vehicles). In particular, electric vehicles have seen a rapid increase in terms of the total number of vehicle registrations and the number of models available between 2010 and 2020 [14]. In the third quarter of 2020, the share increased to 9.9% of total vehicle sales, compared to 3% in the previous year [15].

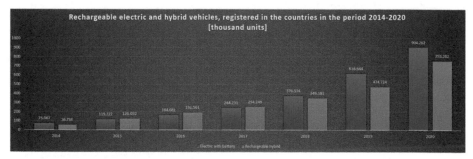

Fig. 1. The situation regarding the registration of electric and hybrid rechargeable vehicles at EU level during the period 2014–2020 [14, 15].

Given the analysis carried out, it turns out that we are only partially prepared for electromobility. Some of the quality indicators of electric vehicles negatively influence the quality. These aspects determine the reluctance of the population towards the electric vehicle, making it unattractive. Because of them, the electric vehicle demonstrates its efficiency, at this time, exclusively in urban environments.

5   Conclusions

The quality of electric vehicles and the services provided is affected by the low autonomy of electric batteries, the deficit and power of charging stations, the weight of the electric vehicles due to additional electrical equipment and the high purchase price. Although the sales of environmentally friendly vehicles have increased at European level, it is necessary to tighten the environmental legislation on pollutants eliminated by internal combustion engines of motor vehicles and we propose the elimination of any cost advantage for heat engine vehicles. At the moment, it is very important to grant eco-bonuses for the purchase of environmentally friendly vehicles. At the level of European states, especially in Eastern European countries, investments are needed in the infrastructure with electricity and maintenance. However, the evolution of sales of green vehicles during the analysis period has evolved significantly. Thus, for battery electric vehicles during the period 2014–2020, the increase was 1,105%, and for rechargeable hybrid vehicles, in the same period, the increase was 1,231%. Quality indicators have been defined by the way in which an electric vehicle satisfies the waiting level of the driver (user or owner), the customer or the passenger (beneficiary of the transport service) within a certain period of time. The period of time can be the whole life cycle of the vehicle, the period of its possession by the owner, a trip on a certain route, a transport of goods, etc. Quality indicators are the totality of a set of attributes that define the electric vehicle. They are defined by the systems, installations and elements that contribute and enhance ergonomics, safety and comfort.

References

1. Faiz, A., Weaver, C.S., Wals, M.P.: Air Pollution from Motor Vehicles: Standards and Technologies for Controlling Emissions, p. 227. World Bank Publications (1996). ISBN 978-0-8213-3444-7
2. David, B.S.: Plug-in Electric Vehicles: What Role for Washington? 1st edn, pp. 2–5. The Brookings Institution (2009). ISBN 978-0-8157-0305-1
3. Dragu, V., Roman, C.V.: Specific aspects of quality in urban public transport. Life and activities in large urban areas. Bucharest, present and future. In: Proceedings of the 7th Edition of the Academic Day of the Romanian Academy of Technical Sciences, Conference Proceedings, Bucharest, 11–12 October, pp 217–225 (2012)
4. Juncker, J.C.: Speech by the President of the European Commission. European Commission, Brussels (2014)
5. Noon, M.: Electric vehicles in urban Europe. EVUE - approaches to e-mobility infrastructure. Urb Act - We connect cities-we build successes. EU-Stakeholder Guide, pp. 1–88. European Regional Development Fund, Brussels (2012)
6. Oprean, I.M.: The Modern Car. Requirements, Restrictions, Solutions. Romanian Academy Publishing, Bucharest (2003)
7. Rodrigues, O.: Electric vehicles in urban Europe. EVUE - approaches to e-mobility infrastructure. Urb Act -We connect cities-We build successes. EU-Stakeholder Guide, pp.10–18. European Regional Development Fund, Brussels (2012)
8. Sepi, M.: The road to wider use of electric vehicles. Off. J. Eur. Union, 11 February, pp. 47–52. EUR-lex Publishing, Issue 2011/C44/08, Brussels (2011)
9. The European Commission: A Clean Planet for All. Brussels, COM 773 final (2018)

10. Toyota. https://www.toyota.ro/world-of-toyota/safety/toyota-safety-sense
11. Toyota RAV 4 Hybrid. User Manual. Published by Toyota Romania (2019)
12. Țîțu, M.: Reliability and Maintenance, pp. 184–186. AGIR Publishing House, Bucharest (2008)
13. WCDE: Brundtland Report. Our Common Future (1986)
14. https://data.consilium.europa.eu/doc/document/ST-6841-2021-INIT/ro/pdf
15. www.acea.be

The Attractiveness of the Railway Transport System - Mobility and the Correlation with Safety and Security

Cătălin-Laurenţiu Bulgariu[1,2](✉)

[1] Faculty of Industrial Engineering and Robotics, Polytechnic University of Bucharest,
313 Splaiul Independenţei, 6 District, Bucharest, Romania
catabulgariu@gmail.com
[2] National Railway Company C.F.R. S.A., 38 Dinicu Golescu Avenue, 1 District, Bucharest,
Romania

Abstract. A transport system must be based on its characteristics which define it and determine its performance and quality. Any failure of the system may affect the safety and security of a society. The rail system is one of the main components of society and one of the basic pillars of its economy. Rail transport is, and will remain, one of the main transport systems in the world, and the future of this system depends on an understanding of the main concepts that define it and on the synergy of its components. Mobility currently faces significant challenges and several factors affect the way people move. These factors are usually linked to the human lifestyle and mobility needs, leading to social constraints and environmental problems, network congestion and undersized transport infrastructure. In this context, the high potential of the rail system within clean and sustainable modes of transport offers the opportunity to increase. The paper summarizes the fundamental concepts that define the interconnections between the attractiveness of the rail transport system, mobility, safety and security. A system perspective on railway requirements in the context of increasing attractiveness is presented in this study. This work provides an overview of the current state and trends of what can be defined as the basis for sustainable and secure mobility.

Keywords: Railway system · Mobility · Attractiveness · Safety and security · Quality

1 Introduction

Due to the problem related to congestion, greenhouse gas emissions, energy consumption, safety and security faced by our society, the role of rail transport is paramount in the policies regarding the development of an environmentally friendly, sustainable and safe transport system. In this context, rail transport has the capacity to respond effectively to the various problems we face and can become one of the main modal choices in passenger transport and freight transport. The promotion of rail transport is a present and future necessity for the healthy development of our society. The transport system

© The Author(s), under exclusive license to Springer Nature Switzerland AG 2022
I. Karabegović et al. (Eds.): NT 2022, LNNS 472, pp. 687–694, 2022.
https://doi.org/10.1007/978-3-031-05230-9_82

of a nation is an important indicator of the degree of development and contributes to a large extent to economic growth and mobility [1]. Efficient management of transport systems in order to ensure socio-economic needs is fundamental and can increase the attractiveness of the services offered. Increasing the attractiveness of rail transport is an essential factor in responding to society's need for mobility and congestion on road transport networks. The rail system must therefore meet users' expectations regarding the provision of a quality, safe and competitive service with other modes of transport. An attractive transport system increases the share of rail transport compared to other transport options. The safety and security of a transport system are two basic and important concepts of increasing the attractiveness of the transport service and responding to users' mobility needs. The correlation between these factors and the need for mobility contributes to the attractiveness of the system. The interrelation of critical factors of a system determines the degree of attractiveness and the satisfaction of mobility needs. Increasing the safety and security of the rail system must be an integral part of the development of national, European transport systems, etc. The rail system is a system vulnerable to threats (bombardments, armed attacks, theft, terrorism) due to the general accessibility of potential attackers to certain parts of the railway infrastructure such as railway stations, railway infrastructure, works of art, traffic control systems, railway installations, etc. [2]. Because of this, the identification of the risks of the rail transport system and of the countermeasures necessary to reduce the risks [3] are indispensable in order to increase safety and security. Most of the times the two concepts of safety and security are assimilated as if they had the same meaning, and therefore in the railway field and beyond, more attention is paid only to the concept of safety, believing that it covers the concept of security as well. However, there are two distinct concepts [4] and it is necessary to increase the attractiveness and quality of the system to treat the two concepts equally and to give due consideration to the prevention of safety and security risks. A correct understanding of the two important factors of the rail system can prevent problems that can affect both persons and goods or the environment and transport

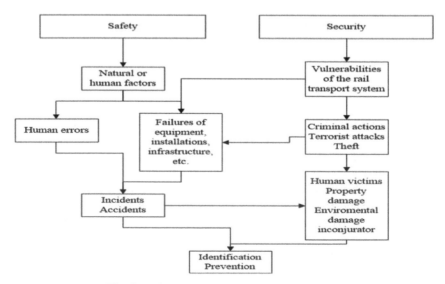

Fig. 1. Safety and security in the railway system

infrastructure. Each transmission system shall take into account the identification of risks, vulnerabilities and how to prevent them (Fig. 1).

2 Safety and Security – The Basis for an Attractive Rail System

Risk factors in the railway system may be defined as actions, happenings or errors which may lead to injury/death to passengers, employees or other persons involved involuntarily due to incidents, accidents, attacks of any kind, theft, and consequently economic and material damage caused. For example, one of the most common accidents in the railway system is derailments, with the effects of damage to infrastructure, rolling stock, cargo, traffic disruptions, delays, etc., which may also lead to human victims or environmental degradation. Understanding these factors and the threats facing the rail system is essential for the development of effective risk prevention strategies [5]. Traffic safety and transport security are among the main concepts determining the quality of the system and its attractiveness. Risk and safety management is an important aspect in any critical industry [6], including the railway sector, in order to improve quality and avoid impacts on all system factors [7]. An essential condition for the identification and management of risks in the railway system is the interrelation of all the actors of the transport system and stakeholders, such as employees, authorities, private operators, etc. Collaboration has an important role to play in defining a risk identification and prevention policy [8]. Hazard identification is the basis of a risk management system and it is very important that the hazard assessment is carried out as thoroughly as possible in order to avoid the omission of potential unidentified hazards which could cause damage to the system [7]. The safety and security of the transport system incorporate the concept of risk, which can be defined as a common characteristic of these two concepts. Risk identification is required in the context of a risk management system [9] (Fig. 2).

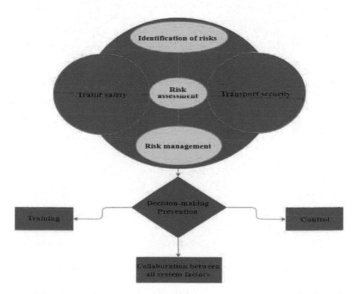

Fig. 2. Proposal for the implementation of a risk management system in the railway system

The risk analysis is a fundamental part of the process within the management system and is determined by the organized use of available information to identify hazards and to estimate the effects to which the transport system is exposed. The hazard identification step is the process prior to identifying the risks to be managed. The purpose of hazard identification is to generate a list of potential risks and events that could have an impact on the achievement of the objectives within the operation activity and beyond. At the risk assessment phase, the risk is estimated and assessed on the basis of the probability and consequences of the hazardous events identified in the hazard identification phase [10]. Rail safety and safety are two important elements, as they relate to people's lives. Therefore, the identification of safety and security risks is vital for an attractive and safe transport system. The attractiveness of the system is reflected in the quality desired and perceived by its users. Rail transport users expect the system to meet their needs and these needs include safety, security, reliability, speed of travel (journey time), comfort, accessibility, etc. [11]. For transport users, safety and security mean social security and is a prerequisite for choosing a mode of transport. If potential users perceive the mode of transport as vulnerable and unsafe, they will avoid using the service. In terms of reliability, passengers expect the service to be provided as expected. Users also expect a certain degree of comfort both in railway stations and in means of transport [12]. For a system to be attractive, the desired quality must be as close as possible to the quality perceived and offered by the system (Fig. 3).

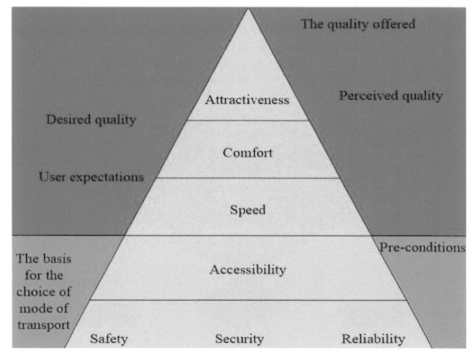

Fig. 3. The pyramid of the conditions of an attractive transport system [12]

The condition that a transport system is attractive is to meet as many user requirements and expectations as possible, at the base of which are safety and security of the system. These requirements may make the modal choices of users different. The supply of the transmission system must be adapted to the demand and the needs underlying it [13].

3 Dynamic Correlations Mobility – Attractiveness

Mobility is a multi-faceted concept used in several adjacent sciences and fields and whose understanding requires extensive and multi-disciplinary examination, from which to result its correlations with the system of activities and with the quality of life [14].

Identifying mobility needs and offers to meet these needs increases the need for a full understanding of the concept. In view of the global concerns to promote a sustainable mobility policy, the role of rail transport is among the main solutions and has a considerable advantage over other modes of transport, with the capacity to meet people's needs and environmental problems.

Promoting this mode of transport is an important factor in achieving a sustainable and attractive transport system. Rail transport is in the process of changing and adapting people's mobility needs. The renaissance of this mode of transport must be based on increasing attractiveness through the adoption of policies for continuous promotion and investment in railway infrastructure.

One of the main identified needs of mobility is the efficient use of travel time [15] and the possibility of the mode of transport to offer this opportunity to users, which influences the modal choice. If the time taken by a user to make the trip can be used for other activities, travel speed may no longer play such a difficult role in weighing decisions on modal choices.

Also another factor influencing user decisions is the reliability of the service. Users want to consume the journey within the specified time period, the certainty of travel time having a considerable impact on the perception of quality of service [16].

In conclusion, it can be stated that in the perception of users, with the advances of technologies and due to behavioral changes, mobility requirements include not only speed of travel but also the efficient use of time and opportunities to perform other activities during the journey [17].

Rail transport has the ability to meet these users' needs by increasing the comfort of the journey, reducing the unusefulness of travel time and improving the conditions which allow this time to be used, in the absence of the large investments needed to increase the speed of travel [17]. The increased attractiveness of the rail transport service is reflected in economic and environmental benefits. Another important factor in promoting an attractive transport system is the social aspect – accessibility – the possibility for all to have access to a range of activities.

Although mobility is often linked only to the concept of traffic, it is much more comprehensive and covers a much wider range of meanings. Understanding the diversity of uses of the concept of mobility is a complex problem that requires thorough deepening and multidisciplinary treatment [14] (Fig. 4).

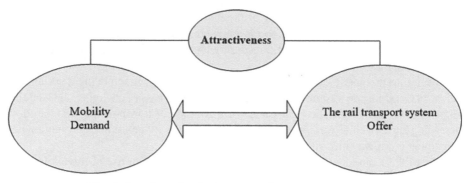

Fig. 4. Interdependencies between rail transport and mobility

In order to increase the attractiveness of rail transport the user must be the focus of attention. Identifying and meeting user requirements determines the choice of mode. The delivery of the desired quality (safety, confidence, comfort, etc.) contributes to the success of the railways and to the achievement of sustainable mobility (Fig. 5).

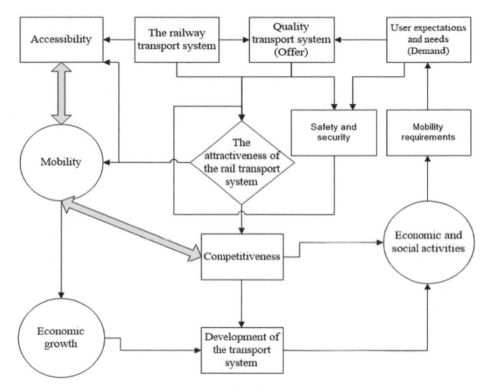

Fig. 5. Dynamic correlation of mobility in rail transport

4 Conclusion

The aim of this work is to identify the drivers of increasing the attractiveness of rail transport and the correlation with the mobility needs of society. An attractive transport system plays an important role in ensuring mobility. The study of mobility in natural and social sciences and associated values is a little explored analysis perspective to date. The main skills of the study are the need to meet the increasing demands of mobility and the need to understand the importance of an environmentally friendly mode of transport. The safety and security of a transport system are also two main elements that define the quality of service and make it attractive. Mobility needs and offers are in a dynamic correlation with accessibility and attractiveness.

References

1. Trepáčová, M., Kurečková, V., Zámečník, P., Řezáč, P.: Advantages and disadvantages of rail transportation as perceived by passengers: a qualitative and quantitative study in the Czech Republic. Trans. Transp. Sci. **11**(3), 52–62 (2020). https://doi.org/10.5507/tots.2020.014
2. Ortiz, D.S., Weatherford, B.A., Greenberg, M.D., Ecola, L.: Improving the Safety and Security of Freight and Passenger Rail in Pennsylvania. RAND Corporation, Santa Monica (2008). https://doi.org/10.7249/TR615

3. Grant, M., Stewart, M.G.: Asystems model for probabilistic risk assessment of improvised explosive device attack. Int. J. Intell. Defence Support Syst. **5**(1), 75–93 (2012). https://doi.org/10.1504/IJIDSS.2012.053664
4. Coppola, P., Silvestri, F.: Gender inequality in safety and security perceptions in railway stations. Sustainability **13**(7) (2021). https://doi.org/10.3390/su13074007
5. Xiang, L., Saat, R.M., Barkan, C.P.: Freight-train derailment rates for railroad safety and risk analysis. Accid. Anal. Prev. **98**, 1–9 (2017). https://doi.org/10.1016/j.aap.2016.09.012
6. Pant, R., Hall, J.W., Blainey, S.P.: Vulnerability assessment framework for interdependent critical infrastructures: case-study for Great Britain's rail network. Eur. J. Transp. Infrastruct. Res. 16(1) 2016. https://doi.org/10.18757/ejtir.2016.16.1.3120
7. Berrado, A., El-Koursi, E.-M., Cherkaoui, A., Khaddour, M.: A framework for risk management in railway sector: application to road-rail level crossings. Open Transp. J. **5**, 34–44 (2011). https://doi.org/10.2174/1874447801105010034
8. Rahmayana, P.E., Purba, H.H.: Risk management in railway during operation and maintenance period: a literature review. Int. J. Eng. Appl. Sci. Technol. **4**(4), 29–35 (2019). https://doi.org/10.33564/IJEAST.2019.v04i04.005
9. Hessami, A.G.: A systems view of railway safety and security. In: Krzysztof, Z. (ed.) Railway Research - Selected Topics on Development, Safety and Technology. InTech (2015). https://doi.org/10.5772/62080
10. Chen, Y.: Improving railway safety risk assessment study. Ph.D. thesis, School of Civil Engineering, University of Birmingham (2013). https://etheses.bham.ac.uk/id/eprint/4465/1/Chen13PhD.pdf
11. Abdul Hamid, N., Tan, P.-L., Mohamad Zali, M.F., HAMAT, U.N., Abd Aziz, N.: Safety and security needs of commuter rail services - travellers' perceptions. J. East. Asia Soc. Transp. Stud. **11**, 1495–1506 (2015). https://doi.org/10.11175/easts.11.1495
12. van Hagen, M., Bron, P.: Enhancing the experience of the train journey: changing the focus from satisfaction to emotional experience of customers. Transp. Res. Procedia **1**(1), 253–263 (2014). https://doi.org/10.1016/j.trpro.2014.07.025
13. van Hagen, M.: How to meet the needs of train passengers? A successful customer segmentation model for public transport. In: European Transport Congress, Noordwijkerhout, Netherlands (2009). http://data.openov.nl/docs/Needscope%20ETC%202009%20def.pdfS
14. Raicu, S., Costescu, D.: Mobilitate. Infrastructuri de trafic. AGIR Bucuresti (2020). ISBN 978-973-720-825-5
15. Lyons, G., Jain, J., Holley, D.: The use of travel time by rail passengers in Great Britain. **41**(1), 107–120 (2007). https://doi.org/10.1016/j.tra.2006.05.012
16. Bates, J., Polak, J., Jones, P., Cook, A.: The valuation of reliability for personal travel. Transp. Res. Part E Logist. Transp. Rev. **37**(2–3), 191–229 (n.d.). https://doi.org/10.1016/S1366-5545(00)00011-9
17. Givoni, M., Banister, D.: Reinventing the wheel - planning the rail network to meet mobility needs of the 21st century. In: Working Paper N° 1036, University of Oxford, Transport Studies Unit (2008). http://www.tsu.ox.ac.uk/

New Trends and Approaches in the Development of Customer Relationship Management

Elma Avdagić-Golub[(⊠)], Amel Kosovac, Alem Čolaković, and Muhamed Begović

Faculty of Traffic and Communication, University of Sarajevo, Zmaja od Bosne 8, Sarajevo, Bosnia and Herzegovina
elma.avdagic@fsk.unsa.ba

Abstract. Today we are talking about saturated service markets characterized by filled distribution channels, intense price competition, and slowed sales growth. A secure tool for successful business in a saturated market are existing customers of the company. A business-focused more on retaining existing customers than attracting new ones involves using the principles of Customer Relationship Management (CRM). CRM implies making key decisions regarding the company's relationship with customers, so with the development of artificial intelligence and data science, this area has become an ideal field for the application of these methods. The level of automation is continuously increasing and will be emphasized in the coming period. By taking advantage of innovative technologies and integrating them into CRM systems, companies can achieve a better market advantage. In this paper, we analyze new trends in customer relationship management that need to be addressed in the coming years. We explore the advantages and disadvantages of new technologies and how they affect the user experience and business revenues of service companies.

Keywords: CRM · Customer · Customer experience · New trends · Data

1 Introduction

New challenges during the previous years and the requirements for high-quality service create a new business environment for service companies. The current dynamics of the environment impose new rules of conduct for market participants, especially those that offer services. Service companies are adapting to these changes to operate successfully. The service industries are setting up new business models, process and technology challenges. The modern business environment is characterized by relatively rapid changes in market conditions and a large amount of information available to consumers and market participants. Successful companies know that performance management processes and the right data flow, from which information and knowledge flow, have a crucial impact on their success.

In the new digital era, where information and knowledge are widely available, evolutions can be seen from large multinational companies to thousands of start-ups. In saturated market conditions, it is necessary to adapt to new changes in the market,

I. Karabegović et al. (Eds.): NT 2022, LNNS 472, pp. 695–703, 2022.
https://doi.org/10.1007/978-3-031-05230-9_83

take advantage of favorable market opportunities and create and develop a base of its customers. A surefire mechanism for this is the application of basic principles of Customer Relationship Management (CRM). In order to increase the quality of customer relations, companies have invested heavily in creating a superior customer relationship management system [1].

Customer relationship management is an approach that focuses on understanding and anticipating customer needs. It is a strategy aimed at attracting and retaining customers, their satisfaction, resulting in the creation of profitability in the long run [2]. Today, CRM is explained as knowledge-based marketing. Therefore, all technologies, especially those in the domain of the Web 2.0 concept, which is in the function of establishing relationships with consumers, collecting and recording information and knowledge about consumers, providing information and support to consumers, and finally building and maintaining relationships with consumers, are called CRM [3]. The authors [4] define a holistic definition of CRM as "a set of business activities supported by both technology and process that is directed by strategy and is designed to improve business performance in an area of customer management".

It is a competitive space, which can be viewed from three key aspects, CRM:

- as Technology:CRM as Technology is designed to help sales professionals manage customer relationships by improving communication, learning more about customer needs, and creating customized solutions for the customer [5]. This means that the CRM is a software product, often hosted in the cloud, and known as the "CRM system".
- as a Strategy: Companies develop strategies to apply the "customer in focus" philosophy across all business processes. CRM is a strategy and business philosophy to which all employees of the company must submit, and which is the principle of individual approach to each client by recognizing and respecting each of his needs.
- as a Process: The established systems contain steps on how to successfully guide the user through the company, from the moment the user requests an offer to use it.

In this paper,the current situation in the CRM development market in terms of technologies, strategies, and processes is analyzed.

2   The Rise of CRM Technology

Companies in the CRM space are integrating innovative technologies and concepts to meet customer requirements. CRM space depends and will increasingly depend on the integrity, reliability, security, and applicability of information systems, and processes, and knowledge in the organization. Software and hardware to support CRM are evolving at a revolutionary rate. These systems have evolved much more than just contact management tools. According to the latest Deloitte CMO report, which examines market behavior affected by the ongoing COVID-19 challenge, marketing experts predict a growing impact of trust relationships, and 29.3% said it will be a top priority for customers [6].

This means that CRM has been evaluated from a simple relationship management tool into a key component of a company (Fig. 1).

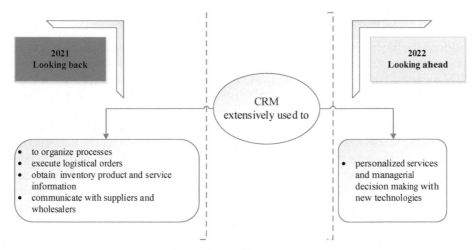

Fig. 1. Evaluation CRM in company

CRM technology will be the single largest revenue area of spending in enterprise software [7]. In such a dynamic environmentto increase efficiency, developed companies are moving from the classic concept of investing, to a new concept of investing in new technologies. The four key fields of technology during the 4.0 Industrial Revolution, that enable to development of perfect CRM systems are IoT, Social, Mobile, Big Data, and Cloud. By reviewing the available literature, we highlight the following trends in the CRM software industry, which the company needs to be aware of if it wants to stay on top of innovation and market competition (Fig. 2).

Fig. 2. CRM trends

2.1 Artificial Intelligence Integration in CRM System

Artificial Intelligence (AI) approaches are appearing at the forefront of research in information retrieval and information filtering systems [8], and are ideal for application in CRM systems, which integrate user data generated from formal and informal interactions, including customers and suppliers. Customer data is becoming huge and complex as it is collected from multiple user touch points set up in the company [9]. The data collected in this way represents both an chance and a challenge. The chance is reflected in the ability to use a wider range of data to improve operational and business performance, as well as create better, more sophisticated, and sometimes a new customer service paradigm. The challenges are to extract useful data from a huge amount of different data so that they have use-value for different departments, interactions, and situations [10]. The application of AI technology in the business domain enables the creation of personalized service and managerial decision-making [11]. The information obtained from the available data through data mining is the basis for directing both reactive and proactive future behavior of the company.

We have recognized three main areas for the application of AI in CRM:

1) Predictive analytics: application of machine learning methods on structured data to predict trends in customer behavior (purchase) [9, 12].
2) Customer segmentation: creating a cluster of customers according to shopping habits and other forms of behavior, thus enabling more productive sales and marketing campaigns. The use of AI provides insight into customer demographics, their preferences, worldviews, interests, allowing companies to fully focus on customers who are likely to buy (use the service) when encouraged in the right way and the right direction [13].
3) Advanced support channels: Another powerful addition of AI to CRM catalogue are advanced support channels including chatbots, video calls, and voice recognition services [14]. These are able to automate responses to customer inquiries, and also to encourage and increase sales outside of business hours. The leading CRM solution providers have already embraced the voice technology and others are expected to follow suit.

A prerequisite for the application of AI in CRM is the organization's technological capabilities [15], all employees must have the skills and expertise to implement, maintain and configure such systems. It is especially important that the management of the service organization constantly reviews whether the service organization has the readiness and capacity to meet customer requirements [16].

A recent report suggests that CRM acts linked with AL will push the global business revenue by up to $1.1 trillion by the end of the year 2021. and the CRM market would pounce to $72.9 in the coming couple of years [17].

2.2 Implications Internet of Things in CRM

Technological changes, digitization and big data have created new services and the growth of the digital economy, which would meet the needs and desires of increasingly demanding users.In order for companies to get as close as possible to their customers, it is necessary for CRM to become woven with IoT technologies [18]. IoT (Internet of Things) is a new paradigm that provides a set of new services for the next wave of technologicalinnovations [19]. According tostatistics [20], there are a few outstanding facts:

a) it is predicted that there will be more than 64B IoT devices worldwide by 2025,
b) IoT will potentially generate $4T to $11T in economic value,
c) 54% of business IoT projects are cost savings, and this is the main driver of revenue,
d) the portable devices market will be worth $1.1 billion by 2022.

IoT technologies enable companies to serve customers in an easier and more proactive way, which was unthinkable in previous years [18]. The connected devices that continuously share data (e.g. knowledge of user behavior, actions, affinity) integrated with smart systems, which use machine learning methods on large amounts of data are a great opportunity for companies to discover deeper knowledge about their customers.We highlight several advantages for the company, which arise from the integration of IoT and CRM technologies:

– Companies can detect a problem with the use of the service/product before the customer reports it because a sensor can be installed in the product and send information to the CRM system about a possible problem with the product. This automates customer service and proactively influences the user experience,
– Application of IoT technologies facilitates the work of customer service representatives,
– Improving business processes from end to end by connecting products, devices, and equipment.

2.3 Advancing Social CRM

The role and daily use of social media have a great impact on CRM as well. Strategy of Social CRM (SCRM) is an advanced version of CRM, enriched with social media, which enables more efficient customer relationship management [21]. The use of social media adds an extra dimension to customer profiling, making the user to whom the service is offered clearer. Previous research confirms that SCRM opens up access to more data sources for companies, which record emotional and behavioral information about customers [22]. This creates better conditions for communication with customers compared to traditional CRM techniques. We can conclude that the benefits which CRM provides to the company are reflected in the following:

– novel support channels,
– access to dataon user behavior, private life, and friends
– segmentation of customers according to the relationships they have on social networks
– discovering new potential users through the database of existing customers.

The challenge in the application of SCRM is the minimum knowledge of the integration of CRM into corporate entrepreneurshipto increase an organization's customer focus [22], which will be certainly the subject of future research. It's interpreted as a complete convergeofmarketing, sales, and service with a large amount of unstructured data, that existingin thestatuses, shares, tweets, comments and likes of over 4 billion social media customers. The challenge is to use all this information at the right time in the right place.

2.4   Advancing Mobile CRM

The development trend of ultra-fast mobile networks and smartphones has caused more and more companies to accept the use of mobile CRM. Mobile CRM (or m-CRM) has been defined as the communication, bilateral orunilateral, that is related to marketing activities via mobilephone to build and maintain relationships betweenthe constomer and the company [23]. This fact represents a strategy companies must takeadvantage of to "collocate their brand in thepockets of consumers" at any time and any place, andcharacterize it by an interactive communication [24]. The global mobile CRM market grew 11% in 2019 to $15 billion worldwide and is expected to grow at an annual rate of 13% until 2029 [25]. Using mobile CRM for sales purposes is not surprising, as companies no longer rely on traditional work ideologies and use mobile applications to improve increase sales targets [26]. Therefore, weselect the following benefits of Mobile CRM:

- Personalization: creating on-site services, using applications that can interactively design a product through the device;
- Increase sales: Attract new customers by applying mobile advertising via RFID, Bluetooth, QR codes or loyalty of newcomers via mobile applications that provide access to products and services from mobile phones
- Increasedproductivity of the workforce and
- Improved customer self service.

The mobile CRM is highly likely to make huge strides, especially in the post-pandemic world, where distributed teams and remote work are common.

2.5   The Effect of New Technologies on Customer Experience

Building strong relationships with customers and understanding their needs or satisfaction distributes the importance of manufacturing service success [27], and largely depends on the associated customer satisfaction and experience. The relationship between customer satisfaction and retention is not yet completely final, some research shows that customer satisfaction has a direct impact on customer retention, loyalty, and competitiveness of the company [28]. The concept of customer satisfaction is a matter of perception of the expected product or service provided [29]. In recent years, a priority of top management in practice is the creation of a superior customer experience (CX) [30]. Research by practitioners constantly believes that it is among the top three priorities of business management around the world, which are considered vital to the long-term success of any company. The reasonthe CX has become crucial in marketing is that customers communicate with companies and their products in many new points of contact, which have emerged as a result of digitalization. A CRM's customer experience is very significant. From the last Salesforce report 84% of customers say theexperience that a company provides is just as important as its products and services, and 87% of consumers believe companies need to offer a more consistent customer experience [18]. The relationship between the customer and the company is more similar to human relationships every day, the company should store previous interactions with the customers, understand what they need, and modify communication with the customer to those needs. It is a process of personalization. Personalization is a secure tool for CX. Not because it's new to the game, but because it's becoming more hyper-focused and individualized, particularly in response to the pandemic.We can conclude that for improved CX the company needs the use of customer data.

2.6 Integration of New Technologies

Through the earlier part of the paper, we wrote about the application of artificial intelligence, social media, IoT technologies, and large amounts of data in CRM systems. However, a new trend in the development of CRM is the integration of all the above into one basic software, which works flawlessly with all the elements together. This includes solutions that integrate: customer service tools, email software, analytics, customer data platforms, and marketing automation. Top customer experiences will incorporatecustomers, the company's enterprise systems, CRM system, Big Data Analytics, Cognitive Computing, and Internet of Things (Fig. 3) [31].

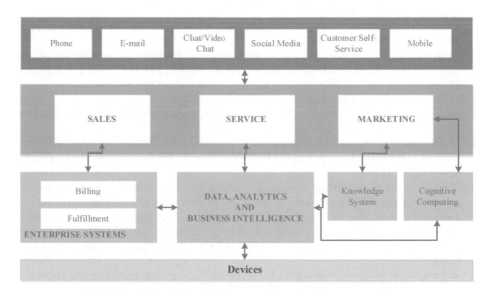

Fig. 3. CRM architecture incorporate with new technologies [29]

3   Conclusion

The philosophy of CRM, as technology, strategy and process have the task to improve the customer experience and optimize various aspects of sales and marketing. If we were to try to make a general prediction for the future development of the use of new technologies in CRM, it would mean that companies will use technology to root out or automate processes that create additional costs. The mechanismfor this is powerful CRM, automated workflow, and connected application stack, where data moves freely between different applications. If companies want to run successfully in the current market conditions, which are much more compressed, more prone to change, enriched with more data sources in the decision-making process, they will have to make sure that their CRM continues to innovate and follow the trends outlined. However, companies need to train their employees to understand how to make the most of such systems and use these new tools to succeed. CRM is now a one-stop solution that helps organizations make data-based decisions to drive hyper sales growth and increase revenue.

References

1. Rekettye, G., Rekettye, G.J.: The changing role of customer experience in the age of industry 4.0. Mark. Menedzsment **54**(1), 17–27 (2020)
2. Soltani, Z., Zareie, B., Milani, F.S., Navimipour, N.J.: The impact of the customer relationship management on the organization performance. J. High Technol. Manag. Res. **29**(2), 237–246 (2018)
3. Dukić, B., Gale, V.: Upravljanje odnosima s potrošačima u funkciji zadržavanja potrošača. Ekon. Vjesn. **38**(2), 583–598 (2016)
4. Richards, E., Jones, K.A.: Customer relationship management: finding value drivers. Ind. Mark. Manag. **37**(2), 120–130 (2008)
5. Agnihotri, R., Trainor, K.J., Itani, O.S., Rodriguez, M.: Examining the role of sales-based CRM technology and social media use on post-sale service behaviors in India. J. Bus. Res. **81**(August), 144–154 (2017)
6. Moorman, C.: Covid-19 and the State of Marketing (2020)
7. Micallef, L.: CRM 101: What is CRM? (2020). https://www2.deloitte.com/mt/en/pages/tec hnology/articles/mt-salesforce-crm-101.html
8. Nilashi, M., Ibrahim, O., Bagherifard, K.: A recommender system based on collaborative filtering using ontology and dimensionality reduction techniques. Expert Syst. Appl. **92**, 507–520 (2018)
9. Dwivedi, Y.K., et al.: Setting the future of digital and social media marketing research: perspectives and research propositions. Int. J. Inf. Manag. **59**(May), 102168 (2021)
10. Avdagic-Golub, E., Begovic, M., Causevic, S., Kosovac, A.: Profiling contact center customers for optimization of call routing using data mining techniques. In: 2021 20th International Symposium INFOTEH-JAHORINA, INFOTEH 2021 - Proceedings, no. March, pp. 17–19 (2021)
11. de Jong, A., de Ruyter, K., Keeling, D.I., Polyakova, A., Ringberg, T.: Key trends in business-to-business services marketing strategies: developing a practice-based research agenda. Ind. Mark. Manag. **93**(December 2020), 1–9 (2021)
12. Duan, Y., Edwards, J.S., Dwivedi, Y.K.: Artificial intelligence for decision making in the era of big data – evolution, challenges and research agenda. Int. J. Inf. Manage. **48**(January), 63–71 (2019)

13. Gupta, S., Leszkiewicz, A., Kumar, V., Bijmolt, T., Potapov, D.: Digital analytics: modeling for insights and new methods. J. Interact. Mark. **51**, 26–43 (2020)
14. Kirkpatrick, K.: AI in contact centers. Commun. ACM **60**(8), 18–19 (2017)
15. Chatterjee, S., Rana, N.P., Tamilmani, K., Sharma, A.: The effect of AI-based CRM on organization performance and competitive advantage: an empirical analysis in the B2B context. Ind. Mark. Manag. **97**(August), 205–219 (2021)
16. Avdagić-Golub, E., Hasković Džubur, A., Memić, B.: Quality management as the basis of business company operations for the purpose of customer satisfaction. Sci. Eng. Technol. **1**(1), 52–58 (2021)
17. Mohanty, S.: Top 10 customer relationship management (CRM) trends for 2021 (2020). https://www.smartkarrot.com/resources/blog/crm-trends/?fbclid=IwAR1xAqIK7O L0Kuej91NHBpuGxr4q42zL2JM95kTdPNpqXqNTSZxuCKC_EuU
18. Fuggle, L.: 20 innovative CRM trends to pay attention to in 2022 (2021). https://blog.hub spot.com/sales/latest-crm-trends
19. Čolaković, A., Hadžialić, M.: Internet of things (IoT): a review of enabling technologies, challenges, and open research issues. Comput. Netw. (2018)
20. Petrov, C.: 49 stunning internet of things statistics 2021 [The Rise of IoT] (2021). https://tec hjury.net/blog/internet-of-things-statistics/#gref
21. Jalal, A.N., Bahari, M., Tarofder, A.K.: Transforming traditional CRM into social CRM: an empirical investigation in Iraqi healthcare industry. Heliyon **7**(5), e06913 (2021)
22. Al-Omoush, K.S., Simón-Moya, V., Al-ma'aitah, M.A., Sendra-García, J.: The determinants of social CRM entrepreneurship: an institutional perspective. J. Bus. Res. **132**(April), 21–31 (2021)
23. Kim, C., Lee, I.S., Wang, T., Mirusmonov, M.: Evaluating effects of mobile CRM on employees' performance. Ind. Manag. Data Syst. **115**(4), 740–764 (2015)
24. San-Martín, S., Jiménez, N.H., López-Catalán, B.: The firms benefits of mobile CRM from the relationship marketing approach and the TOE model. Span. J. Mark. ESIC **20**(1), 18–29 (2016)
25. Hufford, B.: Future of CRM: 6 of the latest CRM trends to look for in 2022 (2021)
26. Mobile CRM Market By Enterprise Size (Large Enterprise, Medium Enterprise, Small Enterprise), Vertical (BFSI, Telecom, Healthcare, Retail, Automotive) & Region - Forecast to 2019 – 2029 (2019)
27. Cristiano, J.J., Liker, J.K., White, C.C.: Key factors in the successful application of quality function deployment (QFD). IEEE Trans. Eng. Manag. **48**(1), 81–95 (2001)
28. Singh, H.: The Importance of Customer Satisfaction in Relation to Customer Loyalty and Retention by Harkiranpal Singh May 2006 UCTI Working Paper. UCTI Working Paper, no. May, p. 6 (2006)
29. Andrade, R., Moazeni, S., Ramirez-Marquez, J.E.: A systems perspective on contact centers and customer service reliability modelling. Syst. Eng. 1–16 (2019)
30. McColl-Kennedy, J.R., et al.: Fresh perspectives on customer experience. J. Serv. Mark. **29**(6–7), 430–435 (2015)
31. Rizvi, M.: Implications of Internet of Things (IoT) for CRM (2017)

Level of Atmospheric Pollution from the Hybrid Vehicle

Gheorghe Neamțu[1] and Aurel Mihail Titu[2]([⊠])

[1] Faculty of Industrial Engineering and Robotics, University Politehnica of Bucharest, Splaiul Independenței nr. 313, 6th District, Bucharest, Romania
[2] "Lucian Blaga" University of Sibiu, 10 Victoriei Street, Sibiu, Romania
mihail.titu@ulbsibiu.ro

Abstract. The scientific paper presents in an elegant and original manner an analysis of the level of pollution with pollutants eliminated in the atmosphere by the thermal engine of hybrid vehicles. The impact on the environment by the pollution with exhaust gases presented in the paper, represents the results of some experiments obtained with measuring equipment, which accurately reproduces and records data of the level of the car pollutants measured at the output of the exhaust installation of the flue gas. The values of the results are graphically processed by means of specific software. The proposed research represents a point of view of the authors that was argued on the basis of the results obtained from tests carried out on mixed routes, using methods of ecological (defensive) driving, but also of the electronic control systems of hybrid vehicle management. At the end of the scientific paper, the results of the research are presented, where we have established the level of pollution and the way in which the hybrid vehicle manifests its friendship with the environment.

Keywords: Hybrid vehicle · Car pollutants · Experimental research · Road traffic · Electronic management system

1 Introduction

Sustainable development in general, as well as the sustainable development of automotive transport [12], has always been a particular concern of researchers. The actual fuel consumption and emissions of pollutants in motor vehicles are directly influenced by the way they are driven [11, 13]. Today, researchers are focusing on making better performing, more efficient and less polluting engines [2, 6], in order to reduce fuel consumption and implicitly pollutant emissions from the combustion of fossil fuels. Since the beginning of the eighth decade, the production of the *clean gas* engine and the limitation of pollutants discharged by the engine into the atmosphere has been defined as a peak requirement [1]. The hybrid vehicle is considered as the interface between the classic vehicle and the electric one. It is a temporary alternative found by researchers to the heat engine vehicle. The hybrid vehicle combines a conventional powertrain with a recoverable energy storage system to achieve better efficiency, lower fuel consumption and lower emissions [10]. In order to establish a level of pollution of the existing

© The Author(s), under exclusive license to Springer Nature Switzerland AG 2022
I. Karabegović et al. (Eds.): NT 2022, LNNS 472, pp. 704–711, 2022.
https://doi.org/10.1007/978-3-031-05230-9_84

gasoline engine on this type of vehicles, we conducted research, applying the method of experimental research [9]. The experimental research was carried out in the extra-urban environment, by going through three times two routes: a mountain route with high difficulty, on a county road and a route with crowded trafficona national road, where it was used the cruising speed dynamic radar control system throughout the range of speed. In order to determine the level of pollutant emissions, a pollutant measuring device [3] was mounted on the hybrid vehicle [7, 8] with the pollution standard Euro 6. The device recorded and stored in its memory the results of pollutants eliminated in the atmosphere by the heat engine. For the accurate, detailed interpretation of the results obtained, two cameras were used that recorded the routes traveled and the data returned by the Electronic Control Unit (ECU). The results obtained were reported to the requirements of Annex no. 13 of the Order of the Minister of Transport no. 2.133/December 8, 2005 [5].

2 Statutory Exhaust Gas Limit Values for Spark-Ignition Motor Vehicles (MAS)

Our country has adapted precisely the domestic legislation to the European one and has regulated by the Order of the Minister of Transport, Construction and Tourism no. 2.133/2005 - RNTR 1, with subsequent additions and amendments [3]. In accordance

Table 1. The legal limit values for exhaust gases resulting from the combustion of gasoline in spark-ignition engines (m.a.s.) [5]

Vehicle type	CO (%)		CO_2 (%)		HC (ppm)		NO (ppm)	
	Low engine speed (rpm)	2,000–3,000 rpm	Low engine speed (rpm)	2,000–3,000 rpm	Low engine speed (rpm)	2,000–3,000 rpm	Low engine speed (rpm)	2,000–3,000 rpm
Vehicle registered until 1986	4,5		Not specified		<1,000		Not specified	
Vehicles registered since 1987	3,5		Not specified				Not specified	
Vehicles with pollution norm EURO 3–4*	0,5	0,3	Not specified		Not specified	<100	Not specified	
Vehicles with pollution norm EURO 5–6**	**0,3**	**0,2**	**Not specified**		**Not specified**	**<100**	**Not specified**	

Note [5]: *For vehicles approved in accordance with the limit values indicated in line A or B of the table in section 5.3.1.4 of Annex I to Directive 70/220/EEC, as amended by Directive 98/69/EC (Euro 3 or Euro 4 motor vehicles of category M_1, N_1, M_2 or N_2).
**For vehicles approved in accordance with Regulation (EC) no. 715/2007 (Euro 5 or Euro 6 motor vehicles of category M_1, N_1, M_2 or N_2).

with the provisions of Annex no. 13 of the regulations, Table 1 shows the legal limit values for the main exhaust gases to spark-ignition motor vehicles (mas).

3   Results and Discussions

3.1   Evolution of Exhaust Gases on the Mountain Route

Figure 1 predicts a general picture of all the exhaust gases emitted into the atmosphere by the motor vehicle on the mountain route, when climbing the ramps and serpentines.

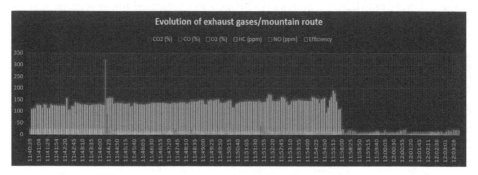

Fig. 1. Evolution of exhaust gases on the mountain route

It is highlighted the continuous, high level of hydrocarbon pollution (yellow color), on more than 2/3 of the route. Their level peaked at up to 160 ppm, above the legal threshold of 100 ppm. They are due to the operation of the heat engine under load, for the propulsion of the vehicle when climbing ramps and serpentines. The graph shows that after climbing the ramps, you reach the plateau area where the hybrid vehicle works more in electric mode. In this case, the level of exhaust gas pollution is reduced or it is even zero.

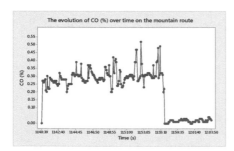

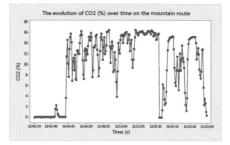

Fig. 2. Evolution of CO on the mountain route.

Fig. 3. Evolution of CO_2 on the mountain route.

Carbon Monoxide (CO), (Fig. 2), is a product of the incomplete combustion of carbon-based fossil fuels and has evolved in the range of [0–0,47]%, well above the legal limit of 0.2%. Towards the end of the route, it returns to normal, below the maximum legal limit of 0.2%. Hence the result that the operation of internal combustion engines at high temperatures and high speed, for a long time, accelerates the oxidation process of the oil, its degradation, thus creating an improper functioning of the engine mechanism. This directly influences fuel consumption and increases carbon monoxide levels. The air temperature admitted in the engine influences the temperature level per cycle, which has an influence both on the spraying of the fuel by the viscosity of the gaseous medium in the combustion chamber at the time of injection, as well as on the burning rate and thus on the formation of the polluting compounds. On the other hand, it is known that as the temperature of the intake air increases, its density decreases and the degree of filling ηv of the engine. Finally, the air mass left in the cylinder at the end of the suction decreases, and in these conditions, we can expect an incomplete combustion of the fuel which leads to a reduction in engine power on the one hand and an increase in its degree of pollution on the other hand [4].

Carbon Dioxide (CO₂), (Fig. 3) evolved in the range [0–16]%. In European and national legislation, there is no stipulated maximum reference value (Table 1). That is why we consider that the maximum value emitted by the engine of the hybrid vehicle are normal emissions, specific to the Euro 6 pollution standard.

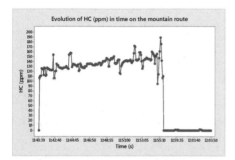

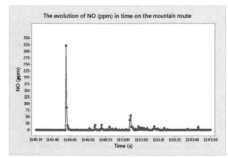

Fig. 4. Evolution of HC on the mountain route.

Fig. 5. Evolution of NO on the mountain route

Hydrocarbons (HC), (Fig. 4), take values in the range [0–189] ppm, well above the legal limit of 100 ppm where, they almost doubled for 18 min. Exceeding the maximum values is due to the high speed and incomplete combustion of gasoline in the cylinders of the heat engine.

Nitrogen Oxides (NO), (Fig. 5), take values in the range [0–321] ppm, having a single appearance, for a short period of time (10–15 s). In European and domestic legislation, there is no stipulated maximum reference value (Table 1). We believe that the maximum value emitted by the hybrid vehicle's engine are normal emissions, specific to the Euro 6 pollution standard. Reaching the maximum value is due to the under-load operation of the heat engine (3,000–4,000 rpm), when climbing ramps and serpentines, specific to the mountain route.

3.2   Evolution of the Exhaust Gases on the Route with Heavy Traffic on Which the Cruising Speed Dynamic Radar Control System was Used Throughout the Speed Range

Figure 2 provides a general picture of all the exhaust gases on the busy traffic route, where the cruising speed dynamic radar control system was used throughout the speed range (the second part of the graph) (Fig. 6).

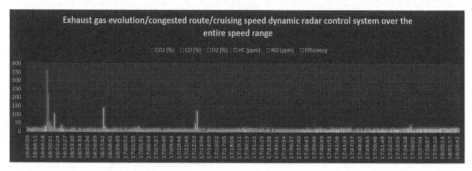

Fig. 6. Evolution of exhaust gases on the route with heavy traffic where the cruising speed dynamic radar control system was used throughout the range of speed.

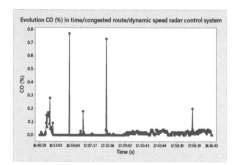

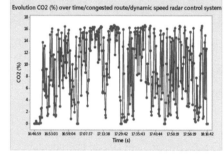

Fig. 7. Evolution of CO on the route with heavy traffic where the cruising speed dynamic radar control system was used throughout the range of speed

Fig. 8. Evolution of CO_2 on the route with heavy traffic where the cruising speed dynamic radar control system was used throughout the range of speed

Carbon Monoxide (CO), (Fig. 7), is a product of the incomplete combustion of carbon-based fossil fuels. In our research, it is kept within the legal limits on almost the entire route. The short exceedances of the legal limit of 0.2%, when it reached the maximum range of 0.77%, respectively 0.73%, appeared in the first part of the route (Fig. 6, the time interval 04:46:59 PM–05:19:29 PM), the hybrid vehicle being driven here by the driver. When using cruising speed dynamic radar control system over the entire speed range, no legal exceedances of this gas were identified (Fig. 6, the time interval 05:29:42 PM–06:02:11 PM). The result is due to the electronic engine and heterogeneous traffic management system that determined the entire route to be covered at an average traffic speed of 40 km/h where, the hybrid vehicle used electric motor traction for propulsion.

Carbon Dioxide (CO₂), (Fig. 8), evolved in the range [0–17]%. In European and domestic legislation, there is no stipulated maximum reference value (Table 1). That is why we consider that the maximum value emitted by the hybrid vehicle's engine are normal emissions, specific to the Euro 6 pollution standard. Given the analyzed data, it results that on almost 70% of the total distance traveled on the entire route, the CO_2 value frequently reached values in the range of [3, 5–16]%. We found that, the pollution with this gas is slightly low compared to the previously analyzed route, where the heat engine operated under load.

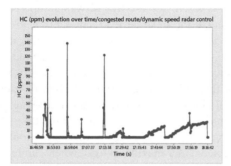

Fig. 9. Evolution of HC on the route with heavy traffic where the cruising speed dynamic radar control system was used throughout the range of speed

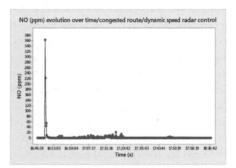

Fig. 10. Evolution of NO on the route with heavy traffic where the cruising speed dynamic radar control system was used throughout the range of speed

Hydrocarbons (HC), (Fig. 9), are kept within the legal limits on almost the entire route. If we analyze the graph of the evolution of hydrocarbons over time, (Fig. 9), we notice that during the first half of its route, when the hybrid vehicle was driven by the driver, they reached for short periods of time (20–40 s), two maximum values, of 139 ppm and 122 ppm, above the legal maximum limit (100 ppm). The exhaustion of the maximum values is due to the high speeds of the heat engine. When completing the second part of the route, when applying the driving electronic management system, the hydrocarbons did not exceed the maximum value of 36 ppm, thus falling below the legal maximum limit. Here, the low hydrocarbon values are exclusively due to the cruising speed dynamic radar control system over the entire speed range, which has ensured proper management of the engine and vehicle controls, substantially reducing hydrocarbon pollution.

Nitrogen Oxides (NO), (Fig. 10), is kept in the range [0–36] ppm throughout the route/itinerary. If we analyze the graph of the evolution of nitrogen oxides over time, (Fig. 5) it is noticed that, during the first part of the route, when the driver acted on the vehicle's controls, they reached for short periods of time (10–20 s), elevated values, in the range [223–363] ppm. When completing the second part of the route, when applying the electronic management system, nitrogen oxides did not exceed the maximum value of 5 ppm. Here, the low values of nitrogen oxides are due to the electronic control system, which has ensured proper management of the engine and vehicle controls, substantially reducing pollution with nitrogen oxides.

4 Conclusions

The level of carbon monoxide pollution is high when going through the mountain route, when the heat engine has been running under maximum load. Here, it increased 2.35 times for 16 min and reached highs of 0.47%, above the legal maximum of 0.2%. When going through the route with heavy traffic on a national road, where the cruising speed dynamic radar control system was also used throughout the speed range, the carbon monoxide pollution was very low, almost non-existent. The phenomenon was due to the busy traffic, where the movement of vehicles was carried out bumper-to-bumper, the average speed of road traffic did not exceed 45 km/h, and the hybrid vehicle operated in electric mode. We found that an overheating of the fresh air load inevitably leads to reduced volumetric efficiency in the process of filling the cylinder. The phenomenon is due to the influence of high temperature and air humidity that affects the principle of burning the heat engine, increasing the quantity of carbon monoxide and carbon dioxide eliminated in the atmosphere. When going through the mixed route, carbon dioxide took the highest values, ranging [0–17]%. Emissions of carbon dioxide and emissions of nitrogen oxide are not regulated by legislation. The hydrocarbons, when going through the mountain route, exceeded the maximum legal limit allowed by the legislation in force, where for 18 min the heat engine operated under maximum load, when climbing ramps. The appearance of maximum values and hydrocarbon pollution may also be due to the use of a low-quality fuel (e.g. gasoline COR 95), or the improper operation of the ignition or fueling systems of the heat engine of the hybrid vehicle, but, in order to be proved, these aspects remain further research directions, which we will investigate and show in another scientific paper. They result in incomplete combustion of gasoline in the heat engine cylinders. The maximum values, above the legal limit, are reached when the ambient temperature influences the temperature of the heat engine operating mode. The phenomenon has a direct influence on the temperature of the compression end which delays the combustion front in the cylinders, resulting in the unburned fuel released into the atmosphere. The phenomenon occurs at high speeds of the heat engine, which operates under load (3,000–4,000 rpm of the crankshaft). Pollution with nitrogen oxides has been reduced on both routes. The phenomenon is due to a good functioning of the catalytic converter of the vehicle. Given the results obtained after the research, we consider that all hybrid vehicles that are equipped with internal combustion engines with the pollution norm Euro 6, emit into the atmosphere carbon monoxide and hydrocarbons above the legal norm when they are overloaded. However, they have the advantage that, at a high ratio of the power and the cylinder capacity of the heat engine, they have a low fuel consumption due to the electric engines that contribute to the traction of the hybrid vehicle, along with the heat engine.

References

1. Grunwald, B.: Theory, Calculation and Construction of Engines for Road Vehicles. Didactic and Pedagogical Publishing House, Bucharest (1980)
2. Feldman, B.: The hybrid automobile and the Atkinson cycle. J. Phys. Teach. **46**(7), 420–422 (2008)

3. Gas analyzer KANE Auto 5-1. http://www.distek.ro/ro/Produs/KANE-AUTO5-1-Analizor-emisii-auto-Clasa-1-OIML-5-gaze--3573. Accessed 11 May 2021
4. Mekki, C., Nagi, M., Experimental study of the dependence of energy performance and the degree of pollution of diesel engine with direct injection at ambient temperature. In: 8th International Conference. Faculty of Engineering, "Constantin Brâncuşi" University, Târgu Jiu, Romania (2002)
5. RTNR-1: Order of the Minister of Transport 2.133, for the approval of the Regulations periodic technical inspection of vehicles registered in Romania, with subsequent completions an amendments. Published in the Official Gazette of Romania. Part I. No. 1.160 of December 21. Annex no. 13 to regulations (2005)
6. Petrescu, F.I., Petrescu, R.V.: Some elements to improve the design of the engine mechanism. In: Proceedings of 8 Naţional Symposium on GTDth, Braşov, Romania, vol. I, pp. 353–358 (2008)
7. Toyota: Toyota Safety Sense. https://www.toyota.ro/world-of-toyota/safety/toyota-safety-sense. Accessed 19 July 2021
8. Toyota RAV 4 Hybrid. User Manual. Published by Toyota Romania (2019)
9. Ţîţu, M., Oprean, C., Boroiu, A.I.: Experimental research applied in increasing the quality of products and services. In: Experimental Data Processing Collection, 684 p. Agir Publishing House, Bucharest (2011). ISBN 978-973-720-362-5
10. Varga, B.O.: Energy efficiency of hybrid or electric powered vehicles for public passenger transport. Habilitation thesis, p. 19. Technical University of Cluj Napoca (2015)
11. Volkswagen: WLTP: New standards for determining consumption. https://www.volkswagen.ro/wltp. Accessed 11 May 2021
12. WCDEL: Brundtland Report. Our common future (1986)
13. Skoda: WLTP: New standards for determining consumption. https://www.skoda.ro/about-skoda/wltp. Accessed 11 May 2021

Multimodal Journey Route Selection as Decision-Making Process

Bia Mandžuka[(✉)], Marinko Jurčević, Krešimir Vidović, and Miroslav Vujić

Faculty of Transport and Traffic Sciences, Univesity of Zagreb, 10000 Zagreb, Croatia
bia.mandzuka@fpz.unizg.hr

Abstract. In recent years, special attention has been paid to the area of travel behavior, perceptions, and preferences for multimodal mobility, but also to the comprehensive facts about what multimodal mobility means for people, for their activities and their daily routines. Travel planning behaviour (choice of travel mode) can be understood as a decision-making process based on accumulated experience as well as developed behavioural patterns during the journey, etc. The paper describes a decision support system for multimodal travel planners based on a behavioural decision-making model.

Keywords: Multimodal travel · User preferences · Decision theory · Decision support system

1 Introduction

Promoting sustainable mobility has been one of the European Commission's priority actions for several years. EU policies and programmes [1, 2] aim to provide seamless mobility and access to key destinations and services for all residents, increase transport safety, reduce harmful emissions and noise, and improve the efficiency and cost-effectiveness of the transport system. The European Union (EU) target for the transport sector is to reduce greenhouse gas (GHG) emissions by at least 60% by 2050 [3]. The EU and its Member States have set a sustainable, energy-efficient approach to mobility as their mobility target. In recent years, driven by the promotion of EU transport policy, multimodal passenger transport systems have become one of the most important tasks in addressing urban mobility [4]. The European Commission has declared 2018 the "Year of Multimodality" and has developed a number of legislative and policy initiatives for better, adequate infrastructure and digital solutions to promote an integrated, multimodal and ultimately sustainable transport system [5].

Establishing sustainable urban mobility is a complex task because sustainable urban mobility should primarily be considered at the individual level, i.e. at the level of the user. The transition from cars to sustainable modes of transport remains a complex behavioral process at the personal level and a societal challenge in general. As passengers need to get from one place to another smoothly, quickly and easily, it is necessary to offer a solution that can compete with the car, which is still the predominant choice.

I. Karabegović et al. (Eds.): NT 2022, LNNS 472, pp. 712–718, 2022.
https://doi.org/10.1007/978-3-031-05230-9_85

According to the ISO taxonomy, Traveler Information is one of the 11 functional areas of ITS. This area includes static and dynamic information about the transport network, pre-trip and on-road information services, and support services that collect, store and manage information for planning transport activities. Pre-trip information services are of great importance to users as they allow to plan a trip from home or from any other location where internet access is available or where information on public transport, time or travel costs is available. In addition to pre-trip information, access to information during the trip is very important [6].

The paper consists of five chapters. A basic description of multimodal travel is given in the second chapter. The third chapter presents the importance and role of multimodal travel planners. The fourth chapter describes the decision-making model, focusing on a behavioral approach. The conclusion presents the main features of this approach and guidelines for further improvements.

2 Multimodal Travel

The concept of multimodal "door-to-door" transport is one of the priority measures in EU transport policy and can solve the transport problems of urban areas. The European Commission, which has declared 2018 as the "Year of Multimodality", has designed a series of legislative and policy initiatives for better, adequate infrastructure and digital solutions to promote an integrated, multimodal, and ultimately sustainable transport system [5].

The multimodal passenger system is the answer that leads to sustainable urban mobility. It includes end-to-end connectivity throughout the journey (service "from anywhere to anywhere"), time-saving (easy change of transport mode) and maximum flexibility in combining transport modes. Such a travel concept includes a combination of available public transport modes from origin to destination (Fig. 1). Urban integration and time-saving are the advantages of developing and introducing such a concept.

Travel planners offer the possibility of unimodal and multimodal travel planning. There are several route options with unimodal travel, but only one mode of public transport. There are several route options with multimodal trips, with each route containing a combination of transport modes according to selected criteria. In certain scenarios of multimodal travel, there are some obstacles such as lack of information for certain modes of transport (transfers, waiting times), lack of personalization of the journey, etc. To achieve such systems' quality and efficient functioning, it is necessary to implement modern technology [7].

The multimodal passenger information system combines a number of different data on which the system is based. These are data on users' destinations, cities, regions, information on the topography of the country and its relation to the transport system, timetables for different modes of transport, schematic maps of transport hubs, schematic maps of transport networks, information on objects of interest (points of interest - POI), information on tariffs and fare zones and, more recently, information on the pollutant emissions of certain modes of transport [6, 8–10].

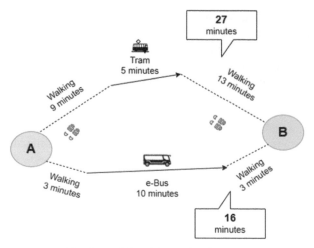

Fig. 1. Multimodal travel options, [7]

3   Multimodal Travel Planners

An appropriate definition of multimodal travel planners was provided by Gentile and Nökel [11]: "These systems are front-end-back-end computer systems, which provide a traveler the best itinerary according to several parameters characterizing an intermodal passenger transport journey. Multimodal travel planners provide better modal integration and more sustainability by enabling travellers to select the most suitable combination of transport modes for the journey and could lead to an increased use of public transport, cycling or walking in urban environment."

Multimodal journey planners are part of urban ITS, which offers a range of passenger transport services. The basic task of the system is to answer the user's question: "How do I get from place A to place B at a certain departure/arrival time and under which conditions? " The system usually contains travel and traffic information such as locations of stops, departure and arrival times of means of transport (which are integrated into the system, e.g. tram, metro, bus, etc.), possibilities to buy tickets, possible incidents, general traffic situation, etc. In contrast to static information systems, dynamic systems enable timely decision-making when choosing means of transport and travel routes [8–10, 12, 13].

Nowadays, multimodal travel planners are mainly based on algorithms that determine the shortest route/shortest travel time. The complexity in building the model that generates options tailored to passengers is reflected in the diversity of their preferences. For this reason, the choice of the most appropriate route can be considered as a multi-criteria problem. An example of the diversity of preferences can be seen in a number of travel scenarios in urban areas. In the case of congestion, passengers can be provided with information and options for other routes, making better use of the available transport infrastructure. It is important to note that the country's economy also benefits from multimodal travel information and planning services, as they provide new business opportunities for service providers and create new jobs in a very dynamic sector. Introducing new transport information and charging systems is an important step towards a better transport system.

Multimodal travel information provides dynamic real-time information integrating several accessible modes of public transport (tram, metro, bus, train, walking, etc.). In this respect, the user creates a customized journey. Previous research has defined multimodal information in three phases: Pre-trip information, on-trip information and post-trip information [6, 8, 9]. Thus, there are different physical locations where a person can be when making a travel decision, and the use of a multimodal information system can change the decision or behaviour pattern during the trip (depending on the information).

In urban areas, multimodal information is of great importance as it increases the share of public transport and other "healthy" ways such as walking and cycling [8, 11]. Multimodal travel information promotes mobility for all user groups, especially for users with disabilities or reduced mobility, by providing information about facilities and assistance at transport hubs. A particular type of user of multimodal travel planners is tourists. They may be guided by additional criteria when selecting a travel route, e.g. choosing a route that includes different attractions (POIs, etc.). The above points to a wide range of criteria and their personalized nature [8].

4 Decision-Making Process

Classical decision theory assumes complete and perfect rationality, whereby rational decision-makers strive to make optimal decisions. In economic theory, the so-called "economic man" (lat. Homo economicus) is an approximation of homo sapiens, who acts and exists in the best way for him and given the possibilities and limits of the environment. He or she knows the relevant aspects of the environment, but even if he or she is not familiar with all aspects, his knowledge of them is pure and coherent. Moreover, he or she is always able to choose the version of the decision that maximizes his or her personal preferences and self-interest (utility). Such an approach has met with numerous criticisms [14, 15]. The homo economicus model is often criticized because there is no perfect knowledge, stability, consistency, or rationality in decision-making.

In this case, the focus is on the user (human) who is faced with the problem of choosing a multimodal travel route. He is not an ideal rational being, but is shaped by his feelings, past travel experiences, emotions, etc. Based on his "travel genome" [16], which is an experienced record of all this, he makes a decision, i.e. the choice of a multimodal travel route. According to [17], the choice of alternative, i.e., decision-making, is an act of compromise between different "I's" within the decision-maker.

The Rational Choice Theory is limited in the human component and does not accurately picture the decision-making process. On the other hand, behavioral decision theory is driven by the decision-maker's cognitive, psychological, and emotional constraints but provides a much more comprehensive picture of the decision-making process. Nobel Prize winner Herbert A. Simon was among the first to recognize the inadequacies of the standard (classical) economic model of decision-making. As a critique and response to the previous classical approach to decision-making, Simon emphasized that decision-making is a complex, dynamic and sequential process that leads to adaptive decision-making. Simon introduced the term "bounded rationality", proving that individuals make rational decisions based on all possible information is not true [18].

People strive for a satisficing, good-enough decision in the decision-making phase, not the "best decision". Satisficing is determined by the level of aspiration of the decision-maker, i.e., the set of minimum requirements that the decision-maker sets. In contrast to the classical "rational" models of decision-making, which do not take into account the deeper psychology or influence on preferences, behavioural economics has offered a solution that considers human decision-making under realistic conditions; where preferences are not stable and are based on the rule of intransitivity [19]. Figure 2 shows the proposal for a multimodal travel route assistance system based on elements of behavioural decision theory.

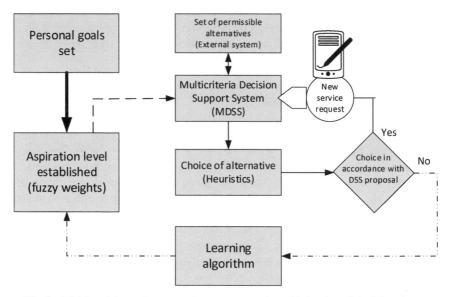

Fig. 2. Multimodal travel route assistance system based behavioural decision theory

5   Conclusion

Accordingly, planners provide travellers with better information when choosing a mode of transport, enabling them to choose the best option for their needs (e.g., in choosing a mode of transport, route, price, travel time and even an option that is less harmful to the environment) and ultimately enabling the successful completion of the journey. The open behavioural model is presented as a sequential, iterative process. In the open behavioral model, the key role is played by aspiration, i.e., the desire and effort to achieve something, and the use of heuristics, which are characteristic of human problem solving and decision making [20].

The system described above can be significantly improved by using real-time information on traffic conditions, possible traffic incidents, congestion, etc. This is made possible using cooperative transport systems [21].In this case, the system becomes significantly more efficient and practical for the end-user - passenger.

Acknowledgements. The research has been supported by the Twinning Open Data Operational project that has received funding from the European Union's Horizon 2020 research and innovation programme under grant agreement No. 857592. In addition, part of the research activities is related to the OJP4Danube project (Coordination mechanisms for multimodal cross-border traveller information network based on OJP for Danube Region) within the Danube Transnational Program.

References

1. European Commission: Sustainable Urban Mobility: European Policy, Practice and Solutions. European Commission, Brussels (2017)
2. Rupprecht Consult - Forschung&Beratung GmbH (ed.): Guidelines for Developing and Implementing a Sustainable Urban Mobility Plan. European Commission (2019)
3. Danish Energy Agency: Energy and Climate Policies beyond 2020 in Europe (Overall and Selected Countries). Danish Energy Agency, Copenhagen (2015). on: https://ens.dk/sites/ens.dk/files/Globalcooperation/eu_energy_and_climate_policy_overview.pdf. Accessed 20 Dec 2021
4. Klug, S. (ed.): Multimodal Personal Mobility. European Commission, Brussels (2013). https://eusmartcities.eu/sites/default/files/201710/Multimodal%20personal%20mobility%20january.pdf. Accessed 2 Jan 2022
5. Civitas: EU Commission declares 2018 the "Year of Multimodality". https://civitas.eu/news/eu-commission-declares-2018-year-multimodality. Accessed 2 Jan 2022
6. Bošnjak, I.: Intelligent Transport Systems I - ITS I. University of Zagreb, Faculty of Transport and Traffic Sciences, Zagreb (2006).(in Croatian)
7. Mandžuka, B.: Personalized Services in Multimodal Journey Planners (Doctoral Qualifying Exam). University of Zagreb, Faculty of Transport and Traffic Sciences, Zagreb (2021)
8. Mandžuka, B., Brčić, D., Škorput, P.: Application of multimodal travel guides for urban and suburban travel. In: Proceedings of the 34th Conference on Transportation Systems with International Participation, Automation in Transportation, Croatia, Dubrovnik, pp. 92–95 (2014). (in Croatian)
9. Mandžuka, B., Brčić, D., Škorput, P.: Application of pre-trip and on-trip information in urban areas. In: Proceedings of the 6th Croatian Road Congress - Via Vita, Croatia, Opatija (2015). (in Croatian)
10. Škorput, P., Mandžuka, B., Vujić, M.: The development of cooperative multimodal travel guide. In: Proceedings of 22nd Telecommunications Forum Telfor (TELFOR), Belgrade, Serbia, pp. 1110–1113 (2014)
11. Gentile, G., Nökel, K. (eds.): Modelling Public Transport Passenger Flows in the Era of Intelligent Transport Systems. Springer, London (2016). https://doi.org/10.1007/978-3-319-25082-3
12. Vujić, M., Skorput, P., Mandžuka, B.: Multimodal route planners in maritime environment. Pomorstvo **29**(1), 1–7 (2015)
13. Vujic, M., Mandzuka, B., Skorput, P.: Multimodal traveller information systems in maritime environment. In: Proceedings of 58th International Symposium ELMAR, Zadar, Croatia, pp. 89–92 (2016)

14. Saad, G.: Behavioral Decision Theory, pp. 1–3. Wiley, Hoboken (2015)
15. Menger, M.: TeorijaodlučivanjaHerberta Simona: Decizionističkipristuporganizaciji. In: PolitičkaMisao, vol. 56, pp. 66–86 (2019). https://doi.org/10.20901/pm.56.2.03
16. Goulias, K.G., Davis, A.W., McBride, E.C.: Introduction and the genome of travel behavior. In: Mapping the Travel Behavior Genome, pp 1–14. Elsevier, Amsterdam (2020)
17. Connolly, T.: Decision Making: Descriptive, Normative and Prescriptive Interactions (Ed. by, Bell, D.E., Raiffa, H., Tversky, A.) Cambridge University Press, New York (1988). (J. Behav. Decis. Making 3, 142–143). https://doi.org/10.1002/bdm.3960030208
18. Redlawsk, D.P., Lau, R.R.: Behavioral Decision-Making. Oxford University Press, Oxford (2013)
19. Brown, R.: Consideration of the origin of Herbert Simon's theory of "satisficing" (1933–1947). Manag. Decis. 42, 1240–1256 (2004). https://doi.org/10.1108/00251740410568944
20. Gigerenzer, G., Gaissmaier, W.: Heuristic decision making. Annu. Rev. Psychol. 62, 451–482 (2011). https://doi.org/10.1146/annurev-psych-120709-145346
21. Mandzuka, S., Skorput, P., Vujic, M.: Architecture of cooperative systems in traffic and transportation. In: Proceedings of 23rd Telecommunications Forum Telfor (TELFOR), Belgrade, Serbia (2015)

Printed by Printforce, the Netherlands